The Correspondence of John Tyndall

VOLUME 14

[illegible] Correspondence of John Tyndall

[illegible]

THE CORRESPONDENCE OF JOHN TYNDALL

VOLUME 14

The Correspondence, October 1873–October 1875

Edited by
Gowan Dawson, Matthew Stanley,
and Matthew Wale

General Editors
Roland Jackson, Bernard Lightman, and Michael S. Reidy

Associate Editors
Michael D. Barton, Erin Grosjean, and Michael Laurentius

UNIVERSITY OF PITTSBURGH PRESS

Published by the University of Pittsburgh Press, Pittsburgh, Pa., 15260

Manufactured in the United States of America
Printed on acid-free paper
10 9 8 7 6 5 4 3 2 1

Cataloging-in-Publication data is available from the Library of Congress

ISBN 13: 978-0-8229-4818-6
ISBN 10: 0-8229-4818-4

Publisher: University of Pittsburgh Press, 7500 Thomas Blvd., 4th floor, Pittsburgh, PA 15260, United States, www.upittpress.org

EU Authorized Representative: Easy Access System Europe, Mustamäe tee 50, 10621 Tallinn, Estonia, gpsr.requests@easproject.com

CONTENTS

ACKNOWLEDGMENTS

Much of the editing of this volume took place during the COVID-19 pandemic of 2020–21, and Tyndall and his correspondents proved amiable companions at that difficult time, even if we, the editors, became envious of their numerous social engagements and foreign trips during our own enforced isolation. Despite the restrictions of the pandemic, we benefited greatly from working with a number of people in the production of the volume. In particular, we are grateful to the general editors of *The Correspondence of John Tyndall*, Roland Jackson, Bernard Lightman, and Michael Reidy; to the wider Tyndall Correspondence Project team, especially Michael D. Barton, Erin Grosjean, and Michael Laurentius; and to James Braund, Ella Coulter, and Nicolas Sanchez-Guerrero for their translations of, respectively, German, French, and Spanish letters.

Like other editors involved in *The Correspondence of John Tyndall*, we owe a huge debt to the army of graduate students and others, on both sides of the Atlantic and over many years, who identified, scanned, and produced initial transcriptions of the letters now presented.

We also thank the many colleagues and friends who generously lent us their expertise in responding to queries on specific issues: Melinda Baldwin, Ruth Barton, Leila Belkora, Bill Brock, Janet Browne, Richard Carter, Anne DeWitt, Pola Durajska, Diarmid Finnegan, Debra Francis, David Hounshell, Frank James, Theresa Levitt, Don Opitz, Greg Radick, Marc Rothenberg, Sally Shuttleworth, James Ungureanu, and Eugene Vydrin. Additionally, Matthew Stanley thanks his wife, Janelle Stanley, and his children, Maya and Zoe, for their great support and patience through many Tyndall anecdotes over dinner.

Finally, Gowan Dawson would like to thank the Leverhulme Trust for the award of a Research Fellowship (RF-2020-085), which enabled him to take the lead role in editing the letters and in the preparation of the front and back matter.

LIST OF ABBREVIATIONS

ANB	*American National Biography*
APS	American Philosophical Society
Ascent of John Tyndall	Roland Jackson, *The Ascent of John Tyndall: Victorian Scientist, Mountaineer, and Public Intellectual* (Oxford: Oxford University Press, 2018)
BAAS	British Association for the Advancement of Science
BBAW, NL. Helmholtz	Archiv der Berlin-Brandenburgischen Akademie der Wissenschaften, Nachlässe Helmholtz
Belfast Address	John Tyndall, *Address Delivered before the British Association Assembled at Belfast, with Additions* (London: Longmans, Green, and Co., 1874)
Belfast Address, 7th thousand	John Tyndall, *Address Delivered before the British Association Assembled at Belfast, with Additions*, 7th thousand (London: Longmans, Green, and Co., 1874)
BL Add MS	British Library additional manuscripts
Brit. Assoc. Rep.	*Reports of the British Association for the Advancement of Science*
CDSB	*Complete Dictionary of Scientific Biography*, 27 vols. (Detroit, MI: Charles Scribner's Sons, 2008)
Cosmos	*Cosmos: Revue encyclopédique hebdomadaire des progrès des sciences et de leurs applications aux arts et à l'industrie*
CS	cable ship
CUL GBR/0012/ MS DAR	Cambridge University Library Darwin Archive
CUL SC	Cambridge University Library Stokes Correspondence
Darwin Correspondence, vol. 22	Frederick Burkhardt et al., eds., *The Correspondence of Charles Darwin*, vol. 22, 1874 (New York: Cambridge University Press, 2013)
Darwin Correspondence, vol. 23	Frederick Burkhardt et al., eds., *The Correspondence of Charles Darwin*, vol. 23, 1875 (New York: Cambridge University Press, 2013)
DCL	doctor of civil law
Faraday Correspondence	Frank A. J. L. James, ed., *The Correspondence of Michael Faraday*, 6 vols. (London: Institution of Electrical Engineers, 1991–2011)

Forms of Water	John Tyndall, *The Forms of Water in Clouds & Rivers, Ice & Glaciers* (London: H. S. King, 1872)
Fors Clavigera	John Ruskin, *Fors Clavigera: Letters to the Workmen and Labourers of Great Britain*, 8 vols. (Orpington: George Allen, 1871–84)
Foster and Huxley Correspondence	William F. Bynum and Caroline Overy, eds., *Michael Foster and Thomas Henry Huxley: Correspondence, 1865–1895* (London: Wellcome Trust Centre for the History of Medicine at UCL, 2009)
Fragments of Science	John Tyndall, *Fragments of Science for Unscientific People: A Series of Detached Essays, Lectures, and Reviews* (London: Longmans, Green, and Co., 1871)
FRS	Fellow of the Royal Society
'Further Papers'	"Further Papers Relative to a Proposal to Substitute Gas for Oil as an Illuminating Power in Lighthouses, Part III," H.L., Sessional Paper [C-1151], *Parliamentary Papers* (1875): 1–66
Glaciers of Savoy	Louis Rendu, *Theory of the Glaciers of Savoy*, trans. Alfred Wills, ed. George Forbes, with supplementary articles by Peter Guthrie Tait and John Ruskin (London: Macmillan, 1874)
Glaciers of the Alps	John Tyndall, *The Glaciers of the Alps* (London: John Murray, 1860)
Hirst Journals	William H. Brock and Roy M. Macleod, eds., *Natural Knowledge in Social Context: The Journals of Thomas Archer Hirst* (London: Mansell, 1980)
HLS	*Historisches Lexikon der Schweiz*
Hours of Exercise in the Alps	John Tyndall, *Hours of Exercise in the Alps*, 1st ed. (London: Longmans, Green, 1871)
IC HP	Imperial College of Science, Technology and Medicine, College Archives, Huxley Papers
Irish Genealogy	Department of Culture, Heritage and the Gaeltacht, Irish Genealogy, https://www.irishgenealogy.ie/
Les Mondes	*Les Mondes: Revue hebdomadaire des sciences et leurs applications aux arts et à l'industrie*
Life and Letters of Huxley	Leonard Huxley, ed., *Life and Letters of Thomas Huxley*, 2 vols. (London: Macmillan, 1900)
Life of Forbes	John Campbell Shairp, Peter Guthrie Tait, and A. Adams-Reilly, eds., *Life and Letters of James David Forbes* (London: Macmillan, 1873)
LLD	legum doctor (doctor of canon and civil law)
LT	Louisa Tyndall, John Tyndall's widow
MP	member of Parliament
M.P. for Russia	W. T. Stead, ed., *The M.P. for Russia: Reminiscences and Correspondence of Madame Olga Novikoff*, 2 vols. (London: Andrew Melrose, 1909)

NDB	*Neue Deutsche Biographie*
ODNB	*Oxford Dictionary of National Biography*
OED	*Oxford English Dictionary*
Phil. Mag.	*Philosophical Magazine*
Phil. Trans.	*Philosophical Transactions of the Royal Society of London*
PMG	*Pall Mall Gazette*
'Report by Professor Tyndall'	"Report by Professor Tyndall to Trinity House on Experiments with regard to Fog Signals," H.C., Sessional Paper [188], Parliamentary Papers 60 (1874): 1–77
RI	Royal Institution of Great Britain
RI MS JT	Royal Institution, London, John Tyndall Papers
Roy. Inst. Proc.	*Proceedings of the Royal Institution of Great Britain*
Roy. Soc. Proc.	*Proceedings of the Royal Society of London*
RS	Royal Society of London
Six Lectures on Light (1873)	John Tyndall, *Six Lectures on Light Delivered in America in 1872—1873* (London: Longmans, Green, 1873)
Six Lectures on Light (1875)	John Tyndall, *Six Lectures on Light: Delivered in America in 1872–1873*, 2nd ed. (London: Longmans, Green, and Co., 1875)
Sound (1867)	John Tyndall, *Sound: A Course of Eight Lectures Delivered at the Royal Institution of Great Britain* (London: Longmans, Green, 1867)
Sound (1875)	John Tyndall, *Sound*, 3rd ed. (London: Longmans, Green, and Co., 1875)
SS	steam ship
THV	Trinity House vessel
TRC	Tennyson Research Centre, Lincoln
Traveller and Plant Collector	Ray Desmond, *Sir Joseph Dalton Hooker: Traveller and Plant Collector* (London: Royal Botanic Gardens, Kew, 1999)
Tyndall Correspondence, vol. 3	Ruth Barton, Jeremiah Rankin, and Michael S. Reidy, eds., *The Correspondence of John Tyndall*, vol. 3, *The Correspondence, January 1850–December 1852* (Pittsburgh: University of Pittsburgh Press, 2017)
Tyndall Correspondence, vol. 6	Michael D. Barton, Janet Browne, Ken Corbett, and Norman McMillan, eds., *The Correspondence of John Tyndall*, vol. 6, *The Correspondence, November 1856–February 1859* (Pittsburgh: University of Pittsburgh Press, 2019)
Tyndall Correspondence, vol. 8	Michael D. Barton, Piers J. Hale, Nathan Kapoor, and Elizabeth Neswald, eds., *The Correspondence of John Tyndall*, vol. 8, *The Correspondence, June 1862–January 1865* (Pittsburgh: Unversity of Pittsburgh Press, 2021)

Tyndall Correspondence, vol. 9 — Michael D. Barton, Iwan Rhys Morus, and James Ungureanu, eds., *The Correspondence, of John Tyndall*, vol. 9, *The Correspondence, February 1865–December 1866* (Pittsburgh: University of Pittsburgh Press, 2022)

Tyndall Correspondence: vol. 10 — Michael D. Barton, Ken Corbett, and Roland Jackson, eds., *The Correspondence of John Tyndall*, vol. 10, *The Correspondence, January 1867–December 1868* (Pittsburgh: University of Pittsburgh Press, 2021)

Tyndall Correspondence, vol. 12 — Anne DeWitt and Kathleen Sheppard, eds., *The Correspondence of John Tyndall*, vol. 12, *The Correspondence, March 1871–May 1872* (Pittsburgh: University of Pittsburgh Press, 2023)

Tyndall Correspondence, vol. 13 — Michael D. Barton, Joseph D. Martin, and Gregory Radick, eds., *The Correspondence of John Tyndall*, vol. 13, *The Correspondence, June 1872–September 1873* (Pittsburgh: University of Pittsburgh Press, 2023)

UCL — University College London

'Vehicle of Sound' — John Tyndall, "On the Atmosphere as a Vehicle of Sound," *Phil. Trans.* 164 (1874): 183–244

Wellcome — Wellcome Library

John Tyndall, stipple engraving by Charles Henry Jeens from a photograph taken in New York, probably in January 1873, by José María Mora. The engraving was included as an unbound portrait in the weekly issue of *Nature* for 20 August 1874—the week of the BAAS meeting in Belfast—accompanying an article by Hermann Helmholtz, "Scientific Worthies. IV.—John Tyndall," *Nature* 10 (1874): 299–302. Tyndall sent the engraving as a present to close friends (see letter 4415), and it is mentioned by several of his correspondents in this volume. *Credit*: Wellcome Collection CC BY 4.0.

INTRODUCTION TO VOLUME 14

In a career regularly punctuated by fierce scientific controversies, the period covered by the 499 letters in the fourteenth volume of *The Correspondence of John Tyndall* was marked by a number of particularly intense and acrimonious disputes. In the two years from October 1873 to the same month in 1875, Tyndall reignited or continued longstanding debates regarding glaciation as well as bacteria germs and spontaneous generation. Still more notably, this volume spans the period of the composition, delivery, and furious reaction to Tyndall's famous—or, more accurately, infamous—Belfast Address. This prestigious lecture, which he delivered as the newly inaugurated president of the British Association for the Advancement of Science (BAAS), has long been heralded as one of the most momentous events of the nineteenth century. A character in George Bernard Shaw's play *Man and Superman* (1903) even declares of it: "Nothing has been right since that speech that Professor Tyndall made at Belfast."[1] The letters in this volume provide a new, and unprecedentedly detailed, account of all aspects of the era-defining address. For Tyndall himself, it afforded a new level of prominence as a public intellectual, and he deployed his position to engage directly with some of the most contentious issues in Victorian society, especially the role of religion in relation to science. But Tyndall's expertise was also required on more practical matters, and the letters in this volume document his extensive role in determining official government policy on urgent questions such as safety at sea and public health. Additionally, they chart a dramatic shift in his personal life, with his initial correspondence with Louisa Hamilton, with whom he had previously communicated only through her family, marking the point where their burgeoning friendship developed into a formal relationship.

Glacial Resurgences

When this volume begins, in October 1873, Tyndall was in the midst of a conflict that had origins stretching back over more than two decades. In the late 1850s and early 1860s, Tyndall had clashed with James David Forbes regarding the question of how glaciers moved. Forbes had proposed that the motion of ice resembled that of a viscous fluid, a theory that gained some currency among his peers during the 1840s. Conversely, Tyndall argued from the mid-1850s onward that glaciers underwent a constant process of thawing and refreezing under high pressure (a phenomenon known as regelation, first described by Tyndall's friend and mentor Michael Faraday). However, what may have started as a case of divergent theories soon developed into a struggle over the basis of scientific authority, with Tyndall mobilizing a network of supporters who would eventually form the X Club and their extended circle.[2] It is unnecessary to recount the dispute in full detail here, but a key turning point in the debate was the allegation of plagiarism leveled at Forbes by Tyndall and Thomas Henry Huxley. It was suggested that the viscous theory was originally the work of the French bishop and man of science Louis Rendu, and although Forbes was never conclusively proven guilty, the hint of wrongdoing was enough to prevent him being awarded the RS's Copley Medal for 1859. Forbes's reputation never fully recovered from this blow, and the anger and resentment felt toward Tyndall by followers of Forbes would not be forgotten.

Although Forbes had been dead for five years by 1873, a coterie of influential supporters ensured that he would continue to trouble Tyndall from beyond the grave. Chief among Forbes's adherents was Peter Guthrie Tait, Forbes's successor as professor of natural philosophy at the University of Edinburgh, who initiated the collation of Forbes's correspondence in *Life of Forbes*, which was published by Macmillan in May 1873. Assisted by two disciples of Forbes, John Campbell Shairp and Anthony Adams-Reilly, Tait and his coeditors took the opportunity to resurrect the glacial controversy, casting aspersions on Tyndall's theories, especially those recently outlined in his *Forms of Water* (1872), and the allegedly underhand manner in which the original dispute had been conducted. This, they hoped, would posthumously vindicate Forbes and repair the damage to his legacy. In July 1873, Tyndall addressed the accusations in an article for the *Contemporary Review* titled "Principal Forbes and His Biographers," and went a step further in having the text reprinted separately as a pamphlet. However, in October 1873 Tyndall was faced with a fresh attack from Forbes's supporters in the form of a pamphlet written by John Ruskin. The influential art critic, who also had a longstanding interest in geology, had been a friend and correspondent of Forbes, and therefore joined the campaign against Tyndall. The pamphlet was the latest in Ruskin's series of "letters" addressed to the "Workmen and Labourers of Great Britain," issued periodically under the title *Fors Clavigera* by Ruskin's assistant George Allen. The letters addressed a wide range of subjects, through which Ruskin hoped to intervene on moral and social issues in the manner of his mentor, Thomas Carlyle. Ruskin accused Tyndall of jealousy and intellectual dishonesty, claiming that he deliberately misrepresented Forbes's theory in order that "he [Tyndall] and his friends may get the credit" of solving the question of glacial motion.[3]

Despite their many differences, Tyndall and Ruskin shared a deep reverence for Carlyle, whose writings had exercised a significant influence over their lives and work. Both men considered Carlyle to be a friend and guide, with Ruskin regarding him as a "second papa," and it is unsurprising that Tyndall's first response to Ruskin's accusations was to write a letter to Carlyle addressed to "My dear General."[4] Unlike Ruskin's rather infantile sobriquet, the military epithet that Tyndall gave to Carlyle demonstrates the self-consciously martial nature of their relationship, with Tyndall playing the role of subordinate to Carlyle's heroic leadership in a knowing reference to the latter's *On Heroes, Hero-Worship, and the Heroic in History* (1840), which Tyndall had first read back in the 1840s.[5] Later, in a letter from May 1875, Tyndall altered his salutation to "Well-beloved Chieftain!" in honor of Carlyle's recently published *The Early Kings of Norway*.[6] Tyndall received mixed advice from his closest friends as to whether Ruskin's attack required a rebuttal. Juliet Pollock counseled him to have "nothing to do with Ruskin," and her husband, William Frederick Pollock, concurred.[7] Conversely, Ellen Lubbock was "in favour of instant war."[8] It was Carlyle's opinion, however, that carried the greatest weight. With regard to Ruskin, Carlyle recommended: "[Y]ou should take no notice of him whatever; such being the one dignified, wise and proper course."[9] "Be it as thou wilt!," Tyndall replied, "I shall do as you advise, because I feel that your advice is the best and noblest that can be given to me."[10] Tyndall clearly treasured Carlyle's "very sweet and noble letter," admitting that it was "precious to me," and sent copies to particularly trusted friends.[11]

When Tyndall wrote to Carlyle regarding Ruskin's attack, he was in Dover, engaged in the practical work that dominated this period of his life. In 1867 Tyndall succeeded Faraday as chief scientific adviser to Trinity House, the authority responsible for the management of lighthouses in England, Wales, and the Channel Islands. In this capacity, he spent much of 1873 conducting experiments to determine the most effective means of warning ships in dense fog, an investigation that naturally led him to take a keen interest in how sound was transmitted through the air. It was an arduous task, requiring Tyndall to be "afloat every day from morning to night," listening to different acoustic signals from aboard a yacht in the English Channel.[12] Tyndall's curiosity was piqued by the varied results of the experiments, and particularly by the significant effect of weather conditions on the distance a sound could travel. An assortment of horns and whistles were tested, and in October 1873 Tyndall began a new series of experiments using a steam-powered siren acquired from the United States with the aid of Joseph Henry, secretary of the

Smithsonian Institution. Tyndall had seen—or rather, heard—the siren in use at lighthouses in New York and New Jersey during his recent American lecture tour, from which he returned in February 1873. The letters in this volume demonstrate the extent to which Tyndall continued to cultivate the transatlantic network he formed during the lecture tour. On 25 October 1873, Tyndall wrote to Henry expressing his satisfaction with the siren, which he considered "unquestionably the most effectual instrument that has come under my notice." As a consequence, he declared his intention to "communicate a paper to the Royal Society on the subject." Tyndall also promised to send an abstract of the paper to the Philosophical Society of Washington, which he regarded as "the best return I can make" for their courtesy during his visit.[13]

With the vagaries of the British climate, it was not until December that Tyndall had the opportunity to test his theories in actual fog. Henrietta Huxley was amused that such undesirable weather was "a Godsend" for Tyndall, while it was nothing but an inconvenience to her.[14] As Tyndall suspected, fog could dampen sound, but given the right conditions, it could also increase the distance over which it was transmitted. This variability posed a huge problem to mariners at a time when shipwrecks were a perpetual threat to global transportation and trade, as those at sea could not rely on acoustic signals to reliably judge how far they were from danger. Tyndall proposed sirens as the most effective solution to this pressing issue, but also recommended further investigation into the use of specially adapted cannons. In addition to his official report to Trinity House, the results of Tyndall's acoustic experiments were read before the RS in February 1874, and later published in the *Philosophical Transactions of the Royal Society of London* (*Phil. Trans.*) under the title "On the Atmosphere as a Vehicle of Sound."[15] Tyndall's correspondence with Gabriel George Stokes, the secretary of the RS, provides a valuable insight into the processes of peer review and revision that were required prior to a paper appearing in the pages of the most prestigious venue for scientific publication.[16] Two readers, one of whom was Stokes himself, compiled reports on the paper and made several suggestions for its improvement. The required revisions mostly involved rearrangement, "chiefly a matter of brain, scissors and paste" as Stokes remarked, particularly in shortening the paper by moving some of the exhaustive details to an appendix.[17] Tyndall complied, albeit wearily.

Any hopes of the glacier controversy dying down while Tyndall focused on his sound experiments were in vain, as Forbes's supporters took a further step in May 1874, publishing a new translation of Louis Rendu's *Théorie des glaciers de Savoie* (1840), again with Macmillan. The book was actually edited by Forbes's son, George Forbes, who wrote an introduction explaining his reasons for bringing Rendu's work to public attention. In short, he hoped to refute the accusations made against his father by circulating the original work that Forbes had been accused of plagiarizing, thereby allowing readers to make a fair judgment. Rendu's book had hitherto not been widely available in Britain, and Forbes *fils* accused Tyndall of misrepresenting the French cleric's writings in order to trump up the charges against Forbes *père*. Also included in the volume were articles by Tait and Ruskin, the former an article written for the *Contemporary Review* but not published by the journal, and the latter an abstract from *Fors Clavigera*. In reply, Tyndall immediately took the offensive, dashing off an article titled "Rendu and his Editors" for the *Contemporary*'s June 1874 issue.

Such a rapid response was, inevitably, rushed and lacking in caution. This gave Tait the upper hand, and he responded, in a private letter that he reserved the right to publish, by asking for the source of a quotation that Tyndall had falsely attributed to him, and noting, with withering disdain, "I cannot remember having used such an expression, nor even having entertained the idea it conveys."[18] Huxley confirmed that Tait was correct, and castigated himself for not having prevented the error: "in reading your proof I took it for granted that the words were Tait's own, but it was very stupid of me not to call your attention specially to the point."[19] Prudently, Huxley cautioned that any further public statement would be inadvisable, and even if Tyndall had wanted to respond, his contretemps with Tait and Ruskin was soon eclipsed by a still more ferocious controversy.

This Horrible Belfast Address

The continuing reverberations of the glacier dispute played a part in the most incendiary incident of Tyndall's entire career. In October 1873 the demands of his "arduous and exhausting . . . Govt work," particularly the experiments on signaling to ships in fog, led Tyndall to consider "giving up the Presidency" of the BAAS.[20] Having declined this prestigious position on three previous occasions, he had finally been persuaded to accept in the summer of 1873, but within weeks was already regretting the decision.[21] His fatigue, however, was not the only consideration, and the BAAS, as Joseph Dalton Hooker confided to Tyndall, was concerned about "their motives for accepting your resignation being misunderstood," a view that Hooker shared. There was no need for him to clarify the cause of this "fear," which Tyndall evidently understood implicitly.[22] After all, his adversaries in the glacier controversy had recently been perpetuating their attacks in Scottish and Irish newspapers, and particularly in Belfast, where Tait had previously been professor of mathematics at Queen's College. The Irish city was also where the BAAS's peripatetic annual meeting would be held the following year, and Tyndall acknowledged ruefully: "If I were to back out they might say that I was afraid to come."[23] He lamented to Hooker, "I fear I am now really committed to it," and soon afterward informed another correspondent that he would remain as the BAAS's incoming president "much against my will."[24]

The main responsibility that Tyndall was grudgingly compelled to undertake was delivering the presidential address at the Belfast meeting in August 1874. Ten months before, Hooker recommended him to "make up your mind to choose a scientific subject, that will demand no great mental strain" and "relieve both your mind and the Association of a load of anxiety."[25] Only two days prior to this particular missive, Hooker had reflected that "it is impossible either officially or unofficially to advise Tyndall," and his counsel about what Tyndall came to call "this horrible Belfast Address" proved no exception.[26]

The address that Tyndall found so onerous, and in which he surveyed more than two thousand years of intellectual history, sparked a storm of controversy that became one of the defining events of the nineteenth century. Frank M. Turner considered it "perhaps the most intense debate of the Victorian conflict of science and religion," and even more recent historians who dispute the validity of this "conflict thesis" concur that Tyndall's presidential address "changed the way modern science came to be perceived by the British public."[27] This was immediately evident to Tyndall's contemporaries, and he himself commended the "sagacity" of Edward Livingston Youmans's view, when reprinting the address in America, that "no scientific paper ever before published has produced so extensive and profound an impression as this."[28] While the address that prompted such hyperbole has, inevitably, been the aspect of Tyndall's career that has received the most scholarly attention, the letters in this volume nonetheless afford numerous new insights into its origin, composition, and reception.[29] The direct, albeit inadvertent, relation to the glacier controversy, which forced Tyndall unwillingly to accept the BAAS's customary presidential duties, was only divulged in private correspondence with close friends, and the discussion of several other features of his subsequent address was similarly confined to the same confidential medium.

The presidential address inaugurated the BAAS's weeklong annual meeting, and was, by tradition, attended by local dignitaries as well as the most eminent members of the scientific establishment. It was, as an aristocratic friend assured Tyndall, "a great position to address all the greatest minds of the country!"[30] His own letters suggest that Tyndall was largely indifferent to the social rank or intellectual prestige of his hearers, and he avowed bluntly to Thomas Archer Hirst: "I do not care what the effect of the Address may be upon the audience." Instead, his principal concern was that "it utters a truth or two which will survive the meeting of the association," a statement combining self-deprecation with a vaunting regard for posterity.[31] Such an aspiration flouted Hooker's advice to avoid the "labour so often bestowed on these

Addresses," and Tyndall actually devoted more than two months of his busy schedule—from early June to mid-August 1874—exclusively to preparing the address.[32] Notably, he spent this time alone, amid the welcome but chilly solitude of the Swiss Alps, and his correspondence with his closest friends, especially Hirst and Huxley, became both more regular and more effusive at this time. These long missives, so different from the customarily curt letters he composed when in London, provide an unprecedentedly precise and detailed account of how the Belfast Address came into being. For instance, while Tyndall publicly acknowledged, in the preface to the published *Address*, that it was "written under some disadvantages," it is only in the letters that it becomes apparent that this referred principally to what Hirst called "that most 'contemptible' of all pains—toothache."[33]

This particular affliction, moreover, had a vital—if rather bathetic—significance for both the composition and the content of the address. It certainly seems to have permeated Tyndall's attitude to writing, and he told Hirst, at a late stage of the process, that he "kept continually gnawing at the address."[34] In fact, he regularly considered that the "Address may possibly be a very poor affair" on account of his dental problems, lamenting that "to do it aright would require perfect health on my part."[35] Tyndall composed the address in three discrete installments, which were sent separately to the printers, Taylor and Francis, back in London. It was when writing the initial two parts, covering the historical roots of the atomic theory and also featuring an imaginary dialogue on the issue of consciousness, that he was most despairing about his teeth. Notwithstanding a minor mishap while swimming in a lake, Tyndall's health had improved by the time he came to the final installment, and he was clearly more sanguine while writing it. As he told Hirst at the end of July: "My work has been slow: but I am now beginning to see the end of it: and I hope to make a somewhat daring but dignified wind-up."[36] It was in this "wind-up" that Tyndall would make some of his most contentious statements, particularly the claim that the "promise and potency of all terrestrial Life" was inherent in "Matter" and did not require a distinct creative act. This was perceived by many critics of the address as an affirmation of both materialism and atheism, dangerous accusations that Tyndall was determined to resist. In the preface to the published *Address*, he conceded that "it is not in hours of clearness and vigour that . . . the doctrine of 'Material Atheism' . . . commends itself to my mind," and "in the presence of stronger and healthier thought it ever dissolves and disappears." As his correspondence during the writing of the address suggests, however, it was actually the return of strength and health, at least to his mouth, that induced Tyndall to compose the provocative peroration in which, tellingly, he reflected that "every meal we eat, and every cup we drink, illustrates the mysterious control of Mind by Matter."[37]

Huxley had earlier advised that Tyndall should be "wise and prudent" in the address, and warned him: "I declare I have horrid misgivings of your kicking over the traces."[38] Huxley's missives are replete with such idioms, and in this case he likened his friend to a horse that evades its harness and becomes uncontrollable. Tyndall himself used the nautical analogy of "steering very close to the wind," although he assured Huxley of his "hope to keep clear of rocks."[39] While Tyndall was no more amenable to Huxley's advice than he had been to Hooker's earlier guidance, he did ask friends to read and comment on the proofs of the respective installments as they became available. Such feedback was an important element of Tyndall's customary compositional method, and he told Hirst: "My first proofs in fact are usually my basis of operations, and the final ones are always very unlike the first."[40] But the incremental process by which the address was written and then printed meant that, beyond Tyndall himself, nobody read the complete text. Both Charles Darwin and Herbert Spencer perused only the pages in which their own work was discussed, while Hirst, himself suffering with ill health, managed just the initial installment.[41] Even Huxley—who had impatiently asked Tyndall, "[S]hall I not see the address? It is tantalizing to hear of your progress and not to know what is in it"—in the end, as Tyndall wrote to Hirst, "read almost, not quite the whole."[42] Crucially, the late completion of the final installment, containing Tyndall's putatively materialistic "wind-up," indicates that this was the part for which he relied solely on his own judgment.

Even without the counsel of his friends, Tyndall's assertion of the potency of matter was carefully qualified by his insistence that all phenomena have their roots in a vaguely understood cosmic life. This is an overlooked aspect of the address that has been recovered in modern scholarship, but it was already evident to at least some of Tyndall's contemporaries.[43] Hirst, who was unable to come to Belfast and finally read the complete address in the suitably sublime setting of a German forest, told his friend: "I dare say some of your audience . . . shook their heads dubiously about your transcendental materialism but it will do them good to ponder it."[44] One of those who might have been expected to shake his head, the anti-Darwinian Duke of Argyll, intuited the same mystical emphasis that Hirst did, and wrote to Tyndall to assure him: "I have taken no part in the outcry about y[ou]r Belfast Address—because I thought it greatly misunderstood—and that its tendency is rather to spiritualise matter, than to materialise Thought."[45] Still more surprisingly, the "great sensation" incited by the address prompted the Indian Hindu divine Protap Chunder Mozoomdar to request a meeting with Tyndall, at which, the former recorded, "his whole nature was glowing with a deep, vague and transcendent sense of the Divine life, beauty and love."[46] After the meeting, Tyndall observed: "I do not think I said anything to our Hindu friend that I should object to hearing proclaimed from the housetops."[47] Inevitably, however, it was in private letters that he could be most candid about his complex personal beliefs.

This was particularly the case with his intense and often intimate correspondence with the Russian socialite Olga Novikoff, to whom he declared of the Belfast Address: "While rejecting the Anthropomorphic notion of superstitious people I said as plainly as words could make it, that inscrutable Power lay behind it all. Here is a mystery that neither you nor I can sweep away."[48] Tyndall could be so open to Novikoff because, as he told her, she had "qualities which seize upon a plain blunt fellow like myself," and there seems to have been a romantic element to their friendship, at least in its initial stages in October and November 1874.[49] Although Novikoff was married, her husband remained in Russia during her lengthy sojourns in London, and, boldly dismissive of bourgeois propriety, she had affairs with several male friends, including William Thomas Stead and Max Nordau.[50] When Novikoff left London in late November to return to Russia, Tyndall appears to have wanted to avoid an emotional farewell, telling her that it was "on the whole best I should not be here."[51] She returned the following October, but by then Tyndall's friendship with Louisa Hamilton had become a tentative courtship, and he was compelled to warn Novikoff: "if I went to you under my present circumstances it would simply beget gossip which neither of us would like."[52] Rumors about Tyndall's personal situation, as a middle-aged bachelor with an evident fondness for female company and a nascent attachment to the considerably younger Louisa, contributed to the controversy over the Belfast Address, especially as materialism was often equated with moral depravity.[53] He joked uneasily, when sending a copy of the "iniquitous Address" to Mary Adair, an earlier object of his amorous affections, that "if it seems the proper fate for it pray put the book in the fire."[54] He was more serious when Novikoff alerted him to a pamphlet that urged his prosecution for blasphemy, assuring her that legally "nothing whatever can come of this petition. England has been long placed beyond any movement of this kind."[55] The impact of such lurid accusations, especially those concerning his personal character and morality, nevertheless compelled Tyndall to consider invoking the law on his own behalf.

In November 1874, Elizabeth Dawson Steuart, an old friend from Tyndall's hometown of Leighlinbridge, wrote to express her "indignation" at an "infamous paragraph in that vile paper."[56] What had provoked such outrage was a report, in the Catholic newspaper the *Tuam Herald*, of a pastoral declaring that "a degenerate Irishman (Tyndall) has selected Irish soil for his platform—and brought the danger home to us by ventilating at our doors his startling theories of . . . Materialism."[57] As a lengthy sequence of letters from

Steuart's family lawyer shows, this charge of being a "degenerate" prompted Tyndall to seriously contemplate bringing an action for libel. After investing considerable sums in obtaining the opinion of a Dublin attorney, it was only the prospect that the "action would go to trial" that finally dissuaded him.[58] Tyndall seems never to have acknowledged this exasperating episode to anyone beyond Steuart and her legal representatives, not even to Hirst or Huxley. Such sensitivity suggests that Tyndall perceived his personal reputation as a potential chink in his intellectual armor, and it was one that critics of the Belfast Address would continue to probe.[59]

Purity and Putrefaction

Reputation and scientific authority were also crucial to another controversy with which Tyndall reengaged at the end of 1874. This was the dispute over spontaneous generation and the germ theory that had been initiated by his Friday Evening Discourse at the RI in January 1870 titled "Dust and Disease."[60] The lecture had provoked an exhaustive response from the pathologist Henry Charlton Bastian, both in a series of letters to the *Times* and his huge tome *The Beginnings of Life* (1872). While Tyndall promoted the conclusions of Louis Pasteur and Joseph Lister, who contended that putrefaction and disease were caused by germs found in dust and dirt, Bastian defended the older idea that bacteria and other primitive organisms appeared continually as a result of natural processes. Notably, Bastian otherwise adhered to the naturalistic assumptions about the universe that Tyndall shared with Huxley, and both had previously offered "fatherly advice as mentors" to the young pathologist.[61] Their breach with Bastian on the origins of microscopic spores, however, was profound and irreparable. In addition to their intellectual differences, James Strick argues, the split was also driven by issues of scientific respectability: Huxley and Tyndall were concerned that Bastian's advocacy of spontaneous generation, with its radical political connotations of life being equated with mere chemical compounds, would contaminate their aspirations to make scientific naturalism gentlemanly and respectable.[62] They consequently traduced Bastian's experiments, not only excluding him from their inner circle but attempting to discredit him as a reliable scientific worker.

In subsequent years the intensity of the dispute abated, although Tyndall still went out of his way to isolate his erstwhile protégé, especially in the confidential medium of private correspondence. In December 1873 Tyndall wrote to Frederick Augustus Porter Barnard, the president of New York's Columbia College who had recently praised aspects of Bastian's experimental work on germs in the American press. He warned Porter that "Bastian's calibre and method of work" could not be trusted, and lambasted his scientific results: "All his more startling ones are to be ascribed to the fact that a man undisciplined in experiment has taken up a subject which requires for its treatment the most consummate experimental tact."[63] The same need to discredit spontaneous generation remained a pressing concern for Tyndall when, in October 1874, an outbreak of typhoid fever in a Lancashire town seemed to confirm the conclusions of William Budd's recent book on the disease. In *Typhoid Fever* (1873), Budd, a physician and epidemiologist, argued in great detail for a "contagion" theory of the fever by which it was spread through direct contact with germs rather than general environmental exposure to miasmas and other noxious vapors.

Tyndall was impressed by Budd's treatise and wrote a long letter to the *Times* explaining the evidence for typhoid germs and their implications for public health. The key point was that rural towns with open cesspools and poor sanitation generally had no typhoid fever—until, that is, someone arrived from another location carrying it, whereupon the disease spread rapidly. In his letter, Tyndall paraphrased Budd's own conclusion: "the living human body is the soil in which the specific poison of typhoid fever breeds and multiplies." He proposed that outbreaks of the disease could be brought under control with disinfectants, but counseled that it must be done "with the precision of a scientific process." If this exacting requirement was adopted, Tyndall insisted that the "plague" could be "instantly stayed," and he rousingly affirmed: "Can it be doubted that with sound medical advisers, backed by an intelligent population . . . rapid destruction of the foe might be accomplished?"[64] As Strick points out, Tyndall worked hard to distinguish precision science (as done by himself) from crude clinical work (as done by physicians such as Bastian). In particular, he attacked the medical press for its resistance to the germ theory, presenting himself as a more scientific, and therefore reliable, authority on whom the government and public health agencies could depend. While Huxley wrote to congratulate Tyndall on the "yeoman's service" that his letter to the *Times* would accomplish, he took the same course he had in the dispute with Tait, gently correcting another of his friend's errors, this time the misattribution of credit for the discovery of the typhoid bacteria.[65]

In the spring of 1875, the Pathological Society held meetings specifically to discuss the germ theory. Bastian was one of the primary speakers and continued to defend spontaneous generation. He accepted that bacteria appeared alongside diseases such as typhoid fever but maintained that they were the result, not the cause, of the disease. As evidence, he pointed to experiments in which putrefaction occurred in material even after it had been boiled, presumably killing any preexisting organisms.[66] Tyndall watched the Pathological Society's debates closely and by "reading a little between whiles" kept abreast of further new developments. He saw "with pleasure the hold which the Germ theory of disease has taken upon thoughtful minds," and declared dismissively: "one need not care to triumph over those who more or less stupidly withstood the theory."[67] By the autumn, he had designed a series of experiments that he felt would, finally and conclusively, disprove spontaneous generation. These were based on Tyndall's earlier work with his "floating matter" box, in which a beam of light was used to detect whether the air within had become optically pure; that is, there was no dust or other material suspended in it. The box was sealed and the light beam showed that the dust within settled after about three days, after becoming trapped on the box's glycerin-coated interior.[68] Tyndall thus convinced himself that he had created a space free of all existing germs, in which he would be able to detect any spontaneously appearing bacteria.

In early September 1875, he placed eight tubes of boiled urine uncovered within the box. Identical ones were placed outside. After a week, the tubes in the box showed no bacterial growth while those exposed to the air were overrun with microbes. Tyndall concluded that the bacteria came from contact with already existing germs carried by dust, and that sterilized material, if kept in a properly clean environment, would not develop any growth. He quickly began repeating the experiment with "infusions" of various materials including turnips, mutton, and haddock, all of which remained pure inside the box but developed bacterial growth outside. After nearly a month of these preliminary experiments, Tyndall wrote triumphantly to Huxley, declaring that he would "reduce to demonstration the practical correctness" of his friend's bold assertion, in his presidential address to the BAAS in 1870, that bacteria germs resided in the atmosphere "in myriads."[69] Huxley was invited to the RI to see the infusions for himself, and was evidently impressed, writing to Tyndall: "I can't tell you how delighted I was with the experiments." He even asked his fellow X Club member to bring to their next meeting, in October 1875, "a little bottle full of fluid containing the Bacteria &c. you have found developed in your infusions." The X Club met every month at St George's Hotel in Mayfair, and if Tyndall brought "a good characteristic specimen" of his putrefying urine-soaked infusions into these plush surroundings, then Huxley promised to "determine the forms with my own microscope and make drawings of them which you can use."[70] Huxley's offer, which was gratefully accepted, reveals Tyndall's enduring hesitance about using microscopy, and in fact, when later relating this incident in public, he exacerbated his own diffidence by suggesting that the request had come from himself rather than Huxley.[71]

Tyndall saw his painstaking experiments as an opportunity to resolve the bitter disputes of the previous five years and to decisively expose Bastian's specious bluster. He told Huxley: "I must live the life of an ascetic, to clear away this Bastian fog. It is amazing what audacity can do in England; and his audacity has powerfully influenced numbers of intelligent people. Without entering into controversy with him I hope to set him in his true light."[72] While the allusion to asceticism indicates the extreme discipline and delicacy needed to perform the experiments, it also suggests the almost holy role Tyndall saw himself as taking in relation to Bastian's imposture. Lorraine Daston and Peter Galison propose that such language was part of a "profoundly moralized vision, of self-command triumphing over the temptations and frailties of flesh and spirit" in which "nineteenth-century objectivity aspired to the self-discipline of saints."[73] But this overlap between ethical and scientific sensibilities is even more pertinent in the loaded language of "purity" and "contamination" in which the debate over spontaneous generation was conducted.[74] In the wake of the furious reaction to the Belfast Address, when Tyndall faced charges of blasphemy and even having created "poisoned waters" with his moral degeneracy, the overdetermined connotations of "purity" would have had a still greater resonance for him.[75]

This righteous, and at times quasi-religious, self-fashioning is evident in several of Tyndall's letters in this volume, and is particularly palpable in an intimate missive he wrote to Novikoff at precisely the time, in October 1875, that he was both working on boiled infusions and worrying about the gossip that their renewed friendship might provoke:

> At present I fear anything and everything that interferes with my work, which demands the most calm and concentrated attention. . . . At the present time I care little for any other danger—But my work must go on without interruption. Pitfalls and enemies are before me and around me, and I am resolved not to be tripped up by the one, nor overcome by the other. But failure can only be avoided by making every part of my investigation unassailable. This God helping me, as old Luther would say, I intend to do, and thus can only be done by solitary thought and severe experiment.[76]

Although this refers specifically to Tyndall's bacterial experiments, he had similarly declared of his presidential address to the BAAS in the previous summer, "I will go to Belfast as Luther did to Worms if necessary—and meet if requisite all the Devils in Hell there."[77] Tyndall had first invoked the sixteenth-century religious reformer several decades earlier, confiding in his private journal in 1850 that the "plea of Martin Luther must be mine 'I cannot otherwise—my God assist me!'"[78] A quarter of a century later, amid the manifold controversies of the tumultuous period covered in this volume, he cultivated such Lutheran austerity and rectitude more assiduously than ever.

Despite what he told Novikoff, Tyndall's experiments were not "done by solitary thought" alone. As he had with the proofs of the Belfast Address, he involved others in his cause, and sent sealed flasks of the boiled liquid to colleagues across the country. These were then opened and exposed to the atmosphere in different environmental conditions, with the resulting bacterial growth demonstrating the ubiquity of airborne germs. Through carefully crafted correspondence, Tyndall recruited the most prestigious members of the scientific elite to cultivate putrefaction and report the results, with Darwin, Hooker, and John Lubbock, as well as many others, all participating in his experiments and implicitly lending their authority to his claims. Without such impressive imprimaturs, Bastian responded with his own investigations that seemed to show bacteria appearing in sterilized environments in exactly the way Tyndall's experiments did not. Tyndall retorted that this was simply more evidence of Bastian's experimental incompetence. After all, dirty tools or cracked glassware could easily be the culprits, although these accusations became harder to maintain once Tyndall himself had trouble with heat-resistant microbes. This situation epitomizes Harry Collins's notion of the "experimenters' regress," in which a lack of consensus on matters of procedure—such as how to sterilize glassware—makes it impossible to resolve this kind of conflict by experiment alone.[79] As such, the resolution of the dispute relied as much on Tyndall's network of authorities, which he cultivated by writing strategically deferential letters to the likes of Darwin, as it did on clear empirical evidence.

Tyndall's correspondence also throws light on a curious incident during his experiments with boiled infusions that has taken on more significance in hindsight: the observation that the rapid die-off of bacterial colonies seemed to be correlated with the growth of *Penicillium* mold. Writing to Huxley in late October 1875, Tyndall described how the mold grew across the top of a tube of turnip infusion, and then how the "white muddiness" of the bacteria quickly fell to the bottom of the tube, leaving nearly clear liquid. He concluded that the "mould has cut off their oxygen and stifled them."[80] Huxley duplicated the phenomenon, but disputed his friend's explanation involving oxygen, though neither of them considered the observation particularly significant. They saw it as an interesting example of a struggle for existence between different kingdoms of nature, but that was all. What Tyndall and Huxley had almost stumbled upon in the brief informal letters sent between themselves was, of course, the antibiotic properties of penicillin, more than half a century before Alexander Fleming's own equally inadvertent discovery of them.[81]

As noted earlier, Tyndall's self-identification with Martin Luther's righteous intransigence helped him endure the numerous controversies, both old and new, that dominated the period of his career covered in this volume. The German reformer was an exemplar of Carlyle's conception of heroism, and Tyndall, in one of his letters to Carlyle, again alluded to him when affirming that he would continue to ignore the provocations of his opponents in the glacier dispute "if it rained Ruskins nine days running."[82] But it was not only in such public intellectual conflicts that Tyndall turned to Luther. In his very first extant letter to Louisa Hamilton, from September 1875, he promised to send her "a scrap of doggerel" that he had composed at Wartburg Castle in central Germany, which Hamilton had herself recently visited. This, Tyndall told her, would "show you how similar our thoughts & reflections have been."[83] Notably, the poem Tyndall promised to send was an homage to Luther, who had been imprisoned in the castle in the 1520s and "whose mighty voice," as Tyndall wrote, "like thunder, / Shook the proud battlements of Rome asunder."[84] Hamilton would soon become Tyndall's wife, only four months after this volume ends in October 1875, after which they enjoyed almost two decades of domestic contentment. While Luther's dauntless example enabled Tyndall to withstand what, in the two years spanned by this volume, were some of the fiercest and most testing controversies of his entire career, it also, as his tentative first letter to Louisa Hamilton indicates, helped initiate a more serene future.

Notes

1. George Bernard Shaw, *Man and Superman: A Comedy and a Philosophy* (London: Archibald Constable, 1903), 164.
2. See Nanna Katrine Lüders Kaalund, "A Frosty Disagreement: John Tyndall, James David Forbes, and the Early Formation of the X-Club," *Annals of Science* 74 (2017): 282–98. See also the introduction to *Tyndall Correspondence*, vol. 6.
3. *Fors Clavigera*, vol. 3 (1873), letter 34, 19.
4. Quoted in Jeffrey L. Spear, *Dreams of an English Eden: Ruskin and His Tradition in Social Criticism* (New York: Columbia University Press, 1984), 86; letter 4185.
5. Tyndall first read Carlyle's *On Heroes* in July 1847; see his journal entry for 18 July 1847 (RI MS JT/2/13a/231).
6. Letter 4602.
7. Letter 4186.
8. Letter 4187.
9. Letter 4188.
10. Letter 4190.
11. Letters 4191 and 4190.
12. Letter 4190.
13. Letter 4201.
14. Letter 4232.
15. Further cited in this volume as "Vehicle of Sound."
16. See Melinda Baldwin, "Tyndall and Stokes: Correspondence, Referee Reports, and the Physical Sciences in Victorian Britain," in *The Age of Scientific Naturalism: John Tyndall and His Contemporaries*, ed. Bernard Lightman and Michael Reidy (London: Pickering and Chatto, 2014), 171–86.
17. Letter 4295.
18. Letter 4355.
19. Letter 4363.
20. Letter 4189.
21. On Tyndall's previous refusals of the position, see letter 4203.

22. Letter 4189.
23. Letter 4202.
24. Letters 4202 and 4203.
25. Letter 4189.
26. Quoted in Ruth Barton, "John Tyndall, Pantheist: A Rereading of the Belfast Address," *Osiris*, 2nd ser., 3 (1987): 114; letter 4341.
27. Frank M. Turner, "John Tyndall and Victorian Scientific Naturalism," in *John Tyndall: Essays on a Natural Philosopher*, ed. W. H. Brock, N. D. McMillan, and R. C. Mollan (Dublin: Dublin Royal Society, 1981), 170; Bernard Lightman, "On Tyndall's Belfast Address, 1874," *BRANCH: Britain, Representation, and Nineteenth-Century History* (2011), https://www.branchcollective.org/?ps_articles=bernard-lightman-on-tyndalls-belfast-address-1874.
28. Letter 4417; "Editor's Table: Professor Tyndall's Address," *Popular Science Monthly* 5 (1874): 746.
29. See, in particular, Barton, "Tyndall, Pantheist"; David N. Livingstone, "Darwinism and Calvinism: The Belfast-Princeton Connection," *Isis* 83 (1992): 408–28; Stephen S. Kim, *John Tyndall's Transcendental Materialism and the Conflict between Religion and Science in Victorian England* (New York: Edwin Mellen, 1996), 43–57, 141–49, and 170–77; Bernard Lightman, "Scientists as Materialists in the Periodical Press: Tyndall's Belfast Address," in *Science Serialized: Representations of the Sciences in Nineteenth-Century Periodicals*, ed. Geoffrey Cantor and Sally Shuttleworth (Cambridge, MA: MIT Press, 2004), 199–237; Matthew Brown, "Darwin at Church: John Tyndall's Belfast Address," in *Evangelicals and Catholics in Nineteenth-Century Ireland*, ed. James H. Murphy (Dublin: Four Courts Press, 2005), 235–46; Gowan Dawson, *Darwin, Literature and Victorian Respectability* (Cambridge: Cambridge University Press, 2007), 82–115; Ursula DeYoung, *A Vision of Modern Science: John Tyndall and the Role of the Scientist in Victorian Culture* (New York: Palgrave Macmillan, 2011), 111–23; Lightman, "Tyndall's Belfast Address"; Ciaran Toal, "Preaching at the British Association for the Advancement of Science: Sermons, Secularization and the Rhetoric of Conflict in the 1870s," *British Journal for the History of Science* 45 (2012): 75–95; Matthew Stanley, *Huxley's Church and Maxwell's Demon: From Theistic Science to Naturalistic Science* (Chicago: University of Chicago Press, 2015), 93–95 and 189–92; *Ascent of John Tyndall*, 327–34; Diarmid Finnegan, "Revisiting Belfast: Tyndall, Science, and the Plurality of Place," in *Geographies of Knowledge: Science, Scale, and Spatiality in the Nineteenth Century*, ed. Robert J. Mayhew and Charles W. J. Withers (Baltimore: Johns Hopkins University Press, 2020), 58–86; and Ian Hesketh, "The Making of John Tyndall's Darwinian Revolution," *Annals of Science* 77 (2020): 524–48.
30. Letter 4271.
31. Letter 4392.
32. Letter 4189.
33. *Belfast Address*, v; letter 4372.
34. Letter 4379.
35. Letter 4370.
36. Letter 4379.
37. *Belfast Address*, 55, viii, and 54.
38. Letter 4363.
39. Letter 4366.
40. Letter 4379.
41. See letters 4386, 4387, and 4392.
42. Letters 4377 and 4392.
43. See, in particular, Barton, "Tyndall, Pantheist"; and Kim, *Transcendental Materialism*.
44. Letter 4394.
45. Letter 4593.

46. Suresh Chunder Bose, *The Life of Protap Chunder Mozoomdar*, 2 vols. (Calcutta: Nababidhan Trust, 1927), 1:46.
47. Letter 4467.
48. Letter 4422.
49. Letter 4446.
50. On Novikoff's affairs with Stead and Nordau, see Stewart J. Brown, *W. T. Stead: Nonconformist and Newspaper Prophet* (Oxford: Oxford University Press, 2019), 33–34; and Michael Stanislawski, *Zionism and the Fin de Siècle: Cosmopolitanism and Nationalism from Nordau to Jabotinsky* (Berkeley: University of California Press, 2001), 44–47.
51. Letter 4492.
52. Letter 4662.
53. See Dawson, *Darwin, Literature and Victorian Respectability*, 105–9.
54. Letter 4529.
55. Letter 4442. The petition in question was C. W. Stokes, *An Inquiry of the Home Secretary as to Whether Professor Tyndall Has Not Subjected Himself to the Penalty of Persons Expressing Blasphemous Opinions* (London: privately printed, 1874).
56. Letter 4493.
57. "The Catholic University of Ireland," *Tuam Herald*, 14 November 1874, 2.
58. Letter 4516.
59. See Dawson, *Darwin, Literature and Victorian Respectability*, 107–9.
60. On 21 January; see J. Tyndall, "On Dust and Disease," *Roy. Inst. Proc.* 6 (1870–72): 1–14.
61. James Strick, "Darwinism and the Origin of Life: The Role of H. C. Bastian in the British Spontaneous Generation Debates, 1868–1873," *Journal of the History of Biology* 32 (1999): 52.
62. See Strick, "Darwinism and the Origin of Life"; and James E. Strick, *Sparks of Life: Darwinism and the Victorian Debates over Spontaneous Generation* (Cambridge, MA: Harvard University Press, 2000).
63. Letter 4229.
64. Letter 4463.
65. Letter 4471; in an additional letter to the *Times* (letter 4472), Tyndall made the correction that Huxley urged.
66. See Henry Charlton Bastian, "An Address on the Germ Theory of Disease," *Lancet* 105 (1875): 501–9.
67. Letter 4620.
68. See *Ascent of John Tyndall*, 342–45.
69. Letter 4655. Huxley had made this assertion in "Address of Thomas Henry Huxley," *Brit. Assoc. Rep. 1870*, lxxxii.
70. Letter 4663.
71. In 1876 Tyndall recounted: "Doubtful of my skill as a microscopist I took specimens . . . and sent them to Prof. Huxley, with a request that he would be good enough to examine them." John Tyndall, "The Optical Deportment of the Atmosphere in Relation to the Phenomena of Putrefaction and Infection," *Phil. Trans.* 166 (1876): 36. On Tyndall's hesitance about microscopes, see *Ascent of John Tyndall*, 265, 343–44, and 367–68.
72. Letter 4667.
73. Lorraine Daston and Peter Galison, "The Image of Objectivity," *Representations* 40 (1992): 82–83.
74. See Strick, *Sparks of Life*, 176.
75. "Catholic University of Ireland," 2.
76. Letter 4668. On Tyndall's self-fashioning, see Bernard Lightman, "Fashioning the Victorian Man of Science: Tyndall's Shifting Strategies," *Journal of Dialectics of Nature* 38 (2015): 5–38.

77. Letter 4384.
78. 7 April 1850, Journal, RI MS/JT/2/13b/484-5.
79. H. M. Collins, *Changing Order: Replication and Induction in Scientific Practice* (Chicago: University of Chicago Press, 1992), 79 and passim. See also Strick, *Sparks of Life*, 175.
80. Letter 4676.
81. See James Friday, "A Microscopic Incident in a Monumental Struggle: Huxley and Antibiosis in 1875," *British Journal for the History of Science* 7 (1974): 61–71.
82. Letter 4185. Tyndall was alluding to Luther's purported comment when warned not to give a sermon in Leipzig because of the presence of the devoutly Catholic Duke George of Saxony: "If God call me to Leipzig, then will I go to Leipzig, though it rain Duke Georges nine days running."
83. Letter 4650.
84. Roland Jackson, Nicola Jackson, and Daniel Brown, eds., *The Poetry of John Tyndall* (London: UCL Press, 2020), 177.

EDITORIAL PRINCIPLES

Our aim is to include all letters to and from John Tyndall that are currently extant. Some of the letters are from published sources, such as letters written by Tyndall to newspapers or journals, but the majority are either the originals or typescripts produced by Louisa Tyndall from originals. One central editorial principle has been to transcribe, wherever possible, from original letters rather than Louisa's typescripts. This is because of Louisa's conscious and unconscious editorial interventions, a point discussed in more detail below. We have also aimed to reproduce, as accurately as possible, the text of the letters as they were written. Spelling mistakes that have appeared in the handwritten letters have therefore been preserved; mistakes appearing where only the typescript letter has survived are silently corrected, as in our judgment they are almost always typographical errors. But, in general, nonstandard spellings (e.g., the use of dialect) have been retained even when in a typescript-only letter (though there is the possibility that the spelling is Louisa's, not her husband's). Where the transcription is based on a typescript transcription by Louisa (or one of her assistants), this will be indicated by "LT Transcript Only."

We have included illustrations that are part of the letter. In cases where it was possible to reprint the original handwritten image, we have done so. In other cases, volume editors have reconstructed drawings because it was not possible to reprint an image from the handwritten letter (due, for example, to deterioration of the original document).

The letters are presented in chronological order of writing. Where more than one letter has the same date, letters from Tyndall (in alphabetical order of addressee) come first, followed by letters to Tyndall (in alphabetical order of writer), with the exception that if the order of writing can be determined, that order takes precedence. Where letters cannot be dated precisely they are placed at the latest likely date of writing. (In some cases if it is possible, but unlikely, that they were written later, notes indicate such possibilities.) To indicate the beginning of a new year, the date of that year has been inserted in a large font before the first letter from that year. Readers should note that although volumes are arranged in chronological order, letters that are discovered from years already covered in published volumes will appear in the final volume of the correspondence.

For non-English-language letters, both the transcription of the original and the English translation are presented. The transcription of the non-English letter appears first, followed by the English translation in a smaller font. Editorial notes are inserted in the English translation. Some foreign letters are in an original hand, while others exist only as transcriptions by Louisa or others who often did not have a full command of the language. In the latter case these letters contain frequent spelling, grammatical, and interpretation errors that do not seem to have been committed by the original author. In these typescript-only cases, and only in these cases, spelling and grammatical errors have been corrected, as have poor interpretations where a better word choice was obvious from the context.

There is a standard format for each letter. The first line lists the author, if it is to Tyndall, or the recipient, if it is from Tyndall; the date; and the letter number. The second line lists any information supplied by the writer in their salutation, such as the place, date, or time of day when the letter was written. (In some cases, though, where the writer put the date at the foot of the letter we have left it there in order to reproduce the text accurately.) The opening salutation, the text of the letter, and closing then follow. Then come the source and editorial notes, the latter indicated in the text by arabic numerals.

Editorial notes are intended to provide the contextualization necessary to understand the contents of each letter. There are basically six types of editorial notes. Many of them are informative notes on persons mentioned in the letters. If a person referred to in a letter is not described in a note, they have been mentioned more than twice in the volume and they will have an entry in the biographical register at the back of the volume. For the rest—those mentioned once or twice—biographical information will appear in the editorial note appended to the letter in which he or she is first mentioned. When there is a second reference to that person a note will refer back to the letter in which he or she was first mentioned. Biographical notes on well-known figures such as Faraday or Huxley stress what that person was doing at the time when the letter was written rather than providing a full overview of the person's life.

A second type of editorial note identifies allusions and quotations. In the case of quotations, where possible, the first edition of the work is cited unless a different edition is mentioned in the letter. If a quote is not in English a translation is given. Glosses on obscure places, abstruse words, and words used in an unfamiliar sense constitute the third type of editorial note. The fourth type provides information about the context of the letter, drawing on such sources as contemporary publications, scholarly sources, Tyndall's personal journals and field notes, and other letters in the *Tyndall Correspondence*.

The fifth type of editorial note is cross-references to other letters in the correspondence, referring to the letter number. The last type is information on significant textual changes that Tyndall made to the original that is thought to be relevant to the meaning of the letter.

There are a series of conventions for indicating when editors have given additional information about the text of a letter:

When Louisa Tyndall has made an annotation or marginal insertion that the editors have deemed important enough to retain, we have included her annotation in an editorial footnote.

When editors other than she have made insertions, they are indicated by unitalicized square brackets; for example, [word].

When certain words are ambiguous and we have had to conjecture their meaning, they are indicated by italicized square brackets; for example, *[*words*]*; when we have considered such words illegible the number of illegible words is indicated in full italicization and square brackets; for example, *[3 words illeg]*.

If a word or series of words are not present or have been destroyed—by inkblots, holes, or mold, for example—they are indicated as missing, italicized and in angular brackets; for example, *<3 words missing>*.

In extremely rare cases where it is clear that Louisa herself has intentionally destroyed a few words, they are indicated as excised, italicized and in angular brackets; for example, *<3 words excised>*.

Finally, a brief word about Louisa Tyndall. She was responsible for collecting together many of Tyndall's letters in order to write a biography of her late husband. This biography was not completed. It is important to recognize that Louisa was not always reliable as a transcriber. She often threw out the manuscript letter when she had made her transcription, but some original manuscripts remain, which indicate her editing processes. Like any transcriber, she was sometimes inaccurate, but corrections show that she proofread her transcriptions. However, she also edited Tyndall's writing to make his letters seem more consistent or grammatically correct. For example, she made punctuation more formal, corrected grammar, and, occasionally, corrected quotations. She consistently corrected Tyndall's usual (deliberate) lowercase for "god" to "God." Her typewriter was unable to insert the superscripts that Tyndall used in abbreviations, or the "&c" symbols that stood for an address at the foot of the letter. (Modern typefaces are unable to deal with all the variant scrawls that the "etc." symbol took and they are all formalized here to "&c.") Louisa also left out details that she considered too private or too salacious for public consumption. She equated postmark and date of writing and overlooked the time taken to write some letters. When she inferred dates from internal evidence, her reasoning was sometimes faulty. Similarly, her identifications of people are sometimes in error. Independent evidence has always been sought for her dates and identifications. Lastly, the patterns of punctuation differ between LT sources and manuscript sources. Tyndall himself used dashes more often than commas, semicolons, and stops, and he used dashes of many different lengths, which are all standardized here to em-dashes. It is often difficult to distinguish between very short dashes and commas or very short dashes and stops. Similarly, capitalization is often ambiguous; distinguishing *S*/*s*, for "science," for example, is a particular problem. Therefore, scholars for whom punctuation and capitalization are significant will need to consult the manuscripts. Users should also be aware that throughout this project the transcribers and editors have generally worked from scans and photographs rather than the original manuscripts.

NOTE ON MONEY

In the nineteenth century the currency used in England was the pound sterling, usually abbreviated to £ (or, occasionally, to *L*). A pound consisted of 20 shillings (abbreviated *s*) and each shilling contained 12 pennies (*d*). A guinea was 21 shillings; that is, £1 1s. A crown was 5 shillings (5s), and half a crown equaled two shillings and sixpence (2s 6d). The smallest coin was a farthing, one quarter of a penny (¼d).

Several forms of representation were in contemporary use. Sometimes the numbers for pounds, shillings, and pence were separated by dots, forward slashes, or spaces; thus, for example, 3 pounds, 4 shillings, and 5 pennies could be written as £3.4.5, £3/4/5, or £3 4 5. A zero could be represented by a dash; for example, 7 shillings was often written as 7/- or just 7/. A half crown was often written as 2/6.

US dollars sometimes feature in this volume. As today, the dollar (abbreviated as $) was a decimal currency, with one hundred cents (abbreviated as ¢) to one dollar. In 1873–75, one British pound was worth approximately 5.5 dollars, which represented a return to a fairly stable exchange rate after historic lows in the value of the dollar during the previous decade.

Other currencies mentioned in this volume are the French franc, Swiss franc, and German thaler and mark. The franc was the national currency in France from 1795 until 1999, when France adopted the Euro. In the mid-nineteenth century, the exchange rate between British and French currencies was approximately 25 francs per pound. A centime was a coin worth one-hundredth of a franc. The franc was adopted as the official currency of the Swiss cantons in 1850, and it was agreed in 1865 that that the Swiss franc have the same value as its French equivalent, which it did until the 1920s. German currency is more complicated, as the mark, introduced as the national currency in 1873 following the country's unification in 1871, was only gradually coming into general use. At the same time, the previous principal Germanic currency, the thaler (formally called a Vereinsthaler) remained legal tender, and in the mid-1870s was still in widespread use, only going out of circulation entirely in 1908. A thaler was a silver coin that had the equivalent worth of three marks, which were made from gold, and in 1873–75 was worth three British shillings (see letter 4419), with the exchange rate approximately 6.6 thalers per pound. The new mark was a decimal currency, with one hundred pfennigs (abbreviated as pf.) to one mark, although lower-denomination pfennig coins were evidently used in conjunction with thalers as well as marks (see letter 4678).

TIMELINE OF JOHN TYNDALL'S LIFE

Year	Event
c. 1822	John Tyndall born at Leighlinbridge, County Carlow, Ireland
c. 1836	Attends John Conwill's National School in Ballinabranna
1839	Begins employment as civil assistant with Ordnance Survey of Ireland in Carlow
1840	Becomes civil assistant in the Ordnance Survey Office at Youghal, Cork County
1841	First appearance in print (poem in *Carlow Sentinel* under pseudonym "W[alter] S[nooks]")
1842	Joins English Ordnance Survey in Preston
1843	Leads written criticism of Ordnance Survey in *Liverpool Mercury*; dismissed November
1844	Home in Ireland, mostly unemployed, until obtaining a position with the firm of Nevins and Lawton, Surveyors, of Manchester
1845–47	Works for Richard Carter, surveyor, of Halifax
1845	Meets Thomas Archer Hirst in Halifax
1847	Begins teaching mathematics and surveying at Queenwood College, Hampshire
1848–50	At University of Marburg, working for PhD with Robert Bunsen and Friedrich Stegmann
1850	Returns to England; meets Michael Faraday and William Francis and attends his first British Association meeting, before returning to Marburg
1851	Moves from Marburg to Berlin (April–June); then returns to Queenwood
1852	Elected fellow of the Royal Society
1852	Friendship developed with Thomas Henry Huxley
1853	First lecture at Royal Institution (RI)
1853	Meets Herbert Spencer
1853	Appointed professor of natural philosophy at RI
1853	Awarded Royal Medal for magnetic work; declines when controversy arises over award
1856	Meets Thomas Carlyle
1857	Climbs Mont Blanc for first time with Hirst
1859	Demonstrates existence of greenhouse gases and climatic implications
1859	Joins Government School of Mines as professor of natural philosophy
1860	*The Glaciers of the Alps* published
1860	Meets John Lubbock and his wife, Ellen Lubbock

1861	First ascent of the Weisshorn
1861	Delivers his first Christmas Lectures at RI
1862	*Mountaineering in 1861* published
1862	Begins to champion Mayer's priority in discovery of conservation of energy
1863	*Heat Considered as a Mode of Motion* published
1864	Awarded Royal Society's Rumford Medal
1864	First meeting of X Club
1865	*On Radiation* published
1865	Engages in public controversy over the efficacy of prayer
1866	Succeeds Michael Faraday as scientific adviser to Trinity House
1867	Faraday dies; Tyndall becomes superintendent of the RI
1867	*Sound: A Course of Eight Lectures* published
1868	Successfully climbs Matterhorn
1868	Discovers the cause of light scattering, to be known as "Tyndall Effect"
1868	*Faraday as a Discoverer* published
1868	"Scientific Materialism" lecture at the British Association
1868	Resigns from Royal School of Mines
1869	Joins Metaphysical Society
1870	"On the Scientific Use of the Imagination" lecture at British Association
1870	*Three Scientific Addresses* published
1870	*Researches on Diamagnetism and Magne-crystallic Action* published
1871	Meets Louis Pasteur for first time while in Paris
1871	*Fragments of Science* published
1871	*Hours of Exercise in the Alps* published
1871	*Light and Electricity* published
1872	Debates over how to measure the efficacy of prayer, aka the "Prayer-Gauge Debate"
1872	*Contributions to Molecular Physics in the Domain of Radiant Heat* published
1872	*The Forms of Water in Clouds and Rivers, Ice and Glaciers* published
1872–73	USA Lecture Tour
1873	*Six Lectures on Light* published
1874	As president of the BAAS, delivers the "Belfast Address"
1876	Marries Louisa Charlotte Hamilton
1877	Develops "Tyndallization" (discontinuous heating and cooling process to destroy heat-resistant spores)
1877	Tyndall and Louisa build summer cottage, Alp Lusgen, at Belalp, northern side of Valais, above Brig
1877	*Fermentation and Its Bearings on Phenomena of Disease* published
1879	*Fragments of Science*, which had gradually expanded, first published in two volumes
1881	*Essays on the Floating Matter of the Air* published
1885	Tyndall and Louisa build Hindhead retreat, Surrey Downs
1887	Resigns from RI
1890	Fight with Gladstone in the *Times* over Irish Home Rule
1892	Hirst dies
1892	*New Fragments* published
1893	Dies at Haslemere, Surrey, accidentally poisoned

TIMELINE OF EVENTS IN JOHN TYNDALL'S LIFE SPECIFIC TO VOLUME 14

Month and year	Biographical details	Notable social, political, and cultural events
October 1873	In Dover for acoustic experiments in relation to fog signaling (8th–1st November) John Ruskin's *Fors Clavigera* criticizes his conduct in glacier controversy; following advice from Thomas Carlyle and Thomas Huxley, does not respond Peter Guthrie Tait's article "Forbes and Dr. Tyndall" rejected by *Contemporary Review* *Nature* refuses to publish his letter to Tait on glacier controversy Considers giving up presidency of BAAS	Financial crisis in United States Girton College opens as Cambridge University's first women's college John Stuart Mill's *Autobiography* published
November 1873	Visits Dover for acoustic experiments in relation to fog signaling (21st–25th)	British troops invade Kingdom of Ashanti in West Africa SS *Ville du Havre* sinks in North Atlantic with loss of 226 lives Formation of Home Rule League in Ireland Herbert Spencer's *The Study of Sociology* published
December 1873	Visits Isle of Wight Conducts acoustic experiments in London's winter fog (9th–13th and 31st) Spends Christmas with Thomas Hirst at the Crystal Palace, Sydenham Begins Christmas Lectures at RI, "On the Motion and Sensation of Sound" (27th and 30th)	Battle of Bocairente (22nd) and Siege of Bilbao escalate Third Carlist War in Spain Death of Louis Agassiz (14th) Hermann Helmholtz awarded Copley medal by RS

Month and year	Biographical details	Notable social, political, and cultural events
January 1874	Completes Christmas Lectures at RI, "On the Motion and Sensation of Sound" (1st, 3rd, 6th, and 8th) Delivers paper, preliminary version of "On the Atmosphere as a Vehicle of Sound," at RS (15th) Delivers Friday Evening Discourse, "On the Acoustic Transparency and Opacity of the Atmosphere," at RI (16th) Publishes letter titled "England and America" in *Daily Telegraph* (19th) Publishes letter titled "Prof. Barrett and Sensitive Flames" in *Nature* (29th) Publishes "Acoustic Transparency and the Opacity of the Atmosphere (I)" in *Nature* (29th)	Marriage of Prince Alfred, second son of Queen Victoria, to Grand Duchess Marie, daughter of Tsar Alexander II (23rd) William Gladstone dissolves Parliament to call a general election (26th) British troops defeat the Ashanti at Battle of Amoafo (31st)
February 1874	Publishes "Acoustic Transparency and the Opacity of the Atmosphere (II)" in *Nature* (5th) Delivers paper, final version of "On the Atmosphere as a Vehicle of Sound," at RS (12th) Begins series of six lectures, titled "The Physical Properties of Gases and Liquids," at RI (17th and 24th) Hector Tyndale visits London	British troops burn Kumasi, tribal capital of Ashanti Agricultural workers' strike begins Gladstone defeated in general election and resigns as prime minister (17th) Benjamin Disraeli becomes prime minister for second time (20th) Verdict in the Tichborne case; Arthur Orton convicted of perjury Expedition of HMS *Challenger* provides evidence for existence of Antarctica
March 1874	Completes series of six lectures, titled "The Physical Properties of Gases and Liquids," at RI (3rd, 10th, 17th, and 24th) Meets Gladstone (4th) Revising "On the Atmosphere as a Vehicle of Sound" for publication, using Stokes's referee's report Travels to Yarmouth to inspect lighthouses at Haisborough Sands (31st)	Peace treaty with the Ashanti affirms freedom of movement for British Gold Coast traders Several battles and continued siege of Bilbao in Third Carlist War in Spain

Month and year	Biographical details	Notable social, political, and cultural events
April 1874	Inspects lighthouses at Haisborough Sands (1st–2nd) Visits Isle of Wight and has to be rescued from a cliff	Mining accident in Cheshire kills fifty-four people Total solar eclipse occurs (16th) Weather maps introduced in the *Times* First exhibition of Impressionist paintings begins in Paris
May 1874	Publication of new edition of Louis Rendu's *Theory of the Glaciers of Savoy* perpetuates glacier controversy; he considers how to respond "Report by Professor Tyndall to Trinity House on Experiments with Regard to Fog Signals" presented to Parliament (21st) Delivers paper, "Further Experiments on the Transmission of Sound," at RS (21st) Delivers paper, "On Some Recent Experiments with a Fireman's Respirator," at RS (21st) Undergoes innovative dental procedure using gold leaf Leaves for Switzerland to write Belfast Address (30th)	Tsar Alexander II makes state visit to Britain First Dorsland Trek of Boer settlers in South Africa begins Siege of Bilbao lifted Sailing vessel *British Admiral* sinks off the coast of Tasmania with loss of 79 lives Levi Strauss receives patent for blue jeans with copper rivets William Crookes's *Researches in the Phenomena of Spiritualism* published
June 1874	Publishes "Rendu and His Editors" in *Contemporary Review* Continues revising "On the Atmosphere as a Vehicle of Sound" for publication Writing Belfast Address; first installment sent to printers Unable to work suffering from toothache	Cavendish Laboratory opens at Cambridge University (16th) Second body discovered in the Battersea Mystery unsolved murders Rules of lawn tennis established Revised edition of Charles Darwin's *The Structure and Distribution of Coral Reefs* published
July 1874	Writing Belfast Address; second and third installments sent to printers Decides to purchase land at Bel Alp in Switzerland	Treaty of Fomena formally concludes war with the Ashanti Fire in Chicago destroys much of the city Sofya Kovalevskaya becomes first woman in Europe to be awarded a doctorate in mathematics First typewriter with cylindrical platen and qwerty keyboard launched in United States

Month and year	Biographical details	Notable social, political, and cultural events
August 1874	Proofs of Belfast Address sent to Hirst, Huxley, Darwin, and Spencer Returns from Switzerland (10th) Publishes "On the Atmosphere as a Vehicle of Sound" in *Phil. Trans.* Leaves for Belfast (17th) Attends BAAS meeting at Belfast (19th–26th), where he assumes presidency Delivers Presidential Address to BAAS (19th) Visits Irish lighthouses for experiments with gas illuminants	Public Worship Regulation Act prohibits ritualism in Church of England services Factory Act introduces fifty-six-hour week and restricts child labor Massacre of freed slaves by members of the White League in Coushatta, Louisiana George Meredith's *Beauchamp's Career* serialized in *Fortnightly Review*
September 1874	Begins friendship with Olga Novikoff Writes preface to Belfast Address (15th) Publishes *Belfast Address* Irish Catholic Church devotes three days to prayers for preservation from infidelity	Typhoon in Hong Kong and South China kills at least seventeen thousand people United States troops defeat Indigenous warriors in Texas at Battle of Palo Duro Canyon (28th) Foundation of Yorkshire College in Leeds Ernst Haeckel's *Anthropogenie* published
October 1874	Meets George Eliot (16th) Submits report to Trinity House on gas as illuminant in lighthouses (16th) Delivers lecture on "Crystalline and Molecular Forces" in Manchester (28th)	Riot in Northampton in support of Charles Bradlaugh's radical atheist candidacy in town's parliamentary by-election London School of Medicine for Women opens Treaty of Bern establishes the General Postal Union to coordinate international postal arrangements
November 1874	Publishes "On the Atmosphere in Relation to Fog-Signalling (I)" in *Contemporary Review* Publishes letters on typhoid fever in the *Times* (9th and 11th) Death of Frances Hooker (13th) Statue of Michael Faraday completed Writes "Preface to the Seventh Thousand" of *Belfast Address*; sends proofs to Novikoff	Massacre of freed slaves by members of the White League during elections in Alabama Sailing vessel *Cospatrick* sinks in south Atlantic with loss of 469 lives James Clerk Maxwell makes clay model visualizing the thermodynamic surface Gladstone's *The Vatican Decrees* published

Month and year	Biographical details	Notable social, political, and cultural events
November 1874	Considers libel action against *Tuam Herald* for printing accusation he is a "degenerate" Novikoff leaves London (23rd) and returns to Russia	Henry Sidgwick's *The Methods of Ethics* published Thomas Hardy's *Far from the Madding Crowd* published
December 1874	Publishes "On the Atmosphere in Relation to Fog-Signalling (II)" in *Contemporary Review* Visits Brighton Publishes *Belfast Address*, 7th thousand Spends Christmas with the Hookers at Kew	A military coup restores the Bourbon monarchy in Spain Transit of Venus observed (9th) Train crash at Shipton-on-Cherwell on the Great Western Railway kills thirty-four people (24th) John William Draper's *A History of the Conflict between Religion and Science* published Anthony Trollope's *Phineas Redux* published Louis Pasteur awarded Copley medal by RS
January 1875	Delivers paper, "On Acoustic Reversibility," at RS (4th) Delivers Friday Evening Discourse, "On Acoustic Reversibility," at RI (15th) Submits further report to Trinity House on gas as illuminant in lighthouses (21st)	Gladstone resigns leadership of Liberal Party (13th) Four-year-old Guangxu becomes emperor of China The Midland Railway abolishes second-class passenger category, leaving only first class and third class Death of Charles Kingsley (23rd) Henry James's *Roderick Hudson* serialized in *Atlantic Monthly*
February 1875	Begins series of six lectures, "Electricity," at RI (11th, 18th, and 25th)	SS *Gothenburg* sinks off the coast of Australia with loss of 102 lives Forced clearance of Indigenous tribes to reservations in Arizona Death of Charles Lyell (22nd)
March 1875	Completes series of six lectures titled "Electricity" at RI (4th, 11th, and 18th) Explosion in RI's chemical laboratory Visits Folkestone with Elizabeth Hamilton	US Congress passes Civil Rights Act prohibiting certain forms of racial discrimination Expedition of HMS *Challenger* measures deepest point of seabed on earth Première of *Trial by Jury*, Gilbert and Sullivan's first comic opera Wilkie Collins's *The Law and the Lady* published

Month and year	Biographical details	Notable social, political, and cultural events
April 1875	Visits Folkestone Death of his cousin John Tyndall (17th)	Franco-German war crisis begins Total solar eclipse occurs (6th) Hot air balloon disaster in Paris (15th) First woman appointed to Poor Law board of guardians in London Snooker invented in India
May 1875	Visits Isle of Wight Spends weekend at home of Joseph Whitworth in Derbyshire (29th–30th)	Britain cooperates with Russia to ease Franco-German war crisis International Bureau of Weights and Measures established in Paris to standardize international systems of measurement SS *Schiller* sinks off the coast of the Isles of Scilly with loss of 311 lives British Arctic Expedition to North Pole sets sail (29th) Carlyle's *The Early Kings of Norway* published
June 1875	Delivers Friday Evening Discourse at RI, "On Whitworth's Planes, Standard Measures, and Guns" (4th) Publishes second edition of *Six Lectures on Light* Publishes third edition of *Sound* Leaves for Switzerland (19th) Measuring glacial motion in the Alps and working on fifth edition of *Fragments of Science*	Alexander Graham Bell first transmits sound electronically (2nd) Serious flooding in France kills more than one thousand people Trollope's *The Way We Live Now* published
July 1875	Joined by the Hamiltons in the Alps Dispute with Joseph Henry over acoustics in relation to fog signaling	Darwin's *Insectivorous Plants* published Public Health Act consolidates local provisions into national code of sanitation Serious flooding in Britain and across Europe Uprising in Bosnia-Herzegovina against Ottoman Turkish rule
August 1875	Returns from Switzerland (*c.* 18th)	Artisans' and Labourers' Dwellings Improvement Act enables clearance of slum housing

Month and year	Biographical details	Notable social, political, and cultural events
August 1875	Attends BAAS meeting at Bristol (25th–1st September), where he relinquishes presidency	Sale of Food and Drugs Act prohibits adulteration Conspiracy and Protection of Property Act permits trade union picketing Society of True Afrikaners secretly formed in South Africa Chemical element gallium discovered spectroscopically by Paul-Émile Lecoq de Boisbaudran Matthew Webb becomes first person to swim English Channel
September 1875	Begins experimenting on airborne germs (10th) First extant letter to Louisa Hamilton (25th)	Joseph Bazalgette's sewerage system for London completed after thirty years First newspaper cartoon strip published in *New York Daily Graphic*
October 1875	Novikoff returns to London (*c.* 9th) Spends weekend with Darwin and Huxley at Down House (16th–17th) Sends friends boiled infusions from his experiments on spontaneous generation to test the effect of air in different locations Attends ceremony at the Guildhall at which Prince Leopold, Queen Victoria's youngest son, is awarded the freedom of the City of London (25th) Visits Shoeburyness for acoustic experiments in relation to fog signaling (26th–29th)	Ottoman Turkish government declares partial bankruptcy, placing its finances in hands of European creditors Theosophical Society founded by Helena Blavatsky Death of Charles Wheatstone (19th)

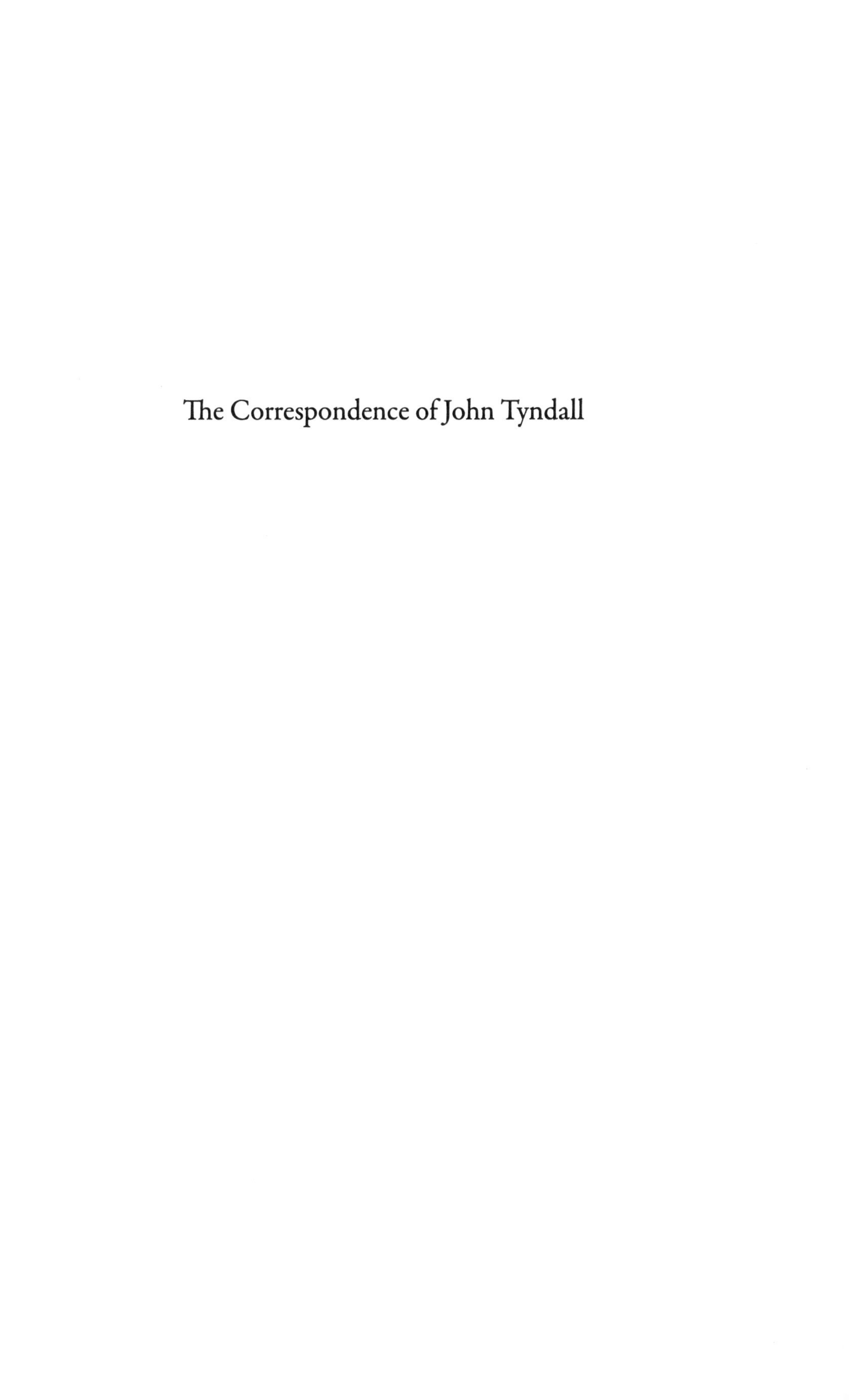

The Correspondence of John Tyndall

1873

To John Richard King[1] 1 October 1873 4182

Royal Institution of Great Britain | 1st. Oct[ober]. 1873

Dear Mr. King

Accept my best thanks for the book[2] and further friendly, and even cordial letter[3] which accompanied it

faithfully yours | John Tyndall

My memory of our day upon the Görner glacier[4] is also a bright one.

Private Collection #2483

---

1. *John Richard King*: John Richard King (1835–1907), vicar of St Peter-in-the-East, Oxford, and a member of the Alpine Club since 1862 (A. L. Mumm, *The Alpine Club Register, 1857–1863* (London: Edward Arnold, 1923), p. 178; 'Notes and News', *Oxford Magazine*, 27 (1909), p. 148).
2. *the book*: possibly R. W. Jelf, *The Thirty Nine Articles of the Church of England Explained*, ed. J. R. King (London: Rivingtons, 1873).
3. *friendly, and even cordial letter*: letter missing.
4. *our day upon the Görner glacier*: possibly 13 August 1858, of which Tyndall later recounted: 'One clergyman especially, with a clear complexion, good digestion, and bad lungs—of free, hearty, and genial manner—made himself extremely pleasant to us all. He appeared to bubble over with enjoyment, and with him and others on the morning of the 13th I walked to the Görner Grat, as it lay on the way to my work. We had a glorious prospect from the summit: indeed the assemblage of mountains, snow, and ice, here within view is perhaps without a rival in the world. I shouldered my axe, and saying "good-bye" moved away from my companions. "Are you going?" exclaimed the clergyman. "Give me one grasp of your hand before we part". This was the signal for a grasp all round; and the hearty human kindness which thus showed itself contributed that day to make my work pleasant to me' (*Glaciers of the Alps*, pp. 133–4). The Görner glacier, and the grat which overlooks it, are on the west side of the Monte Rosa massif close to Zermatt in the Swiss canton of Valais.

To Mr. Harrison[1] [*c.* 4 October 1873][2] 4183

The Lord Warden[3] | Dover Saturday

Dear M^r^. Harrison

I am here at Dover experimenting & I do not know whether I shall be in London on the 6th. Were I sure of being there it would give me pleasure to say 'yes' to your kind invitation[4] I shall probably be away.

Yours very truly | John Tyndall

Huntington Library HARR 1/111.65

---

1. *Mr. Harrison*: not identified.
2. *[c. 4 October 1873]*: Tyndall was in Dover at the same hotel on 9 October (see letter 4185) and refers to an invitation for 'the 6th', which was a Monday, so this letter is likely to have been written on Saturday 4 October.
3. *The Lord Warden*: a large hotel in Dover opened in 1853 and named after Arthur Wellesley (1769–1852), 1st Duke of Wellington, who had held the posts of Lord Warden and Admiral of the Cinque Ports and Constable of Dover.
4. *your kind invitation*: letter missing.

From Mary Egerton 4 October [1873][1] 4184

Mountfield Court, Robertsbridge, Hawkhurst.[2] | Oct[ob]er 4th

My dear Mr Tyndall,

If you are a bad hand at unravelling mysteries, you surely are an excellent hand at making them! Where lay the "Mystery" of my foolish tears, I am at a loss to discover! The case seems to me simple & clear enough, and you have condemned yourself to hear its explanation.

Firstly, then—they were not for jealousy of your friendship with L[ad]y. C. Hamilton,[3] as you seem to suppose!! I ought to be excessively affronted at the notion! but, like yourself, I love "Naturalness", and the frankness that led you to express the thought right out, goes far in my mind towards its condonation.[4] Still I am a little angry, but more vexed, because it shows you do not understand my friendship, or the true pleasure I feel in anything that gives you pleasure.

You have my free permission to fall in love with them all[5] if you like, provided you will tell me! Having once—from circumstances—had a claim upon your confidence,[6] I cannot shake off the desire to keep it, but as to your exclusive friendship!—such a conceited absurdity never entered my head!

But to return to the "mystery". You don't even now understand the intense love I have for those mountains and glaciers![7] And, from the time I first knew them, they have been so associated either with your books[8] or with yourself,[9] that, I confess, the scenery without you, seems to want its finishing touch—the sympathy which completes the delight. Was it then so very unnatural that the thought of a whole month spent among those scenes,[10] with glorious weather, and the very companionship I should have liked best, should have excited a little envy in my mind? The touch of melancholy was added by the consciousness that each year, what with increased expenses, and other ties, there seem more & more difficulties in the way of any foreign excursion; and when one does accomplish going, as last year, the weather may break up, and one may reach the spot one has longed for for years, only to find it an island in an ocean of clouds. Now then, do you understand? And will you promise not to think so meanly of me again?

You ought to see Heathfield[11] in the Autumn tints, the Beech trees must be lovely; it is as pretty a place for the size as I ever saw. (not the house!) I daren't say I should like you to come here too, or you will think I am jealous again! It is about 12 miles from here I think. However, if you have a spare day, you know you will be welcome.

I wish I had seen all your letters to Nature;[12] Could you not send me the rough copy?

We have had such a puzzle over the Rainbow![13] trying to draw the figure with the exact angles, not yours, for I don't quite understand how you mean it to be. Also, we reflected one in a looking glass to perfection, & there were the trees upside down, & the bow in the very identical spot! Of course a sheet of water at a distance would be different, but I certainly think your book[14] wants an explanatory paragraph.[15] I took it just like the man at Schaffhausen.[16]

Goodbye—Believe me ever | Y[ou][r] attached old friend | M F Egerton

What does "Dichroic"[17] mean I wonder! I sent for some ink I saw praised, & behold it turns out purple!

RI MS JT/1/E/53

---

1. *[1873]*: the year is established by reference to Tyndall's letter to *Nature* on the reflection of rainbows (see n. 12).
2. *Mountfield Court, Robertsbridge, Hawkhurst*: the Egerton estate in Sussex.
3. *L[ad]y. C. Hamilton*: Elizabeth Hamilton.
4. *condonation*: the pardoning or remission of an offence or fault, although in English law it referred specifically to the action of a husband or wife in the forgiving of matrimonial infidelity (*OED*).

5. *them all*: the family of Claud and Elizabeth Hamilton, especially their children Louisa, Emma, Mary, and Douglas. The eldest daughter, Louisa, would become Tyndall's wife in February 1876.
6. *from circumstances—had a claim upon your confidence*: possibly relating to the sudden death of Egerton's husband Edward only days after Tyndall had taken them on an expedition to the Oberaletsch Glacier in August 1869 (see n. 9).
7. *those mountains and glaciers*: of the Swiss Alps, particularly in the canton of Valais.
8. *your books*: presumably Tyndall's books on mountaineering and glaciers: *Glaciers of the Alps*; J. Tyndall, *Mountaineering in 1861: A Vacation Tour* (London: Longmans, Green, and Co., 1862); and *Hours of Exercise in the Alps*.
9. *with yourself*: early in their friendship, Tyndall had encouraged Egerton to visit the Swiss Alps, and in August 1869 had accidentally encountered her and her husband Edward in the village of St Niklaus, from where they had travelled, and climbed, together for several days. See *Ascent of John Tyndall*, p. 252.
10. *a whole month spent among those scenes*: Lady Hamilton and her children had joined Tyndall and Thomas Hirst in the Swiss Alps on 11 August 1873, and remained with them until early September (albeit with Hirst leaving on 20 August). Tyndall seems to have spent almost every day of this period with the Hamiltons, and was described by Hirst as 'the life and soul of the party. He takes charge of the party in a most efficient manner' (*Hirst Journals*, p. 1982). See also letters 4145, 4157, and 4164, *Tyndall Correspondence*, vol. 13.
11. *Heathfield*: Heathfield Park in Sussex. The house and estate were owned by the family of Charles Richard Blunt, who had purchased it in 1819, but was occupied by Claud Hamilton and his family.
12. *all your letters to Nature*: in the previous few weeks, Tyndall had published three letters in *Nature*: J. Tyndall, 'Tyndall and Tait', *Nature*, 8 (1873), p. 399; J. Tyndall, 'Reflection of the Rainbow', *Nature*, 8 (1873), pp. 432–3; and J. Tyndall, 'Tait and Tyndall', *Nature*, 8 (1873), p. 431.
13. *such a puzzle over the Rainbow*: Tyndall had recently discussed drawing a figure to see the reflection of rainbows in his letter to *Nature* 'Reflection of the Rainbow'.
14. *your book*: *Six Lectures on Light* (1873).
15. *an explanatory paragraph*: although not adding the explanation Egerton suggested, Tyndall did make some small amendments to the paragraph concerning what he called 'a question often asked—namely, whether a rainbow is ever seen reflected in water' in later editions of his book. See *Six Lectures on Light* (1875), p. 26.
16. *the man at Schaffhausen*: the author of a pseudonymous letter in *Nature*, signed 'Z. X. Y.', who tested—and found wanting—Tyndall's apparent claim in his *Six Lectures on Light* (1873) (on pp. 25–6) that a rainbow's reflection cannot be seen in the surface of a still lake (pp. 25–6). The letter was sent from Schaffhausen, a town in northern Switzerland ('Reflected Rainbows', *Nature*, 8 (1873), pp. 361-2). Peter Guthrie Tait, in one of his own missives to *Nature*, referred to this complaint against *Six Lectures on Light* (1873), and asked: 'What confidence can one have in the accuracy of any statement on a scientific

matter made by the author of it?' (P. G. Tait, 'Tyndall and Forbes', *Nature*, 8 (1873), pp. 381–2, on p. 382).

17. *"Dichroic"*: having the property of showing two different colours when viewed from different angles or in different circumstances (*OED*).

To Thomas Carlyle 9 October 1873 4185

The Lord Warden Hotel[1] | Dover 9th Oct[ober]. 1873

My dear General,[2]

I am here much exercised in the investigation of sound signals[3] intended for warning to the mariner in case of fogs, and from this place I thank you for the addition to the Life of Schiller[4] which, just before I quitted London, reached my hands.

It is a great pleasure to me to have this proof of your vitality—and of your continued interest in your grand old labours.

Last Sunday I was down at Sir John Lubbock's country residence.[5] While her husband and I were out for a ramble Lady Lubbock[6] entered into conversation with a Mr. Allen of Keston,[7] who I believe is in some sense the publisher of Mr. Ruskin.[8] From him she learned many things to my disadvantage, as set forth by Mr. Ruskin, until in the end she told Mr Allen he had better not talk to her in that wise regarding one she knew so well. To shorten matters I am told that Mr. Ruskin has made a bitter attack upon me apropos of Principal Forbes and the glaciers.[9]

This I did not merit at Mr. Ruskin's hands; for over and over again when in private conversation he has been subjected to far graver attacks[10] than he could possibly direct against me, I have defended him.

But I would scorn to influence Mr. Ruskin or any other assailant by considerations such as this. He attacks me on a subject in connexion with which I have only sought to do the right—in connexion with which I have laboured less to exalt myself than to see justice done to others.[11] And this—God helping me—shall be done if it rained Ruskins nine days running.[12]

Always yours affectionately | John Tyndall

He has not had the good manners to send me a copy of his attack.[13]

RI MS JT/1/T/151
RI MS JT/1/TYP/1/192–3

1. *The Lord Warden Hotel*: see letter 4183, n. 3.
2. *My dear General*: Tyndall was deeply influenced by Carlyle's writing, much of which he had read in the 1840s. He considered Carlyle to be his leader, or 'General', a sobriquet

which reflected Carlyle's interest in the values of heroism and martial discipline. Tyndall also used other honorifics relating to Carlyle's work, such as 'Chieftain' (see letter 4602). Despite the familiar greeting, this is the first extant letter from Tyndall to Carlyle since June 1872 (see letter 3775, *Tyndall Correspondence*, vol. 13), and in September 1873 he told a mutual friend 'I have not seen our grand old friend for a long time, the reason in part being that I am frequently so tired at night that I shrink from afflicting him with my dullness' (letter 4181, *Tyndall Correspondence*, vol. 13).

3. *the investigation of sound signals*: Tyndall was engaged in a series of acoustic experiments, as Scientific Adviser for Trinity House (see letter 4187, n. 2), to determine the best method of signaling ships in fog, working from the South Foreland Lighthouse near Dover. In this capacity, he was engaged in testing various methods of warning ships in fog, such as brass horns, steam whistles, and firing cannons. He became particularly interested in the way different climatic conditions altered the audibility of these sounds, which produced conflicting evidence for the efficacy of the different methods (see *Ascent of John Tyndall*, pp. 318–20). For this second series of experiments in October 1873 he used a syren acquired from the United States with the help of Joseph Henry (see letter 4201, n. 2). He reported the results of the experiments to the House of Commons (see 'Report by Professor Tyndall'), and published a paper based on that report (see 'Vehicle of Sound'). The previous year, Tyndall had conducted the first series of experiments at the South Foreland Lighthouse from May to July (see letter 4152, *Tyndall Correspondence*, vol. 13).
4. *the addition to the Life of Schiller*: T. Carlyle, *The Life of Friedrich Schiller: Comprehending an Examination of His Works*, 3rd edn (London: Chapman and Hall, 1873), which contained a 'Supplement of 1872' (pp. 240–78), responding to accounts of the German poet and philosopher (1759–1805) published since the first (1825) and second (1845) editions of Carlyle's book.
5. *Sir John Lubbock's country residence*: High Elms, in Farnborough, Kent, neighbouring Charles Darwin's Down House.
6. *Lady Lubbock*: Ellen Lubbock.
7. *a Mr. Allen of Keston*: George Allen (1832–1907), who lived in the village of Keston in Kent.
8. *in some sense the publisher of M^r. Ruskin*: in 1871 the art critic John Ruskin had established Allen, who was his assistant, as a publisher for his own work, operating from Allen's cottage in Keston.
9. *Principal Forbes and the glaciers*: Tyndall had been embroiled in an acrimonious dispute with James David Forbes regarding the motion of glaciers in the late 1850s (see the Introduction to *Tyndall Correspondence*, vol. 6). The controversy was reignited by Tyndall in *The Forms of Water in Clouds and Rivers, Ice and Glaciers* (London: Henry S. King, 1872), and then intensified by the publication of the *Life of Forbes*, co-edited by John Campbell Shairp, Peter Guthrie Tait, and Anthony Adams-Reilly. Ruskin, who took a keen interest in geology, had corresponded with Forbes and continued to champion his cause against Tyndall.

10. *far graver attacks*: probably a reference to allegations concerning Ruskin's private life. His marriage to Euphemia Gray was rumoured to have been unconsummated and was annulled in 1854. There was also gossip about Ruskin's infatuation with his former student, Rose La Touche, almost thirty years his junior. See also letter 4218, n. 5.
11. *a subject in connexion with . . . justice done to others*: the originality of Forbes's research had been called into question, with Tyndall suggesting that Forbes had plagiarised the work of French geologist Louis Rendu. See J. Hollier and A. Hollier, 'The Glacier Theory of Louis Rendu (1789–1859) and the Forbes-Tyndall Controversy', *Earth Sciences History*, 35 (2016), pp. 346–53.
12. *And this . . . nine days running*: an allusion to the German religious reformer Martin Luther's purported comment when warned not to give a sermon in the city of Leipzig because of the presence of the devoutly Catholic Duke George of Saxony: 'If God call me to Leipzig, then will I go to Leipzig, though it rain Duke Georges nine days running'. Carlyle had himself quoted the same line in several of his works.
13. *a copy of his attack*: letter 34 of *Fors Clavigera* issued in October 1873, in which John Ruskin attacked Tyndall at length. It contained a highly critical account of Tyndall's theories of glacial motion and his dispute with James David Forbes (a friend of Ruskin's) and contended that 'all the ingenuity and plausibility of Professor Tyndall have been employed, since the death of Forbes, to diminish the lustre of his discovery' (*Fors Clavigera*, vol. 3 (1873), letter 34, pp. 1–32, on p. 24). Ruskin's *Fors Clavigera* was a series of ninety-six pamphlets, published monthly by George Allen (see n. 7) between 1871 and 1884. Addressing a wide range of topics, these 'letters' were Ruskin's attempt to communicate his social and moral values to the working classes.

From William Frederick Pollock 13 October 1873 4186

The Athenaeum Club, Pall Mall[1] | Monday | 13 / Oct[ober] / 73

My dear Tyndall

I got your letter[2] this morning—& came here to see the *Fors Clavigera* for October[3]—to which you have called my attention. I have carefully read the passages which have so strangely found their way into such a place, upon glacier theories, & in which your name is introduced—I would most decidedly counsel you to take no notice whatever of Ruskin's attack.[4] The tone is offensive, & the statements, many of them untrue—but I see nothing requiring a personal vindication on your part—nothing which would induce the Court of Queen's Bench to grant a criminal information[5] if the matter was before them. & it comes from a man whose sanity may be doubted[6]—who is a kind of licensed assesser of whomever and whatever happens to strike his wayward fancy the wrong way—& which appears in an obscure pamphlet—with a very small circulation—

Looking at the question from the scientific point of view, it can hardly be thought justifiable to renew a conflict which (but from what I have recently heard, but not seen, has come from Tait)[7] I thought was completely settled as between Forbes & yourself—& with such a champion as Ruskin—You may have to deal with Tait—but as I have not yet had time to read what I am told he has been saying[8]—I cannot know whether that will be certainly necessary—& in that case you had clearly better not be carrying on two controversies at once—one with an opponent of good scientific pretension—the other with a man of none—

At any rate wait & see whether Ruskin's remarks are reprinted in any journal of scientific importance or large circulation—& wait until his next remarks appear, as I observe that he announces his attention is revision to the subject.

My wife[9] has not seen *Fors Clavigera*—but when I read your letter to her at breakfast, she at once said—"Tell him to have nothing to do with Ruskin"—& I felt she must be right—

And so for the present at least, I give you the counsel, so easy to give, & often so hard to take—Do nothing—

We returned on Friday night—We had a fortnight at Fontainebleau[10] which proved not a bad place for a wet fortnight. The soil is always dry—& the beauties of the Forest are unimaginable—& we contrived to find amusement in that place without amusements—of which you will hear in due time—Then came a glorious three weeks of fine weather in Paris—& a great deal of play-going & some society—Delaunay of the Français[11]—is a man of rare good qualities in private life—as well as the first of living actors—& we saw a great deal of him—off the stage—& something of him on. I hope he will be staying with us next year—when you will see him—& I know will like him—Fred & his bride[12] are in the same hotel with us at Paris for ten days—Maurice[13] was with us all the time & did some drawing from the round in the Louvre—Walter[14] joined us for ten days, & came home with us—We are now settled in London—except that I go on Thursday for a couple of days to Cheltenham to see M^rs^. Macready[15] on business.

Yours ever | W. F. Pollock

RI MS JT/1/P/241
RI MS JT/1/TYP/6/2147–8

---

1. *Athenaeum Club, Pall Mall*: the Athenaeum Club, a private members' club located at 107 Pall Mall in London, particularly for those with intellectual interests, and accomplishments in art, science, or literature. The club was founded in 1824, at the instigation of John Wilson Croker. Tyndall became a member in 1860, Thomas Huxley in 1858, Joseph

Hooker in 1851, and Thomas Hirst in 1866. Tyndall was frequently there, often for meals, as it was conveniently close to both the RI and the RS. For Tyndall's election to the club, see *Ascent of John Tyndall*, pp. 142-3. On the Club in general, see M. Wheeler, *The Athenaeum: More Than Just Another London Club* (New Haven: Yale University Press, 2020).

2. *your letter*: letter missing.
3. *the Fors Clavigera for October*: see letter 4185, n. 13. The Athenaeum Club took copies of the monthly parts of *Fors Clavigera* for its library.
4. *Ruskin's attack*: see letter 4185, n. 13.
5. *the Court of Queen's Bench to grant a criminal information*: a formal criminal charge made by the senior court of common law in England, which would be abolished in 1875. The charge Pollock, who was a lawyer, had in mind was clearly that of libel.
6. *a man whose sanity may be doubted*: rumours of Ruskin's mental fragility were fueled by his troubled private life (see letter 4185, n. 10), and in 1878 he would experience the first of several nervous breakdowns.
7. *a conflict which . . . has come from Tait*: Peter Guthrie Tait, who succeeded Forbes as professor of natural philosophy at the University of Edinburgh, had instigated the publication of the *Life of Forbes*, which intensified the Tyndall-Forbes controversy (see letter 4185).
8. *read what I am told he has been saying*: Pollock presumably meant chapters 14 and 15 of *Life of Forbes* (pp. 457–520), which were authored by Tait and contained numerous criticisms of Tyndall's position.
9. *My wife*: Juliet Pollock.
10. *Fontainebleau*: a commune close to Paris known for its spectacular forest and historic chateau.
11. *Delaunay of the Français*: Louis-Arsène Delaunay (1826–1903), a leading French actor with the Comédie-Française, a state-run theatre company in Paris.
12. *Fred and his bride*: Pollock's eldest son, Frederick Pollock, who had married Georgina Harriet Pollock (née Deffell, 1846–1935) in August 1873.
13. *Maurice*: Maurice Emilius Pollock (1857–1932), the third and youngest son of Juliet and William Frederick Pollock. He became a sculptor and married Lydia Helen Roberts (1853–84) in 1880, and Mabel Mary McPherson (1860–1937) in 1889. "Moss" was his family's nickname for him.
14. *Walter*: Walter Herries Pollock (1850–1926), the second son of Juliet and William Frederick Pollock. He was educated at Eton College (1862–65) and was admitted to Trinity College, Cambridge, where he obtained a BA (1871) and MA (1875). He was admitted to the inner temple in 1870 and called to the bar in 1873. Pollock became a poet and journalist, and was editor of the *Saturday Review* (1884–94). He married Emma Jane Pipon in 1876.
15. *M^{rs}. Macready*: Cecile Louise Frederica Macready (née Spencer, 1826–1908), widow of the actor and theatre manager William Macready, who had died on 27 April. Macready was a close friend of the Pollocks, and Pollock was one of his executors, so the business presumably concerned legal matters relating to his estate. Juliet Pollock published *Macready As I Knew Him* (London: Remington and Co., 1885).

To Thomas Archer Hirst 14 October [1873][1] 4187

Trinitas in Unitate[2] | "GALATEA"[3] 14th Oct[ober] | Dover Harbour

My dear Tom

I send you a note from Pollock,[4] which seems sensible. Lady Lubbock[5] will I dare say have written to you: she was exasperated with Ruskin, and was in favour of instant war with him.[6] I shall be staying at the Lord Warden Hotel[7] for some days still. My head has not been in good condition of late, but this beautiful day, and a wholesome luncheon have done it much good. I dine on board the *Galatea* today: the fare is wholesome and the company pleasant.

Ever Dear Tom | Yours affec[tionate]ly | John Tyndall

RI MS JT/1/T/928
RI MS JT/1/HTYP/623

---

1. *[1873]*: the year is established by the reference to Dover, where Tyndall was engaged in a series of acoustic experiments to determine the best method of signaling ships in fog, working from the South Foreland Lighthouse near Dover (see letter 4185, n. 3).
2. *Trinitas in Unitate*: the paper used by Tyndall was headed with the coat of arms of Trinity House, including this Latin motto. It translates as 'Three in One', a reference to the Holy Trinity, from which Trinity House took its name. While Tyndall had assisted Michael Faraday since May 1866 in lighthouse matters (see letter 2423, *Tyndall Correspondence*, vol. 9), in July 1867 he officially took over from Faraday as Scientific Adviser to the Board of Trade (a British government department concerned with making policy and instituting regulations regarding economic activity in Britain and its empire) and Trinity House (the lighthouse authority for England and Wales, established by Royal Charter in 1514) (see letters 2613, 2625, and 2829, *Tyndall Correspondence*, vol. 10). In this role, Tyndall investigated and gave recommendations to Trinity House on matters of lighthouse illumination, and later, sound signaling. He resigned from the position in 1883. On Tyndall and his role with the Board of Trade and Trinity House, see R. M. MacLeod, 'Science and Government in Victorian England: Lighthouse Illumination and the Board of Trade, 1866–1886', *Isis*, 60 (1969), pp. 4–38; S. Courtney, '"A Very Diadem of Light": Exhibitions in Victorian London, the Parliamentary Light and the Shaping of the Trinity House Lighthouses', *British Journal for the History of Science*, 50 (2017), pp. 249–65; and *Ascent of John Tyndall*, pp. 282–5, 316–7, 345–6, 389, 406–13, 420–3, 433–4 and 437.
3. *GALATEA*: THV *Galatea* was a paddle yacht that, having been built in 1868, served Trinity House until 1895. Its name was also part of the printed letterhead on the paper used for this letter. Tyndall conducted his acoustic experiments in the English Channel, close to Dover (see letter 4185, n, 3).

4. *a note from Pollock*: letter 4186.
5. *Lady Lubbock*: Ellen Lubbock.
6. *instant war with him*: see letter 4185.
7. *the Lord Warden Hotel*: see letter 4183, n. 3.

From Thomas Carlyle 16 October 1873 4188

5 Cheyne Row, Chelsea. | 16 October 1873.

Dear Tyndall,

I never yet hear whether you are come home from your Dover investigations,[1] but as I still see nothing of you,[2] fancy that you are not yet in my neighbourhood. The very day before your letter[3] came to me, I had seen, and not till then, the strange wandering, meandering number of Ruskin's *Fors Clavigera*,[4] which contains his ill-natured remarks on your Alpine-Glacier investigations; which naturally seem to me very little called for in that place or any other. But throughout that number, R[uskin]. seems in a specially prickly humour ready to strike out right and left in a more than usually violent manner upon everything that comes above his horizon. Witness that of Sir Charles Adderley,[5] "boiling and eating his mother" instead of burying her;[6] compared with which all he says of you is almost an amenity[7] rather!—Ruskin himself I have not seen for many months, nor heard of except in the most transient way.

As to his criticisms on you and your ideas of glacial movement, my clear opinion is that you should take no notice of him whatever; such being the one dignified, wise and proper course on your part. To no mortal that knows you will it for a moment be credible that you had any private grudge against Professor Forbes's "fame"[8] in that respect, or private favour towards Agassiz[9] or in fact the least purpose in your mind, but that of clearing up the subject and rendering it visible to the bottom, as everybody knows to have been your invariable custom hitherto. Be silent, then, and don't weaken your cause by asking everybody or anybody whether that is not the fact. And go on silently doing the feat, what yet remains of it, as if there were no Ruskin there,—a nine days of raining Ruskins[10] not having yet set in.

I have been here for above a month; visibly not a little smashed and worsened by my painful pilgrimage to northward.[11] Only within the last few days have I begun to feel as if slowly getting back to my old average instead of the better I had counted on. Evidently my time for touring is gone by; "stay where thou art" the rule for me henceforth. Let me hear from you, or still better see you when you return.

Ever gratefully and affectionately yours, | T. Carlyle.

RI MS JT/1/TYP/1/207–8
LT Typescript Only

---

1. *your Dover investigations*: Tyndall was engaged in a series of acoustic experiments to determine the best method of signaling ships in fog, working from the South Foreland Lighthouse near Dover (see letter 4185, n. 3). As Carlyle intuits, Tyndall was away when this letter was written, staying in Dover until the beginning of November.
2. *still see nothing of you*: see letter 4185, n. 2.
3. *your letter*: letter 4185.
4. *the strange wandering, meandering number of Ruskin's Fors Clavigera*: see letter 4185, n. 13.
5. *Sir Charles Adderley*: Charles Bowyer Adderly (1814–1905), a Conservative politician who held various ministerial posts after first becoming an MP in 1841 (*ODNB*).
6. *"boiling and eating his mother" instead of burying her*: at the start of letter 34 of *Fors Clavigera*, Ruskin stated that 'in a recent debate on the treatment of Canada, Sir C. Adderley deprecates the continuance of a debate on a question "purely sentimental". I doubt if Sir C. Adderley knew in the least what was meant by a sentimental question. It is a purely "sentimental question", for instance, whether Sir C. Adderley shall, or shall not, eat his mother, instead of burying her' (*Fors Clavigera*, vol. 3 (1873), letter 34, pp. 1–32, on p. 4). Ruskin was attacking a parliamentary speech made by Adderley regarding Britain's treatment of Canada, which had recently become a single confederation under the Constitution Act of 1867. Adderley, who had played an important part in passing this Act as Undersecretary of State for the Colonies, had suggested that debate over Britain's relationship to its colony was purely sentimental. Ruskin's comments allude to the Biblical narrative of the three-year Siege of Samaria, in which the besieged occupants of the city resorted to cannibalism, with a mother boiling and eating her son (Kings 6:29).
7. *an amenity*: a pleasantry or civility (*OED*).
8. *any private grudge against Professor Forbes's "fame"*: Ruskin claimed of Tyndall: 'To diminish the lustre, observe, is the fatallest wrong; by diminishing its distinctness . . .he still denies, as far as he dares, the essential point of Forbes' discovery; denies it interrogatively, leaving the reader to consider the whole subject as yet open to discussion,—only to be conclusively determined by—Professor Tyndall and his friends' (*Fors Clavigera*, p. 24).
9. *private favour towards Agassiz*: Tyndall had supported Louis Agassiz's claim to priority over Forbes in the publication of work relating to glacial motion. Ruskin referred to the 'spite of Agassiz and his friends' in their treatment of Forbes (*Fors Clavigera*, p. 23).
10. *a nine days of raining Ruskins*: see letter 4185, n. 12.
11. *my painful pilgrimage to northward*: Carlyle had travelled to Scotland in early September to visit the grave of his wife Jane Welsh Carlyle (1801–66) in St Mary's Church, Haddington. After a fortnight away, he recorded that he was 'much broken by the travelling hardships &c' (C. R. Sanders, et al. (eds), *The Collected Letters of Thomas and Jane Welsh Carlyle*, 49 vols (Durham, NC: Duke University Press, 1970–), vol. 48, p. 221). On the death of Jane, see letter 2414, n. 4, *Tyndall Correspondence*, vol. 9.

From Joseph Dalton Hooker 16 October [1873][1] 4189

16th October

My dear Tyndall

I have just returned from Coombebank,[2] having had several earnest conversations with Spottiswoode on the subject of your giving up the Presidency of the Belfast Association.[3]

To this he most strongly objects, on the ground of the consequent embarrassment of the Council, and the fear of their motives for accepting your resignation being misunderstood.[4]

This view is shared by Sharpey[5] and Huxley, and I am convinced of its force. Unfortunately it does not meet your difficulty, which arises from the nature and amount of the Govt work thrust upon you since you accepted the Presidentship, and which you fear may prove a really insuperable obstacle to your undertaking the duties of the latter.

Now it is conceivable that the Govt work may not prove so arduous and exhausting as you anticipate; and whether or no, you should give the Association the benefit of a doubt on this score, and not resign until you feel compelled to do so.

Why not then banish all thought of the Association from your mind, keep your own counsel, get on with your Govt work, and if by Midsummer you find that it is really impossible to prepare an address,[6] then resign?

And as to the Address, make up your mind to choose a scientific subject, that will demand no great mental strain,[7] remembering that it is not for the address that men are chosen Presidents, and that the labour so often bestowed on these Addresses, frightens many a good man from the Chair.

I think that this course would relieve both your mind and the Association of a load of anxiety, and I hope meet your case; if not let us meet again and talk the matter over.

Most sincerely yours | J. D. Hooker.

RI MS JT/1/TYP/8/2721
LT Typescript Only

---

1. *[1873]*: the year is established by reference to the BAAS's presidency (see n. 3).
2. *Coombebank*: William Spottiswoode's house, Coombe Bank, near Sevenoaks, Kent.
3. *your giving up the Presidency of the Belfast Association*: due to the demands of government work, particularly his experiments relating to signaling to ships in fog, Tyndall was considering refusing the presidency of the BAAS, which was to meet in August 1874 in Belfast.

4. *the fear of their motives . . . being misunderstood*: the BAAS's council were concerned their acceptance of Tyndall's resignation might be construed as their bowing to pressure from Peter Guthrie Tait, who had recently revived the controversy over theories of glacial motion between Tyndall and James David Forbes (see letter 4186, n. 7).
5. *Sharpey*: William Sharpey (1802–80), a physiologist who had been secretary of the RS from 1853–72 (*ODNB*).
6. *prepare an address*: it was a traditional responsibility of the incoming president to give an inaugural public address, usually on a topic relating to their field of expertise, at the BAAS's peripatetic annual meeting held at the end of the summer.
7. *make up your mind . . . no great mental strain*: Tyndall would ignore Hooker's advice and spend more than two months—from early June to mid-August 1874—preparing the address, which, covering two thousand years of intellectual history, required extensive reading.

To Thomas Carlyle 17 October 1873 4190

The Lord Warden Hotel[1] | Dover, 17th, Oct[ober], 1873.

My dear Friend,

Be it as thou wilt!—I shall do as you advise,[2] because I feel that your advice is the best and noblest that can be given to me. I am still here,[3] afloat every day from morning till night, and doing good work: Your letter[4] will help me to dislodge all minor irritations from my mind and to bend with wholeness of purpose to the labours I have in hand. Immediately after my arrival I will call to see you. I think you must not give up the notion of quitting London[5] from time to time. I believe the chief problem in your case—and indeed in mine—is rather one of meat and drink[6] than of any thing else. Were our medical men what they ought to be, true "man-healers," I am persuaded that an aliment might be found that would make you practically young again. You are organically immensely strong, save as regards that one point of food. On that score your stomach is fastidious, and hence the need of assiduous search and careful observation to discover the food which it can readily dissolve.

But I must not afflict you with a medical lecture.

I have looked again at your letter. Best thanks for it dear friend—it is very precious to me

Yours ever affectionately | John Tyndall

RI MS JT/1/TYP/1/193
LT Typescript Only

1. *The Lord Warden Hotel*: see letter 4183, n. 3.
2. *do as you advise*: regarding John Ruskin's recently published attacks on Tyndall. Carlyle advised 'take no notice of him whatever' (letter 4188), although Tyndall did not actually follow this advice.
3. *still here*: Tyndall was engaged in a series of acoustic experiments to determine the best method of signaling ships in fog, working from the South Foreland Lighthouse near Dover (see letter 4185, n. 3).
4. *Your letter*: letter 4188.
5. *quitting London*: Carlyle had recently returned to his home at 5 Cheyne Row in Chelsea after a journey to Scotland (see letter 4188, n. 11). Suffering from the privations of travel, Carlyle told Tyndall 'my time for touring is gone by' (letter 4188), and he reflected in his journal that he had 'hurried straight home with the mournful tho*t* "*Empty* House; empty but *quiet*, why sh*d* I ever quit thee ag*n*?"' (C. R. Sanders, et al. (eds), *The Collected Letters of Thomas and Jane Welsh Carlyle*, 49 vols (Durham, NC: Duke University Press, 1970–), vol. 48, p. 221).
6. *the chief problem . . . one of meat and drink*: Carlyle suffered from a chronic, painful stomach condition.

To Thomas Archer Hirst 17 October [1873][1] 4191

17th Oct[ober].

My dear Tom

I have a very sweet and noble letter from Carlyle[2] in which he gives me his clear opinion that I ought to take no manner of notice of Ruskin.[3]

Yours affe[ctionate]ly | John Tyndall

RI MS JT/1/T/930
RI MS JT/1/HTYP/623

1. *[1873]*: the year is established by reference to Thomas Carlyle's letter (letter 4188), which is dated 16 October 1873.
2. *a very sweet and noble letter from Carlyle*: letter 4188.
3. *take no manner of notice of Ruskin*: regarding John Ruskin's recently published attacks on Tyndall and the controversy over glacial movement (see letter 4188).

From Hermann Helmholtz 17 October 1873 4192

BERLIN, 17 October, 1873.

Lieber Freund,

Sie stellen mir in Ihrem Briefe vom 25ten Septb. eine Frage wegen eines 16ten Capitels von „On Heat“. Die Ausgaben des Buches, die ich habe, haben nur 13 und 15 Capitel; also muss dies ein neues sein, was ich noch nicht kenne. In den Auseinandersetzungen, welche die mir bekannten Ausgaben Ihres Buches erhalten, finde ich nichts Anstössiges gegen Magnus. Was den Streit selbst betrifft, so glaube ich können Sie beide mit etwas verschieden gemischten Strahlungen zu thun gehabt haben, und daher sehr verschiedene Absorptionen gefunden. Schon im Sonnenwärmespectrum sind tiefe Absorptionsbänder im Ultraroth, welche anzeigen dass Strahlen von etwas verschiedener Wellenlänge durch die Atmosphäre also wahrscheinlich durch Wasserdampf sehr verschieden afficirt werden. Ich kann mir sonst die beobachteten Unterschiede nicht erklären.

Unsere schottischen Freunde, muss ich zu meinem Bedauern sagen, sind ganz toll vor Nationaleitelkeit. Diese Geschichten mit Forbes sind ja ganz garstig, und ich sehe nicht ein was man gegen ihre Darstellung erwidern will. Weil sie in Schottland von den Gletschern erst durch Forbes gehört haben, glauben sie, die ganze Welt habe erst durch ihn davon gehört. Tait ist ganz verrannt auf wissenschaftliche Boxerei, und leider lässt sich Sir W. Thomson nur allzu stark mitnehmen. Die Hervorhebung von Stokes gegen Kirchhoff bei der Entdeckung der Spectralanalyse ist sehr ungerecht, und bei uns sehr übel aufgenommen. Durch diesen Punct hat Zöllner am meisten für sich günstig gewirkt, viele die das Ganze für verrückt erklären, sagen, Einzelnes darin sei doch ganz vernünftig, und dann berufen sie sich auf diese Geschichte.

Die neuen Angriffe auf Sie in „Nature“ habe ich noch nicht gesehen.

Der Druck von der Übersetzung der „Fragments“ ist lange verspätet worden durch den Streike der Buchdrucker; jetzt geht es schnell vorwärts. Ich beabsichtige eine Vorrede zu schreiben mit Bezug auf Zöllner. Unter den studirt habenden Klassen Deutschlands ist wenig Kenntniss der naturwissenschaftlichen Principien, aber um so mehr heimliche Neigung zur Metaphysik verbreitet, und Schopenhauer hat eine Menge von Anhängern. Während sie sich meistens schämen, ihre metaphysischen Hoffnungen zu bekennen, haben sie doch ihre Freude daran, wenn sie einen Fürsprecher finden, der mit einem gewissen Schein von Sachkenntniss die leitenden Männer der Naturwissenschaft die ihnen sehr unbequem sind, tüchtig zaust. Ausser unserem neusten Popularmetaphysiker Herrn E. v. Hartmann, hat zwar Niemand gewagt, offen für Zöllner Parthei zu nehmen; aber offenbar hat letzterer vielen die sich durch die Naturwissenschaften in den Hintergrund gedrängt sehen, angenehm geschmeichelt.

Ich freue mich, dass es Huxley wieder besser geht; dass B. Jones gestorben, hat auch mich sehr betrübt; ich habe sehr angenehme Tage in seinem Hause verlebt.

Ich war 14 Tage in Pontresina gegen Ende August; dort erwartete man Ihre Ankunft, aber vergebens. Mir und den meinigen geht es gut, nur meine älteste Tochter ist in Folge ihres Wochenbetts leidend, und jetzt mit ihrem Manne für den Winter am Genfer See, in Montreux,

Mit freundschaftlichem Grusse | Ihr | H. Helmholtz.

BERLIN, 17 October, 1873.

Dear friend,

In your letter from Sept[em]b[er]. 25th,[1] you asked me about a 16th chapter of "On Heat".[2] The editions in my possession only have 13 and 15 chapters[3] respectively, and the edition that you mention must therefore be a more recent one that I do not know. I have not found any objectionable remarks against Magnus[4] in the discussions of those editions of your book that are known to me. Concerning the controversy itself,[5] both of you may have dealt with radiations mixed in different proportions and accordingly, you may have found very different values for the absorption. Even the solar spectrum shows deep absorption regions in the ultra red, which shows that the atmosphere, that is, probably water vapour, affects rays with slightly varying wavelengths in very different ways. I cannot explain the observed differences otherwise.

I have to say to my regret that our Scottish friends[6] are quite mad with national pride. These stories about Forbes[7] are quite nasty and I do not see what one could want to reply to their declarations. Since in Scotland they only heard about the glaciers through Forbes, they believe that the entire world had heard about them only through him. Tait is stubbornly attached to scientific quarrels and Sir W. Thomson unfortunately gets drawn into them too easily. Giving priority to Stokes over Kirchhoff[8] on the discovery of spectral analysis is very unfair and is met with great resistance here. Zöllner has profited the most from this issue;[9] many people who declare the whole thing to be crazy say that individual points of it are quite sensible and then refer to this story.

I have not yet seen the new attacks on you that have appeared in "Nature".[10]

The printing of the translation of the "Fragments"[11] has long been delayed due to the book printers' strike;[12] it is now progressing swiftly. I intend to write a preface referring to Zöllner.[13] There is little knowledge of scientific principles among the educated classes in Germany, but all the more a secret inclination towards metaphysics, and Schopenhauer[14] has many followers. While these are usually ashamed to admit their metaphysical hopes in public, they all feel fortunate to have an advocate who gives the impression of knowing his field and is willing to attack the leading figures of science that they find uncomfortable. However, no one, except for our most recent popular metaphysician, Mr. E. v. Hartmann,[15] has dared to publicly side with Zöllner. But Zöllner has apparently been flattering many who feel that science has upstaged them.

I am delighted to hear that Huxley is recovering;[16] I was also very grieved by B. Jones's death;[17] I spent very pleasant days in his home.[18]

I was in Pontresina[19] for 14 days toward the end of August; your arrival was expected there, but in vain. My family and I are fine, only our eldest daughter[20] is ailing following childbirth, but she and her husband[21] have now gone to Lake Geneva, to Montreux, for the winter.

With kind regards | Your | H. Helmholtz.

RI MS JT/1/H/49
LT Typescript Only

---

1. *your letter from Sept[em]b[er]. 25th*: letter 4172, *Tyndall Correspondence*, vol. 13.
2. *you asked me about a 16th chapter of "On Heat"*: Tyndall had written to Helmholtz: 'Vieweg has written to me several times regarding a XVIth chapter to my book on Heat, in which I unfold, and endeavour to solve, the differences between Magnus and myself as regards the action of aqueous vapour upon Radiant Heat. I am almost ashamed to occupy a mind which must be already overladen, with this subject. But I thought you would allow me to say that nothing would grieve me more than to appear captious or critical as regards the work of Magnus; and that if this chapter should appear to you to bear this character I would suppress it. I have confined myself to pure facts and reasonings, and have not omitted to speak with due appreciation of Magnus, still notwithstanding all my efforts the chapter may appear ungenerous, and if so, as I have said, I will abandon the idea of publishing it' (letter 4172, *Tyndall Correspondence*, vol. 13). The publisher Heinrich Rudolf Vieweg was planning a new edition of the German translation of Tyndall's book that would be published as J. Tyndall, *Die Warme Betrachtet als eine Art der Bewegung*, ed. H. Helmholtz and G. Wiedemann, 3rd edn (Braunschweig: Friedrich Vieweg, 1875).
3. *13 and 15 chapters*: the first edition of J. Tyndall, *Heat Considered as a Mode of Motion: Being a Course of Twelve Lectures Delivered at the Royal Institution of Great Britain in the Season of 1862* (London: Longman, Green, Longman, Roberts, and Green, 1863) contained twelve chapters, the second (1865) was expanded to include thirteen, and the fourth edition (1870) had fifteen. Helmholtz's confusion is justified, as there is no subsequent edition containing sixteen chapters.
4. *Magnus*: Heinrich Gustav Magnus (1802–70), a German chemist, physicist, and professor

of technology and physics at the University of Berlin from 1833 until his death. He established a physics laboratory at his home in Berlin, to which he invited young physicists to work, including Tyndall in 1851. Although his early research was primarily in chemistry, by the 1850s he had begun to investigate a range of physical problems, including the absorption of heat by gases. Magnus was one of the founding members of the German Chemical Society (1868). He is perhaps best remembered for demonstrating how a spinning cylinder or sphere curves as it moves through a fluid, such as air, what is referred to as the Magnus effect. Tyndall's series of collected research, *Contributions to Molecular Physics in the Domain of Radiant Heat* (1873) discusses in detail his academic debates with Magnus (*CDSB*; *NDB*).

5. *the controversy itself*: Tyndall had been engaged in a protacted argument with Magnus over priority regarding research into the abosorption of radiant heat by water vapour.
6. *our Scottish friends*: presumably Peter Guthrie Tait, John Campbell Shairp, and Anthony Adams-Reilly. They had intensified the controversy between Tyndall and James David Forbes regarding glacial motion with the publication of their co-edited book *Life of Forbes* (see letters 4185 and 4186).
7. *These stories about Forbes*: see letter 4185, n. 11.
8. *Stokes over Kirchhoff*: working with the German chemist Robert Bunsen, Gustav Kirchhoff had published pioneering work in the field of spectroscopy, through which he was able to analyse the chemical composition of the sun. William Thomson had written to Kirchhoff claiming that George Gabriel Stokes had already published much of the work for which Kirchhoff claimed priority. Although Stokes himself did not pursue these claims, Kirchhoff felt it necessary to defend himself. See I. D. Rae, 'Spectrum Analysis: The Priority Claims of Stokes and Kirchhoff', *Ambix*, 44 (1997), pp. 131–44.
9. *Zöllner has profited the most from this issue*: in his book *Über die Natur der Cometen* (Leipzig: W. Engelmann, 1872), Johann Karl Friedrich Zöllner had used the doubts raised by Thomson's defence of Stokes's priority to question his use of scientific method.
10. *the new attacks on you that have appeared in "Nature"*: P. G. Tait, 'Tyndall and Forbes', *Nature*, 8 (1873), pp. 381–2.
11. *the translation of the "Fragments"*: this translation of *Fragments of Science* would eventually be published as J. Tyndall, *Fragmente aus den Naturwissenschaften: Vorlesungen und Aufsätze*, trans. A. Helmholtz, ed. H. Helmholtz (Braunschweig: Friedrich Vieweg, 1874).
12. *the book printers' strike*: in March 1873 the Union of German Book Printers called more than two thousand of its members out on strike in Leipzig, following the collapse of negotiations over pay with employers; the impact of the strike clearly caused a backlog in printing work that had only just begun to ease.
13. *a preface referring to Zöllner*: H. Helmholtz, 'Vorrede', pp. v–xxv. In it, Helmholtz declared: 'Herr Zöllner möchte die "deductive Methode" welche er selbst in seinen astrophysischen Speculationen befolgt oder wenigstens zu befolgen beabsichtigt, als die urgermanische empfehlen, und Deutschlands geistigen Horizont durch eine chinesische Mauer gegen die inductive Methode des Auslands abschliessen [Mr. Zöllner would like to recommend the

"deductive method", which he himself follows or at least intends to follow in his astrophysical speculations, as the Ur-Germanic one, and close off Germany's intellectual horizon from the inductive method of foreign countries by a Chinese wall]' (p. xxi).

14. *Schopenhauer*: Arthur Schopenhauer (1788–1860), a German philosopher whose works on metaphysics were highly influential in the sciences (*NDB*).
15. *Mr. E. v. Hartmann*: Karl Robert Eduard von Hartmann (1842–1906), a German philosopher whose book *Philosophie des Unbewussten* (Berlin: C. Duncker, 1869) had proved popular across Europe, being translated from the German into French and English (*NDB*).
16. *Huxley is recovering*: Thomas Huxley had been suffering from poor health, both physical and mental, for the last two years. Tyndall was among a large circle of Huxley's friends who collected funds to pay for the health-restoring holiday to the Continent that they reckoned he needed very much (see letters 4028, 4151, and 4154, *Tyndall Correspondence*, vol. 13). For letters from Huxley during his travels in the Mediterranean and Egypt in 1872 when ill, see *Tyndall Correspondence*, vol. 12.
17. *B. Jones's death*: Henry Bence Jones had died on 20 April 1873.
18. *his home*: 84 Brook Street, Grosvenor Square, Mayfair.
19. *Pontresina*: a village in the Swiss canton of Graubünden.
20. *our eldest daughter*: Käthe Branca (née Helmholtz, 1850–78), who had given birth to a daughter, Sophie Branca, on 18 April 1873.
21. *her husband*: Wilhelm Branca (1844–1928).

To Thomas Archer Hirst [18 October 1873][1] 4193

Saturday

My dear Tom.

I send you Ruskin's performance[2] as Lady Lubbock[3] informs me that she cannot get a copy for you. It has been extracted in the Scotsman[4] as I learn today from my friend M[rs]. Rutherford.[5]

I think you would also like to see Carlyle's letter[6] and I therefore send you a copy of it.

Your telegram[7] is here—I am sorry you cannot come

Yours ever | John Tyndall

RI MS JT/1/T/931
RI MS JT/1/HTYP/623

---

1. *[18 October 1873]*: the date is established by reference to John Ruskin's attack on Tyndall (see n. 2) and a letter received from Thomas Carlyle (4188), dated 16 October 1873 (a Thursday). The following Saturday was 18 October.

2. *Ruskin's performance*: see letter 4185, n. 13.
3. *Lady Lubbock*: Ellen Lubbock.
4. *extracted in the Scotsman*: 'Mr Ruskin on Professor Tyndall', *Scotsman*, 14 October 1873, p. 6.
5. *my friend M^rs^. Rutherford*: Elizabeth Rutherford.
6. *Carlyle's letter*: letter 4188.
7. *Your telegram*: letter missing.

To Elizabeth Rutherford 18 October 1873 4194

Royal Institution of Great Britain.[1] | 18th. Oct[ober]. 1873.

My dear Friend

I have just two minutes to say that I am here[2] hard at work: afloat every day from morning to night making observations on Fog signals.[3] No man lends himself more readily to the dissecting knife than Ruskin;[4] and if I willed it I could flay him alive. I may do so yet; but I have had a beautiful letter from Mr Carlyle[5] apropos of his attack on me, which influences me much.[6] I will send you an extract from it not later than tomorrow.[7]

For some days to come a letter addressed to me to the Lord Warden Hotel Dover[8] will reach me.

Yours ever affectionately | John Tyndall.

RI MS JT/1/TYP/3/1022
LT Typescript Only

---

1. *Royal Institution of Great Britain*: although Tyndall was writing on paper with the RI's official letterhead, the letter clearly indicates that he was in Dover at this time.
2. *here*: Dover.
3. *observations on Fog signals*: Tyndall was engaged in a series of acoustic experiments to determine the best method of signaling ships in fog, working from the South Foreland Lighthouse near Dover (see letter 4185, n. 3).
4. *No man lends himself... than Ruskin*: John Ruskin had just published an attack on Tyndall in the latest number of his serial *Fors Clavigera* (see letter 4185, n. 13). Tyndall was considering whether he should respond or not.
5. *a beautiful letter from Mr Carlyle*: letter 4188.
6. *influences me much*: Thomas Carlyle had advised Tyndall to ignore Ruskin.
7. *I will send you . . . than tomorrow*: this was sent in letter 4197.
8. *the Lord Warden Hotel Dover*: see letter 4183, n. 3.

From Millicent Bence Jones 18 October [1873][1] 4195

6 Upp[e]r Wimpole S[tree]t | Oct[ober] 18

Dear Dr Tyndall.

Now that we are for a time at least settled in this house,[2] I must let you know of it in the hope that some day you may have time when any thing brings you into this quarter to come in, if it is only to shake hands again, it will be a sad pleasure to me—& I fear a very sad effort on your part—but we are not likely to meet in any other way as I go no where except as business obliges me, & I see hardly any one—

Hunniball[3] is not here to let you in—& if you are told at the door that I do not see friends will you kindly say that if I am at home I am sure to see you—& should I be out there is not one of the three girls[4] who would not give you a true welcome—We have been of late scattered—two of them went to some relations in Somersetshire but I hope we are not to move again till we find something that will be a future house for them—

When we passed through town a few weeks back you were away, or I should have tried to see you before now—

Believe me always | Your's very truly | M Bence Jones

RI MS JT/1/J/123
RI MS JT/1/TYP/3/740

---

1. *[1873]*: the year is established by the letter evidently being written after the death of Henry Bence Jones on 20 April 1873.
2. *settled in this house*: Bence Jones had moved to this house in Marylebone from 84 Brook Street, Grosvenor Square, Mayfair, after the death of her husband.
3. *Hunniball*: possibly Mary A. Hunniball (b. 1848), who in the 1871 Census was recorded as a servant, living and working within the registration district of Marylebone, the area that Bence Jones had moved to.
4. *the three girls*: Bence Jones's daughters Millicent Mary Beth (1843–1932), Olivia (1845–1911), and Edith Mary (1853–1919). Another daughter, Anabella Mary Bence Jones (1846–1851), died young. Ralph Noel Bence Jones, Olivia's twin brother, died in 1866. There were two other sons, Henry Robert Bence Jones (1844–1912) and Archibald Bence Jones (1856–1937).

To William Frederick Pollock 19 October 1873 4196

59 Montague Square W.[1] | 19 / Oct[ober] / 73

My dear Tyndall,
I found your two notes[2] on coming home from Cheltenham[3] yesterday—but too late for the post. Many thanks to you for sending me the extract from Carlyle's letter.[4] Nothing can be better or better put—and I don't see why the publication in the Scotsman[5] of which your note No. 2. informs me, could alter the <*1 or 2 words missing*>[6] of dignified silence which he recommends—To reply separately to Ruskin, and admit him as an antagonist worthy of your steel in single combat—would be to gratify his wish to provoke you to a conflict.

No—abide by the determination to wait and see what (if any) notice is taken of his attack—and then in due season, and apart from personalities[7] approach the whole subject calmly and deliberately, and as you say—deal with it exhaustively and once for all.

I am glad to hear you are getting good work done,[8] and of a new and interesting kind.

We dined yesterday with "le Jene Menage"[9] at No. 12 Bryanston St.[10] They have a nice little house and everything pretty about them.

Maurice[11] has begun to draw in Poynters school at University College[12]

Yours ever | W. F. Pollock.

RI MS JT/1/TYP/6/2149
LT Typescript Only

---

1. *59 Montague Square W.*: the Pollocks resided at 59 Montagu Square in Marylebone, London, a fashionable London neighbourhood.
2. *your two notes*: letters missing.
3. *Cheltenham*: Pollock had been in the Gloucestershire town visiting the widow of the actor and theatre manager William Macready, for whom he acted as executor (see letter 4186, n. 15).
4. *Carlyle's letter*: letter 4188.
5. *the publication in the Scotsman*: see letter 4193, n. 4.
6. <*1 or 2 words missing*>: there is a blank space here in the typescript of the letter, which may indicate that LT (or one of the typists she employed) could not read the word(s) in the original MS letter.
7. *personalities*: statements or remarks referring to or aimed at a particular person, and usually disparaging or offensive in nature (*OED*).
8. *getting good work done*: Tyndall was engaged in a series of acoustic experiments to

determine the best method of signaling ships in fog, working from the South Foreland Lighthouse near Dover (see letter 4185, n. 3).

9. *"le Jeune Menage"*: 'le Jeune ménage', or 'the young couple' (French), a reference to Pollock's son Frederick and his wife Georgina, who had married in August 1873 (see letter 4186, n. 12).
10. *Bryanston St.*: in Marylebone, London.
11. *Maurice*: Maurice Pollock (see letter 4186, n. 13). He was training to be an artist.
12. *Poynters school at University College*: the Slade School of Art at UCL, where Edward Poynter (1836–1919) was Slade Professor from 1871–5. In March 1870, Tyndall was on a committee to appoint Life Governors to the Slade School of Art, which opened in 1871 (M. Postle, 'The Foundation of the Slade School of Fine Art: Fifty-Nine Letters in the Record Office of University College London', *Volume of the Walpole Society*, 58 (1995/1996), pp. 127–230).

To Elizabeth Rutherford 19 October 1873 4197

Dover | 19th Oct[ober]. 1873.

My dear Friend,

Here is the extract from Mr Carlyle's wise and beautiful letter.[1]

[...][2]

I was strongly minded to make an example of Mr Ruskin, and some of my friends think it ought to be done—others think the reverse, and I am disposed to think with them. He is an ungrateful dog—though perhaps unconscious of his ingratitude. For over and over again, when I have heard in society as grave things laid to his charge, as could well be laid to the charge of any man,[3] I have sought to shelter and defend him—and to accept any interpretation of his conduct rather than the obvious one.

In fact this Forbes affair[4] has aroused many latent antagonisms—which are now showing themselves. As for me I am perfectly safe, for I know the absolute singlemindedness with which I have treated this question. I shall have something more to say upon it. In fact I must place the whole matter in a light that he who runs may read,[5] and then I shall bid good bye to it for ever. I do not think in the long run that the friends of Professor Forbes[6] will have any reason to congratulate themselves on the result of their assault upon me.

Yours ever affectionately | John Tyndall

RI MS JT/1/TYP/3/1023
LT Typescript Only

---

1. *Mr Carlyle's wise and beautiful letter*: letter 4188, regarding John Ruskin's published attack on Tyndall. Tyndall had promised to send extracts of it in letter 4194.
2. *[. . .]*: Tyndall's extracted two parts of letter 4188, the first beginning 'The very day before your letter came to me . . .' and concluding '. . . your alpine and glacier investigations', and the second beginning 'As to his criticisms . . .' and concluding '. . .as if there were no Ruskin there'.
3. *grave things laid . . . to the charge of any man*: probably a reference to allegations concerning Ruskin's private life (see letter 4185, n. 10).
4. *this Forbes affair*: Tyndall's dispute with James David Forbes over theories of glacial motion.
5. *he who runs may read*: Habakkuk 2:2, denoting that something is clearly stated and understandable to all.
6. *the friends of Professor Forbes*: as well as Ruskin, Tyndall presumably meant Peter Guthrie Tait, John Campbell Shairp, and Anthony Adams-Reilly, the co-editors of *Life of Forbes*.

To Mary Egerton — 20 October 1873 — 4198

The Lord Warden Hotel[1] | Dover | 20th. October, 1873

My dear Lady Mary

My days and hours have been filled since I last wrote to you,[2] I start early in the morning and remain all day afloat making my observations.[3] These lines are written in the moments prior to departure. The sea is boisterous this morning, and I know not what awaits me. I have been just looking at a little steamer[4] rolling piteously in the trough of the sea. I am anxious however to vary as much as possible the atmospheric conditions,[5] so though the day may cause me to suffer I hail it as a furtherance to my work.

I have nothing to say to you that you would in the least care to hear, further than this that I am engaged upon a problem of the greatest complexity, and regarding which there is a greater amount of conflicting evidence extant than regarding any other physical question known to me.[6] I hope to show the principal ground of this conflict, to throw some light on the science of the problem, and also to aid in the furtherance of practical measures for the warning and guidance of mariners.

And now I must ask you to give my kindest regards to all my friends at Mountfield.[7]

Ever Yours | John Tyndall

I meant nothing either mysterious or severe by the postscript to which you refer.[8] | Mr Carlyle has written me such a sweet wise letter[9] apropos of

Ruskin.[10] Any number of attacks thus neutralised would be a profit and not a loss.

RI MS JT/1/TYP/1/397
LT Typescript Only

---

1. *The Lord Warden Hotel*: see letter 4183, n. 3.
2. *I last wrote to you*: letter missing.
3. *all day afloat making my observations*: Tyndall was engaged in a series of acoustic experiments to determine the best method of signaling ships in fog, working from the South Foreland Lighthouse near Dover (see letter 4185, n. 3).
4. *a little steamer*: a small steam-powered boat or ship.
5. *to vary . . . atmospheric conditions*: see letter 4185, n. 3.
6. *a problem of the greatest . . . question known to me*: see letter 4185, n. 3.
7. *all my friends at Mountfield*: Mountfield Court in Sussex, the family seat of the Egertons. The friends who were staying there were the Hamilton family (see letter 4200).
8. *the postscript to which you refer*: Tyndall's previous letter is missing.
9. *Mr Carlyle has written me such a sweet wise letter*: letter 4188.
10. *appropos of Ruskin*: John Ruskin had recently published an attack on Tyndall in the controversy over glacial motion (see letter 4185), and Thomas Carlyle had advised Tyndall not to respond.

To Thomas Archer Hirst 21 October 1873 4199

21st Oct[ober] 1873

My dear Tom

Your little note[1] brief as it was, was a pleasure to me—Time was in my life's history when I should have deemed it impossible that I should ever become the object of such attacks[2]—or that I could become hated by any human being as I now am by more than one. I am by nature sensitive to such things; but it is surprising to myself how well I bear these matters now, and how tranquil I feel in the calm confidence that I shall finally overthrow them one and all.

Lady Lubbock is by no means satisfied; for she has seen something else that has roused her indignation. An ignorant writer in the Westminster Review[3] has now stirred her up—I have long known of his existence—but did not think him worth a moment's attention. She evidently thinks I ought to defend myself—and really seems to consider that my pluck needs a stimulus to make me "vindicate my honour"

Well she is true to her friends. But it makes me smile to think that she should deem it necessary to apply this spur to me. My honour shall be vindicated in due time. It is not so dilapidated as to be unable to withstand these assaults upon it until I think it time to return my enemy's fire.

Helmholtz has written me an exceedingly pleasant, and a somewhat surprising letter.[4] He manifestly regards Tait as a scientific bully, and, which surprises me, blames William Thomson for encouraging the mad vanity of Scotchmen. He is going to write a preface to "Fragments of Science"[5] in which he will deal with Zöllner, whom he deems madder than you deem Ruskin.

Yours aff[ectionate][ly] | John Tyndall

RI MS JT/1/T/708
RI MS JT/1/HTYP/624

1. *Your little note*: letter missing.
2. *such attacks*: Tyndall was subject to published attacks by supporters of James David Forbes relating to a controversy over theories of glacial motion.
3. *An ignorant writer in the Westminster Review*: the issue for October contained a critical review of Tyndall's two latest publications, *Six Lectures on Light* (1873) and the pamphlet *Principal Forbes and his Biographers* (London: Longmans, Green, and Co., 1873), the latter of which was Tyndall's response to the attacks made upon him by Peter Guthrie Tait regarding his part in the controversy over glacial motion. The review concluded: 'we can only regret that men like Tait and Tyndall, each great in his own way, should think it worth their while to waste a single moment about quarrels such as those of which this pamphlet, we fear, is only one of further encounters still to come' ('Science', *Westminster Review*, 44 (1873), pp. 487–96, on pp. 488–9). The *Westminster* article was anonymous (as per convention) and the author is not known.
4. *Helmholtz has written . . . a somewhat surprising letter*: letter 4192.
5. *a preface to "Fragments of Science"*: H. Helmholtz, 'Vorrede', in J. Tyndall, *Fragmente aus den Naturwissenschaften: Vorlesungen und Aufsätze*, trans. A. Helmholtz (Braunschweig: Friedrich Vieweg, 1874), pp. v–xxv.

From Mary Egerton 24 October [1873][1] 4200

Mountfield Court, | Robertsbridge, | Hawkhurst.[2] | Oct[ober]. 24th

My dear Mr Tyndall,

Your letter[3] was most welcome. I care about everything but especially I cared to hear of the great scientific interest that attaches to your present investigations.[4] At first sight it appeared as if the question which noise would sound the furthest,[5] might be settled by ears minus brains; and though I was glad you should be inhaling North Sea breezes,[6] I confess I felt disposed to grudge the days & weeks that seemed taken from the dear old R. I.[7] and "original research". But I now find I need not feel this regret, as the two are combined. Pity you cannot bottle up a London fog and send it down, thus equalizing demand and supply! it is rarely that article is wished for so anxiously! & in vain! you must have had some weather to test your sailing prowess this last week!

You scarcely knew when you wrote, how many "friends at Mountfield"[8] your greetings might have included. All the Hamiltons were here![9]

I think I know how to interpret your bitter silence with regard to the subject of which my last letter[10] was full, coupled with certain allusions to a Syren by the Sea,[11] and a fairer Syren by the river,[12] of which I have heard! Dear friend! Why should you have any reserve towards me? Do I not know, better perhaps than most, the flush of feeling which a fresh young life[13] can awaken in your warm heart, till time and wisdom mellow it down! And though the reflected glow from those feelings must naturally fade with them, may I not still "trust"—as you once bade me trust, to find, if not quite "unchanged and unchangeable affection," (my memory is good for some things!) Still such a residue of unaltered dear friends

Ever y[ou]rs | M F Egerton

I need hardly say any inference drawn freely was my own; not prompted

RI MS JT/1/E/55

1. *[1873]*: the year is established by the relation to letter 4198.
2. *Mountfield Court, | Robertsbridge, | Hawkhurst*: the Egerton estate in Sussex.
3. *Your letter:* letter 4198.
4. *your present investigations*: Tyndall was engaged in a series of acoustic experiments to determine the best method of signaling ships in fog, working from the South Foreland Lighthouse near Dover (see letter 4185, n. 3).
5. *which noise would sound the furthest*: see letter 4185, n. 3.
6. *North sea breezes*: the South Foreland Lighthouse overlooks the Strait of Dover. Much of his time was spent aboard a boat from which the various sounds could be monitored.
7. *the dear old R. I.*: i.e. the Royal Institution.
8. *Mountfield*: Mountfield Court in Sussex, the family seat of the Egerton family.
9. *all the Hamiltons*: Claud Hamilton, Elizabeth Hamilton, and their children Louisa, Emma and Douglas. Tyndall had made acquaintance with the family soon after returning from his lecture tour in the United States in February 1873, and had met them during his trip to Switzerland in August. See letters 4145, 4157 and 4164, *Tyndall Correspondence*, vol. 13; and *Ascent of John Tyndall*, p. 323.
10. *my last letter*: letter missing.
11. *Syren by the Sea*: a play on words referencing Tyndall's sound experiments with a syren (see letter 4201).

12. *a fairer Syren by the river*: possibly Louisa Hamilton.
13. *a fresh young life*: Louisa, who was twenty-eight at this time, was twenty-five years Tyndall's junior.

To Joseph Henry 25 October 1873 4201

25th. Oct[o]b[e]r. 1873

My dear Professor Henry.

For the last 3 weeks I have been almost every day afloat making observations upon Fog Signals.[1] It gives me great pleasure to be able so unreservedly to subscribe to your opinion regarding the merits of the Syren.[2] It is unquestionably the most effectual instrument that has come under my notice.

I shall communicate a paper to the Royal Society on the subject;[3] & as soon as it is ready I will send you an abstract of it for communication to the Philosophical Society of Washington. This will be the best return I can make for the courtesy shown to me by the Society during my visit to Washington.[4]

The partisans of Professor Forbes—plus M[r]. Ruskin, who is evidently mad—have made a united assault upon me[5]—I have had a very noble letter from the celebrated Thomas Carlyle,[6] exhorting me to take no heed of them; also a letter from Helmholtz[7] in which he gives very clear and strong expression to his views upon the subject—For the present I will let them have their way—but at some future day may give myself the trouble of putting them to confusion.

Kindest regards to your family—It was exceedingly civil of the Lighthouse Board to lend us the Syren—but I do not think we shall ever let it go back—

Yours ever faithfully | John Tyndall

Smithsonian Institution Archives. RU 26, vol. 139, p. 178

---

1. *observations upon Fog Signals*: see letter 4185, n. 3.
2. *your opinion regarding the merits of the Syren*: in October 1867 at Sandy Hook in Connecticut, Henry compared an air trumpet foghorn developed by Celadon Leeds Daboll with a steam-powered syren developed by Jean Nickolai Henry Adolphus Brown. The latter consisted of a rotating slotted cylinder alternately opening and closing a passageway to either steam or compressed air, and had, Henry concluded, the greater penetrating power. The United States Lighthouse Board, of which Henry was chairman, sent Tyndall the syren to test in his acoustic experiments.
3. *a paper to the Royal Society on the subject*: see 'Vehicle of Sound', which Tyndall read at the RS on 15 January 1874.

4. *my visit to Washington*: Tyndall had visited Washington, DC in December 1872 during his lecture tour of the United States. He delivered a series of lectures on light in Boston, Philadelphia, Baltimore, Washington, DC, New York, Brooklyn, and New Haven, Connecticut. He departed from Liverpool aboard the SS *Russia* on 28 September 1872, and arrived in New York on 9 October. His five months in the United States brought him in contact with numerous notable Americans, dazzled audiences, involvement in a debate on the efficacy of prayer, a side trip to Niagara Falls, the opportunity to set up a fund for pure science research in American universities using his profits from the lecture tour, and a lavish farewell banquet in New York before his departure from American soil aboard the SS *Cuba* on 5 February 1873, reaching London on 19 February. He published the lectures as *Six Lectures on Light* (1873). See Introduction to *Tyndall Correspondence*, vol. 13, as well as the many letters in that volume written during the trip.
5. *partisans of Professor Forbes . . . united assault upon me*: Tyndall was subjected to a series of published attacks from supporters, especially Peter Guthrie Tait and John Ruskin, of the late James David Forbes, with whom Tyndall had been involved in an acrimonious dispute regarding theories of glacial motion.
6. *very noble letter from . . . Thomas Carlyle*: letter 4188.
7. *a letter from Helmholtz*: letter 4192.

To Joseph Dalton Hooker 25 October 1873 4202

Saturday, 25th, Oct[ober], 1873.

Dear Hooker

Supposing I and others do not blackball[1] you, will you tell me when your presidency of the Royal Society[2] begins.

I got your long kind letter[3] at Dover,[4] whither I have come for to day and tomorrow.

I am only too willing to do not only what I deem right, but what my friends deem right as regards Belfast.[5]

I fear I am now really committed to it—for I find that that unscrupulous scientific bully Tait has been republishing in all the newspapers there a slanderous attack of Ruskin's upon me.[6] If I were to back out they might say that I was afraid to come.

Apropos of Ruskin, Carlyle has sent me such a sweet wise letter.[7]

Thank you my dear fellow for all the trouble you have taken regarding Belfast.

Yours ever affectionately | John Tyndall

RI MS JT/1/TYP/8/2722
LT Typescript Only

---

1. *blackball*: record an adverse vote against a candidate for membership of a club or society, or a post within it, by placing a black ball in the ballot box (*OED*).
2. *when your presidency of the Royal Society begins*: Hooker's term as president of the RS began, by tradition, at the annual anniversary meeting, which in 1873 was on 1 December.
3. *your long kind letter*: letter 4189.
4. *at Dover*: Tyndall was engaged in a series of acoustic experiments to determine the best method of signaling ships in fog, working from the South Foreland Lighthouse near Dover (see letter 4185, n. 3).
5. *as regards Belfast*: as president of the BAAS, Tyndall was due to give an address at the meeting to be held in Belfast in August of the following year. Due to the demands of his acoustic experiments, Tyndall had considered stepping down from the presidency, but Hooker and other friends had advised against it (see letter 4189).
6. *Tait has been republishing . . . a slanderous attack of Ruskin's upon me*: Tyndall was subjected to a series of published attacks by Peter Guthrie Tait, John Ruskin, and other supporters of the late James David Forbes, with whom Tyndall had been involved in an acrimonious dispute regarding theories of glacial motion. One of these critiques was contained in the latest number of Ruskin's serial, *Fors Clavigera* (see letter 4185, n. 13), and Tait was engaged in republishing extracts from this work in the newspapers of Belfast, where he had been professor of mathematics at Queen's College in the 1850s. The two Belfast newspapers that printed extracts were: 'Mr. Ruskin on Professor Tyndall', *Belfast News-Letter*, 18 October 1873, p. 3, and 'Mr. Ruskin on Professor Tyndall', *Northern Whig*, 18 October 1873, p. 3.
7. *Apropos of Ruskin . . . a sweet wise letter*: letter 4188.

To Hermann Helmholtz 26 October 1873 4203

26th October 1873.

My dear Helmholtz,

Your letter[1] reached me at Dover where I have been for the last three weeks, afloat almost every day between the coasts of England and France, making observations on fog signals.[2] We have been trying huge air-whistles and steam-whistles, trumpets, guns, and a steam Syren.[3] Worked by steam of 70 or 80 lbs pressure this instrument yields a note of extraordinary power. The observations have quite revolutionised the notions I had been taught regarding the influences which affect the transmission of sound through air, and introduce others which, though hitherto passed over, are of the most potent character.

But I write now simply to thank you for your letter, and to express the gratification which your allusion to this Scotch madness[4] gave me. I shrank from mentioning to you William Thomson's name in connexion with this subject, knowing the intimacy of your friendship with him. With all his brilliancy of intellect he lacks that force of character that Tait possesses, in a coarse and ungentlemanly form it is true, but still strong. Thomson is thus at the mercy of Tait. When I first heard that an attack on me was contemplated, I said I would write to Thomson and appeal to his good sense—but the reply I received[5] was that it would be useless, for he was completely ruled by Tait. I have often felt ashamed of the way in which Kirchhoff has been treated;[6] and no doubt I have added to the list of my offences in Tait's eyes by speaking out so strongly in favour of Kirchhoff in my Lectures on Light.[7]

You know Ruskin by repute. He is thought by some to be going mad, and I have heard it predicted that he will end his days in a lunatic asylum. In an obscure and eccentric way he has been publishing for some years a series of letters to the Working men of England,[8] and in his latest issue[9] he has made a most ferocious attack upon me about this Forbes affair. But I would not waste my breath or your attention upon him were it not to illustrate the character and conduct of Mr Tait. He has taken it upon him to circulate this pamphlet; he has, I am informed, got it inserted into the Scotch newspapers[10]—Much against my will and simply to avoid offending the Council, whose invitation I had already declined three times,[11] I am to preside over the next meeting of the British Association at Belfast. Tait was professor there,[12] and William Thomson I believe was born there.[13] Into the Belfast newspapers therefore Mr Ruskin's assault is thrust.[14] I declare, personally, the matter affects me no more than the dropping of water does granite; and did I need compensation I have it in the noble letters which the subject has brought me—the noblest, wisest, and most affectionate of all coming from Mr Thomas Carlyle[15]—but I do grieve to see English science thus smeared and sullied by its professors. Every conventicle[16] in the land will rejoice to see men of science, whom they regard as a common enemy, thus at loggerheads with each other.

They take care to send me the newspapers where Ruskin has been republished, and from one of them I cut an extract which you may possibly read when you have no other source of amusement. This however is not likely to be the case, and if not throw the extract, without reading, in the fire, and excuse me for inflicting this long letter upon you.

always yours faithfully | John Tyndall.

BBAW, NL. Helmholtz, Nr. 477, BI_22–3
RI MS JT/1/T/491

1. *Your letter*: letter 4192.
2. *at Dover . . . observations on fog signals*: Tyndall was engaged in a series of acoustic

experiments to determine the best method of signaling ships in fog, working from the South Foreland Lighthouse near Dover (see letter 4185, n. 3).

3. *a steam Syren*: a steam-powered syren, lent to Tyndall by the United States Lighthouse Board (see letter 4201, n. 2).
4. *this Scotch madness*: Tyndall was subject to a campaign by supporters of James David Forbes, with whom he had been involved in a controversy relating to theories of glacial motion. Chief among his assailants was the Scottish physicist Peter Guthrie Tait.
5. *the reply I received*: it is unclear from whom Tyndall received this reply, and Helmholtz himself warned about Thomson's closeness to Tait in letter 4192.
6. *the way in which Kirchhoff has been treated*: Gustav Kirchhoff had published pioneering work in the field of spectroscopy, but his claim to priority was disputed by William Thompson (see letter 4192, n. 8).
7. *speaking out so strongly . . . my Lectures on Light*: in *Six Lectures on Light* (1873), Tyndall had stated 'All great discoveries are duly prepared for in two ways: first, by other discoveries which form their prelude; and, secondly, by the sharpening of the enquiring intellect . . . Newton did not rise suddenly from the sea-level of the intellect to his amazing elevation. At the time that he appeared, the table-land of knowledge was already high. He juts, it is true, above the table-land, as a massive peak; still he is supported by it, and a great part of his absolute height is the height of humanity in his time. It is thus with the discoveries of Kirchhoff. Much had been previously accomplished; this he mastered, and then by the force of individual genius went beyond it . . . I do not think that Newton has a surer claim to the discoveries that have made his name immortal, than Kirchhoff has to the credit of gathering up the fragmentary knowledge of his time, of vastly extending it, and of infusing into it the life of great principles' (p. 204).
8. *a series of letters to the Working men of England*: John Ruskin's *Fors Clavigera* (see letter 4185, n. 13).
9. *his latest issue*: letter 34, published in September 1873; vol. 3 (1873), pp. 1–32 (see letter 4185, n. 13).
10. *inserted into the Scotch newspapers*: see letter 4193, n. 4.
11. *avoid offending the Council . . . declined three times*: Tyndall had declined the offer of the Presidency of the BAAS from the association's Council on three previous occasions, and was considering refusing it again but in the end accepted and delivered the Presidential Address at the meeting in Belfast in August 1874.
12. *Tait was professor there*: Peter Guthrie Tait was professor of mathematics at Queen's College, Belfast from 1854–60.
13. *William Thomson I believe was born there:* Thomson was born in 1824 in Belfast, where his father was an instructor of mathematics and engineering at the Royal Belfast Academical Institution.
14. *Into the Belfast newspapers . . . Ruskin's assault is thrust*: see letter 4202, n. 6.
15. *the noblest, wisest . . . coming from Mr Thomas Carlyle*: letter 4188.
16. *conventicle*: a religious meeting or assembly of a private, clandestine or illegal kind (*OED*).

To Thomas Archer Hirst 27 October 1873 4204

Dover[1] Tuesday | (27–10–73)

My dear Tom.

It may be better to keep the proof[2] to yourself. You of course will know better than I can—for you have read it & I have not

Yours aff[ectionatel]y | John Tyndall

RI MS JT/1/T/709
RI MS JT/1/HTYP/624

---

1. *Dover*: see letter 4185, n. 3.
2. *the proof*: possibly of Peter Guthrie Tait's article 'Forbes and Dr. Tyndall', which was rejected by the *Contemporary Review*. It was subsequently published in *Glaciers of Savoy*, pp. 163–98, with a note stating: '*Written (by arrangement with the Publisher) for the "Contemporary Review": of November 1873; but, after having been put in type and corrected for press, considered unsuitable for its pages by the Editor of that Journal*' (p. 163). The *Contemporary*'s editor was James Thomas Knowles, who may have let Tyndall and Hirst see the rejected proof.

From Giovanni Giuseppe Bianconi[1] 28 October 1873 4205

Bologne 28 oct. 1873

Monsieur!

Je prend la liberté de vous faire hommage d'un livre qui vient de paraître, sur la théorie de M. Darwin. L'accueil flatteur que vous avez fait à mes notes sur la chaleur développée par friction des liquides, et sur la flexibilité de la glace, m'engagent à espérer que vous accueillerez favorablement encore ce travail que j'ai consacré à une question de la plus haute importance, et à l'examen logique de quelques points de mécanique animale. Combien de fois j'ai eu envie de donner à mon exposition la lucidité et l'évidence de vos travaux! Mais j'ai la confiance qu'on ne me reprochera pas de n'avoir suivi autant que possible la recherche de la vérité, ni de m'être laissé entrainer par l'imagination dans l'analyse que j'ai fait des mécanismes animales et dans les conclusions qui en découlent.

Agréez M^r^. l'assurance de mon profond respect et de mon estime

J. Jos. Bianconi

P.S. Mon fils vous fait hommage aujourd'hui même d'une note qu'il vient de publier. À M^r. le Prof. J. Tyndall.

Bologna, 28 Oct[ober]. 1873

Sir!

I take the liberty of offering you a recently published book on Mr. Darwin's theory.[2] The flattering welcome that you gave to my notes on heat generated by friction of liquids, and on the flexibility of ice,[3] gives me the hope that you will auspiciously welcome again this work that I have devoted to a very important question, and to the logical examination of some points on animal mechanics. So many times have I wanted to give my presentation the clarity and evidence of your works! But I trust that I will not be blamed for not having followed the search for truth as far as possible, nor for letting myself be carried away by imagination in the analysis that I made on animal mechanisms and in the resulting conclusions.

Please accept, Sir, the assurance of my deepest respect and esteem.

J. Jos. Bianconi

P.S. My son[4] offers you today a note that he has just published.

To Prof. J. Tyndall.

RI MS JT/1/B/98

---

1. *Giovanni Giuseppe Bianconi*: Giovanni Giuseppe Bianconi (1809–78), an Italian zoologist who from 1842 to 1864 was professor of natural history at the University of Bologna.
2. *a recently published book on Mr Darwin's theory*: G. G. Bianconi, *La théorie darwinienne et la création dite indépendante* (Bologna: Nicolas Zanichelli, 1874), although this work was not officially published, in French rather than Bianconi's native Italian, until the following year. Bianconi argued against Darwin's theory of evolution by natural selection, suggesting that the similarity in structure between different species could be explained using mechanical principles.
3. *The flattering welcome . . . the flexibility of ice:* Tyndall had written approvingly of Bianconi's work on the viscosity of ice in *Forms of Water*: 'Still it is undoubted that the glacier moves like a viscous body. The centre flows past the sides, the top flows over the bottom, and the motion through a curved valley corresponds to fluid motion. Mr. Mathews, Mr. Froude, and above all Signor Bianconi, have, moreover, recently made experiments on ice which strikingly illustrate the flexibility of the substance. These experiments merit, and will doubtless receive, full attention at a future time' (p. 163).
4. *My son*: Giovanni Antonio Bianconi, who published a revised Italian edition of his father's book: G. G. Bianconi, *La teoria Darwiniana e la creazione detta indipendente*, trans. G.

A. Bianconi (Bologna: N. Zanichelli, 1875). His recently published note has not been identified.

To Elizabeth Rutherford 30 October 1873 4206

Royal Institution of Great Britain | 30th Oct[ober]. 1873.

My dear Friend,

Very many thanks to you[1]—Had the fire descended, the victim[2] would have been found incombustible. One other deliverance on this head I propose making, and this shall be sufficiently complete to release me from it for ever.

Not only from people here at home, but from the very first heads on Continental Europe I have received justification, approval and support[3] in connexion with this matter.

Yours affectionately | John Tyndall.

You need no assurance from me that Forbes being a Scot has no influence upon me, or such influences as this fact exercises would be distinctly in his favour. Why it was from Walter Scott[4] I got my love of the mountains: and I would venture to say that Burns[5] is one hundred times more closely related to me than he is to his countrymen who assail me.

RI MS JT/1/TYP/3/1024
LT Typescript Only

1. *Very many thanks to you*: the letter to which Tyndall is replying is missing.
2. *the victim*: probably John Ruskin, who had published an attack on Tyndall relating to theories of glacial motion. In Tyndall's previous letter to Rutherford (letter 4197), he had discussed whether to engage with Ruskin, or ignore him. Tyndall seems to suggest he had chosen the latter course.
3. *Not only from people here . . . approval and support*: Tyndall had received advice regarding his course of action from various correspondents, most notably Thomas Carlyle (letter 4188), who had suggested Tyndall should not respond to Ruskin. The reference to the 'very first heads on Continental Europe' probably refers to Hermann Helmholtz, who offered his support in letter 4192.
4. *Walter Scott*: Walter Scott (1771–1832), the popular Scottish novelist and poet (*ODNB*).
5. *Burns*: Robert Burns (1759–96), the Scottish poet (*ODNB*). In October 1866, Tyndall wrote to Mary Anne Coxe of the 'old old days . . . when the poems of Scott and Burns were the delight of my life. When Straths, Bens, Lochs and Corries—purple heather & misty glen—constituted my ideal scenery' (see letter 2487, *Tyndall Correspondence*, vol. 9; also quoted in R. Jackson, N. Jackson, and D. Brown (eds), *The Poetry of John Tyndall* (London: UCL Press, 2020), p. 6).

From Ellen Busk[1] [30 October 1873][2] 4207

32 Harley St[3] | Thursday

Dear Mr. Tyndall

I hear you are in town today[4] & I write a few lines to thank you heartily in the first place for your book[5] which I was delighted to find here on my return awaiting me with a most pleasant welcome—it was so good of you to send it. And now may I say how deeply grieved I have been for all the annoyance you have lately had[6]—anger & indignation being sadly at variance with prudence—I cannot express what one feels at the cowardly base attacks of people wholly unworthy to comprehend the motives that actuate minds like yours[7]—I wont say another word & only these to assure you of sympathy. When you do come to stay more than a few hours in London will you let us have the pleasure of seeing you here—

with kindest regards | Always truly | Ellen Busk

Poor Sir Henry![8]—You will feel his loss I am sure—Will you let us know if by any chance you remain in town and perhaps you w[oul][d]. come & dine?

RI MS JT/1/B/158
RI MS JT/1/TYP/1/174

---

1. *Ellen Busk*: Ellen Busk (née Busk, 1816–90), who married her cousin George Busk (1807–86), the zoologist and palaeontologist (*ODNB*) in 1843, and subsequently became close friends with both Tyndall and Thomas Huxley, even sharing their sceptical approach to theological matters.
2. *[30 October 1873]*: the date is suggested by reference to the death of Henry Holland (see n. 8), which occurred on 27 October 1873, a Monday, with the following Thursday, when this letter seems to have been written, 30 October.
3. *Harley St*: in Marylebone, central London, where the Busks were neighbours of Charles Lyell, although the geologist's wife declined to invite Ellen to social events (see A. Desmond, *Huxley: The Devil's Disciple* (London: Michael Joseph, 1994), pp. 341–2).
4. *in town today*: Tyndall was staying in Dover conducting acoustic experiments in relation to fog signaling, and only returning to London occasionally and briefly (see letter 4185, n. 3).
5. *your book*: probably *Six Lectures on Light* (1873), which had been published in July.
6. *all the annoyance you have lately had*: the acrimonious dispute over the motion of glaciers that had begun in the late 1850s and recently been reignited by Tyndall's *Forms of Water* and then intensified by the publication of the *Life of Forbes*, co-edited by John Campbell Shairp, Peter Guthrie Tait, and Anthony Adams-Reilly.
7. *people wholly unworthy . . . minds like yours*: as well as the editors named in n. 6, particularly Tait, Busk presumably also had in mind John Ruskin, whose letter 34 of *Fors Clavigera*,

issued in October 1873, attacked Tyndall at length for his part in the glacier controversy (see letter 4185, n. 13).

8. *Poor Sir Henry*: Henry Holland (1788–1873), 1st Baronet, an eminent physician and president of the RI from 1865–73, had died on his eighty-fifth birthday on 27 October (*ODNB*).

From Millicent Bence Jones 2 November [1873][1] 4208

6 Upp[e]r Wimpole St[reet] | November 2

Dear Dr Tyndall

Would you kindly at some moment of leisure read the enclosed[2] and tell me what you think of his proposal as to an extract being made of the memoir—& a few copies printed for circulation amongst particular friends—My finest feeling on reading it was "oh no—it never would have been wished"[3]—& I have delayed the answer hoping for an opportunity of consulting you who better than any one that I think would know is what his feelings[4] would have been on the subject would it have come into his thoughts.

The remembrance of him will remain indelibly in the hearts of some & They would not need any such reminder—but if you have at any time heard what would decide one to adopt the suggestion I should be most obliged if you would tell me. & failing that will you give me your own opinion about it—

If such a thing would in any way do honour to him—or if it could gratify those who loved & valued him—most assuredly it should be done—if any friend would undertake it but it would be for their sake not for his—for I feel on that account aversion to it—

You will understand I am sure what I mean by this—& I will abide by what you say either way. Pray forgive my troubling you—I know you will—

Believe me Always | most truly your's | M Bence Jones

A line by post some day is all sufficient

RI MS JT/1/J/124

1. *[1873]*: the year is established by the letter evidently being written after the death of Henry Bence Jones on 20 April 1873.
2. *the enclosed*: enclosure missing.
3. *his proposal as to ... been wished"*: presumably the 'fragmentary notes of autobiography' that Henry Bence Jones 'dictated in the latter weeks of his lifetime, namely in March and April, 1873' (*Henry Bence-Jones, M.D., F.R.S, 1813–1873. An Autobiography with Elucidations at*

Later Dates (London: Crusha, 1929), p. 3). His wife's feeling seems to have been followed, as the notes were not published until 1929.

4. *his feelings*: that is, Henry Bence Jones's feelings.

To Josiah Whitney[1] 6 November 1873 4209

Royal Institution of Great Britain | London 6th Nov[ember]. 1873

My dear Professor Whitney

I return you most special and hearty thanks for your beautiful present[2] which reached my hand half an hour ago. These photographs will adorn my rooms for a long time to come and will always cause me to think with gratefulness on the man who so kindly sent them to me.

Nor have I forgotten my visit to you in Cambridge,[3] nor the patience you evinced in unfolding to me the glories of the region with which your geological labours had rendered you so familiar.[4] Indeed I cherish a bright and pleasant memory of the whole time which I spent at Boston.

The only thing I regret is my inability to devote more time to the cultivation of acquaintances, and friendships on which I should set an especial value. I was hard worked and in poor health at the time. But memory, as somebody says, throws the weeds to the wall and gathers up the flowers of the past.[5] This I do as regards my visit to Boston & Cambridge.

Will you give my best regards and affectionate remembrances to my friends there?

Say to Agassiz when you see him that batteries to no end are opened upon me from the north;[6] but that by and by their fire shall be returned in a manner that will gratify him. Tell him also that the point of my first Reply was omitted in the abstract given in the Popular Science Monthly.[7]

Ruskin, whom many deem mad—mad I should say with overweening conceit; has made a ferocious attack upon me.[8] Grand old Carlyle, Helmholtz and others have written to me appropos of these assaults,[9] in a manner to make amends for any number of diatribes on the part of Ruskin and those he aids.

But I must not bother you more with these personal matters. Thanking you again & again.

Believe me | most faithfully yours | John Tyndall

William Dwight Whitney Family Papers (MS 555). Manuscripts and Archives, Yale University Library

1. *Josiah Whitney*: Josiah Whitney (1819–96), an American geologist and professor of geology at Harvard University. He was also the state geologist for California, in which capacity he led the California Geological Survey from 1860 to 1874 (*ANB*).
2. *your beautiful present*: letter and enclosures missing.
3. *my visit to you in Cambridge*: Tyndall had visited Cambridge, Massachusetts, while staying in Boston in October and November 1872 during the lecture tour of the United States (see letter 4201, n. 4).
4. *the region with . . . rendered you so familiar*: i.e., the Sierra Nevada range of California. A peak in the Sierra Nevada was named Mount Tyndall in 1864 by Whitney's exploring partner Clarence King (see letter 3719, n. 3, *Tyndall Correspondence*, vol. 13).
5. *memory, as somebody says . . . flowers of the past*: not identified.
6. *batteries to no end . . . from the north*: a reference to the recently published attacks on Tyndall by Peter Guthrie Tate and other Scottish supporters of James David Forbes, with whom Tyndall had been embroiled in an acrimonious controversy over theories of glacial motion.
7. *the point of my first Reply . . . in the Popular Science Monthly*: Tyndall's 'Principal Forbes and his Biographers' was reprinted as 'The Glaciers and Their Investigators', *Popular Science Monthly*, 3 (1873), pp. 746–56, but the editor, Edward L. Youmans, removed a key part of the original article and added a note: 'We omit this portion of the discussion, for lack of space' (p. 753). The *Popular Science Monthly* had been founded in 1872, aimed at communicating science to educated but non-specialist readers.
8. *Ruskin . . . made a ferocious attack upon me*: John Ruskin, a friend and supporter of James David Forbes, had published an attack on Tyndall in his serial *Fors Clavigera* (see letter 4185, n. 13)
9. *Carlyle, Helmholtz . . . of these assaults*: see letters 4188 and 4192.

From Mary Egerton 6 November [1873][1] 4210

Mountfield Court, | Robertsbridge, | Hawkhurst.[2] | Nov[embe]r 6th

My dear Mr Tyndall,

Hearing you are in town all this week, we sent you a brace of pheasants; perhaps they ought to have come in time for the festivities of Tuesday![3] L[a][dy] Claud[4] wrote to me about their pleasant afternoon, and moreover knowing how much I should be interested, she most kindly gives me an account of what you told them about those echoes "from nothing" to which you alluded.[5]

What a marvellous power that transparent water vapour[6] is! It will be intensely interesting to ascertain whether the coarser grained fog is capable of conveying the vibrations the vapour rejects. (If so, somebody may write a homily, & say it was made so on purpose for ships to hear fog signals!!) I wonder whether the mechanical violence of a sudden explosion like a cannon, would penetrate the vapoury atmosphere where the steadier syren fails though ear piercing at other times?[7] But strangest of all, that these great waves of sound should be driven back & affected, by their invisible opponents,[8] to howl & groan under the cliff as in a prison house! With what interest I shall look for your reports on the subject! for I conclude you will publish the scientific results, (as well as the practical ones,) in some form?[9]

And now dear friend, before I close there is one thing I have long been wanting to say to you; that is, on reading again the kind letter you wrote me some time ago,[10] I feel I made it a very ungracious & ungrateful reply![11] In my anxiety to defend myself from the implied imputation of a selfishness which would have made me unworthy of your esteem, I hardly at first appreciated the expressions it contained; but believe me—I do value your assurance of "abiding friendship"; and all the more because I cannot help knowing that yours is so much more to me, than mine can ever be to you. I like all the things, & people, you like at least all I know and in my humble way, I take a warmer interest than I can express, in anything I can gather about the subjects of your pursuits &

<1 or more pages missing>

RI MS JT/1/E/58

1. *[1873]*: the year is established by reference to Tyndall's researches on acoustic transparency (see n. 5).
2. *Mountfield Court, | Robertsbridge, | Hawkhurst*: the Egerton estate in Sussex.
3. *the festivities on Tuesday!*: possibly a further meeting with the Hamilton family on 4 November. It is unclear what was being celebrated as it was not a birthday of any of the family.
4. *L[a]dy Claud*: Elizabeth Hamilton.
5. *those echoes "from nothing" to which you alluded*: Tyndall was evidently discussing the aerial echoes heard in perfectly transparent air, and thus without an obvious atmospheric source, which he had been researching in Dover (see letter 4185, n. 3).
6. *transparent water vapour*: Tyndall had been contesting the reigning assumption that optical transparency could be equated to acoustic transparency. This was a critical issue for his research on lighthouses and whether acoustic signalling could be used in circumstances where fog made lights unsuitable.
7. *I wonder whether the mechanical . . . ear piercing at other times?*: Tyndall had been considering both the reports of cannons and a steam-powered syren sent to him from the United States (see letter 4201) as options for fog signals.
8. *their invisible opponents*: 'acoustic clouds' created by water vapour mixing with the air that, although not visible to the eye, prevent sound from passing through (see letter 4261).
9. *your reports on the subject . . . in some form?*: Tyndall published his results in 'Vehicle of Sound'.

10. *the kind letter you wrote me some time ago*: letter missing.
11. *a very ungracious & ungrateful reply*: possibly letter 4200.

To George Gabriel Stokes 8 November 1873 4211

8th Nov[ember]. 1873

My dear Stokes.

I should very much like to test your explanation of the influence of the wind upon sound,[1] and indeed would incur some expense to satisfy myself experimentally upon the point.

Suppose a gun fired on a plane surface and during a stiff breeze—say a wind with a force of 5 or 6 by Beaufort's scale.[2] At what height do you suppose a captive balloon w[oul][d]. be likely to catch the force of the wave,[3] say at a mile to windward of the gun?

Yours faithfully | John Tyndall

RI MS JT/1/T/1403
RI MS JT/1/TYP/4/1485

1. *your explanation of the influence of the wind upon sound*: G. G. Stokes, 'On the Effect of Wind on the Intensity of Sound', *Brit. Assoc. Rep. 1857*, pp. 22–3.
2. *Beaufort's scale*: a measurement of wind speed devised in 1805 by the Irish naval officer and hydrographer Francis Beaufort (1774–1857).
3. *the wave*: sound wave.

From Baldwin Francis Duppa 9 November 1873 4212

Nov[ember]. 9th 1873

My dear Tyndall,[1]

Sometime back I observed a phenomenon gloriously simple, and therefore just as interesting. You will see above a glass which is supposed to be filled with water, or any other liquid. If you take this glass and, filling it with water, and having a candle or any other source of steady light at a proper angle, you will get a reflection of the ray of light as usual. But if you move your eye just so from the centre of the glass, towards its edge, you will find the reflection slowly cut up as into a series of exquisitely dark or light parallel lines.

I leave you to judge of the extreme simplicity & readiness with which this may be produced. The interest of the thing may be judged by its extreme simplicity.

Your old friend | B F Duppa

RI MS JT/1/D/193
RI/JT/TYP/12/3962

1. *My dear Tyndall*: this letter was dictated by Duppa to his wife Adeline Duppa as one of his final acts before his death on 10 November. Adeline then enclosed it with an explanatory letter of her own (letter 4215) which she sent to Edward Frankland, a mutual friend of Duppa and Tyndall, who passed it on to the latter in letter 4217.

From Henrietta Huxley 9 November 1873 4213

4 Marlborough Place[1] | Nov[ember]. 9th 1873.

Dear Brother John,

True and hearty thanks for your affectionate words of sympathy in dear Hal's recovery.[2] "To weep with them that weep", and "to rejoice with them that rejoice"[3] is to be a friend indeed, and that you have been, and have alike in sorrow and in joy clasped hands with me. It is indeed a pleasure to see Hal so like his former self. Still now and then the shadow comes as it did last night. He eats something that does not suit him or perhaps two ounces too much of a simple food and then gets what he calls the "horrors". I suppose that a long illness cannot pass away all at once, and one must expect a return of unpleasant symptoms sometimes. The anniversary of the Royal Society[4] will be a heavy day for the officers,[5] if there is a reception of ladies in the evening. The business begins at 4 o'clock, then there is the dinner lasting till 10, and finally the Soirée. However "che sara sara".[6] By the above you will perceive that age is creeping upon us, and that we begin to love our own fireside. We have just had our usual Sunday night's exhibition; all the children have either played or sang to us, and Ethel,[7] with the wisdom of 7 years, has declared that she is so fond of pigs and cows and sheep that she will marry a farmer—"but praps his lowland won't be well drained and then I couldn't" Deduce from this that the laws of health are taught here!

You are not to forget that you dine here on New Year's Day, if not before. Last year we missed you so.[8]

Ever your affectionate sister | Nettie Huxley.

RI MS JT/1/TYP/9/3024
LT Typescript Only

1. *4 Marlborough Place*: the Huxley's moved to 4 Marlborough Place, Abbey Road N.W. in in London in mid-1872.
2. *dear Hal's recovery*: see letter 4192, n. 16.

3. *"To weep with them that weep", and "to rejoice with them that rejoice"*: Romans 12:15.
4. *The anniversary of the Royal Society*: the RS was founded 28 November 1660, and celebrated its anniversary around this date each year. In 1873, the anniversary meeting was held on 1 December.
5. *the officers*: Huxley had been elected an honorary secretary of the RS, with responsibility for arranging meetings, in 1872 (*Roy. Soc. Proc.*, 22 (1874), p. 12).
6. *"che sara sara"*: anglicized version of Italian phrase 'quel che sarà, sarà' (what will be, will be), used to signifiy fatalism and acceptance of whatever the future might bring.
7. *Ethel*: Ethel Huxley, the youngest of Huxley's seven surviving children.
8. *Last year we missed you so*: Tyndall had been in New York during his lecture tour of the United States (see letter 4201, n. 4).

From Hermann Helmholtz 11 November 1873 4214

Berlin, 11. Novbr 1873

Verehrter Freund

Ihr Telegramm, worin Sie mir Faraday's Bett anbieten, habe ich erhalten und sage Ihnen meinen besten Dank für die sehr freundliche Absicht. Inzwischen weiss ich noch nicht welche Veranlassung dieses Ihr Anerbieten hervorgerufen hat. Hat Ihnen Jemand erzählt, dass ich kommen wolle, oder liegt eine Veranlassung in London vor, die ich noch nicht kenne?

Darf ich bei dieser Gelegenheit noch eine Frage stellen, über die ich für die Abfassung der Vorrede zu den Fragments vergewissert sein möchte, und die ich damit zu entschuldigen bitte? Sind die Nachrichten der Americanischen Zeitungen wahr, dass Sie ein Stipendium für junge Americaner gestiftet haben, welche die Universität Marburg besuchen wollen? Da die Zeitungen es berichteten, hat es nichts auffallendes wenn ich mich darauf beziehe; nur wünsche ich keine falsche oder halb falsche Nachricht benutzen.

Ich kann mich von meinem Erstaunen über Forbes und seine schottischen Freunde immer noch nicht erholen.

Mit vielem Dank und besten Grüssen | Ihr | H. Helmholtz

Berlin, 11th Nov[em]b[e]r 1873

Dear friend

I received your telegram[1] offering me Faraday's bed[2] and thank you sincerely for the friendly intention. Incidentally, I still do not know the reason that motivates your offer. Did someone say I was coming, or is there something going on in London that I do not know of yet?[3]

May I ask you something else on this occasion, and may I beg you to excuse me for it, as it is important that I have certainty before I formulate the preface to the Fragments?[4] Are the notices in the American papers true that you have donated a stipend to assist young Americans who wish to study at the university in Marburg?[5] As the papers mention it, there is nothing remarkable if I refer to it.[6] It is only that I do not want to rely on false or partially false notices.

I still cannot recover from my astonishment over Forbes and his Scottish friends.[7]

With many thanks and the best greetings | Your | H. Helmholtz

RI MS JT/H/50
RI MS JT/H/50–COPY

---

1. *your telegram*: letter missing, but it was reported to have stated simply: 'If you will come over I will make up Faradays own bed for you' (D. Cahan, 'The Awarding of the Copley Medal and the "Discovery" of the Law of Conservation of Energy: Joule, Mayer and Helmholtz Revisited', *Notes and Records: The Royal Society Journal of the History of Science*, 66 (2012), pp. 125–39, on p. 134).
2. *Faraday's bed*: in the private living quarters at the RI where Faraday had resided, and where Tyndall now lived since 25 January 1868.
3. *is there something . . . I do not know of yet?*: Helmholtz had been awarded the RS's most prestigious honour, the Copley Medal (awarded annually in recognition of outstanding research in any branch of science), at the 1 December 1873 Anniversary Meeting for his work in both physics and physiology (*Roy. Soc. Proc.*, 22 (1874), p. 10). However, Tyndall's telegram offering him somewhere to stay in London reached Helmholtz before the news of his prize (see letter 4216); Helmholtz was not able to travel to London (see letter 4223).
4. *the preface to the Fragments*: see letter 4199, n. 5.
5. *notices in the American papers . . . university in Marburg?*: before Tyndall left England for the United States, he decided that the profits from his lecture tour would be donated to establish a fund for pure science research in American universities. In total, $13,033 was donated and invested in securities yielding 7%, through the trustees Tyndall named—Joseph Henry, Hector Tyndale, and Edward L. Youmans—for 'the advancement of theoretic science, and the promotion of original research, especially in the department of physics, in the United States', in the form of Tyndall scholarships to send American students to German universities. This proved difficult, and following the death of Henry and Tyndale, in 1884 the now $32,000 trust was equally divided among Columbia College, Harvard University, and the University of Pennsylvania. Funds from the trust continued to support students well into the twentieth century. See *Ascent of John Tyndall*, pp. 302-3 and 312–5; D. Finnegan, *The Voice of Science: British Scientists on the Lecture Circuit in Gilded Age America* (Pittsburgh: University of Pittsburgh Press, 2021), p. 59; H. S. Miller, *Dollars for Research: Science and Its Patrons in Nineteenth-Century America* (Seattle: University of Washington Press, 1970), pp. 119–24; and K. Sopka, 'An Apostle of Science Visits America: John Tyndall's Journey of 1872–1873', *Physics Teacher*, 10 (1972), pp. 369–75, on p. 74.

6. *I refer to it*: in his preface (see n. 4), Helmholtz commented: 'Uns Deutschen steht Herr Tyndall überdies näher, als viele Andere seiner Landsleute dadurch, dass er einen Theil seiner Studien in Deutschland (hauptsächlich in Marburg) vollendet hat. Seine Liebe für die deutsche Literatur und Wissenschaft bekundet sich immer wieder in seinen Büchern . . . Dieselbe Dankbarkeit documentirt sich in der Stiftung, die er am Schlusse seiner in America mit dem ungeheuersten Beifalle gehaltenen Vorlesungscurse aus dem Ueberschuss seiner Einnahmen gemacht hat [Moreover, Mr. Tyndall is closer to us Germans than many of his compatriots because he completed part of his studies in Germany (mainly in Marburg). His love for German literature and science is expressed again and again in his books . . . The same gratitude is documented in the donation he made from the surplus of his income at the end of his lecture courses held in America with the most tremendous success]' (p. xix).
7. *Forbes and his Scottish friends*: Scottish geologist James David Forbes, with whom Tyndall had been embroiled in a dispute regarding theories of glacial motion. After Forbes's death, his son George Forbes and other supporters including Peter Guthrie Tait had helped reignite the controversy by publishing attacks on Tyndall.

From Adeline Duppa 12 November 1873 4215

Nov[ember]. 12th 1873

Dear Professor Tyndall,

The enclosed letter[1] was written by my dear husband,[2] under peculiar circumstances, which perhaps may make it more valuable to you—It was the last thing he ever signed, or dictated, on Sunday morning last, at 5 o'clock, he woke up with a most violent paroxysm of coughing & shortness of breath, & we thought he would then have died—also at 11. A. M. when the medical man who was present did not expect he could live five minutes. After that he lay in a semi unconscious state till 2.30 P.M. when he suddenly woke, roused himself, & to my utter astonishment dictated the enclosed letter, & gave the clearest directions, as to the diagram I must draw—I know it is very badly done, but I was not prepared to be called on then to use my pencil so I trust you will excuse it, & may be able to make out its meaning.

For some time poor Frank[3] had intended to write to you on the subject, but his weakness had prevented his doing so—I must enclose this to Dr Frankland,[4] not knowing your address, & I trust you will pardon my leaving it open for him to see, as doubtless Frank's last scientific observations will be especially interesting to him—

Very truly yours | A. F. M. Duppa | Budleigh Salterton.[5] R.M.[6]

My husband died at 2.45 on Monday[7] conscious to the very last—he told me to write to you.

RI MS JT/1/D/193
RI MS JT/1/TYP/12/3962

1. *enclosed letter*: letter 4212. The enclosure and this letter were themselves enclosed in a further letter to Edward Frankland (see letter 4217).
2. *my dear husband*: Baldwin Francis Duppa.
3. *poor Frank*: a short form of Francis, her husband's given name.
4. *I must enclose this to Dr Frankland*: Duppa and Frankland were close friends and worked together at the Royal College of Chemistry, having previously done so at the RI.
5. *Budleigh Salterton*: a coastal town in Devon.
6. *R.M.*: Royal Mail.
7. *Monday*: 10 November.

To Hermann Helmholtz 13 November 1873 4216

Royal Institution of Great Britain | 13th. Nov[ember]. 1873

My dear Helmholtz

I thought my telegram[1] would reach you at precisely the same time as the intelligence that the Council of the Royal Society had unanimously awarded you the Copley Medal;[2] which, as you know, is the highest honour the Society has to bestow.

With regard to the other matter[3] I send you a copy of the deed of trust.[4] I did not wish to tie the Trustees[5] too rigidly down, but remembering my own life in Germany,[6] I expressed to them both in conversation and in writing, my conviction that the object I had in view would be best promoted by sending two American pupils to some German University. The interest of the money invested is now sufficient to maintain constantly two such pupils.

I did not limit the thing to Marburg – In fact, though I did not express it in writing, I had chiefly in view the Universities of Heidelberg and Berlin.

I also send you a fragment of a speech delivered just before I quitted New York[7] and in which this subject of education in Germany is referred to.[8]

I think these documents will give you all the information you need.[9]

Let me repeat my invitation if you should be persuaded to come over and receive the Copley Medal in person.[10] It will be handed to you, or to your representative, usually the Foreign Secretary of the Royal Society,[11] on Monday the 1st. of December. The Fellows have their Anniversary dinner[12] afterwards, where your health will be proposed, and where you, if present, will have to reply.

Every faithfully yours | John Tyndall

BBAW, NL. Helmholtz, Nr. 477, BI_24-5
RI MS JT/1/T/492

---

1. *my telegram*: letter missing, but see letter 4214, n. 1.
2. *Copley Medal*: see letter 4214, n. 3.
3. *the other matter*: see letter 4214, n. 5.
4. *a copy of the deed of trust*: see letter 4214, n. 5.
5. *the Trustees*: see letter 4214, n. 5.
6. *my own life in Germany*: Tyndall had spent three years, from October 1848 to June 1850 and October 1850 to April 1851, studying and researching under Robert Bunsen, Hermann Knoblauch, and Christian Gerling at the University of Marburg in Hesse Cassel, Germany, where he was awarded his PhD. His dissertation, 'Die Schraubenfläche mit geneigter Erzeugungslinie und die Bedingungen des Gleichgewichts für solche Schrauben' (1850), was a mathematical analysis of screw surfaces, completed under the direction of Friedrich Stegmann (1813–91), a German mathematician, professor of mathematics at the University of Marburg since 1845.
7. *a fragment of a speech delivered just before I quitted New York*: several days before his departure from the United States, a lavish farewell banquet was given in Tyndall's honour at Delmonico's restaurant in New York on 4 February 1873. Speeches were given by prominent academic and religious figures, and in his own speech, Tyndall stated that 'Hitherto their efforts have been directed to the practical side of science, and this is why I sought in my lectures to show the dependence of practice upon principles. On the ground, then, of mere practical, material utility, pure science ought to be cultivated'. The speeches were published as *Proceedings at the Farewell Banquet to Professor Tyndall given at Delmonico's, New York, February 4, 1873* (New York: D. Appleton, 1873), as well as in the April 1873 issue of *Popular Science Monthly* (the banquet was also reported, with partial transcripts, in 'The Tyndall Dinner', *New York Times*, 5 February 1873, p. 5).
8. *this subject of education in Germany is referred to*: in the speech, Tyndall stated: 'In 1848, wishing to improve myself in science, I went to the University of Marburg . . . I risked this breach in my pursuits, and this expenditure of time and money, not because I had any definite prospect of material profit in view, but because I thought the cultivation of the intellect important—because, moreover, I loved my work, and entertained the sure and certain hope that, armed with knowledge, one can successfully fight one's way through the world. It is with the view of giving others the chance that I then enjoyed that I propose to devote the surplus of the money which you have so generously poured in upon me, to the education of young philosophers in Germany' ('The Tyndall Dinner', *New York Times*, 5 February 1873, p. 5).
9. *the information you need*: Helmholtz used the information in his 'Vorrede' (Preface) to the German edition of Tyndall's *Fragments of Science* (see letter 4199, n. 5), on p. xix.
10. *come over and receive the Copley Medal in person*: see letter 4214, n. 3.
11. *the Foreign Secretary of the Royal Society*: William Hallowes Miller, who served in this role from 1856–73.

12. *The Fellows have their Anniversary dinner*: an annual event, this year on 1 December, commemorating the founding of the RS on 28 November 1660.

From Edward Frankland 13 November 1873 4217

14 Lancaster Gate[1] | Nov[ember]. 13/73

My dear Tyndall

You will no doubt be interested & touched with the enclosed last written words of poor dear Duppa.[2]

Yours sincerely | E. Frankland.

RI MS JT/1/F/55
RI/JT/TYP/12/3962

1. *Lancaster Gate*: in Bayswater, west of the centre of London.
2. *the enclosed last written words of poor dear Duppa*: letter 4212. Baldwin Francis Duppa had died on 10 November (see letter 4215).

From Thomas Henry Huxley 13 November 1873 4218

Council on Education | Kensington Museum. | Science Schools. | Nov[ember]. 13 1873

My dear Tyndall,

I can only hope that Hodgson[1] will not gratify that lunatic[2] by taking any notice of his letter.[3] I heard of his attack upon you[4] but I have not taken the trouble to look at it. Men don't make war on either women or eunuchs[5] and I hope you will let Ruskin have his squall[6] all to himself.

Ever thine | T. H. Huxley.

IC HP 8.159
Typed Transcript Only

1. *Hodgson*: William Ballantyne Hodgson (1815–80), a Scottish education reformer and political economist. In 1854 he lectured at the RI 'On the Importance of the Study of Economic Science as a Branch of Education for all Classes' (*Roy. Inst. Proc.*, 2 (1851–4), p. 420; published separately in 1860 by William Blackwood and Sons), and in 1858 he was appointed an assistant commissioner to the inquiry into primary education. From

1863–8, Hodgson was examiner in political economy at London University. Although he retired in 1870, the following year he was appointed to the new chair of political economy and mercantile law at the University of Edinburgh (*ODNB*).

2. *that lunatic*: John Ruskin.
3. *his letter*: 'Mr. Ruskin and Professor Hodgson', *Scotsman*, 10 November 1873, p. 3. Ruskin complained that Hodgson had misrepresented him in saying that he had 'denounced' the principles of supply and demand. Hodgson did take notice of the letter, and replied in 'Mr. Ruskin and Professor Hodgson', *Scotsman*, 18 November 1873, p. 2.
4. *his attack upon you*: see letter 4185, n. 13.
5. *eunuchs*: Ruskin's failed marriage to Euphemia Gray was allegedly never consummated, and consequently annulled on the grounds of his 'incurable impotency', causing a public scandal (T. Hilton, *John Ruskin* (New Haven: Yale University Press, 2002), p. 200). Huxley was therefore calling Ruskin's masculinity into question, likening him to a eunuch, the castrated men who acted as servants in various ancient and medieval societies.
6. *squall*: a discordant or violent scream (*OED*).

From Edward Frankland 14 November 1873 4219

14 Lancaster Gate[1] | Nov[ember]. 14/73

My dear Tyndall

Your letter[2] is excellent & will be very gratifying to Mrs Duppa, but I return it to you as I have my doubts about the desirableness of your question,[3] at all events at the present moment; & secondly I can tell you much of what you want to know. I visited him near Maidstone about 2 months ago & he there told me when we were alone together how thankful he felt that his scientific pursuits had freed him from religious superstition & enabled him to contemplate the speedy approach of death with calmness & absence of fear. He then knew that he had but a short time to live & from certain remarks in his wife's letters to me I am convinced that he continued in the same frame of mind to the end. She said he was perfectly happy & resigned to everything except leaving her. This was only a few days before his death.

I have not the least notion what Mrs Duppa's feelings in the matter are. She might answer your question with pleasure or it might cause her intense pain. It is so rarely that women attain to philosophical conception of these things.

Yours sincerely | E. Frankland.

Mrs Duppa's address is | Budleigh Salterton | W. Exeter.[4]

RI MS JT/1/F/56

1. *Lancaster Gate*: see letter 4217, n. 1.
2. *Your letter*: Tyndall had drafted a reply to letter 4215 from Adeline Duppa, which had informed him of the death of her husband, Baldwin Francis Duppa. The text of the draft letter was: 'The letter which you have been good enough to send to me, and the circumstances attending its dictation, are profoundly affecting to me. I had a very deep regard for your husband, both as a highminded gentleman, and one purely devoted to his scientific work. Had he not devoted his time to chemistry he certainly would have reached high distinction in physics. I have indeed rarely met such refined experimental skill or more acute powers of observation than he possessed. Curiously enough on yesterday week Dr. Frankland and I were conversing about him; I had not previously known how dangerously ill he was. Would it be too much to ask you to inform me what his notions of religion were. He appears to me to have died as a philosopher with a calm and untroubled mind. As well as a man of science I shall remember him as a gentle, warmhearted friend' (RI MS JT/1/T/430).
3. *your question*: Tyndall was interested by the calm manner in which Duppa had faced his death, so asked his widow about his 'notions of religion' (see n. 2).
4. *Mrs Duppa's address is | Budleigh Salterton | W. Exeter*: it is probable that Tyndall sent a more tactful reply directly to Duppa, but the letter is now missing.

To George Gabriel Stokes 17 November 1873 4220

17th Nov[ember]. 1873

My dear Stokes

Thanks for both your notes:[1] it will certainly be worth while to bring that wonderfully penetrating explanation to an experimental test—am I at liberty to do so if I can?

Will you kindly give me your counsel here? For months I have been experimenting on the transmission of sound through the atmosphere, with reference to fog signals.[2] I aimed at combining throughout the scientific & the practical, and have amassed material which I believe to be of permanent value. The results while adding much that is new, will, I think, clear away a cloud of errors now associated with this subject.

I believe since D^r. Derham's paper in 1708,[3] the Phil. Trans[4] contains no paper which treats of the atmospheric conditions which affect the <u>intensity</u> of sound.

Now Derham's conclusions, which are quoted by Herschel[5] & others, need serious correction & gratification. The Phil. Trans. is, I believe, the proper place to show this: but I am thus circumstanced. I am writing a Report for the Trinity House,[6] which will be presented to the House of Commons. Will this fact interfere with the publication of the results in the Phil. Trans? These Reports are not accessible to scientific men. The paper which I would propose sending to you[7] would not be a copy of the Report; but would be founded on the Report. In some cases however the phraseology of both would be the same.

Supposing you accept the paper would it be correct to send some portion of it in print?—This would save me the trouble of copying.

One word more:—

Let A & B be two sections, say 10 miles apart, and xy a wall, say a hundred feet high, much nearer to B than to A. Suppose guns fired at B and A. The diffracted sound reaches one station and not the other—Which is the station reached?

Yours faithfully | John Tyndall

RI MS JT/1/T/1404
RI MS JT/1/TYP/4/1486

1. *both your notes*: letters missing. They were presumably replies to Tyndall's letter of 8 November 1873 (4211), in which he enquired about Stokes's theory regarding the 'influence of the wind upon sound'.
2. *experimenting on the . . . with reference to fog signals*: Tyndall was testing different methods for warning ships in fog (see letter 4185, n. 3).
3. *D^r. Derham's paper*: W. Derham, 'Experimenta observationes de soni motu', *Phil. Trans.*, 26 (1708), pp. 2–35. In the paper, William Derham (1657–1735), a clergyman and natural philosopher, published the earliest accurate measurement of the speed of sound.
4. *Phil. Trans*: *Philosophical Transactions*, a scientific journal published by the RS since 1665. Stokes was secretary of the RS from 1854–85, during which time he was responsible for refereeing papers for the *Phil. Trans.*
5. *quoted by Herschel*: J. Herschel, 'Treatise on Sound', in *Encyclopædia Metropolitana*, 30 vols (London: B. Fellowes etc., 1817–45), vol. 4 (1830), pp. 747–820, on p. 748.
6. *a Report for the Trinity House*: 'Report by Professor Tyndall'. On Trinity House, see letter 4187, n. 2.
7. *The paper which I would propose sending to you*: 'Vehicle of Sound'.

From George Gabriel Stokes 19 November 1873 4221

Cambridge, | 19th Nov[embe]r. 1873.

My dear Tyndall,

I have consulted with Miller[1] and we both think that the fact of your investigation being in a blue book[2] w[oul]d not prevent you from bringing it before the Royal Society[3] especially as it will to a certain extent be reworked to adapt it to scientific men rather than the general public.

We have no objection to your sending in a printed form such parts as may be common to both reductions. The fact that the substance of the investigation is in print is openly and honestly avowed. That fact must weigh with each individual member of the Council[4] according to his own notions of what ought to be; but we are not such red tapists[5] as to object to a paper merely because it comes in part in a printed instead of a written form.

As to your question[6] about the effect of an interposed wall in the two cases, I have not yet had time to think over it as I have been in a perpetual whirl of engagements of one kind or other. It certainly is not a question I can answer off hand.

Yours sincerely | G. G. Stokes

RI MS JT/1/S/255
RI MS JT/1/TYP/4/1417

---

1. *Miller*: William Hallowes Miller.
2. *a blue book*: Tyndall was preparing a report on his experiments regarding the best method of signaling to ships in fog (see letter 4185, n. 3), which would be presented to the House of Commons as 'Report by Professor Tyndall'. Parliamentary reports were typically issued in blue covers.
3. *bringing it before the Royal Society*: Tyndall wished to publish his experimental findings in the *Phil. Trans.*, but was concerned the paper would be rejected on the grounds that it had already appeared in print as a parliamentary report (see letter 4220). The paper in question was published as 'Vehicle of Sound'.
4. *the Council*: of the RS.
5. *red tapists*: people who adhere strictly or mechanically to official rules and formalities (*OED*).
6. *your question*: Tyndall had posed a question to Stokes regarding the effect of a wall on the diffraction of sound (see letter 4220).

From George Gabriel Stokes 20 November 1873 4222

Cambridge 20th Nov[embe]r 1873

My dear Tyndall,

On thinking over your question[1] it seems to me that, CD being the wall, B the man, A the far station,

the sound would be better heard with gun at B observer at A than with gun at A observer at B

Yours sincerely | G. G. Stokes

RI MS JT/1/S/256
RI MS JT/1/TYP/4/1418

1. *your question*: see letter 4220.

From Hermann Helmholtz 22 November 1873 4223

Berlin 22. Novbr 1873

Verehrter Freund

Ich habe die Benachrichtigung, dass mir die Ehre der Copley Medaille zu Theil geworden sei schon vor Ihrem letzten Briefe empfangen und viel überlegt, ob ich es nicht möglich machen könnte nach London zu gehen, um sie persönlich in Empfang zu nehmen und meine Dankbarkeit für eine so ausserordentliche Anerkennung auszusprechen; aber selbst, wenn es mir gelingen sollte, die Arbeiten von 4 oder 5 Tagen abzuschieben, so würde die schnelle Hin- und Rückreise in Verbindung mit den Festlichkeiten in London mehr werden, als ich meiner Gesundheit zumuthen darf. Haben Sie also meinen besten Dank für Ihre sehr freundliche Einladung, und verzeihen Sie, wenn ich diese für jetzt nicht annehme; hoffentlich später einmal und dann für etwas längere Zeit als zwei Tage.

Ich hatte dasselbe schon an Professor Miller geschrieben, da ich höre, dass diese Briefe bei dem festlichen Dinner vorgelesen werden, werde ich ihm noch einen ostensibleren Brief zu diesem Zwecke schreiben.

Vielen Dank auch für Ihre Zusendung bezüglich der Americanischen Angelegenheit; das genügt vollkommen für meine Zwecke. Es sind übrigens fast immer einige junge Americaner in meinem Laboratorium, und darunter einige, die sehr gut arbeiten, und grossen Eifer zeigen. Vorläufig stecken wir freilich noch in einem elenden provisorischen Local, und der versprochene Neubau ist bisher noch rein negativ, d.h. Löcher in der Erde, die vielleicht später das Fundament aufnehmen sollen. Inzwischen hat unser Finanzminister Anstoss an der Grösse unserer Wohnungen genommen. Darüber muss vielleicht der ganze Plan noch einmal umgearbeitet werden. Ich habe schon oft bereut, dass ich den Ruf an die Universität Cambridge, der mir gleichzeitig mit dem für Berlin in Aussicht gestellt wurde damals nicht angenommen habe. Wenn man nur noch 10 oder höchstens 20 Jahre Arbeitszeit vor sich hat, und Jahr nach Jahr mit einfältigen und plumpen Hindernissen verdorben wird, kann man zuweilen recht muthlos werden.

Im Augenblicke schreibe ich eine kleine Vorrede contra Zöllner zur zweiten Abth. der deutschen Übersetzung von Thomson Tait Natural Philosophy, wo ich nur berühre, was diese direct angeht. Die Hauptsache werde ich in die Vorrede der Fragments bringen

Mit besten Grüssen | Ihr | H. Helmholtz

Berlin 22. Nov[em]b[e]r 1873

Dear friend

I received the notice that I had been given the honour of the Copley Medal[1] already before your last letter[2] arrived and have been thinking much about whether it would be possible to come to London in order to receive it in person and express my gratitude for such an extraordinary recognition; but even if it were possible to push off work for 4 or 5 days, I do not think my health would stand the journey there and back and the social events in London. Please accept my thanks for your very kind invitation and forgive me if I do not accept it at present; I hope to do so at a later date and for more than two days.

I had written the same to Professor Miller,[3] but as I hear that these letters are read at the celebration dinner,[4] I will write a more ostensible letter for this purpose.

Thank you as well for what you sent regarding the American matter;[5] it is quite sufficient for my purposes. By the way, I nearly always have young Americans in my laboratory and among them some who do great work and work with great zeal. Admittedly, we are still stuck in a miserable provisional location and the promised new building is still entirely negative:[6] that is, there are only a few holes in the ground that perhaps one day will host the foundations. In the meantime, the minister of finances[7] has taken offence at the size of our apartments. Because of this it is possible that the whole plan will have to be reworked once again. I have often regretted that I did not accept the chair that was offered to me at Cambridge University[8] at the same time I received the call to Berlin.[9] If one has at most 10 or 20 years of work left and is hindered year after year by fatuous and clumsy obstacles, one can at times lose all courage.

I am writing a small preface against Zöllner for the second part of the German translation of Thomson and Tait's Natural Philosophy[10] where I merely mention what directly concerns it. I will leave the central point to the preface of the Fragments.[11]

With kind regards | Your | H. Helmholtz

RI MS JT/H/51
RI MS JT/H/51–COPY

---

1. *the Copley Medal*: see letter 4214, n. 3.
2. *your last letter*: letter 4216.
3. *Professor Miller*: William Hallowes Miller.
4. *the celebration dinner*: the anniversary dinner, an annual celebration of the RS's founding on 28 November 1660, at which the Copley Medal was awarded. In 1873, the anniversary meeting was held on 1 December.
5. *the American matter*: see letter 4214, n. 5.
6. *the promised new building is still entirely negative*: in July 1870 Helmholtz, then at the University of Heidelberg, had accepted an offer to become professor of physics at the University of Berlin on the condition that the Prussian government provided funds to build a new physics laboratory. These funds had still not been agreed upon by the end of 1870, largely because of the Franco-Prussian war, but Helmholtz had nevertheless come to Berlin in April 1871. The wrangle over the funds was evidently still continuing more than two years later. See D. Cahan, 'The "Imperial Chancellor of the Sciences": Helmholtz Between Science and Politics', *Social Research*, 73 (2006), pp. 1093–128.
7. *the minister of finances*: Otto von Camphausen (1812–96), who had served as minister of finance for the Kingdom of Prussia since 1869 (*NDB*).
8. *the chair that was offered to me at Cambridge University*: on 28 January 1871 William Thomson had written to Helmholtz on behalf of Cambridge University asking him to consider accepting a new professorship of experimental physics and the directorship of the new Cavendish Laboratory.
9. *the call to Berlin*: Helmholtz had declined the offer from Cambridge and finally agreed to accept the alternative professorship at the University of Berlin in April 1871 despite his original conditions still not being met (see n. 6).

10. *a small preface against Zöllner . . . Natural Philosophy*: H. Helmholtz, 'Vorrede zum Zweiten Theile des Ersten Bandes', in W. Thomson and P. G. Tait, *Handbuch der theoretischen Physik*, vol 1, part 2, trans. H. Helmholtz and G. Wertheim (Braunschweig: Friedrich Vieweg, 1874), pp. v–xiv. Helmholtz responded to the criticisms of Johann Karl Friedrich Zöllner, made in his book *Über die Natur der Cometen* (Leipzig: W. Engelmann, 1872), of William Thomson and Peter Guthrie Tait's use of scientific method, particularly in relation to metaphysics and deduction.
11. *the preface of the Fragments*: see letter 4199, n. 5.

From Millicent Bence Jones [1 December 1873][1] 4224

6 Upper Wimpole S[tree]t | Monday

My dear Dr Tyndall

I wish now I had spared you the pain I fear it gave you to look at that portrait[2] without previous intimation! I did write a note but found it impossible to say what I felt—& I gave it up—hoping that my name being on the outside would be sufficient—& I hoped my sending the portrait would say better than words what I wished to express—the great regard I must ever feel for one who loved & was so much beloved by him.

Mr Mitchell[3] kindly had two of the engravings framed one for you & the other for Mr Prescott Hewett[4]—& I was anxious they should both reach before the engraving had been seen by others. Millie[5] took them both to their destinations with a smaller one for Miss Savage[6] from Olivia.[7]

I do not know if it is to you my thanks are owing next to Mr Spottiswoode for his most kind admission to the lectures at the Royal Institution[8]—if so, pray accept my warmest thanks as well as those of my daughter Olivia, on whom he bestowed the same kindness—she is so greatly satisfied by it—& you will believe how much touched I feel by it.

Believe me always | Most truly yours | M Bence Jones

RI MS JT/1/J/128
RI MS JT/1/TYP/3/738

---

1. *[1 December 1873]*: the date is suggested by the publication date of the engraving referred to in the letter (see n. 2). 1 December was a Monday, and Bence Jones seems to have sent the engraving as soon as it was published.
2. *that portrait*: an engraving of Henry Bence Jones based on a chalk drawing of him made by the artist George Richmond in 1865. The engraving was made by Charles Holl and inscribed 'Published Dec 1st, 1873'.

3. *Mr Mitchell*: John Mitchell (*fl.* 1833–74), a bookseller and printer who traded from 33 Old Bond Street, Mayfair, and was the publisher of the engraving of Henry Bence Jones.
4. *Mr Prescott Hewett*: Prescott Hewett (1812–91), a surgeon and anatomical lecturer who had recently become president of the Clinical Society of London (*ODNB*).
5. *Millie*: Bence Jones's eldest daughter Millicent (see letter 4195, n. 4).
6. *Miss Savage*: Alice Savage (1806–1903), daughter of William Savage, the RI's printer and clerk earlier in the nineteenth century. She was unmarried and served as the RI's housekeeper from 1865 to 1900.
7. *Olivia*: Bence Jones's daughter (see letter 4195, n. 4).
8. *the lectures at the Royal Institution*: presumably that year's Christmas lectures for youth at the RI, 'On the Motion and Sensation of Sound', which Tyndall delivered on 27 and 30 December 1873, and 1, 3, 6, and 8 January 1874 (*Roy. Inst. Proc.*, 7 (1873–75), p. 167). The Christmas lectures began in 1827 and still continue. Tyndall delivered them a dozen times over his time with the RI, his first in 1861 (see F. A. J. L. James (ed), *Christmas at the Royal Institution: An Anthology of Lectures* (Hackensack, NJ: World Scientific, 2007)).

To Thomas Archer Hirst 2 December 1873 4225

Tuesday | Dec[ember] 2nd. 1873

My dear Tom.

I have been speaking to Paget[1] & to Fox.[2] It is deemed absolutely necessary that you should overcome all impatience, and conform implicitly to the measures prescribed for you.[3] If this be done nothing, I am informed, is to be feared.

Send me word by Mrs Tunstal[4] how you are.

Yours ever aff[ectionatel]y | John Tyndall

RI MS JT/1/T/710
RI MS JT/1/HTYP/624

1. *Paget*: James Paget (1814–99), an English surgeon and pathologist based at St Bartholomew's Hospital in London (*ODNB*).
2. *Fox*: Wilson Fox.
3. *the measures prescribed for you*: Hirst had been suffering from a number of ailments, including chills and numbness in his legs, since 8 November. In his journal he noted that 'Fox after consultation recommended me to apply warmth to the spine, to sit still at home, keep myself warm and do as little work as possible'. He later recorded that 'Wilson Fox came once a week to see me and in spite of my impatience insisted on perfect stillness and rest' (*Hirst Journals*, pp. 1993 and 1994).

4. *M*[rs] *Tunstal*: Jane Tunstall.

From Jane Barnard 5 December 1873 4226

Barnsbury Villa, | 320, Liverpool Road, | N.[1]

Many thanks dear Dr Tyndall for your kindness in sending the tickets.[2] I have looked among my uncle's papers[3] to see if I could find any letters or notes about Dr Robinson[4] & the Fog Committee,[5] but do not. I am sorry to say the letters & papers are in a very different condition from what he left them. My Aunt[6] remembers that he was vexed about the matter:—she sends her affectionate regards.

Believe me yours | very sincerely | Jane Barnard | D[r] Tyndall | &c &c &c | 5 Dec[ember] 1873

RI MS JT/1/B/51

---

1. *Barnsbury Villa, | 320, Liverpool Road, | N.*: Sarah Faraday and Jane Barnard moved from Hampton Court to Barnsbury Villa in Islington shortly after Michael Faraday's death in August 1867 (Frank A. J. L. James, personal communication).
2. *the tickets*: for Tyndall's Friday Evening Discourse 'On the Acoustic Transparency and Opacity of the Atmosphere', which he delivered at the RI on 16 January 1874 (*Roy. Inst. Proc.*, 7 (1873–75), pp. 169–78). See letter 4252.
3. *my uncle's papers*: those of Michael Faraday.
4. *D*[r] *Robinson*: Thomas Romney Robinson.
5. *the Fog Committee*: at this time, Tyndall was writing a report to Parliament regarding his fog signal experiments (see letter 4220), and his request that Barnard look through her uncle's papers relates to an incident involving Faraday and Robinson over a decade earlier. Robinson was chairman of a committee, appointed in 1861 by the BAAS, with the purpose of advocating for experimental investigation into methods of signaling ships in fog. This committee had addressed a letter to the Board of Trade, which was published in *Brit. Assoc. Rep. 1863*, pp. 105–10. This was passed on to Trinity House, who had consulted Faraday, their official scientific adviser, who agreed that such experiments were desirable, but nevertheless advised against them on the grounds of difficulty and expense. Consequently, Trinity House did not pursue the subject any further at that time. Tyndall, who succeeded Faraday as the scientific adviser to Trinity House (see letter 4187, n. 2), felt it necessary to defend Faraday's position in the conclusion of 'Report by Professor Tyndall'.
6. *My Aunt*: Sarah Faraday.

From Hector Tyndale 5 December 1873 4227

The Arlington.[1] | Washington, D.C. Dec[em]b[e]r 5th 1873

My dear John

I have carried my debt of correspondence to you from the longest Summer to the Shortest Winter day,[2] but one or another matter has happened to engross me and to, temporarily, take possession of my thoughts and feelings to the exclusion of all else. I read your last[3] with interest and am obliged to you for it as well as for the Reply to the Biography of Principal Forbes.[4] I think you have replied to them[5] with great force and quietly as well.

I thank you also for the New edition of your American lectures[6] which I suppose completes your hard labours connected with those lectures. These were so great as to have destroyed any mere pleasure you might have derived from your visit here. I am on a short business visit here and, being near your old quarters,[7] you are rebrought before me and the occupations of these last months being off me, I am free to write you a word of remembrance. I have not yet seen Prof: Henry, and perhaps I may not, but have seen Senator Schurz[8] who remembers you "with pleasure". I see many of your friends in Philadelphia often and they nearly always ask about you—Mr Carey,[9] Prof Lesley and family, Mr Patterson,[10] Mr Childs,[11] Dr Gross Dr McClellan,[12] Dr Morehouse[13] and others. The other evening I met Doctor Gross who asked me very particularly about you and scolded me sharply when I told him that I had not written to you for a long time. He desired to be remembered to you and hoped you would come out again under less pressing demands of work and with more time for society and travel. In fact you have a great many friends and well wishers all over this country. A few days since I saw in the papers the announcement of the death of Herbert Bright (of Philadelphia)[14] who was the young English man with me when I met you on the wharf at Jersey City. He seemed an amiable & kindly man. All my people are now pretty well. Mrs Tyndale has had much neuralgia & Miss Nowlen[15] is never well altogether. You must excuse anything undecipherable in this, as it is a bleak, dark day and the snow fills the air and I can hardly see.

As for myself I am still drifting with the current—with no special course and yet like the sea weeds of Columbus[16] I may bear tidings to some one of a land somewhere. Make my regards to Hirst and Debus of whom I shall be glad to hear. When you write to Gorey,[17] give my kindest regards to all there and especially to your sister.[18]

Affectionately | Hector Tyndale

RI MS JT/1/T/75
RI MS JT/1/TYP/5/1723–4

---

1. *The Arlington*: a grand hotel in Washington, DC that had opened in 1868.
2. *from the longest Summer to the Shortest Winter day*: that is, with the winter solstice—21 December—approaching, Tyndale realized that he had not written since the summer solstice, 21 June. That previous letter is missing.
3. *your last*: probably letter 4121, *Tyndall Correspondence*, vol. 13.
4. *the Reply to the Biography of Principal Forbes*: probably J. Tyndall, *Principal Forbes and his Biographers* (London: Longmans, Green, and Co., 1873), which was a pamphlet reprinting the article of the same name published in the *Contemporary Review*, 22 (1873), pp. 484–508. It was the pamphlet version that Tyndall sent to other correspondents (see letter 4356).
5. *them*: John Campbell Shairp, Peter Guthrie Tait, and Anthony Adams-Reilly, the co-editors of the *Life of Forbes*.
6. *the New edition of your American lectures*: *Six Lectures on Light* (1873). A second edition of the lectures was not published by Longmans in London until 1875, so Tyndale is presumably referring to the first American edition, published in New York by D. Appleton & Co., which contained a different appendix.
7. *your old quarters*: Tyndall had stayed at the Welckerie Hotel when he visited Washington, DC in December 1872 during his American lecture tour (see letter 4201, n. 4).
8. *Senator Schurz*: Carl Schurz (1829–1906), a German-born statesman who had fled across the Atlantic after the failure of the revolutionary uprisings of 1848. He served with Tyndale in the Union Army during the American Civil War (*ANB*).
9. *Mr Carey*: probably Henry Charles Carey (1793–1879), an Irish-American economist, and lifelong resident of Philadelphia, who was a prominent advocate of the American School of economics (*ANB*).
10. *Mr Patterson*: probably Joseph Patterson (1808–87), a Philadelphia merchant who from 1842 was president of Western National Bank in Philadelphia (*The Bankers' Magazine and Statistical Register*, 42 (1887), p. 315).
11. *Mr Childs*: probably George William Childs (1829–94), co-owner of the Philadelphia newspaper the *Public Ledger*.
12. *Dr McClellan*: George McClellan (1849–1913), a Philadelphia-based surgeon and anatomist. He was one of two Philadelphia physicians who cared for George G. Miller (*c.* 1854–72), Tyndall's younger assistant at the RI who joined him on the American lecture tour but succumbed to typhoid fever in Philadelphia on 30 November 1872.
13. *Dr Morehouse*: George Reed Morehouse (1829–1905), a Philadelphia-based physician.
14. *the death of Herbert Bright (of Philadelphia)*: Bright had died on 21 November. In an obituary, he was described as 'a young English engineer of brilliant talents and most versatile powers' ('Current Notes: Obituary', *Industrial Monthly*, 5 (1874), p. 27).
15. *Miss Nowlen*: presumably an unmarried sister of Julia Tyndale.

16. *the sea weeds of Columbus*: in 1492 the presence of weeds, actually algae, in the central North Atlantic had led the explorer Christopher Columbus (1451–1506) to assume that his ships were close to land.
17. *Gorey*: a town in County Wexford, Ireland, where Tyndall's cousin, also named John Tyndall, was a physician.
18. *your sister*: Emily Tyndall.

From John Herschel 6 December 1873 4228

Collingwood[1] | Dec[ember]. 6th. 1873

Dear Sir,

My recent return from India on furlough[2]—which will be of some length—and the near approach of my eldest brothers departure from England,[3] have combined to place me in a position in which the duty devolves especially upon me of taking steps to anticipate the possible event of a publication of my father's letters. By my brother's advice and with the concurrence of my mother[4] and of the other members of the family, I venture to address you with a view to securing your assistance.

It is our wish to form a collection of my father's letters, not so much with a view to publication as to provide against the too probable loss or destruction which takes place with time, as well as to guard against any more fragmentary collection.

I am aware that such applications imply publication in some shape. In the present case I am bound to state that no such intention has been formed; and am conscious therefore that the guarantees which alone I can offer must be less definite than they might otherwise be. I can however assure you, for myself and in behalf of the rest of the family, that any letters which you may entrust to us for transcription shall be returned to you with as little delay as possible, and that the transcripts shall be kept with as scrupulous regard to the nature of their subjects as though they were the writer's own drafts or copies—some of which of course are actually in our hand at this moment.

I am not unmindful of the trouble which we may be giving in making such a request. It is however for you to judge whether it is not a proper one; and to what extent it may be in your power to comply with it. I can do no more than offer our assurance that your response[5] if favourable will be gratefully valued; and, permitting myself to entertain the hope that it will not be otherwise,

I beg to remain | Yours respectfully | J Herschel, Capt[ain]. R.E.

To | Prof. J. Tyndall | L.L.D. F.R.S &c &c

RI MS JT/1/H/118
RI MS JT/1/TYP/2/580

---

1. *Collingwood*: Herschel's home, in Hawkhurst, Kent.
2. *furlough*: a permit or licence giving a member of the armed forces permission to be absent from duty for a certain period of time (*OED*). Herschel was an officer in the Royal Engineers, with the rank of Captain at this time. He had been stationed in India from 1864 to 1873, engaged in the 'Great Trigonometrical Survey', a project that aimed to measure accurately the whole Indian subcontinent.
3. *my eldest brothers departure from England*: William James Herschel (1833–1917), a member of the Indian Civil Service, had presumably returned to Britain upon the death of his father in 1871 but was now returning to India.
4. *my mother*: Margaret Brodie Herschel (née Stewart, 1810–84).
5. *your response*: Tyndall's response is missing, but letter 4364 makes it clear that he acceded to Herschel's request.

To Frederick Barnard 10 December 1873 4229

Royal Institution of Great Britain | London 10th Dec[ember]. 1873

My dear Dr. Barnard.

It is always a pleasure to me to read any thing from your clear and vigorous pen. What you write presents to me a man standing on firm foundations unsurrounded by haze or mystification of any kind. This last paper of yours[1] is excellent and will do good. I am not surprised to find you attaching so much importance to Bastian's work[2]—Still there is not a man of my acquaintance of any scientific weight, and I number among my acquaintances many who know Bastian's calibre and method of work, who attaches any importance to his results. All his more startling ones are to be ascribed to the fact that a man undisciplined in experiment has taken up a subject which requires for its treatment the most consummate experimental tact.

Another part of your paper[3] sobered me a little, if it did not sadden me. A future life is an inheritance in the expectation of which we have been all brought up and which can never be relinquished without pain. But many of us will have to march manfully into the grave leaving this an open question. Carlyle does so,[4] Tennyson does not,[5] but thinks he has a right to immortality—I think Carlyle's position the nobler of the two.

"Because right is right to follow right | In scorn of consequence"[6]

seems to be old Carlyle's motto. It is sometimes annoying to me to observe what little apparent effect this tremendous hope of immortality has on many of those who profess to hold it. I fear they do not all realize the creed they profess to hold.

But I must not bother you with my doubts and hesitations regarding this great theme, but ask you to believe me

in all sincerity | most faithfully yours | John Tyndall

Give my very kindest regards to M^rs^. Barnard.[7]

Pray kindly remember me also to my friends at the College[8]

Columbia University Manuscripts; Box 2 Folder 94; University Archives, Rare Book & Manuscript Library, Columbia University Libraries, Rare Book & Manuscript Library, Columbia University in the City of New York

1. *This last paper of yours*: F. A. P. Barnard, 'The Germ Theory of Disease and its Relations to Hygiene', *Public Health Paper Report*, 1 (1873), pp. 70–87. It discussed theories regarding the origins of life, and addressed the philosophical issues raised by the possibility that living organisms had arisen spontaneously out of non-living matter rather than through an act of God.
2. *attaching so much importance to Bastian's work*: Barnard stated 'one can hardly repress the suspicion that if there be any such thing as spontaneous generation, it is a thing which occurs only under rare and extraordinary conditions, which conditions Dr. Bastian has unintentionally succeeded in establishing, while as a matter of practical importance or daily interest it is as if it were not' (p. 79). Henry Charlton Bastian (1837–1915), a British physician and neurologist (*ODNB*), advocated a controversial theory regarding spontaneous generation, which led to confrontational disputes with numerous scientists and clinicians, including Tyndall, Thomas Huxley, Joseph Lister, and Louis Pasteur, who all supported the increasingly popular germ theory, as illustrated by many of the letters in the forthcoming fifteenth volume of *The Correspondence of John Tyndall*. See also J. Strick, *Sparks of Life: Darwinism and the Victorian Debates over Spontaneous Generation* (Cambridge, MA: Harvard University Press, 2000).
3. *Another part of your paper*: in a section entitled 'Bearing of the Question on the Future of Intelligence', Barnard declared: 'Much as I love truth in the abstract, I love my hope of immortality still more; and if the final outcome of all the boasted discoveries of modern science is to disclose to men that they are more evanescent than the shadow of the swallow's wing upon the lake, it seems to me no better than a heartless mockery to talk of the countless treasures which, along with this withering revelation, she has poured out at their feet. No, if this, after all, is the best that science can give me, give me then, I pray, no more science. Let me live on, in my simple ignorance, as my fathers lived before me, and when I shall at length be summoned to my final repose, let me still be able to fold the drapery of my couch about me, and lie down to pleasant, even though they be deceitful dreams' (p. 80).
4. *Carlyle does so*: Tyndall was perhaps thinking of Thomas Carlyle's statement 'In all ages, those questions of Death and Immortality . . . must, under new forms, anew make their

appearance; ever, from time to time, must the attempt to shape for ourselves some Theorem of the Universe be repeated. And ever unsuccessfully... A region of Doubt, therefore, hovers forever in the background; in Action alone can we have certainty' ('Characteristics', *Edinburgh Review*, 54 (1831), pp. 351–83, on p. 370).

5. *Tennyson does not*: Alfred Tennyson affirmed a hope for personal immortality in several of his poems, most famously in the closing section of the elegy *In Memoriam* (1850).
6. *"Because right is right to follow right | In scorn of consequence"*: a misquotation from Tennyson's poem 'Oenone' (1829). The actual lines are: 'And, because right is right, to follow right | Were wisdom in the scorn of consequence'.
7. *Mrs. Barnard*: Margaret Barnard.
8. *the College*: Columbia College, New York (now Columbia University), where Barnard was President from 1864–89.

To Martha Somerville 10 December 1873 4230

10th December | 1873

My dear Miss Somerville.

Some days ago I received from your publisher a copy of the "Recollections"[1] for which I ask you to accept my heartiest thanks. I have read it through with the greatest interest, and have now a pretty clear image of the life and action of that fine nature whom you have the rare privilege of calling mother.[2] Your own part of that volume has been executed with good taste and great efficacy: the winding up[3] I think particularly fine.

I have been very hard at work since I saw you last, and as a consequence am not so strong a man now as I was when I quitted you at the Riffelberg.[4] I have been trying to get some clue to the caprices of the atmosphere as regards the transmission of sound,[5] and I hope I have succeeded. But the work has made me tired and thin & pale.

The evening after your volume came to me I met Lady Bell[6] and spoke to her about your book. I promised to lend it to her & purpose doing so today. How delighted she will be with your mother's reference to her husband & herself.[7]

I would open all doors to the education of women; but had I known your mother's views on the suffrage question[8] I should have argued the point with her when I had the pleasure & the privilege of seeing her at Naples.[9]

You, I dare say, are now in sunshine; but when I rose this morning it was absolutely impossible to see across Albemarle Street.[10] We are swathed in fog, which has come after I had been vainly seeking for it for months.[11]

I hope you continued to enjoy your life in Switzerland: and I hope that both you and your sister[12] reap the benefit of your tour, and are in the enjoyment of that good health which is the basis of all happiness.

Pray give her my warmest remembrances. | and believe me ever | faithfully yours
John Tyndall

Oxford, Bodleian Library, Dep c.372 (MST–1)

---

1. *a copy of the "Recollections"*: M. Somerville, *Personal Recollections*, ed. M. Somerville (London: John Murray, 1874), an edition of Somerville's late mother's account of her life, together with some of her unpublished papers.
2. *that fine nature . . . privilege of calling mother*: Mary Somerville.
3. *the winding up*: Somerville had added a coda to her mother's recollections, reflecting on her religious beliefs and stating that the 'theories of modern science she welcomed as quite in accordance with her religious opinions' (p. 375).
4. *I quitted you at the Riffelberg*: this was a hotel in the Swiss Alps, near to the mountaineering resort of Zermatt. Tyndall had met Somerville there on 10 August 1873, when Thomas Hirst had recorded in his journal: 'We crossed the Aletsch and the Rieder Alp to the Hotel on the other side where we called upon the two Misses Somerville daughters of the celebrated Mrs Somerville. They were pleasant intelligent ladies great admirers evidently, of Tyndall' (*Hirst Journals*, p. 1982).
5. *the caprices of the atmosphere . . . transmission of sound*: Tyndall had been conducting acoustic experiments to find the most effective way of signaling to ships in fog (see letter 4185, n. 3).
6. *Lady Bell*: Marion Bell (née Shaw, 1787–1876), widow of the anatomist Charles Bell (1774–1842).
7. *your mother's reference to her husband & herself*: in her *Personal Recollections*, Mary Somerville noted: 'What a contrast the refined and amiable Sir Charles Bell formed with Majendie! Majendie and the French school of anatomy made themselves odious by their cruelty, and failed to prove the true anatomy of the brain and nerves, while Sir Charles Bell did succeed, and thus made one of the greatest physiological discoveries of the age without torturing animals, which his gentle and kindly nature abhorred. To Lady Bell I am indebted for a copy of her husband's Life. She is one of my few dear and valued friends who are still alive' (pp. 192–3).
8. *your mother's views on the suffrage question*: in her *Personal Recollections*, Mary Somerville commented: 'I joined in a petition to the Senate of London University, praying that degrees might be granted to women; but it was rejected. I have also frequently signed petitions to Parliament for the Female Suffrage, and have the honour now to be a member of the General Committee for Woman Suffrage in London' (p. 346).
9. *seeing her at Naples*: Tyndall had met Mary Somerville when he was in Naples to visit Pompeii and observe Vesuvius in April 1868. In her *Personal Recollections*, Somerville recalled:

'I owe to Vesuvius the great pleasure of making the acquaintance . . . of Sir John Lubbock, and Professor Tyndall, who had come to Naples on purpose to see the eruption. Unfortunately, Sir John Lubbock and Professor Tyndall were limited for time, that they could only spend one evening with us; but I enjoyed a delightful evening, and had much scientific conversation' (pp. 342–3). See letter 2801, n. 3, *Tyndall Correspondence*, vol. 10.

10. *Albemarle Street*: where the RI is located, in Mayfair.
11. *swathed in fog . . . seeking for it for months*: Tyndall was eager to continue his experiments on the acoustic implications of fog (see letter 4185, n. 3), and later recorded: 'We had . . . been favoured with thunder, hail, rain, and haze, but not with dense fog. All the more anxious was I to turn the recent excellent opportunity to account. On Tuesday, December 9, I therefore telegraphed to the Trinity House, suggesting some gun observations' ('Report by Professor Tyndall', p. 65)
12. *your sister*: Mary Charlotte Somerville.

To Alfred Tennyson 12[?][1] December 1873 4231

12th December 1873

My dear Tennyson

It was very good of you to call upon me, and I should long ago have called upon you did I not know that you were only too sure to be sufficiently visited. I walked over to you on my arrival here[2] yesterday, and rang your bell once. But as nobody came, and as it was past 5oC[lock]. I came away with the purpose of taking my chance of seeing you today.

They tell me, somewhat complainingly, that you do not dine out; and I reply that you show your wisdom in not doing so. How London gets on with its dining out I do not know, for to me the thing is so evil that it is pure sin and wickedness to indulge in it. It may be questioned however whether this is a thing to be regretted. I sometimes say to a friend of mine who can drink beer that if we begin with the lower animals the ox and the ass are formed possessing stomachs capable of dissolving the almost insoluble modicum of nutriment incorporated in grass or straw. And passing from quadrupeds to bipeds we find the Esquimaux[3] no longer able to eat grass, but still to eat blubber and hides. Higher still we find the English navigator[4] demanding at least bread and bacon and beer. Thus we give that the lower the nature the stronger is the gastric juice, so that really the care as regards food necessary on the part of a man like yourself is simply the result of a greater devotion in the scale of being. A common nature can endure bad wine, whereas a pure one is upset by a single glass of the compounds so common at London dinner parties. Let us then conclude, and comfort ourselves in the conclusion, that our fastidiousness of stomach is a thing to be desired, not deprecated; and let those who can consume London dinners not complain if they find us incapable of descending to their level.

I will now end my theory, and will call upon you (but will not bore you by a long visit at about 4:30)

Yours ever | John Tyndall

TRC Letters 6246

1. *12[?]*: although this date is written clearly on the original letter, Tyndall was conducting experiments on the acoustic implications of fog in London until 13 December (see 'Report by Professor Tyndall', p. 65), so he seems to have written the wrong day. He was back in London by 21 December (see letter 4234), and this letter was written the day after Tyndall arrived on the Isle of Wight, so the correct date is presumably sometime in the week commencing 15 December.
2. *here*: presumably the Isle of Wight, where Tennyson had lived, at Farringford House in the village of Freshwater Bay, since 1853. Tennyson also had another residence, Aldworth House, near Haslemere in Surrey, which was built in 1869, but he tended to use this only in the summer to escape tourists visiting Farringford House.
3. *the Esquimaux*: the Indigenous peoples of the Arctic; now considered to be an offensive term.
4. *the English navigator*: a manual labourer engaged in constructing railways, more commonly referred to as a 'navvy'.

From Henrietta Huxley 16 December [1873][1] 4232

4 Marlborough Place | Abbey Road[2] | 16th Dec[ember].

Dear Brother John

Many thanks to you—The young folks will be delighted to have the chance of hearing your Xmas lectures[3]—and I too if I can spare the time.

To think of the fog being a Godsend to you[4]—a blessing in disguise! for us—it was a disguise and disfigurement and anything but a delight.

Hal[5] is fairly well—but—there is no use in disguising it, he is years older than before he fell ill.[6] He goes on ploddingly with his work—but he lacks the old fire—which strength gave him. I wish he could spare himself from evening work, but that is impossible. You will I know keep my report of him to yourself—for I know it does him harm, if it gets about that he is not as strong as ever. The truth is we are all getting old. No one feels this more than your sister.[7] Sometimes when I get into an ungrateful humour—I say—"At least one will have rest when one dies". And yet in my heart I know that I would rather work on—than have that rest. But as years increase, one gets so much more impressed with the sadness of life—and the thought—"Why all this toil? For what"?

You will think this a doleful screed—but my mind has been full of the misery of poor Major Donnelley[8] who has just lost his sweet young wife,[9] and every thing takes colour from this.

Don't forget that you are bound to us for New Year's Day ¼ to 7.

Always with love | affectionately Yours | Nettie Huxley.

RI MS JT/1/TYP/9/3026
LT Typescript Only

---

1. *[1873]*: the year is established by reference to the death of Adeliza Ballantine Donnelly (see n. 9).
2. *4 Marlborough Place | Abbey Road*: see letter 4213, n. 1.
3. *your Xmas lectures*: see letter 4224, n. 8.
4. *the fog being a Godsend to you*: Tyndall had been conducting experiments regarding the effect of fog on the transmission of sound through air, and heavy fog had recently descended on London (see letter 4230, n. 11).
5. *Hal*: Thomas Huxley, Hal being a nickname for Henry, as used by William Shakespeare in *Henry IV* (1598–9).
6. *he fell ill*: Huxley had been suffering from ill health, both physical and mental, for the past two years (see letter 4192, n. 16).
7. *your sister*: i.e. Huxley herself, who referred to Tyndall as 'Brother John'.
8. *poor Major Donnelley*: John Fretcheville Dykes Donnelly (1834–1902), an army officer and promoter of scientific education. In 1859 he became Director of Science in the Department of Science and Art at the South Kensington Museum (*ODNB*).
9. *his sweet young wife*: Adeliza Ballantine Donnelly (née Dykes, 1847–73). She died on 5 December.

From Mary Egerton 18 December [1873][1] 4233

Mountfield Court, | Robertsbridge, | Hawkhurst.[2] | Dec[embe]r. 18th.

Oh! I am So Cross!

So you are going to lecture before Easter,[3] & on a quite new subject, & I have just let the house![4] Perhaps it is as well for my pocket I did not know it three days ago, for looking at it in a practical point of view, it w[oul][d]. be cheaper to go up & down[5] for them than to lose 250 guineas but then the trouble! And Tuesdays too! so that one can't continue them with the Fridays! I only hope they will be the last six before Easter & not the first,[6] so that the fogs may be a little over? (Did you not long for the Syren[7] this last week?)

I rather expected this, for I thought there had not been much preparation for Belfast[8] done this year in Switzerland, but Mr Spottiswoode said so positively there was to be no change, that I thought he must know; so we were planning that you w[oul]^d^. be able to come to us for a Sunday or two, in the comparatively leisure time before Easter; but now I suppose you will want all the Mondays for preparations.

Do you hope by this means to get your Belfast work done so as to have your holiday free next summer? (I do not know what the day of the B. Ass[ociatio]^n^ meeting is to be?)[9] If so I will try to think that is best for you, and will try not to growl & grumble more than I can help! But do write someday & tell me your plans & arrangements, and don't call me your "Excellent" friend! it puts me in mind of what Mr Hirst said, about its being "what country gentlemen call each other at Agricultural dinners!" I am not 'Excellent' a bit, but only your truly attached friend,

M F Egerton.

I was glad to hear an improved account of Mr. Hirst from his niece.[10] I trust it continues. Poor M^r^ De la Rive[11] must be a loss to science as well as to his friends! I wrote to Mrs Sarasin[12] about it.

RI MS JT/1/E/63

1. *[1873]*: the year is established by reference to the death of Auguste Arthur de la Rive (see n. 11).
2. *Mountfield Court, | Robertsbridge, | Hawkhurst*: the Egerton family estate in Sussex.
3. *lecture before Easter*: Tyndall was scheduled to deliver six lectures 'On the Physical Properties of Gases and Liquids' at the RI every Tuesday between 17 February and 24 March 1874 (*Roy. Inst. Proc.*, 7 (1873–75), p. 167).
4. *the house*: 45 Eaton Place, Belgravia, the London residence of Egerton's late husband Edward Christopher Egerton, as opposed to the family's country seat of Mountfield Court.
5. *go up & down*: i.e. travel from Mountfield Court in Sussex to London and back.
6. *the last six before Easter & not the first*: 'Before Easter' was a twelve-week season in the RI's lecture schedule, which in 1874 lasted from 13 January to 28 March; thus Tyndall's lectures were, as Egerton hoped, towards the end of the season.
7. *the Syren*: probably the steam-powered syren Tyndall had been using in his fog signal experiments (see letter 4201, n. 2). However, it might also be an allusive reference to Louisa Hamilton (see letter 4200).
8. *preparation for Belfast*: Tyndall was scheduled to give the presidential address at the annual meeting of the BAAS in Belfast the following year.

9. *what the day of the B. Ass[ociatio]ⁿ meeting is to be*: the meeting began on 19 August 1874, when Tyndall gave his address, and ran for a week.
10. *his niece*: Emily Hirst. Thomas Hirst had been suffering from various ailments (see letter 4225, n. 3).
11. *Poor Mr De la Rive*: Auguste Arthur de la Rive (1801–73), a Swiss physicist who had died on 27 November. He served as professor of natural philosophy at the Academy of Geneva and worked principally on electricity—he was one of the founders of the electrochemical theory of batteries—although he also investigated the specific heats of gases and calculated the temperature of the Earth's crust. His principal book was *Traité d'électricité théoretique et appliquée*, 3 vols (Paris: J.-B. Baillière et H. Baillière, 1854–58) (*HLS*).
12. *Mrs Sarasin*: Albertine Emma Sarasin (née Diodati, 1848–1917), wife of Edouard Sarasin (1843–1917), a Swiss physicist who in 1867 established a private laboratory in which he collaborated with other scientists, including de la Rive, and Jacques-Louis Soret (1827–90) (*HLS*; G. C. Young, 'Edouard Sarasin, (1843–1917)', *Nature*, 100 (1917), p. 28).

To Herbert Spencer [21 December 1873][1] 4234

Sunday night | Athenaeum Club, Pall Mall[2]

My dear Spencer.

I have glanced over your paper,[3] rather than read it critically. It shows the usual penetration; but will you bear with me if I advise you not to publish it as it now stands. Its air is ambitious, and I frankly think that it fails in its aims. If you publish it as a speculation, not as an "explanation", no harm can accrue. But I think harm would accrue if it were published in its present garb.[4]

I often wished to say to you that your chapters on the Persistence of Force &c.[5] were never satisfactory to me. You have taken as your guide a vague & to me I confess altogether unsatisfactory book.[6] The greater part of your volume I consider to be of such transcendent merit; putting one's best thoughts into the clearest language that I feel all the more the transition to the chapters to which I have referred. I expressed I think the opinion to you sometime ago that they ought to be rewritten.

If you have considered how the disturbance of molecules can generate attraction and repulsion at a distance. You ought to state the result of your thought—If you have not thought of this question I think you have omitted the fundamental phenomenon of electricity

I am hard pressed & therefore write briefly. You will excuse my frankness. I certainly should grieve to see any thing with your name attached to it that would give the enemy occasion to triumph.

Yours ever faithfully | John Tyndall

University of London—145.791/94
RI MS JT/1/TYP/3/1031

1. *[21 December 1873]*: the date is suggested by Spencer replying to this letter on 22 December (letter 4235). 21 December was a Sunday.
2. *Athenaeum Club, Pall Mall*: see letter 4186, n. 1.
3. *your paper*: a postscript to Spencer's essay, 'What is Electricity?', which was first published, in November 1864, in the *Reader* and which he was now revising for a new English edition of *Essays: Scientific, Political, and Speculative*, 3 vols (London: Williams and Norgate, 1874). On 12 November, Spencer had written to Edward L. Youmans saying: 'there has dawned on me, after this long delay, an extension of that theory of electricity set forth in the *Reader* and published in the *Essays* [American edition]. I am busy writing a postscript which, when it is in print, I shall submit to Tyndall and other authorities' (D. Duncan, *The Life and Letters of Herbert Spencer* (London: Methuen, 1908), p. 427). Following Tyndall's theory of heat, Spencer argued that electricity was a mode of motion rather than, as was previously thought, a fluid.
4. *published in its present garb*: Spencer did publish the postscript, and added a note to it stating: 'Since the foregoing postscript was put in type, I have received criticisms upon it, oral and written, from several leading electricians and physicists; and I have profited by them to amend parts of the exposition. While I have remained without endorsements of the hypothesis, the objections raised have not been such as to make clear its untenability' ('What is Electricity?', in *Essays*, vol. 3, pp. 191–215, on p. 211).
5. *your chapters on the Persistence of Force &c.*: chapter VIII of Part II 'Laws of the Knowable' of H. Spencer *First Principles* (London: Williams and Norgate, 1862), pp. 251–58. Tyndall probably also has in mind chapter IX of the same part, 'The Correlation and Equivalence of Forces' (pp. 259–85).
6. *a vague & to me . . . unsatisfactory book*: probably W. R. Grove, *On the Correlation of Physical Forces* (London: C. Skipper and East, 1846).

From Herbert Spencer 22 December 1873 4235

38 Queen's Gardens,[1] | 22d Dec[embe]r. 1873

My dear Tyndall,—

Many thanks for your kind note.[2] I wish you had not troubled yourself to the extent of writing: I had quite intended that the thing should stand over till we met.

I quite agree with you as to the undesirableness of publishing this postscript[3] as it stands: indeed I sketched it out with the expectation that criticism would probably oblige me to re-model it. I quite intended (but I see that I must make the intention more clear) to put forth the hypothesis simply as a speculation: apparently having such an amount of congruity with physical principles as made it worth considering—especially in the absence of anything like a satisfactory explanation.

I have had another letter from Clerk Maxwell[4] which considerably startles me by its views about molecular motion. I should like to talk to you about them. They seem to me to differ much from those which I supposed you to hold and which I supposed were held generally.

Thank you for your reminder reflecting the chapter on the "Persistence of Force".[5] I hope to make it worthy of your approval. I am now remodelling it, and the two preceeding chapters.[6]

I hope the effects of the sound-signal report[7] have ceased to echo through your brain.

Ever yours truly | Herbert Spencer

RI MS JT/1/S/103
RI MS JT/1/TYP/3/1118

1. *38 Queen's Gardens*: a boarding house where Spencer had lived while in London since 1866.
2. *your kind note*: letter 4234.
3. *this postscript*: see letter 4234, n. 3.
4. *another letter from Clerk Maxwell*: James Clerk Maxwell wrote to Spencer on 5 and 17 December articulating his views on molecular motions and vibrations (P. M. Harman (ed), *The Scientific Letters and Papers of James Clerk Maxwell*, 3 vols (Cambridge: Cambridge University Press, 1990–2002), vol. 2, pp. 956–61).
5. *the chapter on the "Persistence of Force"*: see letter 4234, n. 5.
6. *I am now remodelling... preceeding chapters*: for H. Spencer, *First Principles*, 3rd edn (London: Williams and Norgate, 1875). The preceding chapters were on the 'Indestructibility of Matter' and the 'Continuity of Motion', and Spencer later asked Tyndall to read the proofs (see letter 4580).
7. *the sound-signal report*: 'Report by Professor Tyndall'.

To Thomas Archer Hirst 24 December 1873 4236

Crystal Palace Hotel[1] | Wednesday | 24–12–73

Dear Tom.

I waited till 2.40. I am obliged to go to town: but shall be back in good time for dinner—at 7 oC[lock] order it—I leave it in your hands. Fowl or mutton as I am stuffed with beef.

Yours aff[ectionatel]y | John
Debus will dine with us

RI MS JT/1/T/711
RI MS JT/1/HTYP/624

1. *Crystal Palace Hotel*: probably the Royal Crystal Palace Hotel, in Sydenham, Kent, which had opened in 1853. Tyndall and Hirst stayed there over Christmas while the latter was convalescing from various ailments (see letter 4225, n. 3).

From Frances Russell 26 December 1873 4237

Pembroke Lodge.[1] | Richmond Park. | Dec[ember]. 26/73

Dear Mr Tyndall

Agatha[2] is extremely obliged to you for y[ou]r kind note[3]—It was a terrible accident,[4] but I am thankful to say she is now nearly recovered, tho' her head is still too weak to allow of more than a mere trifle of read[in]g or writ[in]g, for w[hi]ch reason I write her answer. I think it w[oul]d please you to know how she has longed to get to y[ou]r books—Heat & Light[5]—& how dur[in]g her recovery she delighted in look[in]g at the drawings in them & copy[in]g or trac[in]g them, as next best to read[in]g about them—She is extremely sorry not to be able to attend y[ou]r Lectures on Sound about to begin[6]—

With our best wishes for the New Year, y[ou]rs sincerely | F Russell

We shall always be delighted to see you here—Fri[day]. & Sun[day]. afternoons are the safe times to find us

RI MS JT/1/R/67

1. *Pembroke Lodge*: the Georgian mansion granted by Queen Victoria to Russell's husband, the former Prime Minister Lord John Russell.
2. *Agatha*: Mary Agatha Russell.
3. *y[ou]r kind note*: letter missing.
4. *a terrible accident*: not identified.
5. *y[ou]r books—Heat & Light*: J. Tyndall, *Heat Considered as a Mode of Motion: Being a Course of Twelve Lectures Delivered at the Royal Institution of Great Britain in the Season of 1862* (London: Longman, Green, Longman, Roberts, and Green, 1863) and *Six Lectures on Light* (1873).
6. *y[ou]r Lectures on Sound about to begin*: see letter 4224, n. 8.

From Martha Somerville 27 December 1873 4238

My dear Mr Tyndall

I have been so much pleased with your kind letter[1] & what you say about my mother[2] & the memoirs.[3] Your approval is invaluable to me. I wish the great & interesting studies & discoveries you are engaged in, did not take so much out of you or that you could occasionally give rest to your mind—I fear this is not to be thought of, until the problem[4] is solved to your satisfaction—I hope from what you say this is the case & that previous to entering upon a new subject of inquiry, you will give yourself that rest you so much require. I see you are to preside over the scientific meeting,[5] so that I suppose your Alpine holiday will fall through. Your mentioning the studies you are engaged in upon the transmission of sound[6] &c recall to me the speculations suggested to my mind at a concert lately while listening to some exquisite instrumental music (Mozarts Quatuor in G minor)[7] Nothing is more spiritual than music, mean & coarse passions can hardly if at all be expressed by music alone. Listening to the noble compositions of Mozart, Beethoven, Handel[8] &c &c one loses all consciousness of matter to be absorbed in the purest & most spiritual of delights & yet music depends altogether (at least our perception of it) on matter—the air which vibrates &c &c in waves of melody—shall we in a future state be deprived of this great joy? or is it possible that the melodies & harmonies which charm us may be the expression of something higher & purely spiritual which we will then understand, as we do poetry expressed by words?

After we saw you at Riffell[9] we went to Gruben, St Luc & Zinal[10] where we went to the Roc noir[11] & the cabin of Monntet[12] on the 26th. Sept[embe]r we then crossed the Simplon[13] in lovely weather, but a fortnight later people were obliged to come over on sledges. We spent several weeks delightfully in visiting friends in Lombardy & have lingered on here (on pretext of Cholera at Naples) but now in a few days shall go south. Our address there, in case you should come there is casa Lucchesi, Corso Vitt[orio]. Emmanuele.[14] I need not say, how very happy we shall be if this dream is realised. If we can be of any use to any relations or friends of yours only let us know.

You will scarcely believe that we also have had fogs for 3 days, damp & disagreable, I could hardly see the cupola of the Duomo[15] yesterday, although it is nothing compared to a London fog—My sister[16] desires to be particularly remembered to you. I often think of our meeting amongst the Alps & our walk with you on the Aletsch Glacier!

Believe me | Dear Mr Tyndall | Ever most truly yours | Martha Somerville

27th. Dec[embe]r. 1873 Florence

RI MS JT/1/S/79

1. *your kind letter*: letter 4230.
2. *my mother*: Mary Somerville.
3. *the memoirs*: see letter 4230, n. 1.
4. *the problem*: Somerville may mean the question of the impact of the atmosphere on the transmission of sound, which Tyndall had mentioned in letter 4230. However, as that issue is addressed specifically later in this letter (see n. 6), she could also mean the general scientific questions Tyndall was pursuing rather than a particular piece of research.
5. *preside over the scientific meeting*: Tyndall was due to give the presidential address at the 1874 meeting of the BAAS in Belfast.
6. *transmission of sound*: Tyndall had been conducting acoustic experiments to find the most effective way of signaling to ships in fog (see letter 4185, n. 3).
7. *(Mozarts Quatuor in G minor)*: Piano Quartet in G minor, composed by Wolfgang Amadeus Mozart (1756–91) in 1785. Quatuor is Latin for quartet.
8. *Beethoven, Handel*: Ludwig van Beethoven (1770–1827) and George Frederick Handel (1685–1759), German composers.
9. *we saw you at Riffell*: on 10 August 1873 (see letter 4230, n. 4).
10. *Gruben, St Luc & Zinal*: villages in the Swiss Alps.
11. *Roc noir*: the Grand Roc Noir, a mountain in the French Alps.
12. *Monntet*: a variant spelling of Montet, in the Swiss canton of Fribourg.
13. *the Simplon*: the pass between the Pennine and the Lepontine Alps.
14. *Corso Vitt[orio]. Emmanuele*: a road in Milan, named after the Italian King Victor Emmanuel II (1820–78).
15. *the Duomo*: Milan Cathedral.
16. *my sister*: Mary Charlotte Somerville.

From Mary Egerton 28 December [1873][1] 4239

Englefield House,[2] | Reading. | Dec[embe]r 28th

Dear Mr. Tyndall,

Let it be my turn this year to send you a Christmas greeting. Most heartily would I wish you all happiness for the coming year. Thanks for your letter[3] which was none the less welcome that it was quite uncalled for, & certainly need not have been ushered in with an apology! I conclude Hatfield meant D. Salisbury?[4] You must have had an agreeable visit there if it did not mean too much to politics, for I saw half the government[5] was there that week.

I long to hear some more about your work[6]—when you print any further account of it, you will send me a paper, will you not? or tell me where I can read it. I dare say you can hardly imagine how much interest one can take in the hard facts, though ignorant of so much of the deeper relations that must give them their chief importance & significance to you. I should like to talk it over with you, & all sorts of things besides. I have never had a real good talk with you in my life! Curious it is how human beings are sewn broadcast over the face of the Earth, take root & grow up entangled in the invisible web of circumstance, so that often with those whom one could, if one chose, see any day, one has nothing in common, while those who are associated with all the subjects one most cares about, are almost beyond reach of communication, at least I am always finding it is.—I have just read (for the first time!) your "Rede lecture" on Radiation;[7] it is as beautifully clear as crystal, on your own rock-salt lens!

Another rather different book that has interested me much lately, is Maurice's "Lectures on Conscience".[8] I think you w[oul]d like his noble earnest "Realist" tone, I mean as to the eternal immutable nature of right & wrong & you w[oul]d. also find points of sympathy in his admiration for Fichte[9] and some of his views about the Early philosophers.[10]

The girls told me Sir John Herschel[11] was intensely interested & quite excited about your discoveries. Do not you want to talk them over with him? You should come down by the early train some Saturday to Robertsbridge, walk to Collingwood[12] (5 or 6 miles), & we would fetch you there in the afternoon to spend a Sunday with us & pick up your luggage coming home. The old erect tree has had some Leaden plasters on its wounds, & having stood all these gales without flinching, has proved itself less rotten than we thought it, so when next you act the Gorilla,[13] I shall not be in such terror of the whole thing coming down with you!

We are actually thinking of taking a run to Italy! not quite so rapid as yours last year,[14] but still merely a flying visit; starting about the 19th of Jan[ua]ry., & returning either before or immediately after Easter;[15] so I am particularly glad the lectures are late.[16] Do you give the opening evening one? & on what subjects? There—I had vowed not to friendship left behind, as shall save me from the change of impertinence in letting you perceive the interest I ever shall, because I even must take, in all that interests you for the time being, whether outwardly or inwardly. Don't think I am making too much of this, indeed I am not; but I know you too well to speak of anything of the kind in a merely playful tone. When I saw the two Marys[17] climbing the towers of Bodian Castle[18] together the other day, I could not help wishing you were there, to give a helping hand to each!

Is there no chance of your giving us a day or two when you leave Dover?[19] We are full next week, & I think I know where you will be on the Saturday[20] if you can get away! but after that, if you do not mind the chance of finding us alone, we sh[oul]d be so glad to see you at any time. Christmas I suppose w[oul]d. be out of the question on account of the lectures?[21]

I am so glad the old "Prophet"[22] has sent you words of strengthening and ~~encouraging~~ (that's not exactly the right word!) Do bring his letter[23] when you come. I wish you had mentioned what form this attack of Ruskin's[24] took, & where it was published? I am so glad you do not mean to take any notice of it. I wonder if Tait has sprung out again anywhere else, now Nature is closed to him.[25]—

By the bye, how curious with reference to the Conveyance of Sound, are those "Harmonic Echoes" which Nature has been writing about.[26] I want to hear the one at Bedgebury[27] and now, shall I say "forgive me" I don't think I will, for I don't feel to have done anything that needs it! I only hope I have said nothing you can misinterpret! Read it patiently and believe me, ask any question, because I know you have no time to write, it is bad enough to have written so much! if you ever get this far, forgive the fruits of idleness out sitting. We return home on Wed[nesda]y

Y[ou]rs most truly | M F Egerton

RI MS JT/1/E/64

---

1. *[1873]*: the year is established by various references, especially to John Ruskin's recent attack on Tyndall (see n. 24).
2. *Englefield House*: a country house in Berkshire that was built in 1558 and substantially altered in the 1820s. It was the country seat of the politician Richard Fellowes Benyon and his wife Elizabeth, who were friends of Egerton.
3. *your letter*: letter missing.
4. *Hatfield meant D. Salisbury*: Hatfield House in Hertfordshire was the seat of Robert Arthur Talbot Gascoyne-Cecil, 3rd Marquess of Salisbury, whom Tyndall had visited on multiple occasions. Egerton's 'D.' is presumably an abbreviation of Duke, although this was the incorrect title for Salisbury.
5. *half the government*: this was the Liberal government of William Gladstone, which, having been in power since 1868, would soon lose the general election in February 1874. Although the Marquess of Salisbury was a member of the Conservative opposition, he was friends with many of the aristocratic members of Gladstone's cabinet.
6. *your work*: at this time Tyndall was largely focused on his work with sound, particularly fog signals (see letter 4185, n. 3).
7. *your "Rede lecture" on Radiation*: J. Tyndall, *On Radiation. The "Rede" Lecture Delivered in the Senate-House before the University of Cambridge on Tuesday, May 16, 1865* (London: Longmans, Green, and Co., 1865).
8. *Maurice's "Lectures on Conscience"*: F. D. Maurice, *The Conscience: Lectures on Casuistry* (London: Macmillan, 1868). John Frederick Denison Maurice (1805–72), a liberal Anglican theologian and founder of Christian Socialism (*ODNB*).

9. *his admiration for Fichte*: Maurice discussed the life and ideas of the German philosopher Johann Gottlieb Fichte on pp. 21–5.
10. *the Early philosophers*: Maurice discussed several Greeks philosophers, with a particular emphasis on Socrates, on pp. 10–21 and 167–72.
11. *Sir John Herschel*: unlike his father, who had died in 1871, John Herschel had not been knighted.
12. *Robertsbridge . . . Collingwood*: a village in Sussex, close to Egerton's home at Mountfield Court, and Herschel's home, in Hawkhurst, Kent.
13. *act the Gorilla*: presumably referring to Tyndall's fondness for climbing trees (see *Ascent of John Tyndall*, p. 193).
14. *yours last year*: Tyndall does not appear to have travelled to Italy in 1872 (he did visit Pompeii and Vesuvius in 1868; see letter 4230, n. 9). Thomas Huxley did travel to Naples in 1872, and saw the eruption of Vesuvius that year (see letter 3635, *Tyndall Correspondence*, vol. 12).
15. *Easter*: Easter Sunday was 5 April in 1874.
16. *the lectures are late*: Tyndall's lecture course at the RI was late in the 'Before Easter' season of lectures (see letter 4233, n. 3 and n. 6).
17. *the two Marys*: not identified.
18. *Bodian Castle*: a fourteenth-century castle in East Sussex.
19. *Dover*: Tyndall had been spending a great deal of time in Dover at the South Foreland Lighthouse conducting his sound signaling experiments, although he was no longer staying there (see letter 4185, n. 3).
20. *where you will be on the Saturday*: presumably 3 January 1874; possibly an allusion to a meeting with the Hamilton family, although it would have to be after the Christmas lecture Tyndall was scheduled to give at the RI that day (see letter 4224, n. 8).
21. *the lectures*: see letter 4224, n. 8.
22. *the old "Prophet"*: Thomas Carlyle.
23. *his letter*: letter 4188.
24. *this attack of Ruskin's*: see letter 4185, n. 13.
25. *Nature is closed to him*: Peter Guthrie Tait had published a blistering attack on Tyndall, 'Tyndall and Forbes', *Nature*, 8 (1873), pp. 381–2, which led to a series of escalating exchanges cut off by *Nature*'s editor Norman Lockyer, who noted: 'We feel that we are only consulting the true interests of Science in declining to print further communications on a subject which has assumed somewhat of a personal tone' ('Tait and Tyndall', *Nature*, 8 (1873), p. 431).
26. *those "Harmonic Echoes" which Nature has been writing about*: from August to November 1873, *Nature* published a number of letters discussing locations where echoes seemed to be raised in pitch.
27. *the one at Bedgebury*: Bedgebury Park, an estate in Kent, was the site of the anomalous echo discussed in H. Smith, 'Harmonic Causation and Harmonic Echoes', *Nature*, 8 (1873), pp. 383–4.

From Lucy O'Brien 30 December [1873][1] 4240

Old Church | Limerick | Dec[embe]r. 30th.

My dear Dr. Tyndall,[2]

I don't want you to forget me entirely now. So I send you photographs of myself & also of my husband & the boy.[3] They are not very good, but they will give you at any rate some idea of my belongings.

I wanted to send them to you f[o]r. Xmas but as they came too late I send them f[o]r. New Year's day—

I want you to write to me & tell me how you are—are you very busy? Too busy I fancy—We are quite settled down in Ennis[4] now f[o]r. the next 4 years & I like it much better than I thought I should. Aubrey is adjutant to the Clare Militia & we are amongst all his people in Clare—Now we are in Limerick spending our Xmas Holidays with his Mother.[5]

Papa[6] is living at Sydenham with Uncle C. Wynne[7]—he will come over to us in a few months as soon as Harry's visit to England[8] is over.

You will of course see Harry then. I suppose he will be in England ab[ou]t the end of Jan[uar]y. Good bye now dear Dr. Tyndall

Ever very aff[ectionate]ly. yours | Lucy H V O'Brien

RI MS JT/1/W/112
RI MS JT/1/TYP/5/1867

---

1. *[1873]*: the year is established by reference to a forthcoming visit to England by Henry Le Poer Wynne (see n. 8).
2. *My dear Dr. Tyndall*: this letter was probably enclosed in letter 4246, as O'Brien was unsure of Tyndall's postal address.
3. *my husband & the boy*: Aubrey Stephen Vere O'Brien (1837–98), whom Lucy had married in 1871, and Robert Stephen Vere O'Brien (1872–1936).
4. *Ennis*: the county town of County Clare, Ireland.
5. *his Mother*: Eleanor Jane Lucy Alicia O'Brien (née de Vere, 1811–89).
6. *Papa*: George Wynne.
7. *Uncle C. Wynne*: not identified.
8. *Harry's visit to England*: Harry is the short form of Henry and O'Brien's brother Henry Le Poer Wynne was about to depart from India, where he was the acting Foreign Secretary, for a short visit to Britain in February 1874 (see letter 4246).

To Emily Peel 31 December 1873 4241

Royal Institution of Great Britain | 31st Dec[ember]. /73

Dear Lady Emily,

Fifty times and more during the last few weeks I have thought of writing to you to ask your forgiveness for not answering your last most friendly letter[1] to me. But I cannot allow the year to end without doing so, and wishing you and yours all happiness for many and many a coming year.

Most faithfully yours, | John Tyndall.

RI MS JT/1/TYP/3/969
LT Typescript Only

1. *your last most friendly letter*: letter missing.

From Mary Egerton 31 December [1873][1] 4242

Mountfield Court, | Robertsbridge, | Hawkhurst.[2] | Dec[embe]r 31st

A happy New Year to you, dear friend! Should it bring you as much happiness as I wish you, you will be a fortunate man! I only wish the fates would ever allow me just a tip of "a finger in the pie" in promoting it! How the years do race on! And yet some things seem very long ago; for instance the beginning of our acquaintance,[3] on even the days of the "fat letters", (I wonder if you remember them still!) when, like the Ancient Mariner,[4] a spell seemed laid on me, that I must unburden my soul to you! I fear I taxed your patience sometimes but how good & kind you were, even when delicately administering a gentle snub!

I was thinking the other day what a contrast this Christmas must have seemed to your last, when you were buffetted by the snowstorms of New York![5] On Christmas Day we gathered quite a summer nosegay;[6] Roses, Verbena, Geraniums; but two or three days frost since, have put an end to all that.

I hope Sydenham[7] freshened you up effectively for the lectures?[8] I must ask the Hamiltons to keep me a copy of your notes on programmes; I like to know what is going on. I rather hope for a visit from L[ad]y Claud[9] while they are away; she promised to come, with Douglas to carry her knapsack! She will drive half the way, but a little while ago Louisa & Mary came, & the former literally walked all the way carrying the knapsack with most of her sisters things in it as well! Mary walked three parts of the way! (I wonder if I told you that tale before?) I do delight in them all in their different ways especially of course L[ad][y] Claud! & I think they must like us a little, or they would not be so often planning meetings.

I am going up to the kind Spottiswoodes on the 16th, when I hope I shall see something of you besides the lecture. I am afraid they are not going to include May[10] this time.

I am so longing to hear about the results of the fog;[11] you did not mention whether you went down again to Dover, I hear it extended there, or had any means of experimenting in London. But I should think a Carbon fog[12] would have very different qualities to a water one.

Auf Widersehen![13] Do not trouble yourself to write if you are busy with your lectures; you gave me your good wishes last time.

I was so glad to have a letter from M[r] Hirst himself the other day (about some game) and to find that he was getting out again. His writing had not quite returned to its usual neatness,[14] but I trust hands & all must have very nearly come right again.

Ever y[ou][rs] | M F Egerton

Warm wishes from our party especially May, who says she is chopping out of your acquaintance! She has accidently missed several of our few meetings the last year.

RI MS JT/1/E/67

---

1. *[1873]*: the year is established by various references, especially to Tyndall's researches on signaling in fog (see n. 11).
2. *Mountfield Court, | Robertsbridge, | Hawkhurst*: the Egerton family estate in Sussex.
3. *the beginning of our acquaintance*: their friendship had begun in the summer of 1867.
4. *the Ancient Mariner*: the titular character in Samuel Taylor Coleridge's poem 'The Rime of the Ancient Mariner' (1798), who continually recounts his experience of being cursed for killing an albatross.
5. *the snowstorms of New York*: when Tyndall had lectured in New York in late December 1872 the city had been struck by particularly inclement weather, although the lectures continued despite the heavy snowfall (see letter 4406). On the American lecture tour in general, see letter 4201, n. 4.
6. *a summer nosegay*: a small bunch of flowers.
7. *Sydenham*: Tyndall had spent Christmas at the Royal Crystal Palace Hotel in Sydenham, Kent, with Thomas Hirst (see letter 4236).
8. *the lectures*: see letter 4224, n. 8.
9. *L[ad][y] Claud*: Elizabeth Hamilton.

10. *May*: Mary Alice Egerton (May being a pet name for Mary). It is unclear why she would not be included by the Spottiswoodes.
11. *the results of the fog*: Tyndall had been conducting experiments to determine the most effective means of signaling to ships in fog (see letter 4185, n. 3).
12. *a Carbon fog*: a fog produced by smoke from the burning of coal.
13. *Auf Widersehen*: goodbye (German).
14. *His writing . . . its usual neatness*: Hirst had been suffering from various ailments, including numbness in his hands (see letter 4225, n. 3).

1874

To Thomas Carlyle   1 January 1874   4243

Royal Institution of Great Britain | 1st. January 1874

My dear Friend

I intended to go down last night to wish you as regards both body and mind, that the one may retain its toughness and the other its strength and serenity through many and many a coming year.

I saw on Sunday at Brighton a letter from you which greatly delighted the man to whom it was written, that brave old Captain of Industry Whitworth.[1]

And I see with pleasure that Germany takes due account of her indebtedness to you.[2]

Had I not been bewildered by many things I should have asked your niece[3] whether she would care to come to those lectures to the little people.[4] But she ought to feel sure that she would be welcome to any lectures of mine and act accordingly.

You must allow me to come down to you as soon as the present stress relaxes.

Ever yours affectionately | John Tyndall

National Library of Scotland, Thomas Carlyle Papers (1874), Ms. 1771, f. 1 r&v

---

1. *a letter from you . . . Captain of Industry Whitworth*: Carlyle's letter to the industrialist Joseph Whitworth was made public by the latter and widely reported in the press. It praised Whitworth's plan to supplement his workers' savings by paying them a bonus, and asserted: 'Would to Heaven that all or many of the captains of industry in England had a soul in them such as yours' ('Mr. Carlyle on Modern Work', *Illustrated London News*, 31 January 1874, p. 106). Tyndall seems to have seen the original letter, which was written in late February 1873, probably shown him by Whitworth when they were both in Brighton. See C. R. Sanders, et al. (ed), *The Collected Letters of Thomas and Jane Welsh Carlyle*, 49 vols (Durham, NC: Duke University Press, 1970–), vol. 48, p. 169.

2. *Germany takes due . . . her indebtedness to you*: there were already rumours that Carlyle would be awarded the Prussian Order of Merit, although he did not receive a formal notice of this until February. It was awarded in recognition of his biography *History of Friedrich II of Prussia, Called Frederick the Great*, 6 vols (London: Chapman and Hall, 1858–65), and for his fervent support of Prussia in the recent Franco-Prussian War.
3. *your niece*: Mary Carlyle Aitken (1848–95), with whom Carlyle had lived since 1866.
4. *those lectures to the little people*: see letter 4224, n. 8.

To William Edward Hartpole Lecky[1] 2 January 1874 4244

Royal Institution of Great Britain | 2nd. Jan[uary]. 1874.

My dear Mr Lecky

It would give me particular pleasure to say 'yes' to your friendly invitation;[2] but I am under a bond not to dine out until the heavy labours now pressing upon me are ended on the 16th.[3] Under this bond I have been obliged to decline every invitation that has reached me during the last few weeks.

Trust me I regret this—for you could hardly name a party that I should be more rejoiced to meet

Yours faithfully | John Tyndall

Have you seen Lange's History of Materialism?[4] He often refers to you.[5]

Pray give my warm regards to Count Henri Russell.[6]

RI MS JT/1/TYP/3/835
LT Typescript Only

---

1. *William Edward Hartpole Lecky*: William Edward Hartpole Lecky (1838–1903), an Irish historian and political theorist (*ODNB*).
2. *your friendly invitation*: letter missing.
3. *heavy labours . . . ended on the 16th* : probably the intensive work that Tyndall was putting into his Friday Evening Discourse 'On the Acoustic Transparency and Opacity of the Atmosphere' (see letter 4226, n. 2).
4. *Lange's History of Materialism*: F. A. Lange, *Geschichte des Materialismus und Kritik seiner Bedeutung in der Gegenwart*, 2nd edn, 2 vols (Leipzig: J. Bädeker, 1873–5).
5. *He often refers to you*: vol. 1, pp. 205–10 of Lange's book gave an account of Lecky's *History of European Morals from Augustus to Charlemagne*, 2 vols (London: Longmans, Green, and Co., 1869) and *History of the Rise and Influence of the Spirit of Rationalism in Europe*, 2 vols (London: Longmans, Green, and Co., 1865).
6. *Count Henri Russell*: Henry Patrick Marie, Count Russell-Killough (1834–1909), a French-Irish mountaineer and explorer, known for numerous first ascents in the Pyrenees

(R. Bailey, *The Man Who Married a Mountain: A Journey through the French Pyrenees*, London: Transworld, 2005).

From George Gabriel Stokes 2 January [1874][1] 4245

Cambridge 2nd Jan[uar]y 1873[2]

My dear Tyndall,

Unfortunately I did not make a memorandum of the day you chose, and, trusting to memory, I thought it was the second meeting after Xmas (the 15th)[3] that you chose. I was prevented by illness (I am well now) from being present on the last meeting before Xmas,[4] and I don't know what papers Huxley put on for the 8th.[5] However I will send your abstract to Mattress, to be set up for Thursday.[6]

I think as the paper[7] is not (as yet at least) in our hands it would be more regular to entitle it "preliminary notice" or something to that effect, and let it stand as an independent paper in the Proceedings.[8] When we get the paper the title can be read in due course pro forma,[9] and in lieu of abstract in the Proceedings a reference may be made to the Preliminary Notice. A trifling verbal alteration in one or two places, which can be made on the proof, would be all the change that would be required.

If you meet Huxley you had best ask him what papers he put on. He has left London but returns I believe on Monday.[10]

Yours sincerely | G. G. Stokes

RI MS JT/1/S/254
RI MS JT/1/TYP/4/1416

1. *[1874]*: the year is established by reference to the dates of the meetings of the RS (see n. 3 and n. 5).
2. *1873*: the year given is a slip of the pen common at the start of a new year; as the content of the letter indicates, Stokes evidently meant 1874.
3. *the second meeting after Xmas (the 15th)*: the second of the RS's weekly meetings, held, by tradition, on Thursdays, which took place on 15 January.
4. *the last meeting before Xmas*: probably the ordinary meeting on 27 November 1873, rather than the annual anniversary meeting held on 1 December.
5. *what papers Huxley put on for the 8th*: Thomas Huxley had been elected an honorary secretary of the RS, with responsibility for arranging papers to be read at meetings, in 1872 (*Roy. Soc. Proc.*, 22 (1874), p. 12). Stokes, who had been performing the same role since 1854, could now share and delegate his responsibilities, particularly in relation to papers on the life sciences. The four papers read at the meeting on 8 January 1874 were 'On the Brom-Iodides' by Maxwell Simpson (*Roy. Soc. Proc.*, 22 (1874), p. 51–3); 'Contributions

to the History of the Orcins.—No. IV. On the Iodo-derivatives of the Orcins' by John Stenhouse (pp. 53–5); 'A Memoir on the Transformation of Elliptic Functions' by Arthur Cayley (p. 56); and 'On Electrotorsion' by George Gore (pp. 57–8).

6. *Thursday*: 8 January.
7. *the paper*: the paper's title was soon agreed as 'On the Atmosphere as a Vehicle of Sound' (see letter 4250).
8. *entitle it "preliminary notice" . . . an independent paper in the Proceedings*: the paper was published as 'Preliminary Account of an Investigation on the Transmission of Sound by the Atmosphere', *Roy. Soc. Proc.*, 22 (1874), pp. 58–68.
9. *pro forma*: as a matter of form (Latin).
10. *Monday*: 5 January.

From George Wynne 2 January 1874 4246

The Athenaeum Club, Pall Mall | 2 Jan[uar]y / 74

My dear Tyndall

I do not know why it is I never come across you at the Club.[1] I have as they say in our country[2] been threatening to beat you up at your den[3] but somehow something has always intervened to prevent me. Lucy[4] sent me the enclosed letter[5] to forward to you not being quite certain of your address. I had a telegram from Harry[6] a short time ago to tell me he was running over on privileged leave[7] to see his wife who has never well recovered her confinement, he was to leave about the 5th of this month and I hope he will arrive about the 5th of next month to stop only for a very few weeks. Poor Frank[8] is I fear likely to be again out of work owing to bad times in Germany the iron works where he was engaged erecting blast furnaces has to stop working waiting for better times. I hope he will soon be over to this country. With all the good wishes of the season

Affec[ionately] y[ou]rs | Geo[rge] Wynne

RI MS JT/1/W/99
RI MS JT/1/TYP/5/1868

1. *the Club*: the Athenaeum Club (see letter 4186, n. 1). Wynne became a member of the club in 1858 (*Athenaeum. Rules and Regulations and List of Members. 1862* (London, 1862), p. 93).
2. *our country*: i.e. Ireland.
3. *beat you up at your den*: presumably meaning come to your home or residence.
4. *Lucy*: Lucy O'Brien, Wynne's daughter.
5. *the enclosed letter*: probably letter 4240.
6. *Harry*: Henry Le Poer Wynne, Wynne's eldest son.
7. *running over on privileged leave*: Henry was returning to Britain from India, where he was the acting Foreign Secretary.

8. *Poor Frank*: Francis Wynne, Wynne's youngest son, who was a civil engineer in Germany.

To Thomas Archer Hirst 3 January [1874][1] 4247

3rd Jan[uary]

My dear Tom.

Your letter[2] has been a great comfort to me.

I have not the slightest doubt of your perfect recovery, if only sufficient care be taken.[3]

I am just going in to the lecture room[4] and can write no more.

Rolfe Lubbock[5] is getting on here.

Yours ever affectionately | John Tyndall

RI MS JT/1/T/863

1. *[1874]*: the year is established by reference to Hirst's recovery and Tyndall's Christmas lecture (see n. 3 and n. 4).
2. *Your letter*: letter missing.
3. *your perfect recovery, if only sufficient care be taken*: Hirst had been suffering from a number of ailments, including chills and numbness, since 8 November 1873, and Tyndall had been concerned that he follow the advice of his physician to rest (see letter 4225, n. 3).
4. *the lecture room*: of the RI, where Tyndall was about to give the fourth of his six Christmas Lectures for juvenile audiences on 'The Motion and Sensation of Sound' (see letter 4224, n. 8).
5. *Rolfe Lubbock*: Rolfe Arthur Lubbock (1865–1909), the youngest son of John Lubbock and Tyndall's godson, who was presumably attending the Christmas Lectures.

To Thomas Henry Huxley 5 January 1874 4248

Royal Institution of Great Britain | 5th. Jan[uary],[1] 1874

My dear Huxley

I wished to write you an allocution,[2] but my brain to day is not among those spheral melodies[3] amid which it habitually dwells, and hence I postpone the allocution. Lest you should be doubtful as to the safety of that flattened and desiccated pulp of rags that you have sent me[4] I hereby inform you that it has come safely to hand.

Yours ever | John Tyndall

IC HP 8.163
RI MS JT/1/TYP/9/3031

Typed Transcript Only

---

1. *Jan[uary]*: in the typescript of the letter, this word is given as 'June' and the letter is dated accordingly. However, Tyndall was in Switzerland in early June, and not writing on paper with the RI letterhead (see letter 4353), so it seems more likely that the word is Tyndall's standard abbreviation for January.
2. *allocution*: a formal speech delivering an exhortation or moralizing address (*OED*).
3. *spheral melodies*: the ancient idea, originating with the Ionian mathematician Pythagoras (*c.* 570–*c.* 495 BCE), that the orbital revolutions of celestial bodies emit a form of harmonic or mathematical music.
4. *that flattened . . . you have sent me*: not identified, but possibly a reference to a book or journal article, as pulped rags were widely used in the manufacture of paper.

To Thomas Archer Hirst 7 January [1874][1] 4249

7th Jan[uary]

My dear Tom

I think you have that copy of Ruskins diatribe.[2] If so would you like a good fellow send it to Lady Lubbock?

Yours aff[ectionatel]y. | John.

Hope you go on well[3]

RI MS JT/1/T/864

---

1. *[1874]*: the year is established by the relation to letter 4193 (see n. 2).
2. *that copy of Ruskins diatribe*: *Fors Clavigera*, vol. 3 (1873), letter 34, pp. 1–32. John Ruskin attacked Tyndall for his part in the controversy over glacial movement (see letter 4185, n. 13). Tyndall had sent the copy in letter 4193, in which he told Hirst 'Lady Lubbock informs me that she cannot get a copy for you'. Tyndall was presumably asking Hirst to pass the copy on to Ellen Lubbock, although the reason why she needed it is unclear.
3. *Hope you go on well*: Hirst was staying at the Royal Albion Hotel in Brighton, convalescing from various ailments (see *Hirst Journals*, p. 1993; and letter 4225, n. 3).

To George Gabriel Stokes 7 January [1874][1] 4250

7th Jan[uary].

My dear Stokes—–

Do what you please with the title of the paper.[2]

Yours sincerely | John Tyndall

[The title[3] will be | On the Atmosphere as a Vehicle of Sound.[4] I have already sent it to Mr. White].[5]

CUL SC—add 7656

---

1. *[1874]*: the year is established by the relation to letter 4245.
2. *Do what you please with the title of the paper*: Stokes and Tyndall had been discussing what the paper should be called when it was read at the RS, first in a preliminary version on 15 January and then in a complete form on 12 February, and when it was initially published in the *Roy. Soc. Proc.*, where it appeared as 'Preliminary Account of an Investigation on the Transmission of Sound by the Atmosphere', *Roy. Soc. Proc.*, 22 (1874), pp. 58–68 (see letter 4245).
3. *The title*: it is unclear whether this addendum to the letter is in the hand of Tyndall or Stokes, but it seems likely to be a note written by the latter.
4. *On the Atmosphere as a Vehicle of Sound*: this was the name given to the paper when it was published as 'Vehicle of Sound'.
5. *Mr. White*: Walter White.

From George Gabriel Stokes 8 January 1874 4251

The Royal Society | Burlington House London. W. | 8th Jan[uar]y 1874

My Dear Tyndall,

I am sorry you were disappointed of the 8th.[1] The impression left on my mind as the result of our conversation is still that, though we at first spoke of the 8th you ultimately preferred the 15th, thinking it would be less trouble to make one job of it by taking the R.S. and R.I. lectures on consecutive days.[2] I rather think I am right as I remember being a little surprised at the decision. Perhaps you felt a gush of health and force at the moment which made you like a war-horse sniffing the battle. You are quite right not to do what would interfere with your sleep—.

Yours sincerely | G. G. Stokes

RI MS JT/1/S/257
RI MS JT/1/TYP/4/1419

---

1. *disappointed of the 8th*: Tyndall's preference had been to read his paper 'On the Atmosphere as a Vehicle of Sound' at the RS's first weekly meeting of the year on 8 January; however, Stokes had not made a note of this and instead scheduled the paper for the second meeting on 15 January (see letter 4245).
2. *taking the R.S. and R.I. lectures on consecutive days*: as well as reading his paper at the RS on 15 Janaury, Tyndall was giving a Friday Evening Discourse at the RI on 16 January (see letter 4226, n. 2).

From Jane Barnard 15 January 1874 4252

Barnsbury Villa, | 320, Liverpool Road, | N.[1] | 15 Jan[uary]. 1874

My dear Dr Tyndall

You are very kind in thinking of me in the midst of all your duties. Some weeks ago you gave me some signed tickets[2] & I have ventured to give away two besides the friends I take in my self; I hope I have not taken too much advantage of your liberality—I am looking forward with great pleasure to tomorrow evening.

My Aunt[3] desires her affectionate remembrances; she hopes you are getting a little quiet rest as well as fresh air at Sydenham.[4]

Yours very sincerely | Jane Barnard

Dr Tyndall

RI MS JT/1/B/52

1. *Barnsbury Villa, | 320, Liverpool Road, | N.*: see letter 4226, n. 1.
2. *signed tickets*: for Tyndall's Friday Evening Discourse 'On the Acoustic Transparency and Opacity of the Atmosphere' which he was giving at the RI on 16 January (see letter 4226).
3. *My Aunt*: Sarah Faraday.
4. *Sydenham*: Tyndall had spent Christmas at the Royal Crystal Palace Hotel in Sydenham, Kent, with Thomas Hirst (see letter 4236). Sarah Faraday may have mistakenly thought he was still there.

To the Editor of the *Daily Telegraph*[1] 17 January 1874 4253

SIR—

You have given me a challenge[2] to which I willingly respond. In a speech to which I had the honour of listening just before my departure from America,[3] the Hon[ourable]. William W. Evarts[4] used these words:

> There is a generous and perfect sympathy between the educated men of England and the educated men of the United States. The small matters of difference and political interests which divide these two great countries are nothing to the immense area of uniform and common objects and interests which unite their people.

On the same occasion Dr. John W. Draper, celebrated alike as an historian and a scientific discoverer, concluded a speech in these words:

Nowhere in the world are to be found more imposing political problems than those to be settled here—nowhere a greater need of scientific knowledge. I am not speaking of ourselves alone, but of our Canadian friends on the other side of the St. Lawrence.[5] We must join together in generous emulation of the best that is done in Europe. In her Majesty's representative, Lord Dufferin,[6] they will find an eager appreciation of all that they may do. Together we must try to refute what De Tocqueville has said about us, that communities such as ours can never have a love of pure science.[7] But, whatever may be the glory of our future intellectual life, let us both never forget what we owe to England. Hers is the language which we speak, hers are all out ideas of liberty and law. To her literature, as to a fountain of light, we repair. The torch of science that is shining here was kindled at her midnight lawn.

The President of Cornell University,[8] to which Mr. Goldwin Smith[9] belongs, used, on the same evening, these remarkable words:

We are greatly stirred at times as this fraud or that scoundrel is dragged to light, and there rise cries and moans over the corruption of the times; but, my friends, these frauds and these scoundrels are not the corruption of the times. They are the mere pustules which the body politic throws to the surface. Thank God that there is vitality enough left to throw them to the surface. The disease is below all this, infinitely more wide-spread. What is this disease? I believe that it is, first of all, indifference - indifference to truth as truth; next, scepticism, by which I do not mean inability to believe this or that dogma, but the scepticism which refuses to believe that there is any power in the universe strong enough, large enough, good enough, to make the thorough search for truth safe in every line of investigation; next, infidelity, by which I do not mean want of fidelity to that which underlies all creeds, the idea that the true and the good are one; and finally, materialism, by which I do not mean this or that scientific theory of the universe, but that devotion to the mere husks and rinds of good, that struggle for place and pelf,[10] that faith in mere material comfort and wealth, which eats out of human hearts all patriotism, and which is the very opposite to the spirit which gives energy to scientific achievement . . . I believe that the little army of scientific men furnish a very precious germ from which better ideas may spring; . . . and I trust that love, admiration, and gratitude, between men of science on both sides of the Atlantic, may add new cords and give strength to old cords which unite the heats of the two great English-speaking nations.

On the same occasion, in reference to the question of international amity, I ventured to say this much:

Among the motives which prompted me at the time of accepting your invitation was this: I thought, and friends of mine here thought, that a man withdrawn from the arena of politics, who had been fortunate enough to gain a measure of the goodwill of the American people, might do something towards softening political asperities. I referred to this point in Boston, but my references to it have grown more and more scanty, until, in the three cities last visited, they disappeared altogether. And this not because I had the subject less at heart, but because, as your great countryman Emerson might express it, any reference of the kind would be like the sound of a scythe in December, entirely out of place. During my four months' residence in the United States I have not heard a single whisper hostile to England.

This, Sir, will sufficiently indicate to you my experience of the feeling of the people of the United States towards this country. Either they do not hate us, as alleged; or, if they do, the manner in which they suppressed this feeling, out of consideration for a guest, proves them to be the most courteous of nations.—

I am, Sir, your obedient servant, | JOHN TYNDALL. | Athenaeum Club,[11] Jan[uary]. 17.

Daily Telegraph, 19 January 1874, p. 2.

---

1. *the Editor of the* Daily Telegraph: Edwin Arnold (1832–1904), who had taken on the role in 1873.
2. *a challenge*: this letter was written in response to an article published in the *Daily Telegraph*, 16 January 1874, p. 4. The article reported a speech made a few days earlier by the historian Goldwin Smith (see n. 9) to the Trades Union Congress in Sheffield. Smith had emigrated to the United States to take up a professorship at the newly-established Cornell University in 1868, but then relocated to Toronto, Canada, in 1871. It was Smith's first return to Britain, and during the speech he claimed that 'the Americans hate England'. Tyndall disagreed with the assertion, and wrote this letter, which was published under the heading 'England and America' in the *Daily Telegraph* on 19 January, p. 2.
3. *just before my departure from America*: the speeches Tyndall referred to were given on 4 February 1873 at Delmonico's restaurant in New York (see letter 4216, n. 7).
4. *the Hon[ourable]. William W. Evarts*: William Maxwell Evarts (1818–1901), an American statesman and lawyer (*ANB*).
5. *St. Lawrence*: a river that forms part of the national boundary between Canada and the United States.
6. *Lord Dufferin*: Frederick Temple Hamilton-Temple-Blackwood, 1st Marquess of Dufferin (1826–1902), the Governor General of Canada from 1872–8.
7. *what De Tocqueville has . . . love of pure science*: Alexis de Tocqueville (1805–59), a French political theorist, included a chapter entitled 'Why The Americans are More Addicted to Practical than to Theoretical Science' in his influential book *Democracy in America* (1835–40), in which he stated: 'In America the purely practical part of science is admirably understood, and careful attention is paid to the theoretical portion which is immediately requisite to application . . . But hardly anyone in the United States devotes himself to the essentially theoretical and abstract portion of human knowledge' (A. de Tocqueville, *Democracy in America*, trans. H. Reeve, 3 vols (London: Saunders and Otley, 1835–40), vol. 3, p. 79).
8. *The President of Cornell University*: Andrew Dickson White (1832–1918), an American historian and co-founder of Cornell University, New York. He served as Cornell's first president from 1866–85 (*ANB*).
9. *Mr. Goldwin Smith*: Goldwin Smith (1823–1910), a British historian. He held the regius

professorship of Modern History at Oxford from 1858–66, then held the professorship of English and Constitutional History in the Department of History at Cornell University from 1868–72, though he moved to Toronto, Canada in 1871, travelling back to Cornell periodically to lecture (*ODNB*).

10. *pelf*: money and riches, especially when viewed as a corrupting influence (*OED*).
11. *Athenaeum Club*: see letter 4186, n. 1.

To the Mayor and Town Council of Brighton[1] 17 Janaury 1874 4254

Royal Institution of Great Britain | 17th. Jan[uary]. 1874

Gentlemen

I thank you very much for the honour you have done me in asking me to meet that brave man Sir Samuel Baker.[2]

The qualities that he has shown are highly prized by me, and it is with regret that I find myself unable to meet him

faithfully yours | John Tyndall

Wellcome MS. 7777/17

1. *the Mayor and Town Council of Brighton*: John Leonhardt Bridgen (1814–1903), who served four terms as Mayor of Brighton. The individual members of the Town Council are not identified.
2. *Sir Samuel Baker*: Samuel White Baker (1821–93), an explorer who had recently returned from an expedition to Central Africa, where he had tried to suppress the slave trade. A municipal banquet in Baker's honour was hosted by the Mayor and Town Council of Brighton in the city's Royal Pavillion on 19 January, and it was evidently that to which Tyndall had been invited ('Banquet To Sir Samuel Baker', *Times*, 20 January 1874, p. 12).

From Friederich Albert Lange 19 January 1874 4255

Marburg 19. Jan. 1874

Hochgeehrter Herr!

Auf Ihr sehr freundliches Schreiben vom 9. d. bedaure ich noch nicht mit Bestimmtheit antworten zu können. Das 2. Buch meiner Gesch. des Materialism. sollte im nächsten Sommer erscheinen und der Druck schreitet angeblich rasch voran. Ich kann aber noch nicht wissen, ob ich nicht in den naturwissenschaftlichen Gebieten, zu denen ich jetzt bald komme, auf unvorhergesehen Schwierigkeiten stosse, welche meinen Gang verzögern.

Sie, verehrter Herr, und einige Ihrer Fachgenossen schaffen so viel neuen Stoff, dass ein armer Philosoph nur mühsam folgen kann.

Hochachtungsvoll | Ihr ergebenster | Fr. A. Lange

Marburg 19 Jan[uary]. 1874

Dear Sir!

In response to your very kind letter from the 9th of this month,[1] I regret that I still cannot provide a definite answer. The 2nd book of my History of Materialism[2] should appear next summer and the printing seems to advance quickly. Nevertheless, I am not yet in a position to know whether I will encounter unforeseeable difficulties that will delay my progress once I discuss the natural sciences, a point that I am approaching.

You, admired sir, and some of your peers produce so much that is new that a poor philosopher can follow only with great efforts.

Respectfully | Your humble | Fr. A. Lange

RI MS JT/1/L/22

1. *your very kind letter from the 9th of this month*: letter missing.
2. *The 2nd book of my History of Materialism*: the second volume of F. A. Lange, *Geschichte des Materialismus und Kritik seiner Bedeutung in der Gegenwart*, 2nd edn, 2 vols (Leipzig: J. Bädeker, 1873–5).

To the Editor of *Nature*[1] [*c*. 23][2] January 1874 4256

THOUGH my memory fails to recall the fact, I cannot, with Prof. Barrett's letter[3] before me, refuse to believe that he sent me the paper to which he refers.[4]

Perhaps I ought to have known what Mr. Barrett had been doing before large audiences, but I regret to say that I did not. My excellent assistant, Mr. Cottrell, first heard of Mr. Barrett's experiments from one of my own audience, and steps had been taken to do Mr. Barrett justice before his letter appeared. That act he has anticipated by very ably and very modestly doing justice to himself.

J. Tyndall

Nature, 29 January 1874, p. 241

1. *Editor of* Nature: Norman Lockyer.
2. *[c. 23]*: although the letter was published on 29 January, under the heading 'Prof. Barrett

and Sensitive Flames', Tyndall would likely have written it soon after seeing *Nature*'s previously weekly number, which was published on 22 January (see n. 3).

3. *Prof. Barrett's letter*: William Fletcher Barrett (1844–1925), Tyndall's assistant at the RI from 1863–6. In 1873 he became professor of experimental physics at the Royal College of Science for Ireland. His letter was published on 22 January as 'Dr. Tyndall and Sensitive Flames', *Nature*, 9 (1874), p. 223. Referring to the first of Tyndall's RI Christmas Lectures for juvenile audiences on 'The Motion and Sensation of Sound', which he had given on 27 December 1873 and in which Tyndall demonstrated the reflection of sound by means of a sensitive flame, a gas flame which resonates with sounds or air vibrations, Barrett disputed Tyndall's claim that this phenomenon had 'never before' been made visible. Having first observed the phenomenon during his time as Tyndall's assistant, noting how high-pitched sounds made the flames change size and shape, Barrett stated that he had published a paper on this subject in the *Quarterly Journal of Science* in 1870, and furthermore, had frequently shown his students and 'large audiences' the reflection of sound by a sensitive flame since 1868. See letter 2906, n. 5, *Tyndall Correspondence*, vol. 10.
4. *the paper to which he refers*: W. F. Barrett, 'Light and Sound: An Examination of Their Reputed Analogy', *Quarterly Journal of Science*, 9 (1870), pp. 1–16.

From Jane Barnard 23 January 1874 4257

Friday | 23 Jan[uary] 74

Dear Dr Tyndall

My Aunt[1] thought that as I was passing this way, that I might call & answer your note[2] received this morning.

She feels gratified & honored with Mr Siemens proposal,[3] & seemed very much interested & pleased at the compliments to my uncles name.

She, at the same time said she thought from the letter you enclosed that I & my brother[4] might perhaps have the opportunity of seeing the ship; but I should be very unwilling to put myself forward in any way that might not be suitable & would not wish to be at all prominent in any ceremony.

Mr Siemens mentions the 15th of Feb[ruary]—but that is a Sunday & I suppose that must be a mistake.[5] I am afraid I may be giving you trouble in the matter, but if you think there is any difficulty or impropriety - will you just kindly say so & do nothing more.

I write in case you may be looking for my Aunts reply on your return

Yours very sincerely | Jane Barnard | Dr Tyndall | &c &c &c

RI MS JT/1/B/53

1. *My Aunt*: Sarah Faraday.
2. *your note*: letter missing.

3. *Mr Siemens proposal*: Carl Wilhelm Siemens proposed to name his innovative new telegraph cable-laying ship after Michael Faraday.
4. *my brother*: Frank Barnard.
5. *that is a Sunday & I suppose that must be a mistake*: the sabbath would be an inappropriate day to hold a ceremony to launch a ship, although the Faradays' Sandemanian beliefs did not require strict observance of it. The ceremony to launch the CS *Faraday* was held on Tuesday 17 February ('A New Cable Ship', *Engineer*, 37 (1874), p. 174).

To John Wallace[1] 24 January [1874][2] 4258

Royal Institution of Great Britain | 24th Jan[uar]y.

Dear Sir

Many thanks to you for both the burner[3] and the paper.[4]

Yours truly | John Tyndall

Northumberland Archives SANT/BEQ/4/11/276/B

1. *John Wallace*: a Newcastle-based engineer and inventor who worked for the machine tool manufacturing firm of Tangye Brothers & Rake, to which Tyndall addressed the envelope of this letter.
2. *[1874]*: the envelope is stamped 1874.
3. *the burner*: Wallace was developing a new form of Bunsen burner, a laboratory instrument used to provide a steady, concentrated flame fueled by gas, named after Robert Bunsen. Wallace submitted a patent for his design on 28 September, which was granted on 26 February 1875. See 'Improvements in Gas Burners and in Apparatus Connected Therewith', *English Patents of Inventions*, no. 3318 (1874), pp. 1–5.
4. *the paper*: probably J. Wallace, 'On the Combustion of Coal Gas to Produce Heat', *North of England Institute of Mining Engineers Transactions*, 23 (1873–4), pp. 47–60, which Wallace read at the North of England Institute of Mining Engineers on 11 October 1873.

To Edward Tylor[1] 27 January [1874][2] 4259

Royal Institution of Great Britain | 27th Jan[uar]y

My dear Sir,

I have seen your apparatus[3] tried, and consider it neither more nor less than a stroke of genius.

Yours faithfully | John Tyndall

BL Add MS 50254, f. 70

1. *Edward Tylor*: Edward Burnett Tylor (1832–1917), an English anthropologist who in the 1860s developed the concept of Animism to explain primitive beliefs that spirits animate material objects. He also studied the survival of such primitive ideas in modern religious and metaphysical beliefs, and in November 1872 began attending séances and studying the practices of spiritualists in this light (*ODNB*).
2. *[1874]*: the year is given in the BL's archival listing for the letter, having originally been suggested by its original owner, Richard Max Tylor, the anthropologist's great-nephew. It was presumably based on a postmark on the envelope.
3. *your apparatus*: possibly the 'little apparatus of sprinkled starch & stretched blotting papers' that Tylor brought to a séance held at a 'Miss Kislingbury's' in London on 18 November 1872. The apparatus was intended to record an 'impression' of the raps purportedly made by spirits, but Tylor noted that the 'spirits unequivocally declared their disgust at [the papers], saying they couldn't do anything with them' (quoted in G. W. Stocking Jr, 'Animism in Theory and Practice: E. B. Tylor's Unpublished "Notes on 'Spiritualism'"', *Man*, 6 n.s. (1971), pp. 88–104, on p. 97). Tyndall had previously recorded that when he had attended a séance in the 1860s and 'low knocking was heard from time to time under the table . . . I turned a wine-glass upside down, and placed my ear upon it, as upon a stethoscope. The spirits seemed disconcerted by the act; they lost their playfulness, and did not recover it for a considerable time' ('Science and the Spirits', *Reader*, 4 (1864), pp. 725–6, on p. 725). It is unclear when Tyndall might have seen the apparatus, but séances may have been on his mind at this time as, on the same day as this letter to Tylor, Thomas Huxley recorded his impressions of a séance he had recently attended with Charles Darwin (see *Darwin Correspondence*, vol. 22, pp. 41–5; this is letter 9256 in the online edition of the Darwin correspondence; also published in *Life and Letters of Huxley*, vol. 1, pp. 420–3).

From Mary Egerton [*c*. 27 January 1874][1] 4260

Mountfield Court, | Robertsbridge, | Hawkhurst.[2]

Dear obstinate friend,

I hoped that wretched letter was behind the fire long ago! but lo! it turns up again, to disappoint me, though the signature outside and the cabalistic characters over it, could I have deciphered them, (v. u. J.)? might have warned me there was no inside.[3]

But I should not have written for thus, only I have just heard from L[ad] [y]. Claud[4] that they may have to put off going to London for a little while, on account of this astounding Dissolution.[5] So I thought I would just tell you this, as I said the other night they were going immediately, & it might make a difference about your proposing there. It was only in the bad weather, mind, that she said she had "rather you came in the summer for she w[oul][d] not like you to see the place[6] first under a gloomy aspect"; but Emmy,[7] who was there one of those five days last week, said it was looking lovely. I don't say this only because if you went there it would bring you nearer us,[8] though that might have something to do with it!

Were you not a little bit cross the other night? Something dried me up, & made me feel awfully stupid, and I forgot several things I wanted to talk about. One was your little friend "Amy Lubbock",[9] whom I made acquaintance with at Mrs Spottiswoode's music. I saw such a lovely profile in the corner, that I enquired who it belonged to, and found her as charming as she is pretty. Another was Norwich, where I tried to fancy you listening to the beautiful Anthem![10] (The sermon was not quite a Magee)[11] I talked very coolly about Belfast[12] the other evening, but all the same, I am longing to go, more than I can say! And one ought to see Ireland sometime I think, & it w[oul]^d^ be such an opportunity!

But oh! The "Mrs Grundy's"[13] near home! It is a difficult question how far one sh[oul]^d^ mind being laughed at for pretensions to science one has never made, for it is certainly not that, w[oul]^d^ take me there! I have just a glimmering of a project sometimes, but I dare say it will come to nothing so don't breathe a word.

RI MS JT/1/E/99f

1. *[c. 27 January 1874]*: the date is suggested by reference to the dissolution of Parliament; see n. 4.
2. *Mountfield Court, | Robertsbridge, | Hawkhurst*: the Egerton family estate in Sussex.
3. *that wretched letter . . . was no inside*: not identified, but sending an empty envelope was sometimes a sign of disapprobation (see letter 4446).
4. *L[ad]y. Claud*: Elizabeth Hamilton.
5. *this astounding Dissolution*: William Gladstone's surprise dissolution of Parliament on 26 January in order to call a general election, which was perceived as an unfair attempt to catch the Conservative opposition, whom Egerton supported, off their guard. Claud Hamilton was standing for the constituency of County Tyrone in Ireland, for which he had been an MP since 1839, so presumably had to be there to campaign. He lost his seat in the election on 11 February.
6. *the place*: Heathfield Park in Sussex. The house and estate was owned by the family of Charles Richard Blunt, who had purchased it in 1819, but occupied by the Hamiltons.
7. *Emmy*: Emily Margaret Egerton (1852–94), Egerton's daughter.
8. *it would bring you nearer us*: Heathfield Park was close to Egerton's family seat at Mountfield Court, also in Sussex.
9. *your little friend "Amy Lubbock"*: Amy Harriet Lubbock (1857–1929), daughter of John Lubbock. When introducing her father's evening lecture at the BAAS meeting in Belfast later that year, Tyndall commended the 'beautiful diagrams' and observed that they were the 'production of one of whom he was compelled to speak with a certain degree of reserve and respect, Miss Lubbock', whom he had known since she had 'scarcely attained the age of five or six years, and who had now grown up and become all that those who loved her could desire her to become' ('British Association', *Times*, 24 August 1874, p. 6).

10. *Anthem!*: in Anglican services, an anthem is a musical arrangement sung, in English rather than Latin, by the choir accompanied by an organ, generally with a celebratory tone.
11. Magee: William Connor Magee (1821–91), an Anglican clergyman who was the Bishop of Peterborough (*ODNB*). He had given three celebrated sermons on Christianity in relation to freethought, scepticism, and faith in Norwich Cathedral on 28, 29, and 30 March 1871.
12. *Belfast*: where the BAAS was holding its annual meeting in August, and where Tyndall was due to deliver the presidential address. Egerton did eventually go, although she also expressed doubts about being able to in letter 4306.
13. *"Mrs Grundy's"*: originally a character in Thomas Morton's play *Speed the Plough* (1798), the name had become a shorthand for those with conventional or priggish views. The laughter at Egerton's 'pretensions to science' were presumably prompted by conventional attitudes towards gender, especially among her aristocratic peers.

To George Campbell 29 January 1874 4261

29th Jan[uar]y 1874

Dear Lord Duke.

I send you an incomplete abstract of what I have been doing as regards sound.[1] Since my lecture[2] has been noticed in the papers various corroborations have come to me, but none from so practised an observer as you are; hence your letter[3] is particularly valuable to me.

My position is not that the fog itself helps the sound; but it simply indicates the removal, by condensation, of an agent which is really deadly to the transmission of sound. That agent is aqueous vapour—which mingling with the air renders the atmosphere acoustically flocculent[4] or turbid.[5]

This flocculence or turbidity is quite invisible.

For the last 165 years[6] Scientific men have been going on repeating that fog, falling rain & snow are great deadeners of sound. That they were so was an a priori probability[7]—and this probability rendered a very small amount of evidence sufficient to produce complete conviction.

I suppose I may use your letter, or a portion of it?[8]

Most truly yours | John Tyndall

RI MS JT/1/T/97
RI MS JT/1/TYP/5/72

---

1. *incomplete abstract... as regards sound*: probably Tyndall's paper 'On the Atmosphere as a Vehicle of Sound', which he read to the RS on 15 January 1874 (see letters 4250 and 4251).
2. *my lecture*: presumably the Friday Evening Discourse Tyndall gave at the RI on 16 January, entitled 'On the Acoustic Transparency and Opacity of the Atmosphere' (see letter 4226, n. 2).
3. *your letter*: letter missing, but see n. 8.

4. *flocculent*: having a loosely clumped texture; in chemistry, this relates to the process by which colloidal particles come out of suspension to form an aggregate. Tyndall also discusses his conception of the impact of flocculence on sound in letter 4302.
5. *turbid*: (of a liquid) cloudy, opaque or thick with suspended matter (*OED*).
6. *the last 165 years*: in 1708, the English clergyman and natural philosopher William Derham had published the first reasonably accurate measurement of the speed of sound (see letter 4220, n. 3).
7. *a priori probability*: the Latin phrase 'a priori', which translates as 'from the former', denotes a line of reasoning based on theoretical deduction rather than direct observation.
8. *I suppose I may use your letter, or a portion of it?*: in the published version of his paper, Tyndall observed: 'The Duke of Argyll has also favoured me with the following very interesting account of his own experience. Coming as it does from a disciplined scientific observer, it is particularly valuable. "This fact" (the permeability of fog by sound) "I have long known, from having lived a great part of my life within four miles of the town of Greenock, across the Frith. Ship-building goes on there to a great extent, and the hammering of the caulkers and builders is a sound which I have been in the habit of hearing with every variety of distinctness, or of not hearing at all, according to the state of the atmosphere; and I have always observed on days when the air was very clear, and every mast and spar was distinctly seen, hardly any sound was heard; whereas on thick and foggy days, sometimes so thick that nothing could be seen, every clink of every hammer was audible, and appeared sometimes as close at hand"' ('Vehicle of Sound', p. 219).

To George Gabriel Stokes 29 January [1874][1] 4262

Royal Institution of Great Britain | 29th Jan[uar]y

My dear Stokes,

If you can receive it my paper on the Atmosphere as a Vehicle of Sound[2] shall be in your hands next Thursday.

Would not that be the proper occasion to show the apparatus by which the stoppage of sound by reflection at the limiting surfaces of gases of different densities; & of ordinary air & air saturated with different vapours, has been demonstrated?[3] If you think so I should be ready to show the apparatus and make the demonstrative experiments on Thursday.[4]

If it suit your convenience better of course I am willing to have the paper postponed: I should however wish it off my hands before my lectures here[5] begin.

Whenever it comes on my assistant,[6] who has a short & very beautiful little paper to communicate to the Proceedings,[7] would at the same time make an experiment to illustrate his paper.

very truly yours | John Tyndall

CUL SC—add 7656

1. *[1874]*: the year is established by reference to Tyndall's paper on the atmosphere and acoustics (see n. 2).
2. *my paper on the Atmosphere as a Vehicle of Sound*: Tyndall read a preliminary version of this paper to the RS on 15 January and would deliver a completed version on 5 February.
3. *the apparatus by . . . has been demonstrated*: experimental apparatus through which layers of carbonic acid and coal gas could be observed and thereby used to demonstrate the reflection of sound waves. The apparatus is described, and illustrated, in 'Vehicle of Sound', pp. 202–5 and plate XVIII.
4. *Thursday*: 5 February, the next meeting of the RS.
5. *my lectures here*: see letter 4233, n. 3.
6. *my assistant*: John Cottrell, who devised the experimental apparatus and who, Tyndall declared, was 'eminently skilful in devising apparatus the object of which has been made clear to him' ('Vehicle of Sound', p. 202).
7. *a short & very beautiful . . . to the Proceedings*: J. Cottrell, 'On the Division of a Sound-Wave by a Layer of Flames or Heated Gas into a Reflected and a Transmitted Wave', *Roy. Soc. Proc.*, 22 (1874), pp. 190–1. Tyndall read the paper to the RS on Cottrell's behalf on 12 February.

From George Campbell 30 January 1874 4263

Jan[uary]: 30/74

My Dear Professor Tyndall

Many thanks for y[ou]r letter & paper.[1]

You may make any use you like of my letter.[2]

My notion has always been that as Water conveys sound more rapidly (does it not?) than air, a watery-atmosphere conveys it more rapidly than a dry air.

The molecular condition of the water held in the air must of course be an element in the case. In fogs the water is condensed into separate particles, I suppose, in close—very close—contiguity[3] and it is very conceivable that the impact of sound waves sh[oul]d be transmitted through them with special advantage.

Y[ou]rs tr[ul]y | Argyll

RI MS JT/1/A/100
RI MS JT/1/TYP/1/73

1. *y[ou]r letter & paper*: letter 4261 and 'an incomplete abstract' of what was probably Tyndall's paper 'On the Atmosphere as a Vehicle of Sound', which he read to the RS on 15 January.

2. *make any use you like of my letter*: see letter 4261, n. 8.
3. *contiguity*: the condition of touching or being in contact (*OED*).

From Josiah Parsons Cooke[1] [*c.* late January 1874][2] 4264

<1 or more pages missing>
you would come soon early enough to enjoy a week or two with us then. Poor Agassiz[3] has gone—He was killed by the liberality of his friends who put opportunities in his reach, which he felt he ought not to forego, and in endeavouring to work, after the plainest admonition, he snapped the cord. We are fortunate in having his son Alex[ander]. Agassiz to succeed him, who is not inferior to his Father in genius and will carry on the museum[4] to completion with more direction than the Father ever used; for that was not his strong point.

Youmans wrote me that he sent you a copy of my little book[5] or I should have sent one myself. It is a revision of a course of lectures,[6] which I gave at the Lowell Institute[7] immediately after yours.[8] I hope the book will be more fortunate than the lectures for the great Boston fire[9] occurred soon after the course began and not only interrupted the lectures for a time but also engrossed the attention of our community to such an extent that science could hardly obtain a hearing for some months afterwards—The fire also brought a great deal of anxiety and care to me. I have had to rebuild a good deal of the family property and until within six months have been unable for that reason to do scientific work—Mrs Cooke[10] joins me in kindest regards to yourself—

Very Sincerely Yours | Josiah P. Cooke Jr

RI MS JT/1/C/50

---

1. *Josiah Parsons Cooke*: Josiah Parsons Cooke (1827–94), an American chemist at Harvard University who specialized in the measurement of atomic weights (*ANB*).
2. *[c. late January 1874]*: the date is suggested by reference to Cooke's new book (see n. 5), which was published on 24 January ('D. Appleton & Co.', *Publishers' Weekly*, 5 (1874), p. 73).
3. *Poor Agassiz*: Louis Agassiz died on 14 December 1873 of a stroke.
4. *the museum*: Harvard's Museum of Comparative Zoology, which Agassiz had founded in 1859. Agassiz's son Alexander took over the museum following his father's death in December 1873. See M. P. Winsor, *Reading the Shape of Nature: Comparative Zoology at the Agassiz Museum* (Chicago: University of Chicago Press, 1991).
5. *my little book*: J. P. Cooke, *The New Chemistry* (New York: D. Appleton, 1874), which was published as part of the International Scientific Series, a popular series of scientific books founded by Edward L. Youmans in 1871, with Tyndall, Thomas Huxley, and Herbert

Spencer involved as a 'British Committee'. Tyndall's *Forms of Water* was the first title in the series, which carried Youmans's preface to the whole series, in which he noted: 'Those interested in the series are under many obligations to Prof. Tyndall for his kindness in consenting to furnish its commencing volume' (pp. viii–ix). The series numbered ninety-six titles in all, each with a uniform red cloth binding, published up until 1911 in Britain (Henry S. King), France (Bailliere), Germany (F. A. Brockhaus), Italy (Fratelli Dumolard), Russia (Znanie), and the United States (D. Appleton). See L. Howsam, 'An Experiment with Science for the Nineteenth-Century Book Trade: The International Scientific Series', *British Journal for the History of Science*, 33 (2000), pp. 187–207, and B. Lightman, 'The International Scientific Series and the Communication of Darwinism', *Journal of Cambridge Studies*, 5 (2010), pp. 27–38. See also W. B. Jensen, 'Physical Chemistry before Ostwald: The Textbooks of Josiah Parsons Cooke', *Bulletin for the History of Chemistry*, 36 (2011), pp. 10–21. If Youmans did send the book to Tyndall, his accompanying letter is missing.

6. *a course of lectures*: Cooke's lectures 'aimed to present the modern theories of chemistry to an intelligent but not a professional audience, and to give to the philosophy of the science a logical consistency, by resting it on the law of Avogadro' (Cooke, *New Chemistry*, p. 5). He delivered the lectures at the Lowell Institute in the autumn of 1872.
7. *Lowell Institute*: a Boston organization founded by the philanthropist John Amory Lowell (1798–1881) in 1836, that provided free public lectures on science and the arts to audiences regardless of race or gender.
8. *yours*: As part of his American lecture tour (see letter 4201, n. 4), Tyndall had delivered a series of six lectures on light at the Lowell Institute, between 15 and 25 October 1872.
9. *the great Boston fire*: the fire began on 9 November 1872 and continued into the following day. It remains Boston's largest fire, with thirteen deaths and more than seven hundred buildings destroyed.
10. *Mrs Cooke*: Mary Hinckley Cooke (née Huntington, d. 1911).

To Hermann Helmholtz 10 February 1874 4265

Royal Institution of Great Britain | 10th. Feb[ruar]y 1874

My dear Helmholtz

A young hero-worshipper,[1] who makes you one of the objects of his idolatry, wishes for two minute's conversation with you. He is travelling as Whitworth Scholar[2] on the Continent. I do not know the young man personally, but for his excellent father's[3] sake I give him this note to you.

Believe me | Yours ever faithfully | John Tyndall

BBAW, NL. Helmholtz, Nr. 477, BI_26
RI MS JT/1/T/493

1. *A young hero-worshipper*: not identified, but Tyndall is perhaps invoking Thomas Carlyle's conception of the veneration due to heroes, as articulated in *On Heroes, Hero-Worship, and the Heroic in History* (London: James Fraser, 1841).
2. *Whitworth Scholar*: a three-year scholarship established in 1868 by Joseph Whitworth to fund the academic studies of promising young engineers and chemists. The names of the scholars for each year were published in the annual *Prospectus of Sir Joseph Whitworth's Scholarships for Mechanical Science* (London: Her Majesty's Stationery Office).
3. *father's*: not identified.

From Frederick Arrow[1] 10 February 1874 4266

MY DEAR TYNDALL,—

The enclosed[2] will show how accurately your views[3] have been verified, and I send them on at once without waiting for the details. I think you will be glad to have them, and as soon as I get the report[4] it shall be sent to you. I made up my mind ten days ago that there would be a chance in the light foggy-disposed weather at home, and therefore sent the *Argus*[5] off at an hour's notice, and requested the Fog Committee[6] to keep one member on board. On Friday I was so satisfied that the fog would occur that I send Edwards[7] down to record the observations.

Very truly yours, | FRED. ARROW

Sound (1875), pp. 302–3

1. *Frederick Arrow*: Frederick Arrow (*c.* 1819–75), Deputy Master of Trinity House since 1865, having previously been a commander of several merchant ships ('The Late Sir Frederick Arrow', *Illustrated London News*, 31 July 1875, p. 100).
2. *The enclosed*: letters from Captain William Atkins and Edward Price-Edwards containing accounts of the transmission of sound in fog at sea. They were published in *Sound* (1875), pp. 303–4.
3. *your views*: Tyndall hypothesised that sound carried further in dense fog than in clear conditions, and was proven correct by experimental observation.
4. *the report*: 'Report by Professor Tyndall'.
5. *the Argus*: THV *Argus*, a coastguard boat.
6. *the Fog Committee*: a special committee of the Board of Trade, formed with the aim of addressing issues relating to shipping and fog signals.
7. *Edwards*: Edward Price-Edwards (d. 1904), principal clerk at Trinity House and Arrow's private secretary.

From Mary Agatha Russell 10 February 1874 4267

Pembroke Lodge.[1] | Richmond Park. | February 10. / 74.

Dear Mr Tyndall

I have just received your note & card.[2] Thank you very much indeed for sending it. I shall certainly hope to accept your most kind invitation & am looking forward with great pleasure to the lectures.[3] As I have very seldom been beyond the garden since my accident[4]—I do not feel sure that I shall be able to go so far as London to-morrow (but I hope so) & in case I do not see you I write my thanks at once. We shall be very glad if you are able to come here when you have leisure for such a long journey. With Mama's[5] kindest regards & best thanks, believe me

Yours very sincerely | Agatha Russell.

RI MS JT/1/R/49

1. *Pembroke Lodge*: see letter 4237, n. 1.
2. *your note & card*: letter missing.
3. *the lectures*: probably the series of six lectures 'On the Physical Properties of Gases and Liquids' (see letter 4233, n. 3).
4. *my accident*: not identified, but see letter 4237.
5. *Mama*: Frances Russell.

To James Thomas Knowles 12 February [1874][1] 4268

Royal Institution of Great Britain | 12th. Feb[ruar]y

Dear Mr. Knowles.

On second thought I decide that it will be better for me to employ my spare time in trying to get an account of my sound experiments ready for the Contemporary[2] or some other magazine than to devote it to the preparation of a paper for the Metaphysical.[3]

Yours truly | John Tyndall

Wellcome MS. 7777/43a

1. *[1874]*: the year is established by reference to Tyndall's *Contemporary Review* article (see n. 2).

2. *an account of my . . . ready for the Contemporary*: Tyndall may be referring to 'On the Atmosphere in Relation to Fog-signalling', which was published in two parts in the *Contemporary Review*, which Knowles edited, in November and December 1874. In the articles Tyndall gave an extended account of his experiments with sound-waves and other acoustic phenomena. J. Tyndall, 'On the Atmosphere in Relation to Fog-Signalling', *Contemporary Review*, 24 and 25 (1874), pp. 819–41 and pp. 148–68.
3. *a paper for the Metaphysical*: the Metaphysical Society was an exclusive debating society that Knowles had founded in 1869, and for which he continued to serve as secretary and principal organizer. It met monthly during the social season (November to July) at the Grosvenor Hotel in Mayfair. Tyndall had joined, at Knowles's invitation, in March 1869, but never actually gave one of the short philosophical papers on which the discussion at each meeting was based (*Ascent of John Tyndall*, pp. 217–8). See also C. Marshall, B. Lightman, and R. England (eds), *The Metaphysical Society (1869–1880): Intellectual Life in Mid-Victorian England* (Oxford: Oxford University Press, 2019); and C. Marshall, B. Lightman, and R. England (eds), *The Papers of the Metaphysical Society, 1869–1880* (Oxford: Oxford University Press, 2015).

From Thomas Romney Robinson — 12 February 1874 — 4269

Feb[ruary] 12–74 | Observatory Armagh

Dear Tyndal[1]

As relates to myself there is nothing in the proof[2] which you sent me to which I can take the slightest exception, though it fails to satisfy me that Faraday took the best course.[3] You are quite right in thinking that in the passage which you have quoted from the Report of the Committee it was not my intention to say any thing which would give Faraday pain.[4] In fact it was only a wish to avoid this which prevented me from publishing his report to the Trinity Board and examining the reasons which he assigned to that Body against their taking action on the Report of the Committee.

For had I done so it would, in the then state of public feeling, caused by the wreck of the Anglosaxon,[5] have probably produced a strong impression against him.

As you are acquainted with Faraday's Report[6] you have probably noticed in it an expression of despondency and hopelessness which struck me as singularly unlike the energy of his earlier years.

His statement that "probably none of the researches suggested might lead to a satisfactory result"—and that "as far as we know at present all the propositions fail in giving any assurance of that amount of certainty without which they would be sources of danger rather than of safety"[7]—seems to me quite at variance with the fundamental principles of experimental research.

If we never tried an experiment unless we were certain of succeeding in it we would not experiment at all! And one important object of the proposed researches was to ascertain what methods were not likely to succeed.

As to his suggestion that I should undertake the management of the experiments myself, he forgot that it was impossible for me to leave the duties of my Observatory: And had I been disposed uninvited to recommend to the Trinity Board one of the Committee who was resident in London—I should have been prevented from doing so by their statement that the views entertained by D[r] Faraday were entirely in accordance with their own opinions.

However as the proverb says, "Better late than never"—and though Ten years have been lost I rejoice that they have at last taken up the enquiry which I have no doubt will lead to most satisfactory results; for it is no idle compliment to say they have entrusted it to one who is more likely than any person that I know to work it out well & completely.

I am yours truly | T. R. Robinson

RI MS JT/1/R/33

---

1. *Tyndal*: misspelled by Robinson.
2. *the proof*: presumably of 'Report by Professor Tyndall'.
3. *Faraday took the best course*: Robinson had been the chairman of a committee, appointed in 1861 by the BAAS, with the purpose of advocating for experimental investigation into methods of signaling ships in fog. This committee had submitted a report to the Board of Trade, which was published in *Brit. Assoc. Rep. 1863*, pp. 105–10. This report was passed on to Trinity House, who had consulted Michael Faraday, their official scientific adviser, who agreed that such experiments were desirable, but nevertheless advised against them on the grounds of difficulty and expense. Consequently, Trinity House did not pursue the subject any further at that time. See *Faraday Correspondence*, vol. 6, pp. xxxiv–xxxvii.
4. *You are quite right in . . . give Faraday pain*: after quoting a passage from the report of Robinson's committee which stated 'These opinions Dr. Faraday seems to have formed . . .from a dread of the difficulty, the magnitude, and the expense of the investigation. These, we believe, he exaggerates; but even taking them at his estimate, we think they will not be accepted by the public as a satisfactory excuse for the inertia of this powerful body', Tyndall added: 'Though probably not so intended by Dr. Robinson, these words have been thought to involve a somewhat sharp animadversion' ('Report by Professor Tyndall', p. 72).
5. *wreck of the Anglosaxon*: the SS *Anglo-Saxon*, a steam ship that operated on the Liverpool-Canada route, ran aground on the coast of Newfoundland on 27 April 1863 with great loss of life.
6. *Faraday's Report*: a letter Faraday wrote on 22 July 1863 to Peter Henry Berthon, the secretary of Trinity House, in response to Robinson's request to the Board of Trade for an intensive investigation into fog signaling.

7. *His statement that "probably . . . than of safety"*: the quotations are slight paraphrases of passages from Faraday's letter to Berthon (see n. 6), and Faraday actually wrote: 'it is probable that of the many researches suggested not one in ten, perhaps not one in the whole, would lead to a satisfactory result' and 'as far as we know at present, all the propositions fail in giving any appearance of that amount of certainty without which they would be sources of danger rather than of safety' (*Faraday Correspondence*, vol. 6, p. 317).

From Jane Barnard 14 February 1874 4270

Barnsbury Villa, | 320, Liverpool Road, | N.[1]

My dear Dr Tyndall

Your note[2] came in just before Mr & Mrs Siemens,[3] & we had a most pleasant visit from them. I never doubted your very kind & affectionate feeling towards my Aunt;[4] & your readiness to accompany her & take care of her, if she went to Newcastle[5] is only a fresh proof of it. It is to myself that I feel you are sometimes a little unjust, & I value your good opinion & friendship too much, not to feel hurt when I think you are so: but let bygones be bygones & believe me

Very sincerely yours | Jane Barnard | 14 Feb[ruary]. 1874 | Dr Tyndall | &c &c &c

RI MS JT/1/B/54

---

1. *Barnsbury Villa, | 320, Liverpool Road, | N.*: see letter 4226, n. 1.
2. *Your note*: letter missing.
3. *Mr & Mrs Siemens*: Carl Wilhelm Siemens and his wife Anne Siemens (née Gordon, 1821–1902). The former designed the CS *Faraday*, an innovative telegraph cable-laying ship named after Michael Faraday that was to be launched on 17 February (see letter 4257).
4. *my Aunt*: Sarah Faraday.
5. *Newcastle*: the ship was to be launched from the shipyard of C. Mitchell & Co. on the river Tyne in Newcastle, Northumberland.

From Mary Egerton [*c.* mid-February 1874][1] 4271

<1 or more pages missing>

ourselves, and keep you from wiser talk with Mr. Emmens.[2] I should like to know who deserves the credit of those Capital Valentines![3] I sh[oul]d. give "Science"[4] to Lady Pollock.

You don't know with what intense interest I think of your Belfast address; some topics of which I see glimmering forth from some of the books on your table, of which you allude to.[5] I long to hear it, for I know that whatever views it contains, its tone will be that of Reverence for Truth, and for Goodness also, even where you may not think the two coincide. May the Spirit who lives in all that is great and noble, however expressed, be with you dear friend!

Ever y[ou][rs] | M F Egerton

Don't think me too solemn! It is a great position to address all the greatest minds of the country!

RI MS JT/1/E/98

---

1. *[c. mid-February 1874]*: the mention of Valentines seems to date this letter to shortly after 14 February, while the discussion of Tyndall's reading in relation to writing the Belfast Address establishes the year as 1874.
2. *M[r]. Emmens*: probably Stephen Henry Emmens (1843–*c.* 1900), an industrial chemist, metallurgist, and proposer of non-Newtonian theories of gravity. He was made a member of the RI in 1873.
3. *those Capital Valentines*: not identified, but presumably decorated cards traditionally given on St Valentine's day on 14 February.
4. *"Science"*: possibly a Valentines card with a scientific theme, which would be consistent with Tyndall's flirtatious relationship with Juliet Pollock.
5. *you allude to*: letter missing, although Egerton may refer to one of the books on Tyndall's table, a 'new vol[ume]: of Fichte', in letter 4306.

From Henrietta Huxley 16 February [1874][1] 4272

4 Marlborough Place | Abbey Road N.W.[2] | 16th Feb[ruary]

Dear Brother John

I have been asked to send you the enclosed,[3] which I do not much like doing—the subscription[4] not being on behalf of Scientific folk—and therefore not in your domain.

I have been in bed several days—and write this in my room. My enemy has been severe cold, caught during those few bitter days, which showed itself in rheumatism in the face and head. I got down stairs last evening for a few hours and had a very pleasant time as your American friends say. Our new found American friend Mr Fisk[5]—spending the Sunday evening with us as usual—Together with our dear Sam—Jim's artist cousin.[6]

Only to think of Jess[7] being 16 to day. Do you remember when she was but a few weeks old—your coming into the little drawing room at Waverley Place[8]—and kneeling down beside us—her and me? That seems to have happened in some other land so far off does it seem.

The children[9] have devised a coffee party this afternoon in Mady's studio[10]—and I am to go up and drink coffee with them. My heart is full of gratitude and happiness. The husband well and bright (such a contrast to last year) writing his Aberdeen Address[11]—with all his old fineness and fire. It was so pleasant to sit down and hear him read what he had done and feel that he was himself again.[12]

Now to my actual miseries. I have had no cook for three weeks—and have been suffering from an attack of "helps" "casuals",[13] though I must own to a devoted maid who smooths all—Added to this the new boiler has been so badly set it wabbled about.—That men are at work at it—that not a drop of hot water and little cold is to be had—and that a friend from the country writes to say she will be here on Thursday and you may fancy my feelings! I wish you were at the coffee party this afternoon.

Your affectionate sister | Nettie Huxley

RI MS JT/1/TYP/9/3027
LT Typescript Only

---

1. *[1874]*: the year is established by reference to the sixteenth birthday of Jessica Oriana Huxley, who was born in 1858.
2. *4 Marlborough Place | Abbey Road N.W.*: see letter 4213, n. 1.
3. *the enclosed*: enclosure missing.
4. *the subscription*: not identified.
5. *Mr Fisk*: John Fiske (1842–1901), an American philosopher and advocate of evolution (*ANB*).
6. *our dear Sam—Jim's artist cousin*: Samuel Edmund Waller (1850–1903), a distant cousin of James Henry Huxley (1852–1943), who, as the son of Thomas Huxley's elder brother James Edward Huxley, was known as 'young Jim'. He was supported by the Huxleys, and lodged with them, while studying chemistry at the Royal School of Mines. See A. Desmond, *Huxley: Evolution's High Priest* (London: Michael Joseph, 1997), pp. 5 and 55. Waller worked as an illustrator at the *Graphic* magazine and later in 1874 established his reputation as a genre painter by exhibiting a work called *Soldiers of Fortune* at the Royal Academy. His elder brother Frederick William Waller, who also regularly visited the Huxley's home, would marry Jessica Oriana Huxley (see n. 7) in May 1878.
7. *Jess*: Jessica Oriana Huxley, the Huxleys' eldest daughter.
8. *Waverley Place*: 14 Waverley Place, St. John's Wood, where the Huxley's had lived from 1855–61; from 1861–72 they lived at 26 Abbey Place, St. John's Wood.
9. *The children*: as well as Jessica Oriana, the Huxley's surviving children were Marian, Leonard, Rachel, Henrietta, Henry, and Ethel.
10. *Mady's studio*: Marian had shown artistic promise and was attending drawing classes at the Slade School of Art (see letter 4196, n. 12).

11. *his Aberdeen Address*: Huxley became Lord Rector of Aberdeen University in 1873 (through 1880), having been elected by the students there on 14 December 1872. He delivered his Inaugural Address to the university, titled 'Universities: Actual and Ideal', on 27 February 1874 (P. J. Anderson (ed), *Rectorial Addresses delivered in the Universities of Aberdeen, 1835–1900* (Aberdeen: Aberdeen University, 1902), pp. 170–98, 376–7; Desmond, *Huxley*, pp. 439–40). The address was published as T. H. Huxley, 'Universities: Actual and Ideal', *Contemporary Review*, 23 (1873–4), pp. 657–79; and T. H. Huxley, 'Universities: Actual and Ideal', in *Collected Essays*, 9 vols (London: Macmillan, 1893–4), vol. 3 (1893), pp. 189–233).
12. *himself again*: Huxley had been suffering from ill health, both physical and mental, for the previous two years (see letter 4192, n. 16).
13. *an attack of "helps" "casuals"*: probably servants hired on a temporary, or casual, basis while Huxley sought to hire a new permanent cook.

To Hector Tyndale [18 February 1874][1] 4273

Wednesday

My dear Hector.

No evening dress—We meet[2] in a perfectly informal way—come here[3] at 5.45.

Yours affect[ionatel]y | John

RI MS JT/1/T/1476
RI MS JT/1/TYP/4/1687

---

1. *[18 February 1874]*: the date is suggested by an entry in Thomas Hirst's journal (see n. 2).
2. *We meet*: Hirst recorded in his journal for Wednesday 18 February 1874: 'Dined at Tyndall's to meet General Hector Tyndale who has just come to England' (*Hirst Journals*, p. 1996).
3. *here*: the RI.

From James Joseph Sylvester 19 February 1874 4274

19th Feb[ruary] 1874

My dear Tyndall,

Many thanks for the cheque[1]—I feel really obliged by your kind and cordial good wishes but am perfectly satisfied and do not desire anything more—in fact I have more than once felt that I would rather have received nothing.

I consider the contributions most fair and liberal—

Yours truly | J. J. Sylvester

RI MS JT/1/S/295
RI MS JT/1/TYP/4/1512

1. *the cheque*: presumably a philanthropic contribution, as Sylvester was without a position at this time. On 15 February he told the mathematician Arthur Cayley, 'I am more than half inclined to go out to the Antipodes rather than remain unemployed and living upon charity in England' (quoted in K. H. Parshall, *James Joseph Sylvester: Life and Work in Letters* (Oxford: Oxford University Press, 1998), p. 144).

To Hector Tyndale [20 February 1874][1] 4275

My dear Hector.

Instead of luncheon pray come and dine with me tomorrow at 7. We can then have the whole evening to ourselves.

I will also try to get Debus or Tom.[2]

Yours affect[ionatel]y | <u>John</u>

RI MS JT/1/T/1475
RI MS JT/1/TYP/4/1728

1. *[20 February 1874]*: the date is suggested by an entry in Thomas Hirst's journal (see n. 2).
2. *I will also try to get Debus or Tom*: Hirst recorded in his journal that he dined with Tyndall, Tyndale, and Heinrich Debus on 21 February 1874 (*Hirst Journals*, p. 1996).

From Jean Charles Rodolph Radau[1] 21 February 1874 4276

Paris. 21 février 1874 | 13 rue du Dragon

Monsieur,

Me permettez vous de réclamer de votre bonté un service qu'il vous est facile de me rendre? Il s'agit de certifier en trois lignes que vous m'avez vu à Londres au mois de septembre 1870. J'espère que vous vous rappelez que je vous ai présenté mes respects à l'Institution royale, peu de temps après M. Silbermann. J'ai sollicité un certificat du même genre de M. Wheatstone, qui m'a fait l'honneur de m'inviter à dîner pendant mon séjour à Londres, et qui, je l'espère, voudra bien s'en souvenir. Vous devinez sans peine combien il est important pour moi de pouvoir invoquer de tels témoignages qui établissent qu'au début du siège de Paris je me trouvais à Londres, et je ne vous demande que d'affirmer un fait dont vous êtes certain. J'espère donc, Monsieur, que vous ne refuserez pas de me rendre le témoignage que je sollicite; je vous prie d'excuser mon importunité, et de croire au profond respect avec lequel je suis, Monsieur, votre très humble serviteur.

R. Radau

Paris. 21 February 1874 | 13 rue du Dragon

Sir,

May I ask your kindness for a favour that would be easy to render me? It is to confirm in three lines that you saw me in London in September 1870. I hope you remember that I presented my respect to you at the Royal Institution, a bit after Mr. Silbermann.[2] I have asked a certificate of the same kind from Mr. Wheatstone[3] who made me the honour of inviting me to dine during my stay in London, and who, I hope will be willing to remember it. You must guess without a doubt how important it is for me to be able to call upon such testimonies, which state that at the beginning of the siege of Paris[4] I was in London, and I only ask you to confirm a fact of which you are certain. I hope, then Sir that you will not refuse to send me the testimony I ask for. I beg you to excuse my importunity and to believe in the profound respect with which I am, Sir, your very humble servant.

R. Radau

RI MS JT/1/R/1

---

1. *Jean Charles Rodolph Radau*: Jean Charles Rodolph Radau (1835–1911), a German astronomer and mathematician who lived and worked in Paris (*ANB*).
2. *Mr. Silbermann*: Jean Thiébaut Silbermann (1806–65), a French physicist and physics demonstrator at the Conservatoire des Arts et Métiers in Paris ('Notice biographique sur Jean-Thiébaut Silbermann', *Bulletin de la Société d'histoire naturelle de Colmar*, 6–7 (1865–6), pp. 41–7).
3. *Mr. Wheatstone*: Charles Wheatstone.
4. *the siege of Paris*: the concluding encounter of the Franco-Prussian War, during which Prussian troops captured the French capital. The siege lasted from 19 September 1870 to 28 January 1871. While it is not clear why Radau was required to prove his absence from Paris during this time, he had been born in Prussia and it seems likely that his loyalties would be questioned in the embittered atmosphere of anti-German sentiment that pervaded France in the aftermath of the war.

From Jean-Albert Gauthier-Villars 22 February 1874 4277

Imprimerie et Librairie pour les Mathématiques, les Sciences et la Arts, Quai des Grands Augustin, 55

Paris, le 22 février 1874

Monsieur

J'ai l'honneur de vous adresser le premier exemplaire qui, tout humide encore, sort de mes presses.

Cette nouvelle édition de votre célèbre ouvrage La Chaleur n'est pas, comme vous le savez, aussi parfaite que je l'aurais désiré; j'espère cependant qu'après les corrections faites et l'addition des notes et variantes vous ne jugerez pas cette édition française trop inférieure à l'édition anglaise.

Si vous voulez bien me faire connaître le nombre d'exemplaires que vous désirez recevoir, je m'empresserai de vous les expédier. J'aurai soin de vous envoyer de plus, pour votre Bibliothèque personnelle, un exemplaire convenablement relié.

M. l'Abbé Moigno m'a remis le commencement de la traduction de la Lumière; cette traduction est entièrement écrite de sa main et faite avec soin. Les gravures ont été exécutées avec talent par Mr Dulos, et de mon côté, je donnerai tous mes soins à l'exécution typographique. Aussi j'espère que la traduction de la Lumière ne vous donnera pas les ennuis que vous a occasionnés celle de la Chaleur et que j'ai tout regrettés.

Veuillez agréer, Monsieur, l'assurance de mon bien respectueux dévouement

Gauthier-Villars

M. le Professeur J. Tyndall | Royal Institution | Londres.

Printing House and Bookshop for Mathematics, Sciences and Arts, Quai des Grands Augustin, 55

Paris, February 22 1874

Sir,

I have the honour of sending you the first copy,[1] which still wet, just came out of my press.

This new edition of your famous book Heat is not, as you already know, as perfect as I would have liked; but I hope that after the corrections that were made and the addition of notes and variations, you will not judge this French edition too inferior to the English one.

If you would be kind enough to let me know of the number of copies that you would like to receive, I will hasten to send them to you. Also, I will endeavour to send you, for your personal library, a copy properly bound.

Father Moigno gave me the beginning of the translation of Light.[2] This translation is written entirely by his hand and done with much care. The engravings were done skilfully by Mr Dulos;[3] and as for me, I will take the utmost care in the typographical work. Also, I hope that the translation of Light will not give you as much trouble as Heat[4] had caused you, and about which I deeply regretted.

Please accept, Sir, the assurance of my most respectful devotion.

Gauthier-Villars

Professor J. Tyndall | Royal Institution | London.

RI MS JT/1/V/22

---

1. *the first copy*: of J. Tyndall, *La chaleur: considérée comme un mode de mouvement*, trans. F. Moigno, 2nd edn (Paris: Gauthier-Villars, 1874).
2. *the translation of Light*: Moigno's translation of *Six Lectures on Light* (1873), which was published as J. Tyndall, *La lumière: six leçons faites en Amérique dans l'hiver de 1872–1873*, trans. F. Moigno (Paris: Gauthier-Villars, 1875).
3. *Mr Dulos*: Pierre Edelestand Stanislas Dulos (1820–74), a French engraver.
4. *as much trouble as Heat*: the second French edition of *Heat* was delayed by the Franco-Prussian War, and then problems with the manuscript, which François Moigno claimed was lost. Moigno reworked his translation, but this was based on the third English edition of 1868, which had since been superseded by the revised fourth English edition of 1870. Moigno then had to translate the latest edition, thereby causing a further delay. See F. Moigno, 'Préface du traducteur', in Tyndall, *La chaleur*, 2nd edn, pp. v–ix, on pp. vii–viii.

To Hector Tyndale [*c.* 25 February 1874][1] 4278

Wednesday

My dear Hector

I am <u>extremely</u> glad to find that you have moved[2] near to your friend Reid.[3] I called at Morley's[4] yesterday but found you had gone.

affection[natel]y yours | John Tyndall

I dine somewhere at 7 today—starting from here[5] at 6½ :—

RI MS JT/1/T/1477

---

1. *[c. 25 February 1874]*: the date is suggested by the relation to letters 4279, 4282, and 4283, which show that Tyndale was visiting London the week in which Wednesday was 25 February.
2. *you have moved*: according to a handwritten annotation, presumably by LT, on the typescript of letter 4282 Tyndale had moved to the Langham Hotel. The Langham was in Marylebone, and had opened in 1866. Its manager, James Sanderson, was an American who, like Tyndale, served in the Union army during the Civil War, and the hotel catered to the tastes of a largely American clientele.
3. *your friend Reid*: not identified, but he presumably lived in the Marylebone area of London (see n. 2).
4. *Morley's*: Morley's Hotel on the east side of Trafalgar Square, which was named after its first owner Atkinson Morley (1781–1858). It had opened in 1832, so, unlike the Langham Hotel (see n. 2), it may have not been modern enough for Tyndale's tastes.
5. *here*: presumably the RI.

To Hector Tyndale 26 February 1874 4279

26th Feb[ruar]y | /74

My dear Hector.

I send you gladly two tickets for such friends of yours as may wish to hear tomorrow evening's lecture.[1]

Our Friday meetings are thus constituted. The theatre is divided into two parts one for the Members and managers; one for the Members friends. Each member has the privilege of giving tickets to two friends. Now the limits of our theatre have rendered very stringent laws necessary as regards the members seats. No member can introduce even his own wife or child into those seats—and I am sometimes forced to appear unfriendly through abiding by the Rules of the Institution.

I say this to make it clear to your friend[2] the obstacles that stand in my way to asking him to occupy a members seat. In <u>your</u> case, as my relative, you would enter the theatre with the president[3] and managers, for whom a special seat is reserved. Dress is <u>not</u> necessary.[4]

Yours affectionately | John Tyndall

Please insert the names of your friends on the cards. To get good seats they would have to come at least half an hour before 9 P.M. <u>You</u> will step up to my room & if I am not there kindly await my arrival.

RI MS JT/1/T/1463
RI MS JT/1/TYP/4/1725

1. *tomorrow evening's lecture*: on 27 February the RI's Friday Evening Discourse was 'On Men of Science, Their Nature and Their Nurture', delivered by Francis Galton (*Roy. Inst. Proc.*, 7 (1873–75), pp. 227–36).
2. *your friend*: possibly 'your friend Reid' mentioned in letter 4278.
3. *the president*: Algernon George Percy (1810–99), 6th Duke of Northumberland, who served as the RI's president from 1873–99.
4. *Dress is not necessary*: i.e. formal evening dress.

From William Gladstone 26 February 1874 4280

11 Carlton House Terrace[1] | 26th Feb[ruary]. / 74

Mr Gladstone requests the honour of Mr Tyndall's company[2] at Breakfast on Wednesday next, the 4th March at 10 o'clock.

RI MS JT/1/TYP/1/415
LT Typescript Only

1. *11 Carlton House Terrace*: in St James's, close to Pall Mall; it had been Gladstone's home since 1856.
2. *the honour of Mr Tyndall's company*: Gladstone presumably had more time available, as he had recently stood down as Prime Minister having lost the general election which concluded on 17 February.

To William Gladstone 27 February 1874 4281

Athenaeum Club, Pall Mall[1]

Mr. Tyndall accepts with very great pleasure the invitation[2] with which he has been honoured by Mr. Gladstone.
27th. Feb[ruar]y. 1874

BL Add MS 44784, f. 212

1. *Athenaeum Club, Pall Mall*: see letter 4186, n. 1.
2. *the invitation*: letter 4280.

To Hector Tyndale 27 February [1874][1] 4282

Royal Institution of Great Britain | 27th. Feb[ruar]y

My dear Hector.

Come and have luncheon with me here[2] tomorrow at one o'Clock punctually?

Yours affect[ionate][ly] | John

Dine with me on Thursday next at the Royal Society Club[3] Willis's Rooms[4] 6 P.M. punctually

RI MS JT/1/T/1464
RI MS JT/1/TYP/4/1726

1. *[1874]*: the year is established by the relation to letter 4279, which shows that Tyndale was visiting London at this time.
2. *here*: presumably Tyndall's apartment at the RI.
3. *the Royal Society Club*: an exclusive dining club, formed in 1743, where a select number of Fellows of the RS gathered for dinner every Thursday before the Society's weekly meeting. The current President of the RS acted as President of this club. A second club, the Philosophical Club, was founded in 1847 by William Robert Grove in the aftermath of the reform of the RS's governance. See T. E. Allibone, *The Royal Society and its Dining Clubs* (Oxford: Pergamon Press, 1976); and M. B. Hall, *All Scientists Now: The Royal Society in the Nineteenth Century* (New York: Cambridge University Press, 2002), pp. 4, 82, and 217. See also T. G. Bonney, *Annals of the Philosophical Club of the Royal Society, Written from Its Minute Books* (London: Macmillan, 1919); and A. Geikie, *Annals of the Royal Society Club: The Record of a London Dining-Club in the Eighteenth & Nineteenth Centuries* (London: Macmillan, 1917).
4. *Willis's Rooms*: a suite of assembly rooms in St James's that, since 1781, were named after their longtime manager James Willis (1744–94).

To Hector Tyndale [28 February 1874][1] 4283

Saturday

My dear Hector.

My housekeeper[2] begs one line to say whether you can come to luncheon today at one o' Clock precisely

yours aff[ectionatel][y] | John Tyndall

RI MS JT/1/T/1480
RI MS JT/1/TYP/4/1727

1. *[28 February 1874]*: the date is established by the relation to letter 4282, where the invitation to lunch was first made.
2. *My housekeeper*: Jane Tunstall.

From Robin Allen 28 February 1874 4284

TRINITY-HOUSE,[1] LONDON, E.C.[2] | 28th February, 1874.

Sir,

With reference to the recent fog-signal experiments,[3] and to your concurrence as to the desirability of constructing a gun specially adapted for giving a loud report suitable for fog-signal purposes, I am now directed by the Elder Brethren[4] to forward to you the accompanying copy of correspondence[5] which has passed between the Corporation and the War Office, from which you will perceive that Mr. Cardwell[6] is quite willing that his department should render such assistance as lies in its power, and that a letter has been written to General Adye,[7] giving information as to the system hitherto employed, and the conditions which affect the practical working of the gun. I am thereupon directed to inquire whether any suggestions occur to you arising out of recent experiences at South Foreland[8] which you think desirable to be communicated to General Adye in reference thereto.

I am, Sir, | Your obedient Servant, | ROBIN ALLEN. | PROFESSOR TYNDALL, F.R.S. &c., &c. | *Royal Institution.*

National Archives, MT 10/290. H8476 2523
Typed Transcript Only

1. *TRINITY-HOUSE*: see letter 4187, n. 2.
2. *E.C.*: Eastern Central, a postal district of London.
3. *the recent fog-signal experiments*: Tyndall had been conducting a series of experiments regarding the transmission of sound in order to determine the most effective method of warning ships in fog (see letter 4185, n. 3).
4. *the Elder Brethren*: the governing body of Trinity House.
5. *the accompanying copy of correspondence*: enclosure missing.
6. *Mr. Cardwell*: Edward Cardwell (1813–86), the Secretary of State for War from 1868–74.
7. *General Adye*: John Miller Adye (1819–1900), Director of Artillery and Stores at the War Office from 1870–5.
8. *recent experiences at South Foreland*: see letter 4185, n. 3

From Jean Charles Rodolphe Radau[1] [1 March][2] 1874 4285

Paris. 29 fév. 74 | 13 rue du Dragon

Monsieur,

Permettez-moi de vous remercier du bienveillant accueil que vous avez fait à ma demande; je me souviendrai du service que vous m'avez rendu, et je serais vraiment heureux si je recevais une occasion de vous prouver une reconnaissance. En attendant, veuillez agréer, Monsieur, l'expression de mon profond respect.

R. Radau

Paris. 29 Feb[ruary] 74 | 13 rue du Dragon

Sir,

Allow me to thank you for the welcoming acceptance of my request.[3] I will remember the favour that you have done me, and I would be truly glad if I could get an opportunity to show you my gratitude. In the meantime, please accept, sir, the expression of my sincere respect.

R. Radau

RI MS JT/1/N/1

---

1. *Jean Charles Rodolphe Radau*: see letter 4276, n. 1.
2. *[1 March]*: although Radau clearly dated the letter as 29 February, 1874 was not a leap year and so February only had twenty-eight days; he presumably made an error, and the letter was written on 1 March.
3. *my request*: for a testimonial stating he was in London in September 1870 during the siege of Paris (see letter 4276).

From Edward L. Youmans 1 March 1874 4286

537 Broadway | New York. March 1st 1874

My dear Tyndall,

A year is now gone since you were with us[1] and you must allow me to say that you are very much missed here this winter. You spoiled all the lecturing of this season at any rate. Proctor[2] has been with us, but his deliverances have fallen flat upon the public ear, notwithstanding the sublimity of his themes.

There is general disappointment. He is merely a diffuse and interminable talker without conscience in his work, and with a transparent egotism. He came on crippled from Boston,[3] the Bureau,[4] in whose hands he was, having failed in their arrangements with him by consequence of the business panic[5] and general mismanagement it helped him to get into relation with the Young Christians,[6] who, when they became assured that he would advertise God from the platform,[7] consented to undertake him. He had good houses, and made a little money, of which I was glad, but the lectures disappointed. I was much interested to note the drooping countenances of various of our friends whom you would remember if I enumerated them, and their sad head-shakings at the first lecture[8] as they remarked "this is very different from what we had last winter—wonder if we shall ever have Tyndall again". That work was very effectual, my dear fellow, much more so I suspect than you had any idea of. We thought it hard and inexplicable that you should neglect the grand openings of the most exclusive New York society, in devotion to your work, but you built better than any of us knew. It has been a matter of very pleasing surprise to me to note how vividly last winter's memories are recalled and how they are dwelt upon, notwithstanding our character for transient susceptibility to passing excitements. I saw Frothingham of Brooklyn[9] the other day and his state of mind was quite touching. He had been to hear Proctor's first lecture and he said "it made him sick, and he shouldn't go to the rest." "Don't you think that Tyndall will come to us again? Don't you think he will ever come, I think he will—he must. We never had anything like that in Brooklyn! Don't you think he will want to go to the Rocky Mountains some time". And so he went on in half-despairing, and half-hopeful sententiousness. I was really sorry for him for I could give him no encouragement.

I heard Kingsley the other night on "Westminster Abbey".[10] I liked the admirable art of his awkwardness, and the vigor of his manner, and even enjoyed his horrible intonation: there was such an impressive individuality in his performance. He missed however in one thing, he was quite too sweet on the Americans,[11] and when he wanted one for Westminster Abbey, it was a little too much for his audience. I think he was profoundly sincere, but a little more reserve and discrimination would have been better. Proctor has made an ass of himself in the profusion of his beslathering.[12]

Did you see a copy of the Pen Monthly with an article on original work in science,[13] made up from the banquet proceedings,[14] I meant to have sent it to you.

Mr. Appleton[15] and his family including Willie[16] sail for England on the 7th. Mrs. Appleton[17] has been much broken by the loss of her daughter[18] and they seek a change of scene for a few weeks.

We have been much interested by your late Acoustical work[19] and I thought of reprinting it from Nature[20] but concluded to wait and ascertain if you expected to write it over yourself for future publication.

Mr. Willis[21] has sent all your accounts which contains an omission (I think) of $ *[2–3 words illeg]* Niagara article.[22] I will see about it.

Ever and truly yours | E. L Youmans

RI MS JT/1/Y/2

---

1. *since you were with us*: Tyndall had left the United States, following a five-month lecture tour, on 8 February 1873 (see letter 4201, n. 4).
2. *Proctor*: Richard Anthony Proctor (1837–88), an English astronomer and scientific popularizer, delivered a series of lectures in the United States from October 1873 to April 1874. See D. Finnegan, *The Voice of Science: British Scientists on the Lecture Circuit in Gilded Age America* (Pittsburgh: University of Pittsburgh Press, 2021), pp. 94–129.
3. *crippled from Boston*: Proctor had given a series of six lectures on astronomical topics at the Lowell Institute in Boston, beginning on 21 October 1873.
4. *the Bureau*: probably the Williams Lecture Bureau, based in Boston and owned by Benjamin Webb Williams.
5. *the business panic*: the end of the boom in railroad speculation had led to the failure of a number of American banks in 1873, creating profound economic uncertainty.
6. *the Young Christians*: the Washington, DC chapter of the Young Men's Christian Association. Tyndall had gifted $1500 of his lecture profit to the YMCA (see letter 3975, *Tyndall Correspondence*, vol. 13).
7. *he would advertise God from the platform*: like several other scientific popularizers, Proctor perpetuated the tradition of natural theology in his lectures. See B. Lightman, 'Astronomy for the People: R.A. Proctor and the Popularization of the Victorian Universe', in J. M. van der Meer (ed.), *Facets of Faith and Science. Volume 3: The Role of Beliefs in the Natural Sciences* (Lanham, MD: The Pascal Centre for Advanced Studies in Faith and Science and University Press of America, 1996), pp. 31–45.
8. *the first lecture*: Proctor repeated the same lectures he had given in Boston at Association Hall in New York, beginning on 9 January 1874.
9. *Frothingham of Brooklyn*: Octavius Brooks Frothingham (1822–95), an American clergyman and author (*ANB*).
10. *Kingsley the other night on "Westminster Abbey"*: the British clergyman and novelist Charles Kingsley (1819–75) gave a lecture on this topic at Association Hall in New York on 27 February (published in C. Kingsley, *Lectures Delivered in America in 1874* (Philadelphia: J. H. Coates, 1875), pp. 1–31). His extensive lecture tour of the United States from January to August 1874 had a deleterious impact on Kingsley's health (see letter 4504).
11. *he was quite too sweet on the Americans*: in his lecture, Kingsley commented on the monuments in Westminster Abbey of soldiers who had fought for the United States against the French and Spanish, and he concluded by hoping that 'when an American enters beneath that mighty shade, he may tread on common and ancestral ground as sacred to him as it is to us' (Kingsley, *Lectures Delivered in America in 1874*, p. 31; 'Westminster Abbey', *New York Times*, 28 February 1874, p. 12).

12. *beslathering*: slipping or sliding, here presumably referring to Proctor's hyperbolic flattery of his American hosts when lecturing.
13. *a copy of the Pen Monthly . . . original work in science*: 'On the Value of Original Scientific Research', *Penn Monthly*, 4 (1873), pp. 744–77.
14. *the banquet proceedings*: the speeches given at Delmonico's restaurant in New York on 4 February 1873 to mark the conclusion of Tyndall's lecture tour (see letter 4216, n. 7).
15. *Mr. Appleton*: William Henry Appleton.
16. *Willie*: William Worthen Appleton (1845–1924), Appleton's son.
17. *Mrs. Appleton*: Mary Moody Appleton (née Worthen, 1824–84).
18. *her daughter*: Kate Geary (née Appleton, 1848–73), who had died in Hong Kong on 24 July 1873.
19. *your late Acoustical work*: Tyndall had been conducting experiments regarding the transmission of sound with particular reference to methods of warning ships in fog (see letter 4185, n. 3).
20. *reprinting it from Nature*: presumably J. Tyndall, 'Acoustic Transparency and the Opacity of the Atmosphere', *Nature*, 9 (1874), pp. 251–3 and 267–9, although Youmans seems not to have reprinted it in the *Popular Science Monthly*, of which he was the founding editor.
21. *Mr. Willis*: not identified.
22. *the Niagara article*: J. Tyndall, 'Some Observations on Niagara', *Popular Science Monthly*, 3 (1873), pp. 210–26. Tyndall had visited Niagara Falls during the first week of November 1872, in between his lectures in Boston and Philadelphia (see letter 3738, n. 3, *Tyndall Correspondence*, vol. 13).

To Robin Allen 6 March 1874 4287

Royal Institution of Great Britain | 6th March, 1874.

SIR,

I have read the letter[1] which you have forwarded to me regarding the construction of a gun, with special reference to fog-signalling.[2]

I think it is a wise motion on the part of the Elder Brethren[3] to ascertain the best that can be accomplished in this direction. In some stations the gun will probably offer advantages over other signals, and this being so it is desirable to have the best gun possible.

That the length and bore[4] of the gun have a material influence on the sound produced has been proved by the observations at the South Foreland.[5] In a recent conversation with Sir Joseph Whitworth I learned that he has constructed a gun which could be lengthened or shortened at pleasure,[6] and he kindly expressed his willingness to lend me the gun. Some trial experiments with a gun of this kind ought to be made to determine the influence of length. No doubt, however, the desirability of being able to vary both length and bore, independently and together, will have occurred to the War Department, and, beyond making suggestions which the officers of the department are sure to anticipate, I have no remarks to offer.

The explosive material will of course also enter into the determination of the problem; nor need the amount of light produced by the flash be left out of consideration.

I have the honour to be, Sir, | Your obedient Servant, | JOHN TYNDALL | ROBIN ALLEN, ESQ.

National Archives, MT 10/290. Papers re Fog Signals. H8476 (duplicate: H4380)
Typed Transcript Only

1. *the letter*: letter 4284.
2. *fog-signalling*: in his capacity as scientific adviser to Trinity House (see letter 4187, n. 2), Tyndall had been engaged in a series of experiments to determine the most effective method of warning ships in fog (see letter 4185, n. 3).
3. *Elder Brethren*: the governing body of Trinity House (see letter 4187, n. 2).
4. *bore*: the width of the gun barrel.
5. *the observations at the South Foreland*: Tyndall had conducted his fog-signal experiments from the South Foreland lighthouse near Dover, in Kent.
6. *a gun which could be lengthened or shortened at pleasure*: presumably a development of the muzzle-loading naval gun that Joseph Whitworth had begun manufacturing in the 1860s.

From Mary Egerton 8 March [1874][1] 4288

Mountfield Court, | Robertsbridge, | Hawkhurst. | March 8th.

My dear Mr. Tyndall,

If you have time for a chat after Tuesday's lecture,[2] let me have a line at the door, and we will endeavour to slip upstairs[3] "unknown" to anybody! but I dare say you will be engaged.

How about Mountfield?[4] You will be making your Easter[5] plans soon. I think you might spare us one day of your holidays, & if you cribbed one day from London afterwards, it would never be missed, and that would make two, quite a respectable visit for you![6] If I thought you would come, I would arrange about going to London accordingly, for I do want to have you to myself for a bit; away from London with its hurry & worry, and,—(shall I say it?)—away from L[ad]y Claud[7] with her beauty & attractiveness! By her, I always feel like a farthing rushlight[8] to a "Silber Lamp";[9] or to use a more dignified, (and perhaps truer) comparison, like a Hydrogen flame, to the Lime light![10] When I do see you, your kind cordial manner is just the same as ever, and makes me ashamed of my doubts, but at other times, feeling her fascination, as I do myself, I can hardly persuade myself that your constant meetings, should not blot out the recollection of other friends, except those who have the one unfailing claim upon your memory & affection, that of being friends of quite early days.—Don't call this jealousy! I would not take the least bit of your regard from her, but only keep my own share of a friendship which I treasure more perhaps than I ought to do, because I have so few.

Come then, dear friend, (you see at any rate I have not lost my trust in you, or I should not speak thus!) and if we cannot give you beauty, & grace, & agreeableness, at least you shall find warm hearts & a hearty welcome; and May[11] has got some lovely German songs to sing you. But I did not mean to bother you, for I sh[oul]d not like you to come only because I asked you.

The weather is lovely and the "ruby-budded Limes" are swelling & may be sprinkled with emeralds[12] by the time Easter holidays are ending.

Ever y[ou]rs | M F Egerton

RI MS JT/1/E/22

---

1. *[1874]*: the year is established by reference to Tyndall's lectures given on a Tuesday (see n. 2).
2. *Tuesday's lecture*: the fourth of Tyndall's series of six lectures on 'The Physical Properties of Gases and Liquids', given at the RI on Tuesday 10 March (see letter 4233, n. 3).
3. *upstairs*: presumably to Tyndall's apartment at the RI.
4. *Mountfield*: Mountfield Court in Sussex, the family seat of the Egertons.
5. *Easter*: Easter Sunday was 5 April in 1874.
6. *one day of your holidays . . . visit for you*: Tyndall seems to have come to Mountfield Court on Saturday 29 March, and may have stayed until the following Tuesday (see letter 4301).
7. *L[ad]y Claud*: Elizabeth Hamilton.
8. *farthing rushlight*: a cheap, and only partially effective, form of candle made from the dried pith of the rush plant soaked in animal fat.
9. *"Silber Lamp"*: a domestic light invented in the late 1860s by Albert Marcius Silber (1883–86) in which an inner tube was added to a conventional oil burner, supplying an improved current of air to the flame and thus enhancing its luminosity ('Obituary: Albert Marcius Silber', *Minutes of the Proceedings of the Institution of Civil Engineers*, 89 (1887), pp. 496–8).
10. *a Hydrogen flame, to the Lime light*: while hydrogen burns with a pale blue flame that is barely visible in daylight, the burning of quicklime (calcium oxide) produces an intense illumination.
11. *May*: Mary Alice Egerton (May being a pet name for Mary).
12. *"ruby-budded Limes" . . . sprinkled with emeralds*: an allusion to the line 'A million emeralds

break from the ruby-budded lime' in Alfred Tennyson's poem 'Maud: A Monodrama' (1855), IV.1.1. Limes are trees or bushes of the genus *Tilia*, which are commonly used for ornamental purposes in the gardens of country houses.

From Charles Darwin 9 March [1874][1] 4289

Down, | Beckenham, Kent. | March 9th

My dear Tyndall

If you can attend to balloting at the Athenaeum[2] at the first meeting, on the 16th of this month, will you be so kind as to give your vote (& exert any influence on others which you properly can) in favour of my nephew, Henry Parker.[3] He is a fellow of Oriel,[4] & I can assure you that he is a quite fit member for election from his abilities & in every other respect.

My dear Tyndall | Yours very sincerely | Ch. Darwin

CUL GBR/0012/MS DAR 261.8:33[5]
RI MS JT/1/TYP/9/2818

1. *[1874]*: the year is established by reference to the election of Henry Parker to the Athenaeum, which occurred in March 1874.
2. *balloting at the Athenaeum*: the process by which new members were elected to the exclusive Athenaeum Club. If candidates received any black balls, their applications would be declined. On the Athenaeum Club, see letter 4186, n. 1.
3. *my nephew, Henry Parker*: Henry Parker (1827–92), the son of Darwin's elder sister Marianne.
4. *Oriel*: Oriel College, Oxford.
5. *CUL GBR/0012/MS DAR 261.8:33*: published in *Darwin Correspondence*, vol. 22, pp. 41–5; this is letter 9256 in the online edition of the Darwin correspondence.

From Mary Egerton 16 March [1874][1] 4290

Mountfield Court, | Robertsbridge, | Hawkhurst.[2] | March 16th.

A thousand thanks, kind friend, for your letter;[3] I write this as I may not have an opportunity of saying it, though I am coming up tomorrow, as I have some other business I can combine with the lecture.[4]

I meant no other "fascination" than that you describe; is it not enough?—

Your promise[5] is pleasant to me; as for the time, let it be whenever you are most at liberty—whenever there is least chance of your wishing yourself somewhere else.—You speak of "snatching" the days; I thought you would be sure to take two or three weeks holiday at Easter;[6] after the lectures[7] I am sure you will want it, & in view of your Summer work,[8] and then we might come in at one end or the other.

This is my birthday,[9] so we all took a holiday & spent the morning rambling through the woods getting primroses! I trust I shall never get too old to enjoy Spring flowers, (I don't think you ever will! When is your's[10] I wonder?) What a contrast to the snow of last week!

"It is an ill wind that blows no-good"![11] The endless bother I have had in getting servants, has given me a reason for coming up to town oftener than I sh[oul][d]. perhaps have had the face to do, had it been to spend 4 hours on the railroad for one hour's pleasure! Of course I always arrange my interviews on a Tuesday!

Ever Y[ou][rs]. | M. F. Egerton

RI MS JT/1/E/23

1. *[1874]*: the year is established by reference to Tyndall's lecture 'tomorrow' (see n. 4).
2. *Mountfield Court, | Robertsbridge, | Hawkhurst*: the Egerton family estate in Sussex.
3. *your letter*: letter missing, but presumably a response to letter 4288.
4. *the lecture*: the fifth of Tyndall's six lectures 'On the Physical Properties of Liquids and Gases', given at the RI on Tuesday 17 March (see letter 4233, n. 3).
5. *Your promise*: presumably to visit Egerton over Easter at her home, Mountfield Court, as proposed in letter 4288.
6. *Easter*: Easter Sunday was 5 April in 1874.
7. *after the lectures*: Tyndall's series of six lectures were due to conclude on 24 March (see letter 4233, n. 3).
8. *your Summer work*: in his role as incoming president of the BAAS, Tyndall was due to give an inaugural address at the annual meeting, to be held in Belfast in August. After completing his lectures at the RI before Easter, he intended to spend the summer months in Switzerland writing his address.
9. *my birthday*: Egerton was celebrating her fifty-fifth birthday.
10. *your's*: Tyndall's birthday was 2 August.
11. *"It is an ill wind that blows no-good"*: a variation of the proverb, 'It is an ill wind that blows nobody any good'.

To Frederic William Farrar[1] 17 March [1874][2] 4291

Royal Institution of Great Britain | 17th March

My dear Mr Farrar.

To my regret I see no outlet in the direction of Marlborough.[3] As soon as I shake my lectures[4] off my hands I allow myself a week—the Easter week[5]—on the Isle of Wight—and there like Mahamet[6] I retire to the solitudes to ask myself what I can say to the people of Belfast.[7]

Yours ever | John Tyndall

Wellcome MS. 7777/90

1. *Frederic William Farrar*: Frederick William Farrar (1831–1903), an Anglican clergyman, writer, and philologist. From 1855 to 1870 he was assistant master at Harrow, during which time he wrote three novels and a number of books on philology. Farrar was a supporter of Darwin and adapted his ideas to the development of language during the 1860s, and it was Darwin who nominated Farrar as FRS in 1866. Upon Darwin's death Farrar, by then a canon of Westminster, arranged for his burial at Westminster Abbey. From 1895 until his death Farrar was dean of Canterbury (*ODNB*).
2. *[1874]*: the year is established by reference to Tyndall's preparations for his presidential address to the BAAS (see n. 7).
3. *Marlborough*: Marlborough College in Wiltshire, a boarding school founded in 1843 for the sons of Church of England clergy. Farrar was the headmaster of the school from 1871–6.
4. *my lectures*: the series of six lectures on 'On the Physical Properties of Liquids and Gases' that Tyndall gave at the RI (see letter 4233, n. 3).
5. *the Easter week*: 30 March to 6 April.
6. *like Mahamet*: variant spelling of Muhammad (*c.* 570–632), the founder of Islam. According to Islamic tradition, Muhammad retreated to a cave in mountains near Mecca in order to pray alone, and it was here that he experienced his first divine revelation.
7. *what I can say to the people of Belfast*: in his address as president of the BAAS, to be delivered at the annual meeting being held in Belfast in August. The presidential address was, by tradition, open to the public of the city hosting the meeting.

To William Spottiswoode 17 March [1874][1] 4292

17th March.

My dear Spottiswoode

Mr. Griffith[2] no doubt has informed you of our proposed meeting[3] on Saturday.[4]

I send you documents[5] which will let you know our present state of the question.[6] I should like to send these also to Williamson[7] who has his own opinions on this matter.

It is a thing that ought to be wiped completely out. I do not know at the present moment, save through conversations the merits of the case.

Would you like a good fellow come on Saturday prepared to throw what light you can upon the question.

I am thinking of asking Lockyer to join us at 3 P.M. This would afford time for a free discussion previously; but he ought to have an opportunity of stating his case.

The matter, as far as I can see, is not worth the stir it has created; but I think we ought to finish it once for all.

Yours ever | John Tyndall

RI MS JT/1/T/1345

---

1. *[1874]*: the year is established by the involvement in BAAS business of both Tyndall and Alexander William Williamson, as incoming and outgoing presidents.
2. *M^r^. Griffith*: George Griffith.
3. *our proposed meeting*: presumably of the Council of the BAAS, or at least its principal members. It is unclear if the meeting took place, as the minutes of Council meetings for this period are not included in the BAAS archive at the Bodleian Library, University of Oxford, Dep. B.A.A.S. 1–453.
4. *Saturday*: 21 March.
5. *documents*: enclosure missing.
6. *the question*: presumably concerning accusations made against Norman Lockyer. The precise nature of the issue is not clear, and nothing relating to any potential dispute involving Lockyer and the BAAS at this time is recorded in A. J. Meadows, *Science and Controversy: A Biography of Sir Norman Lockyer* (London: Macmillan, 1972). However, Tyndall soon after called Lockyer 'a man whose conceit has rendered him intolerable to his best friends' (see letter 4300).
7. *Williamson*: Alexander William Williamson, who was the incumbent president of the BAAS, due to hand over to Tyndall in August.

From Mary Egerton 18 March [1874][1] 4293

Mountfield Court, | Robertsbridge, | Hawkhurst.[2] | March 18th.

My dear friend,

Why, in the name of all that's stupid! was I to be possessed by a dumb demon yesterday[3] to spoil half the pleasure of my chat with you! I could feel my conversation coming out in jerks, like the turns of a rusty cog wheel, & every now & then, the thing I most wanted to say, sticking fast altogether! It must have been a bitter consciousness, that made me allude to your remark about "not judging from appearances"!

People talk of "hearts overflowing"! the fuller mine is the more obstinately it keeps shut; I think the valve must open the wrong way! I can then only talk with my fingers, (like the deaf & dumb!) Here are a few violets, all I could find, though on this lovely day I expected quantities. I hope they will be sweeter than the bunch I so ungraciously bequeathed to you yesterday! I can't think why I did it, for they were not worth having!

May's[4] cold is much better & I hope will be all right to go up on Tuesday, for having had so many, we must get the last lecture![5] We shall miss them so after Easter![6]

A.W.S. | v.I.t.F.[7] (Can you read that?)

M. F. Egerton

I forgot to tell you my birthday present from the girls,[8] "The Proceedings of the Royal Inst[itutio]ⁿ" from 1861![9] They found it in a sale list. I think it is quite delightful to have; with Faraday's lectures[10] &c! I suppose that is your first, that is in the 1st vol:[11] Don't trouble to write.

RI MS JT/1/E/25

1. *[1874]*: the year is established be the relation to letter 4290.
2. *Mountfield Court, | Robertsbridge, | Hawkhurst*: the Egerton family estate in Sussex.
3. *yesterday*: Egerton had attended the fifth of Tyndall's six lectures 'On the Physical Properties of Liquids and Gases', given at the RI on Tuesday 17 March (see letter 4290).
4. *May*: Mary Alice Egerton (May being a pet name for Mary).
5. *the last lecture*: Tyndall gave the last of his series of six lectures on 24 March.
6. *Easter*: Easter Sunday was 5 April in 1874.
7. *A.W.S. | v.I.t.F.*: possibly abbreviations of the German expressions 'Auf Wiedersehen' ('Goodbye'), albeit with the second word split into two parts (perhaps reflecting Egerton's lack of proficiency in German), and 'von Ihrer treuen Freundin' ('from your loyal friend').
8. *the girls*: presumably Egerton's four daughters Mary Alice, Charlotte (1849–1926), Emily Margaret (see letter 4260, n. 7) and Georgina Renira (1854–1930).
9. *"Proceedings of the Royal Inst[itutio]ⁿ" from 1861*: the *Roy. Inst. Proc.* were published in yearly parts, which were then bound together into larger volumes. The *Roy. Inst. Proc.* for 1861 (part 11) were collected in volume 3 (1858–62), pp. 295–403. It seems Egerton was given the part for 1861, which had not been bound into a volume.
10. *Faraday's lectures*: the part for 1861 contained two Friday Evening Discourses by Michael Faraday: 'On Platinum' and 'On Mr. Warren De la Rue's Photographic Eclipse Results', *Roy. Inst. Proc.*, 3 (1858–62), pp. 321–2 and 362–6.
11. *your first . . . in the 1st vol*: J. Tyndall, 'On the Influence of Material Aggregation of Force', *Roy. Inst. Proc.*, 1 (1851–54), pp. 254–9. This was the first lecture Tyndall delivered at the RI, on 11 February 1853 (a Friday Evening Discourse).

From Robert G. H. Kean[1] 19 March 1874 4294

Lynchburg Virginia. | March 19th 1874

Professor John Tyndall

Sir,

I have just read with great interest your lecture of January 16th[2] copied by Littell's Living Age, from 'Nature',[3] on the "Acoustic transparency and opacity of the atmosphere."

The remarkable facts you mention induce me to state to you a fact which I have occasionally mentioned, but always when I was not well known, with the apprehension that my veracity would be questioned—It made a strong impression on me at the time—but was an insolvable mystery until your discourse gave a possible solution.

On the afternoon of June 28th 1862 I rode in company with Gen[era]1 G. W. Randolph,[4] then Secretary of War, of the Confederate States, to Price's home[5] about nine miles from Richmond. The evening before Gen[era]1 Lee[6] had begun his attack on McClelland's army,[7] by crossing the Chickahominy[8] about four miles above Price's, and driving in the right wing of McClelland's (the R.S) army. The battle of Gaines' farm[9] was fought the afternoon to which I refer. The valley of the Chickahominy is about one and a half miles wide from hilltop to hilltop—Price's is on one hilltop—that nearest Richmond. Gaines' farm just opposite is on the other, reaching back on a plateau to Cold Harbour.[10]

Looking across the valley I saw a good deal of the battle—Lee's right resting in the valley, the Federal left doing the same. My line of vision was nearly in the line of the lines of battle—I saw the advance of the Confederates—their repulse two or three hours—the final rout of the Federal forces, and in the gray of the evening the triumphant pursuit by the Confederates.

I distinctly saw the musket fire of both lines—The smoke—individual discharges,—The flash of the guns—I saw batteries of artillery on both sides come into action, and fire rapidly—several field batteries on each side were plainly in sight. Many men were hid by the timber[11] which bounded the range of vision.

Yet looking for near two hours, from about 5 to 7. p.m. on a midsummer's afternoon, at a battle in which at least 50,000 men were actually engaged, and doubtless at least 100 pieces of field artillery, through an atmosphere, optically as limpid as possible, not a single sound of the battle, was audible to General Randolph and myself. I remarked it to him at the time, as astonishing—

The cannonade of that very battle was distinctly heard at Amherst Court house,[12] 100. west of Richmond, as I have been most credibly informed. Between me and the battle was the deep broad valley of the Chickahominy partly a swamp, shaded from the declining sun by the hills and forests in the west (my side)—

Part of the valley on each side of the swamp was cleared—some in cultivation, some not—How were conditions capable of producing several belts of air, of varying density—and varying in the amount of watery vapor, arranged like laminae,[13] at right angles to the acoustic waves as they came from the battlefield to me—The direction of the valley is nearly due East & West. The point where the cannonade was heard 100 miles off was directly in the line of the valley, which however, is a short one—less than 20 miles beyond the field—

It occurred to me that this incident might interest you—I owe you thanks for the possible solution—

Respectfully | your ob[edien]t serv[an]t, R. G. H. Kean

RI MS JT/1/K/3
Sound (1875), pp. 274–6

---

1. *Robert G. H. Kean*: Robert Garlick Hill Kean (1828–98), an American lawyer who had served as an officer in the Confederate States Army during the American Civil War.
2. *your lecture of January 16th*: 'On the Acoustic Transparency and Opacity of the Atmosphere', which Tyndall had delivered at the RI on 16 January (*Roy. Inst. Proc.*, 7 (1873–75), pp. 169–78).
3. *copied by Littell's Living Age, from 'Nature'*: J. Tyndall, 'The Acoustic Transparency and Opacity of the Atmosphere', *Littell's Living Age*, 120 (1874), pp. 692–7 and 815–8, which was reprinted from *Nature*, 9 (29 January and 5 February 1874), pp. 251–3 and 267–9.
4. *Gen[era]l G. W. Randolph*: George Wythe Randolph (1818–67), an American Confederate general and briefly the Confederacy's Secretary of War (*ANB*).
5. *Price's home*: probably Price's farm, the site of a Confederate army camp in Virginia.
6. *Gen[era]l Lee*: Robert Edward Lee (1807–70), the commander of the Confederate States Army (*ANB*).
7. *McClelland's army*: the Army of the Potomac, led by George Brinton McClellan (1826–85). During the campaign to which this letter refers, this was the primary Union force operating in the Eastern theatre of the American Civil War.
8. *the Chickahominy*: a river flowing through the east of Virginia.
9. *The battle of Gaines' farm*: 27 June 1862, also known as the Battle of Chickahominy River. The owner of the farm was William F. Gaines (1804–74).
10. *Cold Harbour*: two rural crossroads in Virginia, the site of the Battle of Cold Harbor in 1864, which was essentially fought over the same ground as the Battle of Gaines' Mill.
11. *the timber*: i.e. trees.
12. *Amherst Court house*: in the town of Amherst, Virginia.
13. *laminae*: thin layers (*OED*).

From George Gabriel Stokes 19 March 1874 4295

The Royal Society | Burlington House, London, W. | 19 March 1874.

Sir,

I am desired to return you the thanks of the Royal Society for your Paper on the atmosphere as a vehicle of sound[1] and to inform you that the COMMITTEE OF PAPERS have directed it to be published in the Philosophical Transactions.[2]

I remain, | Sir, | Your obedient Servant, | G. G. Stokes | *Secretary R.S.*

I sent my own report[3] for your perusal.

The other referee[4] remarks: "It seems to me that the paper might with advantage be shortened without any diminution of its scientific value by suppressing the detailed observations made on several days during the experiments at the South Foreland,[5] and by substituting a summary of the results."

A question arose whether you should be asked to curtail the paper as to some of the details, but ultimately it was decided to leave the matter entirely to yourself.[6]

Should you feel disposed to adopt my suggestion of a rearrangement, so as to throw the details into an appendix, which would take some thought but little writing, being chiefly a matter of brain, scissors and paste, it would be well to use a couple of copies of the report for the purpose, so as to leave the original paper intact as a record, making marks on it to guide the printer.

I anticipated your note[7] about a map, which I think is much wanted. I don't think the delineation of the actual forms of the trumpets and syren[8] is of much moment. Should there be room for them on the same plate with the map on a reduced scale they might come in, but it would be better not to reduce the map very greatly. You can arrange with Mr White[9] about that as you may think best.

It was suggested that a reference should be given to the blue book,[10] as some might like to purchase that who would not go to the expense of a Part of the Phil. Trans.[11]

Dr. Tyndall F.R.S.

RI MS JT/1/S/258
RI MS JT/1/TYP/4/1420

---

1. *your Paper on the atmosphere as a vehicle of sound*: this paper was first read to the RS in preliminary form on 15 January, with a complete version read on 12 February.
2. *the COMMITTEE OF PAPERS . . . in the Philosophical Transactions*: in order for a paper to be published in the *Phil. Trans.*, it was subject to review by two anonymous referees before final approval from the Committee of Papers. Stokes, as an honorary secretary of the RS, was instrumental in formalising this process. The paper was published as 'Vehicle of Sound'.

3. *my own report*: G. G. Stokes, 'Report on a Paper by Dr Tyndall entitled On the Atmosphere as a Vehicle of Sound', RR/7/350, Referees' Reports: volume 7, RS. It was not usual practice for authors to see these handwritten reports, which were primarily intended for internal use by the Committee of Papers. However, despite the RS's decision to publish Tyndall's paper, Stokes had some significant reservations regarding the conclusions drawn from the evidence, and therefore sent his detailed report to Tyndall, allowing these criticisms to be addressed in revisions.
4. *The other referee*: Robert Bellamy Clifton (1836–1921), who had studied under Stokes at Cambridge before becoming professor of experimental natural philosophy at Oxford. The report quoted by Stokes is R. B. Clifton, 'Report on a Paper by Dr Tyndall entitled On the Atmosphere as a Vehicle of Sound', RR/7/351, Referees' Reports: volume 7, RS.
5. *the South Foreland*: a lighthouse in Kent, near Dover. It was here that Tyndall had conducted a number of acoustic experiments in order to determine the best method of signaling to ships in a fog. This research formed the basis of his paper.
6. *A question arose . . . entirely to yourself*: Tyndall told Stokes that he would 'try . . . to shorten' the paper (letter 4296), but actually 'added a few more reasons' on particular points when he came to make revisions (letter 4302).
7. *your note*: letter missing, but it presumably related to maps of the area around the South Foreland lighthouse where the sound experiments were carried out, which were included in the final publication (plates XVII and XIX).
8. *the trumpets and syren*: among the various methods tested by Tyndall in his fog-signaling experiments were various types of trumpet (air horns) and a syren lent to him by the United States Lighthouse Board (see letter 4201).
9. *Mr White*: Walter White.
10. *the blue book*: 'Report by Professor Tyndall'.
11. *the expense of a Part of the Phil. Trans.*: while individual parts of the *Phil. Trans.* varied in price depending on their length, the part containing Tyndall's paper, part 2 for 1874 published in December, cost three pounds. See 'The Philosophical Transactions of the Royal Society', *Phil. Trans.*, 184 (1893), p. 817.

To George Gabriel Stokes 20 March 1874 4296

20th March 1874

My dear Stokes.

I thought a good deal over the points you have urged,[1] and will think over them again and give you the result of my cogitations. Differences of temperature[2] was the first thing that occurred to me.

I had also thought of remodelling the paper; and will try now to shorten it; but it is weary work for a tired brain.

And again I thought there is hardly a day that does not contain something which may be of use to future investigations. The length of the paper mainly arises from the length of time the observations cover.[3]

In the face of previous notions it was necessary to give the conclusions of the paper stability, and thus could hardly be done otherwise than by the heaping up of evidence.

However I will see what can be done to meet your views. You remember my writing to you, and your replying to me.[4] That reply encouraged me to make use of a greater quantity of the report[5] than I should have otherwise introduced. I was also encouraged by looking over past numbers of the Transactions & seeing the character of some of the papers printed there.

You speak lightly of brain scissors & paste[6]—But supposing we cannot get the brain? Well I will see what can be done with mine.

Yours faithfully | John Tyndall

You have done your work[7] promptly and I thank you for this.

RI MS JT/1/T/1405
RI MS JT/1/TYP/4/1487

1. *the points you have urged*: see letter 4295. The points relate to suggested revisions to 'Vehicle of Sound'.
2. *Differences of temperature*: in his referee's report, Stokes expressed some reservations regarding Tyndall's conclusions. Specifically, he asserted: 'I think . . . that Dr Tyndall is not justified by the evidence . . . in assuming so confidently as he has done that difference of humidity rather than temperature is the cause of the heterogeneity which is revealed by its effects. I feel pretty confident that on land at least it must be part the reverse, and am by no means clear that such is not the case even out at sea' (G. G. Stokes, 'Report on a Paper by Dr Tyndall entitled On the Atmosphere as a Vehicle of Sound', RR/7/350, Referees' Reports: volume 7, RS).
3. *the length of time the observations cover*: from 19 May to 25 November 1873.
4. *my writing to you, and your replying to me*: letters 4220 and 4221.
5. *the report*: 'Report by Professor Tyndall'.
6. *brain scissors & paste*: see letter 4295, where Stokes suggested the revisions would 'take some thought but little writing, being chiefly a matter of brain, scissors and paste'.
7. *your work*: of refereeing Tyndall's paper.

From Herbert Spencer [21 March 1874][1] 4297

Saturday.

My dear Tyndall

I am glad to get the annexed,[2] referring to a letter I wrote to King[3] some time ago.

Had I seen you lately, I should have given you the memo. I have made with reference to the controversy between Mill and myself.[4]

The portion of Vol II of the Psychology[5] for §[6]402 to §441 inclusive, gives the argument[7] in full, with such preliminary as is needful for following it.

Ever yours truly | Herbert Spencer

RI MS JT/1/S/186
RI MS JT/1/TYP/3/1119

1. *[21 March 1874]*: the date is established by the letter being written on the reverse of another letter addressed to Spencer, which is dated 20 March 1874. This was a Friday and as Spencer gives the date as Saturday on his letter to Tyndall, it is most likely to be the following day.
2. *the annexed*: the letter on which Spencer had written his message to Tyndall (see n. 1). The letter was from the publishers Henry S. King & Co. and read: 'Dear Sir, | Referring to your letter to us of the 27th ultimo, we have the pleasure to inform you that we have this day received a letter from Mr Bailliere enclosing a cheque (X) which we have apportioned to the authors of the works in the "The International Scientific Series" which are already published in French, in accordance with his instructions. We are dear Sir | Yours faithfully | Henry S. King & Co. | (X) 1440 francs. | 540 frs for Prof. Tyndall | 450 frs for Mr Bagehot | 450 frs for Prof. A. Bain'. On the International Scientific Series, see letter 4264, n. 5.
3. *King*: Henry Samuel King (1817–78), a banker and East India agent who, after gaining experience of the book trade, had set up his own publishing firm.
4. *the memo. I have . . . the controversy between Mill and myself*: Spencer had been involved in a long-running dispute with the philosopher John Stuart Mill (1806–73) on the nature of necessary truths. His 'memo.' on the controversy seems to have been prepared specifically for Tyndall, perhaps as an aid for writing his presidential address to the BAAS at its forthcoming meeting in Belfast. In the address, Tyndall stated: '"The question of an external world", says Mr. J. S. Mill, "is the great battleground of metaphysics". Mr. Mill himself reduces external phenomena to "possibilities of sensation" . . . Mr. Spencer takes another line. With him . . .there is no doubt or question as to the existence of an external world. But he differs from the uneducated, who think that the world really is what consciousness represents it to be. Our states of consciousness are mere *symbols* of an outside entity which produces them and determines the order of their succession, but the real nature of which we can never know' (*Belfast Address*, pp. 56 and 57).
5. *the Psychology*: H. Spencer, *Principles of Psychology*, 2nd edn, 2 vols (London: Williams and Norgate, 1870–2).
6. §: a typographical character, often known as a section sign, generally used to refer to a specific section of a document.
7. *the argument*: the sections Spencer indicated address metaphysics and realism, and how to assess the truth or falsity of a statement. Sections 426 onwards relate to his dispute with Mill.

From George Gabriel Stokes 21 March 1874 4298

Lensfield Cottage Cambridge | 21st March 1874

My dear Tyndall,

The question you raised with me[1] was as to the power of presentation—whether the fact of a good portion of the memoir[2] appearing in print, as about to appear in a blue book,[3] would be an obstacle—and I gave it as my individual opinion that it would not.[4] I thought it right however not to ignore the fact in my report to the Committee,[5] while at the same time stating it as my own opinion that that circumstance should be no obstacle; and as you see from the decision[6] my view prevailed.

What should be the amount of detail presented for publication is altogether a different question,[7] which I never understood you to consult me about, nor could I give an opinion without having seen either the paper—or the report.

Long as the details come,[8] I did not recommend their omission. Similar details of laboratory work would not be suitable for publication in the Phil. Trans.; but these experiments were made with government appliances,[9] so as to be out of the reach of a private individual to report, and that forms a reason for giving the results in full.

But supposing they be so given, I think it would be a great improvement if the paper were altered in the manner I indicated, and would give you a much larger circle of readers. At present the record of all the observed facts and of the trains of reasoning which they successively gave rise to comes nearly in chronological order. If the paper were drawn up giving merely the general results (without the details but with dates) of the observations, in connexion with the trains of reasoning entered into and the conclusions arrived at, and were followed by an appendix giving merely the facts of observation,[10] with any abbreviation in the mode of presenting them that may present itself, recorded chronologically with the date in conspicuous type, then the memoir itself would be very generally read, and would be found very interesting, while the appendix would be read through by only some few who were specially working at the subject. At the same time if in reading the memoir the reader wanted fuller information on any point as to the facts of observation he could at once turn to the part of the appendix where it would be found.

This I think would be the most complete form of publication, and I don't think that with a couple of waste printed copies of the report[11] to operate on the trouble would be anything like what it might appear at first sight. It would involve less trouble to adopt the suggestion of the other referee[12] and omit some of the less important details. They would not be lost as they would be preserved in a blue book, but they would of course be more difficult of access to scientific men, as the blue book would have to be ordered through a bookseller, or consulted in some large library. There are pros and cons in this mode of alteration: it would render the paper more readable for the general scientific public, but for the comparatively few who wanted to go as completely as possible into the question, somewhat less complete.

As this is a case in which questions of priority might arise, it would be well to be more cautious than ever to leave the original MS intact, merely introducing marks to guide the printer.

In any case it would be well to mention the blue book. Some might like to purchase that who would not like to go to the expense of a Part of Phil. Trans.[13]

You would naturally think first of heterogeneity depending on temperature, such being the explanation commonly given of the inferior audibility, in many cases, of sounds by day as compared with night.[14]

In the event of any repetition of the contents, it would be desirable to record specially the character of the day as to cloud and sunshine: also (at least on sunny days and some cloudy days) to take the temperature of the sea at the very surface.

Yours sincerely | G. G. Stokes

RI MS JT/1/S/259
RI MS JT/1/TYP/4/1421

---

1. *The question you raised with me*: in letter 4220.
2. *the memoir*: 'Vehicle of Sound'.
3. *a blue book*: 'Report by Professor Tyndall'.
4. *I gave it as my individual opinion that it would not*: in letter 4221.
5. *my report to the Committee*: G. G. Stokes, 'Report on a Paper by Dr Tyndall entitled On the Atmosphere as a Vehicle of Sound', RR/7/350, Referees' Reports: volume 7, RS. Stokes had submitted this to the RS's Committee of Papers.
6. *the decision*: the Committee of Papers directed Tyndall's paper to be published in the *Phil. Trans.* (see letter 4295).
7. *What should be the . . . altogether a different question*: Stokes considered that in letter 4296 Tyndall had misinterpreted his original answer (in letter 4221) as an answer to this question.
8. *Long as the details come*: see letter 4296, n. 3.
9. *made with government appliances*: Tyndall's experiments on fog-signaling were made under the auspices of, and with the facilities of, Trinity House.
10. *an appendix giving merely the facts of observation*: Tyndall added such an appendix on pp. 232–44 of his paper.
11. *a couple of waste printed copies of the report*: Stokes had told Tyndall that in making revisions to the paper 'it would be well to use a couple of copies of the report for the purpose, so as to leave the original paper intact as a record' (see letter 4295).

12. *the other referee*: Robert Bellamy Clifton (see letter 4295, n. 4).
13. *the expense of a Part of Phil. Trans.*: see letter 4295, n. 11.
14. *heterogeneity depending on . . . compared with night*: in his referee's report, Stokes contended that Tyndall had put too great an emphasis on the importance of humidity on the transmission of sound through air, and argued that differences in temperature, especially those depending on the time of day or night, were of far greater significance. See also letter 4296, n. 2.

From Henrietta Huxley 22 March [1874][1] 4299

4 Marlborough Place | Abbey Road | NW[2] | 22nd March—

Dear Brother John,

Will thou dine with us on the 1st April at 7.30—in order that for once thou mayest consort with wise folk on that day? For thy soul's good & body's comfort. Consent.

From | Thine affec[tiona]te Sister | Nettie Huxley.

Thou wilt meet the pleasant Hoffman[3] and his 5th. wife Nay—I am wrong tis but the 4th[4]—

RI MS JT/1/H/515

1. *[1874]*: the year is suggested by reference to August Wilhelm Hofmann's fourth wife (see n. 4).
2. *4 Marlborough Place | Abbey Road | NW*: see letter 4213, n. 1.
3. *the pleasant Hoffman*: August Wilhelm Hofmann (1818–92), a Prussian chemist who had been director of the Royal College of Chemistry before returning to Germany in 1865 (*NDB*).
4. *his 5th. wife . . . but the 4th*: Hofmann's fourth wife was Bertha Wilhelmine Hofmann (née Tiemann, 1854–1922), whom he married on 11 August 1873.

To Rudolf Clausius 26 March 1874 4300

26th March 1874.

My dear Clausius.

I have many times thought of writing to you during these last weeks and months, but like yourself I have been terribly busy. Your letter[1] my dear friend was a great pleasure to me, and the pleasure was enhanced by Johnny's letter[2] which shows how astonishingly quick children grow towards being men. With regard to Tait[3] really though one is obliged, for the sake of the public, to take notice of him sometimes he is not worth a single thought of either of us. Were he a Newton[4] as a mathematician he would still, as a man, be small, and vulgar. I wrote the most innocent of little books—a boy's book of the glaciers[5]—in which I actually ignored myself and gave all possible credit to Agassiz, Forbes, and others. But because I tried to be just to Agassiz,[6] with whom Forbes was always at war, they attacked me.[7] I replied[8] and nothing more was heard for some time. The editor of Nature, a man whose conceit has rendered him intolerable to his best friends, and from whom I never disguised my opinion of his conceit—I mean Lockyer—then permitted Tait to publish a long and spiteful letter in Nature.[9] I was pressed for time, for I was quitting London; so I wrote a rapid reply to Tait[10] in which I told him and the world what I thought of him. But on reflection believing it did not promote the dignity of science to use strong language—I retracted one or two of the expressions I employed. He has since tried to get an article published in some of the more respectable and influential London magazines, but they have closed their doors against him. Even two magazines with Scotch proprietors[11] have refused to let him deface their pages. There is a mad creature here who writes upon Art.—Ruskin—a beautiful imaginative creature, more a woman than a man, and he wrote a scandalous article against me, and in favour of Forbes, in an obscure magazine of his own.[12] Tait had it copied into the Scotch newspapers,[13] and knowing that I am to preside over the British Association at its next meeting in Belfast,[14] where Tait was professor,[15] and where he found a nice little wife[16] (this is the only good thing that I know regarding the man) he had Ruskin's diatribe published in the Belfast newspapers.[17] He would stoop to any measures or any injustice to gratify his vulgar hatred. I dare say I shall be forced to say something more regarding the Forbes controversy, and if they force me to it, now that Agassiz is dead,[18] and can no longer take care of himself, I shall certainly let the public know the real state of the question.

But I have written too much about this man. A very heavy paper embracing the whole of the investigation on Sound[19] is on the point of being handed in to the Royal Society. I hope when you read it that it will please you. It is curious what a length of time error has been handed down from generation to generation of scientific men in connexion with this subject.

I dare say you have read Helmholtz's reply to Zöllner, appropos of Thomson and Tait[20]—it is strong and dignified. I believe he intends writing a reply regarding me in the preface to a little book of mine[21] which is now on the point of publication. This is far better than that I should reply to Zöllner. Helmholtz still believes him to be mad.

I have just finished my lectures here,[22] and after the completion of a few other duties I will retire early to Switzerland to prepare my Belfast address.[23] It will be a difficult piece of work, and I should like to do it well. I did all in my power to escape the duty: indeed I had declined the invitation to preside three times previously, but this time I could not evade it. Tait's friends may possibly make a row there, but I do not care for them. Huxley, Hooker, Spottiswoode and other strong men are coming over with me.

And now my dear Clausius let me repeat what I said at the beginning, that it gave me, and that it will always give me the greatest pleasure to hear of your welfare. Give my kindest regards to your wife,[24] and my love to the children.[25] To Johnny I must write a special letter.

Yours ever faithfully | John Tyndall

RI MS JT/1/T/206

---

1. *Your letter*: letter missing.
2. *Johnny's letter*: letter missing; Johnny is Clausius's son, Rudolf John Clausius.
3. *With regard to Tait*: Peter Guthrie Tait had helped to revive an acrimonious dispute regarding theories of glacial motion in which Tyndall had been involved with James David Forbes in the late 1850s and early 1860s.
4. *Newton*: Isaac Newton (1642–1727), the iconic exemplar of mathematical genius, was an English natural philosopher, mathematician, and Lucasian Professor at Cambridge University noted for many important discoveries in the physical sciences. He was buried in Westminster Abbey (*ODNB*; *CDSB*).
5. *the most innocent of little books—a boy's book of the glaciers*: *Forms of Water*.
6. *be just to Agassiz*: in the 1840s Louis Agassiz and Forbes both made similar measurements of the motion of glaciers, resulting in a dispute over priority of discovery which remained unresolved. In his book, Tyndall remarked that instruments allowing accurate measurement were 'introduced almost simultaneously by M. Agassiz on the glacier of the Unteraar, and by Professor Forbes on the Mer de Glace', although he implied that Agassiz had been first (p. 60).
7. *they attacked me*: in *Life of Forbes*, pp. 457–520.
8. *I replied*: in J. Tyndall, 'Principal Forbes and His Biographers', *Contemporary Review*, 22 (1873), pp. 484–508.
9. *a long and spiteful letter in Nature*: see letter 4239, n. 25.
10. *a rapid reply to Tait*: J. Tyndall, 'Tyndall and Tait', *Nature*, 8 (1873), p. 399.
11. *two magazines with Scotch proprietors*: the *Contemporary Review*, published by Alexander Strahan (1833–1918) and *Macmillan's Magazine*, published by Alexander Macmillan (see letter 4324).
12. *a scandalous article . . . in an obscure magazine of his own*: *Fors Clavigera*, vol. 3 (1873), letter 34, pp. 1–32. See letter 4185, n. 13.
13. *Tait had it copied into the Scotch newspapers*: 'Mr. Ruskin on Professor Tyndall', *Scotsman*, 14 October 1873, p. 6.
14. *I am to preside over the British Association . . . in Belfast*: Tyndall was due to become the president of the BAAS, and give the customary inaugural address, at its annual meeting in Belfast in August.
15. *where Tait was professor*: he was professor of mathematics at Queen's College, Belfast from 1854–60.
16. *a nice little wife*: Margaret Archer Tait (née Porter, 1839–1926).
17. *Ruskin's diatribe published in the Belfast newspapers*: see letter 4202, n. 6.
18. *now that Agassiz is dead*: Louis Agassiz had died on 14 December 1873.

19. *A very heavy paper embracing . . . the investigation on Sound*: 'Vehicle of Sound'.
20. *Helmholtz's reply to Zöllner, appropos of Thomson and Tait*: see letter 4223, n. 10.
21. *the preface to a little book of mine*: see letter 4199, n. 5.
22. *my lectures here*: 'On the Physical Properties of Liquids and Gases', a series of six lectures that Tyndall had given at the RI between 17 February and 24 March (see letter 4233, n. 3).
23. *my Belfast address*: his presidential address to the BAAS (see n. 14).
24. *your wife*: Adelheid Clausius.
25. *the children*: Clausius and his wife had six children: Mathilde (1861–1907), Helene (1863–1919), Rudolf John, Hedwig (1866–1946), Alfred (1868–1950), and Else (1872–1954).

From Mary Egerton [27 March 1874][1] 4301

Mountfield Court, | Robertsbridge, | Hawkhurst.[2] | Thursday

I was sitting under those very limes[3] when your Telegram[4] came! It was as if you had just risen out of the ground in answer to my thought, which always associates you with the spot!

We shall be delighted to see you on Saturday;[5] my brother in law Charles Wynn[6] & his daughter[7] are coming that day. Could you not get off at 10.30, and I would send to Etchingham[8] at 12.30, and you w[oul]d be here to luncheon? "No rose without a thorn"![9] I have got a lame foot at the present moment, and if it don't get well I shall not be able to take a good walk with you, which I had quite set my heart upon in this nice dry weather! If I can't, you must really come again before you go back to London! Why should you not at any rate? You can get here by the coach line;[10] and so you could by the bye on Tuesday to Yarmouth[11] if you could stay till then?

If you can lay your hand upon that letter of Carlyle's,[12] (about the Tait Controversy)[13] do bring it with you. You once asked me whether you had not shown it to me? or I sh[oul]d. not venture to ask; but it w[oul]d. interest me much.

There is something else I

<*1 or more pages missing*>

RI MS JT/1/E/76

---

1. *[27 March 1874]*: the date is suggested by reference to Tyndall's trip to Yarmouth on the following Tuesday (see n. 11).
2. *Mountfield Court, | Robertsbridge, | Hawkhurst*: the Egerton family estate in Sussex.
3. *those very limes*: see letter 4288, n. 12.
4. *your Telegram*: letter missing.
5. *Saturday*: 29 March.
6. *my brother in law Charles Wynn*: Charles Watkin Williams-Wynn (1822–96), a Welsh Conservative politician who was married to Egerton's sister Annora Charlotte Williams-Wynn.

7. *his daughter*: Mary Williams-Wynn (*c.* 1850–1951).
8. *Etchingham*: a village in Sussex with a railway station.
9. *"No rose without a thorn"*: a proverb, meaning no positive situation is without a negative aspect.
10. *the coach line*: presumably a regular horse-driven coach service between the railway station at Etchingham and a stop close to Mountfield Court.
11. *on Tuesday to Yarmouth*: Tyndall travelled to Yarmouth on the Norfolk coast on Tuesday 31 March to inspect the lighthouses at Haisborough Sands (see letter 4438).
12. *that letter of Carlyle's*: letter 4188.
13. *the Tait Controversy*: the controversy with Peter Guthrie Tait over Tyndall's treatment, in the late 1850s, of David James Forbes's claims to priority in his theories of glacial motion.

To George Gabriel Stokes 6 April [1874][1] 4302

6th April

My dear Stokes

I have, as you suggested, preserved the old memoir intact,[2] so that you may at once see any thing additional that has been introduced into the new one.

I have added a few more reasons, which really occurred to me long before the old memoir was written, for holding that the "flocculence" is due to the irregular admixture of air & aqueous vapour.[3] I go out of town tomorrow for a few days & I will write you a line on this subject when I return.

Yours faithfully | John Tyndall

A few new observations on fog have been added: but they can be signalised in a note.

CUL SC—add 7656

---

1. *[1874]*: the year is established by the relation to letter 4298.
2. *as you suggested, preserved the old memoir intact*: in letter 4298 Stokes told Tyndall 'As this is a case in which questions of priority might arise, it would be well to be more cautious than ever to leave the original MS intact, merely introducing marks to guide the printer'. The memoir is the original version of 'Vehicle of Sound', before Tyndall implemented the suggestions made by Stokes.
3. *the "flocculence" is due . . . air & aqueous vapour*: flocculence is a chemical process by which colloidal particles come out of suspension to form an aggregate, thereby stifling the transmission of sound. In the published version of his paper, Tyndall observed that the 'atmosphere, according to my explanation, was rendered acoustically flocculent by wreaths and streaks of mixed air and vapour' (p. 206). Tyndall also discussed his conception of the impact of flocculence on sound in letter 4261.

From Herbert Spencer 7 April 1874 4303

38 Queen's Gardens,[1] | Bayswater, W., | 7 April 1874.

My dear Tyndall,

I have arranged with others[2] for the dinner on the 14th. The hour will be 7 1/2 exactly. Dress interdicted,[3] & whoever comes late takes the consequences.

Ever yours | Herbert Spencer

RI MS JT/1/S/104
RI MS JT/1/TYP/3/1120

---

1. *38 Queen's Gardens*: a boarding house where Spencer had lived while in London since 1866.
2. *others*: not identified, and it does not seem to have been a meeting of the X Club, which usually met on the first Thursday of the month (14 April was a Tuesday). According to his journal, Thomas Hirst did not attend, and did not seem aware of this dinner (*Hirst Journals*, p. 1999). On the X Club, see letter 4363, n. 13.
3. *Dress interdicted*: i.e. formal dress is prohibited, suggesting this was a casual dinner.

From Grace Atkinson Ellis[1] 12 April [1874][2] 4304

Dear Dr. Tyndall,

As we steam down the channel[3] on our homebound-voyage[4] in the Cuba[5] I want to send you a line to say how much my brother[6] and myself enjoyed through your kindness while in London, and hope that if you return to America[7] you will count on us as among those who will warmly welcome you and if at any time you wish to introduce to us any one whom you think proper you will call upon us to do what we can for them in Boston. I will send you my address & hope you will not forget us.

I am, | with best regards | Yours most truly | Grace A. Ellis | April 12th | "Cuba"

Mrs. Ellis | Care of J. L. Little Esq.[8] | Boston Mass.[9]

RI MS JT/1/E/102

---

1. *Grace Atkinson Ellis*: Grace Atkinson Ellis (née Little, later Oliver, 1844–99), an American writer and women's rights activist ('Mrs Grace A. Oliver Dead. Woman of Rare Executive Ability and Noted for Her Charities', *Boston Globe*, 22 May 1899, p. 5). Her first husband was John Harvard Ellis (1841–70), the only child of the American minister and historian George Edward Ellis (1814–94) and his first wife Elizabeth Bruce Ellis (née Eager, m. 1840, d. 1842).
2. *[1874]*: the year is established by reference to Ellis having been in London (see n. 4).

3. *the channel*: presumably either the Queen's or Crosby Channels, which come out from the docks at Liverpool (see n. 5).
4. *our homebound-voyage*: from Britain to the United States. An account of her life later noted that 'in 1874 Mrs. Ellis spent a season in London, Eng., where she enjoyed the best literary society of that metropolis' (F. E. Willard and M. A. Livermore (eds), *A Woman of the Century: Fourteen Hundred-Seventy Biographical Sketches . . . of Leading American Women in All Walks of Life* (Buffalo, NY: Moulton, 1893), p. 547).
5. *the Cuba*: SS *Cuba*, a transatlantic steamer built in 1864 that served on the Cunard Line from Liverpool to New York until 1876. It was the same ship that Tyndall travelled on for his return to England in February 1873 from his American lecture tour. He travelled to New York aboard the SS *Russia*, also of the Cunard Line, in September 1872 (see letter 4201, n. 4).
6. *my brother*: Ellis had five brothers, James (b. *c.* 1846), John (b. *c.* 1848), Arthur (b. *c.* 1852), Philip (b. *c.* 1857) and David Little (b. *c.* 1860), although it is unclear which of them had travelled with her.
7. *return to America*: following Tyndall's lecture tour of the United States in 1872–3 (see n. 5). Tyndall never travelled again to the United States, though he had the desire to visit western mountain ranges.
8. *J. L. Little Esq.*: James Lorell Little (b. *c.* 1810), Ellis's father and a prominent Boston merchant.
9. *Mass.*: Massachusetts.

To William Spottiswoode 16 April [1874][1] 4305

16th April

My dear Spottiswoode.

It was ten minutes past 2 when your circular[2] reached me; just as I was starting to see M^{rs}. Faraday. I cannot attend the meeting,[3] but I shall be glad to belong to the Musical Society.[4]

Yours ever | John Tyndall

RI MS JT/1/T/1351

1. *[1874]*: the year is established by reference to the circular for the Musical Society (see n. 2).
2. *your circular*: a circular letter that Spottiswoode sent 'to some leading Members of the Musical and Scientific World' which ran: '50 Grosvenor Place, *8th April*, 1874. Dear Sir,—It has been suggested by several leading persons interested both in the theory and practice of Music, that the formation of a Society, similar in the main features of its organisation to existing Learned Societies, would be a great public benefit. Such a Musical Society might comprise among its members the foremost Musicians, theoretical as well as practical, of the day; the principal Patrons of Art; and also those Scientific men whose

researches have been directed to the science of Acoustics and to kindred inquiries. Its periodical meetings might be devoted partly to the reading of Papers upon the history, the principles, and the criticism of Music; partly to the illustration of such Papers by actual performance; and partly to the exhibition and discussion of experiments relating to theory and construction of musical instruments, or to the principles and combination of musical sounds. With a view to ascertain the opinions of persons interested in these subjects, and to attempt a more precise definition of the objects and constitution of such a Society, it is proposed to hold a meeting here, at which your presence is requested, on Thursday, April 16th, at 2.30 p.m. I am, Dear Sir, yours faithfully, W. SPOTTISWOODE' ('Origin of the Musical Society', *Proceedings of the Musical Association*, 1 (1874–5), pp. iii–iv, on p. iii).

3. *the meeting*: as stated in the circular, this was to be held at 2.30pm on 16 April at Spottiswoode's residence in Mayfair.
4. *I shall be glad to belong to the Musical Society*: Tyndall became one of the Society's original members, and was appointed, along with Spottiswoode, one of its five vice-presidents; the president was the composer Frederick Ouseley (1825–89). See 'Officers, &c.', *Proceedings of the Musical Association*, 1 (1874–5), pp. vii–ix. Tyndall served on the Society's council until 1885.

From Mary Egerton 19 April [1874][1] 4306

April 19th

My dear Mr. Tyndall,

I thought Mrs Spottiswoode showed great discrimination in her dinner arrangements! Of course L[ad]^y Claud[2] was disposed of by her work and it was so jolly (don't be shocked at the word!) to have a chat with you! I hope you did not pine for L[ad]^y Stanley of Alderby?[3] You shall have forgotten our conversation but never mind! You asked whether Fichte really spoke of a personal immortality? Here is one passage, there are a hundred others, not so particularly I find in the "Seligen Leben",[4] of which however the whole tendency is in that direction, for a life in the unseen & eternal now, "knows" as he says "no death",[5] but in the others, "Vocation of Man"[6] &c. Of course I do not ask you to believe a thing because another man, however noble, has believed it; I know how your independent spirit recoils from the bare idea of "imitation"! But that a man like Fichte, quite independent of all theological bias, should have found in the needs of his own moral and intellectual nature, a deep conviction of the truth of a life beyond this, surely shows the idea not to be one which the Philosopher is bound to crush out, when it arises, as arise it must, in every heart full of tenderness & human sympathy?

That sense of eternalness in love & affection, and those very intellectual cravings after more light, of which you were speaking, do they not point to their own fulfillment? Every other creature that we know of, appears to attain to a state of equilibrium with its circumstances, and then rests content. When these change, a fresh series of variation or development sets in, we are told, till the balance is again restored.[7] But man is ever seeking, never satisfied; he "never is, but always to be blessed"![8] And the nobler the mind the less is it satisfied with its present attainments. Why this? Unless a fuller, higher being, were in store for him? I know I have mixed up the race & the individual, but interchange them as you will, I maintain the same consideration, (I will not call it argument,) applies to either or both!

I think you too much identify the future life[9] with the "vulgar" doctrine of extraneous rewards & punishments.[10] The sense in which I should understand receiving according to our works, it would be simply that he who by high thoughts & noble aims has raised his soul here, will continue to rise higher, with fewer limitations, hereafter; that the unselfish & loving shall live a yet fuller & more loving life, while those who have debased themselves by sin or mean triviality here, would fall lower & lower as long as they have neither wish nor will to rise.

Do take that new vol[ume]: of Fichte[11] with you to your seclusion,[12] and let it shed an insensible spiritual aroma over your mind, when you plunge into the philosophic depths, as I expect you will. I am so looking forward to that address![13] If I cannot hear it,[14] I can read it at all events! But I do long to hear it!

Now I fear you will think I am "a backslider",[15] falling into my old sins! so goodbye!

I trust you have had a breath of country air this most cloudy day. We hope to come up Friday next.

Ever yours | M F E

RI MS JT/1/E/96

1. *[1874]*: the year is established by reference to the Belfast Address.
2. *L[ad]y Claud*: Elizabeth Hamilton.
3. *L[ad]y Stanley of Alderby*: Henrietta Maria Stanley (née Dillon-Lee, 1807–95), Baroness Stanley of Alderley, a society hostess and campaigner for female education (*ODNB*).
4. *"Seligen Leben"*: J. G. Fichte, *Die Anweisung zum seligen Leben* (Berlin: Realschul Buchhandlung, 1806).
5. *a life in the unseen . . . as he says "no death"*: Egerton seems to be paraphrasing discrete passages from J. G. Fichte, *The Way Towards the Blessed Life*, trans. W. Smith (London: John Chapman, 1849), which claims that the 'True Life lives in the . . . Eternal and Unchangeable . . . there is no real death'. To this, Fichte added: 'There is, however, an Apparent Life, and this is the mixture of life and death' (pp. 7 and 4).
6. *"Vocation of Man"*: J. G. Fichte, *The Vocation of Man*, trans. W. Smith (London: John Chapman, 1848), a translation of *Die Bestimmung des Menschen* (Berlin: Vossische Buchhandlung, 1800).
7. *Every other creature . . . again restored*: Egerton seems to be drawing on Herbert Spencer's conception of the equilibrium between an organism and its environmental circumstances,

as articulated in *The Principles of Biology*, 2 vols (London: Williams and Norgate, 1864). More specifically, however, she may be alluding to William Woods Smyth's recent theistic interpretation of Spencer's understanding of evolution, *The Bible and the Doctrine of Evolution* (London: H. K. Lewis, 1873), which proposed of animals that 'the creature is in *Equilibrium* with its circumstances', but then, like Egerton, contended that 'Man is not in equilibrium with his environment', which prompts him to 'wise meditation' on the divine order of the universe (pp. 28 and 85–6).

8. *"never <u>is</u>, but always <u>to be</u> blessed"!*: from the first epistle of Alexander Pope's poem *Essay on Man* (1733–4), l. 96.
9. *the future life*: in the Belfast Address, Tyndall observed that Lucretius 'has no rewards in a future life to offer', and also noted that Pierre Gassendi 'thought that the gods neither rewarded nor punished, and adored them purely in consequence of their completeness; here we see, says Gassendi, the reverence of the child instead of the fear of the slave' ('Address of John Tyndall', *Brit. Assoc. Rep. 1874*, pp. lxvi–xcvii, on pp. lxix and lxxvii).
10. *the "vulgar" doctrine of extraneous rewards & punishments*: the Old Testament doctrine that divine justice rewards good deeds or works, and punishes transgressions externally from the individual, rather than moral qualities being part of personal character.
11. *that new vol[ume]: of Fichte*: *Johann Gottlieb Fichte's Popular Works: The Nature of the Scholar; The Vocation of Man; The Doctrine of Religion*, trans. W. Smith (London: Trübner, 1873). Tyndall cites the second of these works, giving the original German title *Bestimmung des Menschen*, as the basis of a passage in the Belfast Address: 'Fichte, having first by the inexorable logic of his understanding proved himself to be a mere link in that chain of eternal causation which holds so rigidly in Nature, violently broke the chain by making Nature, and all that it inherits, an apparition of his own mind' ('Address of John Tyndall', p. xciii). Tyndall's personal library contained earlier editions of two of William Smith's translations of Fichte's books, *The Characteristics of the Present Age*, trans. W. Smith (London: John Chapman, 1847) and *The Vocation of the Scholar*, trans. W. Smith (London: John Chapman, 1847), but the new compilation of Smith's translations produced by Trübner & Co. would have been more convenient to take with him to Switzerland. See M. DeArce, N. McMillan, M. Nevin and C. Flahavan, 'What Tyndall Read: Provenance, Contents and Significance of the Proby Bequest in the Carlow County Council Library', *Carloviana*, 60 (2011), pp. 134–48, on p. 144.
12. *your seclusion*: at Bel Alp in Switzerland, where Tyndall was going alone to write his presidential address to the BAAS. He often stayed at the Bel Alp Hotel on the Bel Alp (now Belalp), above Brieg (now Brig) on the northern side of the Swiss canton of Valais. In 1877, Tyndall and Louisa built their alpine cottage Alp Lusgen on the Bel Alp, which was a 4.5 hour ascent from Brieg (K. Baedeker, *Switzerland and the Adjacent Portions of Italy, Savoy, and the Tyrol: Handbook for Travellers*, 3rd edn (Coblenz: Karl Baedeker, 1867), p. 256. On Tyndall and the Bel Alp, see T. Braham, 'John Tyndall (1820–1893) and Belalp', *Alpine Journal*, 98 (1993), pp. 193–8).
13. *that address*: Tyndall's presidential address to the BAAS.
14. *If I cannot hear it*: it seems that Egerton did attend the BAAS meeting in Belfast and heard Tyndall's address (see letter 4392).

15. *"a backslider"*: someone who falls away from an adopted course, especially of religious faith or practice; an apostate, renegade (*OED*).

From Clara Wiedemann 19 April 1874 4307

Carlsbad, d. 19tn April 74.

Hochgeehrter Herr Professor!

Mit großer Freude u mit vielem Interesse habe ich heute meine Aufgabe zu Ende gebracht u Ihre interessanten Vorträge dem deutschen Publikum zugänglich gemacht. Ich hoffe, dass alle die, die sie im deutschen lesen werden, eben so viel Freude daran haben, als ich u Ihnen eben so gern in das geheimnisvolle Reich des Lichtes folgen. Nun stehe ich aber am Schlusssatz u weiß keine Deutung u keine deutsche Redeform zu finden, wie ich den Ausspruch: cast your bread on the waters wiedergeben soll. Wollen Sie so freundlich sein, u mir schreiben, wie Sie es verstanden sehen wollen.

Was nun die Anhänge anbelangt, so meint mein Mann, dass zuerst die Polemik zwischen Young u Brougham, wie Sie es schon selbst erwähnten, für das größere, deutsche Publikum ein nicht so großes Interesse, wie in England haben würde u es daher wohl für die deutsche Ausgabe zweckmäßiger sein würde, sie fortzulassen. Wünschen Sie, Henry's wegen, dass der Anhang: Water Crystallisation pag. 257 auch in die deutsche Ausgabe übersetzt? Die übrigen Anhänge sind fertig übersetzt. Sie sind wohl so freundlich, mir nach Leipzig, wohin wir in einigen Tagen zurückkehren, Ihre gütige Antwort zu schicken. Wir sind einige Wochen hier in Carlsbad zu einer frühen Kur gewesen u da war mir Ihr Buch eine eben so liebe, wie belehrende Beschäftigung. Und ich bin Frau Helmholtz sehr dankbar gewesen, daß mir dieses Mal die Arbeit allein blieb, u ich von Anfang bis zu Ende mich gründlich hineindenken u hineinleben konnte u mit gleichem Interesse den ersten, wie den letzten Satz schreiben durfte. Ich kann mir wohl denken, dass Ihnen die Amerikaner durch alle Vorlesungen mit warmer Begeisterung gefolgt sind u Sie ihnen am Schluß so unbefangen die Wahrheit haben sagen können.

Wann werden wir Sie denn einmal in Leipzig sehen, hochverehrter Herr Professor? Der Umweg nach der Schweiz ist doch so groß nicht.

Mein Mann lässt sich Ihnen angelegentlichst empfehlen.

Ihre | ergebene Cl. Wiedemann

Carlsbad, 19th April 74.

Highly honoured Professor!

It is with great joy and much interest that I have today brought my task[1] to an end, and made your interesting lectures accessible for the German audience. I hope that all those who will read it in German take as much pleasure in it as I and will equally gladly follow you into the mysterious realm of light. I am now faced, however, with the final sentence and cannot find an interpretation and a German expression to render the phrase: cast your bread on the waters.[2] Would you be so kind to write to me how you would like it to be understood.

As far as the appendices are concerned, my husband[3] thinks that the polemics between Young and Brougham,[4] as you had mentioned yourself, would not be of such great interest for a German audience as in England, and it would therefore be more to the purpose for the German edition to omit it. Do you wish, for Henry's sake,[5] that the appendix: Water Crystallisation pag. 257 be translated in the German edition? The remaining appendices are completely translated. Please be so kind to send your reply to Leipzig, whither we will return in a few days. We have been here in Carlsbad[6] for an early course of treatments and your book was a dear and instructive occupation for me here. And I was also very grateful to Mrs Helmholtz that this time the work was all mine[7] and that I could try to understand and penetrate it thoroughly from beginning to end and was able to write the first sentence with the same interest as the last. I can easily imagine that the Americans have followed all your lectures with warm enthusiasm and that at the end you were able to unselfconsciously tell them the truth.

When will we be able to see you in Leipzig, honoured Professor? After all, the detour on the way to Switzerland[8] is not that substantial.

My husband sends his sincerest regards.

Your | devoted Cl. Wiedemann

RI MS JT/1/W/65–65

---

1. *my task*: translating Tyndall's *Six Lectures on Light* (1873) into German, which was published as J. Tyndall, *Das Licht: Sechs Vorlesungen gehalten in Amerika im Winter 1872–1873*, ed. G. Wiedemann (Braunschweig: Friedrich Vieweg, 1876).
2. *the phrase: cast your bread on the waters*: answering the question 'What is the use of your work?', Tyndall concluded the final lecture in *Light* by stating: 'If you cast your bread thus upon the waters, then be assured it will return to you, though it may be after many days' (p. 226). He drew upon Ecclesiastes 11:1, 'Cast thy bread upon the waters: for thou shalt find it after many days', which relates to unselfish actions, done without certainty of reward. Wiedemann translated it as 'Lass Dein Brod über das Wasser fahren, so wirst Du es finden auf lange Zeit', and added a footnoted reference to the biblical passage 'Pred. Sal., Cap. XI, Vers 1' (p. 247).
3. *my husband*: Gustav Wiedemann.
4. *the polemics between Young and Brougham*: Thomas Young had established the wave theory of light, which challenged Isaac Newton's theory of light as being emitted as particles. However, Young's theories were attacked by Henry Peter Brougham (1778–1868) in the *Edinburgh Review*. Tyndall added an appendix to *Six Lectures on Light* (1873) containing 'Lord Brougham's Articles on Dr. Thomas Young in the "Edinburgh Review"' and 'Dr. Young's Reply to the Animadversions of the Edinburgh Reviewers' (pp. 227–55). This was not included in the German edition, nor in a French translation (see letter 4350).

5. *for Henry's sake*: another appendix in the *Six Lectures on Light* (1873), entitled 'Water Crystallisation', reprinted a letter on the subject written to Tyndall by Joseph Henry (p. 257). The appendix was not included in the German edition.
6. *Carlsbad*: a popular spa resort in the south-east of Germany (now Karlovy Vary in Czechia), attracting wealthy visitors who wished to be treated for various ailments.
7. *the work was all mine*: although Clara Wiedemann usually translated in collaboration with Anna Helmholtz, *Das Licht* was entirely her own work.
8. *on the way to Switzerland*: Tyndall intended to spend the early summer in Switzerland.

From Henrietta Huxley 21 April [1874][1] 4308

4 Marlborough Place | Abbey Road N.W.[2] | 21st April

Dear Brother John

Will you come and dine with us on the 4th May at 1/4 to 7. Hal's birthday?[3] I hope you will.

I see by the "Times" that you were lost in the Isle of Wight[4] and were at last discovered stranded on a cliff 'Twixt earth and heaven'[5] It was well you were with a party of friends[6] who missed you.

Take warning and do not do such things when you are quite alone.

With love | Yours affectionately | Nettie Huxley.

RI MS JT/1/TYP/9/2982

LT Typescript Only

---

1. *[1874]*: the year is established by reference to the report in the *Times* (see n. 3).
2. *4 Marlborough Place | Abbey Road N.W.*: see letter 4213, n. 1.
3. *Hal's birthday*: Thomas Huxley's forty-ninth birthday.
4. *by the "Times" that you were lost in the Isle of Wight*: 'News in Brief', *Times*, 21 April 1874, p. 9. The report related how Tyndall was 'found up a cliff, where he had climbed in search of geological specimens'. With the aid of a rope, he was 'rescued uninjured'.
5. *'Twixt earth and heaven'*: from Lord Byron's *Werner, A Tragedy* (1823), III.ii.9.
6. *a party of friends*: according to the *Times*, this was 'Sir John Lubbock's party', with whom Tyndall was staying at Freshwater.

To Henrietta Huxley 21 April [1874][1] 4309

Royal Institution of Great Britain | 21st April

My dear Sister.[2]

Calculate on me for the 4th of May.[3]

I want to tell you a pleasant conversation I had last night with Jodrell.[4] He

& a couple more[5] wish to send Hal[6] with Grant Duff[7] to India, taking charge of his duties here and of all necessities ghostly & bodily there!

It is us not living a godly life when the wholly godless are so well cared for.[8]

Your loving brother | John

IC HP 1.118

---

1. *[1874]*: the year is established by the relation to letters 4308 and 4310.
2. *Sister*: Tyndall's affectionate name for Henrietta, who referred to him as 'Brother John'.
3. *the 4th of May*: the date of a dinner to celebrate Thomas Huxley's forty-ninth birthday (see letter 4308).
4. *Jodrell*: Thomas Jodrell Phillips-Jodrell (1807–89), a barrister who donated money for scientific research, at whose house in Stratton Street, Mayfair, Tyndall regularly dined (see letter 4446, n. 5).
5. *a couple more*: the instigator of the plan was seemingly the politician and businessman William Rathbone (1819–1902) (see n. 7).
6. *Hal*: Thomas Huxley, Hal being a nickname for Henry, as used by William Shakespeare in *Henry IV* (1598–9).
7. *Grant Duff*: Mountstuart Elphinstone Grant Duff (1829–1906), who had served as the Under-Secretary of State for India in William Gladstone's Liberal government. Having left the role following Gladstone's defeat in the general election in February, Grant Duff was preparing to visit India for the first time. He later wrote to Huxley's son Leonard: 'You know, I daresay, that Mr. William Rathbone, then M.P. for Liverpool, once proposed to your father to be the companion of my first Indian journey in 1874–5, he, William Rathbone, paying all your father's expenses. Mr. Rathbone made this proposal when he found that Lubbock, with whom I travelled a great deal at that period of my life, was unable to go with me to India. How I wish your father had said "Yes". My journey, as it was, turned out most instructive and delightful; but to have lived five months with a man of his extraordinary gifts would have been indeed a rare piece of good fortune' (*Life and Letters of Huxley*, vol. 1, p. 355).
8. *It is us not . . . well cared for*: seemingly a response to Henrietta's advice to 'Take warning and do not do such things when you are quite alone' after Tyndall had got lost on a cliff in the Isle of Wight (see letter 4308).

From Henrietta Huxley 21 April 1874 4310

4 Marlborough Place[1] | 21st April 1874

"A pleasant conversation"![2]

But not to me. It would not be pleasant that Hal should go to India.[3] It will be bad enough if he must go to America next year.[4] I carry the weight of the thought about with me day and night. And why should he go to India? You speak in riddles to both of us. Will you explain and release my mind from vain imaginings?

Affectionately Yours | Nettie Huxley.

I need not say that if it is for Hal's good that he should go to India I am dumb.

RI MS JT/1/TYP/9/3028
LT Typescript Only

---

1. *4 Marlborough Place*: see letter 4213, n. 1.
2. *"A pleasant conversation"!*: see letter 4309.
3. *Hal should go to India*: a plan for Thomas Huxley to accompany the politician Mountstuart Elphinstone Grant Duff on a five-month tour of India that Tyndall outlined in letter 4309.
4. *go to America next year*: after many delays, Huxley did not depart for his lecture tour of the United States until July 1876, and was accompanied on the trip by Henrietta. On Huxley's lecture tour in the United States, see D. Finnegan, *The Voice of Science: British Scientists on the Lecture Circuit in Gilded Age America* (Pittsburgh: University of Pittsburgh Press, 2021), pp. 64–93.

To Robin Allen 22 April 1874 4311

Royal Institution of Great Britain | 22nd April, 1874.

SIR,

I have read the descriptions, and inspected the drawings of the proposed fog-signal gun.[1]

The thoroughness with which the problem has been thought out, and the completeness of the arrangements proposed to meet the object in view, are in the highest degree satisfactory.

Hitherto the nature of the gun employed has made the minimum interval of firing fifteen minutes. With far less trouble, and with far more safety, the shots with the proposed gun may succeed each other at intervals of five minutes. Were the diminution even of this interval desirable, it is perfectly practicable.

As regards the use of the gun as a fog-signal, I consider the proposition of the War Department a very decided step in advance.

I, therefore, recommend the acceptance of the Department's offer to construct a gun of the proposed description for the sum mentioned.

I do not know the comparative cost of bronze and steel, or iron guns, but I assume that the bronze will be dearest, and that it is chosen in the present instance in consequence of the superior loudness of the report of the bronze gun.

Now the report of a gun, as affecting an observer close at hand, is made up of two factors; the sound due to the shock of the air by the violently exploding gas, and the sound derived from the vibrations of the gun, which, to some extent, rings like a bell. This latter, I apprehend, will disappear at considerable distances, but nobody is entitled to dogmatise on this point. It ought to be decided by actual experiment.[2] Hence it appears to me desirable to have a second gun constructed of steel, or of iron, in all other respects similar to the bronze gun, and that a comparative trial be made at Woolwich,[3] or elsewhere, to determine once for all the question whether, as regards sound at a distance, anything gained by the employment of the more expensive metal.*

I have the honour to be, Sir, | Your obedient Servant, | JOHN TYNDALL. | ROBIN ALLEN, ESQ.

* If the experiment here recommended should have been already made by competent observers, I, of course, should bow to their decision.

National Archives, MT 10/290. Papers re Fog Signals. H8476 (duplicate: H4380)
Typed Transcript Only

---

1. *the proposed fog-signal gun*: in his capacity as scientific adviser to Trinity House, Tyndall had been engaged in a series of experiments to determine the most effective method of warning ships in fog (see letter 4185, n. 3). The plans for a specially adapted gun designed for this purpose were first discussed in letters 4284 and 4287.
2. *actual experiment*: Tyndall later described the experiments undertaken in 'Recent Experiments on Fog-Signals', *Roy. Soc. Proc.*, 27 (1878), pp. 245–58.
3. *a comparative trial be made at Woolwich*: at the Royal Arsenal; the comparative trial is discussed in letters 4513 and 4514.

From Margaret Welsh[1] 23 April 1874 4312

5 Cheyne Row | Chelsea | 23 April 1874

Dear Sir,

Mr Carlyle bids me thank you cordially for your kind offer to come to take him to the Deanery[2] on Saturday.[3] As yet Mr Carlyle has not heard anything of an invitation[4] thither for that day; & even should there be a proposal to that effect, he says he fears he would not be able to accept. So he bids me say, he would not like you to have the trouble (solely on that account at least) of coming so far, unless you should hear further from him before Saturday.

I am, dear Sir, | Yours truly, | Margaret Welsh

RI MS JT/1/W/20

---

1. *Margaret Welsh*: Margaret Welsh (née Kissock, 1803–88), a cousin of Jane Welsh Carlyle, Thomas Carlyle's late wife. On Jane's death, see letter 2414, n. 4, *Tyndall Correspondence*, vol. 9.

2. *the Deanery*: the residence, in the grounds of Westminster Abbey, of Arthur Penrhyn Stanley, the Dean of Westminster. Carlyle visited the Deanery in March 1869 and possibly November 1876, while Tyndall was a regular visitor. See A. A. Adrian, 'Dean Stanley's Report of Conversations with Carlyle', *Victorian Studies*, 1 (1957), pp. 72–4; and *Ascent of John Tyndall*, p. 209 and passim.
3. *Saturday*: 25 April.
4. *As yet Mr Carlyle has not heard anything of an invitation*: there is no record of an invitation being received by Carlyle. See *The Collected Letters of Thomas and Jane Welsh Carlyle*, ed. C. R. Sanders et al., 49 vols (Durham, NC: Duke University Press, 1970–), vol. 49, p. 92.

From Julius Robert Mayer 28 April 1874 4313

Hochverehrter Herr und Freund!

Mit beifolgendem bin ich so frei, Ihnen ein Exemplar der so eben erschienen zweiten Auflage meiner „Mechanik der Wärme" zuzusenden und um wohlwollende Aufnahme der kleinen Gabe zu bitten. Niemals werde ich aufhören daran zu denken, wie vielen Dank ich Ihen, dem edlen Gönner meiner Bestrebungen, schulde. Haben Sie die Güte mich auch den übrigen Herren der Royal Society bestens zu empfehlen und genehmigen Sie den Ausdruck vollkommenster Hochachtung, mit welcher ich verharre

Euer Hochwohgeboren | ergebenster Diener, | J. R. Mayer

Heilbronn, | 28 April 1874

Highly respected Sir and friend!

I take the liberty of enclosing a copy of the newly-published second edition of my "Mechanics of Heat",[1] and ask that you kindly accept this small present. I will never forget how much I am indebted to you, the noble patron of my endeavours.[2] Please be so kind to also give my regards to the other gentlemen of the Royal Society and accept the expression of my greatest respect, with which I remain

Dear Sir, your | most humble servant, J. R. Mayer

Heilbronn, | April 28, 1874

RI MS JT/1/M/87
RI MS JT/1/TYP/7/2536

---

1. *second edition of my "Mechanics of Heat"*: J. R. Mayer, *Die Mechanik der Wärme in gesammelten Schriften*, 2nd edn (Stuttgart: J. G. Cotta, 1874).
2. *the noble patron of my endeavours*: Tyndall had championed Mayer's cause in a dispute over scientific priority with James Prescott Joule. Both Mayer and Joule claimed precedence for demonstrating the mechanical equivalence of heat, the notion that motion and heat are mutually interchangable. See *Tyndall Correspondence*, vol. 8, pp. xvii–xvi.

From William George Valentin 30 April 1874 4314

Royal College of Chemistry | South Kensington | April 30th. 1874

Professor J. Tyndall. L.L.D. F.R.S. &c. &c.

Sir.

I beg to report to you on the experiments conducted by me at the Haisbro' High Lighthouse[1] on Wednesday Thursday and Friday, April 1st, 2nd, & 3rd.

I put my self in communication with Mr. Wigham and arranged with him (1) that a small 100 inch bar Bunsen Photometer[2] should be fixed for testing the illuminating power of the gas, likewise a large 36 ft bar Bunsen Photometer, accurately divided into feet and inches, for measuring the Light given by the new gas Burner; (2) that this powerful Light from the Wigham house should be tested against a light greater than that produced by one or two Sperm candles[3] viz against a bar Burner (3) that all necessary provision should be made for forming the bar as visibly as possible and in conditions similar to those pertaining to the burner in the Cupola of the lighthouse.

From the intimation I received from you I prepared for examining (1) into the state of the gas works at Haisbro' Light House which in accordance with my recommendations of last summer had since been put into thorough repair and to which a Station master had been added for assisting and controlling the manufacture of the Gas; (2) for ascertaining the illuminating capacity of the different power of the Wigham burner: (3) for controlling on land the experiments you and the Light Committee of the Trinity House[4] were proposing to conduct at sea on the nights of Wednesday and Thursday on board the Galatea Steam yacht:[5] (4) for reporting on the comparative consumption of Gas by the Wigham burner and the Colza oil[6] by the new 4 and 6 wick Trinity Lamps, and on the respective cost of the two sources of light.

I arrived at Haisbro' Lighthouse on the morning of Wednesday April 1st. where I met Mr. Wigham and Mr. Douglas and was pleased to find that my instructions had been carried out in a most intelligent and careful manner by Mr. Wigham and that—everything necessary to conduct the—experiments successfully had been—provided for. On your arrival soon after in company with the Gentlemen of the Light Committee,[7] I had the satisfaction to find that the Visiting Committee were satisfied with the improved state of the gas Works and entirely approved of the lighting apparatus which I had placed in the Basement of the Lighthouse Tower in readiness for the experiments.

I had given instructions that a quantity of gas should be made and stored up in the gas holder sufficient to last for the days' experiments. I at once proceeded to ascertain the illuminating power of the gas most carefully by means of the small 100 inch rod photometer. A fish-tail gas burner[8] delivering 4. C. F.[9] of gas and giving a fine spreading flame was employed. Its light was measured with great facility against 2 sperm candles weighed whilst burning, and the consumption of sperm was connected to the standard employed in this Country, viz the sperm candles consuming 120 grains per hour.

The following were the results obtained :—

Consumption of of gas 4. C. F. per hour

Illuminating Power 27.24 candles

To facilitate the photometric measurement of such powerful Lights as are produced by the different powers of the Wigham burner M^r^. Wigham had at my desire provided a rather more powerful test burner and a second photometer rod had been placed at an angle to the first for the transferment of the primary photometric standard.[10] In this manner the Light of the 4. C. F. fish tail burner could be readily compared with that of a burner similar in construction to the large Wigham burner, but consisting of only 10 fish-tail (No. 2.) gas jets, burning without chimney. This burner was found to give 2.61 times the light of the fish-tail or 2.61 x 27.24 = 71.09 Sperm candles. It was thus possible to retain the familiar unit of photometry viz; the sperm candle consuming 120 grains of Sperm per hour, and yet to avoid the difficulty of directly—comparing most powerful lights with so insignificant a light as that disclosed by one, or at best, two Sperm Candles.

Against this intermediary burner giving in round numbers the light of 71 sperm candles I had fitted the Wigham burner. The two were separated by a photometer rod, 36 ft. in length, graduated to indicate directly the photogenic result by multiplying the figures marked on the rod with 71.09, as well as in feet and inches. The results read off by the latter scale had merely to be squared, and the proportional photogenic expression to be found by dividing one square number into the gases, the absolute by multiplying, as before, the quotient thus found with 71.09. Hence the two readings checked each other, and at the same time—verified the divisions made on the large photometer rod. The disc used was about 1 foot in diameter and was fitted in a readily moveable blackened wooden box, about 3 feet in length. Various designs and traceries offset delicacy on the usual paper disc had been prepared which facilitated the photometric work considerably. In fact, every precaution had been taken to accurately measure this formidable light. The burner had been placed a few feet above the floor of the tower of the Lighthouse. A short mica[11] and an iron chimney, about 10 ft. in length, and discharging into the upper part of the tower, conveyed away the noxious products of combustion and created a powerful draught, such as is necessary to lights placed in the cupola of lighthouses. A governor[12]—regulated the supply of the gas. A large and delicately constructed meter, having a drum capacity of 5 cubic feet,—indicated the consumption of gas by the revolutions of a minute hand as well as by the permanently registering hands: the consumption was controlled by—independent observers, by both means, and the readings were found to be remarkably concordant. As far as foresight could provide nothing had, in fact, been left undone to secure good photometric measurements. It gives me pleasure to acknowledge the very intelligent and painstaking manner in which Mr. Wigham had carried out all my directions.

Description of Gas Burner. Before recording the results of the photometrical testings it may be well to describe , as far as can be done, without drawings, the construction of the burner from data—collected by myself and information received from Mr Wigham.

The gas burner devised and perfected by Mr. Wigham consists of a number of fishtail gas jets arranged in concentric rings, and set in narrow upright metal tubes, about 5" in height, fixed in a flat circular hollow metal chamber of cast brass. This chamber is connected from the bottom with the gas supply by means of a mercurial lute or valve, and can be lifted off with the greatest ease, in case of accident when an ordinary oil lamp used in dioptric[13] apparatus may be substituted for it. 28 of these gas jets set in three concentric rings, respectively of 4, 8, 76 constitute what Mr. Wigham terms his 1st power burner. The admission of the gas to the metal chamber is controlled by micrometer cock, placed on the standard gas pipe, leading to the mercurial cup. As the pressure under which the gas is supplied to the burner is necessarily great (2½ inches water pressure), it has been found conducive to a higher photogenic effect to check the flow of the issuing gas by the well known contrivance of having a small lava-cone,[14] bored with five holes in the lower part of the narrow upright delivery tubes, and a lava-tipped fishtail burner at the top. Lights of considerably increased illuminating power are obtained by Mr. Wigham by severally adding to his first power burner concentric ring chambers, made in two halves, containing an additional number of 20 fish-tail jets and which can be connected with the utmost facility, in the manner already described, with a separate gas supply for each chamber. It is obvious that when the different burner powers are once adjusted, a move of the micrometer cock turns on the gas, or shuts it off almost instantaneously from one or the several powers either partially or entirely, and that this gas burner is, therefore preeminently suitable for producing signal or flashing lights.

The burner which I examined at Haisbro' is constructed for lights of 6 different powers: viz

1st.	power or	28	jets of NO.	1.	fish tail burner
2nd.	"	48	"	2	"
3rd.	"	68	"	2	"
4th.	"	88	"	5	"
5th.	"	108	"	5	"
6th.	"	148	"	5	"

The respective diameters of the circles, are 3½", 5½", 7", 8¾", 10½" and 12¼". The openings in the lava-cones within the tubes are somewhat larger for the higher power Lights.

To produce the necessary combustion different sized moveable chimneys made of thin transparent mica plates, riveted together, are employed. There terminal chimneys can be readily attached to a similarly constructed stationary flue of uniform width and height, which is in its turn connected with an iron flue, about 11 feet in height. This conveys the products of the combustion into the open air; a damper in the iron flue regulates the draught. The following are the dimensions of the different terminal chimneys.

	Width		Height
	below	above	
1st. Power	3½"	4½"	8½"
2nd. "	4½"	6½"	7½"
3rd. "	D°.	D°.	D°.
4th. "	8½	9¾	5½
5th. "	D°.	D°.	D°.
6th. "	D°.	D°.	D°.

The stationary mica flue is 10" wide and 16" high; from these dimensions of the terminal chimneys it follows that the different sized flames must become especially contracted and that varying amounts of atmospheric air are supplied thro' the chimneys to the flame, which must cause increased or decreased combustion. This is actually borne out by the corresponding results of the photometric testings of the different powers as will be shewn below. The distance from the top of the fish-tail burners to the inlet of the terminal mica place chimneys varies with the different powers, viz—3½"—4½"—5½"—6½—& 7½ respectively. It appears to me that the shape and intensity of the flame which is primarily dependent upon the size of the burners is greatly influenced also by the height at which the chimneys are placed from the burner.

Photogenic Results. Each power was tested by me photometrically with the gas which had previously given me the light of 27.24 candles for every 4. C. F. consumed. 10 observations were taken during each series of experiments. Both Mr. Wigham and his representative Mr. Edmundson,[15] and Mr. Douglass had ample opportunity given to them to satisfy themselves of the correctness of the photometric readings and of the consumption of gas per burner power. The testings were carried out without the least hitch and passed off to the satisfaction of everybody concerned in it.

I subjoin in a tabular form the results I obtained:—

Table shewing Photometrical Results Obtained with the Wigham Burner.

Power of Burner	Illuminating power in Sperm Candles consuming 120 grains per hour—mean of 10 observations	Hourly consumption of gas in Cubic Feet.	Illuminating Power, produced By 1 Cub. Foot Of Gas	Illuminating Power, produced By 1 Fishtail Jet.	Cubit Feet Of gas, consumed Per Fishtail Jet.
1st. Power, or 28 Jet Burners	429.6 Candles	57.4 C.7.	8.36 Candles	15.34 Candles	1.835 C.7.
2d. Power 48 "	832 "	93.2 "	8.92 "	17.33 "	1.94 "
3rd Power 68 "	1253.18 "	146.3 "	8.57 "	18.43 "	2.15 "
4th Power 88 "	2408 "	244 "	9.87 "	27.36 "	2.77 "
5th Power 108 "	2923.4 "	308 "	9.43 "	27.07 "	2.85 "
6th Power 148 "	3136 "	420 "	7.46 "	21.19 "	2.83 "

It follows from these figures (1) that when gas is burnt in large quantities by this powerful Burner it produces proportionally more light than can be obtained by burning it in an ordinary fish-tail, fatburning or argand burner;[16] for power 4 ex. gr. gave nearly 10 candle light for every cubic foot of gas consumed, equal to an illuminating Power of close upon 40 candles, instead of 27.24 for 4. C. 7. of gas. In fact the least perfect combustion,—owing in all probability to an insufficient supply of atmospheric air—as exemplified by Power 6, viz 7.46 Candle Power for every Cubic Foot of Gas consumed, gives still higher results than that produced by the test fishtail burner I employed. It is clear, then that there is a considerable advantage in burning gas from powerful burners. (2) That a better photogenic effect is produced by Power 2 & 3, than by Power 1. likewise by Power 4 & 5, than by Power 6. This is owing, in my opinion, to the greater indraught of air, dependent upon the terminal mica chimneys: Powers No 2–4 are in fact more economical than Powers 3, 5, or 6.

Cost of Maintaining the Wigham Burner. In order to arrive at a sound and fair estimate of manufacture after satisfying the visiting Committee that the gas Worker has been placed by Mr. Wigham in thorough—working order, I came to an understanding with Mr. Douglass and Mr. Wigham that a fortnightly return of the Gas maker, according to a Tabular Scheme (comp[are]: Appendix C) which I drew up should form the basis of my calculations. I have his return now before me, signed by Mr. Kelsey the principal Light Keeper.[17] From the figures—contained therein it follows, that on an average 10.500 Cubic Feet of Gas were obtained from one ton of Cannel Coal.[18] In addition to the Coke and Tar obtained there were used 8 cwt[19] of Banking (breeze) Coal[20] for effecting the distillation of 1 Ton 9 cwt and 1 qr.[21] of Cannel Coal. The price of Marquis of Lothian's[22] Cannel Coal, incl: of delivery at Haisbro'

Lighthouse, amounts to £2.17.6. per Ton, that of the common banking or breeze coal to 201–per Ton. The Wages paid to one gas maker are 181–per week. The Lime[23] necessary for purifying the gas, equal to 5 per cent of the coal carbonized. Coals at Haisbro' 191–per chaldron (36 bushels).[24] The interest on £1685.6.9 capital (—These figures are based upon an official Statement I received from the secretary of the Trinity House.[25] They include, however, some incidental expenses, which would probably not occur again in establishing another gas Station)—or money disbursed for plant and other expenses in establishing the light House as a gas station amounts to £67.8.3 The cost of manufacturing gas may thus be readily arrived at, as follows:—

	Burner Powers					
	1	2	3	4	5	6
Hourly consumption of Gas in Cub[ic]. Feet	51.4	93.2	146.3	244	308	420
Yearly consumption of Gas during 4,412 Hours in Cub[ic]. Feet	226777,	411198,	645475,	1076528,	1358896,	1953040
Tons of Cannel Coal required to be carbonized	21.6,	39.16,	61.47,	102.52,	129.42,	186
Cost of Cannel Coal at 57/6 per Ton	£62.2,	£112.11.8,	£176.14.6,	£294.15,	£372.16.6,	£534.15.
Tons of Breeze coal required (2 qr per day for 2 daily charges of 1½ cut: each per retort)	Tons, cut; qrs 0.	2	2	D°.,	D°.,	–
Cost of Breeze Coal at 20/– per Ton	£9. 2. 6.	D°.,	D°.,	–	–	–
Cost of Lime used for purifiers	£1.0.7,		£1.17,		£2.18	
Wages at 18/– per week	£46.16.	D°.,	D°.,			
Interest	£67.8.3					

Repairs & Renewal nil—(I have not been able to obtain any return of the expense incurred on the score of repairs and renewal since the establishment of the Haisbro' Gas Station and abstain therefore from giving any figures, as the condition of a small gas Work are so widely different for those of large workers—). The Cost of a 1st. Power Gas Burner, giving a light nearly one fourth superior to a good four wick first class oil Lamp (see House of Commons Paper No 3 of Session 1873. p: 46) would now be £186.9.4, or 10.14d per hour: that of the 2nd. Power would be £237.15.5 or 12.9d p. hour, and that of the 3d. Power, giving 3.8 times the light of a good 1st. class 4 wick oil Lamp would amount to £302.19.3 or 16½d per hour. *[in left margin next to this paragraph]* 10.14 These are decimal points—not shillings & pence

Mr. Douglass and Mr. Wigham are of opinion that an hourly—consumption of 120 C.F. of Gas, all the year round, taking into account all kinds of weather, clear and foggy, as well as the consumption for Cottages etc: will be about what should be calculated upon.

The cost of such a light would be as follows:—

Cannel Coal,	50.42 tons at 57/6	£144.19.0
Breeze Coal,	2 qrs per day	9.2.6
Lime for purifiers		2.7.10
Wages		46.16.0
Interest		67.8.3
Repairs and Renewal		nil
		£270.13.7

or 14³/4d per hour.

Comparative Cost of oil and gas Burners, used at Haisbro' Low and High Lighthouse—

Trials were also carried on simultaneously with the oil lights at the Lower Light House which it had been agreed upon by the Visiting Committee should be controlled by Mr. Wigham who consequently assisted on both Evenings in measuring out the Colza oil for the 1st. order four wick oil Lamp used on the Evening of April 1st. as well as for Mr. Douglass' powerful new 6 wick oil Lamp used on the Evening of April 2nd. From experimental data supplied to me by Mr. Wigham regarding the consumption of oil, and by Mr Douglass regarding the cost of maintaining the Lamps in other respects, I have been enabled to collect the following facts:—

I. Four wick Trinity Oil Lamp. Hourly consumption of Colza on the night of April 1st. = 34.78 fluid ounzes. Annual cost at 2/10d p: gallon (inclus[ion]. of delivery at Haisbro') 4412 hours = 959 gallons. £135.18.0

One years supply of wicks	2..1.0
Do. Chimney glasses	1.16.0
Interest on Capital laid out in Plant &c &c = £84 at 4%	3.10.0
Repairs and Renewals*	2.10.0
	145.15.0

Comparative cost of the two Lights:—

	Oil Light	Gas Light
	£145.15.	*186.9.4
Comparative Illuminating Power	328 candles	429 candles

---

*It should be remembered that no repairs & renewals were taken into account in making out the Estimate for the Gas light because I had no data to guide me. They would probably be much higher, however.

Cost of Lights of signal intensity	£145.15.		£142.12.
Proportional Cost	100	:	97.8

or a Saving of 2.2 P.C. by using Gas.

II. New Six Wick Trinity Oil Lamp.

Hourly consumption of Colza on the night of April 2nd. = 83.7 fluid ounzes (Comp[are]: Hours of Luminous Paper No. 3 of Session: 1873 p: 46. where the consumption is stated to be 87.6 ounzes).

Annual Cost 2/10d p: Gallon (incl: delivery) for 4412 hours = 2308 Gallons	£326.19.4
One years supply of Wicks	£5.1.0.
D°. D°. Chimney Glasses	3.10.0
Interest on Capital laid out in plant &c &c =£106. at 4. p. C.	4.11.7
Repairs and Renewal	2.10.0
	342.11.11

Comparative cost of the Two Lights	Oil Light	Gas Light. Power II
	£342.11.11	£237.15.5
Comparative Illuminating Power	722* candles	832 candles

*Probably too high as the consumption of oil was only 83.7 fluid ounces, instead of 87.6 as given in Parl[iamentary]: Paper already quoted

Cost of Lights of Equal Intensity	£342.11.11		£206.6.
Proportional Cost	100	:	60.2

or a saving of nearly 40. P[er].C[ent]. by using Gas.

Burridge

It is obvious then that the New 6 wick Colza oil Lamp does not burn so economically as the 4 wick Burner, not as the IId. Power Wigham gas Burner, for two reasons viz:—

(1) Whilst the consumption of oil in the two Lamps is as 100 : 240, the illuminating Power is only as 100 to 220.

(2) because the Gas can be burnt more economically from the 2nd. power Wigham Burner than from No. 1. and because it can be manufactured more cheaply on an increased scale.

This proportion would, however, be different for the same size Lamps, burning Paraffin instead of Colza Oil.

———

I employed part of Friday April 3rd to draw up Forms for controlling the manufacture of Gas as well as its consumption. The consumption with Mr. Douglass and Mr. Wigham I discussed and draughted a more concise Regulation for the guidance of the Gas Maker and Lighthouse Keeper at the upper Light House than those which I found in existence. I have since received from Mr. Wigham also a draft of Regulations by which the Light Keeper is to be guided in manipulating the used Gas Burner. This I had drawn up to prevent, in future, any misunderstanding as to ¾ cock or full cock.[26]

I append these Papers marked respectively:—A. B. C. D. & E.

It was but natural that a fear should be entertained, that the powerful new Gas Light might injure the valuable and delicately constructed Light House apparatus. In accordance with the desire expressed by the Visiting Committee Mr. Douglass was good enough to note the temperature of the Lantern in the High Lighthouse. He had handed to me the following observations:—

Night of April. 1st.:—

Mean temperature of Lantern outside the Lenticular[27] Apparatus:—560.F (between 8.30 & 11 o'Clock)

Temperature inside the Lenticular Apparatus:—

48 1st Burner	850. F
68 "	950
88 "	1070
108 "	1240
148 "	about 1320

Night of April 2d.

Mean temperature of Lantern outside Lenticular Apparatus:—620 F.

Temperature inside the Lenticular Apparatus

28 1st burner	820 F.
48 " "	890 "
68 " "	980 "
88 " "	1260 "
148 " "	above 1320 "

Mr. Wigham performed the like Service for me on the night of April 2nd. at the Lower Light House. I quote his observations:—

Temperature of Lantern inside—Lenticular Apparatus—

at 8 oClock	P.M.	830F.
9 "	"	Do.
10 "	"	850
12 "	"	840
2 "	"	830
4 "	"	840
5.30 "	"	820

Temperature of Lantern outside Lenticular Apparatus

at 8.	o'Clock	P.M.	620
9	"	"	620
10	"	"	630
12	"	"	640
1	"	"	640
5.30	"	"	620

As these observations are in harmony with what you observed during the trial of the Wigham Lamp at Howth Bailey Lighthouse[28] in 1869, any fear on the score of injury to the dioptric Apparatus may be confidently dismissed.

It will appear from the searching investigation to which the new Burner was subjected that it is especially adapted for High Power Lights and for Signal and flashing Lights. Its greatest flexibility of Photogenic Power constitutes, in fact, its chief advantage. Being a new method of lighting up Light Houses and requiring new appliances, it is obvious that the manufacture of the gas and the management of the Lights should, for some time to come, be specially looked after, and to this end I would strongly recommend to your notice the system of checks by monthly returns to which I devised conjointly with M^r^. Douglass and M^r^. Wigham and which is set forth in the paper which I append.

I have the honor &c &c | (S[igne]^d^.) | W^m^. Valentin | (Principal Demonstrator | of Practical Chemistry)[29]

National Archives, MT 10/220. Illumination of lighthouses by gas. H7863

---

1. *Haisbro' High Lighthouse*: one of two lighthouses, the high and low, at Haisborough (pronounced Haisbro) Sands on the Norfolk coast, which were opened in 1791. Tyndall had inspected them in April 1873 to determine whether oil or coal gas-burning lights were the most suitable for lighthouses.
2. *bar Bunsen Photometer*: an instrument designed by Robert Bunsen for comparing the luminosity of a candle and a gas flame. The two light sources are fixed at either end of a graduated bar, allowing the observer to measure the distance at which the illuminating effect is equal, from which the luminosity can be calculated.
3. *Sperm candles*: made from spermaceti, a waxy substance that naturally occurs in the head cavities of sperm whales. This substance was prized for burning brightly without odour, and was used to produce candles of a standard photometric value. At this time, 'candle-power' was the unit of measurement for light intensity, defined by the light produced from a pure spermaceti candle of a standardised weight, burning at a set rate.
4. *the Light Committee of the Trinity House*: a committee with specific responsibility for the maintenance of lighthouses.
5. *the Galatea Steam yacht*: see letter 4187, n. 3.

6. *Colza oil*: derived from the seeds of rapeseed, it was considerably cheaper than sperm oil and widely used for domestic lighting before the adoption of coal gas.
7. *the Gentlemen of the Light Committee*: John Sydney Webb (1816–98), who was the chairman of the committee, Edward Parry Nisbet (1810–99), and Richard Collinson (1811–83) (see letter 4438).
8. *A fish-tail gas burner*: a gas burner that produces a spreading flame shaped like the tail of a fish.
9. *C. F.*: cubic feet, the standard measurement for gas volume.
10. *photometric standard*: i.e., the fixed value against which the luminosity of the new gas lamp was compared.
11. *mica*: mineral with a number of industrial applications due to its unique properties.
12. *A governor*: a valve that provides a steady flow of a fluid.
13. *dioptric*: assisting vision by refracting and focusing light (*OED*).
14. *lava-cone*: 'lava' was another name applied to soapstone, a type of rock with various uses, as it is heat resistant and soft enough to carve easily.
15. *M^r^. Edmundson*: Wigham was the owner of J. Edmundson & Co., a flourishing business in Dublin that provided gas plants designed by himself. He had taken on responsibility for the company upon the death of his brother-in-law Joshua Edmundson (d. 1848), which had left Wigham responsible for his sister, who was Edmundson's widow, and her children. It is therefore likely that the Mr. Edmundson referred to here is one of Wigham's nephews acting as an apprentice or junior partner in the business.
16. *argand burner*: an Argand lamp, named after its inventor, the Swiss physicist and chemist Ami Argand (1750–1803). It was brighter and burned more efficiently than earlier oil lamps.
17. *M^r^. Kelsey the principal Light Keeper*: William Kelsey, who served in this role from 1871–7.
18. *Cannel Coal*: a type of bituminous coal, also known as candle coal, used in the production of coal gas. This gas was used for lighting as it burned brightly.
19. *cwt*: abbreviation for hundredweight, an imperial unit of weight or mass. There are twenty hundredweights in an imperial ton.
20. *Banking (breeze) Coal*: a particularly fine, granulated form of coke, which is a fuel produced by heating coal in the absence of oxygen.
21. *qr.*: abbreviation of a quarter, an imperial unit of weight or mass. A quarter of a hundredweight.
22. *The Marquis of Lothian*: Schomberg Henry Kerr (1833–1900), 9th Marquess of Lothian, who owned coal mines in the area surrounding Edinburgh.
23. *Lime*: calcium oxide.
24. *per chaldron (36 bushels)*: a chaldron was a unit for the measurement of coal in volume, though it was by no means standardised. A 'London chaldron' was defined as thirty-six bushels (another imperial unit of measurement for the volume of dry goods).
25. *the secretary of the Trinity House*: Robin Allen.
26. *¾ cock or full cock*: the flow of gas, and therefore the brightness of the flame, was regulated by a stopcock, a valve that the lighthouse keeper could manipulate manually.
27. *Lenticular*: having the double convex form of a lens (*OED*).
28. *Howth Bailey Lighthouse*: the Baily Lighthouse on the Howth Head, in the bay of Dublin

on the east coast of Ireland. It had originally been built in 1667, and was rebuilt in 1814. It was here, in 1869, that Wigham's newly-patented gas-burning light was first installed as an experiment, and proved successful.

29. *(Principal Demonstrator | of Practical Chemistry)*: at the Royal School of Mines and Science Training Schools, South Kensington.

To Joseph Henry 1 May 1874 4315

British Association for the Advancement of Science | 22 Albemarle Street, London, W. | 1st May, 1874

Sir,

We are directed by the British Association for the Advancement of Science, to announce to you that the Forty Fourth Meeting of the Association is appointed to be held at Belfast on Wednesday the 19th of August, 1874, under the Presidency of Professor Tyndall, D.C.L., L.L.D, F.R.S., and to express the earnest desire of the members of the Association to be honoured by your presence on the occasion of this its second visit to Belfast.

The Officers of the Association hope to be supported on this, as on many previous occasions, by the personal assistance and written contributions of the Philosophers[1] of other Countries, and they gladly undertake to make preparation for the convenient reception of those distant friends and Associates who may honour them by accepting this invitation and giving notice of their intention to be present at the Meeting.

We have the honour to be, | Sir, | Your obedient Servants, | John Tyndall President Elect. | Douglas Galton[2] | M. Foster[3]} General Secretaries. | George Griffith, Assist[ant]. Gen[era]l Secretary

Professor Henry, | Washington, U.S.

Smithsonian Institution Archives. RU 26, vol. 141, p. 210

1. *Philosophers*: those who pursue natural philosophy, denoting science in its broadest sense.
2. *Douglas Galton*: Douglas Strutt Galton (1822–99), a British engineer, served as the general secretary of the BAAS from 1871–95, and in 1895 became its president. A graduate of Royal Military Academy, Woolwich, Galton was commissioned second lieutenant in the Royal Engineers in 1840. He joined the British Ordnance Survey in 1846 and the following year was appointed secretary of the railway commission. In 1862 he was appointed Assistant Permanent Under-secretary for War and in 1869 director of public works and buildings. He was elected FRS in 1859, and joined the BAAS in 1860. Galton was active in the efforts of the Society for the Aid to Sick and Wounded in War (now the Red Cross) to assist the wounded of the Franco-Prussian War. The biostatistician and eugenicist Francis Galton was his cousin (*ODNB*).
3. *M. Foster*: Michael Foster (1836–1907), a British physiologist and politician, served as general

secretary to the BAAS from 1872–6. Foster studied medicine at UCL before working for several years as a ship's surgeon and in private practice. He began teaching medicine and physiology in 1867 and worked as a demonstrator for Thomas Huxley. In 1869 he became Huxley's successor as Fullerian Professor of Physiology at the RI. He was named Professor of Practical Physiology at Trinity College, Cambridge in 1870, where his evolution-based research program came to be known as the Cambridge School of Physiology. Foster was instrumental in establishing physiology as a profession in Britain (*ODNB*).

From James Nicholas Douglass 1 May 1874 4316

Engineer's Office. | Trinity House.[1] London, E.C.[2] | 1st May 1874

Dear Dr Tyndall,

I have at last succeeded in getting a few of the Argand gas Burners[3] made that I showed you in June last and I now in part fulfillment of the promise then made send you two fitted as reading lamps for which you will find them well adapted the flame being exceedingly sturdy. I shall be very glad to send more if they will be of any service to the Institution[4] at the same time I have no desire to enroll my name on the list of Innovators of Gas Burners, all I have done is to improve upon the work of others and with the same accessories.

Yours very truly | James N Douglass

RI MS AD/10/C/04/C/03/1

1. *Trinity House*: see letter 4187, n. 2.
2. *E.C.*: Eastern Central, a postal district of London.
3. *the Argand gas Burners*: the original Argand lamp, designed by the French physicist and chemist Ami Argand (see letter 4314, n. 16), utilized a circular wick that permitted air to be drawn through the centre of the flame, thus ensuring more efficient combustion. Douglass developed a gas-burning lamp that employed a similar principle, with concentric rings of flame.
4. *the Institution*: the RI.

From Hector Tyndale 1 May 1874 4317

Philadelphia May 1st 1874

My dear John

I arrived safely at home[1] and found all well. A day or two since a friend of mine sent me a letter to him from a correspondent, a Broker, in New-York, who knew nothing of my being a questioner in the matter—and I give you a quotation from that letter. You remember we spoke on this matter[2] in London[3] and I said I would write you on the subject.

"To day a friend called upon the Secretary of the Road,[4] who reports that 'There is no market value here for the bonds of the Missouri, Kansas, and Texas R[ail]. Road,[5] but in Amsterdam they are quoted 42 to 43, which is equivalent in our currency (paper) to about 48½. About February 1873 they were worth about 85 currency. The highest point they are quoted and known to have been sold at is about 93 currency."

The two or three last annual reports he could not give exactly, but on the close of the last fiscal year, March 31, 1874, the report is as follows. 'Approximate. The Gross earnings for fiscal year, ending March 31/74—$3.500.000. Operating expenses will not run over 55 per cent—which, on say $3.220.000. of 6% bonds and $14.180.000. of 7% bonds, will pay interest and leave a small surplus. The word approximate is used because the returns for the month of March/74 are not all in' These figures I think your friend can rely on. You must be aware that all these outside things have only a quotable value here, which varies every day 1. 2. 5. or 10% according to the amount wanted to buy or sell. Should you want to sell, let me know and I will get a broker, who knows the market in such securities, to transact the business. It is pretty much like buying or selling some of the Southern State or Municipal Bonds—one day some one wants to buy, the next some one wants to sell and prices vary accordingly".

Another friend of my own, a Banker, told me that he "would not have purchased securities of the M. K. and T. RR. Co:[6] as they were of a fluctuating and uncertain character, but, that if he held them, he would not sell them now"; but "that if ever they reached anything like original cost he would sell"—solely on the ground of their uncertainty and the distant locality of the Road—which is not much known here.

Some days since a friend of mine—my Lawyer here[7]—asked me to give a letter introducing his Nephew to you. This is a very usual American habit. Except that I am under compliments and a sense of service to my Lawyer, Mr Spencer, I should have refused. As it is, I have taken upon myself to give a letter[8] for you to his Nephew, Doctor George H. Horne,[9] a Physician of this City, who is about to go to England for the first time, starting about a week hence. I ask your kind attention to him when he calls upon you, which he probably will do within a few weeks, from receipt of this. I cannot ask any attention à l'Americaine[10] for him, but if you can introduce him to men of Science, especially in his own particular study (outside of medicine) which is Entomology, I will feel obliged to you. He would, no doubt, like to meet some celebrities. Dr Horne is a member of our "American Phil[osophical]: So[ciety]: of Philadelphia"[11] and I knew him there as a fellow member and, better, as the nephew of my friend, to whom as said I am somewhat indebted. Dr. Horne has published a number of papers,[12] entomological, and is known in Europe as here. He is a gentlemanly and well connected young man of earnest character. Give my regards to Hirst & Debus who I hope are well.

Affectionately yours | Hector Tyndale

RI MS JT/1/T/76
RI MS JT/1/TYP/5/1729–30

1. *arrived safely at home*: from Tyndale's visit to Britain (see n. 3).
2. *this matter*: presumably the potential risks and rewards of investing in American railway company stocks. Tyndall's lecture tour of the United States in 1872–3 made a surplus of more than $13,000, which was invested to establish a fund for pure science research in American universities (see letter 4214, n. 5). Tyndale was instructed to 'carefully invest it in permanent securities' in order create an endowment for a scholarship fund for physics students ('Professor Tyndall's Deed of Trust', *Popular Science Monthly*, 3 (1873), pp. 100–1, on p. 101).
3. *in London*: Tyndale visited Tyndall in London in late February (see letters 4278, 4279, and 4282).
4. *the Secretary of the Road*: not identified.
5. *Missouri, Kansas, and Texas R[ail]. Road*: the Missouri-Kansas-Texas Railroad Company, which was incorporated in 1870 with the aim of building a railway to connect various military bases on the Western frontier, for which purpose it received various government land grants.
6. *M. K. and T. RR. Co*: see n. 5.
7. *a friend of mine—my Lawyer here*: not identified.
8. *a letter*: letter missing.
9. Doctor George H. Horne: George Henry Horn (1840–97), an American physician and entomologist (*ANB*).
10. à l'Americaine: in the American manner (French).
11. *"American Phil[osophical]: So[ciety]: of Philadelphia"*: the oldest learned society in the United States, founded in 1743.
12. *a number of papers*: Horn had published more than fifty scientific papers at this time; his most recent was 'Descriptions of New Species of United States Coleoptera', *Transactions of the American Entomological Society*, 5 (1874), pp. 20–43.

To Emily Tennyson[1] 2 May [1874][2] 4318

2nd May

My Dear Mrs. Tennyson,

Might I ask you to give yourself the trouble to put that little paper[3] in an envelope and send it to me? It is so simple, & so important that I think I will publish it.

Yours ever faithfully | John Tyndall

You made our visit to the Island[4] very bright.

TRC Letters 6247

1. *Emily Tennyson*: Emily Tennyson (née Sellwood, 1813–96), wife of the poet Alfred Tennyson.
2. *[1874]*: the year is given in the archival listing of the TRC, presumably based on a postmark on the envelope.
3. *that little paper*: not identified.
4. *our visit to the Island*: the Isle of Wight, where the Tennysons had lived, at Farringford House in the village of Freshwater Bay, since 1853. Tyndall had presumably visited them in April, when staying on the island with John Lubbock and his family (see letter 4308). He had also made an abortive visit to them in December 1873 (see letter 4231).

From James Coxe 3 May 1874 4319

Kinellan[1] | Edinburgh | May 3. 1874

My dear Dr Tyndall,

I have delayed for a day or two replying to your note[2] of the 29th ult.[3] that I might have time for consideration. And I now advise you not to think of replying to Tait.[4] Perhaps you have seen the article in the Scotsman[5] which refers to the exhibition he made of himself[6] on the occasion you refer to, but in case you have not I send you a copy. Scarcely one of the young men present would attach the smallest value to what he said, and fewer still would understand his allusions. The students regard such addresses simply as sources of fun, and they enjoy them accordingly, but I doubt if a single one of them bestow a thought upon the matter after it is over. For you to take any notice of it[7] would be to cause Tait to dance with joy, in the belief that he had found a vulnerable slit in your armour. To treat him with silent contempt is, I have not the smallest hesitation, your proper course. It is a saying which I have repeatedly heard, that to mention the name of Tyndall in Tait's presence is like fluttering a red rag before a bull. He loses all self command and self respect, and at once sets off in a furious headlong career, with no control over thoughts or language. To try to stop him would merely show that he had succeeded in galling you. So I hope you will not think of it. If you ever dream of replying to him you must choose a more worthy occasion

My wife[8] is looking forward with pleasure to the arrival of the Philadelphian portrait.[9] She possesses a good many likenesses of you and cherishes them all not even excepting that of Vanity Fair!![10] I hope we shall be lucky enough to find you in Albemarle Street[11] when we pass through London some weeks hence. Always most

truly Yours | J. Coxe

My love and blessing—Likeness <u>not</u> come | M.A.C.[12]

RI MS JT/1/TYP/1/299
LT Typescript Only

1. *Kinellan*: Kinellan House in Murrayfield, an area to the west of Edinburgh.
2. *your note*: letter missing.
3. *ult.*: abbreviation of ultimo, last month (Latin).
4. *Tait*: Peter Guthrie Tait.
5. *the article in the Scotsman*: '[Editorial]', *Scotsman*, 24 April 1874, p. 4.
6. *the exhibition he made of himself*: in an address to the University of Edinburgh on 22 April, Peter Guthrie Tait made unusually personal and derogatory remarks about Robert Lowe (1811–92)—who had been Chancellor of the Exchequer and then Home Secretary in William Gladstone's government until its defeat in the general election in February 1874—albeit without naming him directly. Tait referred to him as an 'ex-Minister' whose taxation policies had 'aggravate[d] the miseries of numbers of the industrious poor'. He also accused Lowe of being 'ready, for the sake of a sorry play upon words, to aggravate the miseries of . . .the industrious poor' ('University of Edinburgh', *Scotsman*, 23 April 1874, p. 3). The editorial from the same newspaper that Coxe sent Tyndall explained this last comment thus: 'That is to say, Mr. Lowe's object in proposing the Match-tax was not to obtain revenue from a superabundant article from which revenue is successfully obtained in other countries, but only to perpetuate a Latin pun (*ex luce, lucellum* [out of light, a little profit]), which object he regarded as so gigantic as to render of no moment the inflicting of new miseries on the miserable'. See also letter 4355, n. 2.
7. *to take any notice of it*: despite Coxe's advice, Tyndall quoted from the *Scotman*'s editorial in 'Rendu and His Editors', *Contemporary Review* 24 (1874), pp. 135–48, on p. 144 (see letter 4355).
8. *my wife*: Mary Anne Coxe.
9. *the Philadelphian portrait*: a photographic portrait of Tyndall taken by Frederick Gutekunst at his studio at 712 Arch Street, Philadelphia. It was taken in January 1873 during Tyndall's lecture tour of the United States, and features as the frontispiece to *Tyndall Correspondence*, vol. 13.
10. *that of Vanity Fair!!*: [A. Cecioni], 'Men of the Day, No. 43: Professor John Tyndall, FRS', *Vanity Fair*, 7 (1872), p. 111. This was a caricature of Tyndall.
11. *Albemarle Street*: the RI was located at 21 Albemarle Street.
12. *Likeness <u>not</u> come | M.A.C.*: presumably a postscript written by Coxe's wife referring to the anticipated arrival of Tyndall's portrait.

From George Wynne 3 May 1874 4320

Sydenham[1] | 3/5/74

My dear Tyndall,

I must not let you hear of the step I am about taking first thro' the newspapers, you know the happy domestic life I had first with my wife[2] and afterwards with my daughter,[3] she is now married and temporarily settled in the far west of Ireland and my three boys[4] are in India & Germany and I am alone in the world and very lonely. If you take all this into consideration you will not be so surprised as you otherwise might be to hear that I am going again to seek happiness in married life. The Lady who is kind enough to take me is a Miss Darrah[5] the daughter of a Colonel Darrah[6] many years dead. Her mother and one married sister[7] is all that remains of her family. I have known Miss D. very intimately a great many years and the only fault I know in her is being a great deal younger than I am. I am sure I shall have your best wishes for our happiness. Lucy quite acquiesces in the step I am taking, as do also the members of my own family as well as of my wife's. I go down to Hastings to-morrow where Miss D. is staying and we are to be married on Tuesday[8] and go at once abroad. As soon as the season is sufficiently advanced we shall go to Switzerland where I hope we may meet you.[9] Till then farewell

most truly yours | Geo Wynne

RI MS JT/1/W/100
RI MS JT/1/TYP/5/1869

1. *Sydenham*: Wynne was 'living' here 'with Uncle C. Wynne' (see letter 4240).
2. *my wife*: Anne Wynne (née Osborne, 1808–64).
3. *my daughter*: Lucy O'Brien.
4. *my three boys*: Henry Le Poer Wynne, Edward Toler Wynne (1837–89), and Francis George Wynne.
5. *Miss Darrah*: Henrietta Jane Darrah (1841–1917).
6. *Colonel Darrah*: Nicholas Lawson Darrah (1788–1851).
7. *Her mother and one married sister*: Jane Darrah (née Luck, *c.* 1819–81) and Susan Catherine Darrah (b. 1837; married name not identified).
8. *Tuesday*: 12 May.
9. *to Switzerland where I hope we may meet you*: Tyndall intended to spend the early summer in Switzerland.

From Robert G. H. Kean[1] 4 May 1874 4321

Lynchburg, Va.[2] | May 4th 1874

Professor John Tyndall

Dear Sir,

Your very courteous letter of April 14th[3] was received a few days ago. Immediately on getting it, I wrote to Mr. Robert M Brown,[4] of Amherst

Court-House[5] a lawyer (barrister) of high character and intelligence, and requested him to make inquiry among persons of character, whether the cannonade of the first battle of Cold Harbor,[6] July 27th, 1862 was heard there, for I could not remember who my informant was, though I did remember as I wrote you, that it was stated by a highly credible source. Mr. Brown's answer is before me. He says "you can assure Prof. Tyndall that this cannonade we distinctly heard, and remarked on by myself and the other inmates of my household, and was ascribed at the time to the stillness of the atmosphere and the peculiar situation of Amherst Court-House where I then (as now) resided, which, although at least 100 miles in a straight line from the battlefield is upon a tributary of James River." His supposition was that the sound had followed the valley of the river.

Mr Brown's letter assists my memory in regard to the generation of the wind. Independently of his statement I would have stated that so far as my memory serves me, the wind, if there was any, was slight. About its direction I cannot speak with certainty but my impression,—(a vague and unreliable one), is, that the smoke of the guns drifted up the Chickahominy[7]—which is West, or in the direction of Amherst C[our]t. Ho[use]. and this would have been diagonally across my line of vision. But I cannot affirm distinctly on this point. I suppose my distance from the nearest field battery engaged in the fight was from 2 to 2 ½ miles in an air line.

Rifled guns in Earthworks on the crest where I stood, were firing upon those I saw and could not hear—a circumstance which assists my statement of the distance an estimate partly by the eye, and partly from my having rode over it a day or two later, from the same point where I stood on the evening of the battle, to the scene of conflict.

If you see cause you are at liberty to make such use of my letters as you please.[8] And it may serve abroad to give it some additional credit (if you should use it) to mention, that in the files of the Treasurer's office,[9] will be found evidence that I had occasion as Rector of the University of Virginia to transmit the acknowledgments of that institution to Her Majesty's Government, for the gift to the Library of the V[irgini]a University of the publications of her Record Office—made in 1873.[10]

I have the honor to be | Respectfully | Your ob[edien]t. Serv[an]t. | R. G. H. Kean

RI MS JT/1/K/4

1. *Robert G. H. Kean*: see letter 4294, n. 1.
2. *Va.*: Virginia.
3. *Your very courteous letter of April 14th*: letter missing.
4. *Mr. Robert M Brown*: Robert Meriwether Brown (1814–94).

5. *Amherst Court-House*: in the town of Amherst, Virginia.
6. *the first battle of Cold Harbor*: the Battle of Gaines' Mill (see letter 4294, n. 9 and n. 10).
7. *Chickahominy*: a river flowing through the east of Virginia.
8. *make such use of my letters as you please*: Tyndall published Kean's previous letter (letter 4294) in *Sound* (1875), pp. 274–6, and added a note to it stating 'I learn from a subsequent letter that during the battle the air was still.—JT' (p. 276).
9. *the Treasurer's office*: the Virginia Department of the Treasury, responsible for the state's finances.
10. *the gift to the Library . . . made in 1873*: in 1873 the British government gifted the University of Virginia more than two hundred volumes of the publications of the Record Commissions, mainly consisting of historical records relevant to the state.

From Mary Anne Coxe 8 May [1874][1] 4322

Friday May 8th

It[2] has come!

And admirable—1000 thanks—

When I say admirable I mean it is you in repose—an attitude one seldom sees in the living man and in any picture of you, one would look in vain for "the light that never yet on sea or shore—the incarnation of the Poet's dream"[3] that the fire of genius gives your living face.

Still, this is by much the best representation I have of you, I am very glad to have it, and thank you exceedingly. It was only last night I got it—it shall go on Monday to be framed, I can't part with it sooner.

We all join in warmest Salaam,[4] and with special thanks from me, believe me now and always

affectionately Yours | M. A. C.

RI MS JT/1/TYP/1/1/300
LT Typescript Only

---

1. *[1874]*: the year is established by the relation to letter 4319.
2. *It*: a photographic portrait of Tyndall taken in January 1873 by Frederick Gutekunst which Coxe, who collected likenesses of Tyndall, had been impatiently awaiting (see letter 4319, n. 9).
3. *"the light that never . . . the Poet's dream"*: a paraphrase of lines from William Wordsworth's poem 'Elegiac Stanzas, Suggested by a Picture of Peele Castle in a Storm' (1807). The original lines are: 'The light that never was, on sea or land | The consecration, and the Poet's dream',ll. 15–6.
4. *Salaam*: greetings (Arabic).

From James Coxe 10 May 1874 4323

Kinellan[1] | Edinburgh | May 10th 1874.

My dear Dr Tyndall,

I have the book[2] but am unable to see in it any cause for wrath. I have therefore given it to a friend[3] who has the means of ascertaining what precise portions are objected to, to get them pointed out, and then we shall see how the land lies and the wind blows. In the meantime I am completely at sea and at fault. Of course my friend gets the information as for himself.

I understand that Tait wrote a scurrilous article for the Fortnightly Review,[4] which was put in print, and afterwards refused publication by MacMillan.[5] If you happen to know anything about it, I should be glad to learn what prevented the publication.

My wife[6] greets you lovingly. She would explain the delay in the photograph[7] reaching her. She thinks it very good, but that it does not give your bright and animated expression. But what photo could? I wish she were in stronger health. However I hope we may get away for a change before this month is out, and trust it will set her up.

most truly Yours | J. Coxe.

RI MS JT/1/TYP/1/301
LT Typescript Only

---

1. *Kinellan*: Kinellan House in Murrayfield, an area to the west of Edinburgh.
2. *the book*: probably *Life of Forbes*. Coxe and Tyndall were corresponding about Peter Guthrie Tait's role in reviving the dispute over James David Forbes's claims to scientific priority on theories of glacial motion (see letter 4319), and Coxe, as letter 4324 suggests, was not yet aware of the other book relating to the controversy that Tait had contributed to, *Glaciers of Savoy*.
3. *a friend*: not identified.
4. *a scurrilous article for the Fortnightly Review*: Tait actually submitted this article, entitled 'Forbes and Dr. Tyndall', to the *Contemporary Review*, rather than its rival the *Fortnightly Review*. The article was not published in either journal, and was instead included in the new translation of *Glaciers of Savoy*, pp. 163–98, with a note stating: '*Written (by arrangement with the Publisher) for the "Contemporary Review": of November 1873; but, after having been put in type and corrected for press, considered unsuitable for its pages by the Editor of that Journal*' (p. 163).
5. *MacMillan*: Alexander Macmillan, although he was not involved with either the *Fortnightly Review* or the *Contemporary Review*. As Tyndall explains in letter 4324, *Macmillan's Magazine*, of which Macmillan was the publisher, also declined Tait's article.

6. *My wife*: Mary Anne Coxe.
7. *the photograph*: see letter 4319, n. 9.

To James Coxe 12 May [1874][1] 4324

Royal Institution of Great Britain | 12th. May.

My dear Sir James

The article[2] I am informed was written for the "Contemporary Review" but refused admission by the editor.[3] It was then presented to Macmillan,[4] and also I am informed refused admission in his Magazine,[5] though he is the publisher of the Life of Principal Forbes.[6] I suppose as a matter of compromise Macmillan agreed to publish, and has published, quite recently a book at the expense and risk, I am informed, of Messrs Forbes and Tait, or of Forbes alone, in which the article is incorporated[7]—whether in its original form or not I do not know.

If you care to have the book, which I have not yet read, I will send it to you.

And to complete your knowledge I send you two copies of a pamphlet which I wrote in reply to their attack upon me.[8]

I do most earnestly hope that the coming warm weather will restore my friend[9] to health and energy.

Ever yours most truly | John Tyndall.

You are a German Scholar and will understand the following words of Helmholtz, the intimate friend of Sir William Thomson

[...][10]

This is not for the public, as I do not wish to involve Helmholtz in these discussions. I could show you similar language from Clausius.[11] In fact your professor[12] has rendered himself notorious throughout the world.

RI MS JT/1/TYP/1/302
LT Typescript Only

---

1. *[1874]*: the year is established by the relation to letter 4323.
2. *The article*: 'Forbes and Dr. Tyndall', written by Peter Guthrie Tait (see letter 4323, n. 4).
3. *the editor*: James Thomas Knowles, who edited the *Contemporary Review* from 1870–7. Knowles may have let Tyndall see the proof of Tait's rejected article (see letter 4204, n. 2).
4. *Macmillan*: Alexander Macmillan.
5. *his Magazine*: *Macmillan's Magazine*.
6. *the Life of Principal Forbes*: *Life of Forbes*. It was this book that had intensified the controversy between Tyndall and Forbes's supporters.

7. *a book at the expense . . . article is incorporated*: *Glaciers of Savoy*. Tait's article was included at pp. 163–98. As Tyndall suggested, the translation was published at the expense of Forbes's son, George.
8. *a pamphlet which I wrote . . . attack upon me*: J. Tyndall, *Principal Forbes and his Biographers* (London: Longmans, Green, and Co., 1873), which was a reprint of the article of the same name published in the *Contemporary Review*, 22 (1873), pp. 484–508.
9. *my friend*: Mary Anne Coxe (see letter 4323).
10. *[. . .]*: Tyndall extracted a passage from letter 4192 in the original German, which, in translation, begins with: 'I have to say to my regret that our Scottish friends . . .', and concludes with '. . . Sir W. Thomson unfortunately gets drawn into them too easily'.
11. *similar language from Clausius*: letter missing, although it was presumably a reply to letter 4300.
12. *your professor*: i.e. Tait, who was professor of natural philosophy at the University of Edinburgh, where Coxe had gained his medical degree in 1835.

To Thomas Henry Huxley 14 May [1874][1] 4325

Royal Institution of Great Britain | 14th. May.

My dear Huxley

Macmillan[2] is sure to have sent you a copy of Rendu's Glaciers of Savoy[x] with Tait's Article.[3] I wish you would consider what ought to be done with it. Perhaps Hirst and you could lay your heads together and discover the best plan, or whether any plan is needed. I have been shaky for the last two or three weeks and run into the country for a breath of air to day.

You know of course that Tait has flooded Belfast with Ruskin's diatribe.[4]

I can smash them, but it is an awful waste of time. Still it appears to me <u>something</u> must be done.

Yours ever | John Tyndall

(x) because you are mentioned by name and otherwise referred to in it.[5]

RI MS JT/1/TYP/9/3030
Typed Transcript Only

---

1. *[1874]*: the year is established by the relation to letter 4327.
2. *Macmillan*: Alexander Macmillan.
3. *a copy of Rendu's Glaciers of Savoy[x] with Tait's article*: *Glaciers of Savoy*. Peter Guthrie Tait's article 'Forbes and Dr. Tyndall' was included at pp. 163–98.
4. *Tait has flooded Belfast with Ruskin's diatribe*: John Ruskin had attacked Tyndall for his part in the dispute over glacial movement in *Fors Clavigera*, vol. 3 (1873), letter 34, pp.

1–32 (see letter 4185, n. 13). Tait had then published extracts from this in two Belfast newspapers (see letter 4202, n. 6).

5. *you are mentioned . . . otherwise referred to in it*: Huxley was mentioned by name twice (*Glaciers of Savoy*, pp. 10 and 11).

To Thomas Henry Huxley — 14 May [1874][1] — 4326

Royal Institution of Great Britain | 14th. May.

My dear Huxley,

I sent a short note[2] down to you to day which missed you.

I hope you are not too opposed to do a little bit of business for me.

Macmillan[3] doubtless sent you the book by Tait & co.[4]

What is to be done with it? Knowles[5] is willing that either you or I should review it. But if the review is to be published in the June number he would require copy on the 20th.

Unluckily before the book reached me I had pledged myself to go into the country—which I need as I am very shaky—I am pledged to a friend[6] to go to day. I think while away I could strike off something—but do you think it ought to be done?

Yours ever | John Tyndall

IC HP 8.161
Typed Transcript Only

1. *[1874]*: the year is established by the relation to letter 4327.
2. *a short note*: letter 4325.
3. *Macmillan*: Alexander Macmillan.
4. *the book by Tait & co*: *Glaciers of Savoy*.
5. *Knowles*: James Thomas Knowles, editor of the *Contemporary Review*.
6. *a friend*: not identified.

From Thomas Henry Huxley — 14 May 1874 — 4327

May 14th 1874.

My dear Tyndall,

I have been at the Royal Commission[1] all the morning so that I had hardly read your first note[2] when your second[3] arrived.

Macmillan[4] sent me the book[5] and I just glanced at it on my return from Cambridge.[6] What is to be done about it is a matter on which I wanted to have a talk with you. But I should say, do nothing in a hurry, and I would not bother about it at present. A good answer to appear some time in July would be effective. I cannot dream of attempting anything serious at present, as I have Royal Commission work, Examinations for the Department[7] and a course of lectures to schoolmasters[8] here, all on my hands in the course of the next two months, to say nothing of the Belfast Lecture.[9] But I suppose that sooner or later I must publish the document laid before the Royal Society[10] and perhaps republish the article in the Westminster.[11]

But as I said before I do not think that there is any hurry and we ought to think the matter well over before anything is done.

When will you be back that we may have a caucus?

Go and get freshened up and we will smite the Amalekites[12] even to the gates of Edinbro'

Ever Yours faithfully | T. H. Huxley.

IC HP 8.162
RI MS JT/1/TYP/9/3031
Typed Transcript Only

1. *the Royal Commission*: the Royal Commission on Scientific Instruction and the Advancement of Science, which sat from 1870 to 1875.
2. *your first note*: letter 4325.
3. *your second*: letter 4326.
4. *Macmillan*: Alexander Macmillan.
5. *the book*: *Glaciers of Savoy*.
6. *return from Cambridge*: Huxley had gone to Cambridge on 9 May to visit Michael Foster and make arrangements for the marking of examination papers (*Foster and Huxley Correspondence*, p. 48).
7. *Examinations for the Department*: the Science and Art Department, based at South Kensington; Huxley was responsible for arranging the marking of 'about 7000 papers' on biology (p. 47).
8. *a course of lectures to schoolmasters here*: Huxley's regular course of lectures to schoolmasters at the Normal School of Science in South Kensington, which he gave every summer.
9. *the Belfast lecture*: 'On the Hypothesis that Animals are Automata, and Its History', which Huxley delivered as the evening lecture on Monday 24 August, at 8:30pm in Ulster Hall, at the BAAS meeting in Belfast (*Brit. Assoc. Rep. 1874*, p. lxv).
10. *the document laid before the Royal Society*: possibly J. Tyndall and T. H. Huxley, 'On the Structure and Motion of Glaciers', *Phil. Trans.*, 147 (1857), pp. 327–46, which Huxley may have been considering republishing in a more accessible form.
11. *republish the article in the Westminster*: presumably [T. H. Huxley], 'Glaciers and Glacier Theories', *Westminster Review*, 11 (1857), pp. 418–44, although it does not seem to have been republished at this time.

12. *smite the Amalekites*: the mortal enemies of the Israelites, whose smiting of them is related in 1 Samuel 15:3.

From Eduardo Lozano[1] 18 May 1874 4328

Madrid 18 de mayo 1874.

Sr. D. Juan Tyndall.

Muy Sr. mio y de toda mi consideracion: Muchas veces he comenzado esta carta y otras tantas arrojaba la pluma, porque nada me autoriza para molestar á persona desconocida distrayéndola de graves ocupaciones, aun cuando por mi parte me viera impulsado por el mas sincero afecto y profundo respeto hacia el eminente Profesor, honra de su Patria, que reune á las excelentes dotes de explorador infatigable una clemencia y sencillez desconocidas hasta ahora en la exposicion de las verdades científicas.

Dispensad si no consultando otra razon que mi entusiasmo he incurrido acaso en vuestro desagrado al proponeros como Socio honorario de la Sociedad de Profesores españoles que se cree honrada con tan esclarecido nombre.

Repito de nuevo useis para conmigo de la indulgencia propia del Maestro pues me juzgaria dichoso si os dignais considerarme como vuestro mas humilde discípulo y entusiasta admirador,

S.S.Z.S.M.B. | Eduardo Lozano

Su casa en Madrid. | Rosas, Mercado, 2 *[1 word illeg]*.

Madrid, 18 May 1874.

Mr. Dr. John Tyndall.

Dear and highly esteemed Sir: I have begun writing this letter many times and so many more I have put down the quill, for nothing authorises me to molest a man whom I do not know, and to distract him from serious duties, even if I have always been moved by the most sincere affection and the most profound respect towards the outstanding Professor, an honour to his fatherland, who adds mercy and simplicity to his excellent gift as a devoted researcher in a manner that was hitherto unknown among students of scientific truths.

Please forgive me if, moved by nothing other than my enthusiasm, I may have importuned you by proposing you as an honorary member of the Society of Spanish Professors,[2] a society that will be honoured by counting on such an illustrious name.

I reiterate, please show me the indulgence characteristic of the master, for I would be pleased if you were so kind to see me as your most humble disciple and enthusiastic admirer,

Your most loyal and humble servant | Eduardo Lozano
His home in Madrid. | Rosas, Mercado, 2 *[1 word illeg]*.

RI MS JT/1/L/39

1. *Eduardo Lozano*: Eduardo Lozano (1844–1927), a Spanish physicist and chemist who held professorships in both Madrid and Barcelona.
2. *the Society of Spanish Professors*: not identified.

To Frederick Barnard 20 May [1874][1] 4329

20th May

My dear Dr. Barnard

The writer of the enclosed[2] is a famous Alpine man.[3] He was chosen first President of the Alpine Club.[4] He is to write the article for the Encyclo[pædia]. Britt[annica].[5]—He was formerly under secretary for our Colonies,[6] and is in the highest sense an accomplished man.

You may deem it a triumph to secure his cooperation—you do not I think want the article[7] in a hurry.

Yours ever | In great haste | John Tyndall

Columbia University Manuscripts; MS#1918 (Barnard Family Papers), Box 2; University Archives, Rare Book & Manuscript Library, Columbia University Libraries, Rare Book & Manuscript Library, Columbia University in the City of New York

1. *[1874]*: the year is suggested by reference to securing a writer for *Johnson's New Universal Cyclopædia*, which would begin publication in 1875 (see n. 7).
2. *the enclosed*: enclosure missing.
3. *a famous Alpine man*: John Ball (1818–89), an Irish politician and glaciologist (*ODNB*).
4. *first President of the Alpine Club*: Ball was the president of the Alpine Club from its foundation in 1857 until 1860.
5. *the article for the Encyclo[pædia]. Britt[annica]*: Ball contributed the entry for 'Glacier' to the *Encyclopædia Britannica*, 9th edn, 25 vols (Edinburgh: Adam and Charles Black, 1875–89), vol. 10, pp. 626–31.
6. *formerly under secretary for our Colonies*: in 1855 Ball was appointed Assistant Under-Secretary of State in the Colonial Department, holding the post until 1857.
7. *the article*: Ball contributed the entry for 'Glacier' to *Johnson's New Universal Cyclopædia*, 4 vols (New York: A. J. Johnson, 1875–8), vol. 2, pp. 557–62, which Barnard co-edited with Arnold Guyot.

To Charles Cecil Trevor 21 May [1874][1] 4330

21st. May

My dear Mr. Trevor.

Would it put the Board of Trade[2] to any inconvenience if I deferred my Report on Haisbro'[3] till after my return from Switzerland?[4]

The works are in good order & the gas is burning efficiently, but the Report will be fuller & better if postponed.

Yours faithfully | John Tyndall.

I find the question complicated by references to the last referees[5] who are extinct.

National Archives, MT 10/220. Illumination of lighthouses by gas. H6071

1. *[1874]*: Trevor annotated this letter 'I have replied no | 21 May 1874'.
2. *Board of Trade*: the short name, formalised in 1861, of The Lords of the Committee of the Privy Council Appointed for the Consideration of All Matters Relating to Trade, which was founded in 1622 and by the mid-nineteenth century had assumed an advisory capacity in relation to government policy on trade and economics.
3. *my Report on Haisbro'*: letter 4438. As chief scientific adviser to Trinity House (see letter 4187, n. 2), Tyndall had inspected the two lighthouses at Haisborough (pronounced Haisbro) Sands on the Norfolk coast in April 1873 to determine whether oil or coal gas-burning lights were the most suitable for lighthouses, and the report gave his conclusions in favour of the latter.
4. *return from Switzerland*: Tyndall would return from Switzerland, where he spent June and July, on 10 August.
5. *the last referees*: Robert Hogarth Patterson (1821–86) and John Sampson Peirce (d. 1904), who had been appointed to the Board of Metropolitan Gas Referees, under the aegis of the Board of Trade, after it was formed in 1868. They were replaced in August 1872 over a conflict of interests, with Tyndall being one of the replacements (see letter 4482, n. 3).

To Hector Tyndale 21 May [1874][1] 4331

21st May

My dear Hector

I have just time to thank you for your valuable letter[2] & to say that any friend of yours[3] will be always welcome to me.

This time I fear I shall be in the Alps[4] when your friend arrives: I am going there next week.

always yours aff[ectionate][ly] | John Tyndall

RI MS JT/1/T/1471
RI MS JT/1/TYP/4/1731

1. *[1874]*: the year is established by the relation to letter 4317.
2. *your valuable letter*: letter 4317.
3. *any friend of yours*: George Henry Horn (see letter 4317, n. 9), an American entomologist whom Tyndale had requested Tyndall make the acquaintance of. He was visiting London for the first time and wished to be introduced to other men of science who shared his interests.
4. *in the Alps*: Tyndall spent June and July in Switzerland, returning on 10 August.

From Gustav Wiedemann 21 May 1874 4332

Leipzig 21 Mai 74

Lieber Tyndall!

Alle Correcturen usf, die Sie in Ihren letzten Zeilen an meine Frau erwähnt haben, sollen in Ihrem „light" bestens besorgt werden. Ebenso werde ich, wie bei Ihren früheren Werken, die Correcturbogen vor dem Druck genau durchlesen, so dass das Werk möglichst Ihren Intentionen entspricht. Ich habe es so ziemlich genau durchgelesen und mit Freuden gesehen, dass es sich als vollkommen ebenbürtiges Drittes sich dem Schall und der Wärme anreiht und so in vortrefflicher Weise den Cyclus von physikalischen Vorlesungen vervollständigt. Wollen Sie nicht einmal die allgemeine Physik, die so interessant zu machen ist, und die Electricitätslehre vornehmen? Zu letzterer stehe ich Ihnen mit allem Rath gern zu Diensten, wenn Sie ihn je brauchen sollten.—Mein Buch ist im Manuscript fertig; im Druck fehlen noch etwa 8 Bogen, die hoffentlich bis Ende Juli in Ihren Händen sein werden.

Im Juli wird meine Frau nach Scheveningen gehen und ich will ihr Anfang August folgen. Ich hatte nun wirklich sehr grosse Lust, nach Belfast zu kommen; hoffentlich kommt nichts dazwischen und ich kann Sie nach so langer Zeit wieder einmal, und zwar gleich als Präsident begrüssen.—Wann werden Sie von London aufbrechen? Sind Sie wohl so freundlich, mir vorher etwa anzugeben, wie es sich mit Belfast verhält; ob man vorher Quartier bestellen muss usf, so wäre ich Ihnen sehr dankbar. Vielleicht machen Sie nach der Versammlung noch eine kleine Exkursion nach Killarney o dglen. Ich begleitete Sie dann sehr gern. Wurde meine Frau in Belfast nicht ungelegen kommen? Damen sind ja meist bei solchen Versammlungen nicht am rechten Ort. Event. wurde ich sie zu bereden suchen, mich zu begleiten; ich fürchte indess, sie scheut die Fahrt von Holyhead nach Dublin. Es giebt ja aber wohl noch eine directere kurzere Seefahrt nach Belfast von Schottland aus.

Beste Grüsse von meiner Frau.

In treuer Ergebenheit | stets | Ihr | G. Wiedemann

Leipzig 21 May 74

Dear Tyndall!

All the corrections and so forth that you mentioned in your latest lines to my wife[1] shall be taken care of in your "Light"[2] in the best manner. Likewise, I shall read the proof pages through carefully before printing, as with your earlier works, so that the work will live up to your intentions as much as possible. I have thus read it through quite closely and seen with delight that it follows on from Sound and Heat[3] as a completely equal third part, and thus completes the cycle of physical lectures in excellent fashion. Will you not be going to work on general physics—which is so interesting to do—and the theory of electricity some time? For the latter, I shall gladly be of assistance to you with any advice, if you should ever need it.—My book[4] is completed in manuscript; in print, there are still some 8 sheets missing, which will hopefully be in your hands by the end of July.

In July, my wife will be going to Scheveningen,[5] and I shall follow her at the beginning of August. I would really very much like to come to Belfast[6] now; hopefully nothing will come up and I shall be able to greet you once again after such a long time, and indeed as president too.—When will you be setting out from London?[7] If you would possibly be so kind as to give me some indication beforehand about how the situation is with Belfast; whether one has to book accommodation beforehand and so forth, then I should be very grateful to you. Perhaps you were going to go on a short excursion to Killarney[8] after the meeting or something like that. I should be very happy to accompany you then. Would my wife not be inconvenient in Belfast? Ladies are of course more often than not out of place at such meetings. Perhaps I would try to convince her to accompany me; I fear, however, that she will shy away from the trip from Holyhead[9] to Dublin. There is of course, though, probably another more direct, shorter sea trip to Belfast from Scotland.

Best regards from my wife.

In loyal devotion | always | Your | G. Wiedemann

RI MS JT/1/W/54–54

---

1. *your latest lines to my wife*: letter missing, but it was clearly a reply to letter 4307.
2. *your "Light"*: J. Tyndall, *Das Licht: Sechs Vorlesungen gehalten in Amerika im Winter 1872–1873*, ed. G. Wiedemann (Braunschweig: Friedrich Vieweg, 1876), which Clara Wiedemann was translating.
3. *Sound and Heat*: J. Tyndall, *Der Schall: Acht Vorlesungen gehalten in der Royal Institution von Grossbritannien*, ed. H. Helmholtz and G. Wiedemann (Braunschweig: Friedrich Vieweg, 1869) and J. Tyndall, *Die Warme Betrachtet als eine Art der Bewegung*, ed. H. Helmholtz and G. Wiedemann (Braunschweig: Friedrich Vieweg, 1867), both of which Clara Wiedemann had also translated, in collaboration with Anna Helmholtz.

4. *My book*: probably G. Wiedemann, *Die Lehre vom Galvanismus und Elektromagnetismus*, 2nd edn, 2 vols (Braunschweig: Friedrich Vieweg, 1874).
5. *Scheveningen*: a coastal resort in the Netherlands.
6. *Belfast*: the location of the annual meeting of the BAAS in August, where Tyndall was to give the presidential address.
7. *When will you be setting out from London?*: Tyndall left London for Belfast on 17 August.
8. *Killarney*: a town in south-west Ireland with surrounding mountains and three lakes that Tyndall had first visited in 1860 and whose physical features he discussed in several subsequent publications (see *Ascent of John Tyndall*, p. 151).
9. *Holyhead*: a town in Wales serving as the major ferry port to Ireland.

To Thomas Archer Hirst 23 May [1874][1] 4333

23rd May

Do like a good dear fellow send me a line of telegram counselling what you think wise regarding this article.[2] I wish Huxley & yourself could have laid your heads together about it.

The time is so tremendously brief that this I fear cannot now be done.

I want the telegram as I may find it necessary to see the editor of the Contemporary.[3]

Yours aff[ectionatel]y | John

RI MS JT/1/T/903
RI MS JT/1/HTYP/626/3

1. *[1874]*: the year is established by the relation to letter 4325, where Tyndall told Thomas Huxley, 'Perhaps Hirst and you could lay your heads together and discover the best plan'.
2. *this article*: J. Tyndall, 'Rendu and His Editors', *Contemporary Review*, 24 (1874), pp. 135–48, which would be published in June. It was a response to *Glaciers of Savoy*, which attacked Tyndall for his role in a dispute over theories of glacial motion with James David Forbes in the late 1850s.
3. *the editor of the Contemporary*: James Thomas Knowles.

To Rudolf Clausius 24 May 1874 4334

Royal Institution of Great Britain | 24th May | 74

My dear Clausius.

You have probably by this time received an official invitation to the meeting of the British Association at Belfast.[1] You know that I am to preside there, and you also know the delight with which I should hail your presence at the meeting.

It occurs about the middle of August. You would be lodged free, and well taken care of in every way.

I do not hope much from this invitation[2]—I know you would come if you could conveniently do so, and I shall be content whatever your decision may be—Still it would be a great pleasure to me to see you.

Love to Johnny[3] and kindest regards to M^rs^. Clausius[4] & the rest of the children.[5]

Ever yours | John Tyndall

RI MS JT/1/T/235

---

1. *the meeting of the British Association at Belfast*: the BAAS's annual meeting from 19 to 26 August, where Tyndall was to give the presidential address.
2. *I do not hope much from this invitation*: Clausius declined it in letter 4343.
3. *Johnny*: Rudolf John Clausius.
4. *M^rs^. Clausius*: Adelheid Clausius.
5. *the rest of the children*: Clausius and his wife had five other children (see letter 4300, n. 25).

To Julius Robert Mayer — 24 May [1874][1] — 4335

Royal Institution of Great Britain | 24th May

My dear Friend

Your book[2] is not yet come, but it gave me the liveliest pleasure to receive your letter.[3] I am extremely pleased to find a second edition of your essays called for. Had I time, and I hope to have it someday, I would put the book into the best possible English, for in this Country also it would be sure to have numerous readers. I knew from the first that I could not be wrong in asserting your claims,[4] and the whole scientific world now ratifies what I have done. My scotch friends[5] are still attacking me; but it is now about Forbes and Rendu. They will gain as little by their attack as they did in your case.

Nothing that they can say can in the least lessen the pleasure I feel in thinking that I was able to lend a hand in lifting you to that high position which you ought to have occupied fifteen years before it was by general consent assigned to you.

It will always give me pleasure to receive a letter from you, and to hear of your welfare in all things.

Believe me | Yours ever faithfully | John Tyndall

It would be too heavy a journey for you to visit Belfast where I am to preside over the British Association in August.[6] If you thought of coming[7] you would be warmly welcomed, and lodged free of expense.

RI MS JT/1/T/1068
RI MS JT/1/T/1068–COPY

1. *[1874]*: the year is established by reference to the BAAS meeting in Belfast (see n. 6).
2. *Your book*: see letter 4313, n. 1.
3. *your letter*: letter 4313.
4. *asserting your claims*: see letter 4313, n. 2.
5. *My Scotch friends*: supporters of James David Forbes, including his son George Forbes and Peter Guthrie Tait, who had published *Glaciers of Savoy*, which attacked Tyndall for his role in a dispute over theories of glacial motion in the late 1850s.
6. *Belfast where I am . . . British Association in August*: the annual meeting of the BAAS held in Belfast from 19 to 26 August, where Tyndall was to give the presidential address.
7. *If you thought of coming*: Mayer declined the invitation (see letter 4346).

To Emil du Bois-Reymond 26 May [1874][1] 4336

Royal Institution of Great Britain | 26th May

My dear Dubois.

Though I have not the faintest hope of drawing you to Belfast to the Meeting of the British Association in August,[2] I cannot help saying that to see you there would be a pleasure not easy to describe.

The Giant's Causeway[3] & some grand coast scenery are at hand.

The steamers between Holyhead[4] & Dublin are of the most splendid character.

It comes back to me that you and I were there at the meeting in 1852.[5] And that in the first instance you stayed in the Plough Hotel.[6]

How oddly those trifles—I mean the recollection of the inn—house themselves in the brain! I certainly have not thought of it for 20 years.

You would be lodged & cared for in every way, and Madame[7] also if you could persuade her to come.

Yours ever | John Tyndall

RI MS JT/1/T/423
RI MS JT/1/TYP/7/2441

1. *[1874]*: the year is established by reference to the BAAS meeting in Belfast (see n. 2).
2. *the Meeting of the British Association in August*: held 19 to 26 August, where Tyndall was to give the presidential address.
3. *The Giant's Causeway*: an area of considerable geological interest on the coast of north-east Ireland, made up of some 40,000 interlocking basalt columns.
4. *Holyhead*: see letter 4332, n. 9.

5. *the meeting in 1852*: the twenty-second meeting of the BAAS also took place in Belfast, in September 1852 (*Brit. Assoc. Rep. 1852*).
6. *the Plough Hotel*: at 7 Cornmarket, Belfast.
7. *Madame*: Jeannette du Bois-Reymond (née Claude, 1833–1911).

To Hermann Helmholtz 26 May [1874][1] 4337

Royal Institution of Great Britain | 26th. May

My dear Helmholtz,

Probably by this time the official invitation from the British Association[2] has reached you. I have little hope of attracting you, but it would be an immense pleasure to me to see you at Belfast.

You would be lodged and cared for in every way; and Mrs. Helmholtz[3] also, if she could be persuaded to come.

I accepted this presidency after refusing it three several times.

The boats between Holyhead[4] and Dublin are large, and not given to tossing.

Yours ever | John Tyndall

BBAW NL Helmholtz, Nr. 477, BI_28
RI MS JT/1/T/505

---

1. *[1874]*: the year is established by reference to the BAAS meeting in Belfast (see n. 2).
2. *the official invitation from the British Association*: for the annual meeting held in Belfast from 19 to 26 August, where Tyndall was to give the presidential address.
3. *Mrs. Helmholtz*: Anna Helmholtz.
4. *Holyhead*: see letter 4332, n. 9.

To Alfred Mayer 26 May 1874 4338

Royal Institution of Great Britain | 26th May 1874

My dear Professor Mayer.

I have been just writing a letter to your celebrated namesake Mayer of Heilbronn,[1] and now on the verge of my departure for the Alps[2] I write a line to you.

I wish what you are pleased to call my 'kindness' could have been far greater than it was. But my position here in London so splits up my time as to render me but poorly competent to show the kindness that I should like to show to my friends. Nevertheless I believe you enjoyed your stay[3] enough amongst us, and this was highly gratifying to me.

Many thanks to you for the photographs.[4]

I am glad to learn that you are in harness and doing such good work.[5]

Pray remember me kindly to your wife,[6] and to President and M^{rs}. Morton.[7] And I charge you particularly to present my best remembrances to M^{rs}. Stephens[8]—I have still a vivid recollection of my pleasant visit to her house.[9]

By the way the expression of an old man's[10] countenance whom I saw on that occasion is still present to me. He wore the dress of a church dignitary: and a more beneficent and saintly countenance I do not think that I have seen.

Goodbye | Yours very faithfully | John Tyndall

Tyndall, John (1820–1893); Hyatt and Mayer Collection, C0076, Manuscripts Division, Department of Special Collections, Princeton University Library

---

1. *a letter to your . . . Mayer of Heilbronn*: letter 4335, to Julius Robert Mayer.
2. *my departure for the Alps*: Tyndall seems to have left on 30 May (see letter 4347).
3. *your stay*: a biographical notice of Mayer later recorded: 'In 1873 he visited England and was most kindly received by such eminent men of science as Tyndall, Wheatstone, De la Rue, Strutt (afterward Lord Rayleigh), Spottiswoode, Ellis, Bosanquet, and Lord Ross' (A. G. Mayer and R. S. Woodward, 'Alfred Marshall Mayer, 1836–1897', *National Academy of Sciences Biographical Memoirs*, 8 (1916), pp. 243–72, on p. 260).
4. *the photographs*: not identified.
5. *such good work*: at this time Mayer was conducting research on acoustics and the duration of sonorous sensations (see letter 4454), as well as on new methods for investigating the oscillatory nature of electrical discharges.
6. *your wife*: Maria Louisa Mayer (née Snowden, 1843–1934), who was Mayer's second wife, his first having died in 1869.
7. *President and M^{rs}. Morton*: Henry Morton (1836–1902), who served as the president of the Stevens Institute of Technology in Hoboken, New Jersey from 1870–1902, and his wife Clara Whiting Morton (née Dodge, 1837–1901) (*ANB*).
8. *M^{rs}. Stephens*: Martha Bayard Stevens (née Dod, 1831–99), a philanthropist and executor of the will of her late husband Edwin Augustus Stevens (1795–1868), which provided for the establishment of the Stevens Institute of Technology, established in 1870 (*ANB*).
9. *my pleasant visit to her house*: on 18 December 1872, when Tyndall had visited New Jersey during his lecture tour of the United States and attended what he called a 'gathering at Hoboken' (letter 3939, *Tyndall Correspondence*, vol. 13). It was later recorded of Stevens: 'On all occasions when opportunity offered, she was quick to extend hospitality to distinguished visitors . . . and such men as Professor Tyndall . . . have enjoyed entertainments of the most delightful character at her beautiful mansion on Castle Point, overlooking the bay and city of New York' (F. De Ronde Furman, *A History of the Stevens Institute of Technology* (Hoboken: Stevens Institute of Technology, 1905), p. 148).
10. *an old man*: Benjamin Bosworth Smith (1794–1884). In a letter from 26 December 1872 Tyndall related how at the 'gathering at Hoboken . . . I met various church dignitaries. Among them the Bishop of America—a most charming patriarchal old man' (letter 3939,

Tyndall Correspondence, vol. 13). Smith was the Presiding Bishop of the American Episcopal Church, to which Stevens belonged, and in this capacity was known as the Bishop of the United States.

From William Henry Appleton 27 May [1874][1] 4339

York. May 27.

Dear Prof. Tyndall.

I read your note[2] respecting that very unfortunate investment.[3] I thought I was doing great things for you and that you would ever remember your visit to America even in the material advantage this would give to you. It shows again how unwise it is to give advice in money investments. I will on my return make the change you propose. I will take to myself the bonds &c and invest the sum in 5½% as you suggest. I will make up the interest on the whole sum at 7% as if no investment had been made, crediting the amt paid you. This would be equitable.

I shall regret very much not seeing you before you return.[4] Willie[5] has received a note from Mr Vincent[6] desiring to know what we want for sketch of Priestley's life.[7] We must leave it for your judgment. We want dates for a good strong article for the science monthly.[8] We saw Baillière[9] in Paris and he is very much pleased with the 'International'[10] sold 2000 each—It ought to encourage authors, as he says translations will appear in Italy and Russia, as well as in other countries, paying the same copyright to the author—wishing you a healthful and pleasant journey.

Believe me | Very sincerely | W. H. Appleton

RI MS JT/1/T/1470
RI MS JT/1/TYP/5/1733

---

1. *[1874]*: the year is established by reference to Appleton's visit to Britain (see n. 4).
2. *your note*: letter missing.
3. *that very unfortunate investment*: Tyndall's lecture tour of the United States in 1872–3 made a surplus of $13,033, which was invested to establish a fund for pure science research in American universities (see letter 4214, n. 5). Overall, the investments made a good return, but this 'very unfortunate' one may relate to the purchase of American railway company stocks, which were extremely volatile at this time; Tyndale wrote to Tyndall on this matter in letter 4317.
4. *before you return*: Appleton and his family were spending time in Britain following the death of his daughter (see letter 4286). Tyndall was about to travel to Switzerland, and would not return until 10 August.
5. *Willie*: William Worthen Appleton (see letter 4286, n. 16).

6. *Mr Vincent*: probably Benjamin Vincent, the RI's librarian.
7. *sketch of Priestley's life*: 'Sketch of the Life of Dr. Priestley', *Popular Science Monthly*, 5 (1874), pp. 480–92, which was published in August. The author was anonymous.
8. *the science monthly*: *Popular Science Monthly*.
9. *Baillière*: Henri Paul Baillière (1840–1905), a French publisher.
10. *the 'International'*: see letter 4264, n. 5.

From Martha Somerville [27 May 1874][1] 4340

Hotel Bedford Rue de l'Arcade | Paris

Dear Mr Tyndall

I have sent you two MSS of my mothers[2] about which I spoke to you in Switz[erlan][d] last summer[3] & you proposed asking Mr Hirst to look over.[4] I thought I sh[oul][d] take them to you myself & we came so far on our way, but on arriving in Paris a fortnight ago we found a telegram announcing the sudden death of my sister in law,[5] with whom we w[oul][d] have stayed in London,[6] & in consequence we have given up all thoughts of England for this year. We are waiting here for a little as we shall probably have some lawyers papers to sign &c &c then go to Switzerland or the Tyrol according as my sister[7] likes for her health. I fear there is little chance of your coming abroad as you have to preside over the meeting at Dublin.[8] I am very sorry not to go to London & see so many friends—you amongst the number—but it cannot be helped. My sister joins me in kind regards

Believe me | Dear Mr Tyndall | most truly yours | Martha Somerville | Wednesday

RI MS JT/1/T/713
RI MS JT/1/HTYP/627

---

1. *[27 May 1874]*: the date is established by reference to the death of Agnes Greig on 10 May (see n. 5), which Somerville heard about 'a fortnight ago', and the letter being written on a Wednesday, as Somerville indicates in the closing salutation, which 27 May was. This date is also suggested by an annotation to letter 4345.
2. *two MSS of my mothers*: probably a 'volume on the form and rotation of the earth and planets' and a 'work of 246 pages on curves and surfaces of the second and higher orders', both written in the early 1830s, of which Mary Sommerville later reflected: 'Had these two manuscripts been published at that time, they might have been of use; I do not remember why they were laid aside, and forgotten till I found them years afterwards among my papers' (M. Somerville, *Personal Recollections*, ed. M. Somerville (London: John Murray, 1874), pp. 201–2). Neither of the manuscripts was subsequently published, and they

remain in the Mary Somerville Collection, Bodleian Library, University of Oxford, MS Dep. b. 207, MSAU2–7 and MSAU2–8.

3. *in Switz[erlan]d last summer*: on 10 August 1873 (see letter 4230, n. 4).
4. *asking Mr Hirst to look over*: Tyndall did this (see letter 4345).
5. *my sister in law*: Agnes Greig (née Graham, 1807–74), who had married Somerville's half-brother Woronzow Greig in 1837. She died on 10 May.
6. *with whom we w[oul]d have stayed in London*: at Greig's home, 5 Cranley Place, Onslow Square, South Kensington.
7. *my sister*: Mary Charlotte Somerville.
8. *the meeting at Dublin*: a mistaken reference to the BAAS meeting in Belfast, where Tyndall was to be the president.

To George Gabriel Stokes 28 May [1874][1] 4341

28th May

My Dear Stokes

This horrible Belfast Address[2] drives me off on Saturday,[3] for in this Babylon[4] I can get nothing done.

You may have some remarks to make on the section about aqueous vapour of my paper.[5] I wish it could have got into your hands before my departure.

My address in Switzerland for some time will be

Bel Alp | Brieg | Canton de Valais[6]

Yours sincerely | John Tyndall

I have given the date of receipt as that of the notice in the Proceedings: but alter it as you like.[7]

CUL SC—add 7656

1. *[1874]*: the year is established by reference to the Belfast Address (see n. 2).
2. *This horrible Belfast Address*: Tyndall's presidential address to the BAAS at its annual meeting in Belfast in August.
3. *Saturday*: Tyndall left London for Switzerland on 30 May, and returned on 10 August.
4. *this Babylon*: i.e. London, which had been compared to the ancient Mesopotamian metropolis since the eighteenth century. For example, Benjamin Disraeli (1804–81), Prime Minister for nine months in 1868 and from February 1874 to April 1880 (*ODNB*), called London 'a modern Babylon' in his novel *Tancred: or, the New Crusade*, 3 vols (London: Henry Colburn, 1847), vol. 3, p. 77.
5. *the section about aqueous vapour of my paper*: 'Vehicle of Sound'. The section was '12. Action of Fog. Observations in London' (pp. 209–14).
6. *Bel Alp | Brieg | Canton de Valais*: see letter 4306, n. 12.

7. *the date of receipt . . . alter it as you like*: the date given in the *Phil. Trans.* was 5 February, but in the *Roy. Soc. Proc.* the paper was recorded as having been 'Received January 1st, 1874' (J. Tyndall, 'Preliminary Account of an Investigation on the Transmission of Sound by the Atmosphere', *Roy. Soc. Proc.*, 22 (1874), pp. 58–68, on p. 58). Stokes gives his reasons for the alteration in letter 4357.

To Hector Tyndale 28 May [1874][1] 4342

28th May

My dear Hector

On the point of my starting for Switzerland[2] I send you the enclosed[3] just received from Appleton. I think it only just that you should know how he deals with me.

Yours affect[ionatel]y | John

RI MS JT/1/T/1470
RI MS JT/1/TYP/4/1732

1. *[1874]*: the year is established by the relation to letter 4339.
2. *starting for Switzerland*: Tyndall left London for Switzerland on 30 May, and returned on 10 August.
3. *the enclosed*: letter 4339.

From Rudolf Clausius 28 May 1874 4343

Bonn, 28 May 74.

Lieber Tyndall,

Es war früher nicht nur mein grosser Wunsch, sondern auch meine bestimmte Absicht, wenn Du einmal Präsident der British Association sein würdest, dann dazu nach England zu kommen. Als Du mit Hirst hier in Bonn warst, und wir eine Partie nach dem Drachenfels, nach Rolandseck und dem Roderberg machten, habe ich es auch mit Hirst besprochen, dass ich bei dieser Gelegenheit mit ihm zusammentreffen wollte. Jetzt bin ich aber nicht mehr so unabhängig in der Ausführung derartiger Beschlüsse wie damals. Durch mein Knieleiden bin ich an Allem verhindert, was mit irgend welchen körperlichen Anstrengungen verbunden ist. Ich muss daher grössere Reisen und Festlichkeiten, die wenn auch in angenehmer Weise, so doch immer ermüdend und anstrengend sind, ganz vermeiden. Unter diesen Umständen wirst Du es mir wohl nicht übel nehmen, wenn ich nicht nach Belfast komme. Du kannst Dich darauf verlassen dass es mir selbst ausserordentlich leid thut, dass ich darauf verzichten muss, bei dieser schönen Veranlassung mit Dir zusammen zu sein, und Deine Rede zu hören.

Sei so gut, auch Hirst zu sagen, weshalb ich unserer Verabredung nicht entsprechen kann.

Mit herzlichem Danke für Deine freundliche Einladung und besten Grüssen von meiner Frau, mir und allen Kindern, besonders Johnny,

Dein | Clausius.

Bonn, 28 May 74.

Dear Tyndall,

It was not only my great wish earlier, but also my definite intention, that if you were going to be the President of the British Association one day, I would then come to England for that. When you were here in Bonn with Hirst, and we went on a trip to the Drachenfels, to Rolandseck, and to the Roderberg,[1] I also discussed it with Hirst that I wanted to meet with him on this occasion. Now, though, I am no longer as independent in carrying out decisions of that sort as I was then. Due to my knee complaint,[2] I am prevented from doing anything which involves any physical exertion at all. I must therefore completely avoid any lengthy trips and any festivities which, even though enjoyable, are always still so tiring and strenuous. Under these circumstances, you will probably not hold it against me if I do not come to Belfast. You can be sure that I am personally extraordinarily sorry that I must forego being together with you on this fine occasion and hearing your speech.

Be so good as to tell Hirst also why I cannot keep our appointment.

With sincere thanks for your kind invitation[3] and best regards from my wife,[4] myself, and all the children,[5] especially Johnny,[6]

Your | Clausius.

RI MS JT/1/TYP/7/2330

1. *here in Bonn with Hirst . . . to the Roderberg*: Thomas Hirst recorded in his journal for 8 July 1869: 'We had a pleasant day with Clausius on the Drachenfels and Rolandseck. We passed the evening at this house' (*Hirst Journals*, p. 1849). Drachenfels and Roderberg are hills, and Rolandseck a village, in the north-western Rhineland.
2. *my knee complaint*: Clausius injured his knee as a volunteer ambulance officer at the Battle of Gravelotte on 18 August 1870 during the Franco-Prussian War.
3. *your kind invitation*: to the meeting of the BAAS in Belfast from 19 to 26 August, where Tyndall was to be the president (see letter 4334).
4. *my wife*: Adelheid Clausius.
5. *the children*: Clausius and his wife had six children (see letter 4300, n. 25).
6. *Johnny*: Rudolf John Clausius.

To Thomas Archer Hirst 29 May 1874 4344

29th May / 74

My dear Tom.

Pray fill up this for such sum as you may need.[1]

Your dinner[2] was wonderfully successful.

Goodbye. Love to both of you.[3] | John Tyndall

RI MS JT/1/T/714
RI MS JT/1/HTYP/627

---

1. *Pray fill up this for such sum as you may need*: the enclosure is missing, but would have been a blank cheque for the London Joint Stock Bank (see letter 4374).
2. *Your dinner*: Hirst recorded in his journal for 28 May 1874: 'Gave my first dinner party. Present Admiral and Lady Key, Capt. and Mrs Campbell, Mr and Mrs Huxley, Mr and Mrs Spottiswoode, Captain Knowles (President of the Mess) and Tyndall. It passed off very comfortably' (*Hirst Journals*, p. 2000).
3. *both of you*: Tyndall presumably refers to Hirst's niece and housekeeper Emily Hirst, who acted as hostess at his dinners.

To Thomas Archer Hirst [*c.* 29 May 1874][1] 4345

My dear Tom.

This[2] is from Miss Somerville—you will come here[3] & look up the manuscript[4]—they want some opinion as to the desirability of publishing it.

And I shall want you to look at my letters & intercept the dividend warrants[5] when I am away.

John.

RI MS JT/1/T/713
RI MS JT/1/HTYP/627

---

1. *[c. 29 May 1874]*: this is the date given in the typescript of the letter, although it was not recorded in the original letter, to which 'May 27 / 74' has been added in a hand other than Tyndall's. However, this was the date of the letter from Martha Somerville that was enclosed with it (letter 4340), and as Somerville sent the letter from Paris it is unlikely that Tyndall received it on the day it was written. As such, 29 May seems likely as the date it was forwarded to Hirst, apparently after he had already sent letter 4344 to the same recipient.
2. *This*: letter 4340.
3. *here*: presumably the RI, to which Hirst was to come without Tyndall, who was about to leave for Switzerland, being present.

4. *the manuscript*: see letter 4340, n. 2.
5. *the dividend warrants*: the documentary order or authority on which a shareholder receives his dividend (*OED*).

From Julius Robert Mayer 29 May 1874 4346

Hochverehrter Herr und Freund!

Aus Ihrem werthen Schreiben vom 24 d. M., welches mir gestern zugekommen, ersehe ich mit grossem Bedauren, dass das Buch, welches ich gleichzeitig mit meinem letzten Briefe unter Kreuzband an Sie abgehen liess, nicht angekommen ist, wesshalb ich mich beeile, Ihnen ein zweites Exemplar zu schicken, das hoffentlich richtig in Ihre Hände kommen wird.—Es wäre allerdings grosse Ehre für mich, wenn Sie Zeit finden könnten, die genannte Schrift ins Englische zu übersetzen; vielleicht sind Sie so gütig, vor der Hand wenigstens eine kritische Anzeige in eines Ihrer geschätzten wissenschaftlichen Organe zu liefern, was mir sicher schon zum grossen Nutzen gereichen würde.—

Wenn die allgemeine Anerkennung meiner Schriften auch etwas spät erfolgt ist, so ist sie doch Dank Ihrer unvergesslichen Bemühungen, erfolgt, und es hat mich diess vorzugsweise ermuthigt, auf dem einmal betretenen Wege weiter fortzuschreiten. Wohl denen die einen so trefflichen Vertheiger und Fürsprecher finden, wie Sie mir einer geworden sind!—Für Ihre gütige Einladung nach Belfast zu kommen, bin ich Ihnen sehr verbunden, werde aber bei meinem vorgerückten Alter und meiner Unkentniss der englischen Sprache wohl keinen Gebrauch davon machen können.

Mit der Bitte, mir Ihr unschätzbares Wohlwollen auch ferner zu erhalten, zeichne ich hochachtungsvoll

Ihr | dankbar ergebenster Diener | J. R. Mayer

Heilbronn, | 29 Mai 1874.

Highly respected Sir and friend!

I read with great regret in your letter from the 24th of this month,[1] which I received yesterday, that the book[2] that I sent to you by book-post[3] at the same time as my last letter[4] has not arrived. I therefore hasten to send you a second copy, which I hope will reach your hands safely.—It would be a great honour for me if you were able to find the time to translate this document into English;[5] you may perhaps at least be so kind as to write a critical review and send it to one of your valued scientific publications, which in itself would certainly be of great help to me.—

Even if the general recognition of my writings came somewhat late, it nevertheless resulted from your unforgettable efforts[6] and greatly encouraged me to progress further on the chosen path. Blessed are those who find such an excellent defender and advocate, as you became for me!—I am very grateful for your kind invitation to Belfast,[7] but I will not be able to take advantage of it given my advanced age and my lack of knowledge of the English language.

With the request that you may maintain your invaluable goodwill towards me in the future, I respectfully sign

Your | grateful and most humble servant | J. R. Mayer

Heilbronn, | May 29, 1874.

Stadtarchiv Heilbronn, D032–155 No. 15
RI MS JT/1/M/88
RI MS JT/1/TYP/7/2537

1. *your letter from the 24th of this month*: letter 4335.
2. *the book*: see letter 4313, n. 1.
3. *book-post*: a special postal service by which books and printed matter other than newspapers were conveyed at reduced rates of postage.
4. *my last letter*: letter 4313.
5. *translate this document into English*: Tyndall suggested he might do this in letter 4335, although it seems that he was never able to.
6. *your unforgettable efforts*: in the early 1860s Tyndall had helped bring to prominence Mayers's overlooked work on the mechanical equivalence of heat (see *Tyndall Correspondence*, vol. 8, pp. xvii–xvi). Shortly after this letter, in June, Tyndall publicly reaffirmed his view that 'By his unaided genius, Dr. Robert Julius Mayer, of Heilbronn in Germany, reached the heart of a generalization, which the professional hierarchy of science in his day had failed to reach, and which in its later developments ranks as high as the principle of gravitation. For this great Bahnbrecher I sought recognition' (J. Tyndall, 'Rendu and His Editors', *Contemporary Review*, 24 (1874), pp. 135–48, on p. 137).
7. *your kind invitation to Belfast*: to the meeting of the BAAS in Belfast from 19 to 26 August, where Tyndall was to be the president (see letter 4335).

To Thomas Carlyle — 30 May [1874][1] — 4347

Royal Institution of Great Britain | 30th May

My dear Friend

It is a cause of great regret to me to leave London without seeing you. I am forced to go to get work done[2] which I should willingly have avoided, and which cannot be done in this Babylon.[3] I had set my heart on coming down to you[4] on Wednesday[5] when I thought I should be free; but I was not free.

In the Contemporary Review you will, if you take the least interest in the matter, see a short article of mine,[6] in which a subject regarding which you wrote to me some months ago[7] is introduced. It will tell its own tale. I should never have alluded to the subject[8] were it not for the manner in which it has been handled by others.[9] This is all I have to say. Goodbye, and may your summer be a pleasant one.

Yours ever affectionately | John Tyndall | 30th May. 5 O'Clock in the morning.

RI MS JT/1/T/164
RI MS JT/1/TYP/1/193–4

1. *[1874]*: the year is established by relation to letter 4188.
2. *to get work done*: Tyndall left London for Switzerland on 30 May in order to write the presidential address for the BAAS meeting at Belfast in August.
3. *this Babylon*: see letter 4341, n. 4.
4. *coming down to you*: presumably to Carlyle's home at 5 Cheyne Row, Chelsea, which was south of central London, hence 'coming down'.
5. *Wednesday*: 27 May.
6. *a short article of mine*: see letter 4333, n. 2.
7. *you wrote to me some months ago*: letter 4188.
8. *the subject*: the controversy with James David Forbes over glacial motion, which, having begun in the late 1850s, had recently been revived.
9. *others*: in his article for the *Contemporary Review*, Tyndall named Peter Guthrie Tait, John Ruskin, and Alfred Wills.

From Gustav Kirchhoff 30 May 1874 4348

Heidelberg 30 Mai 1874

Verehrter Freund!

Ich danke Ihnen herzlich für die freundliche Einladung, die Sie mir haben zu Theil werden lassen, der British Association unter Ihrem Präsidium in diesem August beizuwohnen. Aber—wie Sie es selber vermuthen—ich muss die Einladung trotz ihrer Freundlichkeit ablehnen. Zwar ist mein Fuss, der lange Jahre mich verhindert hat, so weite Reisen zu unternehmen, jetzt soweit hergestellt; aber allerlei andere Umstände nöthigen mich, in den Herbstferien in meiner Heimath oder deren Nähe zu bleiben; und so muss ich auf das grosse Vergnügen verzichten, das ich sonst daran haben würde, Sie bei der gedachten Gelegenheit zu sehn.

Mit herzlichen Grüssen | Ihr | G. Kirchhoff

Heidelberg 30 May 1874

Dear friend!

I sincerely thank you for the kind invitation[1] you sent me to be present at the British Association under your chairmanship this August.[2] But—as you assume yourself—I have to decline the invitation despite your kindness. Admittedly, my foot, which has prevented me from taking long journeys for many years, is now mostly restored;[3] but several other circumstances force me to stay in my home or nearby during the autumn holidays; and so I have to forgo the great pleasure, that I would otherwise have had, of seeing you at the occasion in question.

With warmest regards | your | G. Kirchhoff

RI MS JT/1/K/10

1. *the kind invitation*: letter missing.
2. *the British Association under your chairmanship this August*: the annual meeting of the BAAS at Belfast from 19 to 26 August, where Tyndall was to be the president.
3. *my foot . . . mostly restored*: following an accident that injured his foot in 1868, Kirchhoff was forced to walk with crutches or use a wheelchair. Although it lessened, the ailment never healed fully.

From Emil du Bois-Reymond 31 May 1874 4349

Berlin, W., 17 Victoria Str. | May 31, '74

My dear Tyndall,

Your kind letter and invitation[1] have given me great pleasure by affording me a sign that you have not forgotten me or given me up on account of my long and in my own eyes impardonable silence. On several occasions I was bound in friendship to write to you and I did not. You sent me I do not remember how many valuable books, I remained silent. You came back from America[2] crowned with success, and I did not congratulate you. Poor Bence Jones died,[3] and I did not write you the word: Serrons les rangs.[4] But if I did not write in reality, I intended to do so an infinite number of times. You were constantly present to my mind. But I am pressed for time in a manner which most likely I cannot depict to you more strongly than by saying: like you. As long as I have the least working-power left there is always something more needful to do than the most urgent of letters. But you know all this as well as I can tell it. As to your present invitation, I am sorry it is impossible for me to accept it. I cannot well leave at that time for all sorts of reasons. Otherwise it would have been "good fun" to meet again at Belfast after 22 years, you holding the chair instead of Mr. Robinson.[5] How much I should like to hear you deliver your Address. Altogether I should like so much to see England again, although I can hardly fancy myself there without my friend and support Bence Jones.[6] It cannot, however, be this year. Among the reasons which prevent me from attending the meeting there is foremost the obligation of working out in detail the plan for the fittings of my great laboratory,[7] which is rapidly *[growing]* now, after so many years annoyances and toil. It will be by far the most spacious, sumptuous, glorious place for scientific work which was ever conceived of. Ludwigs celebrated laboratory[8] dwindles into utter insignificance when compared with my palace of Physiology. His building could just as well be used as an inn, a hospital, a school, a factory or a German dwelling-house. My place is organised throughout like the inside of a ship. I sometimes feel a little anxiety at incurring the responsibility of such an enterprise, whose scientific success of course is to a great extent a matter of chance, viz. depending on the good luck in finding the proper men for assisting me in its direction, the officers of the ship. But would it not have been quite as heavy a responsibility not to build such a laboratory, when the money was there, what may not always be the case?

Believe me, my dear Tyndall, | Yours faithfully, | E du Bois Reymond.

RI MS JT/1/D/151
RI MS JT/1/TYP/7/2442

---

1. *Your kind letter and invitation*: letter missing; the invitation was to the meeting of the BAAS in Belfast from 19 to 26 August, where Tyndall was to be the president.
2. *You came back from America*: Tyndall returned from a lecture tour of the United States in February 1873 (see letter 4201, n. 4).
3. *Poor Bence Jones died*: Henry Bence Jones had died on 20 April 1873.
4. *Serrons les rangs*: let's close the ranks (French).
5. *meet again at Belfast after 22 years . . . Mr. Robinson*: the BAAS had last met in Belfast in 1852, but du Bois-Reymond's memory seems confused, as Thomas Romney Robinson had been a vice-president at the meeting, whereas the president was Edward Sabine.
6. *without my friend and support Bence Jones*: du Bois-Reymond and Bence Jones had visited each other in Berlin and London since 1851, and remained close friends until the latter's death.
7. *my great laboratory*: the Berlin Physiological Institute, founded in 1877.
8. *Ludwigs celebrated laboratory*: Carl Ludwig's (1816–95) laboratory at the Physiological Institute in Leipzig, founded in 1865.

From François Moigno 2 June 1874 4350

Les Mondes | Revue Hebdomadaire | Des Sciences et de leurs applications aux arts et à l'industrie | par | M. l'Abbé Moigno | Abonnement : | Paris un an 25f | Départements 30f | Bureaux | 32, rue du Dragon, 32

St. Denis 2 rue de Strasbourg | Paris, le 2 juin 1874

Mon cher Mr Tyndall

J'ai entièrement achevé la traduction de la Lumière, et je vous prie de vous unir à moi pour obtenir de Mr Gauthier-Villars que l'impression soit achevée et le livre lancé avant la fin de l'année scolaire, avant le premier août, pour qu'il puisse être donné en prix.

Le chapitre Résumé et Conclusion a une couleur un peu locale ou américaine. Je crois cependant qu'il fera bon effet.

Mais pensez vous que la discussion entre Lord Brougham et Young (Thomas) ait beaucoup d'intérêt hors de l'Angleterre. Désirez vous qu'elle soit publiée dans l'Edition française? Tout est prêt pour cela, un mot je vous en prie à ce sujet.

Je suis toujours bien embarrassé quand je me vois forcé de traduire les mots Illustrate Discoverer Blackness, et j'aurais bien envie d'adopter définitivement illustrer, au lieu de mettre en évidence; découvreur au lieu de inventeur; on n'invente pas une vérité on la découvre; noirceur au lieu de noir.

Donnez moi s'il vous plaît votre sentiment à ce sujet. Noirceur en français se prend toujours au *[1 word illeg]*.

Je suis bien heureux d'apprendre que vous présiderez la Réunion de Belfast. Pourquoi faut-il que je ne puisse pas y aller. Avec quel plaisir, assisté de l'un de vous je ferais devant le bel auditoire ma revue illustrée par *[1 word illeg]* tableaux photographiques sur verre transparent. Ce serait une brillante soirée. Je viens de *[1 word illeg]*, comme vous l'avez vu peut-être dans les grandes villes de ma chère Bretagne, avec un succès fou. Chacun de ces tableaux devient pour les diverses classes de la société l'occasion d'un enseignement utile et agréable. Mais c'est un beau rêve, parlez en cependant à vos collègues du conseil. Je n'ai jamais tant travaillé, ne travaille avec plus de facilité et de bonheur, si vous me voyez vous serrez effrayé de mon ardeur infatigable. Et cependant je suis *[1 word illeg]*, c'est à dire que je devais me condamner à une vie de mollusque.

A mes conférences illustrées de St. Denis ville très industrielle et très ouvrière j'ai jusqu'à 2400 auditeurs dans une salle immense de bals publiques.

Votre Lumière est charmante, c'est la Vision Intuitive des phénomènes et de leur cause.

Tout à vous d'esprit et de cœur, d'estime, d'affection et d'admiration sincères | F. Moigno

Les Mondes | Weekly Review | Of the sciences and their applications to the arts and industry | by | Father Moigno | Yearly Subscriptions: | Paris for a year 25f, Departments 30f | Offices | 32, rue du Dragon, 32

St. Denis 2 rue de Strasbourg. | Paris, June 2 1874

My dear Mr Tyndall,

I have fully completed the translation of Light,[1] and I beg you to join me in order to get confirmation from Mr Gauthier-Villars that the printing will be completed and the book will be launched before the end of the school year, before the first of August, so that it can be given as a prize.

The chapter Summary and Conclusion has a bit of a local American tone, but I think it will give a good impression.

But do you think that the discussion between Lord Brougham and Young (Thomas)[2] is of interest to those outside England? Would you like it to be published in the French Edition? Everything is ready for that, please give me a word about this.

I never know what to do when I see myself being forced to translate the words Illustrate, Discoverer, Blackness. I would like to use illustrate instead of demonstrate; discoverer instead of inventor; one does not invent the truth, one discovers it; Darkness instead of black.

Please give me your thoughts on the matter. Darkness, in French, is always taken as *[1 word illeg]*.

I am very happy to hear that you will preside over the Meeting of Belfast.[3] Why does it have to be so that I am not able to attend it? With what pleasure, with the help of one of you, I would have showed the marvellous audience my review, illustrated by *[1 word illeg]* photographic images on transparent glass.[4] It would be a brilliant evening. I just *[1 word illeg]*, as you probably have seen it in the large cities of my dear Brittany with insane success. Each of these tables becomes for the various classes of the society an opportunity for a useful and enjoyable education. But it is a nice dream; speak about it to your colleagues on the council. I never worked so hard, nor work with greater ease and happiness, if you see me you would be afraid of my tirelessness. However, I am *[1 word illeg]*, in other words I had to condemn myself to the life of a mollusc.

At my illustrated conferences in the very industrial and worker-oriented city of St. Denis,[5] I have audiences up to 2400 in a huge public ballroom.

Your Light is charming; it is the Intuitive Vision of phenomena and their causes.

Sincerely yours in mind and heart, with respect, affection and admiration | F. Moigno

RI MS JT/1/M/125

1. *the translation of Light*: see letter 4277, n. 2.
2. *the discussion between Lord Brougham and Young (Thomas)*: see letter 4307, n. 4.
3. *the Meeting of Belfast*: the annual meeting of the BAAS in Belfast from 19–26 August 1874, where Tyndall was to be the president.
4. *photographic images on transparent glass*: slides made using the wet collodion method which Moigno had been collecting for his 'salon photographique' since the 1850s, and which he had begun projecting during lectures in 1872. See R. Fox, *The Savant and the State: Science and Cultural Politics in Nineteenth-Century France* (Baltimore: Johns Hopkins University Press, 2012), p. 210.

5. *St. Denis*: the historic city of Saint-Denis was fast becoming an industrial suburb of northern Paris. Moigno had been appointed a canon of the collegiate chapter of the city's basilica in 1873, and the success of his lectures, which took a providential view of nature, soon attracted the attention of Saint-Denis's growing number of secularists, whose protests meant the lectures finally had to be abandoned. See Fox, *Savant and the State*, pp. 210–11.

From Robert Bunsen 3 June 1874 4351

Mein theuerster Freund,

Ihre freundlichen Zeilen vom 24. Mai haben den von mir so lange und oft gehegten Wunsch und Vorsatz, Sie einmal wieder in England aufzusuchen, auf das Lebhafteste angeregt, aber ich fürchte die Anstrengung und Aufregung einer so weiten Reise zu sehr, um nicht selbst meine liebsten Wünsche den Geboten meines vorgerückten Alters zum Opfer zu bringen. Ich habe schon seit mehreren Jahren keine weitere Reise gemacht und werde mich auch in diesem Herbste damit begnügen müssen, in einer schweizerischen Sommerfrische einige Wochen der Erholung zu leben.

So schmerzlicher es für mich ist, dieser Nothwendigkeit nachzugeben, um so herzlicher werde ich Ihrer in den Tagen der Versammlung in alter treuer Freundschaft gedenken.

Von ganzem Herzen, mein theuerster Tyndall,

Ihr | R Bunsen | Heidelberg 3 Juni 1874

My dearest friend,

Your friendly lines from the 24th of May[1] have most vividly stirred my long and often cherished wish and resolution to once again visit you in England, but I fear the exertion and excitement of such a long journey too much to not sacrifice even my dearest wishes to the dictates of my advanced age. I have not made any long journeys for several years already, and this autumn as well I will have to be content with a few weeks' recuperation in a Swiss summer retreat.

As painful as it is to me to give in to this necessity, the more sincerely will I remember you in old loyal friendship during the days of the conference.[2]

With all my heart, my dearest Tyndall,

Your | R Bunsen | Heidelberg 3 June 1874

RI MS JT/1/B/151

---

1. *Your friendly lines from the 24th of May*: letter missing, although Tyndall had clearly invited Bunsen, under whom he had studied in Germany in the 1850s, to attend the BAAS meeting in Belfast at which Tyndall would give the presidential address.
2. *the days of the conference*: the Belfast meeting of the BAAS was held from 19 to 26 August.

From James Coxe 4 June 1874 4352

Kinellan[1] | Edinburgh | June 4 1874.

My dear Dr Tyndall,

In spite of the warning of old Hudibras that those who in quarrels interpose must often wipe a bloody nose,[2] I sent Russel[3] my views of the Forbes controversy.[4] But more than a fortnight has elapsed and he has made no sign. Whether he will print or not I cannot say, but I do not wish to delay longer in telling you frankly the conclusions at which I arrive. In the first place then, I think you have been, as we say in Scotland, too mealy-mouthed, you have spoken too mildly of Forbes, and have given him credit for honour and rectitude more than he deserves.[5] He himself and his friends place James Forbes on a pinnacle, and no man must venture to differ from him. He is the soul of honour and not a breath must tarnish his fair fame. His Champion, John Ruskin, comes forward with "terrible words" to crush you, but John Ruskin paints him as a very sneak.[6] From John Ruskin's account, Forbes goes to the Aar Gletscher, sees what Agassiz is doing, hears all he has got to say, picks up the questions at issue, and goes off on the sly to test them in a more scientific manner than had occurred to Agassiz.[7] My idea is that Forbes felt he had done a dirty thing, and hence his sensitiveness to all criticism. I think if you had kicked him openly and heartily, and told him the reason why, there would not have been the same cry of Mad dog that is raised whenever any one ventures to differ from him. You are in the position of a man fighting with one hand tied. But as things stand I would not counsel you to take any notice of the last book of Forbes (Son) and Tait.[8] Posterity will judge between you. As for Ruskin, I can only say that if I were in Forbes' shoes I would exclaim—"Save me from my friends"![9]

Very truly Yours | J. Coxe.

RI MS JT/1/TYP/1/303
LT Typescript Only

---

1. *Kinellan*: Kinellan House in Murrayfield, an area to the west of Edinburgh.
2. *the warning of old Hudibras... bloody nose*: an allusion to Samuel Colvil's 'Mock Poem' *The Whiggs Supplication, or, The Scotch-Hudibras* (1681), which asserts: 'And many times we see aggressors, | Who trouble others mens reposes, | Gain nothing else but bloody noses' (Part 2,11. 692–4).
3. *Russel*: Alexander Russel (1814–76), editor of the *Scotsman* newspaper from 1848 until his death (*ODNB*).
4. *my views of the Forbes controversy*: Coxe's views appeared as 'Forbes *Versus* Tyndall', *Scotsman*, 2 July 1874, p. 2.

5. *you have spoken too mildly of Forbes . . . than he deserves*: in 'Rendu and His Editors', *Contemporary Review*, 24 (1874), pp. 135–48, Tyndall states: 'It is not, and never was, my design to charge Principal Forbes with conscious wrong' (p. 143).
6. *His Champion, John Ruskin . . . a very sneak*: in letter 34 of *Fors Clavigera*, Ruskin attacked Tyndall at length and contended that 'all the ingenuity and plausibility of Professor Tyndall have been employed, since the death of Forbes, to diminish the lustre of his discovery'. However, Ruskin also conceded that he was not 'prepared altogether to justify Forbes in his method of proceeding, except on the terms of battle which men of science have laid down for themselves' (*Fors Clavigera*, vol. 3 (1873), letter 34, pp. 1–32, on pp. 24 and 23). See also letter 4185, n. 13.
7. *From John Ruskin's account . . . to Agassiz*: see letter 4378, n. 11 and n. 12.
8. *the last book of Forbes (Son) and Tait*: *Glaciers of Savoy*.
9. *"Save me from my friends"!*: part of a saying attributed to the seventeenth-century French military commander Claude Louis Hector de Villars (1653–1734): 'God save me from my friends, I can protect myself from my enemies'.

To Thomas Archer Hirst 5 June 1874 4353

5th. June 1874

Myriad groups of these violets[1] over the Bel Alp bursting through the dead brown grass left behind by the retreat of the winter snow—no eye to see their beauty—for even the sheep & goats have not yet reached this height. I am the only dweller (visitor) in the place; but the very solitude saved me from loneliness.

Gentians and other flowers are also bursting through the sod: but the rhododendron shrubs look as if life had for ever departed; and still in a week or two the slopes will bloom with them.

Grand showers marching over the opposite mountains: Guns of Brieg, called "Katzkopfe"[2]—little things firing perhaps ¼th of powder: heard loudly here in celebration after "lord God's day" fete de Dieu.[3]

The sound of the cataracts rises and falls as acoustic clouds[4] pass between them & me I think I have noticed fluctuations in the song of the lark due to the same cause.

I am pondering—that is all, but something will come out of it.

ever aff[ectionate]ly | John

My dear Tom

When I wrote the foregoing two hours ago I did not imagine I should have a word more to say to you: but a whole bundle of letters have come to me, and among them the enclosed.[5] I think the best plan will be first to ascertain the address of Mr. Crotch's[6] friends and then to forward them this letter.

It records a case of true kindness and sympathy: the writer is Mrs. Lesley, wife of Professor Peter Lesley of Philadelphia, a friend of Emerson's, and a man who took a foremost part in organizing my American Campaign.[7]

Yours ever affec[tionately]. | John.

Love to Lilly.[8]

Ruhmkorff[9] is coming to Belfast. Clausius lame, and deeply sorrowful.[10] Kirchhoff overpowered with work.[11]

BL Add MS 63092, ff. 86–7
RI MS JT/1/HTYP/628

---

1. *these violets*: Tyndall enclosed the flowers in the letter (see letter 4368).
2. *Guns of Brieg, called "Katzkopfe"*: small fireworks, whose German nickname means cat heads, that derived from the Swiss town of Brieg (now spelt Brig) close to where Tyndall was staying at Bel Alp (see letter 4306, n. 12).
3. *fete de Dieu*: the Catholic Feast of La Fête-Dieu (Corpus Christi), or the Blessed Sacrament of the Body and Blood of Christ, is celebrated on the Thursday following the Trinity (sixty days after Easter), which in 1874 was 4 June.
4. *acoustic clouds*: in July 1873 Tyndall discovered that clouds of water vapour in the air could reflect echoes, altering high and low pitch sounds. He named the phenomenon '*an acoustic cloud*' in J. Tyndall, 'On the Acoustic Transparency and Opacity of the Atmosphere', *Roy. Soc. Proc.*, 7 (1874), pp. 169–78, on p. 61.
5. *the enclosed*: enclosure missing.
6. *Mr. Crotch*: George Robert Crotch (1842–74), a British entomologist who in 1873 had become the assistant of Louis Agassiz at the Museum of Comparative Zoology at Harvard, which Agassiz had founded in 1859.
7. *Mrs. Lesley, wife of Professor Peter Lesley . . . my American Campaign*: Susan Inches Lesley (née Lyman, 1823–1904) had married the geologist J. Peter Lesley in 1849, and spent her married life working on behalf of several philanthropic causes. Her husband, as secretary of the APS, had in September 1871 arranged a letter inviting Tyndall to come to the United States with the endorsement of numerous eminent Americans (see letter 3510, *Tyndall Correspondence*, vol. 12). As Hector Tyndale subsequently told Tyndall, 'Prof. Lesley . . . has done all the work & arranged it . . . he wrote that invitation & appended the names'. At Tyndale's behest, Lesley asked the philosopher and poet Ralph Waldo Emerson for permission to add his name to the letter, as what Tyndale told Tyndall was 'an additional gratification to you' (letter 3519, *Tyndall Correspondence*, vol. 12). In the early summer of 1874, the Lesleys were caring for Crotch, who was then in the final throes of tuberculosis, at their home in Philadelphia. He died on 16 June.
8. *Lilly*: Emily Hirst.
9. *Ruhmkorff*: Heinrich Ruhmkorff (1803–77), a German instrument maker (*ANB*).
10. *Clausius lame, and deeply sorrowful*: see letter 4343.
11. *Kirchhoff overpowered with work*: see letter 4348.

From Alexander Agassiz 10 June 1874 4354

Cambridge[1] June 10/74

My dear Mr. Tyndall.

I have this moment seen Rendu's Glaciers of Savoy edited by Geo[rge]. Forbes.[2] What can be done? I am utterly at a loss how to act. Were I in England I should cowhide[3] Ruskin and have at least some satisfaction. I must take more special notice of this than a mere article such as I wrote for the Nation[4] when Forbes's Life and Letters[5] came out. I shall put in a word or two in the Nation[6] now, and what I should like to do is to address you a letter on this subject,[7] and having submitted it to you ask your permission to have it printed, and I shall then give it as extensive a circulation as possible.[8] I had no idea that a scientific man—at least I supposed Tait to be one—could fall so low as to employ a scientific charlatan to write on his side (Ruskin), and then hold him up as a great mogul.[9] It is too contemptable that such sensational stuff[10] should be tacked on to Rendu's Essay. Perhaps you would prefer I should address my letter to Huxley who has had something less to do in the glacier matter, but I should like to address it to you, as you have been the only champion father has had in all this controversy. I am glad that the Geologists have at least had the good sense to steer clear of this infamous attack.

Let me hear from you at your leisure. Meanwhile in the intervals of my Museum[11] and Penikese duties[12] get my materials together.

Believe me | Yours very truly | Alex. Agassiz

IC HP 6.143

---

1. *Cambridge*: the city in Massachusetts.
2. *Rendu's Glaciers of . . . Geo[rge]. Forbes*: *Glaciers of Savoy*.
3. *cowhide*: flog with a cowhide, a strong whip made of the raw or dressed hide of a cow (*OED*).
4. *a mere article such as I wrote for the Nation*: [A. Agassiz], 'Life and Letters of Principal Forbes', *Nation*, 16 (1873), pp. 369–71. The *Nation* was a weekly journal published in New York by E. L. Godkin & Co.
5. *Forbes's Life and Letters*: *Life of Forbes*.
6. *a word or two in the Nation*: [A. Agassiz], 'Theory of the Glaciers of Savoy', *Nation*, 19 (1874), pp. 30–31.
7. *a letter on this subject*: letter 4378.
8. *have it printed . . . circulation as possible*: although it was sent to Tyndall, there is no record of Agassiz's letter subsequently being printed and circulated more widely.
9. *hold him up as a great mogul*: a mogul, originally the name of the leaders of a South Asian Muslim dynasty, was an important, influential or dominant person (*OED*), so Agassiz is presumably

referring to Peter Guthrie Tait's praise for John Ruskin's intervention in the glacier controversy: 'another actor has appeared upon the scene:—and with tremendous effect. The terrible words of Mr. Ruskin . . . will reach myriads of intelligent readers besides those who could otherwise be expected to interest themselves in a question involving (however remotely) scientific issues. Mr. Ruskin's admirable command of language, his clearness, impartiality, acuteness, and his exemplary firmness in declaring truth and doing justice, leave nothing to be desired' (P. G. Tait, 'Forbes and Dr. Tyndall', in *Glaciers of Savoy*, pp. 163–98, on p. 164).

10. *such sensational stuff*: Tait, 'Forbes and Dr. Tyndall' (see n. 9); and J. Ruskin, 'Extract from "Fors Clavigera"', in *Glaciers of Savoy*, pp. 199–210.
11. *my Museum*: see letter 4264, n. 4.
12. *Penikese duties*: the Anderson School of Natural History on Penikese Island off the coast of Massachusetts. The school was founded by Louis Agassiz in July 1873, but following his death five months later was run by his son. The school closed in 1875 following a fire. On this school, considered a forerunner of the Marine Biological Laboratory at Woods Hole, see S. G. Kohlstedt, *Teaching Children Science: Hands-On Nature Study in North America, 1890–1930* (Chicago: University of Chicago Press, 2010), pp. 20–22.

From Peter Guthrie Tait 10 June 1874 4355

University of Edinburgh | 10/6/74.

Sir,

I observe, in the number of the Contemporary Review just published,[1] several matters to which I may possibly, on another occasion, take objection. But, for the moment, I confine myself to the following passage which appears at p. 144:—

> "What he here says, coupled with what he had said before, regarding my ignorance, is a mere feeble copy of his description of Mr Lowe[2]—a man 'compounded in about equal proportions of fiend and fool'; and such repetition is unworthy of Professor Tait."

From this I gather that you attribute to me the words you mark as a quotation.

As I cannot remember having used such an expression, nor even having entertained the idea it conveys,[3] I have to request that you will at once refer me to the passage of my writings or reported addresses from which your quotation was made.

I reserve to myself the right of publishing this letter and any correspondence to which it may lead.

I am, Sir, | Your obedient Servant | P. G. Tait. | Dr Tyndall, F.R.S.

RI MS JT/1/TYP/9/3033
LT Typescript Only

1. *the number of the Contemporary Review just published*: referring to Tyndall's article 'Rendu and His Editors' (see letter 4333, n. 2).
2. *Mr Lowe*: Robert Lowe (1811–92), who had been Chancellor of the Exchequer and then Home Secretary in William Gladstone's government until its defeat in the general election in February 1874. In an address to the University of Edinburgh on 22 April 1874, Tait criticised Lowe, albeit without naming him directly, as an 'ex-Minister' whose taxation policies had 'aggravate[d] the miseries of numbers of the industrious poor'. Tait also objected to Lowe's calls for the centralized control of Scotland's four universities, and cuttingly remarked that he 'had his reward' at the election, and it was 'to be hoped that such a "Professor" will never again be in a position even to attempt to impress his mark upon Scottish, or upon any other, Universities' ('University of Edinburgh', *Scotsman*, 23 April 1874, p. 3). See also letter 4319, n. 6.
3. *I cannot remember . . . the idea it conveys*: in August 1874 the *Contemporary Review* printed a note clarifying where the quotation had come from: 'Professor Tyndall writes to us to correct a passage in his article published the June number of this REVIEW. Referring to Professor Tait as describing Mr Lowe to be a man "compounded in about equal proportions of fiend and fool"—Professor Tyndall was quoting the words of an article in the *Scotsman* of April 24th. The writer of that article sets forth certain charges of Prof. Tait against Mr. Lowe, and adds, "The Professor might have taken some less circuitous, though he could not have taken a plainer, way of intimating that the late Home Secretary is compounded of about equal proportions of fiend and fool". Professor Tyndall quoted these concluding words in his own article, and wishes to say that "they must be understood as the compendium of the *Scotsman*, and not as the *verbatim* utterance of Professor Tait". *Editor of* CONTEMPORARY REVIEW' ('Note', *Contemporary Review*, 24 (1874), p. 502). The particular charge against Lowe that was described in this way in the *Scotsman* was Tait's accusation that the former Chancellor of the Exchequer had been 'ready, for the sake of a sorry play upon words, to aggravate the miseries of numbers of the industrious poor', which the *Scotsman* explained thus: 'That is to say, Mr. Lowe's object in proposing the Match-tax was not to obtain revenue from a superabundant article from which revenue is successfully obtained in other countries, but only to perpetuate a Latin pun (*ex luce, lucellum* [out of light, a little profit]), which object he regarded as so gigantic as to render of no moment the inflicting of new miseries on the miserable' ('[Editorial]', *Scotsman*, 24 April 1874, p. 4).

To Thomas Henry Huxley 11 June 1874 4356

Bel Alp, Brieg, Switzerland.[1] | 11th. June, 1874.

My dear Huxley

Here is a copy of a letter[2] which reached me yesterday from a very influential man in Edinburgh.[3]

[. . .][4]

I know the writer of the above to be a longheaded[5] sagacious man, and therefore placed the whole matter before him, I sent him the little pamphlet,[6] which I sent to you in Switzerland[7] together with the Forms of Water,[8] and this is the result. When he wrote he had not seen the last article in the contemporary.[9]

Now oddly enough in my first treatment of the matter I <u>had</u> "kicked Forbes heartily,"[10] but on reconsideration I thought for the sake of our common vocation that it would be better to soften matters. This he[11] did not appreciate—If they press one further I think—with your sage permission—I should like to try my hand at this old analysis and give the world a picture of the man's doings as they then appeared to me. It would be exceedingly interesting to determine how far God Almighty, to whom he constantly prayed was able to neutralize the molecular motion set up in the gray brain-matter with which his sinful mother Nature had endowed him. In fact a psychological analysis of more than ordinary interest could be founded on the study of the character of Forbes.

I am here trying to work. But I have not yet got into the proper train—I have not yet got even my average power of sleeping. Still I feel myself getting stronger, and though I should like the strength to come more rapidly I have no doubt that a measure of it sufficient to carry me over Belfast will come at last. I have been alone here for several days. And the Alps which later on will swarm with life of men, beasts, and daddy longlegs, are wonderfully lonely. And still they are crowded with the loveliest flowers, violets "blue as thine eyes"[12] scattered like patches of the blue heavens over the hills. And all this beauty exists year after year without an eye to see it. If God Almighty works by "design" he does it rather awkwardly. But I must not blaspheme lest I should lose your good opinion—Quitting God and his doings let me return to the humanities—Give my love to your wife[13] and children,[14] and believe me

ever and always Yours | John Tyndall

IC HP 8.164
RI MS JT/1/TYP/9/3032
Typed Transcript Only

---

1. *Bel Alp, Brieg, Switzerland*: see letter 4306, n. 12.
2. *a copy of a letter*: letter 4352.
3. *a very influential man in Edinburgh*: James Coxe.
4. *[. . .]*: Tyndall copied the text of letter 4352 with the exception of the opening and closing salutations.
5. *longheaded*: discerning, shrewd (*OED*).
6. *I sent him the little pamphlet*: see letter 4324. The pamphlet was J. Tyndall, *Principal Forbes and his Biographers* (London: Longmans, Green, and Co., 1873), which was a reprint of the article of the same name published in the *Contemporary Review*, 22 (1873), pp. 484–508.

7. *I sent to you in Switzerland*: see letter 4154, *Tyndall Correspondence*, vol. 13, in which Huxley, who was staying in the Swiss town of Axenstein overlooking Lake Lucerne, acknowledged receipt of 'copies of your booklet' on 24 August 1873.
8. *Forms of Water*: *Forms of Water*.
9. *the last article in the contemporary*: see letter 4333, n. 2.
10. *in my first treatment of the matter I had "kicked Forbes heartily"*: Tyndall prefaced his article in the *Contemporary* by regretting that the 'attitude of a controversialist is once more forced upon me'. He nevertheless made several imputations against Forbes's character, and suggested that 'many a bitter drop fell from the pen of Principal Forbes into the lives of those whom he opposed' (J. Tyndall, 'Principal Forbes and his Biographers', pp. 485 and 506). After reading the pamphlet version of Tyndall's article in August 1873, Huxley had enthused: 'I think that nothing could be better than the tone you have adopted. I did not suspect that you had such a shot in your locker . . . It is a deadly thrust' (letter 4154, *Tyndall Correspondence*, vol. 13).
11. *he*: i.e. Coxe.
12. *violets "blue as thine eyes"*: Tyndall possibly has in mind James Russell Lowell's paired poems 'A Song' and 'Rosaline' (1842), the first of which begins 'Violet! sweet violet! | Thine eyes are full of tears', while the second line of the latter recounts 'Thine eyes were blue'.
13. *your wife*: Henrietta Huxley
14. *children*: the Huxley's seven surviving children were Jessica Oriana, Marian, Leonard, Rachel, Henrietta, Henry, and Ethel.

From George Gabriel Stokes 11 June 1874 4357

Copy | R.S.[1] Burlington House | 11 June '74

My Dear Tyndall,

I was very busy when y[ou]r. revise[2] reached me & c[oul]d not well examine it while you were still in the country. So I must write my remarks to Switzerland.[3] I will first write ab[ou]t. §[4] 10 w[hic]h is the first point.

My report[5] was sent you that you might adopt such suggestions therein contained as you might agree with. It is not desirable that you sh[oul]d. enter into controversy against arguments that you see no reason to agree with in an unpublished paper so that the public are not in a position to weigh fully what is to be said on both sides. Besides it is almost impossible that in the absence of the content brief extracts sh[oul]d fail to be more or less altered in meaning.[6]

Thus by ordinary circumstances (α)[7] I meant circumstances present to casual observers where in 99 cases out of 100 the sounds came overlaid. You have not yet returned my report but I remember writing that on land at least I thought it must be mainly temperature. Over sea I left it doubtful tho' inclined even there to temperature rather than moisture.[8] On this point my mind was open to be much influenced by the results of observations of surface (say down to a fraction of an inch) temperature of the ocean under conditions like those of July 3.[9]

(β) It is useless to shelter oneself under the authority of a great name in a matter in which once a doubt is started, there is a direct appeal to simple arithmetic. Accepting (γ) I ask how much vapour must be present, supposing the temp[eratu]re 60° F, to lighten the air as much as it w[oul]d be lightened by a dilatation of 1/10th? Arithmetic answers between 14 & 15 times as much as c[oul]d exist in a state of saturation! It is clear to me that Herschel neglected to make some arithmetical jotting as to vapour, unclear to what he made as to temper[atu]re or he w[oul]d not have written quite as he did[10]

(δ) This proves absolutely nothing as to the relative importance of temp[eratu]re & vapour. The ascending currents of lighter air were at the same time warmer & more moist. Concomitants must not be assumed to be causes. If, w[hic]h. is true, the lighter air were more moist, non sequitur, that the moisture was the chief cause of the lightness

(ε) By contact with warmed earth—yes—but not necessarily differently warmed. Be the warming ever so uniform ascending currents must be produced, because an arrangement with bottom strata of air naturally hotter than those above w[oul]d be essentially unreliable.

(ζ) From the sea certainly, but that, whether the surface temp[eratu]re be or be not uniform. In the one case the place of ascending currents w[oul]d be casual, in the other it w[oul]d be determined by the hotter spots.

(η) You have misconstrued the sentence quoted, in a manner w[hic]h. I admit the grammar permits, & it is only the spirit & drift of the argument that forbids. You are therefore arguing against a $\sqrt{-1}$ opponent.[11] If arguing for the honour of England as compared with France I have to say "If England's army be but small she is powerful by sea". I sh[oul]d. not mean that the military weakness was the cause or condition of the naval strength. Just so here my meaning was if on the one hand, as I concede, the warming of the water causes a greater supply of moisture I maintain on the other hand that it causes a greater supply of heat. Of course by communication of sensible heat. I never dreamt of making the vapour the cause of the heating of the air.

X (θ) The need of mottling[12] if such there were w[oul]d be absolutely the same in ascending moist air as in ascending heated air. In truth it is not needed for either (see δ)

What follows is argument directed against what through misapprehension of the intended construction of a sentence you supposed me to mean, but which really nobody maintains

(ι) The explanation of the effect of a local cloud is absolutely the same whether we suppose the uprush produced by warmth or by moisture or rather (since both are or may be really in operation) whether we supposed the one or the other have the lion's share in the result—It contributes therefore nothing towards a settlement of the question.

I have just (in my room at Cambridge)[13] tried an experiment bearing on the question. The room being at 60° I put warm water into a hip basin inserted a thermometer & noted the temp[eratu]re when the last trace of cloud disappeared; it was ab[ou]t. 92°. Doubtless on the large scale the difference between the tempe[ratu]re of sea (surface) & air must be a good deal less or visible clouds (like what you saw over the Serpentine)[14] w[oul]d have been formed [say] sea temp ≯ air temp. +15°

On the other hand the sea temp on July 3 must have been above the dew point or say (air—15°) or there w[oul]d have been fog. But between limits like

air temp ≠ 15°

(probably both really closer to air) the fact of the difference of sea temp. & air temp. w[oul]d not be revealed by a visible phenomenon & we must appeal to the therm[omete]r.

It is desirable to avoid discussion of the subject in the paper of the Phil. Trans. just about to be published[15] for the reason among others that half an hour's experim[en]t. in taking surface temp[eratu]re. (skin temperatures of course at the bows not stern) under conditions similar to those of July 3 w[oul]d do more to clean up the question than half a days discussion on à priori ground.

Please return § 10, of w[hic]h. I had the revise you sent to press, keeping the revise of that, in the form in w[hic]h. you w[oul]d like it to stand. Also to answer the q[uestion]. in page 200. I first thought you had omitted "question" but afterwards that you really meant to write echo in, meaning echo coming inwards from over the sea (or whatever other quarter)

See p. 206 I read "Air passed thro' perchloride of tin & sent into the tunnel produced exceedingly dense fumes. The action in the sounds waves was very strong." & in p 217 "The fumes of perchloride of tin tho' of extraordinary density exerted no sensible effect upon the sound"[16] Why those opposed results?

There is a little difficulty ab[ou]t. the proper date to put.[17]

We agreed that the first communication should stand as a "preliminary account" with an independent date. I enclose what I propose,[18] which please return as I have not a copy saying if you are satisfied.

Y[ou]rs sincerely | G. G. Stokes

CUL SC—add 7656

---

1. *R.S.*: the Royal Society.
2. *y[ou]r. revise*: the revised version of 'Vehicle of Sound'; the letter in which Tyndall sent it is missing, although he alerted Stokes to its imminent arrival in letter 4341.
3. *write my remarks to Switzerland*: Tyndall had given his address as 'Bel Alp | Brieg | Canton de Valais' in letter 4341. On the Bel Alp, see letter 4306, n. 12.

4. §: a typographical character, often known as a section sign, generally used to refer to a specific section of a document.
5. *My report*: G. G. Stokes, 'Report on a Paper by Dr Tyndall entitled On the Atmosphere as a Vehicle of Sound', RR/7/350, Referees' Reports: volume 7, RS. This was sent to Tyndall in letter 4295.
6. *It is not desirable . . . in meaning*: Stokes seems to object to Tyndall quoting from and disputing his own unpublished referee's report in the revised version of 'Vehicle of Sound'; none of the quotations from the report or Tyndall's criticisms of them appeared in the published version of the paper.
7. α: Stokes uses the letters of the Greek alphabet, from Alpha to Iota, to indicate the different parts of his analysis.
8. *I remember writing . . . rather than moisture*: in his referee's report Stokes had written: 'I think . . . that Dr Tyndall is not justified by the evidence . . . in assuming so confidently as he has done that difference of humidity rather than temperature is the cause of the heterogeneity which is revealed by its effects. I feel pretty confident that on land at least it must be part the reverse, and am by no means clear that such is not the case even out at sea' (RR/7/350).
9. *conditions like those of July 3*: on this date in 1873 Tyndall had conducted experiments at the South Foreland lighthouse near Dover (see letter 4185, n. 3). The meteorological conditions, as he recorded in the published version of 'Vehicle of Sound', were such as to ensure 'a calm sea' and 'an optically clear atmosphere' (pp. 195 and 209).
10. *written quite as he did*: presumably in J. Herschel, 'Treatise on Sound', in *Encyclopædia Metropolitana*, 30 vols (London: B. Fellowes etc., 1817–45), vol. 4 (1830), pp. 747–820, where, on p. 766, Herschel discusses the 'propagation of Sound in vapours'. In the published version of 'Vehicle of Sound', Tyndall noted that 'SIR JOHN HERSCHEL gives the following account of ARAGO'S observation' of thunder in '*Essay on Sound*', but he did not mention Herschel's discussion of vapours (pp. 199–200).
11. *a $\sqrt{-1}$ opponent*: a square root of negative, or imaginary, opponent.
12. *mottling*: used figuratively to indicate an unequal distribution of heat and moisture in the atmosphere, which, according to Tyndall, contributed to acoustic opacity.
13. *my room at Cambridge*: in Pembroke College, where Stokes was a fellow.
14. *the Serpentine*: a lake in Hyde Park, London. Tyndall conducted experiments there on 10 and 11 December 1873, the results of which he recorded in 'Vehicle of Sound' (pp. 209–14).
15. *the paper of the Phil. Trans. just about to be published*: see n. 2.
16. *"Air passed thro' . . . upon the sound"*: the respective passages appeared on pp. 214 and 215 of the published version of 'Vehicle of Sound'.
17. *a little difficulty ab[ou]t. the proper date to put*: see letter 4341, n. 7.
18. *what I propose*: presumably to give the date received as 5 February 1874, rather than 1 January.

To Thomas Henry Huxley 15 June 1874 4358

Bel Alp, Brieg[1] | 15th. June, 1874. | A foot of snow all round.

My dear Huxley

Tait has sent me the accompanying letter.[2]

Along with it I received another from Edinburgh[3] warning me against entering into any correspondence with him.

I cannot allow him to bother me now, and I shall therefore take no notice of the letter.

The words he refers to are quoted from the Scotsman Article of April the 24th.[4]

Something may occur—I hope not—to render it advisable to say a word; and I know you will say it if necessary.[5]

Yours ever | John Tyndall

IC HP 8.166
RI MS JT/1/TYP/9/3033
Typed Transcript Only

1. *Bel Alp, Brieg*: see letter 4306, n. 12.
2. *the accompanying letter*: letter 4355.
3. *another from Edinburgh*: letter 4352, from James Coxe.
4. *the Scotsman Article of April the 24th*: '[Editorial]', *Scotsman*, 24 April 1874, p. 4.
5. *Something may occur . . . say it if necessary*: in fact, it was Tyndall who found it necessary to clarify the quotation he had attributed to Peter Guthrie Tait, printing a correction in the *Contemporary Review* (see letter 4355, n. 3) and writing a letter to Tait (letter 4365) that he asked Huxley to forward (see letter 4366).

To George Gabriel Stokes 17 June [1874][1] 4359

Bel Alp 17th June.

My dear Stokes.

In my desire to get off my letter yesterday[2] I omitted to refer to one of your questions: and I am not sure that I am now in a position to make my answer clear. You will see that the first experiments on fumes were made in the "Tunnel";[3] and that they often produced an effect upon the sound waves; but this was due entirely to heat, and to the way in which the fumes were introduced into the tunnel. It was to correct the results thus obtained that the cupboard was resorted to.

I apprehend therefore that one of the results you mention was a tunnel result.

I am very much obliged to you for the trouble you have taken both in the Report, & in your last letter.[4] Let me frankly say however that there is an expression in your paragraph "β"[5] that I would rather see not there. However I dare say no sharpness on your part was intended.

Yours faithfully | John Tyndall

M^r^. Cottrell might be able to answer any question arising in future.

Professor Stokes D.C.L. | Lensfield Cottage | Cambridge

RI MS JT/1/T/1423
RI MS JT/1/TYP/4/1488

---

1. *[1874]*: the year is established by the relation to letter 4357.
2. *my letter yesterday*: letter missing.
3. *experiments on fumes were made in the "Tunnel"*: Tyndall responds to a question asked by Stokes in letter 4357 concerning experiments on the effect on sound of the fumes from perchloride of tin, as described in 'Vehicle of Sound' (pp. 214 and 215). 'Tunnel' refers to the experimental device used to demonstrate the effect in the laboratory.
4. *your last letter*: letter 4357.
5. *an expression in your paragraph "β"*: the passage that offended Tyndall was 'It is useless to shelter oneself under the authority of a great name in a matter in which once a doubt is started, there is a direct appeal to simple arithmetic' (see letter 4362).

From Edward Sabine 19 June 1874 4360

13 Ashley Place[1] | June 19th 74

Dear Tyndall,

I called at the Institution[2] yesterday, to return the Volume of American Scenery[3] which you were so obliging as to lend me some weeks ago, and with it a view of the Queens College at Belfast,[4] which you may like to see, and which you can return when you return from the Continent, where I hear you have been for the last 3 weeks, but are expected in Albemarle S[tree]^t^ shortly.[5] I will beg you to return the Belfast College at your convenience, as I set some value on it, in remembrance of the meeting, in 1852.[6]

always sincerely yours, | Edward Sabine | D^r^. Tyndall.

RI MS JT/1/S/28
RI MS JT/1/TYP/4/1339

---

1. *Ashley Place*: in Westminster.
2. *the Institution*: the RI.

3. *the Volume of American Scenery*: possibly the recently published second volume of W. C. Bryant, *Picturesque America*, 2 vols (New York: D. Appleton, 1872–4). Alternatively, it may be one of the volumes of N. P. Willis, *American Scenery*, 2 vols (London: George Virtue, 1840).
4. *the Queens College at Belfast*: having been founded in 1845, the college hosted sessions during the 1852 meeting of the BAAS, and would again for the 1874 meeting.
5. *expected in Albemarle S[tree]t shortly*: in fact, Tyndall did not return to the RI, in Albemarle Street, until 10 August.
6. *the meeting, in 1852*: the last time the BAAS had met in Belfast, with Sabine serving as president (*Brit. Assoc. Rep. 1852*).

To William Frederick Pollock 20 June 1874 4361

20 June 1874.

Many thanks to you my dear Pollock for the bright sparkle of your friendship which gleamed upon me to-day. It was very pleasant to receive it.[1] I have been working at a tangled skein[2] ever since I came here, and am slowly unravelling it. I roam, and think, and read, as I roam. Sometimes halting under the shade of rocks, amid violets and gentians, soothed by the sound of cascades, whose mellow voices are heard everywhere. Sometimes I abandon my work and resolutely face the hills; bring the drops down my brow and then halt and work again. I am getting slowly stronger; and my grasp of my subject is becoming firmer: but whether it is to bring me weal or woe I know not.

I have not seen the article[3] since its publication. I am glad you think it reads well. Tait has written to me[4] asking for some explanation—but I have not replied to him—nor give him the least attention till my Belfast work[5] is done.

Tell Fred[6] that I had his pleasant letter:[7] I am grieved to hear of the new mishap.[8] But I hope he now sees the end and issue of all mishaps in health and strength.

With best affection to you and yours, wishing you prosperity in your work, peace in your heart, and happiness in your home.

Believe me | always your friend | John Tyndall.

Sir Fred[eric]k. Pollock Bart. | 57 Montague Square | W.[9]

RI MS JT/1/TYP/6/2150
LT Typescript Only

---

1. *receive it*: letter missing.
2. *working at a tangled skein*: his presidential address to the BAAS at Belfast. A skein is a length of thread or yarn, wound into a loose knot (*OED*).

3. *the article*: see letter 4333, n. 2.
4. *Tait has written to me*: letter 4355.
5. *my Belfast work*: writing the presidential address that Tyndall would deliver at the meeting of the BAAS in Belfast in August.
6. *Fred*: Frederick Pollock.
7. *his pleasant letter*: letter missing.
8. *the new mishap*: not identified.
9. *59 Montague Square | W.*: the Pollocks resided at 59 Montagu Square in Marylebone, London, a fashionable London neighbourhood.

From George Gabriel Stokes 21 June 1874 4362

Lensfield Cottage Cambridge | 21st June 1874

My Dear Tyndall,

I am much concerned that anything in my letter[1] should have given you pain.[2] Fortunately I kept a copy for reference in case of need, as I am here so far off, & I have referred to β.[3] I will ask you to draw your pen thro' the words "shelter oneself under" and substitute "invoke", and then I think it will express my meaning without anything liable to cause pain.

In a recondite matter where there are a number of pros & cons to weigh, one naturally <u>and rightly</u> pays great deference to the opinion of a high authority. What I meant was that that does not apply to the present case in which <u>once a doubt is started</u> there is a direct appeal to the same inexorable authority to which doubtless Herschel himself would have appealed had the doubt occurred to him, I mean to arithmetical calculation. Were there reason to think that Herschel wrote as he did[4] <u>after having made the numerical calculation</u>, then it would be reasonable to cite his authority; but everything tends to show that he neglected to make it.

As to perchloride of tin, I thought of temperature—of the cold of evaporation if air were pressed through it—, but I hardly thought that would be enough (the liquid apparently not being externally volatile). Not having heard on this point I sent the paper to press yesterday morning, to take its chance on this point, but I will write to M[r] Mattress to send that part of the paper (if not too late) to M[r] Cotterill, who probably knows the actual experiment, to make it clear if it be not so already.

Believe me | Yours sincerely | G. G. Stokes

CUL SC—add 7656

1. *my letter*: letter 4357.
2. *given you pain*: see letter 4359, n. 5.

3. β: in letter 4357 Stokes used the letters of the Greek alphabet, from Alpha to Iota, to indicate the different parts of his analysis, so β (for Beta) refers to the letter's second section.
4. *Herschel wrote as he did*: see letter 4357, n. 10.

From Thomas Henry Huxley 24 June 1874 4363

4 Marlborough Place | London N.W.[1] | June 24 1874.

My dear Tyndall,

Your letters[2] and the belligerent enclosure[3] in the last have come to hand. I have been looking at the article in the "Contemporary"[4] and I am afraid that Tait has a case against you, for the context may certainly be fairly taken to imply that you quote the phrase about Lowe as Tait's, whereas I suppose they are really the Scotsman's compendious epitome of the sense of what Tait said about Lowe.[5] In reading your proof I took it for granted that the words were Tait's own, but it was very stupid of me not to call your attention specially to the point. As you are abroad I do not know that it is needful for you to take notice of Tait's letter, at present; but I think that you will have to explain that the quotation was from the "Scotsman" and not from him sooner or later. You have a perfect right to hold no communication with him if you think fit, but the explanation should be given somewhere. I will take care of his letter till your return.

I quite agree with your Scotch friend[6] in his estimate of Forbes, and if he were alive and the controversy beginning, I should say draw your picture in your best sepia or lamp black.[7] But I have been thinking over this matter a good deal since I received your letter, and my verdict is, leave that tempting piece of portraiture alone.

The world is neither wise nor just, but it makes up for all its folly and injustice by being damnably sentimental, and the more severely true your portrait might be, the more loud would be the outcry against it. I should say publish a new edition of your "Glaciers of the Alps"[8] make a clear historical statement of all the facts showing Forbes's relations to Rendu and Agassiz,[9] and leave the matter to the judgments of your contemporaries. That will sink in and remain when all the hurly-burly is over.

I wonder if that Address is begun, and if you are going to be as wise and prudent as I was at Liverpool.[10] When I think of the temptation I resisted on that occasion, like Clive when he was charged with peculation, "I marvel at my own forbearance!"[11] Let my example be a burning and a shining light to you. I declare I have horrid misgivings of your kicking over the traces.[12]

The 'x' comes off on Saturday next,[13] so let your ears burn for we shall be talking about you. I have just begun my lectures to Schoolmasters[14] and I wish they were over, though I am very well, on the whole.

Griffith[15] wrote to ask for the title of my lecture at Belfast[16] and I had to tell him I did not know yet. I shall not begin to think of it till the middle of July[17] when these lectures are over.

The wife[18] would send her love but she is gone to Kew[19] to one of Hooker's receptions taking Miss Jewsbury[20] who is staying with us. I was to have gone to the College of Physicians dinner[21] to-night, but I was so weary when I got home that I made up my mind to send an excuse. And then came the thought that I had not written to you.

Ever yours sincerely | T. H. Huxley.

IC HP 8.168
RI MS JT/1/TYP/9/3034–5
Typed Transcript Only

1. *4 Marlborough Place | London N.W.*: see letter 4213, n. 1.
2. *Your letters*: letters 4356 and 4358.
3. *the belligerent enclosure*: letter 4355.
4. *the article in the "Contemporary"*: see letter 4333, n. 2.
5. *Tait has a case against you . . . said about Lowe*: see letter 4355.
6. *your Scotch friend*: James Coxe.
7. *your best sepia or lamp black*: sepia is a pigment of a rich brown colour prepared from the inky secretion of cuttlefish, while lamp black is a pigment of almost pure carbon made by collecting the soot produced by burning oil or gas (*OED*), so Huxley means the darkest colours possible.
8. *your "Glaciers of the Alps"*: J. Tyndall, *The Glaciers of the Alps* (London: John Murray, 1860). Tyndall did not follow Huxley's advice, and a new edition of *Glaciers of the Alps* was only published in 1896, after Tyndall's death. In a 'Prefatory Note', Louisa Tyndall explained that the book was reprinted 'at the suggestion of my husband's Publishers . . . as the older work is still frequently asked for' (J. Tyndall, *The Glaciers of the Alps*, 2nd edn (London: John Murray, 1896), p. ix).
9. *Forbes's relations to Rendu and Agassiz*: James David Forbes stood accused of borrowing elements of his theory of glacial motion from Louis Rendu, and of having been beaten by Louis Agassiz in a dispute over priority in measuring the motion of glaciers. See letter 4185, n. 11.
10. *as wise and prudent as I was at Liverpool*: Huxley's presidential address to the BAAS at Liverpool in 1870, dealing with the generation of life from non-living material (spontaneous generation), was commended in the *Athenæum* for being 'written in the best taste, devoid of any sentences obnoxious to the most sensitive audience' ('British Association', *Athenæum*, no. 2238 (1870), pp. 371–8, on p. 373). The address was published as 'Address of Thomas Henry Huxley', *Brit. Assoc. Rep. 1870*, pp. lxxiii–lxxxix; and 'Address of Thomas Henry Huxley, LL.D., F.R.S., President', *Nature*, 2 (1870), pp. 400–6.
11. *Clive when he was charged . . . my own forbearance!"*: Robert Clive (1725–74) was subject to

parliamentary inquiries in 1772 and 1773 over whether his huge earnings in India had been made at the expense of the East India Company and the government. Clive responded that he was astonished at his own moderation, considering the riches available to him.

12. *kicking over the traces*: (of a horse) to get a leg over the traces—or harness—so as to kick more freely and vigorously; thus, figuratively, to throw off the usual restraints (*OED*).
13. *The 'x' comes off on Saturday next*: the meeting of the X Club that was due to take place on Saturday 27 June was cancelled, seemingly because not enough members were able to attend. Thomas Hirst, who was ill, reflected in his journal for 28 June on the 'X party having been given up' (*Hirst Journals*, p. 2001). In 1864 Tyndall joined George Busk, Edward Frankland, Thomas Hirst, Joseph Hooker, Thomas Huxley, John Lubbock, Herbert Spencer, and William Spottiswoode as members of the X Club, a monthly dining club in which they discussed the promotion of science in British society (see R. Barton, *The X Club: Power and Authority in Victorian Science* (Chicago: University of Chicago Press, 2018)).
14. *my lectures to Schoolmasters*: see letter 4327, n. 8.
15. *Griffith*: George Griffith.
16. *the title of my lecture at Belfast*: see letter 4327, n. 9.
17. *I shall not begin to think of it till the middle of July*: on 22 July Huxley was 'thinking of taking Development for the subject of my evening lecture' (letter 4377), and it seems only to have been in the following month that he finally decided on his topic, telling Michael Foster on 12 August: 'I have chosen "Animal Automation" and its history—about the most completely scabreux topic I could have selected' (*Foster and Huxley Correspondence*, p. 51).
18. *The wife*: Henrietta Huxley.
19. *Kew*: either at the Royal Botanic Gardens, Kew, where Hooker had been Assistant Director (1855–65) and Director (1865–85), or his nearby residence at 49 Kew Green.
20. *Miss Jewsbury*: Geraldine Jewsbury (1812–80), a novelist.
21. *the College of Physicians dinner*: not identified, although the Royal College of Physicians, founded in 1518, held regular dinners for its members at its headquarters in Pall Mall East.

From John Herschel 25 June 1874 4364

21 Sumner Place.[1] | June 25th. 74.

My dear Sir,

I have the pleasure to return you my father's letters.[2] I am greatly indebted to you for hunting them up, for in some respects they are more interesting than any other set I have seen. The frequent reclamations[3] which they contain are especially remarkable. Taken in conjunction with many other scattered instances of the same kind, these denote a phase of character which is at first sight so at variance with the rest that it compels a reconsideration. I do not pretend to have mastered so large a subject, but one must theorize to retain a perception of acquired facts. My theory however is hardly fit to be cast into form yet. Enough that I range this fact along with tenacity of purpose; independence approaching to an assertion of individuality; justice to all alike, self included. So firm a retention of what was his own, from its first manifestation in 1819, and through all the intervening years till 1870, coexisting with a reticence such as is declared in the letter to yourself dated July 21, 1861,[4] (and such as I trace elsewhere) and with a resolute self abnegation in other directions, is a characteristic of importance. I cannot help some curiosity as to the judgment of others on such a point.

Yours very truly | J. Herschel
Prof. Tyndall.

RI MS JT/1/TYP/2/581
LT Typescript Only

---

1. *21 Sumner Place*: Herschel's London home, in South Kensington.
2. *my father's letters*: Tyndall had corresponded with Herschel's father, John Frederick William Herschel, since November 1851 (see letter 566, *Tyndall Correspondence*, vol. 3), and while it is unclear precisely which letters were enclosed, Herschel had requested Tyndall's assistance with his family's 'wish to form a collection of my father's letters' in letter 4228, asking him to provide correspondence 'for transcription' but without the intention of publication.
3. *reclamations*: the action of claiming something back or of reasserting a right (*OED*).
4. *reticence such as is declared in the letter to yourself dated July 21, 1861*: see letter 1796, Diarmid Finnegan, Roland Jackson, and Nanna Katrine Lüders Kaalund, eds., *The Correspondence of John Tyndall*, vol. 7, *The Correspondence, March 1859–May 1862* (Pittsburgh: University of Pittsburgh Press, 2019). Herschel seems to be referring to his father's declaration, in the letter, of his 'principle . . . of never citing my own works—but in some such oblique form of expression'.

To Peter Guthrie Tait [27][1] June 1874 4365

Bel Alp[2] | 29th June | 1874.

Sir,

The passage, with its context, from which the words quoted by me[3] were taken, is this:—"The Professor [yourself][4] might have taken some less circuitous, though he could not have taken a plainer way of intimating that the late Home Secretary[5] is compounded of about equal proportions of fiend and fool."*

I have the honour to be | Sir | Your obedient Servant | John Tyndall
Professor Tait | &c. &c. &c.
*Scotsman 24th April 1874.[6]

IC HP 1.127

---

1. *[27]*: Tyndall seems to have post-dated this letter to the following Monday, 29 June, as it was to be forwarded by Thomas Huxley once he received it as an enclosure to letter 4366, which was written on Saturday 27 June.
2. *Bel Alp*: see letter 4306, n. 12.
3. *the words quoted by me*: in J. Tyndall, 'Rendu and His Editors', *Contemporary Review*, 24 (1874), pp. 135–48, on p. 144 (see letter 4355, n. 3).
4. *[yourself]*: this clarification was inserted, above the written line, by Tyndall himself, seemingly as an afterthought.
5. *the late Home Secretary*: Robert Lowe (see letter 4319, n. 6; and letter 4355, n. 2).
6. *Scotsman 24th April 1874*: see letter 4358, n. 4.

To Thomas Henry Huxley [27 June 1874][1] 4366

Saturday. | a horrible day here:[2] I hope the X[3] have it better.

Dear Hal,

Should you think it desirable to forward this note to Tait[4] put a stamp on it for me like a good fellow & have it posted. It might perhaps be as well to await his attack, which is sure to come. However you will decide.

Alexander Agassiz has written to me a terribly indignant letter[5]—He is going in at them hammer and tongs.[6] I send you his letter[7] & have told him that I am quite willing that he sh[oul]d write to me.

Yours ever | John Tyndall

I am steering very close to the wind,[8] but hope to keep clear of rocks. I want you to read it[9] before it is delivered.

IC HP 1.126
Transcript Only

---

1. *[27 June 1874]*: the date is established by the relation to letter 4365 (albeit this seems to have been post-dated) and to the proposed meeting of the X Club on Saturday 27 June (see n. 3).
2. *here*: Bel Alp in the Swiss Alps, where Tyndall had been since the beginning of the month (see letter 4306, n. 12).
3. *the X*: this meeting of the X Club, which was due to take place on Saturday 27 June, was cancelled (see letter 4363, n. 13). On the X Club, see letter 4363, n. 13.
4. *this note to Tait*: letter 4365. It is unclear whether Huxley forwarded it.
5. *a terribly indignant letter*: letter 4354.

6. *hammer and tongs*: with might and main, like a blacksmith showering his blows on the iron taken with tongs from the forge-fire (*OED*).
7. *I send you his letter*: Agassiz's letter was actually enclosed in letter 4369.
8. *steering very close to the wind*: a variation of the nautical expression 'sailing close to the wind', meaning to come near to transgression of a law or a received moral principle (*OED*); Tyndall was seemingly referring to the potentially controversial elements of his presidential address to the BAAS (see n. 9).
9. *it*: Tyndall's presidential address to the BAAS, the proofs of which were being printed in installments. Tyndall later reported that 'Huxley has read almost, not quite the whole, and approves' (letter 4392).

From Joseph Henry 29 June 1874 4367

June 29. '74

My dear Professor;

Accompanying this I send you copies of several letters[1] relative to an application of the Tyndall Fund[2] the originals of which are on file in the Institution,[3] and you need not, therefore, return the copies, which are as follows;

1. Letter from myself to Gen[era]l. Tyndale and Prof. Youmans.
2. Letters from Prof. C. S. Lyman[4] in relation to Mr. Hastings.[5]
3. Letter from Prof. Youmans.
4. Letter from Gen[era]l. Tyndale.

As I have stated in my letter to Gen[era]l. Tyndale, while on the one hand I think it important the fund should be increased[6] on the other it should be producing fruit in the way you originally contemplated[7] since time is an important element in the progress of the world and assistance timely rendered to a single individual may produce results that will induce others to follow your example in the definite present without waiting for the uncertain future.

I have read with much interest the account of your experiments on Fog Signals[8] and hope to have an opportunity during my coming vacation to verify some of the results.[9] Mr. Brown the patentee of the Sirene,[10] has been getting up a series of modifications of apparatus, at the expense of the Board[11] under my authority for increasing the efficiency of the instrument if possible. In testing the various projects he has employed the effect produced by the vibration of membranes as indicated by the motion of sand structured upon them. It will be part of my duty during this summer to examine these results and to recommend their rejection or adoption as the case may be.

I am very ever truly | yours | Joseph Henry

Prof. John Tyndall, | &c &c, | London.

Smithsonian Institution Archives. RU 33, vol. 39, pp. 182–5

---

1. *several letters*: enclosures missing.
2. *the Tyndall Fund*: Tyndall's lecture tour of the United States in 1872–3 made a surplus of $13,033, which was invested to establish a fund for pure science research in the United States (see letter 4214, n. 5).
3. *the Institution*: the Smithsonian Institution in Washington, DC.
4. *Prof. C. S. Lyman*: Chester Smith Lyman (1814–90), professor of astronomy and physics at Yale's Sheffield School of Science.
5. *Mr. Hastings*: Charles Sheldon Hastings.
6. *the fund should be increased*: the original plan was to 'appropriate the interest of the fund' to make grants to support American students of physics, but with the proviso that 'If in the course of any year the whole amount of the interest which accrues from the fund be not expended in the manner before mentioned, the surplus may be added to the principal' ('Professor Tyndall's Deed of Trust', *Popular Science Monthly*, 3 (1873), pp. 100–01, on p. 101). However, Henry seems to be proposing to spend money from the principal fund (see letter 4390).
7. *producing fruit in the way you originally contemplated*: Tyndall had proposed that the fund should be used for the 'advancement of theoretic science, and the promotion of original research, especially in the department of physics, in the United States' (p. 100).
8. *the account of your experiments on Fog Signals*: probably J. Tyndall, 'Preliminary Account of an Investigation on the Transmission of Sound by the Atmosphere', *Roy. Soc. Proc.*, 22 (1874), pp. 58–68.
9. *an opportunity during . . . verify some of the results*: Henry conducted these tests on 25 August and 1 and 24 September at Sandy Hook, Connecticut. See *Annual Report of the Light-House Board of the United States* (Washington, DC: Government Printing Office, 1874), pp. 111–7.
10. *Mr. Brown the patentee of the Sirene*: Jean Nickolai Henry Adolphus Brown (1823–75), a Belgian-born engineer whose patented steam siren consisted of a rotating slotted cylinder alternately opening and closing a passageway to either steam or compressed air. The rotation ensured the desired frequency, and the cylinder sat in the throat of a large horn.
11. *the Board*: the United States Lighthouse Board, of which Henry had been a member since its foundation in 1852 and was now the chairman.

To Heinrich Debus 1 July 1874 4368

Bel Alp, Brieg[1] | 1st July 1874

My dear Heinrich

I deem myself a good boy for thinking of writing to you while the thought of writing to many other friends has not yet occurred to me. Or at least you ought to think me a good boy, whatever my own thoughts on the matter may be. I have been here now between 3 and 4 weeks, with very variable weather; for the most part in the highest degree reprehensible, and particularly so as it has prevented me from acquiring that physical strength and elasticity which are so necessary to strong brain-work. I ought perhaps to have made myself strong in the first instance, but I feared I should thereby superinduce a certain joyousness and inflexibility of brain, so I attacked my work directly I came here; alternating it of course with moderate mountain rambles. I was getting very strong, but suddenly; I suppose through incautious exposure to rain and a chilly atmosphere, I caught cold, and for the 3rd time in my life my teeth began to annoy me.[2] I could not press them together without pain. This is now gone, but a swelling remains in my right under jaw, and I doubt not a boil, or little abscess is there which will, I suppose, break by and by. All this has shorn me of vigour: still I have lost neither heart nor hope.

I am taking a range over the notions that men have entertained regarding the universe: keeping close throughout to the atomic theory:[3] and briefly contrasting other modes of thought with it. Already I have sent the first instalment of the address[4] to the printer.[5] But the next, which I have not yet begun, will be the most difficult.[6] I have been obliged to read a great deal, and this with the subsequent pondering, and the effort to distil the real meaning and essence out of what I read, makes the work slow. I should like much to have the advantage of your opinion when the paper is in proof, and therefore should feel thankful if you would let me know where you are likely to be at the end of July or during the early days of August.

I sent Tom[7] some flowers, and a few lines[8] soon after my arrival here. I dare say poor Tom has enough on his hands—otherwise he would have written to me before now. Tell me how he is. During this chill I have felt symptoms at my finger ends something like what he describes[9]—but I wish his symptoms were as transitory as mine. That they would be transitory if he could give himself healthy rest and nutrition I am quite convinced.

We have had two snow falls since I came here. The last one a day or two ago—killing the insects in multitudes; rendering the fleshy gentians[10] prostrate—but the tough & elastic violets look fresh immediately the snow has disappeared. Thus far happily I have been able to have a room with a fire in it to work in. Perhaps it was the delivery of this room to a lady who had fallen with a horse into a stream,[11] that precipitated my cold.

But as I have said, I am by no means out of heart. This headache & gum-ache, will soon disappear, and then work will progress gaily.

Ever dear Heins Yours | John Tyndall

RI MS JT/1/T/270
RI MS JT/1/TYP/7/2375–6

1. *Bel Alp, Brieg*: see letter 4206, n. 12.
2. *my teeth began to annoy me*: in letter 4370, Tyndall stated that he had been diagnosed with 'an inflammation of the covering membrane of the tooth'.
3. *the atomic theory*: Tyndall began the Belfast Address by detailing the ancient theory, first expounded by Democritus, Epicurus, and Lucretius, that atoms in empty space, combining in accordance with mechanical laws, were the materials from which all things are constructed. This opening section of the Address concluded: 'In our day there are secessions from the theory, but it still stands firm . . . In fact, it may be doubted whether, wanting this fundamental conception, a theory of the material universe is capable of scientific statement' ('Address of John Tyndall', *Brit. Assoc. Rep. 1874*, pp. lxvi–xcviii, on p. lxxvii).
4. *the first instalment of the address*: pp. lxvi–lxxviii (see n. 2).
5. *the printer*: Taylor and Francis, Red Lion Court, London. See W. H. Brock and A. J. Meadows, *The Lamp of Learning: Two Centuries of Publishing at Taylor & Francis* (London: Taylor and Francis, 1998).
6. *the next . . . will be the most difficult*: the next section of the Belfast Address, pp. lxxix–lxxxi, contained an imaginary dialogue between the eighteenth-century theologian Joseph Butler and a disciple of Lucretius on the issue of consciousness.
7. *Tom*: Thomas Hirst.
8. *a few lines*: letter 4353.
9. *symptoms at my finger ends something like what he describes*: Hirst had suffered from numbness and chills in his legs from 8 November 1873 (see letter 4225, n. 3), and on 7 January recorded in his journal that while his legs now got better his 'hands grew worse. The little fingers first lost sensation and grew stiff and useless. This gradually extended to the other fingers and at last whenever I attempted to close my fist painful tinges crept up the sides of my fingers. This grew worse and worse and even now it clings to me and renders writing difficult' (*Hirst Journals*, p. 1994).
10. *gentians*: plants common to mountainous regions, which typically have blue trumpet-shaped flowers.
11. *a lady who had fallen with a horse into a stream*: not identified.

To Thomas Henry Huxley 1 July 1874 4369

1st. July 1874.

My dear Huxley

I packed up the last letter[1] without enclosing that of Alex[ander]. Agassiz.[2] This I now send to you. I have told him that I do not at all object to his writing to me, and that when his letter[3] comes, you and I will lay our heads together and see what is the best use that can be made of it.

A final exposition will be necessary after the Belfast meeting[4] Agassiz will probably send me a letter prior to the meeting so that if any maneuvers be made there, we can if we like, show that there are two sides to the question, and that uninformed people had better suspend their judgment. I do not know whether you have read that brochure of Agassiz's[5] to which I referred in the last Contemporary Article.[6] It might be published in full. Taking all things into account, and seeing no escape from a final onset there is I think no use in being in a hurry to reply to Tait's question to me.[7] I daresay Knowles would insert a couple of lines for me in the next number of the Contemporary;[8] and these lines I will send for your approval either today or tomorrow.

In the Address[9] my first aim will be to be true to myself, and to avoid as far as it is possible to do so, causing the least pain to others. I dare say some parts of the Address will not be palatable; but I hope on the whole to send it forth with the stamp of fairness upon it. But it cannot come up to the standard which I should like it to reach; for I am slow in pulling things together, and the barbarous weather with which we have been afflicted here[10] has given me a chill, and a spice of neuralgia,[11] which have been very considerable drawbacks, too. The chill however is now surmounted, so that I hope to have plain sailing for the next three weeks.

I have been obliged to read up a good deal: which added to the necessary pondering afterwards has made my work slow: the first instalment of the Address[12] has however been sent off to the printer.[13] I shall feel a great mastery of it when it is in print before me.

I should like to know Spencer's habitat towards the end of this month.[14] If you see him would you ask him to send his address to the R.I.?

ever yours | John Tyndall

It was Knowles, not I, who cut-out the "voluble amateur": he made other excisions also[15]

IC HP 1.128

---

1. *the last letter*: letter 4366.
2. *that of Alex[ander]. Agassiz*: letter 4354.
3. *his letter*: letter 4378.
4. *the Belfast meeting*: of the BAAS, held from 19 to 26 August.
5. *that brochure of Agassiz's*: L. Agassiz, [*Lettre circulaire au sujet de la controverse avec James D. Forbes*] (Neuchatel: privately printed, 1842).
6. *the last Contemporary Article*: J. Tyndall, 'Rendu and His Editors', *Contemporary Review*, 24 (1874), pp. 135–48. Tyndall refers to the 'very rare brochure published by Agassiz, in evident affliction of mind, in 1842' on p. 140.

7. *Tait's question to me*: in letter 4355, Peter Guthrie Tait asked Tyndall to tell him the source of a quotation that was wrongly attributed to him in 'Rendu and His Editors'.
8. *Knowles would insert . . . the Contemporary*: see letter 4355, n. 3.
9. *the Address*: the presidential address that Tyndall would deliver at the Belfast meeting of the BAAS.
10. *here*: Bel Alp in the Swiss Alps, where Tyndall had been since the beginning of June (see letter 4306, n. 12).
11. *a spice of neuralgia*: a pain in the nerves of Tyndall's face, brought on by a problem with his teeth (see letter 4370).
12. *the first instalment of the Address*: pp. lxvi–lxxviii of 'Address of John Tyndall', *Brit. Assoc. Rep. 1874*, pp. lxvi–xcviii.
13. *the printer*: Taylor and Francis (see letter 4368, n. 5).
14. *Spencer's habitat towards the end of this month*: Herbert Spencer would be staying at Ardtornish, an estate in the Highlands of Scotland (see letters 4383 and 4386).
15. *It was Knowles . . . other excisions also*: James Thomas Knowles was the editor of the *Contemporary Review*, so these are presumably editorial revisions to 'Rendu and His Editors', aimed at rendering it less controversial. Tyndall's original choice of words, 'voluble amateur', almost certainly referred to the art critic John Ruskin, and may have been excised by Knowles from a passage in the article, following a quotation from Ruskin criticizing Tyndall's position in the dispute over glaciers, that was published as 'If these "terrible" words be *true* words, why was it left to an amateur to utter them? . . . To these and other observations of Mr. Ruskin I offer no reply' ('Rendu and His Editors', p. 147).

To Thomas Archer Hirst 5 July 1874 4370

5th July 1874

My dear Tom

Your note[1] has come to me & it has been pleasant to receive it.

My life thus far in this place[2] can hardly be considered prosperous: I needed all my thoughts but for 10 whole days it slipped away from me & left me an intellectual cripple. During our severe weather I was incautious, went on the glacier, drank its water, got hot and cold, was rendered moist by rain after having been warm—Still did nothing that I could dream would have seriously affected me. Curiously enough the teeth all began to feel tender: then a concentration of uneasiness in one of them: then a swelling and a long continued, contemptible nagging pain. I understood for the first time your desire to escape from the Alps and to get into sunny Italy.[3]

The bad weather is now gone, and the sun has recovered great power: I am improving. A surgeon here says it was an inflammation of the covering membrane of the tooth that I suffered from. There's a little hole in it; into which Fletcher[4] forced a little gold leaf before I came away—I thought at the time that he did it rather hastily. Sometimes I ask myself whether it would not have been better had I trusted dear old mother Nature, and never put myself into the hands of a dentist at all.

My work has been utterly broken for the last 12 days, but I must now tackle to it with a vengeance. Still it is not work that can be forced: I intended it to be well-pondered work—and such work is not executed on the immediate command of the will. I must lower my ideal—and indeed the Address[5] may possibly be a very poor affair. To do it aright would require perfect health on my part, and perfect health is a thing that does not always come when we call.

That I am selfishly taking up all my space with myself and my affairs, to the exclusion of what is far more important—the state of your health. Your letter is cheerful, and I trust your next one will be more so. It will be necessary to ask you to do something for me—that is to read the Address when it is put into shape. And therefore I shall need to know your address when I return. In fact it would be kind of you, after you have made up your mind as to what your movements are to be, to let me know all about them.

I have quite a library of books around me here: and the difficulty is to shake myself loose from their trammels after I have made myself acquainted with what they contain. I think this must be done to produce anything possessing independent vitality. However you know the difficulties of my work, and I need not dwell upon them.

It amuses me to find Huxley expressing anxiety lest I should kick over the traces,[6] while you have not uttered an anxious word.

Alexander Agassiz has written me a most indignant letter,[7] a copy of which I sent to Huxley.[8] It was apropos of Rendu & his quartette of editors.[9]

My old guide at the Niagara Falls[10] has been covering himself with glory. He saved a man in the Rapids, a little way above the Falls.[11]

I wish Heins[12] would not be "sententious". There is just a little twist in his mind that he ought to have got the better of long ago. Indeed he ought not to have allowed it to appear there at all. Still it is a small matter in so genuine a man.

Now I think I have said as much to you as you will care to read. Should you see Ramsey[13] tell him that a geological survey of Pennsylvania has been voted:[14] and that his friend Lesley, with a salary of 35000 dollars a year, is to have the direction.[15] Say that Lesley will write to him soon.

I have sent the part of your note which refers to her letter to Mrs. Lesley.[16]

Give my love to Lilly;[17] I have no doubt as to her helpfulness & her devotion.

Yours affectiona[te]ly | John

BL Add MS 63092, ff. 86–7

---

1. *Your note*: letter missing.
2. *this place*: Tyndall was spending the summer at Bel Alp in the Swiss Alps (see letter 4306, n. 12).
3. *your desire to escape . . . get into sunny Italy*: when staying with Tyndall in the Swiss Alps in August 1873, Hirst had received an invitation to Genoa in Italy. He recorded in his journal for 19 August that, after much deliberation, 'I came to the conclusion that I must depart at once . . . Accordingly I wrote a few hurried lines to Tyndall to say good bye and went away sorrowfully' (*Hirst Journals*, p. 1982).
4. *Fletcher*: possibly Thomas Fletcher (1840–1903), a dentist based in Warrington, Yorkshire, who developed new dental treatments using gold plugs, and had recently published 'How to Use Adhesive Gold Successfully', *British Journal of Dental Science*, 17 (1874), pp. 314–17.
5. *the Address*: the presidential address to the BAAS at Belfast.
6. *Huxley expressing anxiety . . . kick over the traces*: see letter 4363, where Huxley stated: 'I declare I have horrid misgivings of your kicking over the traces' (meaning to throw off the usual restraints).
7. *Alexander Agassiz has written me a most indignant letter*: letter 4354.
8. *a copy of which I sent to Huxley*: in letter 4369.
9. *quartette of editors*: George Forbes, Alfred Wills, John Ruskin, and Peter Guthrie Tait, although it was only the first of these who had actually edited *Glaciers of Savoy*; Wills was the translator, and Ruskin and Tait provided supplementary essays.
10. *My old guide at the Niagara Falls*: Thomas Conroy (*c.* 1848–1917).
11. *He saved a man . . . above the Falls*: on 1 June 1874, Conroy saved William McCullough, an elderly painter who had fallen while working on a bridge near the rapids at Horse Shoe Falls ('The Hero of Niagara', *Daily Inter Ocean*, 4 June 1874, p. 8).
12. *Heins*: Heinrich Debus. It is not clear what he was being 'sententious' about.
13. *Ramsey*: Andrew Crombie Ramsay (1814–91), a Scottish geologist. He held the chair of geology at the School of Mines from 1851–72 (the Royal School of Mines from 1863), and in 1872 succeeded Roderick Murchison as director of the Geological Survey. Ramsay argued that glaciers formed some lake basins. He accompanied Tyndall to the Alps in 1858 (*ODNB*).
14. *a geological survey of Pennsylvania has been voted*: a proposal for a second Geological Survey of Pennsylvania was approved by Congress in June 1874.
15. *Lesley . . . is to have the direction*: J. Peter Lesley remained the director of the Geological Survey of Pennsylvania from 1874–96.
16. *her letter to Mrs. Lesley*: unclear, but presumably relating to Susan Inches Lesley's care for the dying entomologist George Robert Crotch, discussed in letter 4353.
17. *Lilly*: Emily Hirst.

From Joseph Henry 8 July 1874 4371

Smithsonian Institution, | July 8th 1874

My dear sir:

I write now to introduce to your attention the bearer of this note[1] Mr. Elisha Gray—one of the most ingenious practical electricians of this country. He has discovered a fact in regard to the transmission of musical impressions[2] which you will probably find of interest.

With my best wishes for your continued health and power of scientific investigation.

Very truly | Yours ever | Joseph Henry

Professor John Tyndall | Royal Institution

Smithsonian Institution Archives. RU 33, vol. 39, p. 756

---

1. *the bearer of this note*: Elisha Gray brought Henry's letter to Tyndall at the RI during his trip to Europe in August and September (see letter 4426).
2. *a fact in regard . . . musical impressions*: in early 1874, Gray discovered that composite tones could be sent as 'vibratory currents' through a telegraph wire, enabling the transmission of music.

From Thomas Archer Hirst 12 July 1874 4372

Royal Naval College. | Greenwich | July 12th 1874.

My dear John—

I have your letter of the 5th of July[1] before me, and since its arrival came the note about Mrs Somerville's M.S.,[2] which I will attend to. I was very very sorry to hear what had befallen you; how you had lost 12 days valuable time, and been afflicted, moreover, with that most "contemptible" of all pains—toothache. I hope all that is over now, and that for the rest of your stay bodily strength and mental clearness will increase day by day. I too have had a sorry time of it since I last wrote to you. It was in bed, in fact, that I wrote my last letter,[3] but expecting soon to be up again I said nothing about it. This is how it happened. On Sunday June 14th the day being very sunny although the east wind was keen. I hired a horse and rode over to High Elms[4] and back,—a ride probably of 26 miles—I enjoyed it, though on returning home tolerably quickly, I felt considerable fatigue about the loins, but I attributed it to the new character of the exercise. I expected to be stiff for the next day or two, but that over I anticipated a week's better health. I was disappointed. The stiffness came sure enough; but with it, on the following Wednesday, far worse symptoms, irritation of the bladder, and passage of blood with the urine. Of course I went to bed at once, indeed I was unable to stand up the pain was so great; Wilson Fox came, and the bleeding was soon stopped, but the local pain and irritation diminished very slowly. I was in bed a week. The college examinations[5] were proceeding during all this time, but happily I had prepared every thing beforehand, and I could give all necessary directions from my bed room. During the last week of June I was able to move about the College again, but I felt that the mischief, whatever it was, had not been repaired. On Saturday the 4th of July (after the departure of all the students) the crisis came.[6] You remember that terrible attack of yours at Zermatt[7]—well on that Saturday I had three of precisely the same character. Each time until they stupefied me with opium I was writhing on the floor or in bed with dreadful pain. It was a fearful day. Armstrong,[8] our medical man here, was in attendance all day, and Wilson Fox came down again. They were quite agreed as to the cause; it was the passage of angular crystals from the Kidney to the Bladder[9]—crystals that had probably been shaken loose and started by my long ride to High Elms. During the last week I have been gradually and satisfactorily recovering, and now I am free from all pain and local irritation. I have watched very carefully for proofs of the correctness of Wilson Fox's theory; and have been successful in finding them. During the past week several very small crystals (pronounced by Wills[10] who has examined them under the microscope to be Oxalate of Lime) have passed in the urine. I feel quite sure now that in your case Bence Jones was quite right.[11] I could not help contrasting your activity—however, the day after your attack, with my prostration for nearly a week after mine.

The experiences of the last three weeks, together with the fact that my hands are as bad as ever, have brought us all to a decision as to what I must do. I ought to have gone to the Britannia at Dartmouth,[12] and to the Dockyards to superintend examinations this week, but Admiral Key has put his veto on it, and the Admiralty have on his application made other arrangements. I am finishing up my reports on the recent College examinations, and am then to go away and be perfectly undisturbed until mid-September. The worst is yet to be said, however,—I am afraid I must give up Belfast. I dread the bustle of the meeting,[13] I dread the damp climate of Ireland. I must have perfect rest, and it must be in a dryer climate than this of ours. I think I shall go to Germany, at first perhaps to some of the German Watering places. I must see you, however, and I must, if possible, intercept you on your way home. Please let me know therefore, as soon as you can do so, when you propose to leave the Bel Alp, and how you propose to travel home. I am greatly broken now and

I see clearly that everything depends upon whether I can, during the next six weeks (or two months at most) turn over a new leaf—Write soon. Should I have left your letter will be forwarded

Ever yours affectionately | Tom.—

P.S. My early friend, poor Roby Pridie,[14] is dead. Heart disease carried him off at 49 years of age.

RI MS JT/1/H/256

---

1. *your letter of the 5th of July*: letter 4370.
2. *the note about Mrs Somerville's M.S.*: letter 4345; it must have been delayed in the post, as Tyndall had sent the note on 29 May.
3. *my last letter*: letter missing.
4. *High Elms*: see letter 4185, n. 5.
5. *college examinations*: the summer exams of the Royal Naval College, Greenwich, which ran from 20 to 30 June.
6. *On Saturday the 4th . . . the crisis came*: in the entry for this date in his journal, Hirst recorded: 'On attempting to come down stairs and go out violent pain came on. With difficulty I got up stairs on the first floor again and there flung myself on the bed. I could not lie still there but rolled on the floor writhing with pain and frightening poor Lilly greatly' (*Hirst Journals*, p. 2001).
7. *that terrible attack of yours at Zermatt*: there do not seem to be any records of Tyndall suffering an attack similar to that which Hirst described, in the Swiss mountain resort of Zermatt or elsewhere. Tyndall generally associated his time in the Swiss Alps with what he called 'the blessedness of perfect health', and the only incident of poor health that he recorded occurred in late August 1869 when he gashed his shin while climbing, and, as he later recounted, 'inflammation set in, pus appeared, and in trying to dislodge it I poisoned the wound. It became worse and worse; erysipelas set in, and at last it became evident that I might lose my foot or something more important' (*Hours of Exercise in the Alps*, pp. 314 and 317). Henry Bence Jones later referred to this, in a letter to Tyndall, as the 'attack . . . you had in your leg' (letter 3954, *Tyndall Correspondence*, vol. 13), although the circumstances of it do not otherwise correspond with Hirst's account of his own attack.
8. *Armstrong*: George William Armstrong (b. 1841), medical officer of both the Royal Naval College and the Royal Hospital School in Greenwich.
9. *the passage of angular crystals from the Kidney to the Bladder*: Hirst was suffering from kidney stones in the ureter.
10. *Wills*: Thomas Wills (1850–79), demonstrator in chemistry at the Royal Naval College.
11. *in your case Bence Jones was quite right*: not identified (see n. 7).
12. *the Britannia at Dartmouth*: HMS *Britannia*, a former warship built in 1820 and now moored in the river Dart, was used from 1863 to 1905 as the Royal Naval College at the port of Dartmouth in Devon.

13. *the meeting*: of the BAAS at Belfast from 19 to 26 August.
14. *Roby Pridie*: William Roby Pridie (1825–74), Hirst's friend from his time working as a railway surveyor in Halifax, who died in the Yorkshire town on 29 June.

To Thomas Henry Huxley 15 July 1874 4373

Bel Alp, Brieg[1] | 15th. July, 1874.

My dear Huxley

I have to day sent off the second instalment of my address,[2] and in the gap between it and the third I will insert a letter or two. I have been rather unlucky here. The snow and cold weather chilled me to the marrow before I was hardened; bothered my teeth and for nearly a fortnight left me hardly any working power. And I have been unlucky again in bathing. The day before yesterday in quitting a lake I cut my instep against a stone, and am now laid up with it. This will however soon pas—at least I hope so.

But I confess to you that I am far more anxious about your condition, than about my own; for I fear that after your London labour,[3] the labour of this lecture[4] will press terribly upon you. I wish to heaven it could be transferred to other shoulders. About Hirst I have also poor accounts,[5] and I must endeavour to dissuade him from going to Belfast.

A man named Foster[6] who met you some time ago at Michael Foster's in Cambridge has brought me some account of you. They sent me the list of the great men who assembled there to eat on the occasion of the opening of the new laboratory:[7] but your name was not amongst them, and from this I inferred either occupation or weariness. It is an affliction to me to think that I should just now add to either.

In the Scotsman for July 2nd. is an article on my side of the Forbes controversy.[8] It is long, strong and able. So that if you send that scrap to Knowles[9] the reference to the Scotsman in the concluding paragraph will have to be omitted.

My address will not be what I should like it to be, through want of time and strength. It is most unlucky that the meeting falls so early.[10] I have taken very little exercise since I came here and now I am debarred from taking any for some days. However I hope to be able to keep them profitably employed as listeners for an hour or two.

I have read again those early articles of yours on the Origin of Species:[11] they are very fine, and must have been most opportune. Spencer's Psychology[12] is also on the whole, a noble piece of work. I wish I had time to boil it down: but the wretch is <u>tough</u>.

I wish I could get rid of the uncomfortable idea that I have drawn upon you at a time when your friend and brother ought to be anxious to spare you every labour.

Yours ever | John Tyndall

1 P.M. Have just seen the Swiss Times:[13] am intensely disgusted to find that while I was brooding over the calamities possibly consequent on your lending me a hand that you have been at the Derby Statue[14] and are to make an oration apropos of the Priestley Statue in Birmingham[15] on the 1st Aug[ust]!!!

RI MS JT/1/TYP/9/3036
LT Typescript Only

---

1. *Bel Alp, Brieg*: see letter 4306, n. 12.
2. *the second instalment of my address*: pp. lxxix–lxxxi of 'Address of John Tyndall', *Brit. Assoc. Rep. 1874*, pp. lxvi–xcvii, which Tyndall sent to the printers Taylor and Francis in London (see letter 4368, n. 5).
3. *your London labour*: presumably the 'lectures to Schoolmasters' that Huxley mentioned in letter 4363, with the 'wish they were over'.
4. *the labour of this lecture*: the lecture Huxley was to give at the BAAS meeting (see letter 4327, n. 9), which, as he said in letter 4363, he would 'not begin to think of it till the middle of July'.
5. *About Hirst I have also poor accounts*: see letter 4372.
6. *A man named Foster*: not identified.
7. *the opening of the new laboratory*: the Cavendish Laboratory in Cambridge, the opening of which was marked, on 16 June, by a dinner at Trinity College.
8. *In the Scotsman . . . Forbes controversy*: 'Forbes *Versus* Tyndall', *Scotsman*, 2 July 1874, p. 2. The anonymous article was written by James Coxe (see letter 4352).
9. *that scrap to Knowles*: Knowles was the editor of the *Contemporary Review*, and the 'scrap' was a note of clarification to Tyndall's article 'Rendu and His Editors' that the *Contemporary* published in August (see letter 4355, n. 3). In letter 4369 Tyndall told Huxley that he would request Knowles to 'insert a couple of lines for me in the next number of the Contemporary; and these lines I will send for your approval either today or tomorrow'. It is unclear how the 'scrap' containing these amended lines was transmitted to Huxley.
10. *the meeting falls so early*: the 1874 meeting of the BAAS, beginning on 19 August, was earlier than the 1873 meeting in Bradford, which had begun on 17 September, but it was no earlier than most previous meetings, with, for example, the 1872 meeting in Brighton having begun on 14 August.
11. *those early articles of yours on the Origin of Species*: presumably Huxley's three reviews of Charles Darwin's *On the Origin of Species*: 'The Darwinian Hypothesis', *Times*, 26 December 1859, p. 8; 'Time and Life: Mr. Darwin's "Origin of Species"', *Macmillan's Magazine*, 1 (1859), pp. 142–8; and 'The Origin of Species', *Westminster Review*, 17 (1860), pp. 541–70.

12. *Spencer's Psychology*: H. Spencer, *The Principles of Psychology*, 2 vols (London: Longman, Brown, Green and Longmans, 1855).
13. *the Swiss Times*: an English-language newspaper issued by the Genevan publisher S. H. Bigland between 1869 and 1874.
14. *the Derby Statue*: a statue of Edward George Geoffrey Smith-Stanley (1799–1869), 14th Earl of Derby, was unveiled in Parliament Square, London on 11 July.
15. *an oration apropos of the Priestley Statue in Birmingham*: on 1 August, Huxley unveiled a marble statue of Joseph Priestley in the city where the natural philosopher had spent much of his career before being forced to flee from religious rioters. Huxley had then delivered an address celebrating Priestley's scientific achievements, which was published as 'Joseph Priestley', *Macmillan's Magazine*, 30 (1874), pp. 473–85.

To Thomas Archer Hirst 16 July 1874 4374

16th. July 1874

My dear Tom

Your letter[1] has just come. I am greatly grieved to hear of your illness. Belfast is not to be thought of. Sooner than allow you to run the risk of going there I would throw up the Presidency.[2]

Did I not leave you a blank cheque?[3] Remember to draw heavily on the Joint Stock Bank:[4] There are ample resources there at your disposal.

I do not know my day of return. I have got two instalments of manuscript[5] off my hands: the second went off yesterday. I am told the first is in proof though the proof has not yet reached me. I am very reluctant to let it go into your hands before I cut it up,[6] which I am sure to do. But under the circumstances I think I must ask you to get a proof from M^r^. Mattress (Taylor & Francis) read it over & give me your general counsel regarding it.

My difficulty about the time of my return arises from my inability to say how long the last part of the MS will occupy me. But I should say that I shall certainly be in London during the first days of August—not later than the 5th.[7]

Give my love to Lilly[8]—I know that she has done her duty in taking care of you.

Ever my dear Tom affectiona[te]ly yours | John.

To my amazement I find Huxley is going to deliver an address in Birmingham on the 1st!![9] They have sent me an invitation.[10]

BL Add MS 63092, ff. 99–100

1. *Your letter*: letter 4372.
2. *the Presidency*: of the BAAS at Belfast.

3. *a blank cheque*: see letter 4344.
4. *the Joint Stock Bank*: a type of bank that issues stock and makes its shareholders liable for the company's debts. Tyndall had been depositing his earnings in the London Joint Stock Bank, which was founded in 1836, since the mid-1850s (see *Ascent of John Tyndall*, pp. 106 and 112).
5. *two instalments of manuscript*: pp. lxvi–lxxviii and pp. lxxix–lxxxi of 'Address of John Tyndall', *Brit. Assoc. Rep. 1874*, pp. lxvi–xcvii, which Tyndall sent to the printers Taylor and Francis in London (see letter 4368, n. 5).
6. *cut it up*: Tyndall would tell Hirst, in letter 4379, that his 'first proofs in fact are usually my basis of operations, and the final ones are always very unlike the first'.
7. *I shall certainly be . . . not later than the 5th.*: in fact, Tyndall did not return to London until 10 August.
8. *Lilly*: Emily Hirst.
9. *Huxley is going to deliver an address in Birmingham on the 1st*: see letter 4373, n. 15.
10. *an invitation*: letter missing.

From George Wynne 17 July 1874 4375

Vispach[1] | 17 July /74

My dear Tyndall,

I slept last might at St Nicklaus on my way from Zermatt and I met a gentleman there who had just come down from Bellalp[2] who told me you were there, it would have added to my wish to have gone there which was very strong to know I should meet you, but my whole stay at Zermatt was spoiled by an attack of diarrhea which in the absence of any medicine I found it very difficult to get rid of, it has weakened me a good deal and as I do not feel myself up to another Alpine climb I am making my way with my wife[3] to Andermatt[4] (Hotel Bellevue) where we propose resting ourselves for a week and then going on to St Moritz. I suppose you have learned through the Times[5] the great sorrow that has come upon me in the death of my dear son Harry[6] who died at Calcutta on the 4th May of cholera it has been a great blow to me and to all my family and knowing him as you did I am sure I shall have your full sympathy. He was so acclimatized to India and the climate appeared to agree so well with him that I never anticipated anything happening to stop his brilliant career, and the blow was in consequence the more stunning. How little we thought when we were all lunching together just a few months ago[7] in your room that he so full of life and energy was so soon to be cut off, it is however an inexpressible happiness to me that I saw him so shortly before his death and found him as affectionate and loving as ever. I have had letters from the Viceroy, from Lord Salisbury & the Bishop of Calcutta[8] all speaking of him in the highest terms and of his loss to the

Public Service. He met his death with the greatest calmness, he had four or five days after the attack made his will and settled his worldly affairs. They had great hopes at one time that he would have recovered but the functions of his liver and other organs were so deranged that they would not act and he succumbed. Edward who is in India is terribly cut up and I feel most anxious for his return but his term of service will not be up until another year. Frank's occupation in Germany has gone and he is now in London looking out for employment, Lucy[9] is still at Ennis and I am sorry to say not very strong she has two children a boy and a girl. Is there any chance of seeing you at St Moritz but I suppose your stay in Switzerland will be curtailed by your duties at the British Ass[ociatio]^n.[10] I hope however your stay in Switzerland will be sufficiently protracted to enable you to recover your London dissipation. Farewell now old friend

Ever affect[ionate]^ly y[ou]rs | Geo Wynne

RI MS JT/1/W/101
RI MS JT/1/TYP/5/1870–1

1. *Vispach*: a town (now Visp) at the junction of the Rhône and Vispach Valleys in the Swiss Alps; it was the transport staging post for going up to Zermatt.
2. *Bellalp*: the Bel Alp (see letter 4306, n. 12).
3. *my wife*: Henrietta Jane Wynne (see letter 4320, n. 5).
4. *Andermatt*: a village in the Ursern Valley in the Swiss Alps.
5. *learned through the Times*: 'Deaths', *Times*, 12 May 1874, p. 1.
6. *my dear son Harry*: Henry Le Poer Wynne.
7. *lunching together just a few months ago*: presumably in February, when Henry Le Poer Wynne returned to Britain for two weeks on leave (see letter 4246).
8. *the Viceroy, from Lord Salisbury & the Bishop of Calcutta*: Thomas George Baring (1826–1904), 1st Earl of Northbrook, was Viceroy of India from 1872–6; Robert Arthur Talbot Gascoyne-Cecil, 3rd Marquess of Salisbury, was the Secretary of State for India from 1874–8; and Robert Milman (1816–76) was the Anglican Bishop of Calcutta from 1867–76.
9. *Lucy*: Lucy O'Brien.
10. *your duties at the British Ass[ociatio]^n.*: Tyndall was to be president of the BAAS at the annual meeting at Belfast from 19 to 26 August.

From Edward Sabine — 18 July [1874][1] — 4376

13 Ashley Place[2] | July 18th

Dear Tyndall,

Be sure you return me the Belfast College[3] either before or after your meeting[4] has taken place. It belongs to the Kew observatory[5] where its frame awaits it, being, in the mean time in charge of my clerks at Kew.

Sincerely yours | Edward Sabine | D^r. Tyndall.

RI MS JT/1/S/40
RI MS JT/1/TYP/4/1340

1. *[1874]*: the year is established by the relation to letter 4360.
2. *Ashley Place*: in Westminster.
3. *the Belfast College*: a picture of Queens College, Belfast (see letter 4360).
4. *your meeting*: the meeting of the BAAS in Belfast from 19 to 26 August, at which Tyndall would be president.
5. *the Kew observatory*: the old King's Observatory in the grounds of the former Richmond Palace, which, since 1842, had been administered by the BAAS, and then, from 1871, the RS. See L. MacDonald, *Kew Observatory and the Evolution of Victorian Science, 1840–1910* (Pittsburgh: University of Pittsburgh Press, 2018).

From Thomas Henry Huxley 22 July 1874 4377

4 Marlborough Place | London, N.W.[1] | July 22 1874.

My dear Tyndall,

I hope you have been taking more care of your instep than you did of your leg in old times.[2] Don't try mortifying the flesh again.

I was uncommonly amused at your disgustful wind up[3] after writing me such a compassionate letter. I am as jolly as a sandboy[4] so long as I live on a minimum and drink no alcohol and as vigorous as ever I was in my life. But a late dinner wakes up my demoniac colon and gives me a fit of blue devils[5] with physical precision.

Don't believe that I am at all the places in which the newspapers put me. For example, I was not at the Lord Mayor's dinner[6] last night. As for Lord Derby's statue[7] I went to get a lesson in the art of statue unveiling. I help to pay Dizzie's[8] salary so I don't see why I shouldn't get a wrinkle[9] from that artful dodger.[10]

I plead guilty to having accepted the Birmingham invitation.[11] I thought they deserved to be encouraged for having asked a man of science to do the job instead of some noble swell; and moreover Satan whispered that it would be a good opportunity for a little ventilation of wickedness. I cannot say however that I can work myself up into much enthusiasm for the dry old Unitarian[12] who did not go very deep into anything. But I think I may make him a good peg whereon to hang a discourse on the tendencies of modern thought.[13]

I was not at the Cambridge pow-wow;[14] not out of prudence but because I was not asked. I suppose that decent respect towards a secretary of the Royal Society was not strong enough to outweigh University objections to the incumbent of that office. It is well for me that I expect nothing from Oxford or Cambridge—having burnt my ships so far as they are concerned long ago.[15]

I sent your note on to Knowles[16] as soon as it arrived, but I have heard nothing from him. I wrote to him again to-night to say that he had better let me see it in proof if he is going to print it. I am right glad you find anything worth reading again in my old papers.[17] I stand by the view I took of the origin of species now as much as ever.

Shall I not see the address?[18] It is tantalizing to hear of your progress and not to know what is in it.

I am thinking of taking Development for the subject of my evening lecture,[19] the concrete facts made out in the last thirty years without reference to Evolution. If people see that it is Evolution, that is Nature's fault[20] and not mine.

We are all flourishing and send our love.

Ever yours faithfully | T. H. Huxley.

IC HP 8.165
RI MS JT/1/TYP/9/3037–8
Typed Transcript Only

---

1. *4 Marlborough Place | London, N.W.*: see letter 4213, n. 1.
2. *your leg in old times*: a reference to the time in August 1869 when Tyndall gashed his shin while climbing and allowed it to become infected (see letter 4372, n. 7).
3. *your disgustful wind up*: the postscript to letter 4373.
4. *jolly as a sandboy*: sandboys were street hawkers in early nineteenth-century London who sold sand, and the proverbial expression, meaning extremely happy or carefree, was first recorded by the journalist Pierce Egan in 1821: 'LOGIC . . . appeared as happy as a *sand-boy*, who had unexpectedly met with good luck in disposing of his hampers full of the above household commodity in a short time, which had given him a holiday' (P. Egan, *Life in London* (London: Sherwood, Neely and Jones, 1821), p. 289).
5. *blue devils*: low spirits, despondency (*OED*).
6. *the Lord Mayor's dinner*: the Lord Mayor of London, Andrew Lusk (1810–1909), hosted a dinner at the Mansion House for government ministers on 21 July.
7. *Lord Derby's statue*: see letter 4373, n. 14.
8. *Dizzie*: Benjamin Disraeli (see letter 4341, n. 4) had unveiled the statue of Lord Derby.
9. *wrinkle*: a piece or item of useful information, knowledge, or advice (*OED*).
10. *artful dodger*: a character, who is a skilled and cunning pickpocket, in Charles Dickens's novel *Oliver Twist* (1837–8).
11. *the Birmingham invitation*: see letter 4373, n. 15.
12. *the dry old Unitarian*: Joseph Priestley.
13. *a discourse on the tendencies of modern thought*: Huxley published his address as 'Joseph Priestley', *Macmillan's Magazine*, 30 (1874), pp. 473–85.
14. *the Cambridge pow-wow*: see letter 4373, n. 7.
15. *I expect nothing from Oxford or Cambridge . . . long ago*: Huxley had long made clear his animosity to the Anglican dominance and social privilege of the ancient universities of

Oxford and Cambridge. He would later decline Oxford's Linacre Chair of Zoology when it was offered to him in 1881.

16. *your note on to Knowles*: see letter 4373, n. 9.
17. *my old papers*: see letter 4373, n. 11.
18. *Shall I not see the address?*: the presidential address to the BAAS at Belfast. Huxley must soon have been sent proofs, as Tyndall later reported that 'Huxley has read almost, not quite the whole, and approves' (see letter 4392).
19. *my evening lecture*: Huxley gave the evening lecture on Monday 24 August at the BAAS meeting in Belfast (see letter 4327, n. 9).
20. *Nature's fault*: that is, the concept of evolution would be clear in any facts of nature regardless of whether Huxley intentionally pointed them out.

From Alexander Agassiz 25 July 1874 4378

CAMBRIDGE, MASS., July 25, 1874

MY DEAR PROFESSOR,—

I find it somewhat difficult to treat the articles appended to Rendu's Glaciers of Savoy,[1] for the authors have given us nothing new to answer in the way of evidence, and have reduced the matter to a warfare of vituperation in which I am not inclined to join. You have treated the subject of the measurements so clearly in your article in the Contemporary Review,[2] that I need hardly take it up again. Happily you have my father's own letter,[3] which I read now for the first time, where the facts are very plainly stated. I do not know what Mr. Forbes means when he says, in his Travels in the Alps,[4] "The measurements of Mr. Agassiz were made at my suggestion, and were chiefly pursued by methods specifically indicated by me to him in 1841." My father's methods of measurements were his lines of stakes,[5] and the fact that his experiments in 1841 were repetitions of those tried unsuccessfully in 1840, is evidence enough that they were not suggested by Mr. Forbes. Indeed, Forbes himself shows that he was aware of Agassiz's projected trigonometrical survey of the glacier by his derisive account of the company he found assembled at the Hotel des Neuchâtelois. "His force," he says, "consisted of a paid *surveyor*, a paid draughtsman, a chemist, a geologist, a trumpeter."[6] Why a paid surveyor, except for the express purpose of making the survey which my father had already projected, and which he partly accomplished that summer? Beyond this I can only restate what has been already told before and is loudly proclaimed by the biographers of Forbes[7] himself, namely, that, knowing from my father almost to a day when he intended to examine the stakes which had been recording for him the movement of the glacier of the Aar during a twelvemonth, Forbes made his measurements upon another glacier at that very time, and despatched his results to the Edinburgh Philosophical Journal,[8] hoping to be in advance with the publication of the facts. He thus attempted to destroy, as far as it was in his power to do so, the value of an investigation the whole course of which had been told to him without a shadow of reserve or disguise, and which he knew to be just touching its culmination.

As you have pointed out, my father's letter in the Comptes Rendus[9] gave him the priority after all; but the question here is not simply one of priority, nor whether Forbes did the work better and quicker than Agassiz, but whether he was justified in doing it at all, at least, without giving a hint of his intention to the band of workers who had so confidingly shown him, the summer before, their whole plan of action for the future and all the results of their work in the past. It is a curious fact that the biographers of Forbes, in their desire to secure for him the priority he so much coveted, say harsher things of him than his adversaries, they indeed congratulate themselves upon the celerity and dexterity with which he carried out his plan. "The race," they say (see page 197 of Tait's article),[10] "not unfrequently *is* to the swift." This is specially the case when but one party is running. Mr. Ruskin, while he tells us that he is not prepared "altogether to justify Forbes in his method of proceeding," adds, "except on the terms of battle which men of science have laid down for themselves."[11] I deny that honorable men of science are actuated by such motives. Mr. Ruskin's illustration of "the nugget," given in this connection, which strikes me, by the way, as singularly infelicitous for Mr. Forbes, might be pressed to a sharper conclusion, when it is remembered that the "old pitsman" showed his guest and supposed friend where he hoped to find the nugget, and told him the day and hour on which he meant to dig for it.[12]

I say nothing here of my father's methods as compared with those of Forbes. They were at least the best that he had at his command, and had been faithfully and laboriously pursued. It was because he felt their defects and his own ignorance of the purely physical side of the phenomena, that he invited Forbes to his cabin on the Aar, hoping to associate in his work a physicist of such rare ability and attainments, who could command means and appliances which he and his companions, as geologists, did not possess. The result you know. Mr. Forbes, as he himself tells us, hearing all they had to say, reserved his opinion on glacial theories and investigations until the following summer, when he had established his rival station, having (see page 273 of his biography)[13] "resolved that he would open a campaign by himself."

He seems to have broken his usual reserve only on one subject, and this brings me to the vexed question of the blue bands.[14] After a careful review of all the documents, I confess myself still less able than before to understand Mr. Forbes's pretensions about them. Unquestionably a conversation took place between him and my father respecting them, the very day of their arrival on the glacier. I have heard my father speak of it. He said that, as often happens on the glacier where the surface phenomena are so much influenced by effects of light and shade, recent rains, etc., etc., the appearance of the blue bands differed somewhat on that day from their ordinary aspect. He said himself that he expressed surprise and a special interest in these new features. I do not in the least question that Mr. Forbes thought he was pointing out a phenomenon wholly unobserved before; but that it was so, even setting aside my father's assertion to the contrary, seems to me impossible. Of course he had known of Guyot's paper on the subject at Porrentruy.[15] It was indeed read in the same session with his own "Notice sur les Glaciers."[16] It is possible that it was not in his mind at the moment of his talk with Mr. Forbes, though when he recalled the conversation to which such importance was afterwards given, he believed that he had spoken of it. Even were it otherwise, it would not be strange if, among the mass of facts under observation by the same band of workers from 1837 to 1841, many subjects the importance of which might not at once be fully understood should have been taken up, partially investigated, dropped again, and for the time forgotten. There was no spirit of competition among them. They were patiently accumulating their facts, in no haste to publish, but earnestly trying to solve the questions under investigation. My father and Mr. Guyot especially were as intimate as brothers from their youth to the day of my father's death. In their glacial investigations, if not always associated, they always had reference to each other's work, and it is hardly probable that they should not have discussed a subject on which Mr. Guyot had prepared a paper for a scientific meeting where my father was present. Forbes accused Agassiz of wronging Guyot[17] as much as himself in omitting to mention either the one or the other in his letter to Humboldt.[18] A brief examination of the facts will show how unfounded is this charge. My father was simply telling Humboldt in a private letter the result of his summer's campaign of 1841. As a part of this result he says, "Le fait le plus nouveau que j'ai remarqué, c'est la présence dans la masse de la glace des rubans verticaux de glace bleue."[19] Now what were the details of this investigation, and what part had Forbes in them? In a walk over the glacier *on the first day of his arrival*, he had a conversation about certain appearances which had been reported upon three years before by Guyot, and could therefore hardly have been absolutely new to Agassiz and his companions. Mr. Forbes says he insisted that the blue bands penetrated into the ice more deeply than my father supposed. However this may have been, their depth remained an open question, and continued to be a subject of doubt and discussion among the different members of the party.

Mr. Forbes left early in the season, and after his departure the work of the summer campaign on the glacier went on as before, and the question of the depth of the blue bands came up again. My father, in order to settle the question, caused himself to be lowered into one of the so-called wells in the glacier. This is an old story to you, but you may like to use this letter where the facts are less familiar, and you will pardon me for recalling them. In order to accomplish his purpose, Agassiz was obliged to turn aside the stream which flowed into the well, causing a new bed to be dug for it. This done, he had a strong tripod erected over the opening, and was then lowered by a rope into the aperture; his friend Escher[20] lying flat on the ground with his ear at the edge of the precipice to listen for any warning cry, and directing the descent, which went on prosperously and without accident to a depth of eighty feet. There he encountered an unforeseen difficulty, on account of a wall of ice dividing the well into two compartments. He tried first the larger one, but finding it split again into several narrow tunnels, he caused himself to be raised sufficiently to enter the smaller, and again descended without obstacle. Wholly engrossed in watching the blue bands visible even at that depth in the glittering walls of ice, he was only aroused to the presence of approaching danger by the sudden plunge of his feet into water. His shout of distress was heard, and his friends drew him up, though not without great difficulty, from a depth of one hundred and twenty-five feet, the most serious peril of the ascent being the points of the huge stalactites of ice, between which he had to steer his way. It does not seem strange to me, that, after this singular voyage into the interior of the glacier, my father should have said to Humboldt, in his summary of the season's work, "Le fait le plus nouveau que *j'ai* remarqué, c'est la présence dans la masse de la glace des rubans verticaux,"[21] etc., etc. Neither does it seem strange to me that, in this short and simple record, condensed into a single sentence of his attempt to ascertain the presence and appearance of the blue bands within the body of the glacier, he should not have remembered a casual conversation with Mr. Forbes about them, nor have recalled Mr. Guyot's paper of three years before, treating the external aspect of the same phenomenon. That his expression, "J'ai remarqué,"[22] should, under the circumstances, have caused such commotion in Forbes's mind, and have been so often quoted by his friends as a thing Agassiz had no right to say,[23] remains quite inexplicable to me.

I had hoped to add a concise and connected sketch of my father's glacial work, but I am obliged to defer it, and will only add that I (as did my father also most deeply) have always appreciated the generous stand you have taken in this controversy, without regard to personal annoyances and unjust imputations on yourself, and that I am truly grateful to you for it.

Believe me always, | Yours very truly, | ALEXANDER AGASSIZ.

To PROFESSOR JOHN TYNDALL, | *Royal Institution, London.*

RI MS JT/4/7b

---

1. *the articles appended to Rendu's Glaciers of Savoy*: P. G. Tait, 'Forbes and Dr. Tyndall', and J. Ruskin, 'Extract from "Fors Clavigera"', in *Glaciers of Savoy*, pp. 163–98 and pp. 199–210.

2. *your article in the Contemporary Review*: J. Tyndall, 'Rendu and His Editors', *Contemporary Review*, 24 (1874), pp. 135–48, in which Tyndall asserted: 'I had stated in the "Glaciers of the Alps", and in this Review, that some very important measurements made by Agassiz in 1841 and 1842, by which the differential motion of a glacier was demonstrated, had been ignored in all the writings of Principal Forbes . . . Prof. George Forbes now charges me with forgetfulness of the fact that it was his father who suggested to M. Agassiz the measurements he made; meaning thereby, I suppose, to intimate that his father was not called upon to recognize measurements which were the result of his own instruction. I would, however, ask Mr. Forbes to consider whether I, while endeavouring to hold the balance fairly between contending claims, should have been justified in accepting his father's assertion and ignoring the diametrically opposite assertion of Agassiz?' (pp. 139–40).
3. *my father's own letter*: the original letter, probably from November or December 1859, is missing, but in 'Rendu and His Editors', Tyndall quoted a long passage from it, noting: 'in 1859 . . . I wrote to M. Agassiz . . . I will give the pith of his reply, which, as just intimated, has lain beside me unpublished for fifteen years'. The passage from the letter Tyndall quoted was: '"It was not until after my second visit to the Aar in the winter of 1840–41 that I felt myself prepared for a systematic experimental investigation of the glacier; and I then went up, not with the hope of solving all the problems in one year, but with the view of laying the basis of a solution. The fact that I staked a series of poles across the whole width of the glacier, to a depth which left them standing to the following year, and that I then went up with an experienced engineer to make a minute map of the entire surface of the glacier, which was executed, will show that I had laid my plans for a successful survey of glacier phenomena before Prof. Forbes had, for the first time, set his foot upon the glaciers with a view to studying them. When I invited him to spend some time with me upon the glacier in 1841, I hoped to receive some valuable hints for my investigations from a physicist of so high a standing as his. But he never suggested anything to me, while I showed him everything I had been doing, explained all my difficulties, and the devices with which I proposed to overcome them. That Professor Forbes reached the Mer de Glace in 1842, a few weeks before I went up the Glacier of the Aar, only gave him the opportunity of making a few days' observations at a time when I had already gained an annual average. *That Professor Forbes knew in 1841 of my intention to make this experiment I can affirm the more positively as he saw the iron bars with which I intended to bore the holes, and which had been carried up the glacier before he reached the Grimsel.* That I was going to use instruments of precision in these measurements he must have understood, since I repeatedly mentioned my purpose of making a trigonometrical survey of the glacier the following year. Whether I at any time mentioned the theodolite I cannot remember now. But I am sure that he never suggested anything to me. Allow me one more remark. Everybody knows that I am a naturalist, and not a physicist. My interest in the glaciers arose from a desire to learn something of the mammoth of Siberia, after I had become convinced by Charpentier that the glaciers of Switzerland were much more extensive in earlier times than now. It struck me that there might be some connection between the burial of these gigantic mammalia in the arctic regions and the wider range of glaciers in Switzerland; I am one of those who

believe, as you expressed it in your short and characteristic speech at Geneva, that 'Nature is One', and so I was led to study the accumulations of ice without the necessary preparation. This you cannot fail to perceive in reading the accounts of my successive attempts, and for this, I hope, some allowance will hereafter be made"' (pp. 140 and 140–1). It was presumably this extract from the letter that Alexander Agassiz now read for the first time.

4. *Travels in the Alps*: J. D. Forbes, *Travels Through the Alps of Savoy and Other Parts of the Pennine Chain* (Edinburgh: Adam and Charles Black, 1843). The quoted sentence, however, does not appear in the book, or in its 1845 and 1855 (abridged) editions.
5. *his lines of stakes*: Agassiz had been unable to ascertain the motion of a series of stakes fixed in 1840 on the glaciers at the source of the Aare river as they had not been sunk deeply enough into the ice. He returned the next winter and, as he recounted to Tyndall, 'I staked a series of poles across the whole width of the glacier, to a depth which left them standing to the following year, and . . . I then went up with an experienced engineer to make a minute map of the entire surface of the glacier' (Tyndall, 'Rendu and His Editors', pp. 140–1).
6. *"His force . . . a geologist, a trumpeter"*: the quotation comes from a letter Forbes wrote to his sister on 9 October 1842, which was published in *Life of Forbes* (pp. 294–5).
7. *the biographers of Forbes*: John Campbell Shairp, Peter Guthrie Tait, and Anthony Adams-Reilly.
8. *despatched his results to the Edinburgh Philosophical Journal*: J. D. Forbes, 'On a Remarkable Structure Observed by the Author in the Ice of Glaciers', *Edinburgh New Philosophical Journal*, 32 (1842), pp. 84–91.
9. *my father's letter in the Comptes Rendus*: L. Agassiz, 'Observations sur les glaciers', *Comptes rendus des séances de l'Académie des sciences*, 13 (1841), pp. 818–20.
10. *Tait's article*: 'Forbes and Dr. Tyndall' (see n. 1). However, Agassiz is mistaken in attributing the quotation to Tait or Forbes's supporters, as it actually comes from an article in the *Westminster Review* that was anonymously authored by Thomas Huxley. See 'Glaciers and Glacier Theories', *Westminster Review*, 67 (1857), pp. 418–44, on p. 427.
11. *Mr. Ruskin . . . for themselves"*: the quotation is from 'Extract from "Fors Clavigera"' (see n. 1), p. 202.
12. *Mr. Ruskin's illustration . . . to dig for it*: John Ruskin drew an analogy between Forbes and a newcomer to a mine who strikes lucky by discovering a nugget of copper, much to the chagrin of an 'old pitsman' who has dug the same seam without success for ten years (pp. 202–3).
13. *his biography*: *Life of Forbes*.
14. *the vexed question of the blue bands*: in 'Rendu and His Editors', Tyndall stated of this issue: 'In walking up the glacier of the Aar with Agassiz, Prof. Forbes observed blue veins running through the ice. Agassiz had noticed the grooves answering to them on the surface, but he had not studied them, and in all likelihood he blundered in his conversation about them with his acute and physically-cultured guest. They followed these veins subsequently together for several days, and, after the departure of Forbes, Agassiz traced them to a depth of a hundred and twenty feet. Humboldt, I am informed, had been instrumental in getting him pecuniary aid for his researches, and to Humboldt, after the glacier campaign of 1841 had ended, he addressed a *private* note, mentioning among other things his having seen the

veins. I make no attempt at excusing his omission of the name of Forbes from this note; but, taking every thing into account, the sin of omission does not seem very heinous' (p. 146).

15. *Guyot's paper on the subject at Porrentruy*: the paper given by Arnold Guyot (1807–84) on glacier motion and structure at a meeting of the Société géologique de France in the Swiss municipality of Porrentruy in September 1838 was not published.
16. *his own "Notice sur les Glaciers"*: L. Agassiz, 'Observations sur le glaciers', *Bulletin de la Société géologique de France*, 9 (1838), pp. 443–50. Agassiz seems to be confusing the title of this paper with his father's subsequent 'Notice sur les glaciers', in É. Desor, *Nouvelles excursions et séjours dans les glaciers et les hautes régions des Alpes* (Neuchatel: J.-J. Kissling, 1844), pp. 1–14.
17. *Forbes accused Agassiz of wronging Guyot*: in J. D. Forbes, 'Historical Remarks on the First Discovery of the Real Structure of Glacier Ice', *Edinburgh New Philosophical Journal*, 34 (1843), pp. 133–53, on pp. 144–51.
18. *his letters to Humboldt*: Agassiz's 'Observations sur les glaciers' in the *Comptes rendus* was based on a letter he had sent to the Prussian polymath Alexander von Humboldt (1769–1859).
19. *"Le fait le plus . . . de glace bleue"*: Agassiz, 'Observations sur les glaciers', p. 819. The quotation translates as: 'The newest fact that I have noticed in the mass of ice is the presence of vertical ribbons of blue ice' (French).
20. *his friend Escher*: Arnold Escher von der Linth (1807–72), a Swiss geologist (*HLS*).
21. *"Le fait le plus . . .des rubans verticaux"*: see n. 14.
22. *"J'ai remarqué"*: I have noticed (French).
23. *such commotion in Forbes's mind . . . no right to say*: Forbes said that the expression was as 'articulate as it was unfounded' in 'Historical Remarks on the First Discovery of the Real Structure of Glacier Ice' (p. 142), and his remark was reprinted in *Life of Forbes* (p. 552).

To Thomas Archer Hirst 30 July 1874 4379

Bel Alp[1] | July 30th/74

My dear Tom

I fear much that the reading of my proof[2] will entail upon you a good deal of needless labour. My first proofs in fact are usually my basis of operations, and the final ones are always very unlike the first. Hence if you are taking much trouble—and I shall be rejoiced to find that you are not—it will probably be in part in reference to periods[3] that I should have modified myself.

I have had one or two scraps of intelligence about you; but am hungering to hear from yourself.

My work has been slow: but I am now beginning to see the end of it: and I hope to make a somewhat daring but dignified wind-up.[4] I have had hardly an excursion since I came here, but have kept continually gnawing at the address.

I rise usually very early—often before 5—almost always before 6: go to a chalet[5] with a bit of bread. Eat it and drink two or three cups of milk fresh from the cow—I find it very good. Still my sleep throughout has been bad—but somehow or other I do not seem to require much sleep.

I had thought of starting on Monday.[6] I was to reach London by the 5th. But it is probable now that I shall wait till Tuesday or Wednesday so as to complete the address before I return.

We have had very unsteady weather and today we are clasped by dense fog and abominable drizzle. It is also very cold—Much as I should like to have you here I should not like to see you here today.

The printers[7] have done their duty thus far manfully; and have been prompt with the printing and correction of proofs. You will let me know in your letter all your plans but if you have not written before this reaches you then I trust to open a line from you awaiting my arrival in London.

Give my love to Lilly.[8] I suppose she goes with you to the continent. Give my kind regards also to Heinrich[9] whose hand I hope to shake very soon.

I see the Saturday Review has been mildly backing Forbes:[10] this is what I expected from former articles.[11] I am not sure that the last word is said yet.

In this little world there is not much to talk about. There is the tinkle of the cowbells and the lowing of the cows: but the beasts themselves are hidden in the fog.

I have been measuring a piece of land with a view to its purchase—A Fischel[12] is 7x a square of 75 feet the side. I have been having a Fischel measured; and am to have other perpetuity for 500 francs: but I do not think I shall commit myself to the purchase until I know what the entire house will cost. It would be pleasant to have ones own fireside on such a day as this.

It will be a great joy to me if I see you before your departure for Germany.[13] I shall certainly be in London before the end of next week.

With best affection | always my dear Tom yours | John Tyndall | Thursday

BL Add MS 63092, ff. 92–3

---

1. *Bel Alp*: see letter 4306, n. 12.
2. *my proof*: of 'Address of John Tyndall', *Brit. Assoc. Rep. 1874*, pp. lxvi–xcvii, of which Hirst seems to have received the initial instalment, pp. lxvi–lxxviii (see letter 4374).
3. *periods*: rhetorical or ornamental language (*OED*).
4. *a somewhat daring but dignified wind-up*: the peroration to his presidential address to the BAAS, in which Tyndall would make some of his most controversial statements.
5. *chalet*: a house or cabin of a type traditional in the Swiss and French Alps, typically made of wood and having wide, overhanging eaves, originally used for shelter and for the manufacture of butter and cheese by herders bringing cattle to mountain pastures during the summer months (*OED*).
6. *Monday*: 3 August.

7. *The printers*: Taylor and Francis (see letter 4368, n. 5).
8. *Lilly*: Emily Hirst.
9. *Heinrich*: Heinrich Debus.
10. *the Saturday Review has been mildly backing Forbes*: on 25 July the *Saturday Review* published an anonymous review of *Glaciers of Savoy*, which proposed that 'it can hardly be possible for the merits of each worker to be measured with a degree of precision in which all will agree' and urged that 'there can be no gain from prosecuting any further the controversy . . . and we would fain see it brought to an end by a friendly shaking of hands between the partisans of Forbes and Rendu' ('Rendu on Glaciers', *Saturday Review*, 38 (1874), pp. 121–3, on p. 123).
11. *what I expected from former articles*: Tyndall probably had in mind 'Life and Letters of Professor Forbes', *Saturday Review*, 35 (1873), pp. 791–3, a favourable review of *Life of Forbes*.
12. *Fischel*: a traditional unit of measurement, roughly comprising the amount of farm land that a man could plough in a single day, used in the Swiss canton of Valais.
13. *your departure for Germany*: Hirst planned to travel to Wiesbaden at the end of July (see letter 4388).

To Heinrich Debus 2 August 1874 4380

My Dear Heins

Before I quit Switzerland I will jot down the data required by M^r^. Lockyer.[1] And you shall have it in ample time for publication, if it be thought worth publication.[2]

<u>You</u> are 'naughty', not I.[3] Remember how pressed I am to make a decent finish to my address.[4] I think it will pass muster now that the whole is before me. But you must read it. It has been heavy work & has taken up all my time. But by continually gnawing at it I hope at last to have put it into a form which will not cast discredit on my friends.

With warmest wishes | Ever your friend | John Tyndall

I shall be at home at the end of this week—At least I hope so.[5] | 2nd Aug[ust]. 1874.

RI MS JT/1/T/271
RI MS JT/1/TYP/7/2376

---

1. *the data required by Mr. Lockyer*: Norman Lockyer was the editor of *Nature*, so this was presumably data required for a contribution to the scientific journal.
2. *thought worth publication*: there are no contributions attributed to Debus in *Nature* in 1874 or 1875, so Lockyer may have considered the data not worthy of publication.
3. *<u>You</u> are 'naughty', not I*: a running joke about their slowness in writing to each other (see letter 4368).
4. *my address*: the presidential address for the BAAS meeting in Belfast.
5. *I shall be at home . . . least I hope so*: Tyndall's return to London was slightly later, on 10 August.

To Charles Darwin 5 August 1874 4381

My Dear Darwin,

Inasmuch as I have taken the liberty of mentioning you in my Address to the British Association to be delivered at Belfast, I thought I might ask you to glance over that portion of the address which relates to you,[1] so that I may be sure that I have stated nothing wrong I have therefore asked Taylor & Francis[2] to forward you a proof It is not finally corrected, and some portions of it I have not at all seen; but I think you will be able to see whether any error has crept in If you would return it to me at the Royal Institution as soon as you possibly can I should feel very much obliged to you

Yours ever | John Tyndall | Bel Alp[3] | 5th Aug[ust] 1874

I hope to be in London on M[onda]y or T[uesda]y next[4]

CUL GBR/0012/MS DAR 106:C16[5]
RI MS JT/1/TYP/9/2832

---

1. *that portion of the address which relates to you*: the section of 'Address of John Tyndall', *Brit. Assoc. Rep. 1874*, pp. lxvi–xcviii, which discussed Darwin was pp. lxxxiii–lxxxvii.
2. *Taylor & Francis*: see letter 4368, n. 5.
3. *Bel Alp*: see letter 4306, n. 12.
4. *M[onda]y or T[uesda]y next*: 10 or 11 August.
5. *CUL GBR/0012/MS DAR 106:C16*: published in *Darwin Correspondence*, vol. 22, pp. 402–3; this is letter 9587 in the online edition of the Darwin correspondence.

To Frederick Barnard 6 August 1874 4382

Bel Alp[1] | 6th. Aug[ust]. 1874

My dear Dr. Barnard.

Your note[2] has reached me in my Alpine home, where I have been incessantly at work for several weeks. I have not forgotten Faraday; but until this association business[3] is over I am perfectly crippled for every thing else. During the month of September I will apply myself to the work you require,[4] and before that month ends it shall be on its way towards you. I do not remember the length you mentioned. Perhaps when this reaches you you will do me the favour of naming the number of words the article ought to embrace.

With kindest regards to Mrs. Barnard.[5]

Believe me | ever yours faithfully | John Tyndall

Today is Thursday, and on Saturday I quit this perch with a view to being in London on Monday.[6]

Columbia University Manuscripts; Box 2 Folder 93; University Archives, Rare Book & Manuscript Library, Columbia University Libraries, Rare Book & Manuscript Library, Columbia University in the City of New York

1. *Bel Alp*: see letter 4306, n. 12.
2. *Your note*: letter missing.
3. *this association business*: writing the presidential address which Tyndall was to deliver at the annual meeting of the BAAS in Belfast on 19 August.
4. *the work you require*: Tyndall contributed the entry on Michael Faraday to *Johnson's New Universal Cyclopædia*, 4 vols (New York: A. J. Johnson, 1875–8), vol. 2, pp. 26–7, which Barnard co-edited with Arnold Guyot.
5. *M^rs^. Barnard*: Margaret Barnard.
6. *Monday*: 10 August.

From Herbert Spencer 8 August 1874 4383

Ardtornish.[1] | Morvern. N.W. | 8 Aug[ust] 74

My dear Tyndall

I fear I cannot give you trustworth information on the point you raise. Spaldings paper, read at the Brit[ish]. Ass[ociation]., and afterwards published in Macmillans,[2] does not, as far as I remember, make any reference to Lady Amberley.[3] I am aware that during the next year, she aided in a further prosecution of his experiments; but at the time that her name appeared in connexion with the matter I was fully acquainted with the leading facts as ascertained by Mr. Spalding, and, as I supposed, by him only. My impression was (but it is one produced by the indirect evidence that when first I knew him he was living in town, and not with them, I believe) that he became tutor to their son,[4] at Ravenscroft, after he had published his essential results.

But instead of trusting to my impression it would be best to inquire of him direct. His, address, is I suppose, still

care of Viscount Amberley[5] | Ravenscroft | W. Monmouth

The result of the experiments was to show that the chick did not immediately run about and pick up grains: there was some lapse of time before it began (probably due to the exhaustion of escaping from the egg). But when it began, it succeeded at once—certainly without any tuition.[6]

I hope you are in vigor, but your hand does not show it. It gives one the notion of over-fatigue

Good bye till we meet at Belfast[7]—.

ever yours truly | Herbert Spencer

RI MS JT/1/S/105
RI MS JT/1/TYP/3/1121

1. *Ardtornish*: an estate in the Highlands of Scotland.
2. *Spaldings paper . . . in Macmillans*: Douglas Spalding (1841–77) had originally given his paper on 'Instinct. With Original Observations on Young Animals' at the 1872 meeting of the BAAS in Brighton, before it was published, under the same title, in *Macmillan's Magazine*, 27 (1873), pp. 282–93.
3. *Lady Amberley*: Katharine Louisa Russell (1842–74), Viscountess Amberley. Lady Amberley, with her husband's consent, had a sexual relationship with Spalding, seemingly out of pity for his tuberculosis and celibacy.
4. *their son*: John Francis Stanley Russell (1865–1931).
5. *Viscount Amberley*: John Russell (1842–76).
6. *without any tuition*: Tyndall appended a note to the presidential address to the BAAS he was just then completing, observing of the idea that the consciousness of space is inborn: 'experiments have been . . . made by Mr. Spalding, aided, I believe, in some of his observations by the accomplished and deeply lamented Lady Amberly; and they seem to prove conclusively that the chick does not need a single moment's tuition to enable it to stand, run, govern the muscles of its eyes, and to peck' ('Address of John Tyndall', *Brit. Assoc. Rep. 1874*, pp. lxvi–xcvii, on p. xciv).
7. *at Belfast*: for the annual meeting of the BAAS from 19 to 26 August.

To Heinrich Debus [10 August 1874][1] 4384

Royal Institution of Great Britain | Monday m[ornin]g.

My dear Heins

While I have your address I will write and thank you for your letter.[2] It contains wise advice. I will go to Belfast as Luther did to Worms[3] if necessary—and meet if requisite all the Devils in Hell[4] there.

These things ruffle me less and less, and by and by they will not ruffle me at all.

Yours ever | John Tyndall

RI MS JT/5/11/1340
RI MS JT/1/TYP/7/2375

1. *[10 August 1874]*: the date is suggested by this being the only Monday when Tyndall was in London ahead of the Belfast meeting of the BAAS other than 17 August, which was the day he travelled with Debus to Belfast. It seems unlikely that Tyndall would have written this letter that morning, just ahead of seeing Debus.
2. *your letter*: letter missing.

3. *Luther did to Worms*: in 1521 the German religious reformer Martin Luther (1483–1546), having been excommunicated by Pope Leo X, was called to appear before the Holy Roman Emperor Charles V in the city of Worms. He refused to recant or rescind his opposition to several Catholic practices and teachings, proclaiming, according to tradition, 'Here I stand. I cannot do otherwise. God help me, Amen!'. On 7 April 1850, Tyndall had written in his journal that the 'plea of Martin Luther must be mine "I cannot otherwise—my God assist me!"' (RI MS JT/2/13b/485).
4. *all the Devils in Hell*: possibly an allusion to the dictate regarding personal accountability of the seventeenth-century Puritan theologian Richard Baxter (1615–91): 'If . . . all the Devils in Hell were combined against you, they could not destroy you without yourselves, nor make you sin but by your own consent' (*A Call to the Unconverted to Turn and Live* (London: Nevil Simmons, 1658), p. 253).

From Charles Darwin 11 August [1874][1] 4385

Bassett, Southampton | Aug[ust]. 11th

My dear Tyndall

You will see that I am away from home & your note[2] has been forwarded to me here;[3] but I am sorry to say that I have not received the proofs.[4] These w[oul]d. be apt to be delayed one post[5] in being sent to Down[6] & another in being forwarded here.—I will return them to R[oyal]. Institution by next post, after receiving them if they ever come here.

I most heartily wish you well through your hard week at Belfast. I cannot even fancy surviving such a week of excitement.[7]

Yours very truly | Ch. Darwin

CUL GBR/0012/MS DAR 261.8:20[8]
RI MS JT/1/TYP/9/2833

1. *[1874]*: the year is established by the relation to letter 4381.
2. *your note*: letter 4381.
3. *here*: Darwin was visiting his eldest son William Erasmus Darwin (1839–1914) in Southampton.
4. *the proofs*: to the section of 'Address of John Tyndall', *Brit. Assoc. Rep. 1874*, pp. lxvi–xcviii, in which Tyndall discussed Darwin: pp. lxxxiii–lxxxvii.
5. *one post*: i.e. postal delivery.
6. *Down*: Down House, Darwin's home in Downe, Kent.
7. *such a week of excitement*: the meeting of the BAAS in Belfast from 19 to 26 August.
8. *CUL GBR/0012/MS DAR 261.8:20*: published in *Darwin Correspondence*, vol. 22, pp. 413–4; this is letter 9597 in the online edition of the Darwin correspondence.

From Herbert Spencer 11 August 1874 4386

Ardtornish[1] | Morvern | N.W. | 11 Aug[ust] 74

My dear Tyndall

Your note of the 5th[2] reached me here to-day. This will give you some idea of the postal remoteness of this region, and the ground of my fear that I shall be unable to return you the proof[3] in time to be of any service. There were only three posts a week here, and it takes a week to get a reply from London when one writes there.

I have not got the proof from Taylor & Francis[4] yet, and now I cannot have it till Thursday[5] at 1 o'clock. I cannot send it away till Friday morning, and it will not be delivered till Monday in London. The only chance is that I may be able to send it by a steamer (the Staffa-boat)[6] passing down the Sound of Mull to Oban: If I can do this it will go to Glasgow on Friday & will reach you on Saturday. But I fear that even then it will be too late.

However it will probably be no matter. After all the care and thought you have taken you may very well dispense with criticism.

ever yours truly | Herbert Spencer

RI MS JT/1/S/106
RI MS JT/1/TYP/3/1121

---

1. *Ardtornish*: an estate in the Highlands of Scotland.
2. *Your note of the 5th*: letter missing.
3. *the proof*: presumably the section of of 'Address of John Tyndall', *Brit. Assoc. Rep. 1874*, pp. lxvi–xcviii, which discussed Spencer: pp. lxxxviii–xcv.
4. *Taylor & Francis*: see letter 4368, n. 5.
5. *Thursday*: 13 August.
6. *the Staffa-boat*: a regular boat service which, departing from Oban, made a complete circuit of the Isle of Mull off the west coast of Scotland, stopping at Tobermory, Staffa and Iona.

From Charles Darwin 12 August [1874][1] 4387

Bassett, Southampton | Aug[ust]. 12 | (9. A.M.)

My dear Tyndall,

The sheets[2] have just arrived & I return them by the first post, which leaves this place[3] in ½ an hour. I have read over the whole about my work,[4] & have been most deeply gratified by what you say. As far as a rather hasty reading suffices, I have not one word of criticism to make. It all seems to me excellent, & as clear as light. I shall be most anxious to read the first & last part,[5] as I can clearly see that the whole will interest me in the highest degree.

It is a grand subject—

In great haste | Yours very cordially | Ch. Darwin

CUL GBR/0012/MS DAR 261.8:21[6]
RI MS JT/1/TYP/9/2834

---

1. *[1874]*: the year is established by the relation to letters 4381 and 4385.
2. *The sheets*: the proofs of 'Address of John Tyndall', *Brit. Assoc. Rep. 1874*, pp. lxvi–xcviii (see letter 4381).
3. *this place*: Darwin was visiting his eldest son William Erasmus Darwin (1839–1914) in Southampton.
4. *the whole about my work*: the section discussing Darwin was pp. lxxxiii–lxxxvii.
5. *the first & last part*: Tyndall wrote the address in discrete sections, which were sent to the printers, Taylor and Francis (see letter 4368, n. 5), in separate batches. Darwin presumably refers to pp. lxvi–lxxviii and xci–xcvii.
6. *CUL GBR/0012/MS DAR 261.8:21*: published in *Darwin Correspondence*, vol. 22, pp. 414–5; this is letter 9599 in the online edition of the Darwin correspondence.

From Thomas Archer Hirst 12 August 1874 4388

Bad-Schwalbach | im Taunus[1] | Aug[u]st 12th 1874

My dear John—

This day week you will be entering on your duties of President of the British Association and I shall not be there to enjoy your success I had always looked forward to being at your side when this event happened and it is no small disappointment to me to be compelled to remain so far away. Full of misfortunes and disappointments as the past year has been the greatest to me at the moment is the consciousness that I have been perfectly useless to you as far as the preparation of your address and the transaction of your business as President are concerned. That you will get through your work triumphantly without me I know perfectly well, but I might have removed some of the worry of it, and above all I might have witnessed your triumph.

I want you to receive this letter from me before you start for Belfast in order that no feeling of uncertainty as to how I am getting on can possibly disturb your thoughts. I have felt better[2] ever since I left Greenwich. It was an immense relief to me to break away from the endless little worries connected with College and other examinations and reports;[3] and the journey here by Dover, Calais, Brussels Cologne the Rhine and Wiesbaden[4] was a most wholesome change of occupation. I enjoyed and am still enjoying the sense of freedom from all manner of engagements[5] the consciousness that I can do just what I please, or nothing at all, that I can go where I like, and remain where I please is itself a source of enjoyment to me. The local disorders[6] that troubled me so much lately at Greenwich have almost disappeared were it not that my hands remain obstinately as bad as ever they were and that occasionally my legs ache in an unaccountable way I should say that I ailed nothing. It was at the suggestion of Sir Henry Thompson[7] and with the approval of Wilson Fox that I came here. The climate is dry the air fresh, but not too cold, and the country around hilly and extremely beautiful. Yesterday I walked about 14 miles over hill & dale through forests and corn fields with comparatively little fatigue. When I left Greenwich it was with difficulty I could drag my legs up to the observatory hill. I drink a glass or two of water from the Stahlbrunnen[8] daily and bathe every other day. The water and the baths are very like those of S^{t}. Moritz but the climate is a little more genial and the company at my Hotel,[9] especially the English, much more entertaining. Although I spend the greater part of every day out of doors and alone I never feel lonely. My geometry is always a pleasant occupation for me I forgot to send you a copy of my paper on "Correlation".[10] It is the sole bit of work which has left my hands during the past year, that I can look back on with entire satisfaction. It is to be reprinted, in English, in the Italian Annali di Matematica.[11] And now goodbye. I hope you will have good health at Belfast and that all will go well. Write me a line when you can escape from

<*1 or more pages missing*>.

RI MS JT/H/257

---

1. *Bad-Schwalbach | im Taunus*: a spa town in the Taunus mountain range in western Germany.
2. *felt better*: Hirst had been suffering from kidney stones in the ureter (see letter 4372).
3. *College and other examinations and reports*: the summer exams of the Royal Naval Colleges in Greenwich and Dartmouth, the latter of which Hirst was prevented from superintending on health grounds (see letter 4372).
4. *the journey here by Dover . . . and Wiesbaden*: Hirst left London for Dover on 29 July, travelling by train, and ship across the Channel, until he reached Cologne, from where he took a steamer down the Rhine to Wiesbaden, arriving there on 1 August.
5. *the sense of freedom from all manner of engagements*: Hirst recorded in his journal that 'I enjoyed my trip in a very quiet way. I did not speak to any one more than two words' (*Hirst Journals*, p. 2002).
6. *local disorders*: see n. 2.
7. *Sir Henry Thompson*: Henry Thompson (1820–1904), consulting surgeon at University College Hospital.

8. *Stahlbrunnen*: steel well (German).
9. *my Hotel*: Hirst stayed at the Victoria Hotel, which had opened in 1857.
10. *my paper on "Correlation"*: T. A. Hirst, 'On the Correlation of Two Planes', *Proceedings of the London Mathematical Society*, 5 (1874), pp. 40–70.
11. *reprinted, in English, in the Italian Annali di Matematica*: T. A. Hirst, 'On the Correlation of Two Planes', *Annali di Matematica Pura ed Applicata*, 6 (1874), pp. 260–97.

To Joseph Henry 13 August 1874 4389

ATLANTIC CABLE MESSAGE. | THE WESTERN UNION TELEGRAPH COMPANY.

To Professor Henry | Washington | *Received at* Wash'n D.C. | 11.10 am Aug[ust] 13th *1874*

Help Hastings[1]
Tyndall

Smithsonian Institution Archives. RU 26: Smithsonian Institution Office of the Secretary, Correspondence, 1863–1879 (incoming correspondence), Microfilm Reel 148

---

1. *Help Hastings*: Charles Sheldon Hastings was an applicant to the Tyndall Fund, which used the surplus earnings from Tyndall's lecture tour of the United States in 1872 and 1873 to support American students of physics, and Henry had sent Tyndall a copy of a letter in support of his application at the end of June (see letter 4367). Tyndall approved the application, giving his reasons in letter 4390. On the fund in general, see letter 4214, n. 5.

To Joseph Henry 13 August 1874 4390

13th Aug[ust]. 1874

My dear Professor Henry:

I did not receive your letter[1] until my return from the Alps,[2] for it being rather heavy[3] the porters did not forward it to me. To incur no further delay I at once sent a telegram[4] asking you to help Mr Hastings.[5] The case is just such as we desire to have and to help. You know the matter entirely rests in your hands. With regard to the retention of the fund until the interest amounts to 1.000 dollars,[6] I thought this was your proposition; and to it, as to every proposition of yours regarding the fund I give my prompt consent. But it is never necessary to ask my opinion. The trustees[7] have the entire management of the fund in their hands and what seemeth good to them will also seem good to me.

But I am rejoiced to find the first application to be to so worthy an object as rendering assistance to Mr Hastings. I think I remember him as an earnest young fellow who called upon me at the hotel.[8] At all events, the letter of my excellent friend Lyman[9] makes the thought that the fund is to be applied to the advantage of one so highly recommended especially gratifying to me.

I dare say the Trinity House Report[10] is by this time in your hands. I have desired the printers[11] to send you a paper on the subject which has been just published in the Philosophical Transactions,[12] and in which the subject is illustrated by reference to some striking experiments

With kindest regards to Mrs Henry and your daughters,[13]

Believe me, always yours, | John Tyndall

Smithsonian Institution Archives. RU 33, vol. 40, pp. 441–3

1. *your letter*: letter 4367.
2. *my return from the Alps*: Tyndall returned from Switzerland on 10 August.
3. *rather heavy*: the letter contained copies of several letters relating to an application to the Tyndall Fund (see letter 4214, n. 5).
4. *a telegram*: letter 4389.
5. *Mr Hastings*: Charles Hastings.
6. *retention of the fund . . . interest amounts to 1.000 dollars*: see letter 4367, n. 6.
7. *The trustees*: Henry, Hector Tyndale, and Edward L. Youmans.
8. *called upon me at the hotel*: in Newhaven (now New Haven), Connecticut, where Tyndall had lectured on 22 and 23 January 1873 (see letter 4201, n. 4).
9. *Lyman*: see letter 4367, n. 4.
10. *the Trinity House Report*: 'Report by Professor Tyndall'.
11. *the printers*: Taylor and Francis (see letter 4368, n. 5).
12. *a paper on the subject . . . Philosophical Transactions*: 'Vehicle of Sound'.
13. *your daughters*: Mary Anna Henry, Helen Louisa Henry, and Caroline Henry.

From S. E. Wilson[1] 14 August 1874 4391

Professor Tyndall LL.D. F.R.S. &c. | President, "British Association,"

Sir,

I have just been reading a very interesting sketch of your life and work in one of our popular magazines,[2] apropos of your coming Presidency over the British Association in Belfast, next week. I admire exceedingly the talent and brilliant parts which have raised you to your present honorable position, to occupy which you may justly feel proud.

As an ardent lover of science, and being deeply interested in the successes of her sons in all their varied walks and developments of human knowledge, I cannot help regretting the peculiar phases of thought and opinion occasionally put forth by such eminent men as yourself. I refer now to the unhappy conflicts which occur between the investigators of human knowledge and its sources, from a merely scientific standpoint, and the investigations of the same from a super-human standpoint, and aided by the valuable light of that sacred repository of Truth, the Word of God. Bear with one who is zealous for Scientific and also Scriptural truth, while I give expression to some thoughts which I trust may not be unworthy of your philosophical consideration. These have been suggested by the views you express regarding "phenomena", ("Frag[ments of]. Science" p. 31.) "One by one nat. phen[omena]. have been ass[ociat]e[d] with their prox[imate]. causes, and the idea of distinct personal volition, mixing itself in the economy of nature, is retreating more and more,"[3] [only in the minds of those who as yet do not see far enough in the wonder-working ways of the Omnipotent Creator]. Again you say "Science asserts that without a disturbance of nat. law, quite as serious as the stoppage of an eclipse, or of Niagara turning back in its currents, no act of humiliation or prayer to the Deity could call one shower from heaven, or deflect one beam of the Sun."[4] According to your view, there never has been an event in the external world due to the exercise of any other force than the undirected operation of physical causes—"Nothing has ever intimated that nature has been crossed by spontaneous action, or that a state of things existed at any time which could not be rigorously deduced from the preceding state."[5] Now spontaneous action is antithetical to necessary. It is therefore free action, that of intelligence and will, such as you display in writing or delivering your lectures. You assert that all effects in nature must be referred to blind unintelligent causes. You would preclude the possibility of miracles. May I hope that it may be assumed, in estimating the grasp of your profound researches, that you acknowledge the verity and integrity of Scripture history, and if so, surely you must confess the undoubted evidence of the spontaneous action of the God-man, working in his own world, and appealing to the reason & judgment of his Auditors and followers, 1874 years ago. Would it be a miracle, if, instead of Professor Tyndall coming over from London next week to fill the chair of the B.A.[6] and deliver his address in Belfast he should remain in the Royal Institution and command the aid of the telegraph, and dispatch every word of his paper to a friend at the other end of the electronic wire on the platform of the B[ritish]. Ass[ociatio][n] who would read it aloud to the brilliant and learned audience? Such a feat would have been considered miraculous or magical some years ago, and would seem so still to an unlearned and untutored mind, as e.g. of a rustic who never heard of the art of telegraphy—but to your audience in Belfast, who had faith in your existence, and were assured that you were communicating with them, there would be nothing wonderful or impossible in the supposed case. It is then, dear Sir, "a thing incredible" with you that the Master Mind of the Universe, the Presiding Governor over all worlds, the God "in whom we live and move and have our being",[7] should not send his message of revelation, in all its details to his rational creature, Man; unfolding his will and enforcing his laws; and, at the same time, showing betimes his power & might, in wondrous acts & signs, which we call miracles, just because they are so far above and beyond our finite & feeble activities. Would you venture to assert that when Christ said to the raging sea, "Peace, be still,"[8] the only cause or efficient antecedent was not his will—or when he said to the leprous man "I will, be thou clean",[9] calm and healing immediately resulting in both cases, there was not a spontaneous action of the God-man crossing the ordinary course of nature, or that it was not his personal direct volition which accomplished these phenomena. To say that these facts never occurred, simply because according to the ephemeral theory of the hour, they could not occur, is the infinite of folly. Let me not slight these remarkable facts which are the best authenticated in all history. You and other saváns are prone to think that there is no other evidence of truth than the testimony of the senses. But the reason has its intentions, the moral nature its à priori judgments, the religions consciousness its immediate apprehensions, which are absolutely infallible, and of paramount authority. A man might as well try to emancipate himself from the operations of nature's laws, as from the authority of moral law, or his responsibility to God. When therefore men of science advance theories opposed to these fundamental convictions, they are like the wild seabirds, in a dark Winter night, dashing recklessly against the lantern of yonder light-house, standing out amid the buffeting of the waves, and falling stricken at its base. Take the following illustration:—

When you go to Belfast, next week, you may visit one of those monster spinning mills or weaving factories,[10] which are objects of interest in that town. You approach the building—you admire its size, its structural proportions, and its great capacity for a certain work. You enter in, and examine minutely all the mechanical arrangements; wheels and bands, cranks & pinions, flying shafts and whirring spindles, an endless variety of motion communicated vertically and horizontally—wheels within wheels "from floor to roof", all working with amazing precision. But whence the power that impels all this complicated medley of machinery? You go into the engine-room, and there the "giant with iron arms," is by continuous inspirations from the boiler, drawing out a mighty force which he is extending and exerting through every joint and crank & pinion over the entire factory. But how is all this life-like apparatus started and sustained and controlled? Do you expect to go into that busy mill, and see no presiding genius there?—is there no mind amid the whirr of wheels and the groans of compressed steam, which by its emancipation has wrought such wonders? Impossible!! You find at once the shrewd engineer, watching most intently his machine, in all its movements, guiding, regulating and controlling its actions by his levers, handles & valves—He can do this spontaneously—and you admire the simplicity and ease of his governing power. He tells you how carefully he has constructed each part of the beautiful machinery, and he proves to you by repeated manipulations, that as his mind suggests and directs, his hand can readily accomplish any movement to bring about any result desired, spontaneously. You next accompany the spinning-master through all the maze of spindles and looms, and he shows how each part of the machinery is adapted to every other part, to do a special work – he communicates and cuts off motion at pleasure, spontaneously. He shows his perfect control over all the mechanisms, and finally exhibits the material fabric, yarn or cloth, so deftly and ingeniously manufactured in that indistinct edifice—

So far you have an interesting experience of science & art combined in practical work. Now let us apply it.—We find ourselves in the midst of nature's glorious temple; we gaze around us in this fair world, and behold all that is wonderful and beautiful, and intricate and grand & useful in the works of creation. Whether we contemplate the face of nature—the profundities of earth & sea, or gauge the heights of the starry heavens, we behold order, design, and the most marvellous adaptation of part to part, joint to joint—and a regularity of movement, a succession and development of results infinitely above and beyond what we saw in the factory—Will you venture to deny the grand conclusion that over all these existing phenomena, there is a presiding mind, a lordly Creator, who made, and sustains and controls all his works, spontaneously, by his sovereign will, and at his pleasure? Is it reasonable and easily comprehensible that man, with his finite powers, can originate, control, and either continue or suspend the varied & complex movements of his machine, and yet will it be denied by the philosopher, to man's Creator, as a sentient, powerful and all-wise Being, to Control and regulate every part of this world's phenomena, changes and modes of being and working, and actions more easy to Him, than the forth-putting of power by his impotent creature. The God who made the world and orders all its wondrous phenomena, made the mind of the philosopher, who investigates their mysteries, and demands the submission of the intellect, the homage of the heart; and the humble acknowledgement of a superintending reign by Him who is "Wonderful in counsel and excellent in working!"[11]

With reverence, then, earth's temple tread,
  Its outer and its inner shrine;
There is an active God o'er head,
  A special Providence divine.
That Christ who ruled the raging wave,
  And calmed the storm on Galilee,

Has come this fallen world to save,
To give men light that they might see.
He gave the sun his power to shine,
He gave you power those beams to scan,
Your mind, the noble part divine,
The pride of philosophic man!
Then while you search earth's secrets here,
And strive their wonders to unfold,
While analyzing you brighter sphere,
That all its phases may be told;
Bow to the Mind that made the globe,
Who dwells enthroned in bliss above,
Clothed with the light as with a robe,
"Whose nature and whose name is love."[12]

I trust you will excuse the length of this imperfect dissertation, amid the bustle of preparation for your coming conference. I trust you will have a prosperous and successful meeting, and that in the various learned discussions there will arise nothing to mar the harmonious consistency which should exist between the true philosophic study of the works of nature, and the reverent study of the revelations of nature's God.

I am, Sir, | your obedient servant | S. E. Wilson.

Parsonstown,[13] | Ireland. | August 14, 1874.

RI MS JT/1/W/82

---

1. *S. E. Wilson*: not identified, although apparently a woman and probably a member of the Irish Methodist Church (see n. 12).
2. *a very interesting sketch . . . our popular magazines*: not identified.
3. *"phenomena", ("Frag[ments of]. Science" p. 31.) . . . more and more"*: J. Tyndall, 'Reflections on Prayer and Natural Law', in *Fragments of Science*, pp. 31–7, on pp. 31–2.
4. *"Science asserts . . . one beam of the Sun"*: another quotation from 'Reflections on Prayer and Natural Law' (p. 36), although this time the original wording has been slightly altered.
5. *"Nothing has ever intimated . . . preceding state"*: J. Tyndall, 'Miracles and Special Providences', in *Fragments of Science*, pp. 41–68, on pp. 63–4.
6. *B.A.*: the BAAS.
7. *"in whom we live . . . have our being"*: Acts 17:28.
8. *Christ said to the raging sea, "Peace, be still"*: Mark 4:39.
9. *he said to the leprous man "I will, be thou clean"*: Matthew 8:3.
10. *monster spinning mills or weaving factories*: the spinning of flax and weaving of linen were integral to the economy of Belfast, with the York Street Mill, founded in 1830, reputed to be both the largest spinning and weaving mill in the world.
11. *"Wonderful in counsel and excellent in working!"*: Isaiah 28:29.
12. *"Whose nature and whose name is love"*: the expression is from John Wesley's letter to Miss

A—from 15 January 1767 (*The Works of the Rev. John Wesley*, 14 vols (London: John Mason, 1829–31), vol. 12, p. 342), suggesting that Wilson was a member of the Irish Methodist Church.

13. *Parsonstown*: probably the town (now Birr) in King's County (now County Offaly), although there are also five townlands (small geographical regions) with this name in Ireland.

To Thomas Archer Hirst 15 August [1874][1] 4392

15th Aug[ust].

My dear Tom.

Your note[2] to my great gratification reached me yesterday. Of course it would have been a delight to me to have you at my side—of course you could have helped me in many ways. Indeed instead of a working man at the association[3] you would have made me an indolent man: but do not allow this "disappointment" to press for one moment on your thoughts. I have got the really hard work fairly under my feet. I do not care what the effect of the Address[4] may be upon the audience, but I have the firmest confidence that you will like it, and that it utters a truth or two which will survive the meeting of the association.

You had read but a portion of the address.[5] I also analysed Spencer's doctrines,[6] and wound up with some views of my own. I worked at it long, and got my own head tolerably clear regarding it, and I do not therefore fear its failure.

This moment I have given Mr Mattress directions to send you a copy to the Poste Restante[7] Baden Baden. The papers are already infesting Red Lion Court[8]—wishing to have copies in advance.

Huxley has read almost, not quite the whole, and approves[9]—Darwin has read all that relates to himself,[10] he had it not long enough to do more, and also sends a note of approbation.

Huxley, Debus and myself start on Monday morning;[11] pull up at Salt Hill near Kingstown,[12] where we remain in a charming hotel until Wednesday, when we run on to Belfast.

I am tired but tough, and have now no fears inasmuch as the Address comes quite as near my ideal as I expected. I read your note to Spottiswoode yesterday: he tells me the Hamiltons are going: Lady Mary[13] is going, and a great many besides. And the Belfast people are also showing warm Kindness & cordiality: proving that they have not wholly set their hearts upon home rule.[14]

And now my dear Tom get strong—your letter breathes of health; though I shall not be satisfied as long as these aches continue. The hands I fear will be a longer affair; but even they will slowly mend—You could not expect them to do so in so short a time. I was three weeks in the Alps before I began to be really conscious of returning strength.

Ever affectionately yours | John

RI MS JT/1/T/917

---

1. *[1874]*: the year is established by the relation to letter 4388.
2. *Your note*: letter 4388.
3. *the association*: the BAAS.
4. *the Address*: Tyndall's presidential address to the BAAS meeting at Belfast.
5. *read but a portion of the address*: Hirst seems to have read the first instalment of the proofs of 'Address of John Tyndall', *Brit. Assoc. Rep. 1874*, pp. lxvi–xcviii, which were pp. lxvi–lxxvii.
6. *analysed Spencer's doctrines*: on pp. lxxxviii–xcv.
7. *Poste Restante*: Remainder Post (French), a postal service by which letters are kept for an agreed period until collected by the addressee.
8. *Red Lion Court*: on Fleet Street in London, the location of the printers Taylor and Francis (see letter 4368, n. 5).
9. *Huxley has read almost . . . and approves*: on 12 August Huxley told Michael Foster of Tyndall's presidential address: 'I have read the inaugural and have suggested one or two omissions. I think it's very well done but Lord knows what will be the effect. I wish he had taken another line but having taken it the thing is to do what is done well and never mind language' (*Foster and Huxley Correspondence*, p. 51).
10. *Darwin has read all that relates to himself*: pp. lxxxiii–lxxxvii (see letter 4387).
11. *Monday morning*: 17 August.
12. *Salt Hill near Kingstown*: on the Irish coast near Dublin; the 'charming hotel' is the Salthill Hotel, which, having been built in 1843, was extensively refurbished in the 1860s (see letter 4393). Kingstown is now called Dún Laoghaire.
13. *Lady Mary*: Mary Egerton.
14. *home rule*: self-government for Ireland. The issue was becoming increasingly prominent, as the Home Rule League had made major gains in the British general election of February 1874.

To Elizabeth Dawson Steuart 15 August [1874][1] 4393

Royal Institution of Great Britain | 15th Aug[ust].

My dear Mrs Steuart,

The fact of my going to Ireland recalls to my recollection that since I enclosed to you a small cheque for Edward Hayden[2] you have not written to me. This is a long time ago.

I start on Monday next[3] with my friend Mr Huxley; halt at Kingstown, or rather at the Salt Hill Hotel near Kingstown, on Monday and Tuesday night and proceed to Belfast on the Wednesday. It is a hard and an arduous piece of work, which I should willingly have avoided; but it was to be done, and in the completion of my address,[4] the last page of which is now through the press the weightiest work is accomplished.

I had unsteady weather in the Alps this year. But I was much employed within doors. A week before I came away two charming daughters of the late Lord Dunraven[5] came there. Lady Edith and Lady Emily Quin.[6] I knew their father very well, and with them I had a few pleasant excursions which brightened the last days of my visit.

I also met a fine gentlemanly fellow, son of Brown of Brown's Hill,[7] at the Bel Alp. He told me he had exhausted his purse and worn out his boots, and I have been upbraiding myself ever since for not offering to share my purse with him!

If you care about writing to me, a letter addressed to me

Reception Room | British Association | Belfast

will find me

always yours | John Tyndall.

RI MS JT/1/TYP/10/3412
LT Typescript Only

---

1. *[1874]*: the year is established by the address at the bottom of the letter, at the Belfast meeting of the BAAS.
2. *a small cheque for Edward Hayden*: possibly the Edward Hayden recorded as living on Leighlinbridge Bridge Street in 1852 in *The General Valuation of Rateable Property in Ireland* (Dublin: General Valuation Office, 1847–64).
3. *Monday next*: 17 August.
4. *my address*: Tyndall's presidential address to the BAAS at Belfast.
5. *the late Lord Dunraven*: Edwin Richard Wyndham-Quin (1812–71), 3rd Earl of Dunraven and Mount-Earl (*ODNB*).
6. *Lady Edith and Lady Emily Quin*: Edith Wyndham-Quin (1848–85) and Emily Anna Wyndham-Quin (1848–1940).
7. *Brown of Brown's Hill*: Robert Clayton Browne (1799–1888), who lived at Browne's Hill in Clonmelsh, County Carlow, Ireland. He had three sons, William Clayton Browne (1835–1907), Charles Henry Browne (1836–89), and Robert Clayton Browne (1839–1906), although it is unclear which one Tyndall met.

From Thomas Archer Hirst 20 August 1874 4394

Baden Baden[1] | Aug[u]st 20th 1874

My dear John

Your letter[2] and address[3] arrived yesterday and the latter occupied my thoughts all day to the complete exclusion of Geometry. I took it with me in my days ramble and enjoyed it thoroughly.[4] Of course your ideas were on the main well known to me but the development you gave to them, the incidental matters you brought in, and the whole tone of the address delighted me greatly. I expected a good deal and I have not been disappointed I dare say some of your audience last night shook their heads dubiously about your transcendental materialism[5] but it will do them good to ponder it and depend upon it your address will be pondered long after to-day. I feel light hearted this morning at the thought that your greatest effort is over and has certainly been a successful one; all the rest will be mere routine, tiresome to you perhaps but otherwise of minor importance. I improve steadily and so far without relapse. This place suits me very well. I ramble over the wooded hills all day and in the evening smoke my cigar placidly and listen to excellent music in the grounds of the "Conversations Haus" formerly Kinsaal.[6] In a day or two I shall go on to Wildbad (Württemberg)[7] and explore another part of the Schwartz Wald.[8] Poste Restante[9] will always find me

Ever yours affectionately | Tom

Please give the enclosed[10] to Sabine

RI MS JT/H/258

1. *Baden Baden*: a spa town in the Black Forest, in southwest Germany.
2. *Your letter*: letter 4392.
3. *address*: the copy of 'Address of John Tyndall', *Brit. Assoc. Rep. 1874*, pp. lxvi–xcvii, that Tyndall had requested Taylor and Francis (see letter 4368, n. 5) to send Hirst (see letter 4392).
4. *I took it with me . . . enjoyed it thoroughly*: Hirst recorded in his journal for 19 August: 'Visited the Altes Schloss and Ebersteinburg. It was a charming walk. On my way I read Tyndall's address which is to be delivered at Belfast this evening. It is exceedingly good' (*Hirst Journals*, p. 2004).
5. *your transcendental materialism*: as Hirst intuits, Tyndall's alleged materialism in the Belfast Address was carefully qualified by his insistence that all phenomena have their roots in a vaguely understood cosmic life. See R. Barton, 'John Tyndall, Pantheist: A Rereading of the Belfast Address', *Osiris*, 3 (1987), pp. 111–34; and S. S. Kim, *John Tyndall's Transcendental Materialism and the Conflict Between Religion and Science in Victorian England* (New York: Edwin Mellen, 1996).
6. *"Conversations Haus" formerly Kinsaal*: the conversation house in Baden-Baden, a complex that enabled visitors to the spa to promenade even in bad weather, was opened in 1824.
7. *In a day or two . . . Wildbad (Württemberg)*: Hirst arrived in the spa town of Wildbad on 24 August, staying until the end of the month.
8. *Schwartz Wald*: Black Forest (German).
9. *Poste Restante*: see letter 4392, n. 7.
10. *the enclosed*: enclosure missing.

To Joseph Henry 22 August 1874 4395

BRITISH ASSOCIATION. | BELFAST. 1874. | 22nd. Aug[ust].

My dear Professor Henry,

I am sure I may bespeak your kindness for my friend Mr. Bonamy Price[1] the distinguished Professor of Political Economy in the University of Oxford.

With kindest regard to M[rs]. Henry & the young ladies[2]

believe me | always yours | John Tyndall

Professor Joseph Henry | Smithsonian Institution | Washington D.C.

Henry E. Huntington Library ("The Huntington") John Tyndall to Joseph Henry, 22 August 1874, mssRH 3980, box 45, William Jones Rhees Papers, The Huntington Library, San Marino, California

1. *Mr. Bonamy Price*: Bonamy Price (1807–88), who in 1868 had been elected Drummond Professor of Political Economy at Oxford (*ODNB*).
2. *the young ladies*: see letter 4390, n. 13.

From FitzWilliam Sargent[1] 22 August [1874][2] 4396

Professor Tyndall | Hotel Imbert, Beuzeval, Calvados,[3] France | Aug[ust] 22

Dear Sir:

I have had the pleasure of meeting you several times, on the last occasion at the Bel-Alp, and although I suppose you will scarcely remember my name, yet I take the liberty of writing to you these few lines.

I have just read an imperfect report[4] of your most interesting address to the British Association, and I should like immensely to have a full report of it, if there is to be any such Report.[5] Will you have the kindness (when you have a spare moment) to inform me where I can procure such a copy? And will you also please to inform me where a copy of the paper by Prof. Huxley[6] can be had, which contains the exhaustive & clear exposition of the "Natural Selection", and "Origin of the Species", theories?

Trusting that you will excuse the liberty I have taken in thus addressing you, I am, my dear Sir,

Very respectfully yours, | F. W. Sargent

RI MS JT/1/S/54

1. *FitzWilliam Sargent*: FitzWilliam Sargent (1820–89), a physician from Philadelphia who, with his wife and children, travelled to Europe in 1854, settling in Paris. They thereafter regularly visited mountain resorts in Switzerland. His eldest child was the painter John Singer Sargent (1856–1925), who Tyndall may have met alongside his father.
2. *[1874]*: the year is established by reference to Tyndall's presidential address to the BAAS in Belfast (see n. 5).
3. *Hotel Imbert, Beuzeval, Calvados*: the Hôtel de la Mer-Imbert was built in 1867 in the coastal resort of Beuzeval-les-Bains in the Calvados department of Normandy.
4. *an imperfect report*: Tyndall's presidential address was delivered on 19 August, so the report was presumably in a newspaper.
5. *such Report*: two official versions of the address were published: *Belfast Address* and 'Address of John Tyndall', *Brit. Assoc. Rep. 1874*, pp. lxvi–xcvii.
6. *the paper by Prof. Huxley*: in the address, Tyndall had said that Darwin's work had 'needed an expounder; and it found one in Mr. Huxley. I know nothing more admirable in the way of scientific exposition than those early articles of his on the origin of species' (p. lxxiv). On the articles Tyndall had in mind, see letter 4373, n. 11.

From Mary Anne Coxe [23 August 1874][1] 4397

Kinellan,[2] Sunday

My dear Friend—If humble M. A. C. may still venture so to call the great man who once told her to call him "John Tyndall!"

We[3] only returned last night after a two months wandering—and find this paper[4] which I send along with this (marked in two places)

I need not say with what delight and admiration—and pride in you, we read the Address.[5]

We knew you would make a grand appearance—and did not you do it? "Rather"!

I wonder if you ever got the Scotsman which came out some weeks ago (after being weeks in their hands)[6] James desired 6 copies to be sent to you, and he now bids me to tell you that he would have put far more verve "intult",[7] but had he put any friendly animus in it the Scotsman would simply not have put it in, and he wished the mean detractors[8] to be condemned out of their own mouths. I like the article very much, seeing that he was thus with held from putting more warmth in to it. Say if you have got it, if not I shall try still to get a copy—but it is extremely difficult to do after the time has gone by.

I should like very much to know if you were pleased with it?

I write in frantic haste only returned a few hours and every thing out of its place—painters here reigning supreme.

Good bye | I am thine in | admiration | M. A. C.

RI MS JT/1/TYP/1/304
LT Typescript Only

1. *[23 August 1874]*: this date is suggested by Louisa Tyndall's annotation to the MS letter 'August 22nd or thereabouts', with the 'Sunday' given on the letter confirming the precise date.
2. *Kinellan*: Kinellan House in Murrayfield, an area to the west of Edinburgh.
3. *We*: including her husband James Coxe.
4. *this paper*: enclosure missing, but probably 'The British Association', *Scotsman*, 20 August 1874, p. 5.
5. *the Address*: Tyndall's presidential address to the BAAS at Belfast.
6. *the Scotsman . . . weeks in their hands*: [J. Coxe], 'Forbes *Versus* Tyndall', *Scotsman*, 2 July 1874, p. 2. The article was first sent to the *Scotsman* newspaper in mid-May (see letter 4352).
7. *"intult"*: in to it (Scottish dialect).
8. *the mean detractors*: George Forbes, Peter Guthrie Tait, and John Ruskin (see letter 4352).

To Thomas Archer Hirst 26 August 1874 4398

BRITISH ASSOCIATION. | BELFAST. 1874. | 26th August 1874

My dear Tom

This day ends my labours here and from now on I shall be on the move. If I except the theologians all passed off in the most harmonious manner. Lubbock's lecture[1] was charming, and I never saw Huxley in greater force.[2] I think my address[3] had emptied some of the benches; but the greatness of the audience remained faithful, and Huxley won a triumph.[4] But you can form no notion of the religious agitation. Every pulpit in Belfast thundered at me. Even the Roman Catholics who are usually wise enough to let such things alone came down upon me.[5] But with all this Sound I remain unspent; and though private denunciations & criticisms have reached me, private encouragement of a high kind has not been wanting. I halt at Kingstown for a day or two to see some lighthouse experiments[6] & then proceed homewards. I hope you continue to improve. Your last letter[7] cheered Lilly[8] immensely. She was very sad, but the good account you gave of yourself has abolished all that. And now I must halt, for the people are crowding round me.

always affectionately | John.

RI MS JT/1/T/715

1. *Lubbock's lecture*: John Lubbock gave the evening lecture on 21 August, at 8:30pm in Ulster Hall, on 'Common Wild Flowers Considered in Relation to Insects' (*Brit. Assoc.*

Rep. 1874, p. lxv). This was published as J. Lubbock, 'Common Wild Flowers Considered in Relation to Insects', *Nature*, 10 (1874), pp. 402–6.

2. *Huxley in greater force*: Thomas Huxley's evening lecture on 24 August was 'On the Hypothesis that Animals are Automata, and Its History' (see letter 4327, n. 9).
3. *my address*: the presidential address, given on 19 August.
4. *Huxley won a triumph*: in a letter to his wife Henrietta, Huxley recorded of his evening lecture (see n. 2): 'I am glad to say it was a complete success. I never was in better voice in my life, and I spoke for an hour and a half without notes, the people listening still as mice . . . and though I spoke my mind with very great plainness I never had a warmer reception' (*Life and Letters of Huxley*, vol. 1, p. 414).
5. *Even the Roman Catholics . . . upon me*: in the days following the presidential address on 19 August, the *Ulster Examiner*, Belfast's principal Catholic newspaper, ran a series of leading articles criticizing Tyndall's scientific conclusions. The articles were probably written by the editor, Michael Cahill, curate of St Patrick's Church.
6. *Kingstown . . . lighthouse experiments*: Kingstown (now Dún Laoghaire) had two lighthouses, on the east (built 1847) and west (built 1852) sides of its harbour.
7. *Your last letter*: letter 4388; Tyndall, away in Belfast, had presumably not yet received letter 4394, in which Hirst said much less about his health.
8. *Lilly*: Emily Hirst.

From François Moigno [*c*. late August 1874][1] 4399

Les Mondes | Revue Hebdomadaire | Des Sciences et de leurs applications aux arts et à l'industrie | par | M. l'Abbé Moigno | Rédaction : | 2 rue d'Erfurth | Paris, le 187

Mon cher Mr Tyndall,

J'ai reçu les deux exemplaires de votre adresse, le premier très à temps, le second hier seulement. Je vous dirais volontiers le coeur bien gros

O Tyndal Tyndal, quae te dementia cepit?!

Pourquoi avez vous choisi ce sujet. Ce n'est pas là de la science. La Recherche des origines est interdite à la science, même d'après les lois du Positivisme.

Le spectacle d'une âme à la recherche des origines qu'elle ne trouve pas et qu'elle ne trouvera jamais, s'accrochant dans sa route à tous les novateurs, même et ni moins scientifique et aux moins honorables de tous, comme Jordano Bruno qui fut tour à tour catholique, calviniste, luthérien, anglican, qui fut chassé de partout, et tout cela pour constater que les novateurs pas plus que les orthodoxes n'aboutissent jamais, c'est un bien triste spectacle.

N'importe je publierai une traduction fidèle de votre discours, avec des notes respectueuses mais sévères, dont vous ne serez pas mécontent. Vous savez qu'en fait d'atomisme je suis plus avancé que vous, je vous ai devancé, c'est vous dire assez que je n'ai pas peur des doutes, même les plus avancés.

J'ai même écrit que cette matière, dont vous ne voulez pas qu'on dise du mal, est formée d'atomes simples, bien voisins des esprits, mais qu'à la différence des esprits, elle est inerte!!

A quinze jours donc pour la traduction, je ne ferais pas comme l'Athenaeum Anglais qui a eu peur de vos doctrines, qui pour la première fois peut-être depuis quarante ans n'a pas arrêté l'adresse complète quoique vous soyez peut-être le plus éminent pour la science parmi les Présidents.

J'ai été ravi, infiniment ravi, du portrait que Natur a publié de vous dans la dernière livraison. M. Gauthier Villars a qui je l'ai montré a été plus ravi encore ; je voudrais à tout prix obtenir un cliché Block de ce portrait, pour le publier dans Les Mondes, en tête de la Chaleur, de la Lumière, du Son. Essayez de le faire reproduire en France même en faisant appel à la photographie, ce serait se condamner à un *[1 word illeg]* obtenu de vos amis de Natur qu'ils nous cèdent un cliché, un galvano, nous le paierons ce qu'il faudra. Je compte absolument sur la puissance de votre intervention.

Ah ! Cher et illustre ami, que vous seriez heureux si à tant de science vous pouviez ajouter comme moi beaucoup de foi. Je travaille avec ardeur à finir mon grand ouvrage des Splendeurs de la foi, accord parfait de la Révélation et de la Science, de la foi et de la raison. Vous le lirez, n'est-ce pas ?

Tout à vous d'estime et d'affection sincère, | F. Moigno.

Les Mondes | Weekly Review | Of the sciences and their applications to the arts and industry | by | Father Moigno | Editorial Offices: | Erfurth Street | Paris, the 187

My dear Mr Tyndall,

I have received both copies of your speech,[2] the first early enough, the second only yesterday. And it is with a heavy heart that I would willingly tell you

O Tyndal Tyndal, quæ te dementia cepit?![3]

Why did you chose this subject. This is not science. The Research of origins[4] is forbidden to science, even according to the laws of Positivism.[5]

The sight of a soul in search of the origins it does not find and it will never find, holding on to all the pioneers, even and not as scientific and the least honourable of all, like Giordano Bruno,[6] who was in turn Roman Catholic, Calvinist, Lutherian, Anglican,[7] who was chased away from everywhere, and all that to finally realize that neither the pioneers nor the orthodox ever succeed—this sight is a very sad one.

Whatever, I will publish a faithful translation of your speech, with respectful yet strict notes[8] of which you will not be dissatisfied. You know that I am ahead of you in terms of atomism,[9] I outdistanced you, it goes to say that I am not frightened of doubts even the most advanced ones.

I have even written that this matter, which you do not want to be spoken badly about, is formed of simple atoms, very close to the spirits, but which, to the difference of spirits, is inert!!

Until fifteen days then for the translation![10] I will not act like the English Athenaeum, which was scared of your theories, and, for the first time in forty years, did not state the complete speech[11] although you might be one of the most distinguished amongst the presidents regarding science.

I was delighted, absolutely delighted, by the portrait that Nature published of you in its last issue.[12] Mr Gauthier-Villars, to whom I showed it, was even more delighted; I would like to obtain a cliché block[13] at all costs, to publish in Les Mondes, on the first page of la Chaleur, of la Lumière, of the Son.[14] Any attempt to have it reproduced in France even by calling upon photography would be a condemnation to *[1 word illeg]* obtained of your friends from Nature to let us have a cliché, a galvano,[15] we would pay for it what is required. I absolutely rely on the power of your intervention.

Ah! My dear and illustrious friend, how happy you would be if, to that much science, you could add, like I do, a lot of faith. I am working arduously to finish my major publication of the Splendeurs de la foi,[16] the perfect harmony between Revelation and Science, faith and reason. You will read it, won't you?

All yours with my regards and sincere affection, | F. Moigno

RI MS JT/1/M/137

---

1. *[c. late August 1874]*: the date is suggested by reference to the 'last issue' of *Nature* (see n. 12).
2. *both copies of your speech*: it is unclear precisely which two copies of Tyndall's presidential address to the BAAS at Belfast Moigno received, but as he mentions the *Athenæum* and *Nature* later in the letter, it seems likely that they were 'Literature', *Athenæum*, no. 2443 (1874), pp. 231–3; and 'Inaugural Address of Prof. John Tyndall, D.C.L., LL.D., F.R.S., President', *Nature*, 10 (1874), pp. 309–19.
3. *quæ te dementia cepit?!*: what madness has seized you? (Latin). Moigno is adapting a line, 'Ah, Corydon, Corydon, what madness has seized you?', from Virgil's *Ecologues*, II.viii.69.
4. *The Research of origins*: in his presidential address to the BAAS, Tyndall had controversially asserted: 'By an intellectual necessity I cross the boundary of the experimental evidence, and discern in that Matter which we, in our ignorance of its latent powers, and notwithstanding our professed reverence for its Creator, have hitherto covered with opprobrium, the promise and potency of all terrestrial Life' ('Address of John Tyndall', *Brit. Assoc. Rep. 1874*, pp. lxvi–xcvii, on p. xcii).
5. *the laws of Positivism*: in the Positivist philosophy of the French philosopher August

Comte (1798–1857), the human mind passed through earlier theological and metaphysical stages in which it sought the 'essential nature of beings, the first and final causes (the origin and purpose) of all effects'. Only in the final positivist stage had the mind 'given over the vain search after Absolute notions, the origin and destination of the universe, and the causes of phenomena', and instead focused on the 'study of their laws' and invariable relations (*The Positive Philosophy of Auguste Comte*, trans. H. Martineau, 2 vols (London: Trübner, 1875), vol. 1, p. 2).

6. *Giordano Bruno*: in his presidential address, Tyndall celebrated the Dominican friar and cosmologist as coming close to 'our present line of thought' in his apprehension that 'Nature in her productions does not imitate the technic of man' and instead 'brings . . .forms forth . . .by its own intrinsic force' ('Address of John Tyndall', p. lxxv).
7. *in turn Roman Catholic . . . Anglican*: Bruno spent from 1576 to 1592 in exile from his native Italy because of his heretical beliefs, travelling between several different countries and adapting himself to their respective religious traditions. In his presidential address, Tyndall noted: 'He was accused of heresy and had to fly, seeking refuge in Geneva, Paris, England, and Germany' ('Address of John Tyndall', p. lxxv).
8. *a faithful translation . . . strict notes*: 'Association Britannique pour l'Avancement des Sciences: Discours présidential de M. Tyndall', *Les Mondes*, 35 (1874), pp. 325–97.
9. *in terms of atomism*: Moigno had articulated his adherence to point atomism, which postulates that atoms are made from primary particles which are identical and a single oscillatory law determines their interactions, in 'Recherches de la cause qui maintient les molécules des corps à distance', *Cosmos*, 2 (1853), pp. 371–82.
10. *Until fifteen days then for the translation!*: Moigno's translation was published on 29 October.
11. *the English Athenaeum . . . complete speech*: the *Athenæum*, which usually printed the BAAS president's inaugural address in full, prefaced its abridged account of Tyndall's address by noting that 'disappointment will have been felt by some at the theme chosen by the President', which, rather than providing 'some startling intelligence, fresh from his laboratory', was a 'rather metaphysical dissertation . . . too large for the limits of an address' ('Literature', *Athenæum*, no. 2443 (1874), pp. 231–3, on p. 231 and 233).
12. *the portrait that Nature published of you in its last issue*: in its weekly issue for 20 August, *Nature* included an unbound portrait that accompanied the article H. Helmholtz, 'Scientific Worthies. IV.—John Tyndall', *Nature*, 10 (1874), pp. 299–302. The portrait was a stipple engraving made by Charles Henry Jeens from a photograph of Tyndall taken in New York, probably in January 1873, by José María Mora.
13. *cliché block*: a stereotype or mould, made from either papier-mâché or plaster, of an original printing block that enabled copies to be made.
14. *la Chaleur, of la Lumière, of the Son*: J. Tyndall, *La chaleur, mode de movement*, trans. F. Moigno, 2nd edn (Paris: Gauthier-Villars, 1874); J. Tyndall, *La lumière: six leçons faites en Amérique dans l'hiver de 1872–1873*, trans. F. Moigno (Paris: Gauthier-Villars, 1875); J. Tyndall, *Le son: cours expérimental fait à l'Institution royale*, trans. F. Moigno (Paris:

Gauthier-Villars, 1869). The portrait was not used in any of these publications, and there was no further edition of *Le son*.

15. *galvano*: a metal mould of an original printing block made by the chemical process of electrotyping, which was also known as galvanoplasty.
16. *the Splendeurs de la foi*: F. Moigno, *Les splendeurs de la foi: accord parfait de la révélation et de la science de la foi et de la raison*, 4 vols (Paris: Bureau du journal *Les Mondes*, 1879).

From George Biddell Airy[1] 5 September 1874 4400

Royal Observatory, Greenwich, | London, S.E. | 1874 September 5

My dear Sir

I am much obliged by your present of a copy of your paper on the atmospheric transmission of sound,[2] which I have glanced over with pleasure and hope to read better. The observations which in the first instance strike me as remarkable are those on the non-transmission, accompanied by reflection, of sound from special masses of air.[3] But several popular inaccuracies seem to be corrected for the first time.

My experience of London fogs[4] (vulgo[5] 'London particulars')[6] is not very great. But I have been struck more than once with the clearness with which I heard the sounds of carriages that I could not see.

The variation of audibility of the sounds of Big Ben[7] has often surprised me. In listening for it, perhaps 20 times in succession I cannot hear it at all, and on the 21st time I hear it as distinctly as the clocks at Deptford.[8] (My opportunities of listening are very favourable, because, knowing exactly its error, from galvanic communication,[9] I know precisely when the sound ought to come). Probably some of your new facts may aid to explain it.

I am, my dear Sir, | Yours very truly | G B Airy

Professor Tyndall LLD | &c. &c. &c.

RI MS JT/1/A/56
RI MS JT/1/TYP/1/51
Royal Greenwich Observatory Archives, Papers of George Airy 6/439

---

1. *George Biddell Airy*: George Biddell Airy (1801–92), a British mathematician and astronomer. In 1826 he was appointed Lucasian professor of mathematics at Cambridge University, and in 1828 to the Plumian Chair of astronomy and directorship of the Cambridge University observatory. In 1835 he was appointed Astronomer Royal, a position he held until 1881, and set about improving the instrumentation and research activities of the Royal Observatory, Greenwich. He pursued research on a wide range of topics, including optics,

sound, electricity, calculating the Earth's mean density, and analyzing the inequalities in the motions of the Earth and Venus. From 1872–3, Airy was president of the RS (*ODNB*).

2. *your paper on the atmospheric transmission of sound*: 'Vehicle of Sound', which had been published in August.
3. *the non-transmission . . . special masses of air*: Airy refers to observations in the section of Tyndall's paper entitled 'Experimental Demonstration of the Stoppage of Sound by Aërial Reflection' (pp. 202–5).
4. *London fogs*: Airy alludes to a section of Tyndall's paper entitled 'Action of Fog. Observations in London' (pp. 209–14).
5. *vulgo*: commonly, popularly (*OED*).
6. *'London particulars'*: originally the name of a type of Madeira wine imported through London, by the mid-nineteenth century it had become a colloquial term for the dense fogs that affected the city.
7. *Big Ben*: the nickname for the Great Bell of the striking clock at the Houses of Parliament.
8. *the clocks at Deptford*: the Royal Dockyards at Deptford in east London close to Greenwich hosted many clocks, with that on the clocktower of the Great Storehouse, constructed in the 1720s, the most prominent.
9. *galvanic communication*: Airy had, since the 1840s, used galvanic batteries to send electrical impulses and create time-signals that helped standardize clocks according to the standard time at the Royal Observatory in Greenwich.

From Olga Novikoff 5 September 1874 4401

Ostende.[1] Poste restante.[2] | Septem[ber]. 5/ 74 | Private.

Dear Dr Tyndall,

Pardon the liberty I venture to take, in writing to you; it is, I dare say, too plucky on my part, especially now, when no doubt you get hundreds of letters every day if not every second, expressing infinitely better than a poor "Moscovite" like myself,[3] can do, admiration, gratitude for y[ou]r last address![4] But my superiors are still so very much inferior to you that you probably beat us all, like a legion,[5] without entering into details! I try to find all that is written about y[ou]r last speech, & my conviction becomes stronger & stronger, that even in y[ou]r grand country the number of those who understand you <u>thoroughly</u>, is very limited. Let me therefore tell you that the effect produced by y[ou]r last speech on some of my countrymen & friends is immense! It is refreshing to see a man who has the intelligence to see truth, & the courage to speak it out, inspite of the surrounding cant & humbug those despots both in y[ou]r & my country.

I hope a new era is beginning with you, & I should be happy to be contemporary to y[ou][r] reign! So many different religions have been offered & good naturally accepted by the world how it is that nobody ever follows one, the adherents of which should be compelled to have an opinion of their own & express it honestly & openly.

In countries, w[hic][h] boast with being "orthodox" the opinion prevails, &c, that one cannot live without the belief in immortality. Well look at the Jews (not a very new tribe) & you'll see, how indifferent they were to far-fetched theories of that kind!—Will you pardon my offering you the little here joined pamphlet,[6] published anonymously, w[hic][h] proves what I say?

But I must not intrude upon y[ou][r] time any further. May I confess to you how happy & grateful I should feel, if you offered me the opportunity of seeing you? Are you not going to the Breslau Congress?[7] Don't you pass Ostende? I remain here till the 14th of Sep[tember]. but I may easily go to London afterwards for 2, 3 weeks before returning to Russia. Don't be shocked, please, don't be too shocked with my daring request! After all, it is so natural a wish even in a barbarous foreigner, who has no claims whatever, but whose faith in you is greater than in many recognized scientists!

Olga Novikoff | née Kiréeff

RI MS JT/1/N/17

---

1. *Ostende*: a port city on the coast of Belgium.
2. *Poste restante*: see letter 4392, n. 7.
3. *a poor "Moscovite" like myself*: Novikoff was a Russian who hailed from the capital city, Moscow.
4. *y[ou]r last address!*: Tyndall's presidential address to the BAAS at Belfast, delivered on 19 August.
5. *like a legion*: seemingly an allusion to the lines in John Milton's *Paradise Lost* (1667): 'As each divided legion might have seem'd | A numerous host, in strength each armed hand | A legion', VI.230–2.
6. *the little here joined pamphlet*: probably *Unsterblichkeitslehre nach der Bibel*, a privately printed and anonymous pamphlet in which Novikoff analysed the meaning of the Hebrew word 'Sheol' to better understand ancient Jewish conceptions of the immortality of the soul. She later observed that this 'German pamphlet . . . published, for private circulation' was 'sent . . . to one hundred professors . . . asking their opinion' (O. Novikoff, *Christ or Moses: Which?* (London: Williams and Norgate, 1895), p. ii).
7. *the Breslau Congress*: the forty-seventh annual meeting of the Gesellschaft Deutscher Naturforscher und Ärzte (Society of German Naturalists and Physicians), which was held at various venues across the German city of Breslau (now Wrocław in Poland) from 19 to 22 September.

To Olga Novikoff 7 September [1874][1] 4402

Royal Institution of Great Britain, | 7th Sept[ember].

Dear Madam

Your letter[2] gave me very great pleasure: I thank you much for it. My Address[3] has raised a storm in this country, but that will subside in due time and the atmosphere will be clearer and healthier afterwards.

I am dear madam | Very faithfully Yours | John Tyndall

[Envelope] Madam Olga de Novikoff | 15 Marché aux Herbes | Ostende | Belgium.

[Postmark] LONDON·W | XA | SE 7 | 74

RI MS JT/1/TYP/3/924
John Tyndall Correspondence in the Olga Novikoff Correspondence Collection. Courtesy of Special Collections, Kenneth Spencer Research Library, University of Kansas. MS 30:Q1

1. *[1874]*: the year is established by the relation to letter 4401 and the postmark on the envelope.
2. *Your letter*: letter 4401.
3. *My Address*: Tyndall's presidential address to the BAAS at Belfast, delivered on 19 August.

To William Hugh Spottiswoode 7 September 1874 4403

7th Sept[ember] 1874

Incontestible old Hughie!

Give my love to Papa & Mamma and Cyril;[1] and take the least little crumb of it yourself.

Tell papa I am much obliged to him for his letter.[2]

always my dear Boy | Yours lovingly | J. O. T.[3]

<INFO>RI MS JT/1/T/1291
RI MS JT/1/TYP/3/1263

1. *Papa & Mamma and Cyril*: William Spottiswoode, Eliza Spottiswoode, and Cyril Spottiswoode.
2. *his letter*: letter missing.

3. *J. O. T.*: Jolly Old Tyndall, a nickname by which Tyndall was known to the Spottiswoode children (see letter 3335, Adrian Kirwan and Elizabeth Neswald, eds., *The Correspondence of John Tyndall*, vol. 11, *The Correspondence, January 1869–February 1871* (Pittsburgh: University of Pittsburgh Press, 2022); and letter 3946, *Tyndall Correspondence*, vol. 13).

To Elizabeth Dawson Steuart 7 September [1874][1] 4404

Royal Institution of Great Britain | 7th. Sep[tember].

My dear Friend,

The people that raise this uncandid outcry[2] are not worthy of contradiction. They would roast me, but the time of roasting is happily gone by. You are correct in saying that I am not an Atheist.[3] Though I am far from accepting their crude notions of the Power that rules the Universe.[4]

Your ever | John Tyndall.

RI MS JT/1/TYP/10/3413
LT Typescript Only

---

1. *[1874]*: the year is established by reference to the 'outcry' over Tyndall's Belfast Address (see n. 2).
2. *this uncandid outcry*: over Tyndall's presidential address to the BAAS at Belfast.
3. *I am not an Atheist*: Tyndall would soon after publicly address the accusations of atheism made against him, initially in the preface to first edition of *Belfast Address*, written on 15 September, in which he stated: 'In connexion with the charge of Atheism, I would make one remark . . . Were the religious moods of many of my assailants the only alternative ones, I do not know how strong the claims of the doctrine of "Material Atheism" upon my allegiance might be. Probably they would be very strong. But, as it is, I have noticed during years of self-observation that it is not in hours of clearness and vigour that this doctrine commends itself to my mind; that in the presence of stronger and healthier thought it ever dissolves and disappears, as offering no solution of the mystery in which we dwell, and of which we form a part' (pp. vii–viii). A month later, on 28 October, Tyndall stated in a lecture delivered at the Free Trade Hall in Manchester that the 'profession of that Atheism with which I am sometimes so lightly charged would, in my case, be . . . only slightly preferable to that fierce and distorted Theism which I have had lately reason to known still reigns rampant in some minds as the survival of a more ferocious age' (J. Tyndall, 'Crystalline and Molecular Forces', in *Science Lectures for the People. Sixth Series* (Manchester: John Heywood, 1874), pp. 1–13, on p. 12; this was also published in *Belfast Address*, 7th thousand, pp. 67–83). On Tyndall and religious belief, see G. Cantor, 'John Tyndall's Religion: A Fragment', *Notes and Records: The Royal Society Journal of the History of Science*, 69 (2015), pp. 419–36.

4. *far from accepting . . . rules the Universe*: in *Belfast Address*, 7th thousand, Tyndall, in the revised preface written on 5 December, responded to a Catholic critic by asserting: 'I do not fear the charge of Atheism; nor should I even disavow it, in reference to any definition of the Supreme which he, or his order, would be likely to frame' (p. ix).

To Edward L. Youmans 15 September 1874 4405

September 15, 1874

MY DEAR YOUMANS:

Thanks, many and hearty, for your cordial letter.[1] I have just time to say that before this week ends a revised copy of the Belfast Address[2] shall be on its way to you.

It has caused tremendous commotion. How foolish they are! Their wisdom would have been shown in letting the thing alone, but they are not Wisdom's children.[3]

Cardinal Cullen has just appointed three days of prayer to keep infidelity out of Ireland![4]

Yours ever, | JOHN TYNDALL.

P.S.—I caught a glimpse of Spencer yesterday, and shall dine with him to-morrow. He is flourishing.

J. Fiske (ed), *Edward Livingston Youmans: Interpreter of Science for the People* (New York: D. Appleton, 1894), p. 319

1. *your cordial letter*: letter missing.
2. *a revised copy of the Belfast Address*: *Belfast Address*. Tyndall wrote the preface to the volume on the same day as this letter.
3. *Wisdom's children*: possibly an allusion to the biblical maxim 'wisdom is justified of all her children' (Luke 7:35, also Matthew 11:19).
4. *Cardinal Cullen . . . keep infidelity out of Ireland!*: on 3 September, Paul Cullen (1803–78), the Catholic Archbishop of Dublin, issued a pastoral letter to the clergy of Dublin to be read in the city's churches on the following Sunday (6 September). In the letter, Cullen stated that at a time when 'incessant and perfidious attempts are made to destroy faith . . .I beg you to exhort your flocks to devote three days of this month to prayer for the . . .preservation of Ireland from that infidelity with which philosophy of a false name, under the mask of science, is trying to poison the Catholic people' ('His Eminence Cardinal Cullen', *Freeman's Journal*, 7 September 1874, p. 4). The days of prayer were to be 13, 14, and 20 September, when a relic of the true cross would be put on display at St Mary's Pro-Cathedral in Dublin.

From John William Draper 15 September 1874 4406

University[1] Washington Sq[uare]. | New York | Sept[ember] 15th 1874

My Dear Dr Tyndall,[2]

I am very much obliged to you for your memoir[3] <*x*> what I have <*x*> <*x*> hoping your kindness <*x*> tell you about the reception <*x*> address in America that *[2 words illeg]* <*x*> you & that most severely for it. Your *[1 word illeg]* [was] *[1 word illeg]* & this performance brilliant. The publication of my book[4] brought odium enough on me—so having pressed through the ordeal that you have now to undergo I can sympathize thoroughly with you

It should be a source of unallowed gratification to you to learn that the reception of your address[5] here has been all that you could have meant There is not one of the leading newspapers that does not speak of it with the most profound respect & even with admiration The so called religious journals have to my astonishment lost their vindictiveness They have very little to say

Dont you remember that stormy night when you gave a lecture[6] to an overflowing *[1 word illeg]*. That satisfied me What a change had come over the American [public]. They had a fair excuse for abandoning you[7] <*x*> through thick & thin. When <*x*> over nearly 40 years ago that would never have been an impossibility And now the most influential & *[1 word illeg]* men among us regard your address with approval and *[1 word illeg]* (it is not too strong an expression to use) with affection

I am at this moment printing a book entitled "History of the Conflict between Religion & Science"[8] It was written more than a year ago[9] & will be out in 6 or 8 weeks. I will send <*x*> yet.

When I want <*x*> honored me by quoting <*x*> that it would cost me my <*x*> my social position. I have neither <*x*> that asserted me though I had to *[1 word illeg]* it alone For me there was not a single friendly voice. How different now!

So receive the greeting of a fellow soldier in the cause of truth & believe me

Truly your Sincere friend | WD

Library of Congress, John William Draper Family Papers

---

1. *University*: the University of the City of New York, which was founded in 1831 and would be renamed New York University in 1896. It was located in Washington Square, close to the Greenwich Village area of the city. Draper had taught at the university since 1839, and was professor of chemistry, having recently stood down as president of the medical school after more than twenty years in the role.

2. *My Dear Dr Tyndall*: the MS of Draper's original letter has sustained two large tears that have excised a large number of words, indicated by <*x*>; the letter's contents are reconstructed here as fully as is possible.
3. *your memoir*: presumably a printing of the Belfast Address.
4. *my book*: probably J. W. Draper, *History of the Intellectual Development of Europe* (New York: Harper & Brothers, 1863).
5. *your address*: Tyndall's presidential address to the BAAS at Belfast, delivered on 19 August 1874.
6. *that stormy night when you gave a lecture*: 31 December 1872, when Tyndall gave the last of his series of lectures in New York at the Cooper Union for the Advancement of Science and Art in Manhattan.
7. *a fair excuse for abandoning you*: possibly because of the inclement weather, which had covered New York in thick snow.
8. *printing a book entitled "History of the Conflict between Religion & Science"*: J. W. Draper, *History of the Conflict between Religion and Science* (New York: D. Appleton, 1874).
9. *written more than a year ago*: the book's preface was written in December 1873, and it is unclear what was responsible for the delay. The book contained large amounts of material that had appeared in previous publications, so it does not seem that the delay was due to writing difficulties on Draper's part.

To Thomas Archer Hirst 17 September 1874 4407

17th. Sept[ember]. | 1874

My dear Tom.

I find I am pledged to dine[1] with Hooker today at Kew.[2]

Yours aff[ectionate]ly | John Tyndall

RI MS JT/1/T/924
RI MS JT/1/HTYP/629

1. *pledged to dine*: Hirst recorded in his journal that from 16 to 30 September he was 'at home . . . Dining occasionally at the Senior United Service Club, the Athenaeum being closed for repairs. There I frequently met Tyndall who is being attacked on every side for his Belfast address' (*Hirst Journals*, p. 2006). As such, Tyndall is probably declining an invitation to dine at the Senior United Service Club, which was opposite the Athenaeum Club on Pall Mall (on the Athenaeum, see letter 4186, n. 1).
2. *Kew*: see letter 4363, n. 19.

To Unidentified 17 September [1874][1] 4408

17th Sept[ember].

Dear Sir

A new edition of the Address[2] will be published this week, and I have directed my publishers[3] to send it to you.

Yours faithfully | John Tyndall

BL Add MS 41077, f. 151

---

1. *[1874]*: the year is established by reference to the publication of the Belfast Address (see n. 2).
2. *A new edition of the Address*: *Belfast Address*.
3. *my publishers*: Longmans, Green, & Co., 39 Paternoster Row, London.

To Thomas Archer Hirst [19 September 1874][1] 4409

Saturday

My dear Tom.

I fear I must work during the forenoon of tomorrow, but I will start for Greenwich[2] immediately after luncheon.

Yours aff[ectionate][ly] | John

RI MS JT/1/T/980
RI MS JT/1/HTYP/629

---

1. *[19 September 1874]*: this date, which was a Saturday, has been appended to the MS letter by LT. There is no entry for this date in Hirst's journal, so it cannot be confirmed.
2. *Greenwich*: Hirst was the Director of Studies at the Royal Naval College, Greenwich, from 1873–82.

From Olga Novikoff 19 September [1874][1] 4410

Symonds' Hotel. Brook Str[eet]. | near Bond St. | Saturday. Sept[ember]. 19.

Dear Mr Tyndall,

Y[ou][r] kind note[2] was forwarded to me from Ostende.[3] Many thanks.

I am in despair, & you really must help me! The Times with y[ou][r] address[4] is not to be had by any means! And I do want at least 6 copies of it for my

Russian friends as I want it to be translated. Have you not any spare copies & would you not generously offer them to me? Another request! Can you put aside all y[ou][r] legitimate pride, forget y[ou][r] superiority, & simply call on a simple mortal, who w[ou][ld] have her ears cut, if it could give her the pleasure of y[ou][r] acquaintance. But the day & hour have to be fixed by you!—Don't be indignant please!

Pardon y[ou][rs] truly Olga Novikoff | née Kiréeff

RI MS JT/1/N/40

1. *[1874]*: the year is established by the relation to letters 4401 and 4402.
2. *Y[ou]r kind note*: letter 4402.
3. *Ostende*: a port city on the coast of Belgium where Novikoff had stayed in early September (see letter 4401).
4. *The Times with y[ou]r address*: 'British Association for the Advancement of Science', *Times*, 20 August 1874, p. 4.

To Olga Novikoff 19 September [1874][1] 4411

Royal Institution of Great Britain | 19th. Sep[tember].

Dear Madam

I am just now much entangled:[2] And I never pay visits![3]—But I will make an effort to see you on Monday[4] at 5 P.M.

Yours faithfully | John Tyndall

A new edition of the Address[5] will be printed in a few days.

John Tyndall Correspondence in the Olga Novikoff Correspondence Collection. Courtesy of Special Collections, Kenneth Spencer Research Library, University of Kansas. MS 30:Q2

RI MS JT/3/TYP/3/924
LT Typescript Only

1. *[1874]*: the year is established by reference to the publication of the Belfast Address (see n. 5).
2. *just now much entangled*: Tyndall planned to visit Thomas Hirst in Greenwich the following day (see letter 4409).
3. *I never pay visits*: Tyndall was responding to Novikoff's request for him to 'simply call on a simple mortal' in letter 4410.
4. *Monday*: 21 September. It is not clear if this visit took place, although they certainly seem to have met by 23 September (see letter 4413). Perhaps misremembering these earlier meetings, Novikoff later recalled that she first met Tyndall when Alexander William Kinglake

introduced them in a London park on what she remembered as a 'lovely December day'. She then reflected that the 'acquaintance was made—an acquaintance which will never be forgotten. At that time I generally was "at home" after dinner, from nine to twelve, according to our Russian fashion, by which, without specifying any day, people are allowed to drop in when it pleases them. With me, however, only the elect were invited, and Tyndall naturally belonged to that category. Each time he came was a new pleasure to me, but especially when he was in a poetical mood and recited by heart, which with his strikingly melodious voice and remarkable memory he did to perfection' (*M.P. for Russia*, vol. 1, pp. 150 and 151).

5. *A new edition of the Address*: *Belfast Address*.

From François Moigno 23 September 1874 4412

Les Mondes | Revue Hebdomadaire | Des Sciences et de leurs applications aux arts et à l'industrie | par M. l'Abbé Moigno | Rédaction: | 2, Rue d'Erfurth, 2

Paris, le 23 S[bre] 1874

Cher et illustre ami, j'ai reçu la nouvelle édition de votre Adresse, mais je n'ai reçu ni le portrait, ni le Book of Facts.

Votre petite préface m'a bien attristé.

Cependant j'espère. Est-ce bien de la Théorie matérialiste que vous dites qu'elle se dissout et disparait dans les heures de folie et de santé d'esprit? Mes yeux et mon cœur m'auraient-ils trompé.

Je crois que vous ne serez pas mécontent de moi malgré ma sévérité.

La Lumière avance mais pas assez à mon gré.

J'ai corrigé hier la leçon relative à la Polarisation, quel chef d'œuvre! Quelle lucidité! Quelle instruction comme cela contraste douloureusement avec les nuages de Belfast.

A vous d'esprit et de cœur, d'estime et d'affection sincères.

A vous de toute sympathie la plus vive | l'abbé F. Moigno | Chanoine de St. Denis

Les Mondes | Weekly Review | Of the sciences and their applications to the arts and industry | by Father Moigno | Editorial offices: | 2 Erfurth street, 2

Paris, 23 S[eptem]ber 1874

Dear and illustrious friend, I received the new edition of your Address,[1] but I received neither the portrait[2] nor the Book of Facts.[3]

Your little preface[4] saddened me.

However, I keep hope. Is this really the materialist theory that you say will dissolve and disappear in the hours of madness and sanity of the mind?[5] Would my eyes and heart have deceived me?

I believe that you will not be displeased with me despite my severity.

La Lumière[6] progresses but not sufficiently to my liking.

Yesterday I corrected the lesson on Polarization,[7] what a masterpiece! What insight! Such instruction, as that which contrasted painfully with the clouds of Belfast.

With you in mind and heart, with sincere esteem and affection.

Yours truly with the deepest sympathy | Father F. Moigno | Canon of St. Denis

RI MS JT/1/M/126

---

1. *the new edition of your Address*: *Belfast Address*.
2. *the portrait*: see letter 4399, n. 12.
3. *the Book of Facts*: presumably the abstract of Tyndall's Belfast Address that was published in 'Physical Sciences', in C. W. Vincent (ed), *The Year-Book of Facts in Science and the Arts for 1874* (London: Ward, Locker, and Tyler, 1875), pp. 94–107. Tyndall may have had early access to the proofs of the abstract, as Vincent thanked him for his assistance with the compilation of the volume in its preface.
4. *Your little preface*: the preface to *Belfast Address*, pp. v–viii, which was written on 15 September.
5. *the materialist theory . . . sanity of the mind?*: in the preface, Tyndall stated that 'it is not in hours of clearness and vigour that . . . the doctrine of "Material Atheism" . . . commends itself to my mind', and that 'in the presence of stronger and healthier thought it ever dissolves and disappears' (p. viii).
6. *La Lumière*: letter 4277, n. 2.
7. *the lesson on Polarization*: Tyndall explained the '*polarization of light* . . . by means of a crystal of tourmaline' in the third lecture (*Six Lectures on Light* (1873), p. 115).

From Olga Novikoff [23 September 1874][1] 4413

Private | Symonds' Hotel. Brook Street. London | September. Wednesday.

My dear & charming Majesty:

Would you care to know what Albert Réville (of the Revue des 2 Mondes[2]) says to me of y[ou]r Belfast address? Here it is literally:

"Comme vous, j'ai trouvé ce discours fort remarquable & fort intéressant. On y sent la touche d'un esprit droit et honnête, subtil en même temps et souvent acéré, qui cherche passionnément le vrai, et qui n'aime que lui. Seulement il m'a semblé trop sévère pour les philosophes spéculatifs sans lesquels il n'y aurait jamais eu de philosophie expérimentale. En particulier je ne consentirai jamais à mettre Aristote[3] si bas, et le brave Lucrèce[4] si haut. Je ne sais si Mr Tyndall se rend compte de l'énorme concession qu'il fait au spiritualisme, en reconnaissant que la description pure et simple des phénomènes physiques, producteurs ou déterminateurs, de la sensation, ne rend pas encore compte de cette sensation elle-même, prise au sens subjectif?[5]

N'est-ce pas reconnaître en même temps que le sujet sentant est irrudictible aux explications purement physiques? En somme, j'ai été aiguillonné tout le long de cette lecture par le désir d'aller toujours en avant—ce qui ne m'arrive pas souvent—et je vous remercie infiniment de me l'avais procuré."[6]

The letter has the following funny P.S. "Quel imbécile que le Marquis Ripon."[7]—

Don't forget my prayer about your "Prayer"![8] I think nothing is more humiliating than praying—but with you, one cannot help doing it!—

Y[ou]rs very, very truly & forever! Olga Novikoff

RI MS JT/1/N/32

---

1. *[23 September 1874]*: the date is suggested by the relation to letter 4428 from 8 October, in which Tyndall belatedly responds to Novikoff's 'prayer about your "Prayer"' (see n. 8). Given their exchanges in letters 4421 and 4422, the most likely Wednesday in September for Novikoff to have made this 'prayer', itself seemingly a reminder of a verbal request, is this date.
2. *the Revue des 2 Mondes*: the *Revue des deux Mondes*, an influential monthly journal founded in Paris in 1829 whose name, which translates as *Review of the Two Worlds*, indicated its internationalist aspirations to link the old and new worlds of France and the United States.
3. *Aristote*: French name for the Greek philosopher Aristotle (384–322 BCE), who Tyndall, in the Belfast Address, criticized for the 'sheer natural incapacity which lay at the root of his mistakes'. He also proposed that 'As a physicist, Aristotle displayed what we should consider some of the worst attributes of a modern physical investigator—indistinctness of ideas, confusion of mind, and a confident use of language, which led to the delusive notion that he had really mastered his subject, while he had as yet failed to grasp even the elements of it' (*Belfast Address*, pp. 14–15).
4. *Lucrèce*: French name for the Roman poet and philosopher Titus Lucretius Carus (*c.* 99–*c.* 55 BCE), who Tyndall, in the Belfast Address, praised for his atomistic philosophy, which Tyndall claimed was a precursor of modern physical science. Tyndall also expressed admiration for Lucretius's aspiration for intellectual freedom and the 'destruction of superstition', which Tyndall suggested should be 'deemed a positive good' (p. 8).
5. *pure et simple... sens subjectif?*: in the Belfast Address, Tyndall articulated the putative objections of the eighteenth-century Anglican bishop and theologian Joseph Butler (1692–1752), who, in an imaginary dialogue, is made to say: 'What baffles and bewilders me, is the notion that from those physical tremors things so utterly incongruous with them as sensation,

thought, and emotion can be derived'. The fictive Bishop also tells his Lucretian antagonist 'You cannot satisfy the human understanding in its demand for logical continuity between molecular processes and the phenomena of consciousness. This is a rock on which materialism must inevitably split whenever it pretends to be a complete philosophy of life', and Tyndall concedes: 'I hold the Bishop's reasoning to be unanswerable' (pp. 33–4). Butler served as the Bishop of Bristol and Durham, and as the Dean of St Paul's.

6. *"Comme vous . . . l'avais procure"*: "Like you, I found this speech very outstanding & very interesting. It has the feel of a decent and honest soul, subtle at the same time and often sharp, that passionately seeks the truth and that only loves the truth. Nevertheless, he seemed to me too harsh with speculative philosophers without whom there would have been no experimental philosophy. Specifically, I would never agree to rank Aristotle so low, and the brave Lucretius so high. I do not know if Mr Tyndall realizes how enormous the concession is that he makes to spiritualism, by recognising that the pure and simple description of physical phenomena, producers or determiners, of the sensation, does not yet account for this sensation itself, taken in a subjective sense? That said, is it not a recognition that the sensitive subject cannot be reduced to explanations that are purely physical? In short, I was spurred on throughout this lecture by the desire to always move forward—which I do not feel very often—and I am extremely grateful to you for providing it to me" (French).
7. *"Quel imbécile que le Marquis Ripon"*: "the Marquess Ripon is such an imbecile" (French). George Frederick Samuel Robinson (1827–1909), 1st Marquess of Ripon, had converted to Catholicism in 1873, and was formally baptised at the London Oratory on 4 September 1874 (*ODNB*).
8. *my prayer about your "Prayer"!*: Novikoff seems to have asked Tyndall for details of his contributions to the prayer-gauge debate in 1872 (see letter 4428).

To Gustav Wiedemann 24 September 1874 4414

24th Sept[ember]. 1874

My dear Wiedemann.

Two days ago I desired Longman to send you a copy of the revised Address:[1] you will doubtless find it waiting for you on your return to Leipzig. Vieweg has also had a copy sent to him. Whatever you and he decide regarding it I shall agree to.

I have had letters both from Hamburgh & from Leipzig[2] asking whether it was necessary to ask permission to translate the Report of the Address which appeared in the Academy;[3] I have answered the letter from Hamburg[4] saying that I had no objection to the translating of the articles but warning the writer that the Report in the newspapers was incomplete. You have the complete Report.

Once for all let me explain matters to you regarding your reception at Belfast. A friend of mine[5] who has a beautiful house near that city would have received you and Feddersen[6] with open arms, and you would have fared more sumptuously with him than with me. There were other people also desirous to have you as guests: Knoblauch & his son[7] were received and taken care of in this way. This is quite our custom. Now you preferred being near me, and it was a great pleasure to me to have you near me. The only difference is that you were my guest instead of the guest of one of the gentry in the neighbourhood of Belfast.

I trust that this explanation will tranquilize your mind—You are no man's debtor. I on the contrary am your debtor for giving me what was a real pleasure—the pleasure namely of seeing your face during my year of presidency.

The address has caused great commotion both in England and Ireland. Cardinal Cullen has devoted three days to prayer & fasting,[8] so as to destroy its pernicious influence. It is also to be extinguished by an authoritative reply on the part of the Catholic body.[9]

When you see Helmholtz pray assure him from me that I knew nothing of his being applied to by the editor of Nature to write about me.[10] And that had I any voice in the matter I should have opposed the idea of adding to his great labours a labour of this kind.

With best regards to Mrs. Wiedemann[11]

Believe me always | yours faithfully | John Tyndall

RI MS JT/1/T/1491

1. *the revised Address*: *Belfast Address*.
2. *letters both from Hamburgh & from Leipzig*: letters missing, but the one from Hamburg was possibly from Emil Lehmann (see letter 4429).
3. *the Report of the Address which appeared in the Academy*: 'Science', *Academy*, 6 (1874), pp. 209–17.
4. *answered the letter from Hamburg*: Tyndall's reply is missing.
5. *A friend of mine*: not identified.
6. *Feddersen*: Berend Wilhelm Feddersen (1832–1918), a German physicist who was an independent researcher in Leipzig (*NDB*).
7. *his son*: possibly Rudolf Knoblauch (1861–1926), who became a silk manufacturer.
8. *Cardinal Cullen . . . to prayer & fasting*: see letter 4405, n. 4.
9. *an authoritative reply on the part of the Catholic body*: on the previous day, the *PMG* had reported that 'steps have already been taken towards the early publication on behalf of the Roman Catholic body of an authoritative argument in refutation of the doctrines put forward in the addresses delivered by Professors Tyndall and Huxley at Belfast' ('Professor Tyndall and His Critics', *PMG*, 23 September 1874, p. 8). The body was the Episcopacy of Irish Catholic Archbishops and Bishops, who issued a Pastoral Address on 14 October refuting the 'Materialist theories of to-day which recognise in matter the promise and

potency of every form and quality of life' ('Pastoral Address of the Archbishops and Bishops of Ireland', *Irish Ecclesiastical Record*, 11 (1874), pp. 49–70, on p. 51).

10. *Helmholtz . . . write about me*: Tyndall refers to H. Helmholtz, 'Scientific Worthies. IV.—John Tyndall', *Nature*, 10 (1874), pp. 299–302. The editor of *Nature* was Norman Lockyer; on his request to Helmholtz to write about Tyndall, see A. J. Meadows, *Science and Controversy: A Biography of Sir Norman Lockyer* (London: Macmillan, 1972), p. 34.
11. *M^rs^. Wiedemann*: Clara Wiedemann.

From Elizabeth Dawson Steuart — 24 September 1874 — 4415

Steuart's Lodge | Leighlinbridge | Sep[tember]. 24. 74.

My dear John,

Your truly welcome present[1] has arrived, quite safely, and I lose no time in writing to offer you my best thanks for it. I admire it very much and think the likeness excellent, though there is a sort of stern expression not natural to the usual one. I am very glad to have it, though indeed it is not needed to remind me of one for whom I have so long felt such strong and true affection. I have also to thank you for the copy of address,[2] and am glad you have taken some notice of the disagreeable remarks made on it.[3] Poor old Mrs Tyndall[4] is very thankful for your goodness, which is most acceptable, as she was more in need of assistance than I was aware of. Once more thanking you my dear friend, and with every good and kind wish

believe me, | affectionately yours | E. D. Steuart

I feel sure you will not forget the Vignette.[5]

RI MS JT/1/TYP/10/3414
LT Typescript Only

---

1. *Your truly welcome present*: presumably the engraved portrait made by Charles Henry Jeens from a photograph of Tyndall taken in New York, probably in January 1873, by José María Mora (see letter 4399, n. 12).
2. *the copy of address*: *Belfast Address*.
3. *taken some notice . . . made on it*: Steuart presumably means in the preface to Longmans' edition of the *Belfast Address*, written on 15 September.
4. *Mrs Tyndall*: Dorothea Tyndall.
5. *the Vignette*: possibly the hand-coloured version of the photograph of Tyndall by Mora (see n. 1) that Olga Novikoff seems to have made, which Tyndall referred to as 'The little Vignette' (see letter 4424, n. 8).

From Émile Acollas[1] [25 September 1874][2] 4416

Paris, 25 Rue Mr Le Prince

Monsieur et honoré collègue,

Je vous ai fait adresser un exemplaire de quatre brochures qui, à l'occasion de la science du Droit civil, marquent d'une manière générale ma direction philosophique ; je désire que vous en accueillez l'offre comme un témoignage, bien modeste assurément, mais aussi accusé que possible, des sympathies de mon esprit pour le vôtre.

Déjà votre discours au Congrès de Belfast, malgré certaines réserves que j'ai du faire en le citant, *[1 word illeg]* démontré que nous nous rencontrons dans des points de vue fondamentaux ; votre écrit sur le matérialisme et ses adversaires en Angleterre achève les démonstrations.

J'estime avec vous, monsieur, que l'homme a le besoin permanent d'un idéal, mais que pour le rencontrer cet idéal, il lui suffit de le chercher dans une connaissance de plus en plus approfondie de la nature. Laissez moi subsituter ce mot à l'expression trop dogmatique de la matière ; et des lois qui la gouvernent.

Comme vous aussi je pense, que le besoin le plus immédiat de nos générations c'est de s'affranchir des entraves du passé et du matérialisme pratique du présent, c'est-à-dire d'une part du mysticisme et du fanatisme religieux, et d'autre part, de ce positivisme *[1 word illeg]* philosophique, qui au point de vue spéculatif, est la négation de toute idée, et, au point de vue pratique, la légitimation de tout fait.

C'est là, monsieur et honoré collègue, le bien de nos recherches respectives bien qu'elles s'appliquent à des ordres tout différents.

Veuillez croire à ma cordialité profonde. | Émile Acollas | Professeur de Droit

Paris 25 Rue Mr Le Prince

Sir and honoured colleague,

I sent you a copy of four booklets[3] that, at the occasion of the science of civil law, emphasize my philosophical opinion in general; I would like you to welcome this offer as a token, rather modest of course, but as marked as possible, of my kind feelings towards you.

Your speech at the congress of Belfast, despite some reservations I had while citing it,[4] *[1 word illeg]* already showed that we agree on fundamental points; your writing on materialism and its opponents in England puts an end to the demonstrations.

I think, like you Sir, that mankind has a constant need of an ideal, but in order to meet this ideal, he must simply look for it in a deeper knowledge of nature. Let me substitute this word for the too dogmatic expression of the matter, and the law that governs it.

Like you, I think that the most immediate need of our generation is to free itself from the past impediments and from the practical materialism of the present, namely on the one hand from mysticism and religious fanaticism, and on the other hand, from this *[1 word illeg]* philosophical positivism,[5] which from a speculative point of view, is the negation of all ideas, and from a practical point of view, the legitimacy of all facts.

It is at this point, sir and honoured colleague, the good of our respective research even though they apply to very different domains.

Please believe in my deepest regard. | Émile Acollas | Professor of law

RI MS JT/1/A/2

---

1. *Émile Acollas*: Émile Acollas (1826–91), a professor of jurisprudence at his own private school in Paris.
2. *[25 September 1874]*: this date has been appended to the MS letter, possibly by Tyndall himself.
3. *four booklets*: enclosure missing, but probably *L'Anthropologie et le droit* (Paris: Arnous de Rivière, 1874); *La Science du droit en France au temps présent* (Lausanne: Howard-Deslisle, 1874); *La Philosophie de l'histoire et le droit* (Lausanne: Howard-Deslisle, 1874); and *L'Economie politique et le droit* (Lausanne: Howard-Deslisle, 1874).
4. *Your speech . . . while citing it*: in *La Science du droit en France au temps présent* Acollas noted that 'la gloire poétique de Lucrèce resplendit depuis des siècles, mais ce n'est que de nos jours que la profondeur des ses conceptions en physique a été comprise [the poetic glory of Lucretius has been shining for centuries, but it is only nowadays that the depth of his conceptions in physics has been understood]'. At the end of the sentence he added a footnote which stated: 'Voir le discours par lequel sir John Tyndall a inauguré cette année (1874) la session de l'Association britannique por l'avancement des sciences, discours vraiment magistral et qui serait de tous points admirable si l'auteur, pour ménager les susceptibilitiés religieuses des ses compatriotes, n'eût trouvé bon de se prêter à des concessions que la science ne saurait admettre (*Address delivered before the british association assembled at Belfast*, London) [See the speech in which Sir John Tyndall inaugurated this year (1874) the session of the British Association for the Advancement of Science, a truly masterful speech which would be admirable in all respects if the author, to spare the religious sensibilities of his compatriots, would have seen fit to allow himself concessions that science could not]' (p. 4).
5. *philosophical positivism*: the philosophical theory of Auguste Comte (see letter 4399, n. 5) which proposes that information derived from sensory experience, as interpreted through reason and logic, forms the exclusive source of all certain knowledge.

To William Henry Appleton 28 September 1874 4417

September 28, 1874

My dear Mr. Appleton:

The Address, separately published,[1] is going off with exceeding rapidity. The third thousand was called for in three days.

Yours ever faithfully, | John Tyndall.

P.S.—Being busy just now, I have merely glanced at Dr. Youmans's remarks;[2] but I have seen sufficient to assure me of the sagacity, and, indeed, eminent ability which mark his mind in the treatment of great questions. I often think that had he been less hampered by his ailments in youth[3] he would have made a profound mark on his day and generation. Even as it is, he is doing this.

J. Fiske (ed), *Edward Livingston Youmans: Interpreter of Science for the People* (New York: D. Appleton, 1894), p. 319

---

1. *The Address, separately published*: *Belfast Address*.
2. *Dr. Youmans's remarks*: probably in 'Editor's Table: Professor Tyndall's Address', *Popular Science Monthly*, 5 (1874), pp. 746–8, in which Edward L. Youmans claimed: 'No scientific paper ever before published has produced so extensive and profound an impression as this. The eminent ability of the speaker, the dignity of the occasion, the confessed importance of the subject, and the eloquence and power of the statement, have all concurred to this result' (p. 746).
3. *his ailments in youth*: at the age of thirteen, Youmans contracted ophthalmia, which impaired his vision for the rest of his life. He was practically blind for much of his early adulthood.

From Jane Barnard 28 September 1874 4418

Barnsbury Villa, | 320, Liverpool Road. | N.[1] |
Sep[tember] 28/74 | Monday Even[in]g

Dear Dr Tyndall

I am grieved to say that my dear Aunt[2] is very ill today, so that we must ask you to postpone your visit, to which we were looking forward with much pleasure.

She wrote to you on Saturday[3] I think, & she may have mentioned that she had had a fall on Friday[4] night. Tho' seemingly a very heavy one, she did not appear to suffer much from it till yesterday, when she was very poorly & helpless & today she is quite prostrated both in body & mind. She has had attacks like this before & has rallied from them so that I can only hope she may again.

I am dear Dr Tyndall | Yours very Sincerely | Jane Barnard | 28 Sept 1874

RI MS JT/1/B/55

1. *Barnsbury Villa, | 320, Liverpool Road. | N.*: see letter 4226, n. 1.
2. *my dear Aunt*: Sarah Faraday.
3. *She wrote to you on Saturday*: letter missing.
4. *Friday*: 25 September.

From Karl Hermann Knoblauch 28 September 1874 4419

Halle[1] | Sept[ember] 28th 74

My dear friend,

Returned home I cannot forbear to express you once more my best thanks for that happy and interesting days which issued for me from your kind invitation to Belfast.[2]

I shall never forget the commerce with my old friends and new acquaintances amongst the most celebrated naturalists of England; the splendid hospitality of your countrymen.[3]

The enjoyment produced by lovely landscapes and grand sceneries in the Green Island,[4] the incomparable hours we spent together in your actual home, my dear friend: all these have been far beyond my expectation, when I at first inclined to the journey.

In Greenwich we[5] regretted very much not to be able to make use of your kind recommendation. The observatory[6] was just closed at 2 o'clock when we reached it. The assistants as well as Mr Glaisher had already left it. Besides the latter was not at home, when we called upon him in his lodging.[7]

According to your wish (you will remember at our last evening with you in London), I have ordered a "Töpler Influence Machine"[8] to mechanist Wesselhöft[9] here in Halle. He promised to construct it for 85 Thaler[10] (except the packing and transport) (1 Thaler = 3 shillings) and send it in about 6 or 8 weeks.

My best thanks also for the copy of your excellent address to the British Association,[11] I just received.

Yours for ever most faithfully | Herm Knoblauch

RI MS JT/1/K/25
RI MS JT/1/TYP/7/2509

1. *Halle*: a city in central Germany.
2. *your kind invitation to Belfast*: letter missing.
3. *the splendid hospitality of your countrymen*: while in Belfast, Knoblauch was hosted by a member of the local gentry, as was the custom for BAAS meetings (see letter 4414).
4. *the Green Island*: i.e. Ireland.
5. *we*: Knoblauch travelled with his son (see letter 4414, n. 7).
6. *The observatory*: the Royal Greenwich Observatory, which was founded in 1675.
7. *his lodging*: 20 Dartmouth Hill, Blackheath, London.
8. *a "Töpler Influence Machine"*: an electrostatic induction generator invented by the German physicist August Joseph Ignaz Töpler (1836–1912).
9. *Wesselhöft*: Martin Wesselhöft, a mechanic specializing in electronic and acoustic apparatus, who worked at Jägerplatz 10 in Halle.
10. *Thaler*: a silver coin, formally called a Vereinsthaler, that remained legal tender in the newly-unified Germany even after its official replacement by the mark (see Note on Money).
11. *the copy of your excellent address to the British Association*: *Belfast Address*.

To John William Draper 29 September 1874 4420

29th Sept[ember]. 1874.

My Dear Dr. Draper.

No letter that has reached me for some time has pleased me so much as yours;[1] and no intelligence that I have lately received is more agreeable to me than that which informs me of your intention to give us a "History of the Conflict between Religion and Science".[2]

You stand upon a vantage formed that you never before possessed—few indeed have possessed it. For the mind of this century is wide awake to the opening fight between the ultramontanes[3] & the intelligence of the age. Article after article on the subject appears in the Times,[4] thundering down upon Capel, Manning,[5] and the system which they represent. All over Ireland the church has sounded the alarm, and vials of wrath are poured down upon me from all the Catholic pulpits.[6] As I said to Spencer a few evenings ago it will strengthen mightily all our hands to be able to fall back upon your vast historic knowledge. In fact, we shall smite the enemy hip & thigh.[7]

We had heard from Youmans that you proposed writing this book. I received the intelligence with an inward "hurrah!"

You bring a pleasant, and impressive memory to my mind when you refer to the last night of the year 1872.[8] I had been actually advised to give up the lecture on the assumption that nobody would think of coming out.[9] It was my greatest reward during the whole campaign.

I can well imagine the risks you run in the publication of that book:[10] you have now the reward of courage & you may calculate on the cooperation of strong heart and hard knuckles in the work that you have now in hand.

It shows however the effeminate softness of my nature when I state to you that to hear of the "affection"—or even a small fragment of the affection, of the American public[11]—is to me more pleasant than all their "applause".

Yours ever with the best wishes | John Tyndall

Library of Congress, John William Draper Family Papers

1. *yours*: letter 4406.
2. *a "History of the Conflict between Religion and Science"*: see letter 4406, n. 8.
3. *ultramontanes*: strong adherents or supporters of Papal authority, although the term originally related to representatives of the Roman Catholic Church north of the Alps (i.e. beyond the mountains) as opposed to ecclesiastics in Italy (*OED*).
4. *Article after article . . . in the Times*: the most recent such articles were 'The Pontigny Pilgrimage', *Times*, 4 September 1874, p. 8; and 'Archbishop Manning on Education', *Times*, 22 September 1874, p. 9.
5. *Capel, Manning*: Thomas John Capel (1836–1911), a Catholic priest who in 1873 was made domestic prelate to the Pope and in the following year appointed as rector at the new Catholic University College in Kensington. The latter appointment was made by Henry Edward Manning (1808–92), Archbishop of Westminster and head of the Catholic Church in England.
6. *vials of wrath . . . Catholic pulpits*: see letter 4405, n. 4.
7. *smite the enemy hip & thigh*: paraphrase of 'he [i.e. Samson] smote them hip and thigh with a great slaughter' (Judges 15:8).
8. *the last night of the year 1872*: see letter 4406, n. 6.
9. *nobody would think of coming out*: it was a 'stormy night' (see letter 4406), and New York was covered in thick snow.
10. *that book*: see n. 2.
11. *hear of the "affection" . . . of the American public*: see letter 4406.

From Olga Novikoff 29 [September 1874][1] 4421

Symonds' Hotel | Brook St[reet]. near Bond St[reet]. | Tuesday. 29.

My dear Majesty,

Why did you say, that (what is designated as) y[ou]r "Atheism Disappears & dissolves in the presence of healthier thought, etc . . .[2] Y[ou]r enemies rejoice at what they call y[ou][r] recantation,[3] y[ou][r] enthusiastic friends ...... Well those wait for a word of explanation.

I wonder when you are likely to be honest & keep y[ou][r] word—i.e. concerning poor Symonds' Hotel inhabitants[4]

Y[ou][rs] more than is needed to say | Olga Novikoff | née Kiréeff

RI MS JT/1/N/31

1. *[September 1874]*: the year is established by reference to Tyndall's Belfast Address (see n. 2), and the month by 29 September being a Tuesday.
2. *"Atheism Disappears ... healthier thought, etc.*: in the preface to *Belfast Address*, written on 15 September (see letter 4404, n. 3).
3. *Y[ou][r] enemies rejoice ... y[ou][r] recantation*: a writer in the *Presbyterian Quarterly* later noted that 'this extract from the September preface has been regarded in the light of a recantation ... as proof of a reaction in his own soul from the extreme of that heathen philosophy which denied an intelligent creator and governor of the universe ... Charity to Mr. Tyndall disposes us to take this view of the September preface. It was only the natural recoil ... which every mind, not utterly seared and perverted, must sooner or later experience from the hopeless and dreadful abyss of atheism' (J. W. Mears, 'Theistic Reactions in Modern Speculation', *Presbyterian Quarterly*, 4 (1875), pp. 329–47, on p. 334).
4. *keep y[ou]r word ... Symonds' Hotel inhabitants*: a humorous reference to Tyndall's promises to visit.

To Olga Novikoff 29 September [1874][1] 4422

29th. Sept[ember]

Dear Super Majesty!

I said it because it is a fact[2]—and I like facts. And moreover I do not know a man for whose opinion I would give a half penny who would avow himself an Atheist. While rejecting the Anthropomorphic notion of superstitious people I said as plainly as words could make it, that inscrutable Power lay behind it all.[3] Here is a mystery that neither you nor I can sweep away.

Yours ever | J. Tyndall

I will try and call tomorrow at 5.[4]

RI MS JT/1/T/391

1. *[1874]*: the year is established by the relation to letter 4421.
2. *I said it because it is a fact*: Tyndall is responding to Novikoff's question in letter 4421

about why he had said, in Novikoff's words, 'Atheism Disappears & dissolves in the presence of healthier thought, etc.' in the preface to *Belfast Address* (see letter 4404, n. 3).

3. *I said as plainly as words . . . Power lay behind it all*: in the Belfast Address, Tyndall had stated: 'the whole process of evolution is the manifestation of a Power absolutely inscrutable to the intellect of man. As little in our day as in the days of Job can man by searching find this Power out. Considered fundamentally, then, it is by the operation of an insoluble mystery that life on earth is evolved, species differentiated, and mind unfolded from their prepotent elements in the immeasurable past' (*Belfast Address*, pp. 57–8).
4. *I will try and call tomorrow at 5*: at Symonds' Hotel in Brook Street, Mayfair; Novikoff had made a humorous request for Tyndall to visit in letter 4421.

From Mary Egerton 29 September [1874][1] 4423

Mountfield Court, | Robertsbridge, | Hawkhurst.[2] | Sept[embe]r. 29th

A thousand thanks! Had I thought of its growing into a volume, I should not have returned to ask for it; but I did want to have this address,[3] above all others, from yourself! You have added or restored several most interesting passages,[4] and the beautiful conclusion with those lines of Wordsworth,[5] ought to prove to your assailants that you are not "der Geist der immer verneint,"[6] though your spirits may not run in their grooves.

The last but one paragraph in the preface is to me (I cannot help saying it!), inexpressibly touching! Oh! My dear friend, in those times of "stronger and healthier thought",[7] is not the Spirit of God "bearing witness with your spirit"?[8] Let some say what they will, I shall ever believe that the Spirit of Truth is very near to those who seek Truth, drawing them to seek it in its fullness, Spiritual as well as Intellectual Truth!

But I am forgetting my resolutions, & writing a letter of old times! Forgive it! Once more many thanks;

Ever yours | M. F. Egerton

RI MS JT/1/E/52

---

1. *[1874]*: the year is established by reference to the publication of Tyndall's *Belfast Address*; see n. 3.
2. *Mountfield Court, | Robertsbridge, | Hawkhurst*: the Egerton family estate in Sussex.
3. *this address*: *Belfast Address*.
4. *added or restored several most interesting passages*: in the preface to his published *Belfast Address*, Tyndall noted: 'At the request of my Publishers, strengthened by the expressed desire of many Correspondents, I reprint, with a few slight alterations, this Address. It was written under some disadvantages this year in the Alps, and sent by instalments to the printer. When read

subsequently it proved too long for its purpose, and several of its passages were accordingly struck out. Some of them are here restored' (p. v).

5. *those lines of Wordsworth*: Tyndall closed the published version of his *Belfast Address* with 'words known to all Englishmen, and which may be regarded as a forecast and religious vitalization of the latest and deepest scientific truth,—'For I have learned | To look on nature; not as in the hour | Of thoughtless youth; but hearing oftentimes | The still, sad music of humanity, | Nor harsh nor grating, though of ample power | To chasten and subdue. *And I have felt | A presence that disturbs me with the joy | Of elevated thoughts; a sense sublime | Of something far more deeply interfused, | Whose dwelling is the light of setting suns, | And the round ocean, and the living air, | And the blue sky, and in the mind of man: | A motion and a spirit, that impels | All thinking things, all objects of all thought, | And rolls through all things*' (pp. 64–5). The lines were from 'Lines Composed a Few Miles above Tintern Abbey, on Revisiting the Banks of the Wye during a Tour, July 13, 1798' (1798), ll. 90–104, by William Wordsworth (1770–1850).
6. *"der Geist der immer verneint"*: the Spirit that always denies (German); a misquotation of a line, 'der Geist der stets verneint!', from Johann Wolfgang von Goethe's *Faust. Eine Tragödie* I (1808), l. 1338.
7. *The last but one paragraph . . . "stronger and healthier thought"*: in the penultimate paragraph of the preface to the published *Belfast Address*, Tyndall acknowledged that 'in the presence of stronger and healthier thought . . . the doctrine of "Material Atheism" . . . ever dissolves and disappears' (p. viii).
8. *"bearing witness with your spirit"?*: a paraphrase of 'The Spirit itself beareth witness with our spirit, that we are the children of God' (Romans 8:16).

To Olga Novikoff 1 October 1874 4424

Royal Institution of Great Britain | 1st. Oct[ober]. 1874

My dear Madame Novikoff.

I thank you much for communicating to me M. Reville's opinion of the Address.[1] The opinion of such a man weighs much with me.

I once said, quoting from Emerson, that you can hardly state any truth strongly without apparent injury to some other truth.[2] My remarks on Aristotle[3] were confined to him as a <u>physicist</u> and I had no time to enlarge on his merits in other respects. To one so alive to those merits as M. Reville I may and must have appeared unfair. With regard to "Spiritualists" I do not see the advantage they are to derive from the argument which I have put into the mouth of Bishop Butler.[4] But even were it in their favour I could not shrink from stating the argument, believing that we do not possess the faculties which would enable us to logically step from the physical to the psychical.

With this note I send a book[5] which I would ask you to accept from me. It is a collection of short articles. In the 6th. Article, on "Scientific Materialism"[6] you will see the notion to which M. Reville takes exception more fully carried out.[7]

Yours most faithfully | John Tyndall
The little Vignette[8] is quite exquisite.
[Envelope] Madame de Novikoff | Symonds's Hotel | Brook Street

John Tyndall Correspondence in the Olga Novikoff Correspondence Collection. Courtesy of Special Collections, Kenneth Spencer Research Library, University of Kansas. MS 30:Q3
RI MS JT/1/TYP/3/925

---

1. *M. Reville's opinion of the Address*: see letter 4413.
2. *I once said, quoting from Emerson . . . injury to some other truth*: in his address as president of the Mathematics and Physics section of the BAAS meeting at Norwich on 19 August 1868, Tyndall had said: 'It was the American Emerson, I think, who said that it is hardly possible to state any truth strongly without apparent injury to some other truth' ('Address by Professor Tyndall', *Brit. Assoc. Rep. 1868*, pp. 1–6, on p. 2). The American philosopher and poet Ralph Waldo Emerson had actually said: 'it is the fault of our rhetoric that we cannot strongly state one fact without seeming to belie some other' ('History', in *Essays* (Boston: James Munroe, 1841), pp. 3–33, on p. 32).
3. *My remarks on Aristotle*: see letter 4413, n. 3.
4. *With regard to "Spiritualists" . . . Bishop Butler*: see letter 4413, n. 5.
5. *a book*: *Fragments of Science*.
6. *the 6th. Article, on "Scientific Materialism"*: 'Scope and Limit of Scientific Materialism', *Fragments of Science*, pp. 109–23.
7. *the notion to which M. Reville takes exception more fully carried out*: in 'Scope and Limit of Scientific Materialism', Tyndall maintained that the 'passage from the physics of the brain to the corresponding facts of consciousness is unthinkable . . . we do not possess the intellectual organ, nor apparently any rudiment of the organ, which would enable us to pass, by a process of reasoning, from the one to the other. They appear together, but we do not know why' (p. 121).
8. *The little Vignette*: a vignette is a photographic portrait, showing only the head or the head and shoulders, with the edges of the print shading off into the background (*OED*), so this is perhaps a hand-coloured version of a photograph of Tyndall taken in New York, probably in January 1873, by José María Mora (see letter 4399, n. 12). The photograph is annotated 'Mme Novikoff', suggesting that Novikoff herself may have added the colour using watercolour paints (RI MS JT/8/5047).

From Robin Allen 2 October 1874 4425

Trinity House, | London. E.C.[1] | 2. Oct[obe]r 1874

Dear Dr Tyndall.

I perceive that you have published your Belfast Discourse with notes &c.[2] Would you forgive me if I asked that my shelves might be enriched by an

"Author's Copy".[3] I can assure you that by no one is the noble courage of opinion which produced it held in greater reverence than by myself who regard that spirit as more helpful than most religions to the advent of that millennium the desire for which Bacon seems to have envied as almost too great for a heathen—when Thought shall 'move in Charity Rest in Providence & turn upon the Poles of Truth.'[4]

Very faithfully yours | Robin Allen.

RI MS JT/1/A/84
RI MS JT/1/A/84-COPY

---

1. *E.C.*: Eastern Central, a postal district of London.
2. *your Belfast Discourse with notes &c.*: *Belfast Address*.
3. *an "Author's Copy"*: one of a small number of copies of a book given to its author when it is published, which are often distributed, with an inscription, to friends and professional associates.
4. *'move in Charity Rest in Providence & turn upon the Poles of Truth'*: the full line, from Francis Bacon's essay 'Of Truth' (1625), is: 'Certainly it is heaven upon earth to have a man's mind move in charity, rest in Providence, and turn upon the poles of truth' (*The Essays or Counsels Civil and Moral*, ed. A. Spiers (London: Whittaker, [1625] 1851), pp. 41–3, on p. 43).

To Joseph Henry 6 October 1874 4426

6th Oct[ober] 1874

My dear Professor Henry

Mr. Gray brought me your note,[1] and I was very glad indeed to see your veritable handwriting once more.

Mr. Gray was good enough to leave his instruments[2] in my possession for a day or two. The effect was undoubted and very curious. Still I did not think it had anything to do with vital action; and I was very soon able to produce the sounds in all their intensity without including the human body at all in the circuit.

This Mr. Gray regards as an important step in the practical direction: for it will enable him to reproduce the sounds by purely mechanical means; and will enable him to employ currents of a strength which he dare not employ if the human body were a necessary portion of the apparatus.

During my absence in Switzerland[3] your letter[4] regarding a proposed appropriation of the fund under your trusteeship came here. Being somewhat heavy[5] the porter did not send it on, and consequently I was ignorant of its contents until my return in August. I entirely sympathised with what you proposed to do, and to save time I sent you a telegram[6] to this effect. But American telegrams are expensive; and I therefore made mine short. I hope you understood it.

I see by the papers that you have had some conversations about my sound observations.[7] Trust me that the cause I have assigned—namely reflection by "acoustic clouds"[8] is the true one. Your thoughts[9] had occurred to me. But there is not the slightest doubt as to the correctness of the explanation given; namely that it is through aerial reflection, not refraction that the sounds are stifled.[10]

I will not insist that the acoustic clouds are produced solely by the admixture of aqueous vapor.[11] This certainly is a true cause, as both theory and experiment prove. But so are currents of different temperatures—as both theory & experiment also prove. Both may therefore come into play. But the grand point is that what we have to cope with are these absolutely invisible acoustic clouds produced either by differences of saturation or differences of temperature. In foggy weather happily, though present and distinct from the fog itself, they do not attain a very great density. Hence the optically foggy is acoustically clear; which in many cases, the optically clear & the acoustically foggy go together.

Give my best regards to M^rs^. Henry & the young ladies.[12]

& believe me | always yours | John Tyndall

Smithsonian Institution Archives. RU 33, vol. 157, p. 140

1. *your note*: letter 4371.
2. *his instruments*: probably the organ-like transmitter consisting of eight keys, each activating a single-tone transmitter and tuned to reproduce an octave of the musical scale, that Elisha Gray had built in the summer of 1874. This was connected to a diaphragm receiver which projected the sound, and was known as a 'washbasin receiver' because the diaphragm was originally a metal washbasin. See D. A. Hounshell, 'Elisha Gray and the Telephone: On the Disadvantages of Being an Expert', *Technology and Culture*, 16 (1975), pp. 133–61, on pp. 138–49.
3. *my absence in Switzerland*: Tyndall was away from London from 30 May to 10 August.
4. *your letter*: letter 4367.
5. *somewhat heavy*: see letter 4390, n. 3.
6. *a telegram*: letter 4389.
7. *I see by the papers . . . my sound observations*: Tyndall probably refers to the report of Henry's paper on 'Abnormal Phenomena of Sound' at the meeting of the American Association for the Advancement of Science in Hartford in August 1874 that appeared in *Nature* on 1 October. It was reported that 'Prof. Henry does not exactly accept the deductions recently

made by Prof. Tyndall, having himself observed a large number of similar phenomena, and attributing them to refraction, not absorption, of sound by wind and other causes. Prof. Henry found Tyndall's explanation, that a mixed atmosphere absorbed sound, inadequate to explain the facts' ('The American Association for the Advancement of Science', *Nature*, 10 (1874), pp. 441–4, on p. 443).

8. *"acoustic clouds"*: see letter 4353, n. 4.
9. *Your thoughts*: see n. 7.
10. *through aerial reflection . . . sounds are stifled*: see letter 4400, n. 3.
11. *acoustic clouds are produced solely by the admixture of aqueous vapor*: Tyndall had previously argued that the 'extraordinary opacity [of sound] was proved conclusively to arise from the irregular admixture with the air of the aqueous vapour raised by a powerful sun. This vapour, though perfectly invisible, produced an acoustic cloud impervious to the sound' (J. Tyndall, 'On the Acoustic Transparency and Opacity of the Atmosphere', *Roy. Inst. Proc.*, 7 (1874), pp. 169–78, on p. 171).
12. *the young ladies*: see letter 4390, n. 13.

To Charles Cecil Trevor 7 October 1874 4427

[BOARD OF TRADE | REC[EIVE]D | OCT[ober] 7 1874 | HARBOUR DEPARTMENT] | 7th October. 1874.

Sir,

In the communication with which you honoured me on the 19th of May 1873 (H2677)[1] you made known to me the request of the Board of Trade[2] that I should favour them at my earliest convenience with a Report[3] embracing various points on which the Elder Brethren of the Trinity House required information.

The points specified had reference to the materials, manufacture and cost of the gas used at Haisboro,[4] and its relative value as a lighthouse illuminant.

I beg to say that the Report is now nearly completed; but inasmuch as you also communicate to me the request of the Elder Brethren that I as a Gas Referee[5] should furnish them with a Report on the same subjects[6] I am left in uncertainty as to whom the Report is to be addressed—to the Elder Brethren or to the Board of Trade.

I respectfully await the instructions of the Board on this head.

& have the honour to be Sir | Your obedient servant | John Tyndall | C. Cecil Trevor Esquire | &c. &c. &c.

National Archives, MT 10/220. Illumination of lighthouses by gas. H6071

1. *the communication . . . 19th of May 1873 (H2677)*: letter 4098, *Tyndall Correspondence*, vol. 13.
2. *Board of Trade*: see letter 4330, n. 2.
3. *a Report*: letter 4438.
4. *Haisboro*: two lighthouses, the high and low, at Haisborough (pronounced Haisbro) Sands on the Norfolk coast (also known as Happisburgh Lighthouse), opened in 1791. In the early 1870s the lighthouses were used for experiments on the effectiveness of coal gas as an illuminant, with a small gas works established alongside the high lighthouse.
5. *a Gas Referee*: a member of the Board of Metropolitan Gas Referees, which was created by the City of London Gas Act of 1868. Tyndall had been appointed to the Board in August 1872 (see letter 4482, n. 3).
6. *the request of the Elder Brethren . . . same subjects*: Trevor's letter on behalf of the Board of Trade had stated: 'the Elder Brethren have called the attention of this Board to an arrangement which was come to when this experimental trial of gas at Haisbro' was undertaken, to the effect that the gas referees should be judges upon all points connected with the materials, manufacture, and cost of the gas, and to the fact that down to the present time the gas referees have had every detail of the plant and working under their cognizance and approval' (letter 4098, *Tyndall Correspondence*, vol. 13). The Elder Brethren was the governing body of Trinity House (see letter 4187, n. 2).

To Olga Novikoff 8 October 1874 4428

Royal Institution of Great Britain | 8th. Oct[ober]. 1874

Dear Madame de Novikoff

I have been terribly busy of late; and quite permitted the prayer question[1]—that I think is what you mean—to escape my memory. The two papers with which my name has been connected have been published in the Contemporary Review for July 1871, and for October 1871.[2] I have not copies of these numbers, nor have I the volume of the Review which contains them, otherwise I would send them to you.

I will come to see you as soon as ever I can: probably tomorrow.

Yours most truly | John Tyndall

[Envelope] Madame de Novikoff | Symonds's Hotel | Brook Street | W

John Tyndall Correspondence in the Olga Novikoff Correspondence Collection. Courtesy of Special Collections, Kenneth Spencer Research Library, University of Kansas. MS 30:Q4

RI MS JT/1/TYP/3/925

1. *the prayer question*: see letter 4413.
2. *The two papers . . . for October 1871*: J. Tyndall, 'The "Prayer for the Sick": Hints Towards a Serious Attempt to Estimate Its Value', *Contemporary Review*, 20 (1872), pp. 205–10; and J. Tyndall, 'On Prayer', *Contemporary Review*, 20 (1872), pp. 763–6. While Tyndall correctly remembered the months the papers were published, he mistook the year. Prior to these two 'papers', Tyndall had written about the efficacy of prayer in a chapter entitled 'Reflections' in his *Mountaineering in 1861. A Vacation Tour* (London: Longmans, Green, and Co., 1862), pp. 33–40. As Tyndall biographer Roland Jackson has noted, this was an odd topic to discuss in an otherwise mountaineering-focused book (*Ascent of John Tyndall*, p. 213). This chapter was also published with a different title, 'Thoughts on Prayer and Natural Law', in *Fragments of Science*, pp. 29–37 (on p. 38 is a note explaining what prompted him to write the chapter). In October 1865, Tyndall wrote two letters to the *PMG* in response to the call for prayer against a cholera epidemic (see letters 2314 and 2318, *Tyndall Correspondence*, vol. 9). On Tyndall and the debate over prayer, see *Ascent of John Tyndall*, pp. 212–4, 219–20; N. K. C. Bossoh, 'Scientific Uniformity or "Natural" Divine Action: Shifting the Boundaries of Law in the Nineteenth Century', *Zygon*, 56 (2021), pp. 234–53; S. G. Brush, 'The Prayer Test', *American Scientist*, 62 (1974), pp. 561–3; R. B. Mullin, 'Science, Miracles, and the Prayer-Gauge Debate', in David C. Lindberg and Ronald Numbers (eds), *When Science & Christianity Meet* (Chicago: University of Chicago Press, 2003), pp. 203–24; and R. Ostrander, *The Life of Prayer in a World of Science: Protestants, Prayer, and American Culture, 1870–1930* (New York: Oxford University Press, 2000), pp. 17–34. The 'prayer-gauge debate', as it came to be known, was lively from July 1872 to March 1873. Many of the contributions are collected in J. O. Means (ed), *The Prayer-Gauge Debate* (Boston: Congregational Publishing Society, 1876).

From Emil Lehmann[1] 8 October 1874 4429

Hamburg | Klosterstieg N5. | October 8th 1874

Hochgeehrter Herr!

Beifolgend beehre ich mich, ihnen eine Reihe von Nummern des hamburgische Correspondent zu übersenden, in welchem ich dem deutschen Publikum das Verständniß Ihrer ausgezeichneten, bei Eröffnung des Belfaster Congresses gehaltenen Rede zu vermitteln bemüht gewesen bin.

Wie ich aus den Zeitungen ersehe ist ein selbständiger Abdruck der Rede in ihrer ursprünglichen Fassung von Ihnen veranstaltet worden und würde ich Ihnen sehr dankbar sein, wenn Sie mir diesen Abdruck gütigst zugänglich machen und mir gestatten wollten meine Uebersetzung nach derselben zu ergänzen und in einem Separat-Abdruck erscheinen zu lassen.

Sollten Sie hochgeehrter Herr, wie ich hoffen zu dürfen wage aus meiner Uebersetzung die Ueberzeugung gewinnen, daß es mir gelungen ist den

Sinn Ihrer Rede gänzlich wieder zu geben, so darf ich wohl die fernere Bitte aussprechen, daß Sie mich vorkommendenfalls mit dem Uebersetzungsrecht größerer Arbeiten betreuen möchten, für welches ich das von Ihnen zu fordernde Honorar mit Vergnügen entrichten würde, wie ich dann auch in dem vorliegenden Falle Ihrer geneigten Forderung für das Uebersetzungsrecht entgegensehe. In Erwartung einer gefälligen Antwort verbleibe ich

Hochachtungsvoll | ergebenst | Dr. Emil Lehmann

Herr Professor Tyndall. | London.

Hamburg | Klosterstieg N5. | October 8th 1874

Most respected Sir!

I have the honour of sending you enclosed a number of issues of the Hamburgischer Correspondent,[2] in which I attempted to promote the understanding of your magnificent speech held at the opening of the Belfast congress[3] among the German public.

As I understand from the newspapers, you have arranged for a separate printing of the speech in its original version[4] and I would be grateful to you if you could make a copy accessible to me and to allow me to supplement my translation of the same and publish it as a separate imprint.[5]

Should you, dear Sir, as I dare to hope, be convinced by my translation that I have succeeded in reproducing the meaning of your speech in its entirety, may I express the further request that you might entrust me with the translation rights for any larger works in the future, for which I would gladly pay the honorarium you demand, as I look forward to your kind demand for the translation rights in the present case. In expectation of a favourable response I remain,

Faithfully | devoted | Dr. Emil Lehmann

Mr Professor Tyndall. | London

RI MS JT/1/L/13

1. *Emil Lehmann*: Emil Lehmann (1823–87), a German lawyer, journalist, and translator.
2. *issues of the Hamburgischer Correspondent*: a long-running German newspaper published in Hamburg since 1712. The enclosure is missing, and Lehmann's contributions to the newspaper have not been identified.
3. *Belfast congress*: the meeting of the BAAS, which took place in Belfast from 19 to 26 August, with Tyndall giving the presidential address.
4. *a separate printing of the speech in its original version*: *Belfast Address*.
5. *a separate imprint*: the proposed translation appeared as J. Tyndall, *Der Materialismus in England: Ein Vortrag gehalten in der Versammlung der British Association in Belfast*, trans. E. Lehmann (Berlin: Springer-Verlag, 1875).

To John William Draper 9 October 1874 4430

Royal Institution of Great Britain | 9th Oct[ober]. 1874

My dear Dr. Draper.

I saw the editor of the "Contemporary Review"[1] yesterday & mentioned to him, your forthcoming book.[2]

He thought it highly desirable that some extract from it should appear in the "Contemporary".[3] I wish you would forward me a few, which could serve to give the English public a notion of its character.

Again, I would say that you have not before occupied the vantage ground that you do now.[4]

Always yours | John <u>Tyndall</u>

Library of Congress, John William Draper Family Papers

---

1. *the editor of the "Contemporary Review"*: James Thomas Knowles.
2. *your forthcoming book*: see letter 4406, n. 8.
3. *some extract from it should appear in the "Contemporary"*: the proposed extract from Draper's book did not appear.
4. *Again, I would say . . . that you do now*: Tyndall had made a similar assertion in letter 4420.

From Samuel Clark[1] 9 October 1874 4431

Eaton Bishop Rectory, | Hereford. | Oct[ober]. 9. 1874.

My dear Professor Tyndall,

I am truly grateful to you for your note.[2] Most cordially do I agree with you in your confidence in the ultimate result of the plain utterance of sincere conviction. I believe that every one so uttering himself in purity of life and purpose is seeking for God.[3]

I have this very morning received a copy of an ordination sermon lately preached by an old pupil of mine (Benham,[4] Vicar of Margate) and I have ventured to direct that a copy should be sent to you—you will (of course) not take the trouble to acknowledge it, but I trust that you will just glance at it—

Believe me, | Yours very faithfully | Samuel Clark.

RI MS JT/1/TYP/1/221
LT Typescript Only

1. *Samuel Clark*: Samuel Clark (1810–75), an Anglican clergyman, educationalist, and close associate of the Christian Socialist Frederick Denison Maurice. He was presented with the living of the parish of Eaton Bishop in 1871.
2. *your note*: letter missing.
3. *seeking for God*: on the typescript letter, LT recorded of the MS original that there was a 'scribble on back of this letter in John's hand: All these must eventually revolve themselves in the propagation of energy from life to life'.
4. *Benham*: William Benham (1831–1910), an Anglican clergyman and writer, who was appointed to the vicarage of St John the Baptist Church, Margate in 1872. The ordination sermon does not seem to have been published.

To Elizabeth Dawson Steuart [*c.* early October 1874][1] 4432

Had I chosen to do so I might have brought a fortune with me from America—But I went for another purpose, and left between two and three thousand pounds behind me in the hands of trustees with a view to education.[2] The publication of this in the newspapers[3] has caused many applications to be made to me: the people forgetting that I am a poorer man than if I had not gone to America at all. Applications have come to me from Ireland, England and America. In self defence I must say no.

Young Caleb Tyndall[4] wants a threshing machine and I should like to help him, but I cannot do so just now. His sister Sarah Miller[5] wrote to me some time ago wishing to borrow money. Mr Grogan[6] also wrote to me. But I should go to pieces if I attempted to meet all these requests.

RI MS JT/1/TYP/10/3408
LT Typescript Only

1. *[c. early October 1874]*: this approximate date is suggested by the relation to letters 4434 and 4437, which seem to refer to the letter to Tyndall from Charles James Grogan mentioned in this letter.
2. *two and three thousand pounds . . . a view to education*: Tyndall's lecture tour of the United States in 1872 and 1873 made a surplus of more than $13,000. The 'Tyndall Trust for Original Investigation' was announced in May 1873, with Joseph Henry, Hector Tyndale, and Edward L. Youmans as trustees. See letter 4214, n. 5.
3. *The publication of this in the newspapers*: the announcement was originally made in 'Professor Tyndall's Deed of Trust', *Popular Science Monthly*, 3 (1873), pp. 100–1, which was reprinted in newspapers on both sides of the Atlantic.

4. *Young Caleb Tyndall*: Caleb Tyndall (1837–1914), the son of Tyndall's paternal uncle Caleb Tyndall. Caleb Jr married Mary Anne Williams in 1874, which might explain his desire to improve his options as an agricultural worker with a threshing machine (Irish Genealogy).
5. *Sarah Miller*: Sarah Miller (née Tyndall, 1844–1917), the daughter of Tyndall's paternal uncle Caleb Tyndall. She married David Miller of Bagenalstown in 1869 and had four children, one of whom was, like her father and brother, named Caleb (Irish Genealogy). Before her marriage, Tyndall sent a 'cheque for £50 . . . intended as a dowry' and agreed that she could instead 'employ it in a farm' (letter 2877, *Tyndall Correspondence*, vol. 10). Her letter is missing.
6. *Mr Grogan*: Charles James Grogan (1805–87), the vicar of Dunleckney and thus the clergyman with pastoral responsibilities for Bagenalstown, which was in the Dunleckney parish (*Thom's Irish Almanac and Official Directory, 1874* (Dublin: Alexander Thom, 1874), p. 1067). His letter, which is missing, seems to have asked for financial assistance for the family of Tyndall's late maternal uncle John McAssey (see letters 4434 and 4437).

To Olga Novikoff 12 October [1874][1] 4433

Royal Institution of Great Britain | 12th. Oct[ober]

My dear Madame Novikoff

I greatly regretted my inability to join you yesterday: but if you should be at home[2] to day at 5.30 you may count on my calling

Yours most faithfully | John Tyndall

RI MS JT/1/TYP/3/932
LT Typescript Only

1. *[1874]*: the year is suggested by the relation to letter 4428, in which Tyndall promised 'I will come to see you as soon as ever I can'.
2. *at home*: Novikoff was staying at Symonds' Hotel in Brook Street, Mayfair.

From Elizabeth Dawson Steuart 12 October 1874 4434

Steuart's Lodge | Leighlinbridge | Oct[ober]. 12. 74.

My dear John,

I have heard that it is your intention to pay for the maintenance of a grand-daughter[1] of the late John McAssey of Bagenalstown,[2] and that she is to live with a poor and large family in Carlow, named Walsh, all this may be true, or it may not, but in the event of its being so, I would ask you, for your own sake, to be cautious, lest you might be hereafter annoyed with demands from those people, who may expect you to support <u>them</u> also—a town is a bad place for a young girl who has nothing to do, and many temptations to resist, and I have reason to believe that Sarah Miller,[3] would take charge of her, and treat her well, if the story be true that you are going to settle her somewhere, and I feel sure you would wish to serve her. Poor old Mrs McAssey,[4] is, I hear, dying, but I know nothing further. If I can be of any use in the matter, please employ me. Poor old Mrs Tyndall[5] fell down stairs lately, and is very much injured, keeping her bed and very weak. The money you left me for her, has been of very great use. Sarah is now the only one of her family who helps her at all. Others did, for a while and left off—but for your generosity she would be in a bad way.

With every kind and good wish, | believe me, | Your affectionate old friend, | E. D. Steuart.

RI MS JT/1/TYP/10/3415
LT Typescript Only

1. *the maintenance of a grand-daughter*: not identified, although see letter 4437, n. 5. Steuart presumably heard of Tyndall's apparent intention to maintain her from Charles James Grogan (see letter 4437), who was a friend and local clergyman.
2. *the late John McAssey of Bagenalstown*: Tyndall's uncle on his mother's side. He may be the John McAssey who died on 10 July 1874 in Bagenalstown, aged ninety-one. If so, he was a blacksmith living on The Parade in Bagenalstown, and the recorded cause of death was 'old age' (Irish Genealogy).
3. *Sarah Miller*: see letter 4432, n. 5.
4. *Mrs McAssey*: Mary McAssey (née Smith/Smyth, *c.* 1787–1875), wife of John McAssey.
5. *Mrs Tyndall*: Dorothea Tyndall.

From Charles Cecil Trevor [13 October 1874][1] 4435

Board of Trade S. W. | J. Tyndall Esq F.R.S. | Royal Institution | 21 Albemarle Street | W.

Sir:—

In reply to your l[e]t[te]r of the 7th instant.[2] I am to acquaint you that it will be convenient that your Report on the manufacture, cost, etc of Gas at Haisbro'[3] should be addressed to the Elder Brethren of the Trinity House.[4] The B[oard] of T[rade][5] will if this is done be sure to have the advantage of perusing it.

I am—

National Archives, MT 10/220. H6071 2472

1. *[13 October 1874]*: the date is established by a hand-written addendum to the letter, probably made by Tyndall, stating 'Signed by Mr Trevor | 13.10.74 | 13 Oct.'
2. *your l[e]t[te]r of the 7th instant*: letter 4427; 'instant.' is an abbreviation of instante mense, Latin for this month.
3. *your Report on the manufacture, cost, etc of Gas at Haisbro'*: letter 4438.
4. *Elder Brethren of the Trinity House*: see letter 4187, n. 2.
5. *B[oard] of T[rade]*: see letter 4330, n. 2.

To Olga Novikoff 14 October [1874][1] 4436

14th. Oct[ober].

It is a long time till Friday[2] you will find me terribly stupid beside your two brilliant guests.[3]

(The George Elliots[4] and Kinglake)[5]

With great pleasure | J. T.

RI MS JT/1/TYP/3/939
LT Typescript Only

1. *[1874]*: the year is suggested by Tyndall's regular meetings with Novikoff in the autumn of 1874, and the meetings in the same period that Novikoff had with George Eliot, who invited Novikoff to her home twice in November 1874. See K. McCormack, 'Sundays at the Priory: Olga Novikoff and the Russian Presence', *George Eliot-George Henry Lewes Studies*, 67 (2015), pp. 30–42.
2. *Friday*: 16 October.
3. *your two brilliant guests*: this seems an error on Tyndall's part, as he then lists three guests (see n. 4 and n. 5).
4. *The George Elliots*: Mary Ann Evans (1819–80), who wrote novels under the pseudonym George Eliot, and her partner George Henry Lewes (1817–78) (*ODNB*).
5. *Kinglake*: Alexander William Kinglake.

To Elizabeth Dawson Steuart 14 October [1874][1] 4437

Royal Institution of Great Britain | 14th Oct[ober].

My dear Mrs Steuart,

I thank you heartily for your note[2] and shall be truly glad of your counsel and aid.

What I agreed to do was to provide for Mrs Macassey[3] during her lifetime—nothing more. Still as Mr Grogan[4] describes her daughter[5] as being greatly afflicted with epilepsy I should be sorry to withdraw her support.

I agreed to pay Mrs Macassey £32.0.0 a year. This is the sum that Mr Grogan informed me would be needed to keep her and her daughter in comfort.

After the old woman's death of course the matter would be changed—I should not mind dividing the £40 I used to allow to John Macassey[6] equally between his daughter and Mrs Caleb Tyndall.[7]

I wish to keep out of all further entanglements, and what you propose is, I think, a good means to this end.

If you thought a little more money would be needed I will try to provide it.

Yours ever | John Tyndall.

RI MS JT/1/TYP/10/3416
LT Typescript Only

1. *[1874]*: the year is established by the relation to letter 4434.
2. *your note*: letter 4434.
3. *Mrs Macassey*: Mary McAssey (see letter 4434, n. 4).
4. *Mr Grogan*: Charles James Grogan (see letter 4432, n. 6); his letter is probably the one mentioned in letter 4432.
5. *her daughter*: presumably Sally McAssey, to whom Tyndall was paying twenty pounds a year, sent to Grogan as a biannual 'draft for £10', in the mid-1880s (Grogan to Tyndall, 29 December 1884, RI MS JT/1/TYP/10/3460); this was the amount, dividing the allowance of forty pounds previously given to John McAssey, suggested later in the letter. She may be the mother of the 'grand-daughter of the late John McAssey' mentioned in letter 4434.
6. *John Macassey*: see letter 4434, n. 2. Tyndall began making payments to McAssey at the request of his mother, who was John's sister, giving him 'six or seven and twenty pounds a year' in 1865 (letter 2313, *Tyndall Correspondence*, vol. 9).
7. *Mrs Caleb Tyndall*: Dorothea Tyndall.

To Robin Allen 16 October 1874 4438

Copy | Royal Institution of Great Britain | 16th. October 1874

Sir,

In compliance with the desire of the Board of Trade (H 2677) that I should furnish them with a Report upon the materials, manufacture, and cost of the Gas used at Haisbro',[1] I beg to inform you that soon after the receipt of the letter of the Board[2] I entered upon the desired investigation.

My first care was to secure the services of a gas-analyst of proved trustworthiness, and whose previous experience had rendered him intimately conversant with photometric measurement,[3] and with all the processes employed in the manufacture of Gas. Such an aid I found in Mr William Valentin, principal Demonstrator of Chemistry in the Royal College of Chemistry, South Kensington.

In 1873 Mr Valentin, at my request, went to Haisbro' and made himself acquainted with the condition of the gas-works.

He furnished me with full particulars as to the yield and quality of the gas then produced, and as to its cost of production. It was obvious however from Mr Valentin's Report that no conclusion of any value to the Elder Brethren[4] or the Board of Trade[5] could be drawn from the data then obtained. The flues had been suffered to fall into ruin, and before the desire expressed to me by the Board of Trade could be met, thorough repair was necessary.

If such a state of things could be proved to be a necessary or even probable incident in the manufacture of gas for lighthouses the use of this illuminant would, of course, be at once put an end to; and I thought the onus rested on Mr Wigham to prove that it was not a necessary incident.

After consultation with the Elder Brethren to whom I represented the desirability of avoiding the expenditure of time, money, and good temper in subsequent controversy, and who fully appreciated my motive, it was agreed that the repair of the works should be carried out under Mr Wigham's own superintendence, and that after he had certified their completion, the enquiry demanded by the by the Board of Trade should be made.

That enquiry has now been carried out in as complete and searching a manner as the data at my disposal rendered possible.

In regard to the "materials, manufacture, and, in part, cost of the gas" I have the honour to refer to the Report of Mr Valentin[6] which is herewith transmitted.

In this enquiry the comparative merits of gas and oil were brought into question. The case of the oil was represented by Mr Douglass, the performance of whose lamps was investigated. The case of the Gas was represented by Mr Wigham. It was my desire that everything should be done with the joint concurrence of these two gentlemen. They were therefore present; and I am happy to say that Mr Valentin has efficiently carried out my wishes to make the investigation amicable, and as far as possible, final.

With a view to observations afloat, in accordance with previous arrangement I proceeded to Yarmouth on the 31st of March, and joined Capt[ain]. Webb, Capt[ain]. Nisbet, and Admiral Collinson on board the "Galatea."[7] On Wednesday the 1st of April we all went by land to Haisbro', inspected the gas-works, and made ourselves acquainted with the very admirable arrangements of Mr Valentin for the determination of the quality and consumption of the gas. We afterwards visited the lower Lighthouse, and saw in action the six–wick burner, consuming Colza oil.[8] The shifting of the oil Lamp from the 6-wick to the 4-wick not being deemed practicable, we decided to occupy the first night with a comparison of the 4-wick oil Lamp with the various powers of the gas-burner.

It was arranged that Mr Wigham should satisfy himself as to the quantity of oil consumed, and that Mr Douglass should satisfy himself as to the quantity of gas consumed.

We returned to Yarmouth and went afloat. Our observations were made from two positions, the one four miles, and the other eight miles distant from Haisbro'. Care was taken to make each of these positions equidistant from the two lights.

From both positions the 4-wick lamp and the 28-jet gas burner were first compared together. As regards the volume both lights were alike; as regards intensity a slight advantage might have been possessed by the gas; but they were practically equal in power. This result agrees with that already reported to the Board of Trade as having been observed at Howth Bailey.[9]

The 48-jet burner in both positions yielded a light decidedly superior to the 4-wick Lamp. The 68-jet burner markedly exceeded the 48-jet one, and showed an exceedingly fine light.

The 88, 108, and 148-jet burners were then tried in succession. With regard to the last of these, no advantage is likely to arise from its use that would justify the expenditure of gas necessary to maintain it. In my opinion 108 jets mark the practical upper limit of the Wigham burner.

In the lower powers of the burner the augmentation of light consequent on the addition of twenty jets is, of course, most marked to the eye. The ratio of 48 to 28, for example, is much greater than that of 108 to 88. At Haisbro', however, the advance of light with the higher power appeared to be less distinct than at Howth Bailey.

Ten minutes were allowed for the observation of each burner. During the first five the flame was steady, during the subsequent three minutes it was broken up into flashes of five seconds duration, separated from each other by intervals of darkness of three seconds. These flashes were very effective and quite characteristic. With the 68-jet burner they were exceedingly fine. One peculiarity in the case of the gas must here strike every mind. By the employment of flashes with equal intervals of light and darkness, it is possible to have the intensity of 68 burners concentrated into every flash, with the same expenditure of gas as that required for a 34-jet fixed light. During the intervals of darkness no gas is consumed. No other light, to my knowledge, is capable of being thus saved by the process which confers upon it individuality.

On the night of the first, the programme was carried out with extreme accuracy. In the case of the flashes the turning on and off of the gas was done by hand, and though less perfect than the performance of machinery the work was surprisingly well done. On the 2nd Capt[ain]. Nisbet and I drove to Haisbro', and assured ourselves that everything had proceeded in a satisfactory manner during the previous night. After giving some further instructions we returned to Yarmouth and went afloat.

The wind was very boisterous and as the observations of the preceding night were of precisely the same relative character at the 4-mile and the 8-mile position, we contented ourselves tonight with observations from the nearer Station. The 6-wick Trinity Lamp was burnt on this occasion in the Lower Lighthouse. Compared with the 28-jet burner 6-wick Lamp proved the superior light. Compared with the 48-jet burner, the 6-wick lamp proved to be of practically the same intensity. Mr Brown,[10] the Master of the "Galatea", thought the gas best, and Capt[ain]. Nisbet also thought it possessed a little more body than its rival. But the two lights may be safely regarded as of sensibly equal power. The 6-wick Lamp yielded a very fine light: it is distinctly the most powerful oil Light that I have ever seen. The 4-wick lamp also yielded a pure and beautiful light. Our experience of the higher powers of the gas-burner on the 2nd was precisely the same as it had been on the 1st.

As landward reflectors are not a normal element of dioptric[11] apparatus, I drew the attention of my colleagues to the fact that the reflectors at Haisbro' are exceptionally large, and submitted to them whether a night's observations ought not to be made with the two reflectors suppressed. But the experiments being deemed definite and satisfactory it was thought unnecessary to extend the observations. The results, moreover, are so thoroughly borne out by previous experiments in Ireland[12] that they may safely be relied on.

With regard to the question of cost the comparison will be limited to the 28-jet burner and the 4-wick lamp, and the 48-jet burner and the 6-wick lamp. Though the light is very fine, the inordinate consumption of oil by the 6-wick lamp will probably interfere with its employment. As regards the question of cost I am not in possession of the entire data, and those before me show discrepancies between certain items furnished by the Trinity House,[13] and those furnished by Mr Wigham. The Elder Brethren however desire to work out the question of cost themselves, and in their hands I therefore leave it. But over and above the question of cost the capacity of varying its power and of conferring unmistakable individuality upon a light, which the gas possesses in so eminent a degree, is a point of the highest importance. I have dwelt emphatically upon this point in former reports to the Board of Trade[14] and I venture to employ the same emphasis now. A fixed light a long way off and a ship's light closer at hand may be absolutely indistinguishable; and under such circumstances the coast light virtually ceases to be a guide to the mariner. Cases, I doubt not, will occur to the Elder Brethren in which the absence of all distinctive character on the part of coast-lights proved to be a source of perplexity and danger. This defect will increase in gravity as our coast lights become more numerous and our ships lights more improved. Hence I hold it to be of vital importance that every suitable means of bestowing a character upon a light, which should render its being confounded with any other light impossible, should be taken advantage of. No light hitherto devised, can, in this respect, bear comparison with the gas-light. The method of conferring individuality upon a light by destroying two thirds of the beam with a shade of coloured glass is doubtless necessary in some situations; but it is a necessity to be deplored; and where distinctiveness can be attained by a distribution of the light, not involving its destruction, the method lends itself to common sense.

The gas is perfectly pliant and manageable. At the appointed hour of sunset it flashes out with full power in an instant, whereas the oil lamp requires half an hour's nursing before its maximum power is attained. The Keepers like the gas. I questioned the Assistant Keeper at Haisbro',[15] a most intelligent man, who was at this lighthouse when oil was employed, and who has been there ever since the introduction of the gas. He has never experienced the slightest difficulty in managing it; he never had an accident of any sort with it: in case of manipulation it has the advantage of the oil; while the change from burner to burner in the case of fog may be accomplished in a fraction of a minute. In the 6-wick oil lamp by an ingenious device the three inner wicks may be suppressed, and the three outer ones employed; and in thick weather it is possible to pass from the three wicks to the six wicks. But for the lamp to recover its power after this change a quarter of an hour at least is requisite, while the range of power is far less than that attainable in less than a minute with the gas.

I have received a copy of the brief but lucid report of the Committee[16] with whom I had the honour to act at Haisbro': and notice with satisfaction that the flashing experiments so far pleased them that they recommend further experiments on the flashes. I think the result will prove to the Elder Brethren that in gas we have an illuminant of extraordinary power and of unrivalled individuality. But Haisbro' being a fixed light is not able to bring out all the virtues of the gas. Its applicability to revolving lights, and its power of producing perfectly novel modifications of such lights, have been demonstrated by conclusive experiments at Rock-a-Bill,[17] for an account of which I beg to refer to the Reports which I have already submitted to the Board of Trade.[18] When first honoured by the commands of the Board to go to Ireland and report on the applicability of gas to Lighthouse illumination,[19] I knew nothing of Mr Wigham or his work. But I soon discerned in him a man of ability, energy, and fruitfulness of invention; and I deemed it in the highest degree desirable that these qualities, instead of being overshadowed by discouragement, should have free but prudent scope in the public service. Every recommendation which I have ventured to forward to the Board of Trade regarding him and his work was based upon the careful consideration of antecedent experiments; and I venture to think that every one of them has been justified by the result. The recent investigation at Haisbro' has not altered the conviction that I have long entertained and expressed, that in the introduction of gas a great step has been taken in the art of Lighthouse illumination.

I have the honour to be, | &c | (S[igned]?) John Tyndall | Robin Allen Esq

P.S. I beg here to add an observation of importance, an account of which has just reached me. I may be permitted to introduce it by reference to a passage in my Report to the Board of Trade dated 25th July 1873.

> "Fogs vary in density, and between our densest fog and our clearest air there are infinite gradations. It is not sound reasoning to conclude that because in a dense fog weak lights and strong are alike extinguished, all lights are equally valueless as regards fog. For there are states of the atmosphere when it is of the highest importance that the mariner should see his light, and where if a low power be used he will not see it, while if a high power be used he will. It is in the commonest forms of "thick" weather that the superiority of a strong light would be most evident."
>
> "I have already stated that the Triform[20] flash, which occurs at Rock-a-Bill once in a revolution, is superior to the seven other flashes.
>
> Reasoning alone would lead us to conclude that with a gradually thickening atmosphere, the feebler flashes would disappear first, the strong beam of the Triform making itself useful after the others had ceased to be so. This is proved by observations made from the Hill of Howth,[21] and from the "Kish" Lightship.[22] At all times when the Light is seen one flash in eight is paramount; but on many occasions the seven minor flashes disappeared altogether, the eighth flash alone being visible."

The observation which I have now to cite is a striking confirmation of the one here referred to. It was made by Sir William Thomson from the Salthill Hotel[23] on the 21st of September. Speaking of the performance of the Triform light he says

> "The triform light exhibited from the lower position in the neighbourhood of the chief tower was strikingly superior even to the great fog power of 108 burners, exhibited in the chief tower; so much so, that a heavy thunder-shower which happily chanced to pass during our experiments between the Salthill Hotel and the Lighthouse completely eclipsed the light of the chief tower while the Triform still shone conspicuously through it."[24]

The superiority of the triform over even the 108-jet burner being thus marked, its enormous superiority over the 28-jet burner, or its equivalent 4-wick lamp may be inferred.

With regard to what I have called the 'flexibility' or 'plasticity' of the gas, Sir William Thomson expresses himself thus:—

"On the evening of 21st of September, and in the experimental trial of distinctive eclipses on the evening of the 22nd I was much pleased with the admirable 'plasticity' of the gas illumination, in respect to the system which I have urged of giving distinction to different lights by combinations of short and long eclipses in accordance with the principles of the Morse Alphabet."[25]

The Board of Trade is already in possession of my statement that by proper combinations of flashes every lighthouse might be made to spell its own name.[26]

J. T.

National Archives, MT 10/220. Lighthouses, Illumination by means of gas. H7863

---

1. *the Board of Trade (H 2677) . . . Gas used at Haisbro'*: see letter 4427, nn. 1–4.
2. *the letter of the Board*: letter 4098, *Tyndall Correspondence*, vol. 13.
3. *photometric measurement*: a method of measuring and comparing the intensity of light directly emitted by sources such as candlelight or gaslight.
4. *Elder Brethren*: the governing body of Trinity House (see letter 4187, n. 2).
5. *Board of Trade*: see letter 4330, n. 2
6. *the Report of Mr. Valentin*: letter 4314.
7. *Capt[ain]. Webb . . . on board the "Galatea"*: John Sydney Webb, Edward Parry Nisbet, and Richard Collinson, members of Trinity House's Elder Brethren and its Committee of Light (see letter 4314, n. 7). For THV *Galatea*, see letter 4187, n. 3.
8. *Colza oil*: see letter 4314, n. 6.
9. *Howth Bailey*: see letter 4314, n. 28.
10. *Mr. Brown*: not identified.
11. *dioptric*: assisting vision by refracting and focusing light (*OED*).
12. *previous experiments in Ireland*: at the Baily Lighthouse (see n. 9) and the Wicklow Head Lighthouse, which Tyndall conducted in June 1869.
13. *Trinity House*: see letter 4187, n. 2.
14. *former reports to the Board of Trade*: 'Papers Relative to Proposal to substitute Gas for Oil as Illuminating Power in Lighthouses', H.C., Command Paper [4210], *Parliamentary Papers*, (1869), pp. 1–20, on pp. 16–20; and 'Further Papers Relative to Proposal to substitute Gas for Oil as Illuminating Power in Lighthouses', H.C., Command Paper [C.282], *Parliamentary Papers*, (1871), pp. 1–33, on pp. 27–33.
15. *the Assistant Keeper at Haisbro'*: probably G. Monk, who served in this role, alongside the principal lighthouse keeper William Kelsey, from 1871–7.
16. *the Committee*: the Trinity House Light Committee, which was chaired by Webb (see n. 7).
17. *Rock-a-Bill*: a lighthouse on one of the two islands of Rockabill off the east coast of Ireland, close to Skerries. Construction of its granite tower started in 1855, and it began operating in 1860.

18. *the Reports . . . the Board of Trade*: see n. 14.
19. *When first honoured . . . to Lighthouse illumination*: Tyndall had first inspected Irish lighthouses on behalf of the Board of Trade in June 1869 (see n. 12).
20. *Triform flash*: a flash appearing in three different forms produced by a light with three lenses in tiers devised by Wigham.
21. *the Hill of Howth*: a peninsula northeast of Dublin.
22. *the "Kish" Lightship*: a ship used, since 1811, as a beacon at Kish Bank in the bay of Dublin, where attempts to build a permanent lighthouse had foundered in the 1840s because of the difficult weather conditions.
23. *the Salthill Hotel*: see letter 4392, n. 12.
24. *"The triform light . . . through it"*: Thomson's words are from a letter he wrote to Wigham on 12 October, of which Tyndall evidently had a copy as he later quoted from it more fully in 'Report to the Commissioners of Irish Lights, by Professor Tyndall, of an Inspection of Galley Head Lighthouse', H.C., Sessional Paper [405], *Parliamentary Papers*, 64 (1879), pp. 1–8, on p. 4.
25. *the Morse Alphabet*: the sequences of dots and dashes representing letters of the alphabet devised by the American inventor Samuel Morse (1791–1872).
26. *my statement that . . . be made to spell its own name*: Tyndall later explained of coal-gas lights: 'By a simple automatic apparatus, its flame can be caused to send forth flashes in any required succession, and of any required duration. Long and short flashes might therefore be combined, so as to render the identity of a lighthouse unmistakable. It might, I doubt not, be made to spell its own name' (J. Tyndall, 'A Story of the Lighthouses', *Fortnightly Review*, 44 (1888), pp. 805–28, on p. 808).

To Alfred Mayer 17 October 1874 4439

Royal Institution of Great Britain | 17th Oct[ober]. 1874

Dear Professor Mayer.

I had been just revising my lectures on 'Sound' with a view to a new edition[1]—so your 'criticisms'[2] will arrive just in time—probably I shall have anticipated them.[3]

M^r^. Gray came here[4] and showed me his singular apparatus[5]—After working a little with it, I was able to exclude all vital action, and to obtain the musical sounds with dead matter.[6]

M^r^. Gray considered this a step in the practical direction; for one objection to the use of his instrument was the danger incurred by the operator. Now the whole thing can be accomplished by mechanism.

Believe me | ever faithfully yours | John Tyndall

Will [you] present my complimentary & kind remembrances to M^rs^. Stevens?[7]

Tyndall, John (1820–1893); Hyatt and Mayer Collection, C0076, Manuscripts Division, Department of Special Collections, Princeton University Library

---

1. *my lectures on 'Sound'... a new edition*: *Sound* (1875).
2. *your 'criticisms'*: in A. M. Mayer, 'Researches in Acoustics', *American Journal of Science*, 8 (1874), pp. 241–55. In particular, Mayer noted that 'Helmholtz also dwells with emphasis and with some minuteness on the ... important fact, that the dissonant and consonant effects of beats do not altogether depend on their frequency per second, but also on the position in the musical scale of the sounds producing these beats. Tyndall, in his otherwise admirable little book "On Sound", has overlooked the latter important fact, and while assuming broadly that 33 beats per second—no matter what the pitch of the notes producing them—give the greatest dissonance, while 132 beats per second give consonance, he has made logical deductions from these premises which do not bear the test of experiment or conform to the experiences of musicians ... These criticisms on the book of my friend Professor Tyndall are not given merely for the sake of criticism, but because his eminence as an original investigator, and his great power as a popular teacher of scientific truths, have given an extensive distribution to his writings, and I am sure he will be obliged to any one who, in the proper friendly spirit, will show how his work, written to diffuse scientific knowledge, may be rendered more efficient in accomplishing that object' (pp. 253 and 255).
3. *probably I shall have anticipated them*: in the new edition of *Sound* (1875), Tyndall removed the passage in the book's first edition which stated: 'In our last lecture I operated with a resultant tone produced by 33 vibrations a second. That tone was perfectly smooth and musical; whereas beats which succeed each other at the rate of 33 per second, are pronounced by the disciplined ear of Helmholtz to be in their condition of most intolerable dissonance. Hence the resultant tone referred to could not be produced by the coalescence of beats. When the beats are slower than 33 they are less disagreeable. They may even become pleasant through imitating the trills of the human voice' (*Sound* (1867), pp. 298–9). However, it is not clear whether he did this before or after he saw Mayer's article in the *American Journal of Science*.
4. *M^r^. Gray came here*: Elisha Gray's visit is described in letter 4426.
5. *his singular apparatus*: see letter 4426, n. 2.
6. *to exclude all . . . musical sounds with dead matter*: Gray's instrument was intended to enable musical tones to be transmitted telegraphically on submarine cables; a more detailed account of Tyndall's experiments with it is given in letter 4426.
7. *M^rs^. Stevens*: see 4338, n. 8. Tyndall had corrected the misspelling of her name in his previous letter.

To Olga Novikoff [17 October 1874][1] 4440

Royal Institution of Great Britain

Surely nothing could be "Cosier" then your little dinner party.[2] It quite charmed away a feeling of intense weariness with which I entered your gracious presence

J. Tyndall

RI MS JT/1/TYP/3/939
LT Typescript Only

1. *[17 October 1874]*: the date is suggested by the relation to letter 4436.
2. *your little dinner party*: presumably the dinner party arranged for Friday 16 October in letter 4436. A handwritten note by LT on the typescript of the letter suggests the party took place at Symonds' Hotel in Brook Street, Mayfair, where Novikoff was staying, and, beyond Tyndall, was attended by George Eliot (see letter 4436, n. 4), George Henry Lewes (see letter 4436, n. 4), Alexander William Kinglake, and Abraham Hayward (1801–84), an English essayist and translator (*ODNB*).

To Henry Enfield Roscoe[1] 17 October [1874][2] 4441

Royal Institution of Great Britain | 17th Oct[ober].

My dear Roscoe.

I shall always be glad to see my friend Sir Joseph Whitworth.

Dare I ask you to give me a glass of good dry champagne at dinner? I w[oul][d] not trouble you with the request were it not important for the well being of my brain during the lecture.[3]

The parsons do not "bother" me[4] save through the trouble of opening the various letters & other communications which my sins have brought upon me. My faith is great that time will justify me & put the parsons in the wrong.

Have you a good large screen,[5] or will it be necessary for me to send one down?

Can I address the audience from the side of the Hall?[6]—it looked a long way to the end when I heard Huxley.[7]

I hardly think he was heard at the end.

Yours truly | John Tyndall

The Royal Society of Chemistry—Ex Coll H.E. Roscoe, 1915
H. E. Roscoe, *The Life & Experiences of Sir Henry Enfield Roscoe, D.C.L, LL.D., F.R.S. Written by Himself* (London: Macmillan, 1906)

1. *Henry Enfield Roscoe*: Henry Enfield Roscoe (1833–1915), professor of chemistry at Owens College, Manchester, and founder of the city's Science Lectures for the People series, which he ran from 1866–79 (*ODNB*).
2. *[1874]*: the year is established by reference to Tyndall's contribution to Science Lectures for the People (see n. 3).
3. *the lecture*: Tyndall's lecture on 'Crystalline and Molecular Forces', delivered on 28 October, which inaugurated the sixth series of Science Lectures for the People. The lecture was published as J. Tyndall, 'Crystalline and Molecular Forces', in *Science Lectures for the People. Sixth Series* (Manchester: John Heywood, 1874), pp. 1–13; this was also published in *Belfast Address*, 7th thousand, pp. 67–83.
4. *The parsons do not "bother" me*: after giving his controversial presidential address to the BAAS in August 1874, Tyndall had been subject to numerous allegations of materialism and atheism from members of the clergy (see letter 4404, n. 3).
5. *a good large screen*: Tyndall often used projections to illustrate his lectures, which required a screen.
6. *the Hall*: the Free Trade Hall, Peter Street, Manchester, which was opened in 1856 to commemorate the repeal of the Corn Laws ten years earlier.
7. *when I heard Huxley*: Huxley had given a lecture on 'Coral and Coral Reefs' at the Free Trade Hall, Manchester on 4 November 1870, inaugurating the second series of Science Lectures for the People (*Science Lectures for the People. Second Series* (Manchester: John Heywood, 1870), pp. 3–17).

To Olga Novikoff 19 October [1874][1] 4442

Royal Institution of Great Britain | 19th. Oct[ober].

Your note[2] has just reached me—It arrived while I was in the country.[3] Nothing whatever can come of this petition.[4] England has been long placed beyond any movement of this kind.[5]

I had a capital game of Hockey yesterday with my Godson.[6] He is a brave and beautiful little fellow.

always Yours | John Tyndall

RI MS JT/1/TYP/3/926
LT Typescript Only

1. *[1874]*: the year is established by reference to a petition against Tyndall (see n. 4).
2. *Your note*: letter missing.
3. *in the country*: presumably at High Elms, the country residence of John Lubbock and his family in Farnborough, Kent (see n. 6).
4. *this petition*: C. W. Stokes, *An Inquiry of the Home Secretary as to Whether Professor Tyndall*

Has Not Subjected Himself to the Penalty of Persons Expressing Blasphemous Opinions (London: privately printed, 1874).

5. *England has been . . . movement of this kind*: although the Court of Exchequer Chamber had in 1867 upheld the common laws against blasphemy that were created in the sixteenth century, reaffirming that Christianity was part of the laws of England, it had been agreed in the 1840s that it was only when irreligion involved insults to God that it might be liable to prosecution. This was the case when the secularist George Foote was imprisoned for blasphemy in 1883.
6. *my Godson*: Rolfe Arthur Lubbock (see letter 4247, n. 5).

From Olga Novikoff [21 October 1874][1] 4443

Wednesday. Symonds' H[otel].[2]

Always most welcome—as you may easily imagine—but as you fix no time, I just say, that I will be at home from 3.30 to 7.30.

Professor Alois Riehl's pamphlet on "Moral & Dogma"[3] will be sent to you one of these days. It is, perhaps, rather too condensed, & written as if in fear not to say every thing wanted, but in the whole, I think, deserved the sensation it made in Germany. Martineau's answer to y[ou]r address[4] is exceedingly weak & vague. Y[ou]r notes[5] are real stars: Distinct—but distant, "Clear—but oh, how cold! They seem as if they were intended for print! How shocked you must be with mine! Pardon me! But I always have the impression as if no real feeling—of any kind nature—could exist where caution prevails. You'll exclaim again. "What a woman she is!" Well—not exactly—all women—let them be as plain & old as you like—are inclined to admit what they long for! All women are conceited . . . As to me, I expect nothing, & have claims upon nothing and nobody.—

Ever Y[ou]rs Olga

RI MS JT/1/N/35

---

1. *[21 October 1874]*: the date is suggested by the relation to letter 4444, and that 21 October fell on a Wednesday in 1874.
2. *Symonds' H[otel].*: in Brook Street, Mayfair.
3. *Professor Alois Riehl's pamphlet on "Moral & Dogma"*: A. Riehl, *Moral und Dogma* (Wien: Carl Gerold, 1871).
4. *Martineau's answer to y[ou]r address*: James Martineau criticized Tyndall's Belfast Address in an address at Manchester New College, London on 6 October. It was reported in newspapers, and subsequently published as J. Martineau, *Religion as Affected by Modern Materialism* (London: Williams and Norgate, 1874).
5. *Y[ou]r notes*: not identified.

To Olga Novikoff 21 October [1874][1] 4444

21st. Oct[ober].

Dear Friend.

Forgive me for not mentioning the time.[2] Let me say 5.30. The day's work will be then well nigh ended.

With me it is to those who claim least that I yield most.[3]

Ever yours | J. T.

RI MS JT/1/TYP/3/932
LT Transcript Only[4]

1. *[1874]*: the year is suggested by the relation to letter 4443 and by Tyndall's regular meetings with Novikoff in the autumn of 1874.
2. *Forgive me for not mentioning the time*: in letter 4443 Novikoff told Tyndall 'as you fix no time, I just say, that I will be at home from 3.30 to 7.30'.
3. *With me it is to those who claim least that I yield most*: a response to Novikoff's statement 'As to me, I expect nothing, & have claims upon nothing and nobody' in letter 4443.
4. *LT Transcript Only*: the letter is written in hand, presumably by LT, on a page otherwise containing typescripts of letters.

From Charles Lyell 21 October 1874 4445

73, Harley St, | London, W. | 21. Oct[obe]r. 74

Dear Tyndall

I write to thank you for the copy of the Belfast Address,[1] which you were so kind as to send me. I congratulate you on its success and that of the meeting[2] which has certainly been very great—owing, I have no doubt in a great degree to your having spoken out so freely & fearlessly.

In a paper called the "Enquirer" an eloquent defense has been written on your address[3] to prove that the materialism such as it may assert can by no means be treated as Atheism,[4] but I daresay you have seen this[5] & therefore I need not trouble you with it.

Believe me, with thanks, | most truly yours | Cha Lyell

RI MS JT/1/L/59
RI MS JT/1/TYP/3/860

1. *the copy of the Belfast Address*: Belfast Address.

2. *the meeting*: of the BAAS at Belfast, which was held from 19 to 26 August.
3. *In a paper called the "Enquirer" . . . your address*: T. E. P., 'Professor Tyndall's Inaugural Address', *Inquirer*, 33 (1874), pp. 574–5. The *Inquirer* was the leading Unitarian journal, having been founded in 1842; the author of the article was Thomas Elford Poynting, a Unitarian minister in Manchester who identified his theological position as Higher Pantheism.
4. *prove that the materialism . . . treated as Atheism*: the article asserted that: 'It is not against the idea of GOD Himself that the hostility of science, as represented by the President of the British Association, is directed, but against a form of thought in which men in general have clothed GOD and presented him . . . under the image of . . . an Almighty Artificer, separate from matter . . . But the destruction of this form of thought instead of plunging us into the darkness of Atheism, opens upon us the light of true Theism. It leaves us free to form another far grander and worthier thought of GOD, that of the *Indwelling all-forming and all-sustaining spirit of the Universe*, which it is clear that Dr. TYNDALL recognises under what he calls a Cosmical life' (pp. 574–5).
5. *I daresay you have seen this*: Tyndall had actually written to the *Inquirer*'s editor asking him to 'express to T.E.P. my cordial acknowledgement of his able and sympathetic article on my Belfast Address, which appeared in the Inquirer for Sept. 5th, 1874' (quoted in C. B. Upton, 'Thomas Elford Poynting: In Memoriam', *Theological Review*, 16 (1879), pp. 487–507, on p. 501). The original letter is missing.

To Olga Novikoff 22 October 1874 4446

Royal Institution of Great Britain | 22nd. Oct[ober]. 1874

Be calm! There is nothing whatever wrong. You have sent me no secrets; and if you had, I should have closed my eyes against them. The pamphlet[1] has come & I thank you for it. With it came an empty envelope.[2] Also a note informing me that Mr. Liebreich[3] had called upon you. For this again I thank you. So you see I know nothing that you need care about. I wish I could now sit down & rest myself over Moral & Dogma,[4] but I must postpone that for a day or two. I hardly think a visit tomorrow evening will be possible. If possible I will come—But I have pledged myself to a little party in Stratton Street,[5] and I know not how long they will keep me.

I shall forget you just as little as you will forget me. For you are not one to be readily forgotten. Bright, cultivated, natural, honest. These are qualities which seize upon a plain blunt fellow like myself, and hold a place in my memory. I could say more but I won't.

Yours ever | John Tyndall

Museums and Galleries, City of Bradford MDC
RI MS JT/1/T/387

1. *The pamphlet*: A. Riehl, *Moral und Dogma* (Wien: Carl Gerold, 1871) (see letter 4443).
2. *an empty envelope*: sometimes sent as a sign of disapprobation (see letter 4260).
3. *M^r^. Liebreich*: Richard Liebreich (1830–1917), a German eye surgeon who moved to London in 1870 and was head of ophthalmology at St Thomas's Hospital until 1877 (*NDB*).
4. *Moral & Dogma*: see n. 1.
5. *a little party in Stratton Street*: Stratton Street is in Mayfair, only three streets along from Albemarle Street off Piccadilly. Tyndall later recalled that 'Stratton Street is well known to me. Many a pleasant dinner I had there with my friend Joddrell in days that are gone' (Tyndall to Emily Peel, 26 June 1887, RI MS JT/1/TYP/3/974). Thomas Jodrell Phillips-Jodrell (see letter 4309, n. 4) lived at number 12, which was probably the location of the 'little party'.

To Olga Novikoff 23 October [1874][1] 4447

Royal Institution of Great Britain | 23rd. Oct[ober].

Dear Friend

To my great regret I find we have not the works of Schopenhauer[2] in our library. But I think our librarian[3] might be able to borrow them from the London Library[4] and to have them brought here for your inspection.

It was nearly 11 when I returned last night[5]—and I was very weary.

Yours most faithfully | John Tyndall

RI MS JT/1/TYP/3/928
LT Typescript Only

1. *[1874]*: the year is established by the relation to letter 4446.
2. *Schopenhauer*: Arthur Schopenhauer (1788–1860), a German philosopher whose best-known work was *Die Welt als Wille und Vorstellung* (Leipzig: F. A. Brockhaus, 1818), published in English as *The World as Will and Representation*, which developed Kantian idealism by arguing that we experience the world as representation rather than a thing-in-itself. The sole thing-in-itself is the will, an ultimate inner essence, the human servitude to which is the cause of all suffering (*NDB*). Novikoff had first become acquainted with 'S.'s philosophy of will, being the foundation, the essence of everything' in Russia in the mid-1860s (*M.P. for Russia*, vol. 1, p. 93).
3. *our librarian*: Benjamin Vincent.
4. *the London Library*: a private lending library founded in 1841 and located in St James's Square.
5. *nearly 11 when I returned last night*: Tyndall had been at a 'little party in Stratton Street', probably at the home of his friend Thomas Jodrell Phillips-Jodrell (see letter 4446).

From Thomas Henry Huxley 25 October 1874 4448

Athenaeum Club, Pall Mall[1] | October 25 1874.

My dear Tyndall,

If you will put questions into a triangular postscript,[2] they will get forgotten.

The outer layers of the shells of big Globigerinæ[3] are made up of angular rods like

separate crystals but not the inner layers, which are laminated. As to these outer columns being a calcspar[4] I am not certain, but I know that Sorby[5] has worked at the question and can tell you all about it.

I am glad you think me right about the Fullerian.[6] I advised Spottiswoode to try Ray Lankester[7] who is a very able fellow.

Ever Yours | T. H. Huxley.

Some people can flourish as well as other people.

IC HP 8.166a
RI MS JT/1/TYP/9/3039
Typed Transcript Only

---

1. *Athenaeum Club, Pall Mall*: see letter 4186, n. 1.
2. *a triangular postscript*: Tyndall's letter with a triangular postscript is missing.
3. *Globigerinæ*: a planktonic marine protozoan with a calcareous shell. Huxley hypothesized that these tiny shells built up on the ocean floor over eons to form layers of chalk. Previously, Huxley had discussed Globigerina in his lecture 'On a Piece of Chalk' at the BAAS meeting in Norwich on 26 August 1868 (*Brit. Assoc. Rep. 1868*, p. lvii), published as T. H. Huxley, 'On a Piece of Chalk', *Macmillan's Magazine*, 18 (1868), pp. 396–408. Following this letter, on 29 January 1875, Huxley referred to Globigerina in a Friday Evening Discourse he delivered at the RI, 'On the Recent Work of the "Challenger" Expedition, and its Bearing on Geological Problems' (*Roy. Inst. Proc.*, 7 (1873–75), pp. 354–7, on p. 356).
4. *calcspar*: calcareous spar or rhombohedral crystallized carbonate of lime (*OED*).
5. *Sorby*: Henry Clifton Sorby (1826–1908), a geologist who specialized in using microscopes to study the composition of rocks and minerals (*ODNB*).
6. *the Fullerian*: the Fullerian Chair of Physiology and Comparative Anatomy at the RI,

which, with the equivalent Chair of Chemistry, was named after the politician and philanthropist John Fuller (1757–1834).

7. *Ray Lankester*: Edwin Ray Lankester (1847–1929), who in 1874 was appointed Jodrell Professor of Zoology at UCL. In 1875, Alfred Henry Garrod (1846–79) replaced William Rutherford (1839–99), who had held the Fullerian Chair since 1872.

From Richard Grattan 26 October 1874 4449

Drummin House | Carbury Co[unty]. Kildare | Ireland | 26th. October, 1874.

Dear Sir,

Nothing can be worse than the iniquitous manner in which you and Professor Huxley have been misrepresented and vilified, by the religious fanatics of all sects, on this side of the channel.

I have taken up the subject and will defend you to the utmost. Your opinions have been denounced as immoral and antisocial, in fact you are represented as corruptors of youth, and as persons with whom it is unsafe to hold intercourse.

It would be inconsistent with your characters and position to enter into a controversy with those people. This, however, I will do. But you must, on your part aid me, by carrying out my plan, and forming an Association such as I have suggested.[1]

For the present, I intend merely to obtain the opinions of a few clerics, on the subjects to which I have referred in my letter, I will then publish another edition,[2] with them as an appendix, and with a commentary of my own. In this last, of course, you and Professor Huxley will give me your assistance. I guess we shall puzzle those gentry.

Did you get the Books from Trübner?[3] Do not trouble yourself about Theology. Leave that to me. I send you some copies of my letter to Watts[4] for distribution among your scientific friends. If you can turn to use a greater number tell me and you shall have them.

Perhaps, you might send this letter to Professor Huxley,[5] with my best respects.

I am Dr. Sir | truly Yours | Richard Grattan.

Professor Tyndall | Athenaeum Club[6]

RI MS JT/1/TYP/9/3041
LT Typescript Only

---

1. *an Association such as I have suggested*: not identified, but presumably for the defence of rational Protestant thought.

2. *my letter . . . another edition*: Grattan's letter does not seem to have been formally published, but copies of the new edition appear to have been sent to the press. On 26 December, the *Family Herald* reported: 'RICHARD GRATTAN, M.D., sends us some tracts of a freethinking nature, and "wishes to know the exact limit between Protestant freedom of thought and Protestant rationalism". And of course Doctor GRATTAN defends in great measure Professors Tyndall and Huxley, and says they are not Atheists, and that, in denouncing them as Atheists, the entire Irish press has sinned, and that no one of ordinary intellect is an Atheist, and that he believes that "the most pious amongst us are those who inquire into the nature and attributes of the Creator"' ('To Correspondents', *Family Herald*, 34 (1874), p. 124).
3. *the Books from Trübner*: probably R. Grattan, *Considerations on the Human Mind, Its Present State, and Future Destination* (London: Trübner, 1861); and R. Grattan, *The Right to Think: Addressed to the Young Men of Great Britain and Ireland* (London: Trübner, 1865). They were published by the firm that Nicholas Trübner (1817–84) had founded in the 1850s (*ODNB*).
4. *Watts*: Robert Watts (1820–95), a Presbyterian theologian (*ODNB*), whose sermon criticizing Tyndall's presidential address to the BAAS, preached at Belfast's Fisherwick Place Church on 30 August, was published as *Atomism: Dr. Tyndall's Atomic Theory of the Universe Examined and Refuted* (Belfast: William Mullan, 1874).
5. *send this letter to Professor Huxley*: Tyndall obeyed the request (see letter 4461).
6. *Athenaeum Club*: see letter 4186, n. 1.

To William Spottiswoode 27 October [1874][1] 4450

27th Oct[ober].

My dear Spottiswoode

Liebreich, whom I saw last night, will be ready in March to give us an evening on "The Real and the Ideal in Portraiture"[2]

Ten Kisses to Cyril and love to Hugh.[3]

Ever yours | John Tyndall

Morley has as yet given no sign.[4]

RI MS JT/TYP/2/1262
LT Typescript Only

---

1. *[1874]*: the year is established by reference to Richard Liebreich's Friday Evening Discourse (see n. 2).
2. *Liebreich . . . "The Real and the Ideal in Portraiture"*: Tyndall seems to have been introduced to the German eye surgeon Richard Liebreich (see letter 4446, n. 3) by Olga Novikoff. Liebreich delivered a Friday Evening Discourse with this title at the RI on 19 March 1875

(*Roy. Inst. Proc.*, 7 (1873–75), pp. 430–43). Although Liebreich focussed principally on Greek and Roman sculpture in the discourse, he also 'referred his audience to a collection of sculpture by eminent living artists placed in the Library of the Royal Institution' ('Royal Institution Lectures', *Illustrated London News*, 27 March 1875, p. 333).

3. *Cyril and . . . Hugh*: Cyril Spottiswoode and William Hugh Spottiswoode.
4. *Morley has as yet given no sign*: not identified, but John Morley (1838–1923), editor of the *Fortnightly Review*, had previously given a Friday Evening Discourse, on 12 April 1872 on the topic of 'Rousseau's Influence on European Thought' (*Roy. Inst. Proc.*, 6 (1870–72), pp. 475–6), so it may relate to a request to give another discourse. It is unlikely that Morley would have given a positive 'sign', as he reflected of his previous discourse in a private letter: 'I wish I had not agreed to make an absurd discourse at the Royal Institution. What is the good? The pen is my instrument, rather than the tongue' (F.W. Hirst, *Early Life and Letters of John Morley*, 2 vols (London: Macmillan, 1927), vol. 1, p. 213).

From John William Draper 27 October 1874 4451

University[1] Washington Square | New York | Oct[ober] 27th 1874

My Dear Dr Tyndall

I am very obliged to you for the interest you take in my forthcoming work.[2] Of course as soon as it appears here there will be a typhoon, but it will weather that.

Youmans,[3] to whom I have communicated your letter[4] informs me that the Appletons[5] will send you as soon as they can be struck off proof sheets of the whole book. The book proofs were corrected yesterday. It is of about 400 pages, and is one of the International series[6]

As there may be some delay in your receiving the sheets, I enclose proofs of the Preface[7] They will give you an idea of the scope and management of the book

Your Belfast address[8] continues to be the subject of criticism and of current conversation. It came at a very opportune moment. There has been none of that under tone bitterness that would have been manifested had it appeared ten years sooner. There has been a wonderful change in that respect. I have already told you of the sentiments of personal esteem it has generated toward you[9]

I hope that I too may find that vantage ground of which you speak.[10] Anyhow I have tried to deserve it.

Very truly yours

Library of Congress, John William Draper Family Papers

1. *University*: the University of the City of New York (see letter 4406, n. 1).
2. *my forthcoming work*: J. W. Draper, *History of the Conflict between Religion and Science* (New York: D. Appleton, 1874).
3. *Youmans*: Edward L. Youmans was the science editor for D. Appleton & Co. and had commissioned Draper's book.
4. *your letter*: letter 4430.
5. *the Appletons*: the publishers D. Appleton & Co., 72 Fifth Avenue, New York.
6. *the International series*: Draper's book was volume 77 in the International Scientific Series (see letter 4264, n. 5).
7. *the Preface*: pp. v–xvi, which was dated December 1873.
8. *Your Belfast address*: Tyndall's presidential address to the BAAS at Belfast, delivered on 19 August.
9. *I have already told . . . toward you*: letter 4406.
10. *that vantage ground of which you speak*: letters 4420 and 4430.

From Olga Novikoff 29–30 October [1874][1] 4452

Symonds' Hotel | Brook Street | Thursday night. & |
1. o'C[lock]. A.M. Friday. Oct[ober]. 30.

My dear Friend,

I have received 2 letters w[hic]h might interest you; one is from Albert Réville,—the other from Alois Riehl—both are full of you. Here is Réville: ". . .. . . Je ne connaissais pas l'ouvrage du Mr. T. (Fragments of Science).[2] J'ai grand envie d'en faire le sujet d'un article pour la revue des 2 mondes[3] cet hiver. Vous pouvez assurer Mr Tyndall qu'il sera écrit dans un esprit de sincère admiration pour son talent et son caractère".[4]

<u>Alois Riehl</u>:

Tyndall's Fragments

"haben in gewissem Sinne in mir Epoche gemacht. Sie geben mir den Beweis an die Hand dass die Ergebnisse einer forschenden nicht blos phantasirenden Philosophie schlieslich mich den höchsten Resultaten & Zielen der Naturwissenschaften zusammentreffen. Diese Ubereinstimmung zweier so heterogenen Wissenskreise ist eine stichhaltige Probe auf Wahrheit für beide.

Die ausführliche Anzeige des Buches T.'s[5] wird noch im Laufe dieser Woche erschienen. Ich werde einen Abdruck davon Ihnen zusenden.—Es wundert mich, um wie viel schroffer und stärker in England der Widerstand gegen die neue naturwiss: und philosophische Weltanschauung ist, als in Deutschland. Das haben wir denn doch unserer nationalen Philosophie, ich meine, „der Kritischen“ zu danken. Man hat in Deutschland von Seite der Wissenschaft Du Bois Reymond's Rede über die Grenzen des Naturerkennens,[6] welche, wie ich jetzt sehe—eine bloße Reprise Tyndall'ischer Ideen ist—ob zwar nicht Unrechts—als zu wenig entschieden angriffen."[7]——

—I was going to send you this letter at once.[8] But—hesitation comes over me—you already seem bored to death with me—& I shall give you some time of repose—before reminding you again of

Y[ou]r Olga N.—

RI MS JT/1/N/38

---

1. *[1874]*: the year is established by reference to Alois Riehl's review (see n. 5).
2. *Fragments of Science*: *Fragments of Science.*
3. *article pour la revue des 2 mondes*: Réville's proposed article on *Fragments of Science* did not appear in the *Revue des deux Mondes* in winter 1874, although he contributed an article considering the Belfast Address and Tyndall's career more widely the following March. See A. Réville, 'Les Sciences naturelles et l'orthodoxie en Angleterre', *Revue des deux Mondes*, 8 n.s. (1875), pp. 283–318 (see letter 4535).
4. *"...... Je ne connaissais pas l'ouvrage ... son caractère"*: "...... I did not know M^r^. T.'s work (Fragments of Science). I would very much like to make it the topic of an article for the Revue des Deux Mondes this winter. You can assure Mr Tyndall that it will be written in the spirit of sincere admiration for his talent and his character" (French).
5. *Die ausführliche Anzeige des Buches T.'s*: A. Riehl, 'Feuilleton: Tyndall, *Fragmente aus den Naturwissenschaften*', *Wiener Abendpost*, 2 and 3 November 1874, pp. 2004 and 2012–13. This newspaper was an evening edition of the *Wiener Zeitung.*
6. *Du Bois Reymond's Rede über die Grenzen des Naturerkennens*: on 14 August 1872, Emil du Bois-Reymond delivered a public lecture in Leipzig on the limits of natural knowledge which was published as *Über die Grenzen des Naturerkennens* (Leipzig: Zeit, 1872).
7. *"haben in gewissem Sinne ... entschieden angriffen"*: "were, in a certain sense, epoch-making. They furnished me with the proof that the results of an enquiring—and not only fantasizing—philosophy in the end correspond with the highest results & goals of the natural sciences. This agreement between two so heterogeneous spheres of knowledge is a valid test for the truthfulness of both. The comprehensive review of T.'s book will appear in the course of this week. I will send you a copy of it.—I am surprised at how much harsher and stronger resistance against the new scientific and philosophical worldview is in England than it is in Germany. For this, after all, we have to thank our national philosophy, I mean the 'critical'. In Germany, Du Bois Reymond's speech on the limits of the knowledge of nature, which, as I now see is nothing more than a reprise of Tyndall's ideas was attacked by scientists—although not unjustifiably—for being too indecisive" (German).
8. *I was going to send you ... bored to death with me*: Novikoff presumably wrote this postscript in the early hours of the morning of Friday 30 October, having written the initial

part of the letter the previous evening after spending time with Tyndall at her salon, where his behaviour gave the impression he was bored with her (see letter 4453). The letter could have been sent 'at once' as Tyndall and Novikoff, who were living close to each other in Mayfair, seem to have sent at least some of their letters via servants rather than the regular postal service (see letter 4469, n. 7).

To Olga Novikoff [30] October [1874][1] 4453

Royal Institution of Great Britain | 20th[2] Oct[ober].

My dear Friend,

I thank you exceedingly for the opportunity you have so kindly given me of reading the letter of Professor Riehl.[3] It is one of those solid earnest utterances which give me incomparably greater pleasure than the cheers of the newspapers, rare as those latter may be. I am much gratified to learn that he has thought the Fragments of Science[4] worth reading. It costs time to produce even Fragments; and I should never permitted myself to be deflected even to this slight extent from my original inquiries, did I not think the diffusion of sound science among the community at large to be of unspeakable importance at the present time. I sought in all honesty to make my own mind clear as to the subjects on which I wrote, and as far as in me lay to carry the same clearness to others. But no man is a proper judge of his own work; and it is therefore peculiarly gratifying to find one so eminently competent to form an opinion upon the subject speaking so favourably and indeed so kindly of me and my work, as Professor Riehl has done in this letter.

I was not "brilliant" last night—but every body else was—Serjeant Parry[5] in particular and I enjoyed <u>their</u> brilliancy to which <u>my</u> dullness was a foil.[6]

I am going to day to have a mutton chop with M^rs^. Faraday,[7] but I hope to be able to call to see you tomorrow

Ever truly yours | <u>John Tyndall</u>

John Tyndall Correspondence in the Olga Novikoff Correspondence Collection. Courtesy of Special Collection, Kenneth Spencer Research Library, University of Kansas. MS 30:Q31
RI MS JT/1/TYP/3/950
M.P. for Russia, vol. 1, p. 158[8]

---

1. *[1874]*: the year is established by the relation to letter 4452.
2. *20th*: this appears to be a slip of the pen by Tyndall, as the letter he was replying to (letter 4452) was clearly dated by Novikoff as 30 October.
3. *the letter of Professor Riehl*: see letter 4452.

4. *the Fragments of Science*: Alois Riehl had read the German translation of *Fragments of Science* (see letter 4192, n. 11).
5. *Serjeant Parry*: John Humffreys Parry (1816–80), a barrister whose title, formally Serjeant-at-law, denoted his membership of an elite group of lawyers who enjoyed privileges at the common law courts.
6. *I was not "brilliant" last night . . . my dullness was a foil*: seemingly a response to Novikoff's comment 'you already seem bored to death with me' in letter 4452, an impression that, presumably, had been given by Tyndall's behaviour at her salon earlier that day.
7. *Mrs Faraday*: Sarah Faraday.
8. *M. P. for Russia, vol. 1, p. 158*: this published version incorrectly gives the year as 1878.

From Alfred Mayer 30 October 1874 4454

Hoboken N[ew]. Jersey | US. America. | Oct[ober]. 30. 1874

Dear Professor Tyndall;

I received your letter of Oct[ober]. 17th[1] while I was preparing proof and plates of my paper, on the law of the duration of the residual sonorous sensation,[2] for republication in England.[3] As the 'criticisms'[4] are nought, now that the new edition of your charming little book[5] appears, I have deleted the latter portion of my paper, relating to your account of Helmholtz's labour.[6]

I think that my determination of the law of the duration of sonorous sensations is a real step in progress in Acoustics; and although,—as I have indeed shown—the constants of its mathematical expression may change with different individuals, yet I feel confident that it will always retain a hyperbolic expression.[7]

Its physiological bearings, and its applications to harmony are at once apparent, rendering much of Helmholtz's qualitative work quantitative, and serving to give a clear and concise idea of his great discovery of the physiological cause of musical harmony.[8]

With the highest consideration of regard | I remain | Yours ever faithfully | Alfred M. Mayer

RI MS JT/1/M/76

---

1. *your letter of Oct[ober]. 17th*: letter 4439.
2. *my paper, on the law of the duration of the residual sonorous sensation*: A. M. Mayer, 'Researches in Acoustics', *American Journal of Science*, 8 (1874), pp. 241–55.
3. *republication in England*: Mayer's article was reprinted in Britain as 'Researches in Acoustics', *Phil. Mag.*, 49 (1875), pp. 352–65.
4. *the 'criticisms'*: see letter 4439, n. 2.

5. *the new edition of your charming little book*: *Sound* (1875).
6. *deleted the latter portion ... Helmholtz's labour*: in the *Phil. Mag.*, Mayer removed a lengthy passage which appeared on pp. 253–5 of the original version of his article in the *American Journal of Science*. The passage criticized Tyndall's account of Hermann Helmholtz's work on the dissonant and consonant effects of beats in the initial edition of *Sound* (1867) (see letter 4439, n. 2).
7. *a hyperbolic expression*: that is, the graph of the sensation of a sound should decrease following the shape of a hyperbola. Mayer's data, however, did not quite fit this expectation (see 'Researches in Acoustics', p. 246).
8. *the physiological cause of musical harmony*: in 1857 Helmholtz had delivered a public lecture in Bonn entitled 'Ueber die physiologischen Uraschen der musikalischen Harmonie', and developed his ideas in his book *Die Lehre von den Tonempfindungen als physiologische Grundlage für die Theorie der Musik* (Braunschweig: Friedrich Vieweg, 1863). He proposed that the human body responded to specific wavelengths of sound, with the nerve fibres of the inner ear acting like strings that vibrate sympathetically to sonic events.

From Maria Schroeder[1] 30 October 1874 4455

Höchst geehrter Herr Professor!

In unserem Zeitalter wo sich der große Culturkampf zwischen Kirche & Staat vollzieht, wo die Säulen des alten Glaubens nicht nur zu wanken beginnen sondern, wo die Wissenschaft ihm jede Basis thatsächlich entzogen hat—in dieser Zeit des Übergangs, wo jeder Denkende die heilige Verpflichtung hat, sich an dem geistigen Kamph thätig und handelnd zu betheiligen und der Wahrheit zum Siege zu helfen, sei es mir gestattet einige Worte an Sie, höchst verehrter Herr Professor, zu richten.—

Als Präsident der "British Association for the Advancement of Science" besitzen Sie augenblicklich eine Herrschaft welche Sie zum Mittelpunkt der Aufmerksamkeit aller denkender Geister macht und die Presse, deren gewaltige Macht ihnen ganz zu Gebote steht, trägt in Tausenden von Exemplaren diesseits und jenseits des Oceans, Ihr gesprochenes, wie Ihr geschriebenes Wort in Tausend Geister! Sie halten—ein moderner Petrus—die Schlüssel zum Himmelreich der Ewigen Wahrheit in Ihrer Hand. Jetzt ist der Augenblick gekommen, o, lassen Sie die nach Erkenntniß ringenden Geister, die nach Wahrheit & Klarheit Dürstenden frei und ungehindert einziehen in das Reich des Lichtes; entschleiern Sie ganz das noch halb verdeckte Bild zu "Saïs", unbekümmert darum ob einige schwache Seelen dabei in Ohnmachtfallen; auch hier gilt jenes ewige Naturgesetz „das Individuum muß untergehen damit die Race fortbestehe". Nicht der Kirche in dem heutigen Sinne sondern dem Philosophen und dem Gesetzgeber wird das Hohepriesteramt der Zukunft zufallen. In dem "Heute" wandelt das "Morgen" bereits und darum müssen auch schon jetzt die Männer der Wissenschaft, der reinen Vernunft, die das hohe Ziel der sittlichen Freiheit (das wahrhafte Paradies) Männer, die das lebendig gewordene Princip des Kant'schen Categorischen Imperativus sind die Führerschaft übernehmen für das arme, irrende Menschengeschlecht. Bei dem unbedingten Vertrauen, das Ihnen von Männern aller Fractionen—ja lassen Sie mich hinzufügen—aller Nationen entgegengebracht wird, liegt es unmittelbar in Ihrer Hand Tausende von schwankenden Gemüthern für die Sache der Vernunft & Wahrheit zu gewinnen. Aber dies kann nur geschehen; wenn Sie Ihre persönlichen Ansichten, die Resultate Ihres eigenen wissenschaftlichen Forschens in ihren Folgen mit Bezug auf Religion der Welt offenbaren, welche bei der ausgezeichneten Adresse in Belfast, wohl eine wunderbar klare Geschichte der Philosophie, nicht aber eine Darlegung Ihrer eigenen Anschauungen gegeben hat. Das ganze Religionssystem der sogenannten geoffenbarten Religionen ist unhaltbar geworden für denkende Menschen des 19ten Jahrhunderts. Das Ammenmärchen der Schöpfungsgeschichte, des Sündenfalls und der Erlösung kann niemand befriedigen, der wahrhaft denkt und tief empfindet. Hingegen die Weltentwicklung ohne Anfang und Ende, der Ewigkeitsbegriff in Zeit und Raum, der den Menschen einschließt, hat etwas wahrhaft Erhebendes. Das ewige Streben des Menschengeistes nach Veredlung ist schön und giebt uns das tiefste Gefühl von der Würde des Menschen, von der Weihe seines Berufes. Von dem Philosophen gilt daher das Wort Schiller's, welches er an die Künstler richtete: "Der Menschheit Würde ward in leere Hand gegeben, bewahret sie."—

Lassen Sie mich Ihnen sagen, daß meine Worte, obwohl der Sache geweiht, dictirt worden sind von der höchsten Verehrung für Ihre Person, lassen Sie mich hoffen, daß sie nicht ganz ungehört verhallen mögen und genehmigen Sie, Herr Professor, die Versicherung meiner ausgezeichnetsten Hochachtung & Ergebenheit

Maria Schroeder | London, 30ten Oc. 74.

Randolph House 61[a] Portsdown-Road. | Maida-Hill W. London.—

Most revered Professor!

In our age, where the great culture war between church and state is taking place, where the pillars of the old belief not only begin to wobble, but where science truly has withdrawn its every basis—in this time of transition, where every thinking person has the holy duty to actively participate in this spiritual fight and help truth to victory, may I be allowed to address a few words, most revered Professor, to you.

As president of the "British Association for the Advancement of Science" you currently hold a position which puts you at the centre of attention of all thinking spirits, and the press, whose tremendous power is fully at your service, in thousands of copies on both sides of the ocean carries your spoken as your written word into a thousand minds! A modern St Peter,[2] you hold the keys to the heavenly kingdom of eternal truth in your hand. The time has come now, o, let those minds which struggle for knowledge, which thirst for truth and clarity enter freely and unhindered into the kingdom of light; unveil the as yet half-hidden image at "Sais",[3] uncaring whether a few weak souls shall faint; here also, the eternal principle holds true that "the individual has to perish so that the race may live".[4] Not to the Church in its modern sense, but to the philosopher and to the lawgiver will the Office of High Priest fall in the future. The "today" already contains the "tomorrow" and thus, men of science, of pure reason—the highest goal of moral freedom (the real paradise)—, men who are the incarnation of Kant's principle of the categorical imperative,[5] have to assume the leadership for wretched, mistaken mankind. With the absolute trust that men of all factions—let me add—of all nations are placing in you, it is directly in your hand to win thousands of wavering minds for reason & truth. But this can only happen if you reveal your personal opinions, the results of your own scientific research and their consequences for religion to the world, which during the excellent address in Belfast was given a wonderfully lucid history of philosophy, but not an exposition of your own opinions. The whole religious system of the so-called revealed religions has become untenable for thinking people of the 19th century. The fairy tale of Genesis, the Fall of Man and the Redemption can satisfy nobody who truly thinks and feels deeply. On the contrary, the evolution of the world, without beginning or end, the concept of eternity in time and space, which includes mankind, has something truly uplifting. The eternal quest of the human mind for improvement is beautiful and gives us the deepest feeling of human dignity, the sanctification of his vocation. To the philosopher, therefore, the lines that Schiller[6] addressed to the artists apply: "The dignity of Man into your empty hands is given, Protector be."[7]—

Let me assure you that my words, although consecrated to the cause, were dictated by the highest admiration for your person, let me hope that they may not trail off completely unheard and accept, dear Professor, the assurance of my sincerest respect & devotion

Maria Schroeder | London 30th Oct[ober]. 74.

Randolph House 61a Portsdown-Road. | Maida-Hill W. London.—

RI MS JT/1/S/58

---

1. *Maria Schroeder*: no one of this name, or anything similar, was living at 61a Portsdown-Road, Maida-Hill at the time of either the 1871 or 1881 Censuses, but she may be the Maria Schroeder (b. *c.* 1819) recorded in the 1841 Census as living in Westminster.
2. *St Peter*: St Peter (*c.* 30–*c.* 64 CE), one of Jesus's twelve Apostles and the first leader of the early Christian Church. He was given the keys of heaven by Jesus in Matthew 16:19.
3. *half-hidden image at "Sais"*: the veiled statue of the goddess Isis in the ancient Egyptian city of Sais, which traditionally represented the secrets of nature.

4. *the individual has to perish so that the race may live*: paraphrase of John 11:50, 'it is expedient for us, that one man should die for the people, and that the whole nation perish not'.
5. *Kant's principle of the categorical imperative*: the name given by the German philosopher Immanuel Kant (1724–1804) to an absolute, unconditional requirement that must be obeyed in all circumstances and is justified as an end in itself. Kant formulated the principle in his *Grundlegung zur Metaphysik der Sitten* (Riga: J. F. Hartknoch, 1785).
6. *Schiller*: Friedrich Schiller (1759–1805), a German poet and philosopher (*NDB*).
7. *"The dignity of Man . . . Protector be"*: F. Schiller, 'Die Künstler' (1789),11. 442–3.

From Thomas Henry Huxley 31 October 1874 4456

4 Marlborough Place[1] | October 31 1874.

My dear Tyndall,

I have been unable to see King[2] before to-day when I went to Cornhill[3] and had a long talk with him. He really is not unreasonable but seemed to have been worked up to a boiling point of irritation by the "Evolution" of correspondence with Spencer.[4] He had not the slightest idea of disputing our position with regard to the contributors to the International series,[5] and is quite willing to execute a formal agreement with us on the basis of the old one[6] with these modifications.

King to print 1250 of the first edition accounting for 1000.

This is the actual arrangement at present in consequence of Youmans's modification of the original agreement.[7]

King to print 1250 of all following editions but to pay author £55 for each instead of £50 as at present.

In other words he accounts for 1100 instead of 1000 leaving him 150 instead of 100 in each of these editions as we agreed would be satisfactory when we discussed the matter.

He was quite fair and offered to leave the matter in my hands, but after hearing all he had to say I thought this proposition equitable. He was anxious to get the matter settled without further discussion, and as that course seemed to me very desirable, I agreed to the proposal. If your ambassador has exceeded his powers you must do as other potentates do and dismiss him.

Ever Yours very sincerely | T. H. Huxley.

I hope Manchester[8] went off well.

IC HP 8.167
RI MS JT/1/TYP/9/3040
LT Typescript Only

1. *4 Marlborough Place*: see letter 4213, n. 1,
2. *King*: Henry Samuel King (see letter 4297, n. 3).
3. *Cornhill*: King's publishing firm, Henry S. King & Co., was located at 65 Cornhill, in the City of London.
4. *the "Evolution" of correspondence with Spencer*: presumably a joke concerning the amount of correspondence King was receiving from the evolutionary philosopher Herbert Spencer.
5. *the International series*: see letter 4264, n. 5.
6. *the old one*: the original series agreement, made between Edward L. Youmans and King in October 1871, was for a payment to authors of £50 for each edition of a book, which was defined as a printing of 1250 copies, representing a royalty of 20%. Authors would also receive a 10% royalty from American sales, and 7.5% on sales in Europe.
7. *Youmans's modification of the original agreement*: not identified; this modification is not mentioned in either L. Howsam, 'An Experiment with Science for the Nineteenth-Century Book Trade: The International Scientific Series', *British Journal for the History of Science*, 33 (2000), pp. 187–207; or B. Lightman, 'The International Scientific Series and the Communication of Darwinism', *Journal of Cambridge Studies*, 5 (2010), pp. 27–38.
8. *Manchester*: a lecture (see letter 4441, n. 3).

To Olga Novikoff 1 November [1874][1] 4457

Royal Institution of Great Britain | 1st. Nov[ember].

My dear Friend.

I am going out of town for a few hours to rest my dull and weary head: I will call to see you sometime tomorrow—probably at 5.30.

Thank you many times for the "Neue Wissen"[2] and I beg of you to thank your friend Prof. Frohschammer on my behalf. I shall read all that he has written, for what I have read gives me a high idea of his ability, earnestness, and honesty.

I wish I had some authentic account of the sufferings and persecutions referred to by his friend.

Yours ever | John Tyndall

[Envelope] Madame de Novikoff | Symonds's Hotel | 34 Brook Street

John Tyndall Correspondence in the Olga Novikoff Correspondence Collection. Courtesy of Special Collections, Kenneth Spencer Research Library, University of Kansas. MS 30:Q5

RI MS JT/1/TYP/3/929

1. *[1874]*: the year is established by reference to Jakob Frohschammer, whom Tyndall and Novikoff were discussing at this time.

2. *the "Neue Wissen"*: J. Frohschammer, *Das neue Wissen und der neue Glaube* (Leipzig: F. A. Brockhaus, 1873).

From Olga Novikoff [*c.* 1 November 1874][1] 4458

Second Edition.[2]

Frohschammer begs you to read the article in "Unsere Zeit"[3] w[hic]h w[ou]ld acquaint you with his "struggle for life", or at least—"for truth". He would like to send you 2 of his books, but wants to know whether you care to read them?

Have you seen in the "Neue Wissen & N. Glauben" what he says of you?[4]

Frohschammer has not yet seen y[ou]r Fragments of Science[5] & wants to get them. I could easily send them, but I think—or I feel sure—he w[ou]ld be happier to get them from y[ou]rself. I am in a great hurry & very tired. Pardon my dreadful scrawl.

Olga

Prof. Froschammer's address is | München | 7 Amalienstrasse. | Bavaria.

RI MS JT/1/N/42

1. *[c. 1 November 1874]*: the date is suggested by Novikoff seeming to respond to Tyndall's request for 'some authentic account' of Jakob Frohschammer's 'sufferings and persecutions' in letter 4457, which is dated 1 November. Novikoff seems to be responding by the second post (see n. 2) that day.
2. *Second Edition*: possibly Novikoff's way of indicating that the letter was sent in the second post.
3. *the article in "Unsere Zeit"*: probably 'Die oppositionelle Bewegung in der katholischen Kirche', *Unsere Zeit: Deutsche Revue der Gegenwart*, 8 (1872), pp. 1–23, which contains an account of Frohschammer's career on pp. 12–19.
4. *in the the "Neue Wissen & N. Glauben" what he says of you*: J. Frohschammer, *Das neue Wissen und der neue Glaube* (Leipzig: F. A. Brockhaus, 1873). In the book, whose title translates as *The New Knowledge and the New Belief*, Frohschammer, having discussed the materialism of modern science, stated: 'Mit gleicher Entschiedenheit, in gleichem Sinne sprach sich über diesen Gegenstand der englische naturforscher Tyndall in der britischen naturforscherverfammlung ebenfalls in August 1868 aus. Bei der Wichtigleit der Sache und angesichts der dreisten Behauptung der Bertreter der materialistischen Weltauffssung, daß diese als ficheres refultat der modern naturforschung betrachtet warden müsse, möge auch aus diefer Rede einiges wörtlich hier angefuhrt warden [The English naturalist Tyndall spoke out on this subject with the same determination and in the same sense at the British naturalists' meeting in August 1868. Given the importance of the matter and in view of the bold assertion of the advocates of the materialistic world view that this must be regarded as a reliable result of modern natural research, some of this speech should also be quoted verbatim here]' (pp. 62–3). The speech quoted was Tyndall's address as president

of the Mathematics and Physics section of the BAAS meeting at Norwich on 19 August 1868 (*Brit. Assoc. Rep. 1868*, pp. 1–8 of 'Notices and Abstracts of Miscellaneous Communications to the Sections'), which was also published as 'Scientific Limit of the Imagination' in J. Tyndall, *Use and Limit of the Imagination in Science* (London: Longmans, Green, and Co., 1870), pp. 52–65; and as 'Scope and Limit of Scientific Materialism' in *Fragments of Science*, pp. 107–23.

5. *y[ou]r Fragments of Science*: *Fragments of Science*.

From George Henry Wilson[1] 2 November 1874 4459

Wellington, New. Zealand. | November 2/74.

Dear Sir

Enclosed you will receive three newspaper clips:[2] the contents of which refer to your late address delivered at the Belfast Meeting.[3] &c—The first report in N.Z. Times date Oct[ober]. 16/74[4] gave much pain to very many of your readers here at the Antipodes, and if allowed to pass unchallenged would have done your name an indirect injury. I replied to the implied triumph of bad men and a worse system as you perceive in the letter to the Daily Tribune of Oct 24/74[5]—awkwardly or feebly, it was very probably my best,—it has had the effect of producing from the "Times" a short paragraph—enclosed—disclaiming responsibility[6]—&c. yet you will perceive the evident idea of triumph which is implied in the first notice of your address by the "Times" when it goes on to draw parallels[7]—

Hoping you will excuse my troubling you with this communication

I am dear Sir | Your ob[edien]t serv[an]t | George H. Wilson.

To Professor Tyndall. | &c. &c.

RI MS JT/1/W/76
RI MS JT/1/TYP/5/1793

1. *George Henry Wilson*: George Henry Wilson (1833–1905), a novelist who had emigrated to New Zealand and whose *Ena: Or, The Ancient Maori* (1874) had recently been published in London by Smith, Elder.
2. *three newspaper clips*: the enclosures have been preserved, but are not transcribed as part of the letter. For details of them, see n. 4 and n. 6.
3. *your late address delivered at the Belfast Meeting*: Tyndall's presidential address to the BAAS meeting at Belfast, delivered on 19 August.
4. *report in N.Z. Times date Oct[ober]. 16/74*: '[Editorial]', *New Zealand Times*, 16 October 1874, p. 2. The report claimed that in his presidential address, Tyndall 'charged "religion,

especially Christianity, with being inimical to human progress"' and 'ignored faith altogether, and declared himself a materialist', concluding that: 'we had not thought that the time had yet come when views such as those of Professor Tyndall would be put before, and apparently received by, such a body as the British Association'.

5. *the letter to the Daily Tribune of Oct 24/74*: 'Letters to the Editor', *Tribune* (Wellington), 24 October 1874. Wilson, who signed his letter 'G.H.W.', quoted passages from the report of the *Times* in London of Tyndall's address, and asserted: 'In no single sentence can it be discovered that the Professor was defining religion, or any of its characteristics, most certain it is that he never once accuses Christianity as being "inimical to human progress"'.
6. *from the "Times" a short paragraph . . . disclaiming responsibility*: '[Editorial]', *New Zealand Times*, 23 October 1874, p. 2. The paragraph stated: 'We disclaim responsibility for the briefest of brief summaries of the address of Professor Tyndall . . . We quoted it—not having seen the speech itself . . . as given by a popular Glasgow newspaper, as the essence of Professor Tyndall's address'.
7. *the evident idea of triumph . . . goes on to draw parallels*: the initial report in the *New Zealand Times* had drawn a parallel between Tyndall's putative views in the Belfast Address and those of a 'sect' in Australia 'who do not scruple to avow the doctrine that "Christianity is the invention of the enemy of mankind", and that it is "doing the Devil's work"' ('[Editorial]', *New Zealand Times*, 16 October 1874, p. 2).

From Thomas Woolner[1] 2 November 1874 4460

29, WELBECK STREET,[2] | W. | Nov[ember] 2 '74

My dear Tyndall,

I went down to the Athenaeum[3] tonight hoping to meet you and put your mind at rest with respect to the bust of Bence Jones.[4]—

The mark is what we call a vent[5] in the marble, and has been there ever since it was marble. It will never be different to what it is now; and no circumstance that would not destroy the bust will ever affect it in any way. As the whole bust becomes toned with time it will gradually become less perceptible to the eye, and in all probability eventually disappear, tho' of course, it will always be actually there.—We never think much of vents unless they cross fingers or some delicate part which it would be impossible to work without breaking.—

Believe me | Very truly yours | T. Woolner

Unlike dark spots on veins we rarely see these vents till the last fine cutting.

RI.G1b/4
(also found in Managers' Minutes vol. XIII pp. 37–8 of 7 December 1874)

1. *Thomas Woolner*: Thomas Woolner (1825–92), a British sculptor who in 1848 had been a founder member of the Pre-Raphaelite Brotherhood. He subsequently established a successful career as a sculptor of public monuments (*ODNB*).
2. *WELBECK STREET*: in the West End of London.
3. *the Athenaeum*: the Athenaeum Club (see letter 4186, n. 1).
4. *the bust of Bence Jones*: a sculpture had been commissioned from Woolner to mark the retirement of Henry Bence Jones in March 1873 (see 'Testimonial to Dr. Bence Jones', *Roy. Inst. Proc.*, 7 (1873–75), pp. 57–8 and 92), and it had been installed in the ante-room of the RI in February 1874.
5. *a vent*: an internal flaw in a block of marble that takes the form of an opening or aperture.

To Thomas Henry Huxley 4 November [1874][1] 4461

4th Nov[ember].

Dear Hal

Writer of the enclosed[2] is a man of 85[3]—I suppose slightly mad.

Did you not enjoy the pounding which the Times caused you to administer to me?[4] He[5] evidently thinks that he has caught one of us.

Yours ever | J. Tyndall

IC HP 8.168a
RI MS JT/1/TYP/9/3041
Typed Transcript Only

1. *[1874]*: the year is established by the relation to letter 4449.
2. *the enclosed*: letter 4449 from Richard Grattan.
3. *a man of 85*: Grattan had been born in February 1790, so was in fact 84.
4. *the pounding which the Times caused you to administer to me*: in an editorial in the *Times* on 31 October that was critical of both the Belfast Address and its Catholic detractors, it was noted of Tyndall: 'If he is to be effectually criticized, it should be by the eminent natural philosopher whom the Roman Catholic Prelates associate with him. It is easy to imagine Professor Huxley applying to his colleague's speculations precisely the same kind of criticism he has often bestowed upon such doctrines as the Prelates uphold. "If you like", he might say, "to prolong your vision backward beyond the boundary of experimental evidence, and to exercise your imagination in dreaming what you can never prove, you may find it a relief from the monotony of mere experience, and within certain limits you will do no harm. You please and console yourselves, and you hurt no one else. But the only things I care to assert are confined within this modest horizon of experience, and you must excuse

me troubling myself with your visions, whether idealistic or materialistic, theological or scientific. On either side they are compressible, malleable, indefinite, mere airy visions, given an imaginary reality by language"' ('[Editorial]', *Times*, 31 October 1874, p. 9).

5. *He*: the editor of the *Times* was John Thadeus Delane, but it is unclear whether this particular editorial was written by him.

To Olga Novikoff 5 November 1874 4462

Royal Institution of Great Britain | 5 Nov[ember]. 1874

My dear Friend.

As to the dinner I hope you will not press me. My work at the present time is of so serious a character,[1] that, except on official occasions where I cannot well help myself, I must set my face against dining out. Besides it is probable that on Monday[2] I shall be away from London. The noise of the builders on all sides of me[3] here will compel me to fly away. I shall come of course to night, I hope between 9 & 10.

Yours ever | John Tyndall

[Envelope] Madame de Novikoff | Symonds's Hotel | 34 Brook Street

[Postmark] LONDON·W | C5 | NO 5 | 74

John Tyndall Correspondence in the Olga Novikoff Correspondence Collection. Courtesy of Special Collections, Kenneth Spencer Research Library, University of Kansas. MS 30:Q6

RI MS JT/1/TYP/3/928

1. *My work . . . so serious a character*: presumably the composition of Tyndall's letter to the *Times* on the nature of typhoid and how it might be combatted most effectually, a subject that, he insisted, was of the 'very gravest importance' (see letter 4463).
2. *Monday*: 9 November.
3. *The noise of the builders on all sides of me*: the cause of this noise is unclear, as the substantial building works to expand and modernize the RI's old chemical laboratory, as well creating a new physics laboratory and an additional staircase in the east wing of the building at 20 and 21 Albemarle Street, were completed in December 1872. The work was overseen by Henry Bence Jones (F. A. J. L. James and A. Peers, 'Constructing Space for Science at the Royal Institution of Great Britain', *Physics in Perspective*, 9 (2007), pp. 130–85, on pp. 162–3). Tyndall avoided the worst of the noise during those building works as he had been on his lecture tour of the United States (see letter 4201, n. 4). Following this renovation, William Spottiswoode delivered a Friday Evening Discourse at the RI on 17 January 1873, 'On the Old and New Laboratories at the Royal Institution' (*Roy. Inst. Proc.*, 7 (1873–75), pp. 1–10).

To the Editor of the *Times*[1] 6 November 1874 4463

Sir,—

Some time ago a book, the labour of preparing which, added to the effects of other labours, has, it is to be feared, permanently damaged a superior mind, was placed in my hands at the request of its author. It was a treatise on "Typhoid Fever", by Dr. William Budd, F.R.S., of Clifton.[2] Taken in connexion with the recent revelations concerning the state of Over Darwen,[3] the facts recorded in that book are so remarkable, its logic is so strong, and its conclusions are so momentous, that I venture to ask space for a notice of it in the columns of The Times.

In medicine, as elsewhere, knowledge grows and consolidates through the conflict and sifting of opinions and evidences. With regard to the great class of diseases known as epidemics, which flourish through the transfer from place to place, and from person to person, of a something which continues to exist through its own powers of reproduction, physicians have been long divided in their notions. And with regard to the title of certain diseases to be ranked as epidemic, the opinions of the medical world have been equally divided. On this last question, more especially, theoretic notions may be of the last importance, for they more or less determine the physician's practice, and have, therefore, a direct bearing upon the lives committed to his care.

On hardly any point of medical theory, and the practice flowing from it, has this division of opinion been more distinct than on the question of typhoid fever. The pith of the controversy is this:—Can typhoid fever be generated anew? Is it produced by the decomposition and putrefaction of animal and vegetable substances, or must the matter producing it have had previous contact with an infected body? In other words, for every new case of typhoid fever may we with certainty infer a pre-existing case, of which the new one is merely the propagation or continuation, or are we entitled to conclude that organic matter, which has never been in contact with a typhoid patient, is, in virtue of its own decomposition, capable of starting the fever anew? When we consider that this just sends 15,000 of the inhabitants of these islands yearly to the grave, and causes 150,000 to pass through its protracted miseries, the question here stated assumes the very gravest importance, because our relation to it must determine our mode of attack upon this enemy of mankind.

The position taken by Dr. Budd in reference to this question is one which will render his name memorable in the history of medicine. In the work before us he seeks to prove that the first of the positives just laid down is the true position; that there is no such thing as the spontaneous generation of typhoid fever; that the malady is propagated, as surely as smallpox is propagated through a special virus, by contagion. He begins by developing his evidence on this head; he then fixes the principal seat of the contagious matter in the intestine; he examines the nature of the intestinal affection, the relation of typhoid fever to defective sewerage, the character of the contagious agent, the employment of disinfectants and disinfection. He discusses the so-called "pythogenic" or putrescent theory,[4] and winds up with some remarks on the spontaneous origin of typhoid fever. The book from the beginning to end, is one comprehensive argument, with reference to which it may be said that the facts alleged are of the most conclusive character, while the logic which binds them together is, as far as I can see, simply irresistible.

This is a question which is sure to occupy the attention of legislators as well as physicians, and it is therefore desirable to place it in the clearest untechnical light. Dr. Budd takes his reader to the village of North Tawton,[5] where he was himself born and brought up, and every inhabitant of which was personally known to him. In the village there was no general system of sewers. Round the cottages of those who earned their bread with their hands, and who formed the great bulk of the population, collected various offensive matters. Each cottage, or group of three or four cottages, had a common privy, to which a simple excavation in the ground served as a cesspool. In many cases, hard by the cottage door there was not only an open privy, but a dungheap, where pigs rooted and revelled. For a long period there was much offensive to the nose, but no fever. An inquiry, conducted with the most scrupulous care,[6] showed that for 15 years there had been no severe outbreak of the disorder, and that for nearly ten years there has been only a single case. "For the development of this fever", adds Dr. Budd, "a more specific element was needed than the swine, the dungheaps, or the privies were able to furnish."[7]

That element at length came, and formed a starting-point from which its further progress might be securely followed. On the 12th of July 1839, a case of typhoid fever, doubtless imported from without, occurred in a poor and crowded dwelling, and before the end of November 80 of the inhabitants had suffered from it, a proportion about the same as that now suffering at Over Darwen.[8] The reader will, I trust, bear strictly in mind that the question now before us is whether typhoid is contagious, and he is asked to weigh the answer which facts return to this question. Two sawyers[9] living near the stricken house at North Tawton fell ill and quitted the village for their own homes at Morchard,[10] where no previous case of typhoid fever had been. In two days one of these men took to his bed, and at the end of five weeks he died. Ten days after his death his two children were laid up with the fever. The other sawyer also took to his bed, and when at the worst a friend from a distance came to see him, and assisted to raise him in bed. On the tenth day after this friend was seized with the fever. Before he became convalescent two of his children were struck down, and his brother, who lived at a distance, but who came to see him, also fell a victim. Was this series of events the result of chance, or was it the work of contagion? Let us pursue the inquiry further. On the 20th of August a Mrs. Lee began to droop at North Tawton, and, not knowing what was impending, she visited her brother at Chaffcombe,[11] seven miles off. She was smitten with the fever, and before she became convalescent her sister-in-law, Mrs. Snell, who had nursed her, was attacked and died subsequently. Then came Mr. Snell, then one of the farm apprentices, then a day labourer, then a Miss S., who had come to take charge of the house after Mr. Snell's death; and finally, a group consisting of a servant man, a servant girl, and another young person who had acted as nurse.

The case here submitted to the reader[12] is not one of medical practice but of common evidence, which does not even require a trained scientific mind to weigh it. Let us proceed. A boy who had been smitten at Chaffcombe went to his mother's cottage between Bow[13] and North Tawton. Before he recovered, his mother, who had nursed him, sickened and afterwards died. Two children of the family next door were next attacked, then the sister of the boy who had carried the infection from Chaffcombe. She, in her turn, removed to another place, and became a new focus for the propagation of this disease. Again, to lighten the list of invalids a girl named Mary Gibbings was sent from Chaffcombe to her home at Loosebeare,[14] four miles off. Here she lay ill for several weeks. Before she recovered her father was seized. A farmer who lived across the road, and who visited Gibbings, was next struck down. His case was followed by others under the same roof; and the fever, spreading from this to other houses, became the centre of an epidemic which gradually extended to the whole hamlet.

At the same time, scattered over the country side were some 20 or 30 other hamlets, in each of which were the usual manure yard, the inevitable pigsty, and the same primitive accommodation for human needs. "The same sun shone upon all alike, through month after month of the same fine, dry, autumnal weather. From the soil of all of these hamlets human and other exuviae exhaled into the air the same putrescent compound in about equal abundance. In some of them, indeed, to speak the exact truth, these compounds, if the nose might be trusted—and in this matter there is no better witness—were much more rife. And yet, while at Loosebeare a large proportion of the inhabitants were lying prostrate with fever, in not one of the twenty or thirty similar hamlets was there a single case."[15] There is no confusion of data here; no blur or indistinctness in the observer's vision, no flaw, as far as I can see, in his reasoning. He follows the morbific agent from place to place, sees it planted, developed, shedding its seeds, producing new crops; growing up where it is sown, and there only. Ashpits fail to develope it; putrescence fails to develope it; stench fails to develope it; even the open privy is powerless as long as it is kept free from the discharges of those already attacked. The case of North Tawton is typical; numerous other cases equally conclusive are adduced—among them the foul condition of the Thames in the hot weather of 1858 and 1859, when stench for the first time "rose to the height of a historic event",[16] and when, nevertheless, London, even along the river, enjoyed a singular immunity from fever. It is, I think, impossible for any intelligent reader, and I should say certainly impossible for any man trained to scientific reasoning, to quit Dr. Budd's volume without closing with his conclusion that the living human body is the soil in which the specific poison of typhoid fever breeds and multiplies.[17]

What is the seat of the poison? Dr. Budd is too cautious to shut out the possibility of infection by any of the emanations from a person suffering from the disease. But its special and almost exclusive locus is the diseased intestine. He gives drawings and photographs of the bowel at various stages of the disease;[18] and it is hardly possible to look on these without coming to the conclusion that the whole interior surface of the bowels is the seat of an eruption. The pustules or protuberant patches, called "Peyer's patches",[19] thicken and stand out in relief from the surface of the gut. They feel, to use the words of Chomel,[20] as if a solid and elastic substance had been inserted between the coats of the intestine, while, when a patch is cut through, its texture is seen to be occupied by a yellowish-white cheese-like matter. "This is the peculiar 'typhoid matter' whose presence is typical of the disease, and whose formation and elimination constitute the essence of the intestinal process."[21] Louis[22] has made careful observations as to the duration of the alvine[23] discharge which accompanies typhoid fever, and finds it to be in mild cases 15, and in severe cases 25 days. For this period, therefore, every individual smitten at Over Darwen has been flooding the undrained ground with the poison of this contagious fever. It reaches the drinking-water; it partially dries and floats in the air; it rises mechanically with the gases issuing from cesspools, and thus the pestilence wraps like an atmosphere the entire community.

How could a disease whose characteristics are so severely demonstrable have ever been imagined to be non-contagious? How could such a doctrine be followed out, as it has been, to the destruction of human life? Mainly because practice in cities, where the greatest medical authorities reside, was directly calculated to throw the physician off the scent. The seat of the disease being the intestine, with well-appointed water-closets it is not in the sick-room that the mischief is done, but often at a distance from the sick-room, through the agency of the sewer, which Budd graphically describes as "a direct continuation of the diseased intestine".[24] Hence the mystic power of "sewer gas."[25] Hence the inability of the metropolitan practitioner to trace the disease to its origin. Hence the immunity of undrained country villages as long as the specific poison keeps away, and hence also the localized ravages of the disease in such villages as soon as it appears.

The results point with unmistakable precision to the mode of attack. But this point will be best illustrated by a case out of Dr. Budd's own practice. It occurred in the Convent of the Good Shepherd, at Arno's Court, near Bristol. At no previous time had typhoid fever ever appeared in the Convent. In November, 1863, a girl actually labouring under the disease was admitted into the Reformatory. She was removed by the permission of a medical man, who, in accordance with the prevalent doctrine, held "that typhoid fever was the result of bad drainage, and in no way contagious."[26] When Dr. Budd visited the place, on the 29th of February, 30 young women were in bed with the fever, and within 48 hours 20 more were added to the list. Fifty-six of whom eight died, were struck down before the plague was subdued. The measures adopted were these:—

Flooding all the drains of the place with disinfectants with a view to destroy as far as possible the poison already cast off.

The reception of all discharges from the sick, immediately on their issue from the body, into vessels charged with disinfectants.

The instant immersion of all bed and body linen used by the sick into a disinfecting liquid before its removal from the ward.

Scrupulous ablution and disinfection of the hands of the nurses.

Lastly, the burning or disinfection of all beds occupied by the sick, as soon as vacated by death, convalescence, or otherwise.

All these things were done, not in a loose or slipslop way, but with the precision of a scientific process. It may be said that the plague was instantly stayed; only three certain cases of infection having occurred subsequent to the adoption of these measures. Can it be doubted that with sound medical advisers, backed by an intelligent population, an equally rapid destruction of the foe might be accomplished at Over Darwen?

Were it not that I have already drawn far too heavily upon your space, I might enlarge upon these subjects. I will limit myself to one more point of commanding interest. What is the nature of the typhoid poison? The "yellow typhoid matter" already referred to, Budd describes as made up of nucleated cells.[27] The term "germ-theory" does not, to my knowledge, occur once in the volume, possibly because of the opposition and ridicule which that theory encountered in the English Medical Press.[28] Over and over again Budd speaks of "germs", but it might be imagined that he used the word figuratively. Those who knew him, however, were well aware that this was not the case; and in the early part of the present volume, after describing the calamities incident to typhoid fever, he remarks:—"It is humiliating that issues such as these should be contingent on the powers of an agent so low in the scale of being that the mildew which springs on decaying wood must be considered high in comparison."[29] Four or five years ago I, an outsider, ventured upon this ground of medical theory, for it involved no knowledge of medical practice, but simply a capacity to weigh evidence; and the evidence that epidemic diseases were parasitic appeared to me very strong. On the 9th June, 1871, I ventured to express myself thus:—"With their respective viruses you may plant typhoid fever, scarlatina or smallpox. What are the crops that arise from this husbandry? As surely as a thistle rises from a thistle-seed, as surely as the fig comes from the fig, the grape from the grape, and the thorn from the thorn, so surely does the typhoid virus increase and multiply into typhoid fever, the scarlatina virus into scarlatina, the smallpox virus into smallpox. What is the conclusion that suggests itself here? It is this:—that the thing which we vaguely call a virus is to all intents and purposes a seed; that, excluding the notion of vitality, in the whole range of chymical science you cannot point to an action which illustrates this perfect parallelism with the phenomena of Life—this demonstrated power of self-multiplication and reproduction."[30] It was the clear and powerful writings of William Budd, joined to those of the celebrated Pasteur, that won me to these views. It is partly with a view of stomping at a receptive moment salutary truths upon the public mind, but partly also through the desire of rendering justice to a noble intellect, which has been literally sacrificed to the public good,[31] that I draw attention, not only to the masterly combination of observation and inference exhibited from beginning to end of Dr. Budd's volume, but also to the crowning fact already published in the medical journals, and to which my attention was first drawn by my eminent friend Mr. Simon,[32] that Dr. Klein has recently discovered the very organism which lies at the root of all the mischief,[33] and to the destruction of which medical and sanitary skill will henceforth be directed.

I am, Sir, your obedient servant, | John Tyndall.

Royal Institution, Nov[ember]. 6.

RI MS JT/6/5/3
Times, 9 November 1874, p. 7

---

1. *the Editor of the* Times: John Thadeus Delane, who edited the newspaper from 1841 to 1877. Tyndall's letter was printed under the heading 'Typhoid Fever'.
2. *a treatise on "Typhoid Fever" . . . of Clifton*: W. Budd, *Typhoid Fever: Its Nature, Mode of Spreading, and Prevention* (London: Longmans, Green, and Co., 1873). Budd (1811–80) was a physician and epidemiologist (*ODNB*).
3. *the recent revelations . . . Over Darwen*: the *Times* had recently noted the 'disgracefully defective sanitary arrangements' in the mill town, near Blackburn in Lancashire, and

reported that there was an outbreak of typhoid fever in late October ('The Manufacture of Fever', *Times*, 2 November 1874, p. 4).

4. *the so-called "pythogenic" ... theory*: the 'appellation of "pythogenic fever"' was coined by Charles Murchison in 'Contributions to the Etiology of Continued Fever', *Medico-Chirurgical Transactions*, 41 (1858), pp. 219–306, on p. 221.
5. *North Tawton*: in Devon, on the river Taw. Its inhabitants worked mostly in either the production of serge (a twill fabric) or in agriculture.
6. *An inquiry ... scrupulous care*: the inquiry was conducted by Budd himself.
7. *"For the development ... able to furnish"*: Budd, *Typhoid Fever*, p. 11.
8. *a proportion about the same ... at Over Darwen*: it was recorded of Over Darwen in the *Times* that there were 'at present, in a population of 23,000, more than 1,500 cases of typhoid fever' ('Manufacture of Fever', p. 4).
9. *sawyers*: workmen whose business it is to saw timber (*OED*).
10. *Morchard*: a village in Devon.
11. *Chaffcombe*: a village in Somerset.
12. *The case here submitted to the reader*: all the details are taken from Budd, *Typhoid Fever*, pp. 13–17.
13. *Bow*: a village in Devon.
14. *Loosebeare*: a village in Devon.
15. *"The same sun shone ... was there a single case"*: Budd, *Typhoid Fever*, p. 18.
16. *"rose to the height of a historic event"*: paraphrase of 'Never before, at least, had a stink risen to the height of an historic event', in W. Budd, 'Observations on Typhoid or Intestinal Fever', *British Medical Journal*, 2 (1861), pp. 485–7, on p. 485.
17. *that the living ... breeds and multiplies*: paraphrase of '*The living human body, therefore, is the soil in which this specific poison breeds and multiplies; and that most specific of processes which constitutes the fever itself is the process by which the multiplication is effected*' in Budd, *Typhoid Fever*, p. 37.
18. *drawings and photographs ... stages of the disease*: plates 1–3 between pp. 48–9 of Budd, *Typhoid Fever*.
19. *"Peyer's patches"*: lymphoid follicles named after Johann Conrad Peyer (1653–1712), a Swiss anatomist (*HLS*).
20. *Chomel*: Auguste François Chomel (1788–1858), a French pathologist.
21. *"This is the peculiar 'typhoid matter' ... intestinal process"*: Budd, *Typhoid Fever*, p. 47.
22. *Louis*: Pierre-Charles-Alexandre Louis (1787–1872), a French physician and pathologist.
23. *alvine*: relating to the intestinal tract (*OED*).
24. *"a direct continuation of the diseased intestine"*: Budd, *Typhoid Fever*, p. 42.
25. *"sewer gas"*: atmospheric air mixed with gas formed by the decomposition of sewage (*OED*), which was considered by Murchison and other exponents of pythogenic fever as the cause of contagion.
26. *"that typhoid fever ... in no way contagious"*: Budd, *Typhoid Fever*, p. 133.
27. *The "yellow typhoid matter" ... nucleated cells*: Budd, *Typhoid Fever*, p. 50.

28. *the opposition and ridicule . . . English Medical Press*: the *British Medical Journal*, for instance, noted that in considering the 'truth of the "germ-theory"' it had 'had from the first the invaluable assistance of Dr. Bastian', the theory's leading opponent, and it approvingly printed his dismissive view of 'what in modern times has been termed "The Germ-Theory of Disease". Like homœpathy and phrenology, this theory carried with it a kind of simplicity and attractiveness, which insured its acceptability to the minds of many. Now, however, it seems to rest upon foundations only a little more worthy of consideration than those upon which these other theories are based' ('The Missing Introductory', *British Medical Journal*, 2 (1871), p. 418; and H. C. Bastian, 'Epidemic and Specific Contagious Diseases', *British Medical Journal*, 2 (1871), pp. 400–9, on p. 402).
29. *"It is humiliating . . . considered high in comparison"*: Budd, *Typhoid Fever*, p. 4.
30. *"With their respective viruses . . . and reproduction"*: Tyndall made the statement in a Friday Evening Discourse on 'Dust and Smoke' delivered at the RI on 9 June 1871 (*Roy. Inst. Proc.*, 6 (1870–72), pp. 365–76). The lecture was published as 'Dust and Smoke', *Nature*, 4 (1871), pp. 124–8, with the passage Tyndall quotes here on p. 125.
31. *a noble intellect . . . sacrificed to the public good*: an obituary of Budd in the *British Medical Journal* noted that he was 'smitten by severe disease' in 1873, and that, seven years later, he 'died undoubtedly a martyr to his work, of which the greatest part and the most exhausting was done for the public good' (W. M. Clarke, 'William Budd, MD, FRS: "In Memoriam"', *British Medical Journal*, 1 (1880), pp. 163–6, on pp. 163 and 166).
32. *Mr. Simon*: John Simon (1816–1904), who served as the government's Medical Officer of Health from 1855 to 1876 (*ODNB*).
33. *Dr. Klein has recently discovered . . . the mischief*: Edward Klein's identification of the microphyte responsible for the contagion of enteric disease had been reported to the RS in June 1874 and was published as E. Klein, 'Research on the Smallpox of Sheep', *Roy. Soc. Proc.*, 22 (1874), pp. 388–91. Edward Klein (1844–1925), was a British histologist and bacteriologist (*ODNB*).

From Olga Novikoff [*c.* 6 November 1874][1] 4464

Friday e[ven]ing

My dear Friend,

Read F's letter[2] attentively, then return it to me.—the story about the loss of his eye-sight[3] has been pointed out to me by his religious enemies who saw in it the "punishment of God."[4] I begged F to tell me all about it. Legends are invented even in our <u>truthful</u> days!

In desperate haste but sincerely y[ou]rs Olga

RI MS JT/1/N/39

1. *[c. 6 November 1874]*: the date is suggested by this seeming to be a further response to Tyndall's request for 'some authentic account' of Jakob Frohschammer's 'sufferings and persecutions' in letter 4457, after Novikoff had responded initially in letter 4458. 6 November was the first Friday after Tyndall made the request.
2. *F's letter*: F is Frohschammer, but his letter is missing.
3. *the loss of his eye-sight*: Novikoff later recounted how she had told Tyndall of Frohschammer that 'once, as he was finishing one of his straightforward rebukes to Rome, he was struck with a kind of paralysis which rendered him half blind . . . he still worked on for conscience' sake, as if ignoring all the terrible conditions of a blind man's solitude. Occasionally his housemaid read to him, but it is easy to guess the way in which she must have deciphered some of the terms contained in his scientific books. This, however, often made my poor friend smile, and only added to his collection of amusing anecdotes about his "literary help". Tyndall seemed interested in this very incomplete and fragmentary sketch of mine' (*M.P. for Russia*, vol. 1, p. 153).
4. *the "punishment of God"*: there are several biblical allusions to blindness as a divine punishment, particularly Acts 13:11 and Deuteronomy 28:28, but Novikoff seems to conflate these with the words traditionally attributed to Genghis Khan: 'I am the punishment of God. If you had not committed great sins, God would not have sent a punishment like me upon you'.

To Olga Novikoff 6 November [1874][1] 4465

Royal Institution of Great Britain | 6th Nov[ember].

Dear Friend

I send you five copies of the Manchester lecture;[2] but do not, please, forget to tell your friend[3] that it was rapidly put together & not fully corrected for the press.[4]

There was not the least necessity to write the little note[5] you have just sent to me.

Yours ever | John Tyndall

The Waller Manuscript Collection, Uppsala University, alb–77:258
RI MS JT/1/TYP/3/929

1. *[1874]*: the year is established by reference to the publication of Tyndall's Manchester lecture (see n. 2).
2. *copies of the Manchester lecture*: the lecture on 'Crystalline and Molecular Forces' that Tyndall had delivered on 28 October at Manchester's Free Trade Hall was reprinted in *Belfast Address*, 7th thousand, pp. 67–83, and the 'five copies' Tyndall sends are either advance copies of these pages or the version of the lecture published in *Science Lectures for the People. Sixth Series* (Manchester: John Heywood, 1874), pp. 1–13.
3. *your friend*: probably Jakob Frohschammer.

4. *rapidly put together & not fully corrected for the press*: in the 'Preface to the Seventh Thousand' of *Belfast Address*, 7th thousand, Tyndall noted that the printed version of the lecture was 'my own corrected edition' of 'a verbatim report . . . in the "Manchester Examiner"' (pp. x–xi).
5. *the little note*: probably letter 4464.

From James Coxe 6 November 1874 4466

General Board of Lunacy,[1] | Edinburgh, | Nov[ember]. 6th 1874.

My dear Dr Tyndall,

I thank you much for your kind remembrance of me.[2] For some time you have deservedly been the best abused man in England, and I hope you will long labour in the vocation that has earned you so much notice; that is in bringing down the strong-holds of ignorance and superstition.

My wife[3] is fairly well, and I hope we shall next year again rap at the door of a certain house in Albemarle Street,[4] and be fortunate to find the Spider[5] at home,

Most truly Yours | J. Coxe.

RI MS JT/1/TYP/1/305
LT Typescript Only

1. *General Board of Lunacy*: the organization put in charge of Scottish mental health institutions by the Lunacy (Scotland) Act of 1857; Coxe served as Commissioner in Lunacy on the Board from its foundation until his death in 1878.
2. *your kind remembrance of me*: letter missing.
3. *My wife*: Mary Anne Coxe.
4. *a certain house in Albemarle Street*: the RI, at 21 Albemarle Street.
5. *the Spider*: presumably a nickname for Tyndall, perhaps relating to his prowess in climbing; another friend used 'act the Gorilla' to describe his propensity for climbing trees (see letter 4239).

To Moncure Daniel Conway[1] 7 November [1874][2] 4467

7th. Nov[ember].

My dear Conway.

I do not think I said anything to our Hindu friend[3] that I should object to hearing proclaimed from the housetops—Do with the conversation[4] what seemeth unto you good: and use my name as you please.

Yours ever | John Tyndall

Moncure Daniel Conway Papers, Box 21; University Archives, Rare Book & Manuscript Library, Columbia University Libraries

1. *Moncure Daniel Conway*: Moncure Daniel Conway (1832–1907), an American freethinker who, having become a leading figure in the movement for the abolition of slavery, settled in London in 1863, where he served as minister of the South Place Chapel (*ODNB*).
2. *[1874]*: the year is established by reference to Protap Chunder Mozoomdar's visit to Britain (see n. 4).
3. *our Hindu friend*: Protap Chunder Mozoomdar (1840–1905), a Bengali leader of the Indian Hindu reform movement, the Brahmo Samaj, who endeavoured to show the harmony between the ethics of Hinduism and Christianity.
4. *the conversation*: Conway accompanied Mozoomdar when he visited Britain from May to November 1874 and later recalled that 'He longed to obtain in England some scientific patronage of his Theism, and requested me to arrange for him an interview with Tyndall, my memorandum of which I conclude to insert here. The interview took place at the Royal Institution. We three were alone, and Mozoomdar, after alluding to the high reputation of Tyndall in India, said, "I feel the need for a few axioms of religion". Tyndall said, "Should we call them by such a precise term as axiom?". Mozoomdar suggested that the term principles might be substituted, adding in illustration, "Such as God and the soul". Tyndall in his gracious way said, "It is not possible for me to use those words except with reservations and explanations". Mozoomdar asked in what form possible those ideas or principles could be expressed. Tyndall asked, "Is any form possible or even desirable?". Mozoomdar said, "In India we do stand in need of some form to embody our new religious ideas, for the sake of morality. Young men abound there who are not only parting with their old beliefs, but with their morality at the same time". Tyndall expressed a hope that Mozoomdar might be mistaken, and added, "I cannot believe that there would be any such result if these young men were properly taught in moral principles". Mozoomdar acknowledged that they were not, properly speaking, taught morality at all. Tyndall said, "I cannot believe that any man requires the aid of theology to teach him that an honest man is better than a rogue". Tyndall stated that certain purely moral passages in a work of Fichte had wrought an important effect in himself in early life, and he did not deny that sacred books might stimulate into activity the higher principles in every human mind and heart—where they always exist though sometimes latent unless influenced from without. Mozoomdar said with emphasis, "I feel that religion must conform to science". Tyndall rejoined, "Such religion as that I cannot condemn, but the reverse. In true religion there is a permanent and indestructible element; the forms may frequently have to be abandoned, the essence never. I think it is not wise to mould this fluid element into form, however new. They who see farthest cannot discern the ultimate forms into which this religious sentiment will mould itself". On hearing this Mozoomdar stretched forth his hand to Tyndall, who cordially grasped it' (M. Conway, *My Pilgrimage to the Wise Men of the East* (London: Archibald Constable, 1906), pp. 221–2). Mozoomdar gave his own account of his conversation with

Tyndall, which took place 'soon after the great sensation caused by the latter's address at Belfast', reflecting: 'The impression with which I left him . . . was that his whole nature was glowing with a deep, vague and transcendent sense of the Divine life, beauty and love; but his intellect, self-bound, loyal and logical to its creed, hesitated and failed to grasp or admit the import of that Life upon the origin, growth, facts and laws of being. It is a gross injustice to call him an atheist. "Working in the cold light of the understanding for many years", he said in effect to me as we rose to part, "we here *do* feel the want of the fire and vigour of that Life. It is all but extinct in England. In saying so, and in not accepting it at the hands of those who have it not, I have become unpopular. Let those who have the Life give it unto us. To you, therefore, in the East we look with real hope; life came from those regions once before and it must come again. Take, therefore, my hearty sympathy and good will. And know that the sympathies of men like you are the few crumbs of comfort left to me in my unpopularity"' (S. C. Bose, *The Life of Protap Chunder Mozoomdar*, 2 vols (Calcutta: Nababidhan Trust, 1927), vol. 1, p. 46).

To Robin Allen 9 November 1874 4468

Royal Institution of Great Britain | 9th November, 1874.

SIR,

I shall use my best efforts to be present at such experiments as may be made as to the sound producing power of the projected new gun.[1]

It is doubtless not without consideration that bronze has been fixed upon as the material for the gun.[2] I can hardly imagine any other quality of the metal than its *strength* coming into play in the projected experiments. Here the "Whitworth metal,"[3] which is enormously strong, has, no doubt, been taken into consideration.

I have the honour to be, Sir, | Your obedient Servant, | JOHN TYNDALL. | ROBIN ALLEN, ESQ.

National Archives, MT 10/290. Papers re Fog Signals. H8476 (duplicate: H4380)

---

1. *such experiments . . . the projected new gun*: those conducted at the Royal Arsenal, Woolwich on 26 November, to ascertain the effectiveness of different gun-cylinders in producing acoustic fog signals. Tyndall was able to be present (see letter 4513, n. 3).
2. *bronze has been . . . material for the gun*: see letter 4513, n. 2 and n. 4.
3. *the "Whitworth metal"*: steel that has been compressed in a liquid state to remove air cells, and then, after reheating, compressed by hammering. The resulting metal, produced by the company of Joseph Whitworth, was particularly dense and strong.

To Olga Novikoff 9 November [1874][1] 4469

9th Nov[ember].

Dear Friend.

I this day found the passage to which you referred in Wissen und Glaube.[2]

I have received & read with extreme interest the account of Frohschammers relation to the Vatican.[3] This I shall turn to account[4]

I this morning gave directions to my publisher[5] to forward to Munich a copy of Fragments of Science.[6]

Finally, as your maid came,[7] I was in the act of seeking for the large photograph[8] which I now find pleasure in sending you.

Yours ever | John Tyndall

RI MS JT/1/T/394/2

---

1. *[1874]*: the year is established by reference to Jakob Frohschammer, whom Tyndall and Novikoff were discussing at this time.
2. *the passage to which you referred in Wissen in Glaube*: see letter 4458, n. 4.
3. *the account of Frohschammers relation to the Vatican*: possibly the letter from a friend with which Frohschammer began his preface to *Das Christenthum und die moderne Naturwissenschaft* (Wien: Tendler, 1868), pp. iii–v, but possibly also an account provided by Novikoff (see n. 4).
4. *This I shall turn to account*: in the 'Preface to the Seventh Thousand' of *Belfast Address*, 7th thousand, pp. v–xxxii, Tyndall added a footnote to a point about the Vatican's 'moral assassination' of Catholic writers who defied it, noting: 'See the case of Frohschammer as sketched by a friend in the Preface to "Das Christenthum und die moderne Wissenschaft". His enemies contrived to take his bread, in great part, away, but they failed to subdue him, and not even the Pope's Nuncio could prevent five hundred students of the University of Munich from signing an Address to their Professor' (p. xxiii). Much of this information was actually provided by Novikoff who later recounted of her conversations with Tyndall about Frohschammer: '"Tell me his biography!" he exclaimed. "How did he come to hold these views? Did you not tell me that he took Holy Orders?". "So he did", I explained in reply. "His life has been a very hard one, full of struggle and privations. An orphan depending entirely on a remote relation, who was an old fanatical priest, the boy had no choice . . . But the more he matured in study and meditation, the more he felt that truth was not religiously adhered to by the Holy See . . . he began publishing pamphlets which not only made a stir in Rome, but were severely criticised and all put upon the Index. Frohschammer, naturally, was thus compelled to give up his parish. This deprived him of his living . . . The Holy See then forbade Roman Catholic undergraduates to attend his lectures on philosophy, and as there were hardly any students of any other

persuasion at Munich University this meant ruin to Frohschammer's career as a professor'" (*M.P. for Russia*, vol. 1, pp. 153–4).

5. *my publisher*: Longmans, Green, & Co., 39 Paternoster Row, London.
6. *Fragments of Science*: *Fragments of Science*. Novikoff had provided Frohschammer's address in Munich in letter 4458.
7. *as your maid came*: as they were living close to each other in Mayfair, Tyndall and Novikoff seem to have sent at least some of their letters via servants rather than the regular postal service.
8. *the large photograph*: not identified, but possibly a version of the photograph of Tyndall taken in New York, probably in January 1873, by José María Mora. Novikoff seems to have wanted to send a copy to Frohschammer (see letters 4474 and 4478).

To Louis Pasteur 9 November 1874 4470

9th Nov[ember]. 1874

My dear Mr. Pasteur.

I cannot allow the occasion to pass without expressing to you the pleasure which the recent award of the Royal Society[1] has given me.

That you may live long to enjoy your well-merited honours, and to gather others is my earnest wish.

Faithfully Yours | John Tyndall

Bibliothèque nationale de France, Manuscrits, Papiers Louis Pasteur, NAF 18107

---

1. *the recent award of the Royal Society*: Pasteur was to receive the RS's Copley Medal 'for his researches on Fermentation and on Pebrine'; the formal award was made at the RS's anniversary meeting on 30 November, where the medal was received on Pasteur's behalf by the RS's foreign secretary Alexander William Williamson (*Roy. Soc. Proc.*, 23 (1875), pp. 68–70).

From Thomas Henry Huxley 9 November [1874][1] 4471

4 Marlborough Place[2] | Nov[ember]. 9th.

My dear Tyndall,

Your letter in the Times to-day[3] will do yeoman's service[4] but your reference to Klein at the end[5] does not put what he has done in quite the right relation to what others have effected in the matter.

If you will look at my B.A. address for 1870[6] you will find that Chauveau[7] and Sanderson proved that living microzyms[8] are the efficient agents of contagion. Subsequently Cohn[9] has demonstrated that these minute living particles grow and multiply under suitable conditions, out of the body. And quite recently Klein has made the very important further discovery that, in sheep-pox, the microzyms not only grow and multiply but rise to a higher grade of organization than they were previously known to attain—becoming converted in fact into a mycelium[10] or something analogous to the crust formed by a common green mold.

Klein's discovery[11] is a most important enlargement of our knowledge of the parasitic organism which gives rise to sheep-pox and similar disorders, but the "very organism" was discovered before.

Ever yours very sincerely | T. H. Huxley.

There is method in our Irish friend's madness,[12] but I do not see how we can help him.

IC HP 8.170
RI MS JT/1/TYP/9/3042
Typed Transcript Only

1. *[1874]*: the year is established by reference to Tyndall's letter to the *Times* (see n. 2).
2. *4 Marlborough Place*: see letter 4213, n. 1.
3. *Your letter in the Times to-day*: letter 4463, which was published as J. Tyndall, 'Typhoid Fever', *Times*, 9 November 1874, p. 7.
4. *yeoman's service*: good, efficient or useful service, such as is rendered by a faithful servant of good standing (*OED*).
5. *your reference to Klein at the end*: Tyndall concluded his letter by noting the 'crowning fact already published in the medical journals . . . that Dr. Klein has recently discovered the very organism which lies at the root of all the mischief [i.e. typhoid], and to the destruction of which medical and sanitary skill will henceforth be directed'.
6. *my B.A. address for 1870*: in his presidential address to the BAAS at Liverpool in 1870 (see letter 4363, n. 10), Huxley observed 'does the vaccine matter contain living particles, which have grown and multiplied where they have been planted? The observations of M. Chauveau, extended and confirmed by Dr. Sanderson himself, appear to leave no doubt upon this head' (p. lxxxv).
7. *Chauveau*: Auguste Chauveau (1827–1917), a French pathologist.
8. *microzyms*: i.e. bacteria.
9. *Cohn*: Ferdinand Cohn (1828–98), a German bacteriologist (*NDB*).
10. *mycelium*: branching, thread-like growth characteristic of fungi or fungi-like bacteria.
11. *Klein's discovery*: see letter 4463, n. 33.
12. *our Irish friend's madness*: see letter 4461; the 'Irish friend' was Richard Grattan.

To the Editor of the *Times*[1] 10 November 1874 4472

Sir,—

It is not my intention to enter into any controversy as to the points raised in the letter which you did me the honour to publish,[2] for the mind which does not see the strength of the argument as it now stands would hardly be convinced by controversy. I would, however, thank you for permission to say an additional explanatory word.

It is regarded by some as "a flaw in the argument" that I have not been able to point to the beginning of typhoid fever. Now, if the fever, as alleged, be the produce of an organism, it appears to me that one might be as fairly asked to point out the beginning of mushrooms, in order to prove that they are propagated by seeds.

And as regards the argument that typhoid fever occurs where the contagion cannot be traced, I would oppose to it the enlightened remarks of Sir Thomas Watson,[3] who has stood forth as a light amid the darkness which has surrounded this question:—"The argument would be equally valid in favour of the spontaneous origin of smallpox. There are a thousand unsuspected ways in which the invisible contagion may be conveyed. It may lurk in a hackney coach; you may catch the complaint from your neighbour in an omnibus, in a theatre or at a concert, at church, or in a casual jostling crowd. Your linen may be impregnated with it at the house of a laundress, or your coat may bring it from the workshop of your tailor; nay I have heard it affirmed that the contagion of smallpox has been carried in a letter."[4]

The same eminent authority will help me to express myself upon another point:—"Mind, I neither deny nor doubt that filth, foul air, and the gaseous products of animal and vegetable decomposition are things hurtful to health, or that they are capable, especially where abundant and concentrated, of causing serious disease, and even death. What I do doubt and deny is that of themselves they ever produce a contagious fever."[5]

It would have given me pleasure to dwell upon the excellent experimental observations which preceded those of Dr. Klein, more especially those of Chauveau, Burdon Sanderson, and Cohn;[6] but my reluctance to draw further upon your space prevents me from dwelling upon the historic development of the subject.

I am Sir, your obedient servant, | John Tyndall | November 10.

RI MS JT/6/5/2
Times, 11 November 1874, p. 5

1. *the Editor of the* Times: John Thadeus Delane, who edited the newspaper from 1841 to 1877. Tyndall's letter was printed under the heading 'Typhoid Fever'.
2. *the letter which you did me the honour to publish*: letter 4463, which was published as J. Tyndall, 'Typhoid Fever', *Times*, 9 November 1874, p. 7.
3. *Sir Thomas Watson*: Thomas Watson (1792–1882), professor of principles and practice of medicine at King's College London and president of the Royal College of Physicians from 1862 to 1866 (*ODNB*).
4. *"The argument would . . . carried in a letter"*: T. Watson, *Lectures on the Principles and Practice of Physic*, 5th edn, 2 vols (London: Longmans, Green, and Co., 1871), vol. 1, p. 136. Although this book had first been published in 1843, based on lectures given in 1836 and 1837, Tyndall was presumably quoting from the most recent edition, which had been revised and enlarged.
5. *"Mind, I neither deny . . . a contagious fever"*: Watson, *Lectures*, vol. 2, p. 846.
6. *the excellent experimental observations . . . and Cohn*: this statement reflects the advice of Thomas Huxley in letter 4471. On Chauveau and Cohn, see n. 7 and n. 9 in that letter.

From Gilbert Govi[1] 10 November 1874 4473

Rome le 10 Novembre 1874 | Hôtel de New-York

Monsieur Tyndall,

Je viens de lire dans les Mondes le magnifique discours que vous avez prononcé à l'ouverture des Séances de l'Association Britannique, et j'en suis si heureux, que je me permets de vous en témoigner par cette lettre ma plus vive reconnaissance.

Il est beau, en effet, de voir un homme tel que vous, arrivé à l'apogée de sa gloire, dans un pays éminemment religieux et conservateur, risque sa réputation et son repos pour l'amour de la vérité, et les risques sans réticences, sans détours, sans quelques-uns de ces faux-fuyants, que les habiles savent si bien se ménager.

Moi aussi, M. Tyndall, je lutte depuis bien des années pour le triomphe du vrai, et voila pourquoi j'ai tressailli de joie en me voyant tout-à-coup un allié aussi puissant, et aussi universellement estimé que vous.

La poste va vous apporter, avec cette lettre, un mien discours sur: Les lois de la Nature, composé et publié en 1869, pour la réouverture de cours à l'Université de Turin. Cet écrit, longuement médité, trop condensé peut-être, mais suffisamment clair pour les hommes instruits auxquels il l'adressait, souleva dans le camp religieux un cri général de réprobation. L'abbé Moigno me fit alors l'honneur de me sermonner longuement dans les Mondes (VIe année 1869 T. XIX Janv.–Avril. Pag. 291–300) sans toutefois reproduire mon discours, car il lui était beaucoup plus facile d'en attaquer des phrases, que d'en renvoyer les arguments. Je ne crus pas devoir répondre alors aux insultes de l'abbé, je lui suis même gré d'avoir ergoté sur ma thèse, car il n'y a pas de vérité qui ne finisse par se faire jour, quand on en parle, ne fût-ce que pour en médire, et je vois que j'ai bien fait de m'épargner une polémique inutile, puisque l'Idée a fait son chemin, et qu'elle vient de trouver en vous un défenseur si considérable et si éloquent.

Le parti religieux n'a pas cessé depuis lors de me poursuivre en y employant toutes les ressources de sa stratégie la plus raffinée. On a soulevé à Turin contre moi une partie de la jeunesse, on a étouffé sous le silence tous mes travaux, on m'a empêché d'avoir la Chaire de Physique à l'Université de Rome, on a ôté au ministre *[1 word illeg]* les moyens d'y fonder pour moi une chaire d'Histoire de la Physique, et après m'avoir bercé de vaines promesses pendant près de deux ans, on va, peut-être, me mettre un de ces jours fort poliment à la porte. Cela ne me décourage cependant pas, ni m'empêche de poursuivre ma propagande rationaliste, j'ai vécu pauvre en ma jeunesse, j'y saurai vivre encore à l'avenir, mais je garderai mes convictions, et je lutterai jusqu'à mon dernier jour contre les doctrines absurdes d'un passé qui voudrait ne pas déchoir et qui cherche à se défendre par tous les moyens.

Je vous dis cela, M. Tyndall, uniquement pour vous mettre en garde contre ce qui vous attend, et pour que vous n'en soyez pas pris au dépourvu. De même que les gens de la coterie religieuse parviennent à faire un grand homme de l'esprit le plus médiocre, à force de répéter partout, toujours et sur tous les fronts qu'il n'y a pas de plus haute intelligence que la sienne, de même il leur est loisible de renverser la réputation la mieux établie à force de la calomnier. Mais le temps est un grand justicier et l'on peut s'en rapporter à lui.

Cependant, vous allez avoir à endurer beaucoup d'attaques injustes et d'oublis outrageants—Les notes que l'abbé Moigno a mises à votre discours ne sont que des attaques d'avant postes—Quelque gauches et inhabiles qu'elles puissent vous paraître, elles n'en sont pas moins de Sages et Saintes réputations pour un certain monde, et vous ne devez pas vous attendre à des objections beaucoup plus sérieuses. Les protestants prendront peut-être un ton plus respectueux et moins méprisant à votre endroit . . . ils ne vous traiteront, peut-être pas comme un Érudit de seconde main, mais ils ne vous en attaqueront pas moins sans trêve et sans raison.

Quand on va avoir beaucoup d'ennuis, il est bon d'être assuré qu'on a aussi des amis sur qui peut compter . . . et voila pourquoi j'ai tenu à vous écrire cette lettre et à vous envoyer mon discours.

Veuillez donc croire à mon attachement le plus sincère, et disposez en toute occurrence de

Votre bien dévoué Gilbert Govi | Profess. De Phys. à l'Université de Turin | Directeur provisoire de la Bibliothèque Casanate à Rome

Rome, November 10 1874 | New-York Hotel

Mr Tyndall,

I just read in Les Mondes[2] the magnificent speech that you delivered at the opening of the British Association sessions, and I am so happy about it, that I would like to take the opportunity to show you my deepest gratitude.

It is beautiful, in fact, to see a man like you, who arrived at the verge of his glory, in a highly religious and conservative country, risking his reputation and his rest for the love of truth, and who risks it all without hesitation, directly, without some of those excuses that the skilled can use so well for caution.

Me too, Mr Tyndall, I struggled for many years for the triumph of truth, and that is why I rejoiced when I suddenly saw such a powerful ally and such a universally esteemed one as yourself.

The mail will deliver to you, with this letter, my own speech: The Laws of Nature,[3] composed and published in 1869, for the reopening of the course at the University of Turin. This writing had been given a lot of thought; perhaps it is too brief[4] but it is clear enough for the educated men to whom it is addressed, raised a general cry of condemnation in the religious camp. Father Moigno did me the honour of sermonizing extensively in Les Mondes (VI of the year 1869 T. XIX Janu.–April, Pages 291–300)[5] without duplicating my speech because it was much easier for him to attack sentences rather than reply to the arguments. I did not think that I should respond to the insults from the abbot. I even appreciated that he quibbled about my thesis, because all truth will eventually come out, when we speak about it, even if only to slander. And I see that I did well to save myself from an unnecessary debate, since the idea has made its way onward and found an advocate as great and as eloquent as you.

The religious sector continues to pursue me with all of its resources and with the strongest of its strategies. They have raised against me part of the youth in Turin, all my work has been shrouded in silence, I was prevented from being the Chair of Physics at the University of Rome,[6] the means to fund for me a chair of History of Physics[7] by the minister Scialoja[8] have been removed, and after having deluded me with empty promises for almost two years, one of these days they will perhaps put me politely out the door. This, however, does not discourage me nor does it prevent me from pursuing my rationalist propaganda. I was poor in my youth, and I might have to live like that again in the future, but I will keep my beliefs and I will fight until my last breath against the absurd doctrines of a past that does not wish to fall, and which seeks to defend itself by all means possible.

I am telling you this, Mr Tyndall, only to warn you against what awaits you and so that you will not get caught off guard. Just as people from the religious side manage to make a great man out of a mediocre mind by always repeating everywhere and to everyone, that there is no higher intelligence than his; similarly they are capable of reversing even the most established reputation through slander. But time is a great dispenser of justice and we can rely upon it.

However, you will have to endure a lot of unfair attacks and outrageous omissions. The notes that Father Moigno made about your speech[9] are only preliminary attacks. However clumsy and incompetent they might seem to you, they are of no less than a sagely and saintly reputation to some people, and you should not wait for more serious objections. Perhaps the Protestants will take a more respectful and less contemptuous tone in your city[10] ... perhaps they will not treat you as a second-hand scholar, but they will not attack you any less relentlessly or unreasonably.

When there will be a lot of trouble, it is good to rest assured that there are also friends whom you can count on ... and here is why I wanted to write you this letter and to send you my speech.

So please believe my sincerest commitment, and in any event be sure to make use of your devoted Gilbert Govi | Profess[or] of Phys[ics] at the University of Turin | Interim Director of the Casanate Library[11] in Rome.

RI MS JT/1/G/11

---

1. *Gilbert Govi*: Gilbert Govi (1826–89), an Italian physicist who had been professor of physics at the University of Turin since 1860. He fought in the Sardinian army in both of its wars of independence in 1848 and 1859, and had spent time in France as a political exile.
2. *Les Mondes*: 'Association Britannique pour l'Avancement des Sciences: Discours présidential de M. Tyndall trans. F. Moigno', *Les Mondes*, 35 (1874), pp. 325–97 (see letter 4399).
3. *The Laws of Nature*: G. Govi, *Le leggi della natura* (Torino: Stamperia Reale, 1868).
4. *too brief*: Govi's published speech was 52 pages.
5. *Father Moigno did ... Pages 291–300)*: F. Moigno, 'Philosophie des sciences', *Les Mondes*, 19 (1869), pp. 291–300.
6. *prevented from being the Chair of Physics at the University of Rome*: Pietro Blaserna (1836–1918) had been appointed instead in November 1872 (G. H. B., 'Prof. P. Blaserna', *Nature*, 101 (1918), p. 287).
7. *a chair of History of Physics*: in November 1872, Blaserna, along with chemist Stanislao Cannizzaro (1826–1910) and Paolo Volpicèlli (see letter 4642, n. 1), had written to the University of Rome's rector Filippo Serafini (1831–97) urging him to establish an academic post in the history of physics. Although Serafini agreed, and steps were taken toward the establishment of the post, bureaucratic delays and internal university politics meant that by November 1874 Govi had abandoned his hopes of being appointed. See A. Borrelli and E. Schettino, 'La prima cattedra di storia della fisica in Italia: un'occasione mancata', *Scienza e politica*, 33 (2005), pp. 75–110.
8. *the minister Scialoja*: Antonio Scialoja (1817–77), an Italian politician who was minister of public education in the governments of Giovanni Lanza and Marco Minghetti.
9. *The notes that Father Moigno made about your speech*: the notes by Moigno appended to

his translation of 'Association Britannique pour l'Avancement des Sciences: Discours présidential de M. Tyndall' in *Les Mondes* (see n. 2).

10. *your city*: presumably London, but Govi might mean Belfast.
11. *the Casanate Library*: an historic library in Rome, formally called the Biblioteca Casanatense, named after Girolamo Casanate (1620–1700). It was founded in 1701 and nationalized in 1872, with Govi appointed the interim director in November 1873.

To Olga Novikoff 11 November [1874][1] 4474

Royal Institution of Great Britain | 11th. Nov[ember].

My dear Friend

I will put my name on the photograph[2] and carry it to you tomorrow myself.

Is it possible to get a glimpse of the "Einleitung in die Philosophie"[3] which first brought upon Frohschammer the indignation of the Jesuits?—I should like much to see it. The Jesuits have not heard the last of Frohschammer believe me

Yours ever | John Tyndall

RI MS JT/1/TYP/3/930
LT Typescript Only

1. *[1874]*: the year is established by reference to Jakob Frohschammer, whom Tyndall and Novikoff were discussing at this time.
2. *the photograph*: see letter 4469, n. 8. By putting his name on it, Tyndall presumably meant signing the photograph with his autograph, ahead of it being sent to Frohschammer.
3. *the "Einleitung in die Philosophie"*: J. Frohschammer, *Einleitung in die Philosophie und Grundriß der Metaphysik* (München: J. G. Cotta, 1858).

To Olga Novikoff 12 November [1874][1] 4475

Royal Institution of Great Britain | 12th Nov[ember].

My dear Friend.

Not until last night was I able to read the articles by Professor Riehl,[2] but last night I read them from beginning to end. They are the utterances of a strong, clear and cultivated mind. And surely I have reason to be proud that those scraps of mine[3] should have attracted so much of the attention of such a man.

I will tell you my circumstances, and allow you to decide when I am to come to see you. I could come any time this afternoon between 2.30 and 5.30.

This evening from 8 oC[lock]. until a late hour I shall be in the city at Sion College[4] in the midst of 3 or 400 parsons. So that this renders it impossible for me to call after dinner.[5] But if you wish I will do so tomorrow.

Perhaps you would kindly give me one line stating what you will permit me to do.

Yours ever | John Tyndall

[Envelope] Madame de Novikoff | Symonds's Hotel | 34 Brook Street.

John Tyndall Correspondence in the Olga Novikoff Correspondence Collection. Courtesy of Special Collections, Kenneth Spencer Research Library, University of Kansas. MS 30:Q7
RI MS JT/1/TYP/3/930

1. *[1874]*: the year is established by reference to Alois Riehl's articles (see n. 2).
2. *the articles by Professor Riehl*: see letter 4452, n. 5.
3. *those scraps of mine*: Riehl's articles were a two-part review of the German translation of *Fragments of Science* (see letter 4192, n. 11).
4. *Sion College*: an ecclesiastical college, founded in 1630, for the incumbent clergy of Anglican parishes in the City of London. Although there is no record of Tyndall's visit on this occasion, the *Church Herald* had previously noted that 'persons whose opinions and unbelief are totally at variance with the Christian Religion, have been from time to time invited to Sion College, to propound those opinions and that unbelief before an assembly mainly composed of clergyman of the National Church'. It complained that the meetings in the 'new series for 1873–1874' would make this 'venerable and useful institution . . . a propaganda for anti-Christian principles' ('"The Philosophers" at Sion College', *Church Herald*, 4 (1873), p. 737). See also letter 4561.
5. *impossible for me to call after dinner*: in letter 4474 Tyndall promised to bring Novikoff a photograph, but he was evidently unable to fulfil the obligation and so sent it instead (see letter 4478).

From François Moigno [12 November 1874][1] 4476

Mon cher Monsieur Tyndall.

J'arrive épuisé au terme de la course vertigineuse dans laquelle vous m'avez entraîné; j'aurai cependant encore assez de force pour vous faire entendre une voix amie.

Comment, sur le beau tapis vert des Hautes-Alpes, sous cette belle voûte bleue dont vous nous avez révélé le secret, inondé des douces clartés des neiges éternelles si blanches et si pures, échauffé par ces gloires aux rayons cramoisis dont le soleil couchant couronne les sommets qui le dérobent au regard, avez-vous eu le triste courage d'organiser ce cortége effréné des novateurs de tous les pays, cette danse échevelée des témérités de tous les âges!

Obéissiez-vous à un plan arrêté longtemps d'avance?

Aviez-vous mission de déchaîner la tempête qui, après s'être montrée comme un point noir sur le ciel de Norwick, s'était accentuée de plus en plus à Exeter, à Édimbourg, pour devenir l'ouragan de Belfast?

Je l'ai cru, je j'ai craint, et je n'ai pas osé franchir les deux détroits de la Manche et de la mer d'Irlande, tremblant d'assister à une bataille et non plus à un triomphe.

Je n'avais pas cependant perdu tout espoir d'avoir à redire après vous une de ces brillantes épopées d'observations, de faits, d'expériences que le monde entier admire.

Et voici que vous m'avez condamné à faire avec vous de la métaphysique et de la polémique religieuse. Vous avez prêché l'évolution: j'ai été forcé de prêcher la création, au risque d'importuner et de fatiguer mes chers lecteurs. Vous avez exercé votre apostolat de destruction: j'ai du exercer mon apostolat de conservation.

Mais il en est temps, abandonnez pour toujours ce terrain, qui n'est pas le vôtre.

Je n'ai pas votre talent admirable, votre génie! Mais j'ai plus appris que vous, et l'universalité de mes études, jointe au calme d'une vie abritée par l'autorité, a défendu ma tête bretonne de la cérébration inconsciente.

L'évolution est le rêve scientifique du panthéiste, qui pour s'émanciper de Dieu consent à n'être qu'un nuage léger prêt à se fondre dans l'azur infini du passé, selon vos propres paroles. Vous vous rappelez les angoisses de l'infortuné qui, dans le conte d'Hoffmann, court à la recherche de son reflet, dont il se voit partout dépouillé. Bien plus grandes, croyez-moi, seront les angoisses de la pauvre âme qui a consenti à se laisser absorber par le grand tout de la nature; car, sa personnalité abdiquée, il faudra bien qu'elle la reprenne quand elle se verra face à face avec son premier principe et sa fin dernière, avec Dieu, qu'elle voudra fuir épouvantée, si elle le force de lui apparaître dans sa justice, et qui l'aurait faite éternellement semblable à lui, s'il lui avait permis de lui apparaître dans sa gloire.

F. Moigno.

My dear Mr. Tyndall.

I am exhausted at the end of the breathtaking journey on which you have taken me; however, I still have enough strength to let you hear a friendly voice.

How, on the beautiful green carpet of the Hautes-Alpes, under the beautiful blue vault whose secret you revealed,[2] flooded with the soft radiance of eternal snow so white and so pure, warmed by the glory of crimson rays in which the setting sun crowns the summits that hide it from sight, have you had the sad courage to organise this frantic procession of innovators from all the countries, this wild dance of temerity of all ages![3]

Do you comply with a plan that was defined well in advance?

Was it your mission to unleash the storm which, after showing up as a black spot in the sky of Norwich, had become worse and worse in Exeter, in Edinburgh, and turned into a hurricane in Belfast?[4]

I believed it, I feared it, and I did not dare to cross the two straits of the Channel and the Irish Sea, fearing that I would witness a battle instead of a triumph.

However, I did not give up all hope of retelling one of your brilliantly epic observations, facts, and experiences that the whole world admires.

And now you have condemned me to do metaphysics and religious controversy with you. You preached about evolution: I was forced to preach about creation, at the risk of annoying and boring my dear readers.[5] You have exercised your apostolate of destruction: I had to exercise my apostolate of preservation.

But the time has come, leave forever this field,[6] which is not yours.

I do not have your admirable talent, your genius! But I have learned more than you, and the universality of my studies, together with the peacefulness of a life sheltered by authority, protected my Breton head from unconscious cerebration.[7]

Evolution is the scientific dream of a pantheist,[8] who, to emancipate himself from God, agrees to be like a light cloud ready to fade into the infinite azure of the past, according to your own words.[9] You remember the anguish of the unfortunate who, in Hoffmann's tale,[10] runs off in search of his reflection, of which he is always deprived. There will be much more anguish, believe me, for the poor soul that agrees to become absorbed by the larger whole of nature; because, having abdicated its personality, it will have to take it on again when it will go to face its first principle and its final end, with God, and it would want to flee in terror, if it would force him to appear in front of it in his righteousness, and who would have made it eternally similar to him, if he was allowed to appear in his glory.

F. Moigno.

Les Mondes, 35 (1874), pp. 412–13

1. *[12 November 1874]*: the date is given by the date of publication of this letter in *Les Mondes* (see n. 5).
2. *the beautiful blue vault whose secret you revealed*: in J. Tyndall, 'On the Blue Colour of the Sky, the Polarization of Skylight, and on the Polarization of Light by Cloudy Matter Generally', *Roy. Soc. Proc.*, 17 (1869), pp. 223–33.
3. *this frantic procession . . . all ages!*: in his presidential address to the BAAS at Belfast, Tyndall had traced the history of atomistic scientific thought to its roots in ancient Greece, discussing an array of historical figures including Democritus, Lucretius, Giordano Bruno and Pierre Gassendi.

4. *a black spot in the sky of Norwich . . . a hurricane in Belfast*: Moigno refers to the meetings of the BAAS at Norwich in 1868, Exeter in 1869, Edinburgh in 1871, and Belfast in 1874. At the latter, he suggests, the 'black spot' of incipient materialism in Tyndall's address as president of the Mathematics and Physics section at Norwich has, six years later, become the 'hurricane' of his presidential address. On Tyndall's address in Norwich, see letter 4458, n. 4.
5. *my dear readers*: Moigno's letter was published in his journal for 12 November 1874 (*Les Mondes*, 35 (1874), pp. 412–13).
6. *this field*: presumably of religion and metaphysics.
7. *unconscious cerebration*: a term coined by the prominent English physiologist and devout Unitarian William Benjamin Carpenter (1813–85) in the 1850s to describe the process by which the unconscious mind can spontaneously produce logical conclusions below the plane of consciousness.
8. *a pantheist*: an adherent of the view that God is immanent in or identical with the universe, and that God is everything and everything is God (*OED*). In his presidential address, Tyndall suggested that 'Some form of pantheism was usually adopted by . . . Men of warm feelings and minds open to the elevating impressions produced by nature as a whole, whose satisfaction, therefore, is rather ethical than logical'. He also proposed that Giordano Bruno was 'a "Pantheist", not an "Atheist" or a "Materialist"' ('Address of John Tyndall', *Brit. Assoc. Rep. 1874*, pp. lxvi–xcvii, on pp. lxxvii and xciii). See R. Barton, 'John Tyndall, Pantheist: A Rereading of the Belfast Address', *Osiris*, 3 (1987), pp. 111–34.
9. *a light cloud ready to fade . . . your own words*: a paraphrase of the original ending of Tyndall's presidential address: '. . . when you and I, like streaks of morning cloud, shall have melted into the infinite azure of the past' (p. xcvii).
10. *Hoffmann's tale*: 'Die Geschichte vom verlorene Spiegelbild' (The Story of the Lost Reflection, 1814) by Ernst Theodor Amadeus Hoffmann (1776–1822).

From Louis Pasteur 12 November 1874 4477

Paris le 12 nov. 1874

Mon cher monsieur Tyndall,

Très fier et très heureux de la bonne nouvelle que m'a annoncée la lettre du Président Hooker, je vous remercie bien cordialement de l'empressement que vous avez mis à me la confirmer. Vous ne sauriez croire le prix que nous attachons sur le Continent à la médaille de Copley à cause de l'ancienneté de son origine et des noms des savants illustres auxquels elle a été décernée.

Veuillez agréer l'assurance nouvelle de ma vive sympathie et de mon entier dévouement.

L. Pasteur

>

Paris, 12 Nov[ember]. 1874

My dear Mister Tyndall,

Very proud and very happy of the good news that I received in the letter from President Hooker,[1] I thank you very cordially for the eagerness that you put towards confirming it. You would not believe the price that we place on the Copley Medal on the Continent, because of its ancient origin and the names of the distinguished scholars to whom it was awarded.[2]

Please do accept the new assurance of my heartfelt sympathy and of my entire devotion.

L. Pasteur

RI MS JT/1/P/6

1. *the good news . . . letter from President Hooker*: as president of the RS, Joseph Hooker conferred on Pasteur the Copley Medal for 1874, received on Pasteur's behalf by the RS's foreign secretary Alexander William Williamson. His letter probably contained the eulogy to Pasteur's researches given in his 'Presidential Address', *Roy. Soc. Proc.*, 23 (1874), pp. 49–73, on pp. 68–70. See also letter 4470.
2. *its ancient origin . . . to whom it was awarded*: the Copley Medal, named after the politician Godfrey Copley (1653–1709) and given by the RS for outstanding achievements in scientific research, was first awarded in 1731. Its recipients included Benjamin Franklin (1753) and François Arago (1825), and, more recently, Alexander von Humboldt (1852) and Charles Darwin (1864).

To Olga Novikoff [*c.* 13 November 1874][1] 4478

<*start of letter missing*>

I am pale and wan, but my work must be accomplished, and I have now only five days in which to wind it up.[2]

I send you the photograph[3] with my compliments to your friend[4]

Yours ever | John Tyndall

John Tyndall Correspondence in the Olga Novikoff Correspondence Collection. Courtesy of Special Collection, Kenneth Spencer Research Library, University of Kansas. MS30:Q18

1. *[c. 13 November 1874]*: the date is suggested by the relation to letters 4474 and 4475 (see n. 3).

2. *work must be . . . to wind it up*: probably writing 'Preface to the Seventh Thousand', in *Belfast Address*, 7th thousand, pp. v–xxxii. Tyndall sent Novikoff an incomplete draft five days later, complaining that he was 'ill and weary' (letter 4487), and only completed it in early December (see letter 4501).
3. *the photograph*: see letter 4469, n. 8. Tyndall had promised to himself bring the photograph to Novikoff in letter 4474, but letter 4475 suggests he never made the visit. As such, he seems to have sent it with this letter instead.
4. *your friend*: probably Jakob Frohschammer.

To William Spottiswoode 13 November [1874][1] 4479

13th Nov[ember]

My dear Spottiswoode.

Busy bodies have been occupying themselves about Faraday's statue,[2] and Mr Teniswoode,[3] Foley's executer, has been here today. I desired him to write to you.

The statue is finished.[4] It appears that Mr Foley's will[5] has produced litigation. Chancery proceedings[6] will soon be begun and Mr Teniswoode thinks it would save trouble if the statue could be approved by the Committee[7] and taken away from the studio[8] within the coming fortnight. He wishes to avoid having to apply for an order, which would be necessary if the statue remained in the studio until proceedings had begun.

Mr. Leighton (whom Bence Jones regarded as a busy-body) speaks of the British Museum.[9] Such was one decision—but it was hardly formed when a reconsideration of it was thought necessary. Mrs. Faraday's feelings[10] were thought to be consulted by the decision; but Mrs. Faraday is too sensible to meddle in a matter which it is the province of the subscribers to decide.

I suppose the proper way will be to call the committee together on as early a day as possible.

Ever faithfully yours | John Tyndall

Westminster Abbey, as near Newton[11] as possible is the place for Faraday.[12]

RI MS JT/1/T/1372
RI MS JT/1/TYP/3/1292

---

1. *[1874]*: the year is established by reference to the repercussions from the death of John Henry Foley (see n. 4 and n. 5).
2. *Faraday's statue*: on 21 June 1869, a public meeting at the RI had resolved to erect a monument to Michael Faraday funded by public subscription, with the Irish sculptor John Henry Foley (1818–74) being commissioned to create a full-length marble sculpture.

3. *Mr Teniswoode*: George Francis Teniswood (1820–92), a painter.
4. *The statue is finished*: Foley completed only a sketch model and the head of the statue before his sudden death on 27 August 1874. The remainder of the sculpture was finished by his pupil Thomas Brock (1847–1922).
5. *Mr Foley's will*: Foley's will left his studio and practice to Brock, his casts to the Royal Dublin Society and much of the rest of his property to the Artists' Benevolent Fund.
6. *Chancery proceedings*: these began with a bill of complaint submitted to the Lord Chancellor in the Court of Chancery in London. The court, which had heard civil equity suits since 1558 but had become notorious for bureaucracy and mismanagement, was finally dissolved in 1875.
7. *the Committee*: of the Faraday Memorial Fund, whose members included George Bentham (1800–84), Thomas Huxley, Austen Henry Layard (1817–94), Edward Sabine, Tyndall, and Alexander William Williamson.
8. *the Studio*: John Henry Foley's studio was in Osnaburgh Street, Marylebone, close to Regent's Park.
9. *Mr. Leighton . . . the British Museum*: John Leighton (1822–1912) was an artist and member of the RI who, after several requests to Faraday and Henry Bence Jones, had given a Friday Evening Discourse on Japanese art in May 1863 (*Roy. Inst. Proc.*, 4 (1862–66), pp. 99–108). Despite his only negligible acquaintance with Faraday (and perhaps justifying Bence Jones's barbed opinion of him), Leighton wrote a letter to the *Times* on the Faraday memorial, insisting that he was 'one who as a monitor and friend has cause to venerate the great philosopher'. He recalled 'giving several suggestions' to the 'able sculptor Foley'—with whom he also claimed to be 'intimate'—about the poses in which his purported friend might be portrayed. In the same letter, Leighton proposed that, rather than a more grandiose location such as St Paul's Cathedral, 'in deference to the simple feelings of the great practical philosopher, the British Museum' would be a 'site more appropriate' for the statue ('The Faraday Memorial', *Times*, 13 November 1876, p. 6). Tyndall later reflected in his journal on how 'that extremely unpleasant busybody Leighton . . . has been pushing himself forward with the greatest effrontery since 1851' (21 April 1884, RI MS JT/2/19).
10. *Mrs. Faraday's feelings*: Sarah Faraday had initially been unhappy at using public subscriptions to fund her husband's statue.
11. *Westminster Abbey, as near Newton*: the monument to Isaac Newton (see letter 4300, n. 4) in Westminster Abbey, created in 1730 by the Flemish sculptor John Michael Rysbrack (1694–1770), stands in the Abbey's nave, north of the entrance to the choir.
12. *place for Faraday*: the statue did not end up at Westminster Abbey, nor St. Paul's Cathedral, another location that was proposed, but was placed, and remains, at the RI, which, as William Frederick Pollock noted in his memoir, was very appropriate as this was 'where Faraday's work was done, and in whose service he spent the greater part of his life' (*Personal Remembrances of Sir Frederick Pollock*, 2 vols (London: Macmillan, 1887), vol. 2, p. 207).

From William Henslow Hooker[1] 13 November 1874 4480

Kew, 13 Nov[ember]. 1874.

Dear Dr Tyndall

I am very sorry to have to inform you of our sad loss. My mother[2] expired suddenly this afternoon. She was perfectly well this morning but at about noon she was seized by a fit, and was quite unconscious, she expired within two hours. My father was in town at the time and found her gone when he arrived home.

Believe me, | dear Dr Tyndall, | Yours very aff[ectiona]tely | W. Hooker.

RI MS JT/1/TYP/8/2724
LT Typescript Only

1. *William Henslow Hooker*: William Henslow Hooker (1853–1942), eldest son of Joseph Hooker, who later became an economist.
2. *My mother expired suddenly this afternoon*: Frances Hooker died on 13 November 1874, and was interred in the Hooker family vault at St Anne's Church in Kew on 18 November.

From Daniel Oliver[1] 13 November 1874 4481

Kew, 13 Nov[ember]. 1874.

My dear Sir

You will be deeply grieved to hear of the very sad affliction which has befallen dear Dr Hooker this afternoon.

Mrs Hooker died in his absence (in town) about 3.0'clock after some two hours' illness.[2]

Dr Hooker returned about 4.30. He left her apparently in excellent health this morning.

Very sincerely yours | D. Oliver.

RI MS JT/1/TYP/8/2723
LT Typescript Only

1. *Daniel Oliver*: Daniel Oliver (1830–1916), a botanist and keeper of the herbarium of the Royal Botanic Gardens, Kew from 1864 to 1890 (*ODNB*). Oliver had prepared *Lessons in Elementary Biology* (London: Macmillan, 1864), based upon unpublished material left by

Frances Hooker's father John Stevens Henslow, and illustrated by her sister Anne Henslow Barnard.

2. *Mrs Hooker died . . . two hours' illness*: see letter 4480, n. 2.

To Frederick Alexander McMinn[1] 14 November 1874 4482

17, Buckingham Street, Adelphi,[2] Nov[ember]. 14, 1874.

We[3] have to thank you for your letter of the 2nd of November,[4] giving us the information asked in our letter to you of the 12th of October,[5] in reference to the increase your company[6] are making in their purifying apparatus.[7] We were desirous to obtain complete information on the facts, before reconsidering the maximum of sulphur impurity we have prescribed, and we are happy to express our satisfaction with the measures your company are taking to increase the purity of the gas supplied by them, and our sense of the difficulties which have delayed the execution of those measures.

Having now the facts correctly before us, we have carefully considered your request; and we may say at once that if we were convinced that the retention of the maximum at the present amount would press so hardly on the company as to be likely to subject them to fine through the fact of the accidental non-completion of the new apparatus, we should be disposed to make a relaxation in their favour. But on carefully considering the data before us we cannot come to that conclusion. In the first place, looking at the sulphur returns of last winter, we find that, although the maximum then in operation was 30 grains, the actual amount did not often exceed 25 grains. The principal exception occurred at The Haggerston works[8] in December, and at Fulham[9] during the other half of the same month.

In the following months, owing, no doubt, to increased experience in the management of lime, although the make of gas had scarcely at all diminished, a maximum of 28 grains was never exceeded at Haggerston, and on three occasions only at Fulham. At the St. Pancras works[10] the same maximum was exceeded twice only in each of the four winter months. Since that time two advantages have accrued to the company, which should enable them to supply gas of a somewhat higher degree of purity than last winter—first, the opening of the spacious works at Bromley,[11] which enables them to transfer a portion of their make from works, the demand on which has outgrown the means of purification, to a place where every facility should exist for the reduction of sulphur to the lowest point; and, secondly, the experience gained not only from the experiments specially instituted by the company on the subject (as named in your letter), but from a year's working of a method of purification, which, even where imperfectly applied, reduces the amount of sulphur considerably below 25 grains. A comparison of the returns of the present year with those of the corresponding period of last year shows very much improved results, which (as there has been no increase of purifying apparatus brought into use at the urban stations) we can only attribute to increased knowledge, care, and skill in the mode of purification.

In the month of October, 1874, the maximum amount of sulphur returned was—

	Grains.
At Fulham	17·6
At St. Pancras	19·3
At Haggerston	18·2
At Bromley	15·4

(excepting two days, on which they were quite anomalous, and evidently accidental).

In November the maxima have been—

At Fulham	16·1
At St. Pancras	13·1
At Haggerston	17·1
At Bromley	10·3

The averages have been as follows:—

	Oct., 1874.		Nov., 1874.
Fulham	15·6	...	14·8
St. Pancras	15·4	...	12·0
Haggerston	15·2	...	15·6
Bromley	10·4	...	7·3
General average	14·1		12·4
The average for the corresponding months of last year were	18·2	...	20·5

Hence it is clear that your managers[12] are employing their existing purifiers much more successfully than at the corresponding periods of last year, and although the make is approaching its maximum they are keeping a long way below the prescribed maximum of sulphur. At Bromley, judging by the very low sulphur returns, there can be no reasonable doubt of the sufficiency of the purifying capacity; and at St. Pancras four out of six new purifiers are so far advanced that they may certainly be used before the depth of the winter. These two works yield nearly half the supply.

From these data we cannot doubt the power of your engineers, by skill and care, to keep during the winter within the present maximum under ordinary circumstances. It must be borne in mind that if it should be exceeded, owing to extraordinary contingencies, you have your power of appeal, and we have no doubt that in such a case the chief gas examiner[13] would take into consideration any *bonâ fide*[14] pleas on your part founded on temporary difficulties arising out of the non-completion of the new apparatus.

You must, we are sure, be aware that the raising of the maximum would be viewed with great displeasure by the public, and it is consequently a step we could not resolve on, except under more urgent necessity on the part of the companies then we at present see. The public may reasonably expect a lowering of the amount of sulphur, now that the practicability of keeping it down has been established. And we hope that next year we shall be able to fix a lower maximum than heretofore.

We take this opportunity of expressing our satisfaction at finding that purification by lime has been carried on at Fulham during the summer without a repetition of the former complaint of nuisance to the neighborhood. We were confident, at the time you brought this matter before us, that such a result might be achieved by care and skill on the part of your managers, and we are pleased to find that our anticipations have been realized.

Sulphur Compounds in Gas. Report of the Proceedings before a Committee of the House of Commons on the Crystal Palace District Gas and The Gaslight & Coke Company Bills. Session 1877 (London: William B. King, 1877), pp. 151–2

---

1. *Frederick Alexander McMinn*: Frederick Alexander McMinn (1842–1932), chief engineer of the Imperial Gas-Light and Coke Company's gasworks at Fulham. He had worked at the gasworks since *c.* 1856, and lived at Sandford Manor House in Fulham.
2. *17, Buckingham Street, Adelphi*: the Office of the Metropolitan Gas Referees in central London.
3. *We*: the Metropolitan Gas Referees, who were appointed by the Board of Trade under the City of London Gas Act (1868). In August 1872, Tyndall and Augustus George Vernon Harcourt (1834–1919), a chemist at Oxford University, replaced J. H. Patterson and John Sampson Pierce, while William Pole (1814–1900), an engineer and musician, had replaced Frederick John Evans (1815–85) in 1870. Tyndall served in this position until 1892, Pole until his death in 1900, and Harcourt until 1917 (K. Hutchison, 'The Royal Society and the Foundation of the British Gas Industry', *Notes and Records: The Royal Society Journal of the History of Science*, 39 (1985), pp. 245–70, on p. 265; *Sulphur Compounds in Gas. Report of the Proceedings before a Committee of the House of Commons on the Crystal Palace District Gas and The Gaslight & Coke Company Bills. Session 1877* (London: William B. King, 1877), p. 87).
4. *your letter of the 2nd of November*: letter missing.
5. *our letter to you of the 12th of October*: letter missing.
6. *your company*: the Imperial Gas-Light and Coke Company, which was founded in 1821 and supplied North London with coal gas and coke.
7. *their purifying apparatus*: cast-iron cylinders containing lime suspended in water through which gas is forced at high pressure to remove sulphur.
8. *The Haggerston works*: the gasworks located near the Regents Canal in Haggerston, close to Shoreditch, in East London. They were established in 1823.

9. *Fulham*: the gasworks near to the River Thames at Sands End, Fulham, in West London. They were established in 1824.
10. *the St. Pancras works*: the gasworks located near the near the Regents Canal in St Pancras in North London. They were established in 1824.
11. *the spacious works at Bromley*: the gasworks located near the River Thames at Bow Creek, Bromley-by-Bow, in East London, which were established in 1873 at a cost of £300,000.
12. *your managers*: the managers, or chief engineers, of the individual gasworks.
13. *the chief gas examiner*: Henry Letheby (1816–76), an analytical chemist whom the Board of Trade had appointed Metropolitan Chief Examiner of Gas in 1860 (*ODNB*).
14. *bonâ fide*: in good faith (Latin).

From Olga Novikoff [15 November 1874][1] 4483

Sunday.

Dearest Friend,

On y[ou]r leaving me last night, I went to my bed room, where I found my maid & a nice, bright fire. I sent away my maid, but made the fire still nicer & brighter; then I put myself on the floor, à la turque,[2] stirring at the flames & thinking of you a long, long time, interrupting the stream of my thoughts only on perceiving that I was not thinking of you alone, that I was carried away too much, that I allowed myself to play a certain part in the destiny so precious to me, but so different from mine.—At last the coals were out, the candles as well—& I had to go to bed in perfect darkness! Shall I tell you what some of my anxieties are?? I am afraid, that, in referring to several religious questions, the idea of wounding some of y[ou]r friends' feelings will prevent y[ou]r speaking out y[ou]r mind as well, as you might do. Most anything you utter is heard by all the reading & the thinking world; & if y[ou]r friends can not see how much kindness there is in perfect frankness, let them discover it at last, though slowly. Yes, I entreat you, be entirely y[ou]rself. You could not be better!—Y[ou]r religious enemies took hold of what you said about "y[ou]r healthier & stronger thoughts about God"[3] (as they explain it) & will see a recantation[4] now in any & every thing you'll say!—When I remember what I felt in reading y[ou]r address,[5] I sympathise now more than ever with the great numbers of people, unknown to you but living upon the bread you give them. Let the faith of those unknown but united be complete in you. Truth is superior to all y[ou]r friends, & must go first!—

Pardon my writing you all this. I am nothing but the poor Molière's kitchen maid, who always said her impressions openly when she spoke to her Master.[6]—I shall be curious to know whether you have Frohschammer's "Das Recht der eigenen Überzeugung"[7] or not? I think it's a little gem. As to the Einleitung in der Ph.[8] it is—I think—the weakest thing he ever wrote.

—Why did you speak of "pledges" yesterday? Indeed—I am the very last person in the world, to insist upon my claims even if I ever had any.—What is the prospect I may have for tomorrow? There—value this riddle, & accept the share of my affection you are kind enough to care for.

Y[ou]r Olga

RI MS JT/1/N/41

1. *[15 November 1874]*: the date is suggested by reference to a book by Jakob Frohschammer (see n. 8) which Tyndall had enquired about in letter 4474, dated 11 November. 15 November was the following Sunday.
2. à la *turque*: in the Turkish style (French); sitting cross-legged with the lower parts of the legs folded towards the body.
3. *"y[ou]r healthier & stronger thoughts about God"*: see letter 4421, n. 2.
4. *see a recantation*: see letter 4421, n. 3.
5. *y[ou]r address*: Tyndall's presidential address to the BAAS at Belfast, delivered on 19 August.
6. *the poor Molière's kitchen maid . . . spoke to her Master*: Dorine, the outspoken and shrewd maid in the theatrical comedy *Tartuffe* (1664) by Molière, the stage name of Jean-Baptiste Poquelin (1622–73).
7. *Frohschammer's "Das Recht der eigenen Überzeugung"*: J. Frohschammer, *Das Recht der eigenen Überzeugung* (Leipzig: Fues, 1869). Novikoff later recalled of Tyndall: 'One evening he came and found me alone, or, rather, only in company with several books just received from a great Munich friend of mine,—Professor Frohschammer. To my surprise Tyndall knew nothing either of the books or of the life of that remarkable scholar; he asked me to give him a general idea of both, not an easy task, to be sure. But I marked several passages in the work on *The Right of Independent Conviction* (*Das Recht der eigenen Ueberzeugung*)' (*M.P. for Russia*, vol. 1, p. 151–2).
8. *the Einleitung in der Ph.*: J. Frohschammer, *Einleitung in die Philosophie und Grundriss der Metaphysik, zur Reform der Philosophie* (München: J. G. Cotta, 1858); Tyndall had asked 'Is it possible to get a glimpse of the "Einleitung in die Philosophie"' in letter 4474.

To Olga Novikoff 15 November [1874][1] 4484

15th Nov[ember]

Many thanks dear Friend for your most excellent letter[2]—Trust me—I shall commit no act of weakness unworthy of your friend.[3] I am strong because I stand upon the truth. What I meant to express last night was a desire not to unnecessarily lacerate tender minds. But trust me for the truth as far as I see it.

Yours ever | John Tyndall

APS Mss.509.L56
RI MS JT/1/TYP/3/931

1. *[1874]*: the year is suggested by the relation to letter 4483.
2. *your most excellent letter*: letter 4483.
3. *your friend*: Jakob Frohschammer.

From Thomas Archer Hirst [16 November 1874][1] 4485

My dear John

I called half an hour ago just to say three words

<u>Keep your temper</u>

If I replied at all to the nasty little letter from Beale in this mornings Times[2] I would simply chaff[3] him. Neither he nor it is worth any serious consideration or remonstrance whatever

Ever yours 'Tom'

Poor Morell is pegging away in an insane manner at me in a monthly Educational Journal.[4] I shall let him have his say undisturbed.

RI MS JT/1/H/490

1. *[16 November 1874]*: the date is established by reference to a letter in the *Times* for that day (see n. 2).
2. *the nasty little letter from Beale in this mornings Times*: 'Fever Germs', *Times*, 16 November 1874, p. 10. In his letter, Lionel Smith Beale (1828–1906), professor of pathology at King's College London, entered the debate on the 'nature and origin of fever poisons', contending that they were not of the 'nature of a microscopic fungus originating without', but were instead a 'living particle arising within, and from the living matter of man's body itself'. His advocacy of this zymotic theory was in opposition to the germ theory of disease advanced by Tyndall in his own letters to the *Times* (see letters 4463 and 4472). See P. A. Richmond, 'Some Variant Theories in Opposition to the Germ Theory of Disease', *Journal of the History of Medicine and Allied Sciences*, 9 (1954), pp. 290–303, on pp. 299–301.
3. *chaff*: to banter, rail at, or rally, in a light and non-serious manner, or without anger, but so as to try the good nature or temper of the person 'chaffed' (*OED*).
4. *Poor Morell . . . a monthly Educational Journal*: this seems to have been a dispute over the teaching in schools of Euclid's conceptions of proportion and ratio, for which John Reynell Morell (1821–91) suggested a simplified syllabus. Hirst had discussed proposals for the teaching of Euclid in his presidential address to the Association for the Improvement of Geometrical Teaching, which was published as 'The Geometrical Association', *Monthly Journal of Education*, 1 (1874), pp. 219–21.

To Mary Egerton 18 November [1874][1] 4486

Royal Institution of Great Britain | 18th. Nov[ember].

My dear Lady Mary

I was stunned on Saturday[2] morning by the account of Mrs Hooker's death. I went out to Kew[3] in the afternoon, and on Tuesday[4] I also went out and shared their evening meal, and tried, I hope with success, to lessen the weight of woe oppressing them all. I go this morning to see her buried.[5]

On Friday morning she was perfectly well, strong and vivacious. Hooker left her to come to town. She did not answer the luncheon bell, and was found when looked after on the drawing room floor. She was able to speak feebly. The servants were sent about doctors, and the dear child[6] to whom you refer had to hold her mother in her arms. It was a fearful trial for the child, for there was a good deal of convulsion. She died in two hours.

Hooker was away and no body knew where to send a message to him. He returned knowing nothing of what had happened. But people had been placed at the gates to intercept him and thus the matter was made known to him.

He is a man of the tenderest nature; and he loved his wife deeply—the shock is commensurate. But it pleases me to see that he dwells on the happiness of her life; and I hope with his work, and his children,[7] that he will successfully pull through.

Yours ever | John Tyndall

RI MS JT/1/TYP/1/399
LT Typescript Only

1. *[1874]*: the year is established by reference to the death of Frances Hooker (see letter 4480, n. 2).
2. *Saturday*: 14 November.
3. *Kew*: Hooker's residence at 49 Kew Green.
4. *Tuesday*: 17 November.
5. *see her buried*: see letter 4480, n. 2.
6. *the dear child*: presumably Grace Ellen Hooker (1868–1955) as Egerton specifies that the child was female, and Hooker's other daughter, Harriet Hooker, was twenty at the time, and so unlikely to be regarded as a child.
7. *his children*: Joseph Dalton Hooker's other children were William Henslow Hooker (see letter 4480, n. 1), Charles Paget Hooker (1855–1933), Brian Harvey Hodgson Hooker (1860–1932), and Reginald Hawthorn Hooker (1867–1944).

To Olga Novikoff 18 November [1874][1] 4487

Royal Institution of Great Britain, | 18th. Nov[ember].

Dear Friend

I send you as much of the preface[2] as I have been able to get ready. It is not finally corrected, and I would therefore beg of you not to allow it to fall into any other hands than your own. You will return it to me when you have read it?

Tonight I am entangled—tomorrow night also: the Royal Society meets for the first time[3] and I have pledged myself to be there.[4] Saturday and Sunday[5] I shall be away from London. I could call to see you at 5.30 any day before Saturday and on Friday I do not know that I shall be fettered in the evening.

I am ill and weary—and rendered a little sad by my work today.

Yours ever | John Tyndall

RI MS JT/1/TYP/3/931
LT Typescript Only

1. *[1874]*: the year is established by reference to Tyndall's preface (see n. 2).
2. *the preface*: 'Preface to the Seventh Thousand', in *Belfast Address*, 7th thousand, pp. v–xxxii. The published version of the preface was signed 'Athenæum Club: *December* 5, 1874'; p. xxxii. For Novikoff's recollections of her own contributions to the preface, see letter 4469, n. 4.
3. *the Royal Society meets for the first time*: Tyndall means the first time this season, as the RS had last met on 18 June, before the summer recess.
4. *I have pledged myself to be there*: Tyndall presumably wished to show support for Joseph Hooker. He was the RS's president, but, following the death of his wife on 13 November, the meeting on 19 November was chaired by William Spottiswoode, the treasurer, who proposed a 'Resolution' which was 'unanimously agreed to "That the Royal Society desire to condole with their President for his loss, and to express to him their deep sympathy in his great affliction"' ('*November* 19, 1874', *Roy. Soc. Proc.*, 23 (1874), p. 1). See letter 4480, n. 2.
5. *Saturday and Sunday*: 21 and 22 November.

From Olga Novikoff [18 November 1874][1] 4488

Thursday.[2] 8 o clock P.M.

Hearty thanks. I have read your proofs[3] twice, slowly, each time trying to find fault with you. I do not see any single word to be omitted, but if you want to avoid absurd comments, many explanations have to be added.—However—I have marked the pages and when you call on me tomorrow, or the day after, I shall explain my impressions better than I can do it now.[4]

RI MS JT/1/TYP/3/931
LT Typescript Only

1. *[18 November 1874]*: the date is suggested by the relation to letter 4487 (see n. 2).
2. *Thursday*: this seems to be an error on Novikoff's part, and should instead be Wednesday. This letter is a brief preliminary response to letter 4487, which Tyndall dated 18 November, before the more detailed reply in letter 4490 which was written at 8am on Thursday 19 November.
3. *your proofs*: of 'Preface to the Seventh Thousand', in *Belfast Address*, 7th edn, pp. v–xxxii.
4. *I shall explain . . . than I can do it now*: Novikoff in fact explained her comments the following morning in letter 4490.

To William Spottiswoode 19 November [1874][1] 4489

19th Nov[ember]

My dear Spottiswoode.

Did I tell you that Huxley has promised to give us a short course on the work of the Challenger?[2]

D^r^. Richardson[3] who was here yesterday would I fancy be glad to give you a Friday ev[enin]g.[4] but we must talk about the matter

Yours ever | John Tyndall

RI MS JT/1/T/1370

1. *[1874]*: the year is established by reference to Thomas Huxley's lecture on HMS *Challenger* (see n. 2).
2. *Huxley has promised . . . work of the Challenger*: Huxley delivered a Friday Evening Discourse, 'On the Recent Work of the "Challenger" Expedition, and Its Bearing on Geological Problems', at the RI on 25 January 1875 (*Roy. Inst. Proc.*, 7 (1873–75), pp. 354–7). HMS *Challenger* undertook an expedition of nautical scientific research, sponsored by the RS, from 1873–6.
3. *D^r^. Richardson*: probably the physician and sanitarian Benjamin Ward Richardson (1828–96), who may have spoken with Tyndall about his recent work on typhoid (see letters 4463 and 4472).
4. *a Friday ev[enin]g*: the RI's weekly Friday Evening Discourses, which, since 1825, had

attracted large audiences made up of the RI's members and their guests. Richardson does not seem to have been invited to give one, and only spoke at the RI when it hosted a meeting of the Sanitary Institute in July 1877.

From Olga Novikoff [19 November 1874][1] 4490

Thursday. 8 o'c[lock]. AM.

Dearest friend,

Y[ou][r] answer is admirable! It is merciless (w[hic][h] is only just) but conveys throughout the largeness & generosity of mind, w[hic][h] are y[ou][r] characteristics. A man on his deathbed, w[ou][ld] burst out of laughter, if, instead of an ad profundis[2] they read to him what you say of the Bishop of Manch[ester]:[3] The idea of quoting the passages, omitted at first[4]—is capital. The name of "enlightened heathens"[5] will be accepted by many as a flag, I am sure. I read the proofs twice,[6] slowly each time trying to be as antagonistic to everything you say as possible. According to my impression not one single word ought to be omitted; but, if you want to avoid tedious attacks—great many explanations have to be added. Will you also not refer to the silly interpretations given to the word "scepticism"?[7] A word, the meaning of w[hic]h is clear to mere school boys, & w[hic][h] brilliantly illustrates the ignorance of some of y[ou][r] opponents.—the words "I prayed"—"the world to come" "I avow myself an Atheist in reference to any definition" etc[8] . . . might be—I think—more explained now—in order not to be bothered afterwards.—

In speaking of the extravagances of the religious world, you might give some more instances—it w[ou]l[d] impress more the reader.

—In referring to the persons who share y[ou]r views, I think, You ought to speak with more decisiveness of "the weary souls who would welcome them with gratification & relief".[9] In fact—I am sure you have had already more than one token of sympathies of that sort. It now strikes me whether, on the other hand, it w[ou] l[d] not be better to avoid the word "thousands."[10] However—I have not thought it over & only give you a hasty impression.

What you say of the boundaries of scientific experiments is as clear, as what you say of Genesis is grand![11] It is a great, great treat that you are giving to the world by y[ou][r] straightforward, splendid utterances!—I dare say, you'll smile at my candid & daring criticisms. You know already so well my weak sides that I give myself as I am without any reserve!—

Y[ou][rs] Olga.

RI MS JT/1/N/36

1. *[19 November 1874]*: the date is established by relation to letter 4487, which Tyndall dated 18 November. 19 November was a Thursday.
2. *ad profundis*: into the depths (Latin).
3. *what you say of the Bishop of Manch[ester]*: James Fraser (1818–85), the Anglican Bishop of Manchester from 1870–85 (*ODNB*). In his second preface to the published version of the Belfast Address, Tyndall stated: 'Soon after the delivery of the Belfast Address the able and respected Bishop of Manchester did me the honour of noticing it; and in reference to that notice a brief and, I trust, not uncourteous remark was introduced into my first Preface. Since that time the Bishop's references to me have been very frequent. Assuredly this is to me an unexpected honour. Still a doubt may fairly be entertained whether this incessant speaking before public assemblies on a profoundly emotional subject does not tend to disturb that equilibrium of head and heart which it is always so desirable to preserve—whether, by giving an injurious predominance to the feelings, it does not tend to swathe the intellect in a warm haze, thus making the perception, and consequent rendering of facts, indefinite, if not untrue' ('Preface to the Seventh Thousand', in *Belfast Address*, 7th thousand, pp. v–xxxii, on p. x).
4. *passages, omitted at first*: in his first preface, Tyndall observed that he would 'pass over a recent sermon attributed to the Bishop of Manchester with the remark, that one engaged so much as he is in busy and, I doubt not on the whole, beneficent outward life, is not likely to be among the earliest to discern the more inward and spiritual signs of the times, or to prepare for the condition which they foreshadow' ('Preface', in *Belfast Address*, pp. v–viii, on p. vi). In the second preface, he quoted a passage from 'the "Times" of November 9' with the 'grave, misrepresentation, contained in the Bishop's last reference to me' ('Preface to the Seventh Thousand', p. x).
5. *"enlightened heathens"*: speaking of his religious critics, Tyndall lamented 'the waste of energy on the part of good men over things unworthy, if I might say it without discourtesy, of the attention of enlightened heathens' ('Preface to the Seventh Thousand', p. xiii).
6. *I read the proofs twice*: Novikoff here expands on the brief comments given in letter 4488.
7. *not refer to . . . given to the word "scepticism"*: the passage referring to these interpretations seems to have been omitted from the preface.
8. *the words "I prayed" . . . to any definition" etc*: these precise expressions do not appear in the published 'Preface to the Seventh Thousand', although there might be revised versions of the last two on p. v ('. . .recent remonstrances, appeals, menaces, and judgments—covering not only the world that now is, but that which is to come'), and p. viii ('My critic is very hard upon the avowal in my Preface regarding Atheism').
9. *"the weary souls who . . . gratification & relief"*: a slightly altered version of the line, albeit without any additional decisiveness, appears on p. xvii of 'Preface to the Seventh Thousand'.
10. *better to avoid the word "thousands."*: the word is not used by Tyndall in the published 'Preface to the Seventh Thousand', so was presumably omitted.
11. *What you say of the boundaries . . . Genesis is grand*: Novikoff presumably refers to passages

in the 'Preface to the Seventh Thousand' where Tyndall asserts that 'crossing the boundary of the experimental evidence . . . is the habitual action of the scientific mind—at least of that portion of it which applies itself to physical investigation' (p. xiv), and that 'the Book of Genesis has no voice in scientific questions . . . It is a poem, not a scientific treatise. In the former aspect it is for ever beautiful: in the latter aspect it has been, and it will continue to be, purely obstructive and hurtful' (p. xvi).

To Olga Novikoff 20 November [1874][1] 4491

Royal Institution of Great Britain, | 20th. Nov[ember].

I was just thinking of writing to you—The proofs[2] have come, and most thankful am I for your observations.[3]

I shall take my chance of seeing you this evening at 3 or thereabouts.

Yours ever | John Tyndall

I have not read the work to which you refer.[4]

RI MS JT/1/TYP/3/931
LT Typescript Only

1. *[1874]*: the year is established by the relation to letter 4483.
2. *The proofs*: of 'Preface to the Seventh Thousand', in *Belfast Address*, 7th thousand, pp. v–xxxii, which Novikoff had been reading.
3. *your observations*: see letters 4488 and 4490.
4. *the work to which you refer*: possibly J. Frohschammer, *Das Recht der eigenen Überzeugung* (Leipzig: Fues, 1869), of which Novikoff had said 'I shall be curious to know whether you have Frohschammer's "Das Recht der eigenen Überzeugung" or not?' in letter 4483.

To Olga Novikoff [21 November 1874][1] 4492

Royal Institution of Great Britain. | Saturday.

I will think over the Preface[2] but it has nearly reached its limit.

I do not think that I shall be in London on Monday; and indeed I think it on the whole best I should not be here.[3]

Goodbye and may all fair breezes,[4] and kindly thoughts attend you.

Yours ever | John Tyndall

RI MS JT/1/TYP/3/928
LT Typescript Only

1. *[21 November 1874]*: the date is suggested by the relation to letters 4487 and 4490, and by this letter seeming to have been written on the Saturday before Novikoff's departure from London, which was on Monday 23 November (see n. 3).
2. *Preface:* 'Preface to the Seventh Thousand', in *Belfast Address*, 7th thousand, pp. v–xxxii. Tyndall would presumably think over the detailed suggestions that Novikoff made in letter 4490.
3. *I do not think that . . . best I should not be here*: Novikoff was leaving London to return to Russia on Monday 23 November (see letter 4495); that Tyndall considered it best that he not be there for her departure might suggest how close their relationship had become.
4. *all fair breezes*: for Novikoff's sea journey to Ostend in Belgium, from where she would travel by train to Brussels (see letter 4495).

From Elizabeth Dawson Steuart — 21 November 1874 — 4493

Steuart's Lodge | Leighlinbridge | Nov[ember]. 21st 74.

My dear John,

No words can express my indignation against the infamous paragraph in that vile paper,[1] which has been, (or will be) sent to you, and as I know Caleb is not in favor with you,[2] I write a line to say, that if you think well of taking legal steps in the matter, you will be perfectly safe in the hands of my excellent friend Mr Thorp, than whom a more upright honorable gentleman does not exist. He is very cautious, and never allows a client to go to him, without a good chance of success: in this case his first step would be to lay it before first rate Counsel,[3] and on his opinion to proceed or not. If it be a case of libel the parties deserve punishment richly and it would be only right to give it to them, besides acting as an example to others.

I write in great haste

my dear John | ever sincerely yours | E. D. Steuart.

RI MS JT/1/TYP/10/3417
LT Typescript Only

1. *the infamous paragraph in that vile paper*: 'The Catholic University of Ireland', *Tuam Herald*, 14 November 1874, p. 2. The paragraph printed a pastoral to be 'read at all the Churches throughout the Diocese of Achonry' on Sunday 15 November, issued by Patrick Durcan and Francis Joseph MacCormack, who shared the Bishopric of Achonry. The pastoral stated that 'When a degenerate Irishman (Tyndall) has selected Irish soil for his platform—and brought the danger home to us by ventilating at our doors his startling theories

of Evolution and Materialism—It behoves us, the divinely-constituted guardians of Faith and Morals, to watch closely the interests of Pure Christian Education'. The parishioners of Achonry were also urged to assist in 'warning off the youth of the country from the poisoned waters' by contributing to the funds of the Catholic University of Ireland.

2. *Caleb is not in favor with you*: Caleb Tyndall was Tyndall's paternal uncle, although there had been animosities between Tyndall's father and his brother. Tyndall had recently declined to help 'young Caleb Tyndall', Caleb's son, with the purchase of a 'threshing machine' (see letter 4432), which might have enhanced Steuart's sense that Tyndall would not ask Caleb for advice on appropriate legal representation.
3. *Counsel*: a barrister, who would present the case in a courtroom.

To Olga Novikoff [*c.* before 23 November 1874][1] 4494

What I heard from you about your poor Munich friend,[2] and all I found in his works,[3] has so deeply impressed me that I beg you to forward him this cheque[4] as a little help and as a token of my sympathy.

M.P. for Russia, vol. 1, pp. 153–4

---

1. *[c. before 23 November 1874]*: the approximate date is suggested by Novikoff's comment that Tyndall sent her the cheque 'a little before my departure for Russia' (see n. 4), which was on 23 November.
2. *your poor Munich friend*: Jakob Frohschammer.
3. *all I found in his works*: Tyndall had declared 'I shall read all that he has written' in letter 4457 at the start of November; for details of the works of Frohschammer that he seems to have read, see letters 4457, 4458, 4469, 4474, and 4483.
4. *this cheque*: Novikoff later recalled that Tyndall had sent her a 'cheque of one hundred pounds sterling', saying it was meant 'as a little help and as a token of my sympathy' for Frohschammer. She then recounted: 'I was naturally much touched by such spontaneous generosity, but knowing Frohschammer's independent character I returned the cheque . . . This happened a little before my departure for Russia. My direct route lay via Berlin, but having some misgivings about the part I had played in this transaction I resolved to return home via Munich, in order to see Frohschammer, which I did. I told him about the £100 cheque, and the liberty I took in depriving him of so large a sum. "That is quite impossible! I have never had such a big sum in all my life", interrupted he. No sooner had I finished my story than he exclaimed, "Thank you heartily for having anticipated my feelings. Yes, Tyndall has done me the greatest kindness I could have desired, which I will always remember. You have understood me quite well"' (*M.P. for Russia*, vol. 1, pp. 153–4).

From Olga Novikoff 25 November [1874][1] 4495

Nov[ember]. 25. Wednesday. Bruselles

I am starting off for Munich this afternoon. Shall I not find a letter from you there? This w[ou]ld be really dreadful. I am so sick of my Bruselles swells,[2] & long for Munich where I shall speak about you with Frohschammer.[3] Here they read nothing & know nothing. On my arriving here[4] I sent for letters at the Embassy.[5] My arrival thus became known, & the very day lots of people came, offering luncheons, dinners, boxes, parties, tedious discussions about tedious questions. They were all talking last night about love & devotion &c &c &c. All were shouts—I alone silent like a mute, & amongst all these jokes & silly "witticisms" feeling as if I were drowning & sick. At last somebody exclaimed "But what is the matter with Mme N? We are talking all this time, she alone is not uttering a word". Well I said—The fact is—I have lived all my life not admitting the very existence of love & have therefore no theory about it".—great indignation & commotion . . . as to me—I was glad I could merely become a listener once more.—Oh, do write to me! Don't give me up entirely.—If I had the courage to bother you more with my biography, you w[ou]ld see, that I am not so bad as you think. I am dreadfully alone my nearest connections have other ties, & every thing is avowed to me.—People call certain avowals cynical—but there is at least no comedy & gives me the moral right to feel myself free—

Y[ou][rs] "absolutely" | Olga

RI MS JT/1/N/34

1. *[1874]*: the year is established by reference to Jakob Frohschammer, whom Tyndall and Novikoff were discussing at this time.
2. *swells*: fashionably or stylishly dressed people; hence, people of good social position, or highly distinguished people (*OED*).
3. *Munich where I shall speak about you with Frohschammer*: Novikoff later recalled of the route of her return to Russia: 'My direct route lay *via* Berlin, but . . . I resolved to return home *via* Munich, in order to see Frohschammer, which I did' (*M.P. for Russia*, vol. 1, p. 154). On why she wanted to speak with Frohschammer about Tyndall, see 4494, n. 4.
4. *On my arriving here*: presumably on the evening of 23 November, having travelled from London to Belgium via the sea crossing to Ostend.
5. *the Embassy*: presumably the Russian embassy.

From Elizabeth Dawson Steuart — 25 November 1874 — 4496

Steuart's Lodge, | Leighlinbridge. | Nov[ember]. 25. 74.

My dear John,

I believe there is not the least doubt that the paper[1] is under the patronage of the R.C.[2] Priests, supported by them and the R.C. farming class, in opposition to all that is good, upright, or honorable. I might be able to ascertain all about it only it never ceases raining, and I cannot get out. But any information you may require, it is hard if I do not make it out. How proud I should be if I could help you!

I received the odious paper from you this morning, and shall send it this night to Mr Thorp who is in Dublin[3] and may want it there. I forwarded your last[4] to him also to show your wishes.

ever your true old friend | E. D. S.

RI MS JT/1/TYP/10/3418
LT Typescript Only

1. *the paper*: the *Tuam Herald* (see letter 4493, n. 1).
2. *R.C.*: Roman Catholic.
3. *Mr Thorp who is in Dublin*: Charles Thorp attended the Dublin Commission Court at certain times of the year (and would soon become the sessional crown solicitor for County Carlow in this court), at which time he worked from offices at 60 Lower Dominick Street.
4. *your last*: letter missing.

To James MacLehose[1] — 27 November [1874][2] — 4497

27th Nov[ember].

I am very much obliged to the publisher of Dr. Caird's Lecture,[3] for his courteous remembrance of me.[4] I have heard the lecture spoken of as the ablest utterance hitherto urged against me.

John Tyndall

APS Mss.509.L56

1. *James MacLehose*: James MacLehose (1811–85), a bookseller and publisher based at 61 St Vincent Street, Glasgow, who, in 1871, became publisher to the city's University.

2. *[1874]*: the year is established by reference to John Caird's lecture (see n. 3).
3. *Dr. Caird's Lecture*: J. Caird, *The Unity of the Sciences: A Lecture Delivered at the Opening of the Winter Session of the University of Glasgow* (Glasgow: James MacLehose, 1874). Caird (1820–98) was a Presbyterian theologian who had been appointed as principal of the University of Glasgow in 1873. At the end of the lecture, which was delivered on 3 November, he quoted Tyndall's comments in the Belfast Address that religion was a 'force . . . capable of being guided to noble issues in the region of emotion', before concluding: 'But, if this be the only God to whom science points, and with the notion of whom it proposes to meet and satisfy the infinite aspirations and inextinguishable longings of the spirit of man, the boon it brings is one for which I, for one, cannot pay it the poor tribute of my gratitude. I cannot bow before this blank inscrutability, of whom you help me neither to affirm nor deny anything, and for whom, therefore, I can feel no rational reverence' (p. 37).
4. *his courteous remembrance of me*: the letter which MacLehose presumably sent to Tyndall, along with a copy of Caird's lecture, is missing.

From Harriet Hooker 29 November 1874 4498

Kew | Nov[ember]. 29th 1874.

My dear Dr Tyndall

Will you please be so kind as to come and spend Christmas with us; we shall be so very glad if you can come and stay with us for it; we shall be all at home, but both Papa[1] and I want someone whom we know well to come too, to help to cheer us up,[2] we will try and not make it a very dismal time, so please will you come. Papa would like it so much.

Papa is not very well, he has a little touch of bronchitis, I hope this milder weather will send that away though. Hoping you will send a favourable answer to this

I remain with love from us all, | Yours aff[ectiona]tely | Harriet Hooker.

You know you have only to say which day you can come and your room shall be ready.

RI MS JT/1/TYP/8/2725
LT Typescript Only

---

1. *Papa*: Joseph Hooker.
2. *help to cheer us up*: Frances Hooker, Harriet's mother, had died suddenly on 13 November (see letter 4480, n. 2).

To Joseph Dalton Hooker 30 November [1874][1] 4499

Royal Institution of Great Britain | 30th Nov[ember].

My dear Hooker—

Let me express to you once again the great enjoyment which your address[2] gave me. It was a credit to you and to the Society.[3]

Ever aff[ectionatel]y | John Tyndall

Royal Botanic Gardens, Kew, Joseph Dalton Hooker Collection, vol. 4, 1270–70a

1. *[1874]*: the year is established by reference to Hooker's address (see n. 2).
2. *your address*: Hooker's presidential address to the anniversary meeting of the RS earlier that day, published as 'President's Address', *Roy. Soc. Proc.*, 31 (1874–5), pp. 50–74. Although written by Hooker, the address was actually delivered by William Spottiswoode, probably because Hooker was still recovering after the sudden death of his wife on 13 November (see letter 4480, n. 2).
3. *the Society*: the RS.

From Olga Novikoff [*c.* late November 1874][1] 4500

Let your English readers know him[2] and his works through your kind introduction. Now that everybody is reading your Belfast address, and that you are printing its sixth edition,[3] could you add a few lines about him?[4]

M.P. for Russia, vol. 1, p. 154

1. *[c. late November 1874]*: the date is suggested by Tyndall's reply, letter 4501, which is dated 2 December 1874.
2. *him*: Jakob Frohschammer.
3. *your Belfast address . . . printing its sixth edition*: *Belfast Address*. Novikoff uses edition in the sense of a printing of a thousand copies. The first six printings of the address remained unchanged from the original, with Tyndall making alterations and additions in the seventh thousand (see n. 4).
4. *add a few lines about him*: Tyndall replied in letter 4501, 'I have taken care to refer to Professor Frohschammer', which he did in the 'Preface to the Seventh Thousand' of *Belfast Address*, 7th thousand, pp. v–xxxii, on p. xxiii. On Tyndall's account of Frohschammer in the preface, see letter 4469, n. 4.

To Olga Novikoff 2 December 1874 4501

Royal Institution of Great Britain | 2nd Dec[ember]. 1874

My dear Friend.

I had hoped to be able to send you my preface[1] earlier; but it has been slow work, for my head has not been in the best possible condition. It is now nearly ended and I hope to be able to send it to you within the coming week.

I have taken care to refer to Professor Frohschammer,[2] with whose courage, earnestness and profound knowledge I have been greatly impressed. The preface will no doubt augment my difficulties and bring down upon me a new discharge of wrath.

I hope you will make known to your friend Frohschammer, that the Fragments of Science[3] are really nothing more than Fragments; struck off from time to time in the intervals of harder work. My principal work in life has been scientific enquiry, and what I have done here is embodied in work so purely scientific and technical that it did not occur to me to present them to you when you were here.

The Fragments are rather an indication of my intellectual sympathies than a contribution to the thought of the age. They have had however some influence in England.

I was rejoiced to learn a day or two ago that Mr. Kinglake was quite restored. He had been riding with a friend of mine[4] in the Park.[5]

With regard to Switzerland I have not yet made my plans and you know I am rather a wild wanderer among the heights: For years I have desired to visit Munich, and the knowledge that Professor Frohschammer is to be found there will be a spur to me to do so at no distant day.

Believe me | with best wishes | Most truly Yours | John Tyndall

[Envelope] Madame Olga de Novikoff | Poste Restante[6] | Munich | Bavaria

[Postmark] LONDON·W | 1 | DE 2 | 74

John Tyndall Correspondence in the Olga Novikoff Correspondence Collection. Courtesy of Special Collections, Kenneth Spencer Research Library, University of Kansas. MS 30:Q8

RI MS JT/1/TYP/3/932

1. *my preface*: 'Preface to the Seventh Thousand', in *Belfast Address*, 7th thousand, pp. v–xxxii (see letters 4487 and 4491).
2. *refer to Professor Frohschammer*: see letter 4469, n. 4.
3. *Fragments of Science*: Tyndall had arranged for a copy of *Fragments of Science* to be sent to Jakob Frohschammer in Munich (see letter 4469).

4. *a friend of mine*: not identified.
5. *the Park*: Hyde Park, where Alexander William Kinglake lived at 28 Hyde Park Place.
6. *Poste Restante*: see letter 4392, n. 7.

From Emily Peel 2 December 1874 4502

Drayton Manor, | Tamworth. | Dec[embe]r. 2/74

Dear Professor Tyndall

On my return from Switzerland I have found an author's copy of y[ou]^{r}. address at Belfast.[1] I am much touched by your remembrance of me, for out of sight, out of mind is evidently no rule of yours.—Won't you come here next week & renew your acquaintance with Drayton & its inmates?[2] It would give us great pleasure to see you on any day that might suit you.—

I tried my strength at mountaineering this season in Switzerland—& found out alas! that 10 years will tell upon one's strength. Yet after the ascent of the Buet & the Dent du Midi[3] I found myself getting into fine training & bad weather alone stopped me in my proposed ascent of M^{t}. Blanc. When worthy of the distinction you must enroll me amongst your Alpine Clubbists,[4] if such an honour is ever accorded to one of the weaker sex!

Sir Robert[5] begs to join in best regards.

Yours sincerely | Emily Peel

RI MS JT/1/P/65

---

1. *an author's copy of y[ou]r. address at Belfast*: *Belfast Address*.
2. *Drayton & its inmates*: the Peel family had owned Drayton Manor since 1790, and the Prime Minister Robert Peel had it rebuilt in the early 1840s. Its 'inmates' at this time included Emily's children: Agnes Helen (1860–1964), Victoria Alexandrina Julia (1865–1935), Robert (1867–1925), Evelyn Emily (1869–1960), and Gwendolin Cecilia (b. 1872).
3. *the Buet & the Dent du Midi*: mountains in, respectively, the French and Swiss Alps.
4. *your Alpine Clubbists*: the Alpine Club had been founded in 1857 and its members met at the Metropole Hotel in London. Women were not permitted to join (and would not be until 1974), and instead a Ladies' Alpine Club was created in 1907. On women and alpine mountaineering, see A. C. Colley, *Victorians in the Mountains: Sinking the Sublime* (New York: Routledge, 2010), pp. 101–42.
5. *Sir Robert*: Robert Peel (1822–95), Emily's husband and son of the former Prime Minister (*ODNB*).

To Emily Peel 5 December 1874 4503

Royal Institution of Great Britain | 5th Dec[ember]. /74

Dear Lady Emily,

Assuredly not "out of sight out of mind,"[1] and could I at all manage it nothing would give me greater pleasure than the adding together both of sight and mind by going down to Drayton Manor.[2] But this Belfast affair[3] has got me into so many entanglements and has so invaded my proper work, that its demands upon me now are inexorable.

I am very much delighted with your account of your achievements in Switzerland.[4] The "ten years" appear to have missed their usual action in your case. That you should have accomplished the Buet,[5] and thought of Mont Blanc, is to me wonderful. Well, if ever you climb the Monarch of Mountains[6] I hope I may be there to see.

I wish you from my heart all the good things, ghostly and bodily, of the coming festive season. I spend my Christmas with my friend Hooker, who has had such a calamity in the loss of his wife[7] and helper. He needs sustainment, and I must try what can be done to promote this.

Always yours most truly, | John Tyndall.

RI MS JT/1/TYP/3/970
LT Typescript Only

1. *"out of sight out of mind"*: see letter 4502.
2. *Drayton Manor*: see letter 4502, n. 2.
3. *this Belfast affair*: the controversy provoked by Tyndall's presidential address to the BAAS at Belfast, which he had delivered on 19 August.
4. *your account of your achievements in Switzerland*: in letter 4502.
5. *the Buet*: see letter 4502, n. 3.
6. *the Monarch of Mountains*: Mont Blanc, as described in Lord Byron's *Manfred* (1816), I.i.62.
7. *his wife*: Frances Hooker, who died suddenly on 13 November (see letter 4480, n. 2).

To Mary Egerton 8 December 1874 4504

Royal Institution of Great Britain | 8th. Dec[ember].

My dear Lady Mary

With a hot and tired brain I fly from London this day to Brighton, and stay there, for I have work to do, for a week. Cottrell is coming down to me with the Laboratory notes,[1] so that I may improve the time. Under these circumstances you see how difficult it would be for me to join you on Friday.[2]

It would have given me great pleasure to meet "May"[3] at Kingsley's[4] and I grieve to hear that they are not well.[5] I met him in St. James' St[reet].[6] some time ago, and was much concerned to note his changed appearance—America did not agree with him.[7]

I do not know where Clerk Maxwell's lecture was published[8]—He sent me a separate copy—but I cannot lay my hands on it—otherwise you should have had it.

I am glad you like the paper in the Contemporary[9]—Unhappily you must take me as I am, and the Contemporary only embodies one side of me.

I spend the Christmas with dear Hooker. He needs cheering up and I must try to meet this emergency.[10]

Kindest regards to all. | Ever Yours | John Tyndall | 1874.

RI MS JT/1/TYP/1/398
LT Typescript Only

---

1. *the Laboratory notes*: possibly concerning the 'facility with which sounds pass through textile fabrics', on which Tyndall submitted to the RS on 4 January 1875 an addendum to his paper 'On Acoustic Reversibility'. The addendum gave an 'experimental elucidation of this subject', and Tyndall commented: 'In these experiments my assistant, Mr. Cottrell, has rendered me material assistance' (J. Tyndall, 'On Acoustic Reversibility', *Roy. Soc. Proc.*, 23 (1875), pp. 159–65, on pp. 164 and 165).
2. *Friday*: 11 December.
3. *"May"*: presumably Mary Alice Egerton (May being a pet name for Mary).
4. *Kingsley's*: the home of Charles Kingsley (see letter 4286, n. 10), at 56 Old Church Street, Chelsea.
5. *they are not well*: Kingsley was suffering from pneumonia, and would die on 23 January 1875, while his wife, Frances Eliza Kingsley (1814–91), had a heart attack on 3 December from which she was initially not expected to survive.
6. *St. James' St[reet].*: in Mayfair, directly across Piccadilly from Albemarle Street.
7. *America did not agree with him*: Kingsley had given an extensive lecture tour in the United States from January to August (see letter 4286), but he contracted pleurisy in San Francisco in June and had to be treated for several weeks in Colorado Springs. His wife later recounted that Kingsley looked 'altered and emaciated' in the autumn of 1874 (F. E. Kingsley, *Charles Kingsley, His Letters and Memories of his Life*, 2 vols (London: C. Kegan Paul, 1878), vol. 2, p. 449).
8. *where Clerk Maxwell's lecture was published*: presumably J. C. Maxwell, 'Molecules', *Nature*, 8 (1873), pp. 437–41, which was based on a lecture James Clark Maxwell had delivered on 23 September 1873 at the BAAS meeting at Bradford. Tyndall discussed and criticized Maxwell's lecture in his presidential address in Belfast a year later.

9. *the paper in the Contemporary*: J. Tyndall, 'On the Atmosphere in Relation to Fog-Signalling', *Contemporary Review*, 24 and 25 (1874), pp. 819–41 and pp. 148–68, which was published in November and December.
10. *this emergency*: the death of Frances Hooker, who had died suddenly on 13 November (see letter 4480, n. 2).

From Olga Novikoff 10 December [1874][1] 4505

Thursday night | Decem[ber]. 10 | Novem[ber]. 28.[2]

On my arriving here[3]—going from the railway to my hotel, I stopped at the best library to inquire, whether y[ou]r address[4] had been translated into Russian? They gave me a scientific almonack called "Nature"[5] where it was to be got, & my first occupation on my reaching Moscow tomorrow will be to compare the translation with the original. I also got a translation of y[ou]r delightful Switzerland climbing[6]—w[hic]h made me forget all the fatigue of my journey, & will give it to my boy[7]—who, I hope will deserve y[ou]r kindness one day, as he is a very fine, intelligent, & straightforward boy indeed. & one of the best pupils of the classical Lyceum[8] where he is now, for Greek, Latin & Mathematics. My great wish is to make of him a man of science—if he'll keep what he promises now to be, but is it possible to make anything "one likes one" self from another living being? I wish I had more faith in the importance of education! As yet I only believe in self-education, & this even......

But I must not bother you too much today.—As I told you already
<rest of letter missing>

RI MS JT/1/N/37

1. *[1874]*: the year is established by reference to Tyndall's Belfast Address (see n. 4).
2. *Novem[ber]. 28*: Novikoff also gives the date according to the Julian calendar that was then used in Russia.
3. *here*: possibly Minsk, then a Russian city, which was the main juncture on the railway to Moscow. It was also somewhere from which Moscow could be reached in a day, and Novikoff indicates that she expects to arrive there tomorrow.
4. *y[ou]r address*: Tyndall's presidential address to the BAAS at Belfast, delivered on 19 August.
5. *a scientific almonack called "Nature"*: not identified.
6. *a translation of y[ou]r delightful Switzerland climbing*: probably Альпийские ледники [Alpine Glaciers] (Moskva: Izd. A. I. Glazunova, 1866), a translation of *Glaciers of the Alps*.
7. *my boy*: Alexander Novikoff, Novikoff's only child.
8. *the classical Lyceum*: probably the Imperial Katkov Lyceum in Moscow, which was founded in 1867.

To Eliza Spottiswoode 12 December [1874][1] 4506

Brighton | 12th December

Need I tell my dear friend how profoundly I sympathise with her in her great sorrow?[2]

John Tyndall

I caught the sad intelligence today in one of the newspapers.[3]

RI MS JT/TYP/1/1289

1. *[1874]*: the year is established by reference to the death of Spottiswoode's father (see n. 2).
2. *her great sorrow*: Spottiswoode's father William Urquhart Arbuthnot (b. 1807) died on 11 December.
3. *sad intelligence today in one of the newspapers*: reports of Arbuthnot's death appeared in 'Summary of This Morning's News', *PMG*, 12 December 1874, p. 6; and 'Death of Mr. W. U. Arbuthnot', *Times*, 12 December 1874, p. 9.

From Henrietta Huxley 13 December 1874 4507

4 Marlborough Place | Abbey Road N.W.[1] | 13th Dec[ember] 1874.

My dear Brother John,

Every day since I came home[2]—this autumn—I have mentally written a letter, to say how kind I thought it of you to have the children[3] to dinner—and to talk to them about so many interesting things. But—pressing daily affairs always prevented me from doing more than this. Fast friends I think know all that passes in each other's mind and therefore you will have known what I felt and thought, and needed no letter from me, to tell it you. It is a grand thing when you can always take up your friends like a book—where you last left off with them.

Do not forget to come to us on New Years Day at ¼ to 7. My heart aches when I think of poor Dr. Hooker's Christmas.[4]

Sir Charles Lyell has had a nasty fall head foremost down stairs,[5] but I hear is going on well.

Always | very affectionately Yours | Nettie Huxley

Leonard was 14—on the 11th. He has got some chemical apparatus—and he and Harry[6] spend all their spare moments in experimenting.

RI MS JT/1/TYP/9/3043
LT Typescript Only

1. *4 Marlborough Place | Abbey Road N.W.*: see letter 4213, n. 1.
2. *came home*: presumably from the summer holiday that the Huxley family usually took in August, although it is not clear where the family went in 1874.
3. *the children*: Jessica Oriana, Marian, Leonard, Rachel, Henrietta, Henry, and Ethel. There is no record of when Tyndall had invited them to dinner.
4. *poor Dr. Hooker's Christmas*: Joseph Hooker's wife, Frances Hooker, died suddenly on 13 November (see letter 4480, n. 2).
5. *Sir Charles Lyell . . . head foremost down stairs*: the *Times* later recorded that Lyell's death on 22 February 1875 was 'accelerated by a fall downstairs' and reported the evidence given at the inquest into his death: 'James Moss, the butler, said that on the 9th of December last . . . he was informed by the maid that Sir Charles had fallen downstairs. He found the deceased sitting in a chair, having a bruise on the forehead. He helped him to bed'. Lyell had 'fallen down about ten stone steps', and, according to his physician, sustained a 'severe contusion over the right eye, and a slight dislocation of the left thumb' ('The Late Sir Charles Lyell', *Times*, 27 February 1875, p. 5).
6. *Leonard . . . and Harry*: see n. 3. Harry is the short form of Henry.

To Eliza Jane White[1] 14 December 1874 4508

Royal Institution of Great Britain | 14th December, 1874.

DEAR MADAM,—

I thank you very much for the beautiful lines on nature[2] which you have been good enough to send me.[3]

The "disinheriting"[4] is sure to be made good in some way to him on whom the penalty has fallen because of his loyalty to the truth. ––I am, dear madam, yours faithfully,

JOHN TYNDALL

Derry Journal, 21 December 1874, p. 4

1. *Eliza Jane White*: Eliza Jane White (née Cameron, *fl.* 1862–1901), an Irish poet who advocated republican and feminist views, and would later declare herself an atheist. She published under the pseudonym Ida White, and was married to George White, editor of the *Ballymena Observer*.
2. *the beautiful lines on nature*: White had sent Tyndall a poem entitled 'Nature (From the Tyndall Standpoint)', which was written shortly after Tyndall delivered his presidential

address to the BAAS at Belfast on 19 August. It was published in the *Ballymena Observer*, 19 September 1874, p. 1, and later in I. White, *The Three Banquets and Prison: Poems* (London: Swan, Sonnenschein, 1890), pp. 182–4. White's original letter, to which this letter was a reply, is missing.

3. *good enough to send me*: letter missing.
4. *The "disinheriting"*: in the final stanza of the poem, White wrote, seemingly in the voice of Tyndall addressing nature: 'Then if I am thy work, and thine *alone*, | All *disinherited*, behold I yield | Time, Space, Infinity, and Heaven itself, | My sacred aspirations and my hopes | To earth, to earth! and 'mid their ruins stand | Beneath the stars, with no bright home beyond, | The tenant and the wanderer of a day!'.

From Olga Novikoff 17 December 1874 4509

Moscow. Poste Restante[1] | Decem[ber]. 17/5[2] 74

Dearest Friend,

Could you not tell me whether there is a good biography of Sir Roderick Murchison, or not? I sh[ou]ld like to read it not so much for the sake of the dear, old Baronet, as for that of my gifted friend, Count Keyserling who has travelled in Siberia with M[n] & is probably mentioned also.[3]—Let me tell you what Keyserling says of you after having read y[ou]r Fragments & y[ou]r address:[4]

> "Tyndall est un des grands docteurs de l'avenir. C'est <u>un génie enseignant</u>. L'avenir, où les sciences naturelles pourront, en certaines mesures, satisfaire aux besoins métaphysiques du cœur, est infiniment éloigné mais par un homme, comme lui, on avance dans ce sens. Une religion, pour être puissante, doit régler non seulement nos pensées pendant les catastrophes qui nous accablent, mais dominer les détails de la vie ordinaire. Même maintenant, les naturalistes fourniraient un tribunal plus compétant que ces théologiens pour ce qui concerne la santé physique et intellectuelle. La tournure que prend la science moderne doit être considérée comme un bienfait pour l'humanité. Les doctrines originaires du christianisme sont tout à fait incompatibles avec la science de nos jours. Les zélateurs du classicisme ont tort de déplorer cette direction. Ce sera plutôt l'étude de la nature que le Grec et le Latin—qui renouvellera l'humanité."[5]

I have received 2 letters from Frohschammer in both the same question "Was wissen sie von Professor Tyndall?"[6] is repeated. He'll get a note from me, with the identical question!

I don't like the R[ussia]n translation of y[ou]r B[elfast]: Address.[7] The words are there, but not the same fire, the same vivifying spirit!

Y[ou]rs, more than you can ever care for, Olga

RI MS JT/1/N/18

1. *Poste Restante*: see letter 4392, n. 7.
2. *17/5*: Novikoff gives the date according to both the Gregorian calendar (17) and the Julian calendar (5) then used in Russia.
3. *Count Keyserling . . . probably mentioned also*: Roderick Impey Murchison had conducted geological fieldwork in Russia in 1840 and 1841, on which he was accompanied by Alexander Keyserling as well as the French paleontologist Édouard de Verneuil (1805–73). Keyserling was mentioned regularly in the biography of Murchison that Archibald Geikie (see letter 4512, n. 2) was just then completing, which recorded the view of Murchison and Verneuil that 'in their Russian colleague von Keyserling they found an admirable travelling companion, and one to whose judgment and powers of observation the success of their conjoint work in the empire of the Czar was largely indebted' (A. Geikie, *Life of Sir Roderick I. Murchison*, 2 vols (London: John Murray, 1875), vol. 1, p. 315).
4. *y[ou]r Fragments & y[ou]r address*: *Fragments of Science* and *Belfast Address*.
5. *"Tyndall est . . . qui renouvellera l'humanité"*: Tyndall is one of the greatest doctors of the future. He is a genius teacher. The future where natural science could, in some measure, satisfy the metaphysical needs of the heart is infinitely distant, but with the help of a man like him we move forward in that direction. In order to be powerful, a religion must regulate not only our thoughts during catastrophes that overwhelm us, but it must also dominate the details of ordinary life. Even now, naturalists provide a more proficient tribunal than the theologians regarding physical and mental health. The turn that modern science takes should be considered as a boon to mankind. The doctrines native to Christianity are entirely incompatible with today's science. The zealots of classicism are wrong to deplore that direction. This will rather be the study of nature than that of Greek and Latin—which will renew humanity (French).
6. *"Was wissen sie von Professor Tyndall?"*: what do you know of Professor Tyndall? (German).
7. *the R[ussia]n translation of y[ou]r B[elfast]: Address*: see letter 4505.

From Charles Thorp 17 December 1874 4510

Bagenalstown[1] | 17. Dec[ember]. 1874.

My dear Sir,

Herewith I send you the case I laid before Counsel[2] with his opinion written on it,[3] by which you will find he advised an action for libel to be brought against the Tuam Paper.[4]

From whom did you get the Paper you sent me? Can that person prove having purchased it at the office of the Journal? If so, it would not be necessary to send to purchase another paper—If not, it will be necessary to buy a paper at the office, which might now be difficult particularly for a stranger.

Perhaps whoever sent the paper to you,[5] can prove having obtained it from the Proprietor, or if not, that he could now get another.

Pray let me know this as soon as possible.

If I were to send a stranger there it would create suspicion and might defeat our object.

Let me have back the Case, when I shall, if you so desire it, send instructions to junior Counsel to prepare the Summons and Plaint,[6] for the sooner the action is commenced, if you decide on bringing it, the better.

Faithfully yours | Charles Thorp[7]

Professor Tyndall.

P.S. I also send a letter I got from Counsel with the Case, by which you will see he advises caution—I presume his reason is lest the Tuam paper should publish something in the nature of a withdrawal of the libel as was done by the Carlow paper.[8]

RI MS JT/1/TYP/10/3421
LT Typescript Only

1. *Bagenalstown*: a small town on the River Barrow in County Carlow, Ireland, about two miles south of Leighlinbridge.
2. *Counsel*: see letter 4493, n. 3.
3. *the case I laid before Counsel with his opinion written on it*: enclosure missing.
4. *an action for libel to be brought against the Tuam Paper*: see letter 4493, n. 1.
5. *whoever sent the paper to you*: not identified, although Elizabeth Dawson Steuart seems to have first drawn Tyndall's attention to it in letter 4493.
6. *the Summons and Plaint*: documents by which a legal proceeding is initiated.
7. *Charles Thorp*: LT's typewritten transcript of this and other letters give his surname as Thorpe, but all printed sources, and Steuart in letter 4493, give it as Thorp, so the addition of an 'e' seems an erroneous intervention of LT's part.
8. *the Carlow paper*: not identified, and neither the *Carlow Sentinel* nor the *Carlow Post* seem to have made any mention of the potentially libellous pastoral printed in the *Tuam Herald*.

From Charles Thorp 22 December 1874 4511

Bagenalstown[1] | 22. Dec[ember]. 1874.

My dear Sir,

I thought it better to write to Mr Byrne[2] on the subject of your letter,[3] so that we might have his further advice—and I now enclose you his reply.[4]

Perhaps it is better to issue the Writ,[5] and then write to the Proprietor of the paper,[6] and if he proposes to make such an apology as we may consider sufficient, as it's more than likely he will, the action need not be proceeded with.

This however is as you wish.

There will be great difficulty in being able to purchase one of the papers in which the letter is published at this distance of time,[7] unless it can be done through some person on the spot, as of course a stranger applying at the office of the paper would at once create suspicion, and the object would appear, so that no paper could be obtained in that way. If we could find out some one who has already obtained a paper at the office at which it was issued, it is the only way I see.

Perhaps you may know some one there who can assist in this way?

I shall also try, but I do not know any one living in Tuam.

There is no Queen's College at Tuam, there is one in Galway.[8]

Waiting to hear further from you

Believe me, | Yours faithfully | Charles Thorp

Professor Tyndall.

Many thanks for the Book you sent me[9]—I had already seen a former number of it, lent me by Mrs Steuart, and was much interested in its perusal.

Whoever sent the paper to you would I have no doubt assist.

RI MS JT/1/TYP/10/3423
LT Typescript Only

---

1. *Bagenalstown*: see letter 4510, n. 1.
2. *Mr Byrne*: the Counsel mentioned in letter 4510. There were two barristers of this name in Ireland at this time: John Alexander Byrne (*c.* 1824–1902) of 30 Lower Fitzwilliam Street, Dublin, who had been called to the bar in 1849; and John Ouseley Byrne (*c.* 1846–89) of 2 Leinster Street, Dublin, who was called to the bar in 1868 (*Thom's Irish Almanac and Official Directory* (Dublin: Alexander Thom, 1874), p. 989). It is unclear which was the 'Mr Byrne' consulted by Thorp.
3. *your letter*: letter missing.
4. *enclose you his reply*: letter missing.
5. *the Writ*: a formal written order issued by a court of law. The writ in question was to be issued to initiate a proceeding for libel (see letters 4493, 4496, and 4510).
6. *the Proprietor of the paper*: Richard Kelly (1810–84), who, having founded the *Tuam Herald* in 1837 and then relinquished its ownership to his son Jasper, returned as proprietor in 1866 following Jasper's death.
7. *the papers . . . this distance of time*: the potentially libellous letter was published in the *Tuam Herald* on 14 November (see letter 4493, n. 1).
8. *Queen's College . . . in Galway*: founded in 1849, and part of the Queen's University of Ireland, with other colleges in Belfast and Cork, which from 1850 awarded degrees to students of all religious denominations, especially Catholics.

9. *the Book you sent me*: probably the revised edition of *Belfast Address*, 7th thousand, which was relevant to the libel case. Tyndall had sent Elizabeth Dawson Steuart a copy of the original edition in September (see letter 4415).

To Olga Novikoff 23 December 1874 4512

Royal Institution of Great Britain | 23rd December | 1874.

My dear Friend.

Your letter from Moscow[1] reached me last night. Murchison left all his papers in the hands of M^r^. Geikie,[2] the Director of the Geological Survey of Scotland; he, I doubt not, is now engaged upon the work,[3] but I do not know the time of its probable appearance.

As soon as the preface[4] was ended I gave most distinct directions to the printers[5] to send you a copy to Moscow. I also gave them directions to send a complete copy of the preface, address, and Manchester lecture[6] to Professor Frohschammer. Probably both you and he have by this time received the documents.

I am so much obliged to you for all the kind sayings you communicate to me; they are ample compensation for the virulence which the "religious" world has poured down upon me. Really on reading the religious newspapers I sometimes ask myself, are those people mad, or am I mad. Their delusions appear to me more and more grotesque and horrible.[7] But we must assume that the world is in some sense mad—As Carlyle says, there are 30 millions in England but they are mostly fools[8]—and in dealing with such a mad world we must be patient.

I saw Kinglake last night at the Athenaeum;[9] he looked fresh and well.

We are visited by terribly cold weather here. I like it, but it has proved very fatal to the weak and aged. I went down to Brighton[10] some days ago for a little rest, and was smitten there with a severe bronchial cold: however colds have no great hold upon me, and I am now almost right again.

I mentioned to you the death of M^rs^. Hooker. Well I go out to Kew for the Christmas in order if possible to keep my friend's mind from dwelling too sadly upon his bereavement.[11]

I am greatly pleased to read what you said regarding your boy.[12] To me the love of a mother for her son has always been the most sacred thing on earth.

Please mention to your friend Count Keyserburg how much I appreciate his kind sympathy.[13] You must always bear in mind that the Fragments[14] really are such, and that when they were written I had no idea of ever collecting them together. With all the warm wishes of the Season,

believe me | Ever Yours | John Tyndall

John Tyndall Correspondence in the Olga Novikoff Correspondence Collection. Courtesy of Special Collections, Kenneth Spencer Research Library, University of Kansas. MS 30:Q9
RI MS JT/1/TYP/3/933

1. *Your letter from Moscow*: letter 4509.
2. *Mr Geikie*: Archibald Geikie (1835–1924), who served as director of the Geological Survey for Scotland from its creation in 1867 until 1881 (*ODNB*).
3. *the work*: A. Geikie, *Life of Sir Roderick I. Murchison*, 2 vols (London: John Murray, 1875).
4. *the preface*: 'Preface to the Seventh Thousand', in *Belfast Address*, 7th thousand, pp. v–xxxii.
5. *the printers*: Spottiswoode & Co., New Street Square, London.
6. *Manchester lecture*: see letter 4441, n. 3.
7. *Their delusions . . . more grotesque and horrible*: Tyndall may be referring to the potentially libellous accusations made by Catholic Bishops and reprinted in an Irish newspaper about which he was just then consulting lawyers (see letters 4510 and 4511).
8. *As Carlyle says . . . mostly fools*: Thomas Carlyle had originally said: 'The practice of modern Parliaments, with reporters sitting among them, and twenty-seven millions mostly fools listening to them, fills me with amazement' (T. Carlyle, *Latter-Day Pamphlets* (London: Chapman and Hall, 1850), p. 177). However, the quotation was subsequently regularly adjusted, as Tyndall does here, to take into account the country's increased population.
9. *the Athenaeum*: the Athenaeum Club (see letter 4186, n. 1).
10. *I went down to Brighton*: see letter 4504.
11. *my friend's mind . . . his bereavement*: Joseph Hooker's wife, Frances Hooker, had died suddenly on 13 November (see letter 4480, n. 2).
12. *your boy*: Alexander Novikoff.
13. *his kind sympathy*: see letter 4509.
14. *the Fragments*: *Fragments of Science*.

To Robin Allen 24 December 1874 4513

Royal Institution of Great Britain | 24th December, 1874

Sir,

I have read with interest the lucid Report of Colonel Campbell on the recent experiments with gun-cylinders at Woolwich.[1]

The "order of merit"[2] comes out in a very satisfactory manner; but I would emphasize the statement that the figures recorded indicate the *order* only, and give no notion of comparative intensity. It is to the credit of the observers[3] that differences so slight should be so clearly detected.

The difficulty referred to by Colonel Campbell as to the superiority of the range of cast-iron over that of bronze[4] will, in all probability, be resolved in subsequent experiments. Further experience, indeed, is needed to place the *fact* beyond question;[5] and prior to the establishment of the fact, attempts at explanation would be premature.

I entirely concur in the proposals of Colonel Campbell as regards future experiments with the guns,[6] and find particular pleasure in observing that he is determined to develop to the uttermost of the capabilities of gun-cotton.[7] Here the experiments in free air, and in association with a reflector, cannot fail to be of interest.

I have the honour to be, | Sir, | Your obedient Servant, | JOHN TYNDALL. | Robin Allen, Esq.

National Archives, MT 10/290. Papers re Fog Signals. H8476 (handwritten: H4380)

---

1. *Report of Colonel Campbell . . . gun-cylinders at Woolwich*: 'Fog Signal Guns. Minute 32,498', *Extracts from the Proceedings of the Department of the Director of Artillery*, 12 (1874), pp. 332–3. The experiments, on the effectiveness of different gun-cylinders in producing acoustic fog signals, were conducted on 26 November at the Proof Butts at the Royal Arsenal, Woolwich.
2. *The "order of merit"*: of four types of gun cylinder with a 'charge of 4 oz. of powder and a half-inch wad', whose 'sound was observed at intervals commencing at 100 yards and going as far as 3,000 yards during the first experiment', and 'from a distance of about two miles' in 'the second experiment'. It was reported that 'in the first experiment the cast-iron gun with conical reflector (2) proved the best, the cast-iron with parabolic reflector (3) next, the bronze ditto (4) third, and the plain muzzle gun (1) the worst. In the second experiment this order was maintained' (pp. 332–3).
3. *the observers*: it was reported that 'a Committee of the Elder Brethren of the Trinity Board, accompanied by Professor Tyndall and Mr. Douglass, attended and assisted as observers during the experiments' (p. 332). On Trinity House, see letter 4187, n. 2.
4. *The difficulty . . . cast-iron over that of bronze*: the report noted that 'the increase in distance was proved to be unfavourable to bronze' (p. 333).
5. *Further experience . . . the fact beyond question*: Tyndall later quoted Frederick Campbell's comments in a paper submitted to the RS: 'The result of subsequent trial, as reported by General Campbell, is, "that the sonorous qualities of bronze are greatly superior to those of cast iron at short distances, but that the advantage lies with the baser metal at long ranges"'. Tyndall added a footnote stating: 'General Campbell assigns a true cause for this difference. The ring of the bronze gun represents so much energy withdrawn from the explosive force of the gunpowder. Further experiments would, however, be needed to place the superiority of the cast-iron gun at a distance beyond question' (J. Tyndall, 'Recent Experiments on Fog-Signals', *Roy. Soc. Proc.*, 27 (1878), pp. 245–58, on p. 246).

6. *the proposals . . . future experiments with the guns*: the report stated 'Superintendent Royal Gun Factories proposes to experiment further with 1 and 2 [see n. 2], and also with two new guns, one having a cone longer and broader at the base, and the other a parabolic reflector to match, both to be made of cast-iron' (p. 333).
7. *develop to the uttermost . . . capabilities of gun-cotton*: the report concluded: 'Recommends that a parabolic reflector be made for trial in the focus of which the charge of guncotton could be exploded' (p. 333).

To Robin Allen 25 December 1874 4514

Royal Institution of Great Britain | 25th December, 1874.

Sir,

Would you permit me to supplement my remarks on the Report of Colonel Campbell[1] by the suggestion that an 18-pounder gun (as employed at the Trinity House) and 3 lbs. of powder be fired, and that the equivalent quantity of gun-cotton be determined?[2]

In this way we shall arrive most directly and practically at the comparative cost and efficiency of the two substances.

Your obedient Servant, | JOHN TYNDALL. | ROBIN ALLEN, ESQ.

National Archives, MT 10/290. Papers re Fog Signals. H8476 (handwritten: H4380)

---

1. *my remarks on the Report of Colonel Campbell*: see letter 4513.
2. *the suggestion that . . . gun-cotton be determined*: on 11 January 1875, the Director of Artillery, Royal Arsenal, Woolwich, noted: 'Trinity House, 31/12/74, forwards Dr. Tyndall's observations on the experiments made in November 1874, and request that a comparative trial, as suggested, between an 18-pr. gun and guncotton be embodied in the further experiments' ('Fog Signal Guns. *Minute* 32,645', *Extracts from the Proceedings of the Department of the Director of Artillery*, 13 (1875), pp. 67–70, on p. 67). Tyndall later described the experiments that were made at his suggestion: 'On à priori grounds, then, we are entitled to infer the effectiveness of gun-cotton, while in a great number of comparative experiments, stretching from 1874 to the present time, this inference has been verified in the most conclusive manner . . . On the 22nd of February, 1875, a number of small guns, cast specially for the purpose—some with plain, some with conical, and some with parabolic muzzles, firing 4 ounces of fine-grain powder—were pitted against 4 ounces of gun-cotton, detonated both in the open and in the focus of a parabolic reflector. The sound produced by the gun-cotton, reinforced by the reflector, was unanimously pronounced loudest of all. With equal unanimity, the gun-cotton detonated in free air was placed second in intensity. Though the same charge was used throughout, the guns differed not

among themselves, but none of them came up to the gun-cotton, either with or without the reflector' (J. Tyndall, 'Recent Experiments on Fog-Signals', *Roy. Soc. Proc.*, 27 (1878), pp. 245–58, on pp. 247–8).

From Olga Novikoff 25 December 1874 4515

Russia. Moskau.[1] Poste restante.[2] | Decem[ber]. 25. 74.

You know me too well not to guess my delight in reading y[ou]r last Preface.[3] It is capital. I won't enter into details as you seem over bothered with my letters already. I do not know who sent me the address,[4] as my name on the cover was not in y[ou]r handwriting. Still I may thank you for it most heartily.

More than ever I regret not to be now in England. How I want to learn the way in w[hic]h y[ou]r noble, powerful words have been received in England. Frohschammer is enthusiastic about it—as you'll see it, black on white, very soon. Probably in the Allgem[eine] Zeitung.[5] Y[ou]r kind way of mentioning him,[6] will suffice, I hope, to make him more studied in y[ou]r country, and one of y[ou]r contemporaries will perhaps be induced now to review his works in English. But it w[ou]ld carry me too far were I to express all I hope besides.

Good bye. Be happy. Who ever deserved more than you to be so?

Y[ou]rs ever & entirely | Olga Novikoff | née Kiréeff

P.S. | Kinglake wrote to me the other day, mentioning his meeting you[7] & finding you "very cordial & extremely pleasant."—w[hic]h I easily conceive & admit!

RI MS JT/1/N/19

1. *Moskau*: Moscow (German)
2. *Poste restante*: see letter 4392, n. 7.
3. *y[ou]r last Preface*: 'Preface to the Seventh Thousand', in *Belfast Address*, 7th thousand, pp. v–xxxii.
4. *who sent me the address*: it was the printers Spottiswoode & Co., New Street Square, London (see letter 4512).
5. *Frohschammer is enthusiastic . . . Allgem[eine] Zeitung*: J. Frohschammer, 'Professor John Tyndall und die Freiheit der Wissenschaft in England', *Beilage zur Allgemeine Zeitung*, 9 January 1875, pp. 121–3.
6. *Y[ou]r kind way of mentioning him*: see letter 4469, n. 4.

7. *his meeting you*: Tyndall met Alexander William Kinglake at the Athenaeum Club on 22 December (see letter 4512).

From Charles Thorp 26 December 1874 4516

Bagenalstown[1] | 26. Dec[ember]. 1874.

My dear Sir,

Should the action against the Tuam paper,[2] as advised by Mr Byrne,[3] proceed to a trial, there is no doubt but that your attendance and evidence at the hearing would be necessary. It does not however follow as inevitable that the action would go to trial, as the chances are it would be arranged in the way suggested by Mr Byrne namely, that the paper would publish a sufficient apology and pay the expenses, which I presume would satisfy you.

This however is matter of speculation, and though most probable, is not at all a certainty, and as we should proceed with the action if once commenced, in case the suggested arrangement was not made, (as it would never do to stop short then, which would make matters worse than at present) you should decide before the Writ[4] is issued, whether you would let the Case go to trial, which would involve your attendance here, in the event of no apology being offered.

My own belief is that the apology would be given and that there never would be any occasion for a trial, but as I before said there is no certainty of this.

I have not seen Mrs Steuart since receipt of your last letter,[5] but I know her feelings on the matter, she is very indignant at the libel,[6] and would gladly see the parties[7] punished, but I also know she would rather waive this than cause you additional annoyance, should your having to attend the trial of the Case have that effect.

Perhaps then under all the circumstances it is as well to let the matter drop. Unless the doing so would embolden the parties who appear in arms against you to further aggression

It is purely a matter for your own personal consideration and I feel so confident that you are far more able to decide than I am, I shall leave it to your better judgment after the observations I have made.

Believe me, | Yours very faithfully | Charles Thorp

Professor Tyndall.

RI MS JT/1/TYP/10/3424

LT Typescript Only

1. *Bagenalstown*: see letter 4510, n. 1.
2. *the action against the Tuam paper*: see letters 4493, 4496, 4510, and 4511.
3. *Mr Byrne*: see letter 4511, n. 2.
4. *the Writ*: see letter 4511, n. 5.
5. *your last letter*: letter missing.
6. *very indignant at the libel*: see letter 4493, in which Elizabeth Dawson Steuart stated 'No words can express my indignation against the infamous paragraph in that vile paper'.
7. *the parties*: presumably Richard Kelly (see letter 4511, n. 6), whose newspaper printed the libel, and Patrick Durcan (1790–1875) and Francis Joseph MacCormack (1833–1909), who issued the pastoral letter in which the libel was contained (see letter 4493, n. 1).

From Charles Darwin 27 December 1874 4517

(Private) | Down | Beckenham, Kent. | RAILWAY STATION | ORPINGTON, S.E.R. | Dec[ember] 27 1874

My dear Tyndall

I know how deep an interest you feel in Lady Lubbock, & I believe that you are more likely to be able to influence her than any other person. Mr Birkbeck[1] called here yesterday & said he thought she was dying, as she eats nothing; but I think this is too gloomy a view—

She is attended only by Dr Erasmus Wilson,[2] who, some time ago, said he was puzzled by her case. Mr B. has tried to persuade her to see some other D^r^; but she answered it w[oul]^d^ only send her to her grave so much the sooner. He seemed rather huffed at Sir John,[3] & said it was of no use speaking to him again. Now I believe diet, restriction of stimulants &c, will alone save her. Therefore I further believe that Dr Andrew Clark, who does not trust in physic,[4] but has the great art, of compelling his patients to obey him, would be the best man. If you agree with me, & can do any thing, I am sure you will be anxious to do so. I feel sure that my intervention ought <u>in no way</u> to appear—

I ought to add that Mr B. thought that Sir John is now thoroughly alarmed, but has no influence over her—

yours very sincerely. | Ch. Darwin

CUL GBR/0012/MS DAR 261.8:22[5]
RI MS JT/1/TYP/9/2835

1. *Mr Birkbeck*: probably Robert Birkbeck (1836–1920), John Lubbock's brother-in-law.
2. *Dr. Erasmus Wilson*: William James Erasmus Wilson (1809–84), a surgeon who specialized in dermatology (*ODNB*).

3. *Sir John*: John Lubbock.
4. *physic*: cathartic or purgative medicines (*OED*).
5. *CUL GBR/0012/MS DAR 261.8:22*: published in *Darwin Correspondence*, vol. 22, p. 593; this is letter 9784 in the online edition of the Darwin correspondence. The handwritten letter is in the hand of Emma Darwin, and Charles signed his own name.

From Charles Thorp [*c.* 27–8] December 1874[1] 4518

Bagenalstown[2] | 6. Dec[ember]. 1874

My dear Sir,

Absence in Dublin[3] prevented my sooner acknowledging receipt of your letter and cheque.

Pray accept my thanks for your very liberal payment which is much more than I would have been entitled to for the little trouble I took in the matter. I really would not have thought of making any charge as the case[4] was not to proceed further, but as you have been so kind as to send me the cheque I shall accept it with many thanks.

I think under all circumstances you have decided rightly in not proceeding further in the matter, particularly after the paper withdrawing what was contained in the letter previously published and properly designating it as a vile slander,[5] for though their doing this would not wholly exonerate them from the consequences of publishing so infamous a libel,[6] it would have a great effect in reducing the damages that would otherwise be given.

But besides this, you can well afford to let anything that could appear in such a journal, pass without notice.

I think they have now got such a fright as will make them more cautious what they allow into their paper in future.

I shall however be on the look out and should any thing further appear shall take care to let you know.

Again thanking you for your liberality
Believe me, | very faithfully yours | Charles Thorp
Professor Tyndall.

RI MS JT/1/TYP/10/3419
LT Typescript Only

1. *[c. 27–8]*: LT's typescript gives the date as '6. Dec. 1874', but this seems to be an error as Tyndall was still considering whether to proceed with the libel case when Thorp wrote to him on 26 December in letter 4516. As such, this letter must have been written between 26 and 29 December, when, in letter 4520, Thorp mentions the payment acknowledged here.

2. *Bagenalstown*: see letter 4510, n. 1.
3. *Absence in Dublin*: see letter 4496, n. 3, although Thorp may simply have been paying social visits in the Christmas season.
4. *the case*: for libel (see letters 4493, 4496, 4510, 4511, and 4516).
5. *the paper withdrawing . . . designating it as a vile slander*: it is not clear whether Thorp was suggesting that the withdrawal had actually been made by the *Tuam Herald*, although no such retraction seems to have appeared in the newspaper.
6. *so infamous a libel*: see letter 4493, n. 1.

To Charles Darwin 28 December 1874 4519

Royal Institution of Great Britain | 28th. Dec[ember]. 1874. | 11 P.M.

My dear Darwin.

I have just received & read your letter,[1] and I need not tell you how concerned I feel about it. It was only this morning I had a note from her,[2] informing me that she intended to accompany the Spottiswoodes[3] to Lubbock's lecture,[4] but giving no hint that she was ill. This, however, is like her. She is quite capable of dying without giving any sign.

I will so arrange matters that I may have an hour's earnest conversation with her. I do not know that she will pay any attention to me; but I think if she listens to anybody she will be inclined to listen to me.

I quite think with you that Andrew Clark is the man most likely to give her sound advice, and I shall do my best to induce her to consult him.

Last night I returned from Kew, whither I went on Thursday.[5] Hooker passed the crisis[6] well. On christmas day he had some skating, and without our making any effort, which would assuredly defeat itself, his mind was kept cheerfully occupied throughout—It was a happiness to me to be able to be at his side during this time of trial.

He told me about Mivart,[7] and allowed me to read the correspondence.[8] On à priori grounds, & by an indescribable intuition, I could predict Mivart's act as the natural outflow of his character.

always yours | John Tyndall

CUL GBR/0012/MS DAR 106:C17–18[9]
RI MS JT/1/TYP/9/2836

1. *your letter*: letter 4517.
2. *her*: Ellen Lubbock.
3. *the Spottiswoodes*: William Spottiswoode and his wife Eliza Spottiswoode.

4. *Lubbock's lecture*: presumably John Lubbock's lecture 'On the Relation of English Wild Flowers to Insects', given at the RI on 22 January 1875.
5. *Thursday*: 24 December.
6. *the crisis*: the death of Joseph Hooker's wife, Frances Hooker, who had died suddenly on 13 November (see letter 4480, n. 2).
7. *Mivart*: in an anonymous contribution to the *Quarterly Review*, the Catholic biologist St George Mivart (1827–1900) criticized an article by Darwin's son George and erroneously alleged that he 'speaks in an approving strain of . . . the encouragement of vice in order to check population' ('Primitive Man: Tylor and Lubbock', *Quarterly Review*, 137 (1874), pp. 40–77, on p. 70).
8. *the correspondence*: Darwin had been corresponding with Hooker and Thomas Huxley regarding an appropriate response. See G. Dawson, *Darwin, Literature and Victorian Respectability* (Cambridge: Cambridge University Press, 2007), pp. 77–80; and letters 9757, 9768, 9769, 9770, 9773, 9774, 9776, 9777, 9780, 9785, and 9788, *Darwin Correspondence*, vol. 22 (or in the online edition of the Darwin Correspondence).
9. *CUL GBR/0012/MS DAR 106:C17–18*: published in *Darwin Correspondence*, vol. 22, p. 594; this is letter 9787 in the online edition of the Darwin correspondence.

From Charles Thorp 29 December 1874 4520

Bagenalstown[1] | 29. Dec[ember]. 74

My dear Sir,

I dare say you have decided rightly in allowing that Tuam paper[2] to remain unnoticed—though it certainly is very provoking that such a vile publication should go unpunished. However should it appear again in anything of a respectable journal we can thus take satisfaction,[3] and I shall certainly be on the look out.

I paid Mr Byrne[4] a fee of two guineas on the Case,[5] so if you send me that I shall not make any further charge for myself as the fee you sent me before[6] is quite sufficient compensation for the small trouble I have had. I am only sorry you should have been put to so much expense, as well as the annoyance you must have felt by this vile business.

Believe me, | Yours very faithfully | Charles Thorp

Professor Tyndall.

RI MS JT/1/TYP/10/3425
LT Typescript Only

1. *Bagenalstown*: see letter 4510, n. 1.

2. *that Tuam paper*: the *Tuam Herald* (see letter 4493, n. 1).
3. *take satisfaction*: Thorp seems to mean this in a legal sense, implying that the action for libel could be revived were the same words printed in the *Tuam Herald* to appear in more reputable publications; there is no record that they were.
4. *Mr Byrne*: see letter 4511, n. 2.
5. *the Case*: for libel (see letters 4493, 4496, 4510, 4511, 4516, and 4518).
6. *the fee you sent me before*: see letter 4518.

1875

From Elizabeth Dawson Steuart 1 January 1875 4521

Steuart's Lodge | Leighlin Bridge | 1.1.75.

My dear John,

Your d[ra]ft.,[1] has just arrived, and will be of great help to poor old Mrs Tyndall.[2] Caleb[3] has been kinder to her of late, but his idle family[4] take a great deal from him. I think you are wise in not proceeding against those low people,[5] for it would be a most difficult matter to trace it home, for they would swear anything. I suppose you have written to Mr Thorp, not to proceed.

Could you, without inconvenience, let me know some of the best places in London for buying Scrap Photographs?[6] exclusive of "Ackerman" and "Beal",[7] I get some about this time every year for Albums &c. and they amuse me. A collection comes over to select from.

With many kind wishes | Your affectionate | E. D. S.

RI MS JT/1/TYP/10/3426
LT Typescript Only

1. *Your d[ra]ft:* a cheque Tyndall presumably sent in support of his aunt, although it may also have been to settle an outstanding debt on her account (see letter 4524).
2. *Mrs Tyndall*: Dorothea Tyndall.
3. *Caleb:* Caleb Tyndall, Tyndall's paternal uncle.
4. *his idle family*: Caleb had at least seven surviving children at this time: Emma (1824–1904), Patrick (1828–1901), John (1829–83), James (1831–80), Caleb Jr (see letter 4432, n. 4), Sarah (see letter 4432, n. 5) and William (1844–86). Caleb Jr and Sarah had both recently written to Tyndall asking for financial assistance (see letter 4432), while William and his wife Fanny had earlier received assistance from Tyndall after the deaths of their cows, although they appear to have overvalued the loss (see letters 2828 and 2831, *Tyndall Correspondence*, vol. 10).
5. *not proceeding against those low people*: Tyndall had been considering whether to sue the *Tuam Herald* for libel (see letters 4493, 4496, 4510, 4511, 4516, 4518, and 4520).

6. *Scrap Photographs*: unmounted photographs that could be collected in a scrap-book.
7. *"Ackerman" and "Beal"*: Arthur Ackermann (1830–1914), who ran his family's long-established print-selling business in Regent Street, and Samuel Benoni Beal (1826–75), a picture dealer at 47 St Paul's Churchyard, who traded in photographs as well as fine art.

To George Gabriel Stokes 4 January 1875 4522

4th Jan[uar]y. 1875.

My dear Stokes.

The paper[1] was intended for the proceedings.

The observation to which you refer[2] is this. At Fort Scammel[3] there is a fog signal station. Immediately under the cliff at the back of the Island[4] on which the fort stands—where you are deep in the sound shadow—you do not hear the signals; you do hear them further out at sea.

At A you do not hear the sound from S: at A′ you do. I thought of it when writing my paper—but it does not sufficiently bear upon it. I have made many observations of a like kind myself.

I have written to Mattress to ask him for a proof of the paper when it is printed.

I shall have a couple of pages to add to the MS. already in your hands.

Yours faithfully | John Tyndall

RI MS JT/1/T/1406
RI MS JT/1/TYP/4/1489

1. *paper*: J. Tyndall, 'On Acoustic Reversibility', *Roy. Soc. Proc.*, 23 (1875), pp. 159–65. Tyndall delivered the paper at the RS on 4 January, and also gave it as a Friday Evening Discourse at the RI on 15 January (*Roy. Inst. Proc.*, 7 (1873–75), pp. 344–50).
2. *The observation to which you refer*: if Stokes referred to the observation in a letter, rather than verbally, it is missing.
3. *Fort Scammel:* a fort guarding the harbour in Portland, Maine, USA.
4. *the Island*: House Island.

From John William Draper 4 January 1875 4523

University[1]—Washington Square | New. York | Jan[uary] 4th 1875

My Dear Tyndall.

I received this morning from your printers[2] proof sheets of your Addresses and Prefaces.[3] I have read them with renewed pleasure and heartily thank you for your kindness in having them sent.

~~The warfare against science goes on here very uproariously and about as abusively as it does with you. Its bitterness is however telling against itself~~[4]

In return for that courtesy and as a sample of the facetious phase of the conflict ~~into which I have dropped~~ I send you by this mail a newspaper the Daily Graphic[5]—it may stand as somewhat the parallel of your Punch's poem "Atom the Architect"[6] and may draw a laugh from you as it does from me. This newspaper which has a very large circulation[7] has put out some very indignant articles on my book[8]—full of misrepresentation & vilification.

Of course before you gave your Belfast address you weighed well the consequences as also I did before I printed this book and I suppose we are both ready to say to our assailants as old Anaxarchus did when they were pounding him in the mortar. 'Beat as long as thou wilt on the bag of Anaxarchus, himself thou canst not hurt.'[9]

I told you what a profound sensation your Address created in America[10] and how many friends it made you here—or rather how it brought up your old friends to your support. You will not then be surprised to learn how they are standing by me. In the midst of this storm of abuse—this damning with faint praise by some and cursing by others my book is steadily making its way. On the next day after its publication 3000 copies had been sold. Now nearly 10,000 have been disposed of. The third edition is exhausted and a fourth in press.

More and more impressively do I recognize the great change that has taken place of late years. The social conquests that Science has made have outstripped her philosophical ones. Whatever may be the condition of things on the European continent—we have got the ear of the English speaking race. And it only remains for us to be true to her and true to ourselves to obtain a great victory for posterity.

I may truly say I was under a "religious conviction" that I published the "Intellectual Development".[11] I thought it dire to the age in which I live, and as to my present book I may confess to you what I would hide from others that the infirmities of age are so over coming me—I live a life of physical suffering,[12] that during its composition my chief anxiety was to be spared until I could complete it. Destiny has granted me that, and now I am content. Every day the mail is bringing me a swarm of letters, many from friends but many more from strangers—their burden is "Stand fast! Your book is doing great good. If you need help let me know".

So as the Greek race—runners delivered the torch to each other.[13] I say to you "Stand fast your address is doing great good. If you want help let me know".

And I hope that the friends of Science—will standby you in England as they are standing by me in America. Let us all fight shoulder to shoulder and fighting not for ourselves but for posterity.

Your friend | John W Draper

Professor Tyndall

Library of Congress, John William Draper Family Papers

1. *University*: the University of the City of New York (see letter 4406, n. 1).
2. *your printers*: Spottiswoode & Co., New Street Square, London.
3. *Addresses and Prefaces*: the Belfast Address, 'On Crystals and Molecular Forces', 'Preface to the First Thousand', and 'Preface to the Seventh Thousand' in *Belfast Address*, 7th thousand.
4. ~~*The warfare . . . against itself*~~: these lines were excised with a cross in the original letter, which may indicate that this is an initial draft of the letter that was actually sent to Tyndall.
5. *the Daily Graphic*: enclosure missing, but Draper presumably sent Tyndall a copy of the newspaper's issue for 30 December 1874, the front page of which carried a large cartoon and accompanying poem presenting Draper as an Islamic warrior with a scimitar that bears the words: 'The Conflict between Religion and Science. By Dr. John W. Draper'. He confronts Pope Pius IX, who, seated on the Papal throne in St Peter's, holds aloft a scroll labelled 'Infallibility' ('The Conflict between Religion and Science', *Daily Graphic*, 30 December 1874, n.p.) The *Daily Graphic: An Illustrated Evening Newspaper* was the first daily newspaper in the United States with illustrations, and ran from 1873–89.
6. *your Punch's poem "Atom the Architect"*: a comic poem satirizing Tyndall's lecture in Manchester on 'On Crystalline and Molecular Forces' (see letter 4441, n. 3) published in *Punch*, 7 November 1874, p. 198. Tyndall reprinted it in *Belfast Address*, 7th thousand, p. 68.
7. *a very large circulation:* the *Daily Graphic*'s circulation peaked at 10,000, although the newspaper was not a financial success.
8. *some very indignant articles on my book*: Draper's book was *History of the Conflict between Religion and Science* (New York: D. Appleton, 1874), of which the *Daily Graphic* carried a review 'exposing the demerits of this much-puffed book' that, it was alleged, was 'full of the baldest atheism and most unjust form of irreligion that has ever been printed'. In remarking on Draper's own apparent flaws, the same review also noted: 'As if to emphasize the inherent cowardice of the scientific character, have we not lately seen Professors Tyndall and Huxley crawfish when Gog and Magog hinted their displeasure? Bold as lions in the Alps and at Belfast, they roared like sucking doves when the coteries of London signified their displeasure' ('Religion and Science', *Daily Graphic*, 23 December 1874, pp. 388–9).

9. *old Anaxarchus . . . canst not hurt'*: Anaxarchus (*c.* 380–*c.* 320 BCE) was an ancient Greek philosopher who accompanied Alexander the Great on his conquests. He was considered to be an adherent of the ideas of Democritus and Pyrrho. Diogenes Laertius' *Lives of the Eminent Philosophers* states that he was condemned to death by Nicocreon, the tyrant of Cyprus. The specific translation of his words seems to come from J. Platts, *A New Universal Biography*, 5 vols (London: Sherwood, Jones, 1825), vol. 1, p. 484.
10. *your address created in America*: see letters 4406, 4420, and 4451.
11. *the "Intellectual Development"*: J. W. Draper, *History of the Intellectual Development of Europe* (New York: Harper & Brothers, 1863). Draper had sent Tyndall a copy (see letter 2020, *Tyndall Correspondence*, vol. 8), who used it as one of his sources for the Belfast Address.
12. *a life of physical suffering*: a biographical notice later said of Draper: 'His health, which through life had been generally good, was disturbed during his later years by severe attacks of gravel [i.e. kidney stones], which incapacitated him for journeying. These attacks wore upon him and finally ended his life' (G. F. Barker, 'John William Draper, 1811–1882', *National Academy of Sciences Biographical Memoirs*, 2 (1886), pp. 349–88, on p. 382).
13. *runners delivered the torch*: in the ancient Olympics, the Greeks carried out the practice of lampadedromia, relaying torches from the fire in Olympia to the Acropolis in Athens.

To Elizabeth Dawson Steuart [5 January 1875][1] 4524

Royal Institution of Great Britain | Tuesday.

My dear Mrs Steuart,

I thought really that this man's account[2] was under a pound; otherwise I should never have thought of sending him this money. I am sorry you offered it to him.

But you will understand my feelings when I say that I could not bear any thing of this kind to associate itself with the thought of me and mine. I always knew he had no claims, but I must sweep all tattle[3] which might connect itself with an affair of this kind[4] away from my mother's[5] memory.

Pray therefore pay the man his claim, and explain to him that I was under the impression that his account was under a pound when I sent the cheque. It would annoy me more than I can tell you, to have the thought entertained that I sought to stop the man's importunity by offering him a part of his claim—Do please pay him.

The cheque enclosed.

In terrible haste | Yours ever sincerely | John Tyndall.

If you would tell me what you mean by photographic scraps[6] I will name photographers—or send you what you wish for.

RI MS JT/1/TYP/10/3427
LT Typescript Only

1. *[5 January 1875]*: the date is suggested by the relation to letter 4521, where Steuart asks for help obtaining scrap photographs, which was dated 1 January. The following Tuesday was 5 January.
2. *this man's account*: not identified, but an account is a statement of financial expenditure and receipts relating to a particular period or purpose, with calculation of the balance; or a detailed statement of money due (*OED*).
3. *tattle*: idle or frivolous talk; chatter, gossip (*OED*).
4. *an affair of this kind*: not identified, but as Tyndall was worried about damage to his mother's memory it may relate to the family of his late maternal uncle, John McAssey (see letter 4434, n. 2). Tyndall and Steuart discussed their financial difficulties in letters 4434 and 4437.
5. *my mother*: Sarah Tyndall (née McAssey, 1795–1867). She was descended from a prosperous farming family with lands in Ballybrommell, County Carlow, Ireland, that included in earlier generations Quakers by the name Malone. In 1813 she married John Tyndall senior (1792–1847). After her husband's death she remained in the family cottage at Leighlinbridge, supported financially by Tyndall. Sarah died on 30 April 1867, and Tyndall travelled to Ireland to attend her funeral (see letter 2589, *Tyndall Correspondence*, vol. 10).
6. *photographic scraps*: see letter 4521, n. 6.

From George Gabriel Stokes 5 January 1875 4525

Cambridge. | 5th Jan[uary]. 1875.

My dear Tyndall,

From your account of the American experiment contained in your letter of yesterday,[1] I see that my memory must have somewhat deceived me and I must have mixed up that experiment with the question you proposed to me, and supposed that reciprocal signals entered into the former as they did into the latter. I agree with you that the experiment described in your letter of the 4th has only a very remote bearing on the subject of your paper.[2]

Yours sincerely | G. G. Stokes.

RI MS JT/1/TYP/4/1427
LT Typescript Only

1. *your letter of yesterday:* letter 4522.
2. *your paper*: see letter 4522, n. 1.

From Anna Helmholtz 7 January 1875 4526

45. Königin Augusta-Strasse.[1] | January 7th 75.

Dear Professor Tyndall,

You have so impressed me by your beautiful Christmas gift[2]—and given me so much pleasure that I wanted to thank you for it immediately, but was not very well and put it off—as lazy people are apt to do—I am afraid you would have felt less inclined to cause me any particular pleasure if you had looked at my translation of your Fragments.[3] Nobody can feel its cranks and defects more deeply than I do—but it was rather too difficult for my poor capacity and whenever I thought that I had achieved something remarquable in rendering your thoughts and expressions—my husband[4] was sure to tell me that it was all nonsense or a mistake and that I had to do it again—and as [I] am neither very patient nor very good tempered it did not always make me feel amiably towards you or him!

So you may judge of my "Zerknirschung"[5] as Vieweg made over to me a much larger sum than I was entitled to—and I felt that I was taking away your monnaye[6] besides doing but poor justice to your work—and now you go on heaping coals on my guilty head!—In spite of all these moral reflections I am very proud of my bracelet—not the less so that it is my only one—and send my very best thanks to you for it and your kind letter.

RI MS JT/1/H/36

---

1. *Königin Augusta-Strasse*: in central Berlin, alongside the Landwehr Canal.
2. *Christmas gift:* the associated letter is missing. This letter suggests the gift was a bracelet, and Tyndall had given Anna jewelry before, sending a brooch to thank her for her work on the translation of the third edition of *Heat*.
3. *my translation of your Fragments*: see letter 4192, n. 11. Anna translated many of Tyndall's books into German, also collaborating with Clara Wiedemann and Estelle du Bois-Reymond.
4. *my husband*: Hermann Helmholtz.
5. *"Zerknirschung"*: guilty conscience (German).
6. *monnaye*: presumably Helmholtz's irregular spelling of money.

From Olga Novikoff 7 January 1875 4527

Moscow. Post Restante[1] | January.7.75.

My silence[2] might induce you to think me wiser, better, than I am. The truth is: not a day passed without my thinking twenty times of writing to you. Don't frown, don't be angry! I cannot be answerable for my constancy, as I really cannot help it! I was silent only because I had to spend all my time near my poor boy's[3] bed, who had a severe inflamation of the lungs, w[hic]h he is very prone to. At the last medical consultation I picked up some few Latin words & found new confirmation to what I already suspected myself—that his lungs were weak. I don't expect him to live q[ui] te many years; but Epicurus was quite right—it is only a lack of sound views w[hic] h makes us fear death.[4] The survivors suffer more than the departed. But what a gloomy Xmas letter I am writing! (our Xmas has just begun).[5] Pardon me! I always speak to you as if to my conscience—without any restriction—without any "curly" circumlocutions & *[ambiguity]*. I have received the other day a complete copy of the Preface, address & Manchester lecture.[6] (How I do love y[ou] r few additional lines about G. Bruno!)[7] but again I do not know who the kind sender is. As I want several copies I wrote to different persons about it & do not see exactly who is to be thanked in the present case.—

One thing is certain, if there is one single being on earth who is [in] full possession of his mental faculties—it is you. What Keyserling said about you,[8] is thought by all those who are capable of thinking at all. Frohschammer's review[9] will be a new proof of this, I am sure. The Belfast address, has several translations[10]—neither of w[hic]h pleases me, as they are all incomplete, thanks to the omissions made by the censure.[11] I cannot understand the necessity of a censure in a country, where all the reading people generally live a g[rea]t part of their life abroad, & reading there whatever they like! But enough for today. I'll try not to write to you before I get a new encouragement from y[ou]r part for doing so!

Y[ou]rs ever, Olga N.

P.S. | Who is Mr. Auberon Herbert,[12] who wrote a letter in the Times "On Prayer"???[13]

RI MS JT/1/N/20

---

1. *Poste restante*: see letter 4392, n. 7.
2. *My silence*: Novikoff had last written to Tyndall on 25 December 1874 (see letter 4515).
3. *my poor boy*: Alexander Novikoff.
4. *Epicurus was quite right . . . fear death*: in his Epistle to Menoeceus, the Greek philosopher

argued that while we live, death is not yet present, and when death is present, then we do not exist. As such, death affects neither the dead nor the living, so there is no need to fear it.

5. *our Xmas has just begun*: the Russian Orthodox Church uses the Julian calendar, in which Christmas begins on 7 January.
6. *complete copy of the Preface, address & Manchester lecture*: *Belfast Address*, 7th thousand.
7. *y[ou]r few additional lines about G. Bruno*: Tyndall added an 'Appendix: *Note on Bruno*' to the new edition of the *Belfast Address*, contending that 'Bruno was neither an atheist nor a materialist' (p. 68).
8. *What Keyserling said about you*: see letter 4509.
9. *Frohschammer's review*: see letter 4515, n. 5.
10. *several translations*: see letter 4505.
11. *omissions made by the censure*: Tsarist Russia had a long tradition of state censorship of newspapers and books, and while Tsar Alexander III had lifted some censorship laws in the mid-1850s, they had been reimposed in 1866.
12. *Mr. Auberon Herbert*: Auberon Herbert (1838–1906), a British politician and intellectual disciple of Herbert Spencer (*ODNB*).
13. *a letter in the Times "On Prayer"*: Herbert's letter to the *Times* criticized the superstition of 'prayer with the view of working on God's will', but maintained a faith in a higher 'form of prayer—purer, more spiritual, and more disinterested than those with which we are usually familiar' ('Mr. Auberon Herbert on Prayer', *Times*, 2 January 1875, p. 6).

To John Dougall[1] [*c.* 9 January 1875][2] 4528

Dear Sir

I should like to express to you the pleasure with which I have read your lecture before the Glasgow Chemists and Druggists Association.[3] I do not relinquish the hope that a mind so clear and searching as yours will finally, and after the bad experimental work with which we have been drenched during the past six years has been swept away, see emergent from the chaos that Pharos light[4] which will assuredly be the pride of the physicians of the future—the germ theory of epidemic disease.[5]

Believe me dear Sir | Yours very truly | John Tyndall

RI MS JT/1/TYP/1/331
LT Typescript Only

1. *John Dougall*: John Dougall (1829–1908), medical officer of the burgh of Kinning Park in Glasgow.
2. *[c. 9 January 1875]*: the date is suggested by reference to the publication of Dougall's lecture (see n. 3). Tyndall is likely to have written soon after reading the second part, which was published on 9 January.

3. *your lecture before the Glasgow Chemists and Druggists Association*: Dougall gave a lecture on 'Zymotic Poison' on 23 December 1874 at a meeting of the Glasgow Chemists and Druggists' Association at the city's Andersonian University. It was published in two parts, on 2 and 9 January, as J. Dougall, 'Zymotic Poison', *Pharmaceutical Journal*, 5 (1874–5), pp. 524–6 and 548–50.
4. *that Pharos light*: the famous lighthouse that towered over the ancient city of Alexandria in Egypt, and whose illumination protected shipping from the treacherous island of Pharos.
5. *the germ theory of epidemic disease*: in his lecture Dougall explained: 'Believers in the germ theory hold that zymotic diseases are caused by living vegetable organisms or ferments, or fungus spores, which, entering the body, reproduce themselves . . . each disease having its special ferment or germ'. However, he disagreed with this and instead concluded: 'I consider there is a large balance of evidence proving that bacteria and fungi in dead matter, also the various animal and vegetable parasites of living organisms, are absolutely non-zymotic, the result and not the cause of the morbid conditions of their habitat' (pp. 526 and 549).

To Mary Adair 11 January [1875][1] 4529

11th Jan[uar]y

My dear Miss Adair

Amid the shower of abuse which has descended on me during the last four months I have sometimes asked myself how my sins have affected you. Will you accept a copy of the iniquitous Address[2] from me, in the Preface to which[3] I have sought to clear up my case. If it seems the proper fate for it pray put the book in the fire.

most faithfully Yours | John Tyndall

RI MS JT/1/T/89

1. *[1875]*: the year is established by reference to the publication of Tyndall's Belfast Address (see n. 2).
2. *a copy of the iniquitous Address: Belfast Address*, 7th thousand.
3. *the Preface to which*: 'Preface to the Seventh Thousand', pp. v–xxxii.

From Thomas Henry Huxley 11 January 1875 4530

4 Marlborough Place[1] | Jan[uary]. 11th 1875.

My dear old Shylock,[2]

My argosies[3] have come in and here is all that was written in the bond![4] If you want the pound of flesh[5] too, you know it is at your service, and my

Portia[6] won't raise that pettifogging objection to shedding a little blood into the bargain which that other one[7] did.

Ever yours faithfully | T.H. Huxley.

IC HP 8.171
RI MS JT/1/TYP/9/3044
Typed Transcript Only

1. *4 Marlborough Place*: see letter 4213, n. 1.
2. *Shylock*: the Jewish usurer in William Shakespeare's play *The Merchant of Venice* (*c.* 1600).
3. *argosies*: merchant-vessels of the largest size and burden; especially those of Venice (*OED*).
4. *the bond*: not identified, but possibly relating to payments from the International Scientific Series, as discussed in letter 4456.
5. *the pound of flesh*: in *The Merchant of Venice*, Shylock lends money to Antonio but demands that 'the forfeit | Be nominated for an equal pound | Of your fair flesh, to be cut off and taken | In what part of your body pleaseth me' (I.i.156–9).
6. *my Portia*: presumably Henrietta Huxley.
7. *that pettifogging objection . . . that other one did*: when, in *The Merchant of Venice*, Shylock seeks to claim his forfeit from Antonio, Portia demands of him 'Shed thou no blood, nor cut thou less nor more | But just a pound of flesh. If thou takest more | Or less than a just pound, be it but so much | As makes it light or heavy in the substance | Or the division of the twentieth part | Of one poor scruple—nay, if the scale do turn | But in the estimation of a hair, | Thou diest and all thy goods are confiscate' (IV.i.317–24).

To Thomas Henry Huxley 12 January 1875 4531

Royal Institution of Great Britain. 12th. Jan[uary]. 1875

Dear Hal

I give you a full acquittance—That 5 per cent[1] goes more against my grain than I can tell you—Still I hope to be even with you some day.

Ever Yours | John Tyndall

Do you care about tickets for next Friday?[2]

IC HP 8.172
RI MS JT/1/TYP/9/3044
Typed Transcript Only

1. *a full acquittance—That 5 per cent*: see letter 4530.
2. *tickets for next Friday*: for Tyndall's Friday Evening Discourse 'On Acoustic Reversibility', which he gave at the RI on 15 January (see letter 4522, n. 1).

To Charles Cecil Trevor 12 January 1875 4532

Royal Institution of Great Britain | 12 Jan[uar]y. 1875

My dear Mr. Trevor.

I will most willingly correct the proofs.[1]

There has however been some additional correspondence on the subject with the Trinity House.[2] I should like to know whether this correspondence is to be published—as on that will depend whether a slight alteration and addition (needed to prevent misapprehension) should not be introduced into those proofs.

May I keep these till Saturday? I am in the midst of my preparations for a severe evening on Friday.[3]

faithfully yours | John Tyndall

The appendices A, B, C, D & E were sent to the Trinity House last Saturday.[4]

National Archives, MT 10/220. Lighthouses, Illumination by means of gas. H935

1. *the proofs*: of Tyndall's contribution to 'Further Papers', pp. 53–6.
2. *some additional correspondence on the subject with the Trinity House*: not identified, although it may relate to the issues that Tyndall mentioned in letter 4540.
3. *a severe evening on Friday*: Tyndall's Friday Evening Discourse 'On Acoustic Reversibility', which he gave at the RI on 15 January (see letter 4522, n. 1).
4. *sent to the Trinity House last Saturday*: letter missing. For Trinity House, see letter 4187, n. 2.

From Mary Adair 12 January [1875][1] 4533

Heatherton.[2] Tuesday. | Jan[uar]ry. 12th

Dear Dr Tyndall,

I thank you for your note and gift[3] . . . which I do value. Though I do not in all you say agree with you, yet I am content to attribute much of this to my inferiority of mind—my brain being unequal to such a wide grasp of truth as your's. But I can see that much of the clamour is simply due to the ignorance and willfulness of those who make it; and I fought many battles on y[ou]r behalf last autumn in Scotland. In Nov. I had the pleasure of hearing Mr. Haweis preach on the "Belfast Address"[4] and it was by far the calmest, clearest, and most satisfactory of the many I have heard on the subject. I hope you are strong and well and don't mind the "clattering of tongues"[5]—

Your's very faithfully | H. M. Adair.

RI MS JT/1/A/13

---

1. *[1875]*; the year is established by the relation to letter 4529.
2. *Heatherton*: Heatherton Park in Somerset had been the seat of the Adair family since 1807.
3. *your note and gift:* letter 4529 and *Belfast Address*, 7th thousand.
4. *M^r. Haweis preach on the "Belfast Address"*: the Anglican clergyman Hugh Reginald Haweis (1838–1901) was the incumbent of St James's Church, Marylebone, which was probably where Adair heard him preach. Some of what he said might have been reprinted in Haweis's collection of 'pulpit discourses' *Current Coin* (London: Henry S. King, 1876), in which the chapter on 'Materialism' discusses the Belfast Address, pp. 3–47, on p. 20.
5. *"clattering of tongues"*: possibly a mistaken reference to Marcus Aurelius's warning that 'the acclamations of the multitude are but a clapping of tongues' in his *Meditations*, VI.xvi.

From Rudolf Clausius 13 January 1875 4534

BONN, 13/1/75.

Lieber Tyndall,

Ich habe so lange nichts von Dir gehört, dass ich doch einmal schreiben muss, um eine Nachricht von Dir zu bekommen. Du weisst ich bin ein nachlässiger Briefschreiber. Ohne besondere Veranlassung schreibe ich an Niemand. Aber wenn auch noch so lange Zeit vergangen ist, ohne dass ich geschrieben habe, so kannst Du doch immer sicher sein, dass meine Gesinnungen gegen Dich unwandelbar dieselben sind.

Ich will Dir nun vor Allem recht herzlich Glück zum neuen Jahre wünschen. Besonders wünsche ich dass du Dich bei Deinem Leben voller Arbeit und Aufregung nicht zu sehr anstrengst, sondern Dir Deine Gesundheit kräftig erhältst.

Deinen Vortrag in Belfast für dessen Zusendung ich dir bestens danke, hab ich mit grossem Interesse gelesen. Er hat Dir neben grosser Anerkennung auch manche Anfechtungen zugezogen; aber ich denke, diese Art von Anfechtungen wird Dich nicht sehr schmerzen. Sie sind für dich nur ehrenvoll.

Vor der Versammlung in Belfast hattest Du in der Contemporary Review einen Artikel gegen Tait und seinen Anhang veröffentlicht, der sehr scharf, aber gewiss vollständig verdient war. Ist darauf keine Antwort erfolgt? Ist Tait auf der Belfaster Versammlung gewesen, und wie hat er sich dort benommen?

Ich habe sehr bedauert, nicht hinkommen und Dich in Deinem Ehrenamte zu sehen, und hören zu können. Ich bin aber jetzt mit meiner Gesundheit so bestellt, dass ich, um mich frisch und arbeitsfähig zu erhalten, alle körperlichen Anstrengungen vermeiden muss, und daher an grössere Reisen und das Mitmachen von Festlichkeiten nicht denken darf. Ich habe selbst die Naturforscherversammlungen von denen eine ganz in der Nähe, in Wiesbaden, war, in den letzten Jahren nicht besucht.

Vor einiger Zeit schrieb mir Vieweg in Braunschweig, dass von meiner Sammlung von Abhandlungen über die mechanische Wärmetheorie eine neue Auflage nöthig ist. Ich werde diese Gelegenheit benutzen, um den ganzen Gegenstand mehr im Zusammenhange zu bearbeiten, und eine Art Lehrbuch daraus zu machen. Der Druck wird sehr bald beginnen.

Vor einiger Zeit hatten wir das Vergnügen, Sylvester hier in Bonn zu sehen, und hatten ihn einen Abend bei uns, wo er die auch anwesenden hiesigen Mathematiker durch seine anregenden mathematischen Ideen in eine sehr lebhafte Unterhaltung verwickelte.

Nun, lieber Tyndall, lebe wohl, und nimm auch von meiner Frau und den Kindern, besonders von Johnny, die herzlichsten Glückwünsche zum neuen Jahre, und von uns allen die besten Grüsse,

Dein | Clauius.

Sei so gut, auch Hirst ... zu grüssen

BONN, 13/1/75

Dear Tyndall,

I have not heard from you for such a long time[1] that I have to write once in order to receive a message from you. You know I am a neglectful correspondent. Without a special reason I write to no one. But no matter how much time has passed without my writing, you can still always be sure that my sentiments towards you remain unwaveringly the same.

First of all, I want to cordially wish you the best for the New Year. I especially hope that you will not exert yourself too much in your life filled with work and excitement, but that you keep in robust health.

I have read with great interest your Belfast lecture, for the sending of which I thank you very much.[2] In addition to wide recognition, it also brought you quite a few challenges; but I think that this kind of challenge will not pain you much. They are only honourable for you.

Before the conference in Belfast,[3] you had published an article in the Contemporary Review[4] against Tait and his followers[5] that was very harsh, but certainly entirely deserved. Was there no response to it?[6] Has Tait been at the Belfast conference,[7] and how did he behave there? I very much regretted not being able to come and see and hear you in your honorary office. But my current state of health[8] is such that in order to keep myself fresh and able to work, I must avoid all physical effort and therefore may not think about longer journeys or attending festivities. In recent years, I did not even attend the conferences of natural scientists, one of which was quite close in Wiesbaden.[9]

Some time ago, Vieweg wrote to me in Braunschweig that a new edition of my collection of articles on the mechanical theory of heat was necessary.[10] I will use this opportunity to bring the entire subject more into context and to turn it into some kind of textbook. Printing will begin very soon.

Some time ago we had the pleasure of seeing Sylvester here in Bonn, and had him at ours one evening, where he also engaged the local mathematicians that were present in a very vivid discussion with his stimulating mathematical ideas.[11]

Now, dear Tyndall, farewell, and also accept from my wife and children, especially Johnny,[12] the most sincere best wishes for the New Year, and the warmest greetings from us all,

Your | Clausius.

Be so good, to also . . . greet Hirst.

RI MS JT/1/TYP/7/2331–2
LT Typescript Only

---

1. *such a long time*: seemingly since 24 May 1874; Tyndall's last extant letter was 4334, to which Clausius replied in 4343.
2. *your Belfast lecture, for the . . . you very much*: probably *Belfast Address*, 7th thousand, which was presumably sent by the printers Spottiswoode & Co. rather than by Tyndall himself as Clausius has just complained at the length of time since he last heard from him.
3. *the conference in Belfast*: the meeting of the BAAS from 19 to 26 August 1874.
4. *an article in the Contemporary Review*: see letter 4333, n. 2.
5. *Tait and his followers*: Peter Guthrie Tait, George Forbes, and John Ruskin.
6. *no response to it?*: Tait responded privately in letter 4355.
7. *Has Tait been at the Belfast conference*: Tait seems not to have attended the BAAS meeting in Belfast and instead received satirical commentary on Tyndall's presidential address from James Clerk Maxwell. See C. G. Knott, *Life and Scientific Work of Peter Guthrie Tait* (Cambridge: Cambridge University Press, 1911), pp. 173–4.
8. *current state of health:* see letter 4343, n. 2.
9. *conferences of natural scientists . . . in Wiesbaden*: in September 1873 the Versammlung Deutscher Naturforscher und Ärzte (Assembly of German Naturalists and Physicians) had met in the spa town of Wiesbaden in central Germany.
10. *a new edition . . . mechanical theory of heat*: R. Clausius, *Die mechanische Wärmetheorie*, 2nd edn (Braunschweig: Friedrich Vieweg, 1876).

11. *his stimulating mathematical ideas*: James Joseph Sylvester was at this time likely working on invariant and covariant theory.
12. *Johnny*: Rudolf John Clausius.

From Albert Réville 13 January 1875 4535

Neuville sur Dieppe (Seine-Inférieure) | 13 Janvier 1875

Monsieur,

On ne peut vous lire sans se sentir animé du désir de vous connaître, et pour parler de vous convenablement—comme je compte le faire prochainement dans la Revue des deux Mondes—il faut avoir quelques données précises sur votre personne et votre position. Je suis encore encouragé à tenter la démarche que je fais en ce moment près de vous par une dame distinguée que nous connaissons également, Mme Olga Novikoff. Voici *[en]* peu de mots ce que j'ose vous demander, et si vous voulez bien satisfaire à ma demande, il m'est indifférent que vous le faisiez en anglais ou en français.

Je viens vous demander de me donner une courte notice sur votre date de naissance, le lieu qui vous a vu naître, votre carrière et la position que vous occupez actuellement. Je pousse même l'indiscrétion jusqu'à vous demander la même chose au sujet de votre ami Mr Le professeur Huxley. Car je compte parler aussi de lui dans l'article que je prépare et vous indiquer tous les deux comme les champions au bon droit de la science contre les entraves arbitraires d'une tradition soi-disant religieuse.

J'ai trop d'estime pour vous et votre libéralisme pour craindre d'ajouter que nous n'envisageons pas les choses religieuses tout à fait du même point de vue. Je crois avoir le droit d'être plus affirmatif que vous, mais je vous reconnais complètement celui de ne pas affirmer plus que vous n'êtes persuadé. En fait, et dans la mesure où j'ose me comparer à vous, nous sommes partis de deux points opposés de l'horizon. Je suis terriblement hérétique pour un théologien, et vous avez un sentiment religieux très pur pour un savant naturaliste. Il en résulte selon moi que, toute proportion gardée, nous marchons à la rencontre l'un de l'autre, et cela suffit, n'est-il pas vrai, pour justifier les sympathies réciproques.

Ne connaissant pas votre adresse, j'envoie cette lettre à votre *[éditeur]* qui saura sans doute vous la faire parvenir.

En vous priant d'excuser ma liberté grande et de la mettre, si vous voulez, sur le compte de cette impertinence française qui n'attend pas d'avoir été présentée pour interpeller les gens dont elle désire obtenir quelque chose, j'ai l'honneur de vous exprimer mes sentiments de haute consideration

A. Réville

Neuville sur Dieppe (Lower-Seine) | January 13 1875

Sir,

One cannot read your work without being impelled by a desire to get to know you and to talk of you properly—as I intend to do soon in the Revue des Deux Mondes[1] and I need some accurate facts about yourself and your position. I am still encouraged to approach you, by a distinguished lady who you also know, Mrs Olga Novikoff.[2] Here are a few things that I would like to ask you, and if you would consent to answer, it makes no difference to me if you do so in English or in French.

I am asking you to give me a short note on your date of birth, the place where you were born, your career and the position that you occupy now. I even push the indiscretion to ask you the same thing about your friend Professor Huxley, since I intend on speaking about him as well in the article that I am preparing, and to present the two of you as the rightful champions of science against the arbitrary barriers of a so-called religious tradition.

I have too much respect for you and for your liberalism to fear adding that we do not see religious things quite from the same perspective. I think that I have the right to be more assertive than you, but I recognize your right not to assert more than what you believe in. In fact, and as far as I dare compare myself to you, we came from opposite sides of the horizon. I am terribly heretical for a theologian, and you have a very pure religious feeling for a natural scientist. I think that this results in that, all proportions guarded, we move to meet each other, and that is enough to justify mutual sympathies, is it not?

Not knowing your address, I send this letter to your editor,[3] who will probably send it to you. I beg you to excuse my great freedom, and to put it, if you would like, on the account of this French impertinence, which does not wait to be presented to approach the people from who she wants to obtain something. I have the honour of expressing my feelings of the highest consideration towards you.

A. Réville

RI MS JT/1/R/20
RI MS JT/1/TYP/3/1013

1. *soon in the Revue des Deux Mondes*: A. Réville, 'Les Sciences naturelles et l'orthodoxie en Angleterre', *Revue des deux Mondes*, 8 n.s. (1875), pp. 283–318. On the article's origins, see letter 4452.
2. *encouraged to approach you ... Mrs Olga Novikoff*: Novikoff had told Tyndall about Réville's interest in him (see letters 4413 and 4452).
3. *your editor*: Réville presumably means Tyndall's publisher, and so probably sent the letter to Longmans, Green, & Co., 39 Paternoster Row, London.

To Olga Novikoff 14 January [1875][1] 4536

Royal Institution of Great Britain | 14th. January

Dear Friend

I am in the midst of my preparations for a lecture[2] but I turn aside to write to you. Your letter[3] reached me last night. Accept my cordial sympathy, but you may be too much cast down about your boy.[4] I had twice over, when young, inflammation of the lungs. I could plainly see that the two physicians who attended me had little hopes of my life—One of them said to me when I was recovering "John you will never be secure from attacks of inflammation." Well from that hour to this I have known nothing of the kind. And I hardly think any man in England has tried his lungs or indeed any of his internal organs, more severely than I have tried mine. They have been subjected to the iciest air, and the hardest work of the Alps. This I think ought to cheer you.

Frohschammer has just sent me his friendly review.[5] I will write to him on Saturday or Sunday. My preface[6] appears to have silenced them—they seem somewhat ashamed of themselves. One thing is very noticeable—that while religion in many cases exalts natural nobleness, and strengthens the spirit of self sacrifice, in other cases it has a purely evil influence—making the base ten-fold more base. Some of them lie without remorse to gain their ends with an adversary.

Auberon Herbert is the brother of Lord Carnarvon:[7] a very pure, good, wellmeaning radical, but not an over strong young man.

I see Kinglake from time to time, he seems very well. He has got another instalment of his work[8] off his hands.

You write to me sometimes as if I were ungrateful for your kindness—this is not the case; my seeming neglect and delay arise wholly from my heavy work, and, though you will not believe me, my bad brain. Had nature given me the capacity of resting that brain by sleep, I might have really approached what you suppose me to be. I might have made my mark in the world. But as it is I am like a climber in the mountains who has to depend upon a broken leg. At rare intervals I feel what I might have been had the power that rules human destinies been propitious to me. But I will not complain I have only too much reason to be content.

Goodbye for the present | Ever Yours faithfully | John Tyndall

RI MS JT/1/TYP/3/951
M.P. for Russia, vol. 1, pp. 159–60

1. *[1875]*: the year is established by reference to Tyndall's lecture (see n. 2).
2. *a lecture*: 'On Acoustic Reversibility', given as a Friday Evening Discourse at the RI on 15 January (see letter 4522, n. 1).
3. *Your letter*: letter 4527.
4. *your boy*: Alexander Novikoff.
5. *Frohschammer has just sent me his friendly review:* see letter 4515, n. 5.
6. *My preface*: 'Preface to the Seventh Thousand', in *Belfast Address*, 7th thousand, pp. v–xxxii.
7. *Auberon Herbert is the brother of Lord Carnarvon*: on Herbert, see letter 4527, n. 12. Henry Howard Molyneux Herbert (1831–90), 4th Earl of Carnarvon, was, like his younger brother, a politician (*ODNB*).
8. *another instalment of his work*: volume 5 of A. W. Kinglake, *The Invasion of the Crimea*, 9 vols (Edinburgh: William Blackwood, 1863–87), which was published in 1875.

From Jane Barnard 14 January 1875 4537

Barnsbury Villa, | 320 Liverpool Road, | N.[1] | Thurs Eve[nin]g | 14 Jan[uary]. 1875

My dear Dr Tyndall

I am disappointed to think that I shall miss the great enjoyment of being at the R.I. tomorrow evening[2] for though my dear Aunt[3] is better I cannot leave her, for she needs constant attendance. Today is the first day since her attack[4] that I could say she was herself again.—she is still very weak & entirely confined to bed, but her mind is lively & she can take up matters with something of her old vigour. She was very much pleased at your coming & asked me two or three times,—"what was the word you used as applicable to herself & you?, & to find it in the dictionary for her & tell her the exact meaning (resiliency).

We have been much interested in the sad case of our old housemaid M[rs] Brown.[5] She told us of your kindness to her. Unless she can get some help to get her over the next few months, it will be very grievous for her, her hands are quite tied.

I have had the pleasure of giving one of your tickets to my nephew Clement Reid[6] for tomorrow evening. It was he that you so kindly helped on to the Geological Survey; & we believe it has been a step, satisfactory to all concerned. Hoping this will not trouble you in the midst of your preparations (I know something of the setting up of a F.E.[7] such as yours & with my Aunts affectionate regards

I am dear D[r] Tyndall | yours very Sincerely | Jane Barnard

RI MS JT/1/B/56

1. *Barnsbury Villa, | 320 Liverpool Road, | N.*: see letter 4226, n. 1.
2. *the R.I. tomorrow evening*: for Tyndall's Friday Evening Discourse at the RI 'On Acoustic Reversibility' (see letter 4522, n. 1).
3. *my dear Aunt*: Sarah Faraday.
4. *her attack*: possibly from intestinal failure (see letter 4640). Faraday had also suffered a fall in September 1874 which brought on a bout of illness (see letter 4418).
5. *the sad case of our old housemaid Mrs Brown*: not identified.
6. *Clement Reid*: Clement Reid (1853–1916), a geologist and paleobotanist, largely self-taught. Tyndall helped him get a position with the Geological Survey of Great Britain in 1874.
7. *a F.E.*: a Friday Evening Discourse. As Michael Faraday's niece, Barnard likely had seen preparations for them many times.

To Rudolf Clausius 16 January 1875 4538

16th. January 1875

My dear Clausius,

I was very much rejoiced to receive your last letter.[1] For though I know that you are one of those whose friendship does not shift like the wind, still it is so great a pleasure to me to hear from time to time some account of you from yourself, that I thought your silence very long. I return with all cordiality every kind wish that you have expressed. Indeed I doubt whether you have a friend who desires your welfare and happiness more than I do.

That Belfast Address has brought down upon me a perfect avalanche of abuse. But it has taught me many things. Among others that I am capable of bearing a great deal of abuse without caring for it. I dare say Tait will make his appearance again,[2] but I do not think that I shall waste any more time in discussion with him. He is from nature a coarse man, from whom it is vain to expect a dignified or gentlemanly behaviour.

Hirst is now well. But a few months ago he caused me great anxiety of mind. He was attacked by something that looked very like paralysis.[3] But he has happily shaken it off, and is now in fair health. He has a capital position at Greenwich[4]—a beautiful house and 8000 Thalers[5] a year. So that he only needs good health to render his life a happy one.

It grieves me very much to hear, my dear Clausius, that your health is not strong.

Johnny[6] must now be growing—a big boy. I remember the little man very well, and I remember with great distinctness all the children sleeping so peacefully and so innocently under the care of their loving mother.[7]

It gives me great pleasure to learn that you intend to throw your memoirs into the form of a Lehrbuch.[8] It will be an epoch in the history of a great subject. It will also be of great service to me personally.

I had a long conversation last night with Sylvester about his "Arborescences"[9]—M. Camille Jordan[10] is here and his stimulated Sylvester. I always liked Sylvester much: for 20 years and more we have known each other, and have never had a difference: but his temper is very warm and some of his friends have not been so fortunate as I have been. Give my kindest regards to M^{rs}. Clausius[11] and the children[12] and believe me always

faithfully yours | John Tyndall

RI MS JT/1/T/207

---

1. *your last letter:* letter 4534.
2. *Tait will make his appearance again*: see letter 4534, n. 7.
3. *something that looked very like paralysis*: in the summer of 1874 Hirst suffered from kidney stones in the ureter (see letters 4372 and 4388.)
4. *a capital position at Greenwich*: Hirst was the Director of Studies at the Royal Naval College, Greenwich, from 1873–82.
5. *Thalers*: see letter 4419, n. 10.
6. *Johnny*: Rudolf John Clausius.
7. *their loving mother*: Clausius's wife Adelheid would die on 1 March.
8. *Lehrbuch:* textbook (German). Tyndall presumably meant R. Clausius, *Die mechanische Wärmentheorie*, 2nd edn (Braunschweig: Friedrich Vieweg, 1876), which appeared in English as *The Mechanical Theory of Heat*, trans. W. R. Brown (London: Macmillan, 1879).
9. *"Arborescences"*: in the branch of mathematics that deals with graphs, an arborescence is a particular form of directed graph. Working on this subject, James Joseph Sylvester later noted: 'I have found it a profitable exercise of the imagination, from a philosophical point of view, to build up the conception of an *infinite* arborescence and to dwell on the relations of time and causality which such a concept embodies' (J. J. Sylvester, 'On an Application of the New Atomic Theory to the Graphical Representation of the Invariants and Covariants of Binary Quantics', *American Journal of Mathematics*, 1 (1878), pp. 64–104, on p. 90).
10. *M. Camille Jordan*: Camille Jordan (1838–1922), a French mathematician and pioneer in group theory.
11. *M^{rs}. Clausius*: Adelheid Clausius.
12. *the children*: Clausius and his wife had six children (see letter 4300, n. 25).

From Hector Tyndale 17 January 1875 4539

Philadelphia January 17th 1875

My dear John

I have to thank you for your book "On the Atmosphere as a Vehicle of Sound"[1]—received so long ago and read with much interest—and for a copy of your Belfast Address with the preface to the Seventh Thousand.[2] I had nothing, of any interest, to say and I knew you would be deluged with all sorts of words about the Address. The two prefaces to the Address[3] pleased me greatly, more especially the later one. They demonstrate your good feeling in the debate and desire for the truth.

In this Country, as with you, people's opinions have been divided pretty much according to their theologies, but, on the whole, very largely in favor of the clearness and fairness of your Address. But of all this you know more than myself as your friends & others have no doubt kept you informed.

At a party the other evening, I met Mr Eli K. Price,[4] an old and prominent lawyer and a good deal of a polemic and disputant, and when he mentioned your name I supposed he was about to traverse your Address. But on the contrary he spoke of the fairness and good feeling of it and wound up by asking me to present his kind respects to you—he met you here, at Judge Mitchell's[5] I believe. On several occasions Dr: Gross has asked me to give his kind remembrances—and so many others upon whom you made an impression personally. A few days since I met Rev[erend]: W[ilia]m H. Furness and as usual he spoke of you. As I told him I should report the conversation, although it might make you blush, here is the gist of it. "I have just read Tyndall's preface to his Belfast Address and am struck with it's clear truthfulness, I am more and more pleased with him." "He is my pattern of a Christian man" To which I replied that I had known you more called otherwise quite as frequently. "Yes I know," said he, "but I mean in his single devotion to Truth—in his sincere and perpetual search for Truth and his courage in the search" He said he should like to send you "a little sermon" he had just preached[6] but feared to annoy you. I told him that, no doubt, you would be pleased to get it and that you would read it. And then he desired his kind respects &c sent to you. A short time since I read a work with the singular title of "Deicides".[7] It is an "analysis of the life of Jesus" from a Jewish stand point and treated fairly. For fear you have not seen the book I have copied a legend from it, attributed to the Talmud Babo Meziah[8] by the Author.[9] Enclosed is the copy.[10] I think it one of the grandest and fairest illustrations of earnest, honest, self assertion I ever saw. In some respects it almost equals the grand self defense and supra-egoistic assertion of Job.[11] The defense of enthusiasm—of the God within him. I hope you have not seen the legend before so that I may anticipate your pleasure in another fine illustration of Truth.

Some weeks ago I received kind letters from the young ladies of Dr Tyndall's family—with them came a kind message from Emma. I had not written to any of them for a long time, nor did I go over to see them when in England last Spring. It gave me much pleasure to learn that they had not taken my long silence and neglect illy and so I wrote them. I feel a strong affection for your Sister Emma, 'though I have seen her so seldom. She appears to me an earnest, patient seeker after Truth and of a real womanly, kindly nature. I have a real regard for all of Dr Tyndall's family and trust that no seeming neglect of mine may wound them. Give my love to Hirst and Debus. Mrs Tyndale desires to be remembered to Hirst and sends kindest regards to you.

Affectionately your Cousin | Hector Tyndale

Professor John Tyndall &c

RI MS JT/1/T/77
RI MS JT/1/TYP/5/1734–7

---

1. *your book "On the Atmosphere as a Vehicle of Sound"*: this was not a book, but rather a long article, 'Vehicle of Sound', of which Tyndall presumably sent an offprint.
2. *your Belfast Address with the preface to the Seventh Thousand*: *Belfast Address*, 7th thousand.
3. *The two prefaces to the Address*: 'Preface to the Seventh Thousand', pp. v–xxxii, and 'Preface to the First Thousand', pp. xxxiii–xxxvi.
4. *Mr Eli K. Price:* Eli Kirk Price (1797–1884), a Pennsylvania State Senator and political reformer. He was a member of, and regular contributor to, the APS.
5. *Judge Mitchell*: James Tyndale Mitchell (1834–1915), who was a judge of the Philadelphia District Court.
6. *"a little sermon" he had just preached*: W. H. Furness, *Discourse Delivered January 10, 1875, on the Occasion of the Fiftieth Anniversary of his Ordination* (Philadelphia: Sherman, 1875), which stated: 'The Darwinian law of Natural Selection and the Survival of the Fittest is in all men's minds, and in the material, organized world of plants and animals, we are all coming to consider it demonstrated. As an animal, man must be concluded under that law. In the physical world, as Professor Tyndall tells us, "the weakest must go to the wall". But man is something, a great deal more than an animal. He has an immaterial, moral, intellectual being, for which he has the irresistible testimony of his own consciousness; and as an immaterial being, it is not at the cost of the weak, but it is by helping the weak to live that any individual becomes strong' (pp. 24–5).
7. *"Deicides"*: J. Cohen, *The Deicides: Analysis of the Life of Jesus, and of the Several Phases of the Christian Church in their Relation to Judaism*, trans. A. M. Goldsmid (London: Simpkin, Marshall, 1872).

8. *Talmud Babo Meziah*: the Talmud is the principal text of Rabbinic Judaism, and the Babo Meziah, usually transliterated as Bava Metzia, is the second of the first three Talmudic tractates in the order of Nezikin.
9. *the Author*: Joseph Cohen (1817–99), a French lawyer and journalist.
10. *Enclosed is the copy*: the enclosure is a transcription of folio 59 of the Talmud Bava Metzia taken from pp. 40–1 of Cohen's book.
11. *the grand self defense and supra-egoistic assertion of Job*: Job 26–31.

To Robin Allen 21 January 1875 4540

Royal Institution of Great Britain | 21st January 1875.

SIR,

I HAVE read the report of Mr. Douglass[1] which you have submitted to me; and although it throws upon me unexpected labour at a very busy time, I comply, as far as it is possible for me to do so, with the desire of the Elder Brethren[2] by offering upon the report the following observations.

Firstly, with regard to the 28-jet burner[3] I am not disposed to assume with Mr. Douglass that the difference between the photometric measurements,[4] and the observations at sea, is due to the "vertical height" of the gas flame, rendering its light exfocal.[5]

It would have been gratifying to me had Mr. Douglass mentioned this objection at Haisbro'.[6] For by a process familiar to both of us it would have been easy to cut off the exfocal light and to determine its exact numerical value. After reading the report of Mr. Douglass the best means open to me was to ask the celebrated optician Mr. Grubb, of Dublin,[7] to determine the photometric value of that portion of the 28-jet flame which rises above the height of 4 inches, this being the height hitherto aimed at in the case of the 4-wick Trinity lamp. Mr. Grubb has done this, and he certifies the photometric value of the part of the flame referred to, to be under 4 per cent. of the whole illumination. Mr. Douglass makes the amount 31 per cent.

I expressed my regret at Haisbro' that we were not able to make the two lights reversible; for in this way alone is any peculiarity of the dioptric[8] apparatus, which might affect the one of the other of the flames, to be eliminated. To one peculiarity I expressly drew the attention of the Elder Brethren with whom I had the honour to act, and I proposed experiments with a view to its elimination. But they did not deem such experiments necessary. Nor, in view of the performance of the gas, and its demonstrated power, "plasticity," and handiness in other respects, did I deem them necessary.

The three elements which, in my opinion, are the determining ones in the comparison of gas and oil are *power*, *cost*, and *ease of manipulation*. Instead of discussing these Mr. Douglass fixes attention upon a fourth element, which he calls "superiority as a lighthouse illuminant."[9] My only reference to this phrase shall be made with a view of rendering its meaning perfectly clear. In using it the practical question of cost and actual performance is not in the mind of Mr. Douglass, but the idea that *were the gas pulled down so as to render the two lights photometrically equal*, the better shape of the new Trinity lamp-flame would render it 31 per cent. superior to its rival. It is sufficient to say that this idea can be first experimentally tested after the two lights have been rendered reversible.

There is one point in connexion with the 28-jet burner which in justice to its inventor I must here signalise. Anxious to give the Elder Brethren the maximum of information, Mr. Valentin has stated in a table[10] the cost of the 28-jet burner, supposing it to be pitted *at Haisbro'* against the 4-wick lamp. In that table he debits the small burner[11] with the whole of the gas-maker's wages, and with the whole of the interest on 1,685*l*. 6*s*. 9*d*. Now the gas-maker is there, not to produce gas for a 28-jet burner only, but for a series of fog burners, culminating in one of 108 jets. He manufactures, not 51.4 cubic feet of gas an hour, but an average of 120 cubic feet an hour. Were it, moreover, proposed to erect works for a 28-jet burner, an idea to my knowledge not entertained, a fraction of 1,685*l*. 6*s*. 9*d*. would cover the cost.

In a supplementary sheet herewith sent (now incorporated in the report on the experiments at Haisbro') Mr. Valentin calculates the cost of the 28-jet burner on the basis of debiting it with its proportion of the gas annually consumed. The result is, that taking the flame with its alleged defects of shape—assuming the observations at sea to represent its true value—there is still, in point of economy, a large per-centage in favour of the gas as compared with the oil, while a considerably larger per-centage accrues if we take the photometric results as the correct ones.

We now pass on to the comparison of the 48-jet burner and the 6-wick lamp. Here two of the observers preferred the gas as having the greater "body." But I waive this and permit the two lights to stand as equal to the eye. Mr. Valentin credits the gas with a superiority *as to cost* of 40 per cent. Mr. Douglass credits the oil "as a lighthouse illuminant" with a superiority of 15 per cent. I have already endeavoured to render clear the meaning of this latter phrase. Let us now look at the real practical facts.

The photometer in the present instance makes the 48-jet burner 121 candles better than the 6-wick lamp. Let us regard the light of these candles as exfocal, in the worst sense; that it is thrown not upon the sea, where it might be of use, but upon the sky, where it can be of no use. The cost of the gas, calculated from the best data obtainable by Mr. Valentin, is 237*l*. 15*s*. 5*d*., the cost of the oil is 342*l*. 11*s*. 11*d*. If these numbers are correct then for every 100 pounds spent on gas, even though it should embrace 121 useless candles, 144 pounds will be spent on oil. In these calculations the cost of the gas-maker is included. I have been frequently assured by Mr. Douglass that this cost is unnecessary; and that for a slight augmentation of their pay the light-keepers would gladly undertake the making of the gas. But I would respectfully impress upon the Elder Brethren the fact that no comparison of the kind here made can exhibit the full advantage possessed by the gas. With it we can at once pass from the 48-jet burner of 832 candles to the 88-jet burner of 2,408 candles. And by cutting up the beam into equal intervals of light and darkness, with an expenditure of gas not much in excess of that necessary to feed the 48-jet burner as a fixed light, we can obtain a succession of flashes, or luminous shocks, each possessing nearly three times the power of the 48-jet beam. Nothing of this kind is possible to the oil light.

In the larger powers of the Wigham burner,[12] while showing in clear weather lights of the utmost splendour, a great part of the light is necessarily more or less exfocal. The focus is fixed for the horizon, and every addition to the power of the burner strengthens the beam in this direction. The large powers, however, are not required for the horizon; they are intended for thick and foggy weather when the horizon is utterly unattainable, and when the object is to make the light strong nearer shore.

The glare of these powerful flames upon a fog, more especially if the system of flashing be introduced, cannot fail to be of the utmost service as a guide to the mariner.*

The reference of Mr. Douglass to his experiments at Westminster[13] caused me to ask you to send me the report of those experiments. This you have been kind enough to do. To one only of the questions raised by this report I will here refer, and it is one of fundamental importance in relation to the present question. Mr. Douglass states emphatically his belief in the accuracy of the photometric results at Haisbro'. He considers them "most conclusive." Well those measurements make the 108-jet burner equal to 2,982 candles, while the measurements at Westminster make it equal to only 1,199 candles.

Either then the measurements of Mr. Douglass are inexact, or the gas he operated on must have been of extraordinary poorness.

To this point, doubtless, the Elder Brethren will direct his attention.

Mr. Douglass in his report brings the electric light into prominence.[14] I hope that noble light will ever shine upon the shores of England. Wherever sharp lines of demarcation are needed it has no rival. In such cases, and not in such alone, I should strongly recommend its adoption.

But we do not possess the data that would entitle us to place it above the gas when the maximum power of this illuminant is invoked.

It would give me great satisfaction to have this question decided; and to see the triform gas light[15] pitted, as regards cost, penetrative power, and general handiness and efficiency, against the best electric light which has stood a year's wear and tear upon our coasts. But even granting that the electric light should come out triumphant, the question would still have to be settled between gas and oil.

It would perhaps been well if the gas system had been permitted to develop itself further in Ireland before testing it in this country. When Atlantic mariners have had the advantage of comparing the Fastnet with Galley Head,[16] they will be able to pronounce authoritatively on the merits of the gas station. Meanwhile I think the Elder Brethren may accept my assurance that gas places in their hands a new power, both as regards penetration and distinction; and that great advantage will accrue from its partial intercalation with oil and electricity along our coasts. Let us look for a moment to its established ranges of power. At Haisbro' and Howth Bailey[17] there is a capacity of variation from 28 jets to 108 jets, or from 429 candles to 2,923 candles, which is a ratio of nearly 1:7. In the triform we have a power of variation from 28 jets to 324 jets, or from 429 candles 8,769 candles, which in round numbers is a ratio of 1 to 20. In the triform we dispense with the upper and lower prisms. This makes the 28-jet flame of less power than when the full apparatus is employed, but if thought necessary the addition of a few jets would re-establish the power. In clear weather the 28-jet burner seems amply sufficient for the mariner, while as the weather thickens we can rise by steps, accomplished in a moment, to a photometric intensity twenty times as great. The force is thus at hand when it is needed, while it is husbanded when a clear atmosphere renders its employment unnecessary.

By the simplest mechanical contrivances, moreover, and without any change in the dioptric apparatus, the powerful beams of the gas light can be broken up into flashes of longer or shorter duration, ascending, in the case of fogs, from the mere winking of the 28-jet burner, to the sevenfold shock of the 108-jet one, and to the twenty-fold shock of the triform. By making the intervals of light and darkness equal to each other these powerful flashes are obtained with the expenditure of gas needed to maintain the burner as a fixed light; while by varying the flashes as regards duration and succession the light might be rendered distinctive to the dullest seaman; the possibility of mistaking one light for another, which is now so frequent and disastrous,* being thus reduced to a minimum. Until I read his report I did not imagine that there was any difference of opinion between Mr. Douglass and myself on these important points.

And here I would once more emphasise the advantage likely to accrue from these powerful and sudden flashes in the case of fogs too heavy to permit of the light itself being seen through them. I have operated upon artificial fogs so dense that a stratum of two or three feet in thickness sufficed to cut off totally the view of the electric light. But the fitful *glare* produced upon the fog by the alternate extinction and ignition of the light was most remarkable. In reference to this point I may be permitted to quote a remark added on the 27th of last May to a paper presented to the Royal Society.[18] Here it follows verbatim:—"The more I think of it, and the more I experiment upon it, the more important does this question of flashes appear to me. In one of the sections of the foregoing paper experiments on artificial fogs are described. The densest of these were suddenly and strikingly illuminated throughout by the combustion of half a grain of gunpowder, and of a still smaller quantity of gun-cotton. The cutting off and restoration of the candle light, or the electric light employed to test the density of the fog, produced a similar effect. It is its suddenness that enders the lighting-flash so startlingly vivid through a cloud. A revolving light like the South Stack[19] does not fulfil the necessary conditions. Its revolution is slow and the angular spaces between the beams being filled by laterally scattered light the differential action is practically abolished. At a distance the luminosity, when uniform, may be so feeble as to be unseen, while its sudden extinction and revival would render it sensible."

In my earlier reports to the Board of Trade[20] I defended the system of gas illumination against an opposition based upon mistaken data. Here, as there, I have endeavoured to set forth clearly its merits and capabilities, which in my opinion are beyond dispute. I have neglected no means of testing this conviction in the most thorough manner. The gas system has for years successfully stood the test of experiments for themselves. They declined to do so. At the end of August last I wished the President of the Board of Trade, and the Elder Brethren his companions, who were then in the Kingstown Harbour,[21] to witness the performance of the light; they were unable to do so. Once I had the good fortune to secure the co-operation of Admiral Collinson,[22] and in Kingstown Harbour, with the triform shining at Howth Bailey five or six miles away, I asked him whether he had ever seen so fine a light. "Never, I believe" was his reply. Nor is it to be supposed that the Wigham burner has reached perfection. With due encouragement the burner of ten years hence would probably be as superior to the present one as the 4-wick lamp of to-day is to that of ten years ago.

I have, &c. | (Signed) JOHN TYNDALL.

ROBIN ALLEN, ESQ.

* *Vide* Correspondence at the end of my Report on Fog Signals, House of Commons Paper, No. 188 of Session 1874.

* I am guided here by accounts of wrecks in the newspapers, and by conversations with nautical men; deferring, of course, to the more definite and accurate records kept by the Trinity House and the Board of Trade.

'Further Papers', pp. 53–6

1. *the report of Mr. Douglass*: this had been submitted to Trinity House on 11 December 1874 and was printed in 'Further Papers', pp. 49–52. For Trinity House, see letter 4187, n. 2.
2. *Elder Brethren*: the governing body of Trinity House (see letter 4187, n. 2).
3. *28-jet burner*: John Wigham's gas burning design, which Tyndall and others had assessed as being superior at Haisbro (see letter 4438).
4. *photometric measurements*: measurements comparing the intensity of light directly emitted by sources such as candlelight or gaslight.
5. *exfocal*: not passing through the focus (*OED*). This meant that some of the light produced would not be channeled through the lens and therefore be useless.

6. *Haisbro:* Tyndall had visited the two lighthouse towers off Haisborough Sands with James Nicholas Douglass in April 1874.
7. *Mr. Grubb, of Dublin*: Howard Grubb (1844–1931), an optical instrument maker who traded from the Dublin suburb of Rathmines (*ODNB*).
8. *dioptric*: assisting vision by refracting and focusing light (*OED*).
9. *"superiority as a lighthouse illuminant"*: the quotation, of which this is a slightly amended version, appears on p. 50 of Douglass's report.
10. *a table*: see letter 4314.
11. *small burner:* the gas was produced on-site at the lighthouse in specially-built facilities.
12. *Wigham burner:* the gas burner designed by John Wigham.
13. *his experiments at Westminster*: on the capacity of the two 'Gramme' machines that powered the lights of the Clock Tower at the Houses of Parliament in Westminster.
14. *Mr. Douglass . . . electric light into prominence*: in his report, Douglass stated that 'with the electric light it would appear doubtful whether any limit can be assigned to the power that may be obtained' (p. 51).
15. *triform gas light*: a light devised by Wigham with three lenses in tiers that produced a flash appearing in three different forms.
16. *Fastnet with Galley Head*: two lighthouses off the south coast of Ireland. The former had opened in 1854 and used oil as the source of light; the latter was completed in 1875 but only became operational in 1878 using coal gas.
17. *Howth Bailey*: see letter 4314, n. 28; the customary spelling is Baily.
18. *a paper presented to the Royal Society*: 'Vehicle of Sound'. The added remark that Tyndall quotes is on p. 234.
19. *the South Stack*: a lighthouse off the coast of Anglesey in Wales, built in 1809.
20. *my earlier reports to the Board of Trade*: 'Papers Relative to Proposal to substitute Gas for Oil as Illuminating Power in Lighthouses', H.C., Command Paper [4210], *Parliamentary Papers*, (1869), pp. 1–20, on pp. 16–20; and 'Further Papers Relative to Proposal to substitute Gas for Oil as Illuminating Power in Lighthouses', H.C., Command Paper [C.282], *Parliamentary Papers*, (1871), pp. 1–33, on pp. 27–33.
21. *the Kingstown Harbour*: Tyndall had visited Kingstown (now called Dún Laoghaire) on the Irish coast near Dublin both before and after delivering his presidential address to the BAAS at Belfast on 19 August 1874. On 26 August he told Thomas Hirst that he would 'halt at Kingstown for a day or two to see some lighthouse experiments'; letter 4398.
22. *Admiral Collinson*: see letter 4314, n. 7.

From Albert Réville 21 January 1785 4541

Neuville sur Dieppe (Seine-Inférieure) | 21 Janvier 1875

Cher Monsieur,

Deux mots seulement de réponse à votre aimable lettre du 18 et en attendant les détails que vous voulez bien me promettre à bref délai sur vous et M. Huxley.

Je ne me rappelle pas vous avoir dit que nous envisagions les choses d'un point de vue opposé, je pensais vous avoir dit que, partis de points opposés, nous marchions à la rencontre l'un de l'autre. Il me semble que vous commenciez par poser la thèse matérialiste ou du moins purement expérimentale, et je me réjouis de voir que, sans sortir de l'expérience, vous constatez le fait naturel de "l'impulsion religieuse" dans l'être humain. Pour moi, théologien, je suis parti d'un ensemble d'affirmations religieuses que j'ai modifiées et modifie continuellement à l'école de l'expérience physique, historique, et psychologique, et je suis arrivé à considérer le fait dont vous parlez comme le dernier résidu, mais le résidu inattaquable, de la théorie religieuse.

Voilà dans quel sens je pense que nous marchons à la rencontre l'un de l'autre, guidés, j'ose affirmer, pour vous et pour moi, par l'amour pur de la vérité.

Maintenant je crois pouvoir être de plus un peu plus affirmatif que vous. Sans doute je crois comme vous que toutes les formules religieuses sont susceptibles de changer, et qu'il ne faut jamais l'oublier. Mais au dehors des vérités mathématiques pures, il n'est pas, il me semble, une seule branche d'affirmations humaines dont on n'en puisse dire autant. Toutefois cela doit-il nous condamner à ne rien affirmer de positif ? Voici mon raisonnement : si l'impulsion religieuse dans l'être humaine est naturelle, elle est légitime, logique comme tout ce qu'est naturel. Si elle est légitime, elle doit avoir un objet, et un objet en rapport avec sa tendance essentielle. Cet objet peut sans doute se dérober à nos définitions comme à nos expérimentations directes, mais il est a priori certain qu'il répond, fût-ce d'une manière incompréhensible, aux desiderata essentiels de l'impulsion religieuse. Impulsos pulsante denotar, terminum practicar, là-dessus je m'appuie pour stipuler la légitimité du sentiment chrétien essentiel, celui de l'affinité de l'esprit humain et de l'objet que nous appelons Dieu, affinité qui, dans le symbolisme du langage religieux, se confond avec le sentiment du rapport filial de l'homme et de Dieu. Ce qui suffit, selon moi, d'une part, pour expliquer la genèse historique des religions; de l'autre, pour revendiquer pour mon christianisme épuré les mérites d'élévation, de purification et de consolation que les orthodoxies aujourd'hui dépassées possédaient à des degrés divers.- Je n'ai pas besoin de montrer à un penseur tel que vous quelle liberté pleine et entière je reconnais aux sciences naturelles sans faire brèche à mon principe religieux, ni de vous expliquer pourquoi je suis heureux de voir un représentant aussi distingué et aussi loyal de ces sciences arriver à reconnaitre la réalité du fait qui me sert à moi-même de base fondamentale. Ceux qu'on appelle chez nous matérialistes ne veulent pas reconnaitre cette réalité.

J'ai reçu votre Address 8ème éd. avec la préface. Je la possédais déjà, elle et la préface, grâce à l'aimable attention de Mme de Novikoff. Vous êtes dans son esprit à une hauteur vertigineuse. Je ne vous en ferai pas descendre, soyez en sûr, et merci pour votre charmante manière de répondre à un indiscret tel que moi.

Mes deux mots sont devenus une longue lettre. Il y a des vibrations cérébrales qui sans doute s'engendrent impérieusement l'une l'autre sans qu'on puisse les arrêter. La dernière sera l'expression physique de l'assurance sincère que je vous offre de ma parfaite considération.

A. Réville

Neuville sur Dieppe (Lower-Seine) | January 21 1875

Dear Sir,

Allow me to say a couple of words in reply to your kind letter of the 18th[1] and pending the details that you wanted to give me without delay on you and Mr. Huxley.[2]

I do not remember telling you that we look at things from opposite points of view; I thought that I said that coming from opposite sides, we walk forward to meet each other. It seems to me that you started by posing the materialist thesis, or at least one that is purely experimental, and I am glad to see that without removing the experiment, you state the natural fact of "religious impulse" in human beings. As a theologian, I started from a collection of religious assumptions that I changed and constantly modify at the school of physical, historical and psychological experience, and I came to consider the fact about which you speak as the last remainder, but the unquestionable remainder, of the religious theory.

And this is the way in which I think that we walk towards one another, guided, dare I say, both you and me by the pure love of truth.

Now I think that I can be a little more assertive than you. Without any doubt, I believe like you that all religious formulations are susceptible to change, and that we should never forget about it. But aside of the pure mathematical truths, there is not, it seems, any one branch of human assertions about which we cannot say the same. However, should it condemn us to have nothing positive to assert? There is my reasoning: if the religious impulse in human beings is natural, then it is legitimate and logical like everything else that is natural. If it is legitimate, it must have an object, and an object in accordance with its essential tendency. That object can likely evade our definitions just like our direct experiments, but it is certain a priori that it will respond, even if in an incomprehensible manner, to the main requirements of the religious impulse Impulsus pulsantem denotar terminum praedicare.[3] I lean on this to stipulate the legitimacy of the main Christian sentiment, that of the affinity of the human mind and the object that we call God, an affinity that in the symbolism of religious language merges with the sense of a filial relationship between man and God. Which is enough, I believe, firstly, to explain the historical genesis of religions; and secondly, to claim for my refined Christianity the merits of elevation, of purification and of consolation that the orthodoxies, today outdated, had at various degrees. I do not need to show a scholar such as you what complete and utter freedom I recognize in the natural sciences without breaching my religious principle, nor do I need to explain why I am happy to see such a distinguished and loyal representative of these sciences coming to realize the reality of the fact that serves as a fundamental basis for myself. Those we call materialists do not wish to recognize this reality.

I received your Address of the 8th ed. with the preface.[4] I already owned it, along with the preface, thanks to the kind attention of Mrs Novikoff.[5] She thinks of you very highly. I will not lessen that, be certain, and thank you for your lovely way of responding to a prying person such as me.

My couple of words became a long letter. There are vibrations in the brain that may abruptly beget one another without being able to stop them. Lastly, please accept the physical expression of my sincere wishes and the offer of my highest consideration.

A. Réville

RI MS JT/1/R/21
RI MS JT/1/TYP/3/1014–5

---

1. *your kind letter of the 18th*: letter missing.
2. *the details that . . . on you and Mr. Huxley*: Réville requested these details in letter 4535.
3. *Impulsus pulsantem denotar terminum praedicare*: The impulse to observe who is driving is the impulse to predict the end (Latin). This is an approximate translation as there are errors in Réville's Latin.
4. *your Address of the 8th ed. with the preface*: J. Tyndall, *Address Delivered Before the British Association Assembled at Belfast, With Additions*, 8th thousand (London: Longmans, Green, and Co., 1874), which reprinted 'Preface to the Seventh Thousand', pp. v–xxxii.
5. *the kind attention of Mrs Novikoff*: see letter 4413.

From George Gabriel Stokes 22 January 1875 4542

22nd Jan[uar]y 1875

My dear Tyndall

Will you think over recommendations in physics for the Arctic expedition?[1] To be written on foolscap, on one side only, and sent in by the middle of March. Those named with you are B. Stewart, Thomson, Wheatstone[2] and myself.

Yours sincerely | G. G. Stokes

RI MS JT/1/S/263
RI MS JT/1/TYP/4/1428

---

1. *the Arctic expedition*: the British Arctic Expedition of 1875–6 led by George Strong Nares (bap. 1831–1915), who had also commanded the expedition of HMS *Challenger* until December 1874 (*ODNB*; on HMS *Challenger*, see letter 4489, n. 2). Both the RS and the Royal Geographic Society advised the planning of the expedition. It is not known if

Tyndall recommended any physicists, but the scientific personnel on the expedition were Henry Wemyss Feilden (1838–1921), an army officer and naturalist who served on HMS *Alert*, and Henry Chichester Hart (1847–1908), a botanist who served on HMS *Discovery* (Darwin Correspondence Project).

2. *B. Stewart, Thomson, Wheatstone*: Balfour Stewart (1828–87), a Scottish physicist and meteorologist (*ODNB*), William Thomson, and Charles Wheatstone.

To Charles Cecil Trevor 23 January [1875][1] 4543

23rd. Jan[uar]y.

Dear Mr. Trevor.

Communications with the Trinity House[2] prevented me from sending this[3] in earlier.

I have corrected a small clerical error at the request of Mr. Valentin, and added, in manuscript, a statement which is necessary to prevent misapprehension.

I am quite willing to look at the revise, and return it immediately

Ever yours | John Tyndall

National Archives, MT 10/220. Lighthouses, Illumination by means of gas. H935

1. *[1875]*: the year is established by the relation to letter 4540.
2. *Communications with the Trinity House*: letter 4540. On Trinity House, see letter 4187, n. 2.
3. *this*: the proofs of Tyndall's contribution to 'Further Papers', pp. 53–6 (see letter 4532).

From Olga Novikoff 24 January 1875 4544

Moscow. Poste Restante.[1] 24.1–75. | "Tecum vivire amem," | "Tecum obeam libens" Horatio.[2]

What a funny creature my boy[3] is! (I think, I told you, that he lives with my husband's Parents,[4] whilst I live quite alone, & only spend the day with the Novikoffs.) Well—this m[orn]ing I come to see Sacha (diminutive of Alexander) & the first thing he says, is: "I am glad to see you, mother, now, sit down & tell me something about Professor Tyndall."

—Why about him, Sacha?—

"Because I like him(!!) & besides, you like to speak of him. For today, tell me all he did in America."

The little man is not so stupid, after all! What do you think?

—But let me thank you for y[ou]r friendly note,[5] w[hic]h reached me last night; & the arrival of w[hic]h seemed to me every day less & less probable. Somebody told me kindly once that my misfortunes were chiefly created by my imagination. I know nothing of the origin of things, but it is clear to me is, that pain is painful, whatever it comes from. Correspondence is the next best thing to personal interview, & the distance between dear London & Moscow is so dreadfully great;—Are you not fast asleep? If you resist such a capital soporific as my present letter, you do suffer from sleeplessness!—

Have you actually written to Frohschammer? How delighted he must have been to get y[ou]r letter.[6] Here is a missive from Albert Reville,[7] who is also going to pay you his best compliments literarily (in the Revue des 2 Mondes).[8]

—Kinglake never writes to me without mentioning y[ou]r name. Both he & Charles Villiers are great friends & admirers of y[ou]rs. I was on the point of sending you an enthusiastic letter about you of my friend, Mr A Keyserling, but it is rather a private letter, & those are to be burned at once if propos of Keyserling. He published in English a description of his travels with Murchison & Verneuil in 2 large volumes "Russia & the Oural Mountains".[9] You w[ou]ld greatly oblige me, in letting me know, whether this work is to be had in England, or not? I'm afraid not, as it is rather oldish just as the German work Keyserling published with Krusenstern. "Wissenschaftliche Beobachtungen auf einer Reise in Petschoraland".[10]

What a victory it is if you have actually silenced y[ou]r clerical opponents! I can hardly hope to see them honest enough to confess their defeat. Are they not merely collecting new materials for unfair attacks??—I understand Gladstone's giving up his leadership.[11] Was it not said over & over again that he attacked the Catholics only in order to get numerous Protestant votes? He proves now, how much he cares for them!—It is ½ past 2 A.M.. Good night!

Y[ou]rs ever. Olga.

RI MS JT/1/N/21

1. *Poste Restante*: see letter 4392, n. 7.
2. *"Tecum vivire amem," | "Tecum obeam libens" Horatio*: 'With you I should love to live, with you be ready to die' (Latin). Horace (Quintus Horacius Flaccus; 65–27 BCE), *Odes*, III.ix.24.
3. *my boy*: Alexander Novikoff.
4. *my husband's Parents*: Novikoff's husband was Ivan Novikoff (d. 1890), a general on the General Staff of Grand Duke Nicholas, brother of Tsar Alexander II. According to one of Novikoff's English friends: 'The mother of Ivan Novikoff . . . was a Princess Dolgourouki, the daughter of a Russian poet, Prince Ivan Dolgourouki. She was a Russian matron of the old school, who upheld the ancient Muscovite tradition which made the mother-in-law rather than the wife the mistress of the household' (*M.P. for Russia*, vol. 1, pp. 13).
5. *y[ou]r friendly note*: letter 4536.

6. *y[ou]r letter*: letter missing, but Jakob Frohschammer responded to it in letter 4558.
7. *a missive from Albert Reville*: enclosure missing; this seems to be a letter that Réville wrote to Novikoff, seemingly with comments about Tyndall, rather than letter 4541, which Réville appears to have sent to Tyndall separately (see letter 4561).
8. *his best compliments literarily (in the Revue des 2 Mondes)*: see letter 4535, n. 1.
9. *a description of his travels . . . "Russia & the Oural Mountains"*: R. I. Murchison, E. de Verneuil, and A. von Keyserling, *The Geology of Russia in Europe and the Ural Mountains*, 2 vols (London: John Murray, 1845).
10. *the German work . . . Reise in Petschoraland"*: A. Keyserling and P. T. von Krusenstern, *Wissenschaftliche Beobachtungen auf einer Reise in das Petschora-Land in Jahre 1843* (St Petersburg: Carl Kray, 1846). Paul Theodor von Krusenstern (1809–81) was a Russian polar explorer from the Baltics.
11. *Gladstone's giving up his leadership*: William Gladstone resigned as leader of the Liberal Party on 14 January following Benjamin Disraeli's victory in the previous year's general election.

To William Thomson 29 January 1875 4545

29th. Jan[uar]y. 1875

My dear Thomson.

It would have given me high pleasure to have seen you when you were good enough to call here;[1] and that you were moved to do so is a source of gratification to me.

For however it may harden the character, it is not one of the pleasures of life to live on any other terms than those of friendship with a man like you.

I know so little of what is going on in the world that it was only through the oblique incidence of a note from Dublin,[2] apropos of your visit to Howth Baily,[3] that I was made aware of the existence of Lady Thomson.[4] Let me wish you cordially every happiness, and that the celebrity which surrounds your home may for long years to come be warmed up in your consciousness by the more sacred glow of the affections. Amid many acts of more than ordinary kindness the powers that rule man's destiny have held this one blessing back from me; and have not at the same time withheld the consciousness that it would have been a blessing.[5]

Believe me faithfully yours | John Tyndall

Cambridge University Library, Kelvin Correspondence—Add 7342/T628

1. *here*: the RI.
2. *a note from Dublin*: letter missing.
3. *your visit to Howth Baily*: Tyndall alludes to Thomson's visit in letter 4438; on Howth Baily, see letter 4314, n. 28.

4. *Lady Thomson*: Frances Anna Thomson (née Blandy, 1837–1916), Thomson's second wife, whom he had married on 17 June 1874.
5. *Amid many acts . . . it would have been a blessing*: Tyndall would himself marry Louisa Hamilton exactly thirteen months later, on 29 February 1876, and this might suggest that he was already contemplating matrimony.

To Charles Cecil Trevor 29 January [1875][1] 4546

Royal Institution of Great Britain | 29th. Jan[uar]y

Dear Mr. Trevor.

The accompanying note[2] from M^r^. Douglass has been forwarded to M^r^. Valentin. The data furnished to Mr Valentin[3] subsequent to the experiments at Haisbro differ from the data now forwarded by M^r^. Douglass. This will need a recalculation and a correction of some of the figures. M^r^. Valentin promises that this shall be done by tomorrow. Meanwhile bear in mind that the delay is not to be ascribed to any tardiness on my part in correcting the Report.[4]

Faithfully Yours | John Tyndall

National Archives, MT 10/220. Lighthouses, Illumination by means of gas. H935

1. *[1875]*: the year is established by the relation to letter 4543.
2. *The accompanying note:* letter missing.
3. *The data furnished to Mr Valentin*: probably relating to the cost of the 28-jet burner used at the two lighthouses at Haisborough Sands (see letter 4540).
4. *correcting the Report*: Tyndall had been correcting the proofs of his contribution to 'Further Papers', pp. 53–6.

To Charles Cecil Trevor 30 January [1875][1] 4547

Royal Institution of Great Britain | 30th. Jan[uar]y.

Dear Mr. Trevor.

I now send you the corrected proof.[2]

From a letter addressed by M^r^. Douglass to M^r^. Valentin, which is now before me, it appears that he, Mr Douglass, thinks that M^r^. Valentin derived his erroneous data[3] from M^r^. Wigham: this is not the case—the error, such as it is, is to be ascribed to M^r^. Douglass.

I hope, however, the delay thus arising is not of much importance.

faithfully yours | John Tyndall

National Archives, MT 10/220. Lighthouses, Illumination by means of gas. H935

1. *[1875]*: the year is established by the relation to letter 4546.
2. *the corrected proof*: of Tyndall's contribution to 'Further Papers', pp. 53–6.
3. *his erroneous data*: see letter 4546, n. 3.

To Charles Cecil Trevor 1 February [1875][1] 4548

Royal Institution of Great Britain | 1st. Feb[ruar]y.

Dear Mr. Trevor.

My report[2] was intended to be my final word on the matter of gas illumination.

To make it so I introduced a short paragraph about the glare on fogs; and a note on the gas maker's wages.[3]

M^r^. Allen informs me that a report of M^r^. Douglass[4] & my remarks upon it,[5] made at the invitation of the Elder Brethren[6] are to be published, & he reminds me that some of these remarks would be repetition of the passages in my Report to which I have just referred.

Would it not be therefore better to postpone the working off of the Report until this additional correspondence has been got ready for prep? The omission of the passages referred to from the Report may render explanation unnecessary.

faithfully yours | J. Tyndall

National Archives, MT 10/220. Lighthouses, Illumination by means of gas. H935

1. *[1875]*: the year is established by the relation to letter 4540.
2. *My report*: letter 4540, published as 'Further Papers', pp. 53–6.
3. *the glare on fogs . . . gas maker's wages*: both are discussed in letter 4540.
4. *a report of Mr. Douglass*: see letter 4540, n. 1.
5. *my remarks upon it*: letter 4540.
6. *the Elder Brethren*: the governing body of Trinity House (see letter 4187, n. 2).

From Sarah Faraday 2 February 1875 4549

Barnsbury Villa, | 320, Liverpool Road, | N.[1] | 2nd Feb[ruary]. 1875

Thanks dear Dr Tyndall for your kind and prompt communication[2] concerning Mrs Brown.[3] I am much interested about her, and think the Managers[4] have been very kind.

Yours affectionately | (signed) S. Faraday.

RI MS JT/1/TYP/12/4187
LT Typescript Only

1. *Barnsbury Villa, | 320, Liverpool Road, | N.*: see letter 4226, n. 1.
2. *your kind and prompt communication*: letter missing.
3. *Mrs Brown*: see letter 4537.
4. *the Managers*: of the RI, where Mrs Brown had been a housemaid. The RI's managers at this time were William Bowman (1816–92), William Cavendish, 7th Duke of Devonshire (1808–91), Henry John Codrington (1808–77), Warren De la Rue (1815–89), Thomas Frederick Elliot (1808–80), Francis Galton (1822–1911), William Robert Grove (1811–96), Caesar Henry Hawkins (1798–1884), Alfred Latham (1801–85), Josceline William Percy (1811–81), William Pole (1814–1900), John William Strutt, 3rd Baron Rayleigh (1842–1919), Arthur John Edward Russell (1825–92), Carl Wilhelm Siemens, and Charles Wheatstone.

To Charles Cecil Trevor 5 February [1875][1] 4550

Royal Institution of Great Britain | 5th. Feb[ruar]y

Dear Mr. Trevor,

Pray save me the time required to rummage over the back numbers of the Times by giving me information on the following points.

I distinctly remember reading the account of a ship wreck on the French coast[2] not long ago (say about a year or so) Evidence was taken either at Boulogne or Calais, and it was then stated that the lights of the French coast were sometimes mistaken, & had been in that case mistaken, for those of the English Coast. I want you kindly to give me the reference to that evidence.

I also remember some reference being made to the mistaking of one light for another in the case of a disastrous shipwreck on the American Coast[3] a couple of years ago. Can you refer me to this?

Are no other cases known to you where one light has been mistaken for another? I derived a different impression from consultations with nautical men.

yours faithfully | John Tyndall

Of course in such matters I defer to the Brethren,[4] but I must show cause for the remarks I have made.[5]

National Archives, MT 10/220. Lighthouses, Illumination by means of gas. H935

1. *[1875]*: the year is established by the relation to letter 4540 (see n. 5).
2. *the account of a ship wreck on the French coast*: not identified, and no such report seems to have been published in the *Times* during the period Tyndall mentions.
3. *a disastrous shipwreck on the American Coast*: after the wreck of SS *Atlantic* on 1 April 1873

off the coast of Nova Scotia, with the loss of more than 500 lives, an editorial in the *Times* attributed the accident to 'a sheer blunder in the observation of a Lighthouse', noting that 'another Light was mistaken for the Sambro Light and the vessel's course was thus steered on an entirely false supposition' ('[Editorial]', *Times*, 3 April 1873, p. 9).

4. *the Brethren*: the Elder Brethren of Trinity House (see letter 4187, n. 2).
5. *the remarks I have made*: in letter 4540, where Tyndall commented on the 'possibility of mistaking one light for another, which is now so frequent and disastrous' and stated: 'I am guided here by accounts of wrecks in the newspapers, and by conversations with nautical men; deferring, of course, to the more definite and accurate records kept by the Trinity House and the Board of Trade'.

From Henry Anthony Hammond[1] 5 February 1875 4551

Sundridge House, | Bournemouth. | Feb[ruary]. 5. 1875.

Dear Sir.

Assuming that the enclosed statement[2] is correct *[*(not so)*]*,[3] it cannot fail to be interesting to you to see what such an independent thinker as M[r] Thomas Carlyle thinks of your views.[4]

A similar story is told of M[r] Carlyle meeting four ladies in a train who framed him for his opinion on M[r]. Darwin's ideas as to the Origin of Species.[5] He evaded all answer for some time, and last is said to have replied—"Man—If you will have my opinion—I read. 'Thou madest him a little lower than the angels'—"[6]

Trusting you will excuse what might seem an impertinence, my sending you the enclosed,

I am Sir | Y[ou][r] ob[edien][t] serv[an][t] | Henry A Hammond

You will excuse this accidental tear.

RI MS JT/1/H/27

1. *Henry Anthony Hammond*: Henry Anthony Hammond (1829–1910), an evangelical Christian and member of the Plymouth Brethren since 1848.
2. *the enclosed statement*: enclosure missing, but probably a report of Thomas Carlyle deriding Tyndall's Belfast Address as a 'philosophy fit for dogs' as mentioned in letter 4556.
3. *[(not so)]*: this has been added in a different hand, probably Tyndall's.
4. *what such . . . Carlyle thinks of your views*: letter 4556 suggests that Tyndall soon after wrote to Carlyle's niece Mary Aitken to ask her exactly this.
5. *the Origin of Species*: C. Darwin, *On the Origin of Species* (London: John Murray, 1859).
6. *'Thou madest him a little lower than the angels'*: Hebrews 2:7. The full version of Carlyle's comment is given in letter 4556.

To Robin Allen 8 February 1875 4552

Note.—The opportunity of reading the remarks of the Elder Brethren[1] on the foregoing investigation[2] has been courteously and spontaneously granted me. The concluding paragraph, which makes known the intentions of the Brethren with regard to the future of the gas,[3] is, I think, in accordance with that true conservatism which, while resisting capricious change, does not refuse scope and encouragement to real improvements. This being my opinion, it would be as ungracious as unnecessary on my part to go over their arguments in detail. I will only say that had the Brethren and their adviser[4] worked together on this gas question as long and assiduously as in our researches on Fog-signals,[5] we should, I am persuaded, (possibly by a little yielding on both sides) be as unanimous regarding the former as we now happily are regarding the latter.

One more word. I hold in my hand the letter of a friend,[6] a passenger from Calais to Dover on the 31st of October, describing the extinction by fog of the electric lights at the South Foreland.[7] But the Elder Brethren themselves have cited (page 59, footnote) a still more striking illustration of the same kind. Here, it seems to me, we have the precise conditions necessary for testing the utility of flashes in the case of the electric light. The highly intelligent light-keepers at the South Foreland (and a more steady and intelligent body of men than the light-keepers generally it would be difficult to find) may readily be taught the alternate occultation[8] (which ought to be complete) and opening of the light, and to observe whether the glimmer on the fog is not seen after the light itself has ceased to be visible. Similar observations, though with less ease, might be made at Haisbro'. To be really effectual the change from light to darkness, and from darkness to light, ought to be *sudden*. With this practical suggestion I bring these remarks to an end.

J. T. | 8th February 1875.

'Further Papers', p. 56

---

1. *the remarks of the Elder Brethren*: the governing body of Trinity House (see letter 4187, n. 2); the remarks are contained in a letter from Allen dated 30 January printed in "Further Papers', pp. 57–60.
2. *the foregoing investigation*: letter 4540.
3. *The concluding paragraph . . . future of the gas*: the intention stated was to convert the second light at Haisborough Sands to gas 'so as to make the entire establishment a gas station'. This would also facilitate 'learning more of the value of its higher powers in relation to fog,

and possibly, of its application to intermittent lights; and of advancing its utility generally by improvements in detail' (p. 60).

4. *their adviser*: Tyndall.
5. *our researches on Fog-signals*: according to 'Report by Professor Tyndall', Tyndall began his researches on this question on 19 May 1873, and he was still involved in experiments with gun signals (see letter 4514).
6. *letter of a friend*: letter missing.
7. *the South Foreland*: a lighthouse near Dover used to warn ships approaching the Goodwin Sands, a feature responsible for some 2,000 shipwrecks.
8. *occultation*: cut off from view by something interposed (*OED*).

To Charles Cecil Trevor 8 February [1875][1] 4553

8th. Feb[ruar]y.

Dear Mr. Trevor.

If it sh[oul]d be thought necessary, or desirable, I shall be glad to glance at the corrections & additions here introduced, and to return the proof[2] by the messenger who brings it to me—or by my own.

Yours faithfully | John Tyndall

National Archives, MT 10/220. Lighthouses, Illumination by means of gas. H935

1. *[1875]*: the year is established by reference to Tyndall's report (see n. 2).
2. *the proof*: of Tyndall's 'Further Papers', pp. 53–6.

From Frances Russell 8 February 1875 4554

Pembroke Lodge.[1] | Richmond Park. | Feb[ruary]. 8 /75

Dear Mr Tyndall

When can you give us a Sat[urda]y to Mon[da]y here? or, short of that if you cannot compass it, a Sun[da]y afternoon extends to after dinner—

Yours very sincerely | F Russell

RI MS JT/1/R/68

1. *Pembroke Lodge*: see letter 4237, n. 1.

From William Thomson 8 February 1875 4555

The University, | Glasgow. | Feb[ruary] 8/75

My dear Tyndall

I was excessively busy all last week or I should sooner have written to thank you for your kind letter,[1] and your good wishes for myself and my wife.[2] We were on our way northward to her new house[3] when we made our little visit to Howth Baily[4] and were much interested in all we saw there. I hope when we are next in London to have an opportunity of talking over many things about lighthouses with you.

Cordially wishing you every happiness

I remain yours very truly | William Thomson

RI MS JT/1/T/19
RI MS JT/1/TYP/5/1561

1. *your kind letter*: letter 4545.
2. *my wife*: Frances Anna Thomson (see letter 4545, n. 4).
3. *her new house*: Netherhall House, the baronial mansion that Thomson had recently had built in Largs, Ayrshire.
4. *our little visit to Howth Baily*: see letter 4545, n. 3.

From Mary Carlyle Aitken[1] 9 February 1875 4556

5 Cheyne Row, Chelsea | 9 Feb[ruary]. 1875.

Dear Mr Tyndall,

I need hardly say that your small request[2] gives me no trouble whatever & I hasten to comply with it. I myself heard my uncle use the expression "a philosophy fit for dogs", but it was in reference to Darwin's theory of Evolution <u>not</u> to your Belfast Address. He "the General"[3] bids me to say that he never said or thought anything of the kind in reference to what you have said or written.

As to the "similar story".[4] I daresay you yourself have often heard my Uncle repeat the words "Thou hast made him a little lower than the angels",[5] & add that the Darwinites seemed to say "Thou hast made him a little higher than the tadpoles". The <u>railway train</u> I should think is an invention on the part of the "ladies"[6]—in order to make the effect more dramatic.

I think it is somewhat hard that I am twice addressed by you as "M^rs^

Aitken".[7] The first time I thought it an oversight, but I now conclude that you think my superior years entitle me to level rank.

My uncle sends his best regards & hopes to see you soon.

Yours very truly | Mary Carlyle Aitken

RI MS JT/1/A/65

1. *Mary Carlyle Aitken*: Mary Carlyle Aitken (1848–95), a writer and niece and assistant of Thomas Carlyle, with whom she had lived since 1866.
2. *your small request*: letter missing, but presumably a request to check the account of Carlyle's views of the Belfast Address included in letter 4551.
3. "*the General*": a nickname for Carlyle presumably relating to his views of the importance of heroic martial leadership.
4. *the "similar story"*: see letter 4551.
5. *"Thou hast made him a little lower than the angels"*: Hebrews 2:7.
6. *The railway train . . . the part of the "ladies"*: see letter 4551.
7. *"Mrs Aitken"*: Aitken was at this time unmarried, although she would marry her cousin Alexander Carlyle in 1879.

From Mary Agatha Russell 9 February 1875 4557

Pembroke Lodge.[1] | Richmond Park. | Feb[ruary]. 9. /75.

Dear Professor Tyndall

Your invitation[2] is very delightful, & my brother[3] & I are most grateful to you for your kind thought of us. We shall, I hope, be able to go up to the Royal Institution for your lectures,[4] & we look forward with much pleasure to devoting our Thursday afternoons to the subject of Electricity!—We are very glad that you can come here next Sunday.[5] With our best thanks, believe me,

Yours very sincerely | Agatha Russell.

RI MS JT/1/R/50

1. *Pembroke Lodge*: see letter 4237, n. 1.
2. *Your invitation*: letter missing.
3. *my brother*: Francis Albert Rollo Russell.
4. *your lectures*: Tyndall delivered a series of seven lectures on electricity at the RI on successive Thursdays from 4 February to 18 March (*Roy. Inst. Proc.*, 7 (1873–75), p. 342). Russell did attend them with her brother (see letter 4575).
5. *come here next Sunday*: 14 February; Tyndall had been invited in letter 4554.

From Jakob Frohschammer 11 February 1875 4558

München den 11 Febr. 1875 | Amalienstraße 7/0

Hochgeehrter Herr und College!

Ich sage Ihnen besten Dank für die freundlichen Zeilen, mit denen Sie mich beehrt haben. Es gereicht mir zur Genugthuung und Freude wenn mein Artikel in der Allg[emeine]. Z[ei]t[u]ng einigen Werth für Sie hat, und ich wünsche, daß derselbe Einiges beitragen moege zur Foerderung Ihres Ruhmes und zur Verbreitung Ihrer Werke besonders in Deutschland. Sobald ich durch Frau O. Novikoff naehere Kenntniß von Ihrem Conflicte erhielt, beschloß ich darüber etwas zu schreiben und war nur einige Zeit darüber unschlüßig in welcher Form dieß am besten geschehen koenne. Ich hoffe daß die, welche ich gewählt habe die beste sei, wenigstens in Deutschland, wenn auch allerdings mein Artikel in England kaum dem Publikum ganz oder theilweise zur Kenntniß kommen wird.

Ihre Lage ist allerdings, wie Sie sagen, den Gegnern gegenüber viel günstiger als die meinige seit 15 Jahren war und noch ist in diesem Bayern, das von den Jesuiten durch Jahrhunderte hindurch beherrscht ward und deßen geistige Kraefte so sehr gelaehmt wurden, daß es kaum noch moeglich ist, sie wieder zu wecken und zum Aufschwung zu bringen. Volk und Jugend sind noch fast ganz vom Ultramontanismus beherrscht und der Liberalismus selbst ist ohne Energie und Entschiedenheit und kann am allerwenigsten durch den bei uns sehr lahmen Altkatholicismus eine Foerderung finden. Aller Wahrscheinlichkeit nach wird bei der in diesem Jahre stattfindenden Neuwahl für die Kammer der Abgeordneten der Ultramontanismus entschieden siegen und dann werden wir allem Anschein nach auch ein ultramontanes Ministerium bekommen. Für mich wird dieß in so fern von Bedeutung sein, als dadurch meine Stellung an der Universitaet in Gefahr kommen wird. Indeß wird dieß an meinem Verhalten nichts aendern.

Auch in so fern sind Sie in beßerer Lage als ich, weil Sie, wie ich glaube, nicht von sog. Freunden Zugleich mit der Sache der Wissenschaft verrathen worden sind, wie es leider mir geschah. Eine Erfahrung, die mir bitterer war als die Maßregeln, die Papst und Bischoefe gegen mich ergriffen.

Es wird mich freuen, wenn ich auch in der Folge Kunde von Ihrer Thätigkeit und Ihrem Conflicte erhalte. Es waere überhaupt sehr zeitgemaeß, wenn auch die Vertreter des Rechtes der Wissenschaft in Anbetracht der Solidaritaet ihrer Intereßen in naehere Verbindung treten und sich gegenseitig unterstützen und foerdern würden, um ihre Seite siegreich zu vertheidigen—nachdem die Gegner der freien Forschung, die verschiedenen Orthodoxien, wenn auch unter sich sehr uneinig, doch der Wissenschaft gegenüber einig sind. Insbesondere in der Kathol[ische]. Kirche ist die internationale

Verbindung und Gemeinschaft im Kampfe für ihre Intereßen gegen Staat und Wissenschaft sehr entschieden. Ich meinerseits bin stets bereit zu solcher internationaler Gemeinschaft den Gegnern gegenüber. Mit den liberalen franzoes. Schriftstellern laeßst sich vorlaeufig an keine Verbindung denken oder nur sehr ausnahmsweise. Aber zwischen England und Deutschland ist sie wohl moeglich.

Mit der Versicherung vorzüglicher Hochachtung | Ihr | ergebenster | J. Frohschammer

P.S. Die 2 Nummern d[er]. Quarterly Review habe ich gestern richtig erhalten und sage Ihnen verbindlichsten Dank dafür.

Munich, 11 Febr[uary]. 1875 | Amalienstraße 7/0

Dearest Sir and Colleague!

I thank you very much for the friendly lines[1] with which you have honoured me. It would give me satisfaction and joy should my article in the Allg[emeine]. Z[ei]t[u]ng[2] have some value for you, and I hope that the same may contribute much to promote your reputation and the distribution of your work especially in Germany. As soon as I received more information about your conflict through Mrs. O. Novikoff, I decided to write something about it and was only undecided for some time in what form this could best be done. I hope that the one I have chosen may be best, at least in Germany, although my article will hardly be noticed, as a whole or in part, by the English audience.

Your situation against the enemies, however, is, as you say, much more favourable than mine has been for 15 years[3] and still is in this Bavaria that for centuries was dominated by the Jesuits and whose intellectual power was so crippled that it is almost impossible to reawaken and restore it. People and youth are still almost entirely dominated by ultramontanism,[4] and liberalism itself is without energy and decisiveness and can find support least of all from the very paralysed Old Catholic Church here.[5] In all likelihood, ultramontanism will have a decided victory during the election of the chamber of deputies this year[6] and it appears that we will then also get an ultramontanist ministry. For me, this will be of significance in so far as my position at the university[7] will be endangered. However, this will not change anything in my behaviour.

You are also in a better position than me because I believe that you—together with the matter of science—were not betrayed by so-called friends, as unfortunately has happened to me. An experience that was bitterer to me than the measures that pope and bishops took against me.

It would make me happy if I also received subsequent communications about your work and your conflict. As a matter of fact, it would generally be timely if, considering the solidarity of their interests, representatives of the law also entered into closer contact with science and supported and promoted each other in order to victoriously defend their side—seeing that the enemies of free research, the various orthodoxies, albeit disunited amongst themselves, are united against science. In the Cathol[ic]. Church in particular, international solidarity and unity in the struggle for its interests against state and science are very determined. I for my part am always ready for such international fellowship against the enemies. For now, there is no way to consider a connection with the liberal French writers, or only very exceptionally. But between England and Germany it should be possible.

With the assurance of my highest regards | Your | most humble | J. Frohschammer

P.S. Yesterday I have received the 2 issues of t[he]. Quarterly Review[8] and I sincerely thank you for them.

RI MS JT/1/F/67

1. *the friendly lines*: letter missing.
2. *my article in the Allg[emeine]. Z[ei]t[u]ng*: see letter 4515, n. 5.
3. *mine has been for 15 years*: see letter 4469, n. 4.
4. *ultramontanism*: from the Latin for beyond the mountains, or north of the Alps, this was a position within the Roman Catholic Church that placed strong emphasis on papal authority.
5. *the very paralysed Old Catholic Church here*: a schismatic group in Bavaria led by Ignaz von Döllinger (1799–1890) that separated from the Roman Catholic Church after rejecting papal authority and insisting that authority should be supreme in the Church. Frohschammer, who declined to join them, felt that their liberalism did not go far enough.
6. *ultramontanism will have . . . chamber of deputies this year*: while the election to the second chamber of the Bayerische Ständeversammlung (Bavarian States General) in July 1875 was indeed won by the Ultramontanes, their majority over the Liberals was only two (see letter 4625).
7. *my position at the university*: as professor of philosophy at Ludwig-Maximilians-Universität in Munich.
8. *the 2 issues of t[he]. Quarterly Review*: possibly those for July (no. 273) and October (no. 274) 1874 containing an anonymous attack by the Catholic biologist St George Mivart on George Darwin and the latter's response: 'Primitive Man: Tylor and Lubbock', *Quarterly Review*, 137 (1874), pp. 40–77, and 'Note Upon the Article "Primitive Man: Tylor and Lubbock" in No. 273', *Quarterly Review*, 137 (1874), pp. 587–89. Charles Darwin had discussed the matter with Joseph Hooker and Thomas Huxley (see letter 4519, n. 8).

To Elizabeth Dawson Steuart 12 February [1875][1] 4559

Royal Institution of Great Britain | 12th Feb[ruary].

My dear Mrs Steuart,

I am very sorry that I made such a bad shot with the photographs.[2] I did not know before you wrote to me of the existence of such things as scrap photographs.[3] So I made enquiry, found out a place and went straight there and bought them. Pray give those you do not need to some of your young friends—or throw them in the fire.

I will make further enquiries and if I learn any thing about the thing[4] I will let you know.

Yours ever | John Tyndall.

RI MS JT/1/TYP/10/3428
LT Typescript Only

1. *[1875]*: the year is established by the relation to letters 4521 and 4524.
2. *such a bad shot with the photographs*: Tyndall was responding to Steuart's request in letter 4521.
3. *scrap photographs*: see letter 4521, n. 6.
4. *the thing*: not identified, but seemingly businesses in London selling scrap photographs. Alternatively, it may relate to the payment of 'this man's account' discussed in letter 4524.

To Hermann Helmholtz 16 February [1875][1] 4560

Royal Institution of Great Britain | 16th. Feb[ruary].

My dear Helmholtz,

Let me thank you for the strong and dignified reply to Zöllner which you have been good enough to send me.[2]

Your mode of getting Thomson out of difficulty as regards the transport of germs is exceedingly good. The possible blowing away and subsequent subsidence of Superficial air germs rescues the hypothesis from destruction.[3]

I am truly glad to find you taking a definite public stand in the matter of Kirchhoff[4]—It is especially necessary at the present time. I wish Stokes himself would disclaim the pretensions put forward on his behalf.[5]

It is a questionable compliment to him to suppose him possessed of such a discovery and unable for years to know the worth of it.

ever faithfully yours | John Tyndall.

RI MS JT/1/TYP/2/504
LT Typescript Only

1. *[1875]*: the year is established by reference to Helmholtz's preface (see n. 2).
2. *the strong and dignified reply . . . good enough to send me*: see letter 4223, n. 10.
3. *Your mode of getting Thomson . . . hypothesis from destruction*: Thomson proposed that organic germs might be conveyed through space by meteors, explaining how they originated on earth, but Zöllner claimed that the heat generated by the meteors' entry through the earth's terrestrial atmosphere would destroy the germs. Helmholtz responded that the

germs might be blown away at the highest strata of the earth's atmosphere, before too much heat had been generated, and then float to earth. See Helmholtz, 'Vorrede', p. xi.

4. *the matter of Kirchhoff*: Gustav Kirchhoff's claim to priority in the discovery of spectrum analysis in the late 1850s. In a footnote in his 'Vorrede', Helmholtz declared: 'Auf dem Gebiete der persönlichen Fragen muss ich bezüglich der die Principien der Spectralanalyse betreffenden Prioritätsreclamation, mit welcher Herr W. Thomson für Herrn Stokes gegen Herrn Kirchhoff aufgetreten ist, mich auf die Seite des Letztgenannten stellen in voller Anerkennung der Gründe, die er selbst geltend gemacht hat [In the field of personal questions, I have to side with the latter [Zöllner] on the priority complaint concerning the principles of spectral analysis, with which Mr. W. Thomson stood up for Mr. Stokes against Mr. Kirchhoff, in full recognition of the reasons he himself asserted]' (p. xiii).
5. *I wish Stokes . . . on his behalf*: Thomson claimed that George Gabriel Stokes had taught him the principles of spectrum analysis in their conversations no later than 1852, although Stokes did not publish anything on the matter at the time and himself disclaimed priority in favour of Kirchhoff. Thomson, though, continued to insist vigorously that Stokes had priority, and that is what Tyndall wishes Stokes would disclaim. See I. D. Rae, 'Spectrum Analysis: The Priority Claims of Stokes and Kirchhoff', *Ambix*, 44 (1997), pp. 131–44.

To Olga Novikoff 19 February 1875 4561

Royal Institution of Great Britain | 19th. Feb[ruary]. 1875.

Dear Friend

Every thing, as far as I am concerned, is serene in the religious atmosphere. All hostile sounds have subsided, and the enemy permits me to pursue in peace my ordinary labours.

Hence I had nothing that you would care to hear to communicate to you, and therefore I delayed writing. I have had excellent letters from M. M. Reville and Frohschammer.[1] The latter writes somewhat despondingly: he foresees the coming ultramontane triumph in Bavaria. He thinks it may affect his position there; but it does not affect his resolution. If he be a bachelor he may well defy the malice of his enemies; but it is become a serious matter when the interests of a family are involved. I will try to keep myself acquainted with whatever changes may occur in Munich to modify the position of Frohschammer.

But I have been too hasty in saying that the storm had quite subsided. You probably do not know the editor of the Quarterly Review.[2] Well, for some months this gentleman has been feeling the pulse of society to ascertain whether it would be safe to attack me. For if unsafe he has too great a regard for himself to attempt it. I am told that he has resolved to annihilate me; from which I infer that he found a sufficient amount of sympathy and support from those whom he consulted. He is one of those warriors who always fight on the side of the big battalions. I know nothing of the tone of his article, though he has been boasting about it in various places and ways. If he behave ill, I shall lay upon his back a thong which he will consider the reverse of pleasant.

I met Mr Villiers a few days ago at Lady Stanley's.[3] He was very vigorous. He had a discussion with Sir James Kay Shuttleworth[4] after dinner, in which he showed uncommon energy. Your friend Kinglake I frequently see, and he is flourishing, though his Inkerman has not passed unscathed through the hands of the Reviewers.[5]

I am now plunged in my lectures[6] and to my great horror the Messrs Longman's[7] have called upon me to prepare new editions of all my books.[8] Thus you see my hands are very full.

I may possibly give the parsons a sermon at Sion College[9] after Easter. If I should do so I will let you know all about it.

Good bye and with very many thanks for your last letter,

Believe me | Ever Yours | John Tyndall

RI MS JT/1/TYP/3/934
LT Typescript Only

---

1. *excellent letters from M. M. Reville and Frohschammer*: letters 4541 and 4558.
2. *the editor of the Quarterly Review*: William Smith (1813–93), who edited the Tory *Quarterly Review* from 1867 to 1893 (*ODNB*). However, no such article by Smith was published in the *Quarterly* or anywhere else, and Tyndall may instead mean its Whig rival the *Edinburgh Review*, whose editor Henry Reeve (1813–95) anonymously reviewed the Belfast Address in its January 1875 number. Reeve contended that Tyndall 'mistook his duty when he plunged, at Belfast, into the atoms of Democritus' and other similar ancient doctrines, and he exclaimed: 'Are we to return to Paganism or something behind Paganism—to the flux of Heraclitus, the *voûs* of Anaxagoras, or the atoms of Democritus—are we to take our morals from Epicurus and our gods from Lucretius?' ([H. Reeve], 'Mill's *Essays on Theism*', *Edinburgh Review*, 141 (1875), pp. 1–31, on pp. 6 and 3).
3. *Lady Stanley*: Mary Stanley (née Sackville-West, 1824–1900), Countess of Derby, a society hostess and wife of Edward Henry Stanley (1826–93), 15th Earl of Derby (*ODNB*).
4. *Sir James Kay Shuttleworth*: James Kay Shuttleworth (1804–77), a politician and educational reformer (*ODNB*).
5. *his Inkerman . . . hands of the Reviewers*: the fifth volume of A. W. Kinglake, *The Invasion of the Crimea*, 9 vols (Edinburgh: William Blackwood, 1863–87) was published in 1875 and entitled 'The Battle of Inkerman'. Its account of the decisive victory of the British and French over Russia in November 1854 was criticized for, among other things, besmirching the reputation of the French troops, with a review in the *Academy* commenting that 'it is to be hoped that our gallant neighbours will not accept the statements advanced in this book as in any degree representing the opinions or feelings of their late allies' (G. Chesney, 'Kinglake's Battle of Inkerman', *Academy*, 7 (1875), pp. 181–2, on p. 182).
6. *my lectures*: Tyndall's lecture course on electricity at the RI (see letter 4557, n. 4).

7. *Messrs Longman's*: Thomas and William Longman, proprietors of the publishers Longmans, Green, & Co.
8. *new editions of all my books*: *Fragments of Science: A Series of Detached Essays, Lectures, and Reviews*, 5th edn (London: Longmans, Green, and Co., 1876); *Heat: A Mode of Motion*, 5th edn (London: Longmans, Green, and Co., 1875); *Six Lectures on Light* (1875); *Sound* (1875).
9. *Sion College*: see letter 4475, n. 4. Tyndall had previously told Novikoff that he was going there in November 1874 (see letter 4475).

To Charles Cecil Trevor 24 February [1875][1] 4562

Royal Institution of Great Britain | 24th. Feb[ruar]y.

My dear Mr. Trevor

Many thanks to you for the Blue book[2] & for your compliance with my request regarding Sir W[illia]m. Thomson.[3]

I see the printer[4] has allowed an important word to drop out of his type. On page 55, 18th line from bottom for the words "with the expenditure of gas" we ought to have "with *[such]*, half the expenditure of gas".[5] It is rather an unlikely omission.

Yours faithfully | John Tyndall

This just caught my eye. I hope it is the only omission.

National Archives, MT 10/211. Lighthouses, Illumination by means of gas. H1650

1. *[1875]*: the year is established by the relation to letter 4547.
2. *the Blue book*: 'Further Papers'. Tyndall's contribution, of which he had sent corrected proofs in letter 4547, was on pp. 53–6. Blue books were volumes, characteristically bound in blue, containing official British government publications (*OED*).
3. *my request regarding Sir W[illia]m. Thomson*: not identified, but possibly relating to Thomson's trip to the lighthouse at Howth Baily (see letters 4438, 4545, and 4555).
4. *the printer*: the report was printed by the firm of George Edward Briscoe Eyre (1840–1922) and William Spottiswoode, printers to the Queen.
5. *the words "with the expenditure . . . half the expenditure of gas"*: Trevor added a handwritten note to the letter stating: 'Queen printers informed | 24 Feb | & also the 8 recipients of the Paper', although the line remains uncorrected in the parliamentary record (p. 55).

From Olga Novikoff 25 February [1875][1] 4563

Moscow. February 25th

Dear Friend,

It is a very bad feeling to be envious of a friend. I dare say it is, but had you seen Frohschammer's letter, in w[hic]h he mentions y[ou]rs[2] you w[ou]ld understand my envying him!—

I wish you could let me know, whether y[ou]r lectures at the B. Associ[atio]n[3] are going to be published or not?[4] A friend of mine gave me such a very interesting account of the first lecture, that, as the Germans say, it really "schmeckt nach mehr."[5]—of proposed problems & discoveries! A Professor Strumpell[6] from Leipzig writes, that a young Italian called Masso[7] has just invented an instrument w[hic]h is likely to solve many curious questions in experimental psychology. It is an apparatus w[hic]h notes with wonderful accuracy the continually varying expansion & contractions of the muscles according to the greater or lesser amount of blood w[hic]h they contain in the moment. It is found, for instance, that in sleep the arm is slightly swollen, that if the sleep is disturbed, the arm diminishes in bulk, that if he is awakened, it diminishes still more; that if he directs his attention to something that interests him, there is a still further contraction. The blood leaves the arm and flows probably to the brain, and thus the amount of intellectual effort may be determined by the bulk of the arm!—

I wonder how far my translation[8] may have interested you? No easy thing, indeed, to interest you anyhow!—Do not forget or give me up altogether.

Ever y[ou]rs Olga

RI MS JT/1/N/26

1. *[1875]*: the year is established by the relation to letter 4564.
2. *y[ou]rs*: letter missing, but Frohschammer replied to it in letter 4558.
3. *your lectures at the B. Associ[atio]n*: this seems an error on Novikoff's part and she instead means the RI, where Tyndall was then delivering a series of lectures on electricity (see letter 4557, n. 4).
4. *to be published or not*: see letter 4564.
5. *"schmeckt nach mehr"*: tastes like more (German), a colloquialism expressing a desire for more of something, especially food.
6. *Professor Strumpell*: Ludwig von Strümpell (1812–99), a professor of philosophy at Leipzig University (*NDB*).
7. *a young Italian called Masso*: Angelo Mosso (1846–1910), a physiologist who developed a 'human circulation balance' which indirectly measured intellectual activity.

8. *my translation*: not identified.

To Olga Novikoff 1 March [1875][1] 4564

Royal Institution of Great Britain | 1st. March

Dear Friend.

Your brief little note[2] came to me this morning, and I write at once to answer your question regarding the lectures.[3] They are not to be published: for I have not the time necessary to organize them But I am having notes of the course printed, and as soon as it is ended I will prepare a special copy of the notes and experiments, and send it to you.

That is a capital case of correlation[4] that you describe And so linked together are both classes of phenomena,[5] that wild as the thing may appear it is at bottom by no means unphilosophical.

We laid poor Lyell in Westminster Abbey on Saturday.[6]

We had a lecture here on Friday from a gentleman who appears to be well acquainted with Russia—Mr Ralston of the British Museum.[7] He was good enough to say that I have some friends in Russia—at all events I believe that I have one.

Goodbye | Ever Yours | John Tyndall

John Tyndall Correspondence in the Olga Novikoff Correspondence Collection. Courtesy of Special Collections, Kenneth Spencer Research Library, University of Kansas. MS 30:Q10
RI MS JT/1/TYP/3/935

---

1. *[1875]*: the year is established by reference to the funeral of Charles Lyell (see n. 6).
2. *Your brief little note*: letter 4563.
3. *the lectures*: Tyndall's lecture course on electricity at the RI (see letter 4557, n. 4).
4. *a capital case of correlation*: the physiological measurements of Angelo Mosso that Novikoff referred to in letter 4563.
5. *both classes of phenomena*: mental activity and blood flow.
6. *We laid poor Lyell in Westminster Abbey on Saturday*: Charles Lyell, who had been ailing since an accident on 9 December 1874 (see letter 4507), died on 22 February and was buried in the north aisle of the nave of Westminster Abbey five days later ('The Late Sir Charles Lyell', *Times*, 24 February 1875, p. 5; 'Obituary. Sir Charles Lyell', *Sunday Times*, 28 February 1875, p. 8).
7. *a lecture here on Friday . . . Mr Ralston of the British Museum*: William Ralston Shedden-Ralston (1828–89) worked in the printed-book department of the British Museum, specializing in Russian literature (*ODNB*). On 26 February he delivered a Friday Evening Discourse at the RI entitled 'On Popular Tales: Their Origin and Meaning' (*Roy. Inst. Proc.*, 7 (1873–75), pp. 378–83).

From Olga Novikoff 2 March 1875 4565

March. 2. 75. | Moscow. Poste Restante.[1]

Dearest Friend,

How am I to thank for y[ou]r kind, charming & most clever letter of Feb. 19?[2] Y[ou]r time is so precious, that I feel ashamed of myself for being so happy in y[ou]r sacrificing sometimes a small part of it to me. But does shame annihilate happiness?

Along with y[ou]rs came a missive from C. P. Villiers, in w[hic]h he is quite enthousiastic about you, "considered from a social point of view". "How agreeably he lowers himself to the level of a London table, putting every body at their ease, & never intruding the truth upon them, unless required."—

You actually possess a peculiar talent to charm even those, who w[ou]ld like to hate y[ou]r opinions. I think I spoke to you of my 2 brothers.[3] I do not know anybody more true, more kind, more generous, than they are & besides—they love me, as nobody could ever do—I suppose. But there is a positive gulf between their views & mine, w[hic]h in former days—was most painful to me & to them & greatly diverted those who listened to our hot discussions. I read sometimes their orthodox lucubrations—they read mine or those I recommend. On becoming acquainted with y[ou]r "address & Preface"[4] my closest brother[5] could not help exclaiming "what an attractive man he must be". though he regretted "you made such a little distinction between Xdom & Xians."[6] hoping thus to save the first in sacrificing the second!—

I am anxious to see Réville's article on y[ou]r works.[7] The hostile intentions of the Contemp[orary]: Review[8] were unknown to me, but the attacks of the Edinbro' Review (I think December 74)[9] struck even the Ultra "Orthodox"[10] as being too unfair, too much beside the mark, & antiquated.

As to Frohschammer, being a priest, quite alone in the world, & used to extreme poverty, has no fear of material losses. But you may greatly help him as you have done already,[11] in making him known & appreciated in England. It w[ou]ld be a great thing, if somebody ever wrote a review of his works! Few people live upon the hope of justice being done to them after their death. Frohschammer has been all his life underrated, & that wounds his feelings more than he likes to avow even to himself. In this respect, you stand much higher than he; but his life has been a very trying one, & who can say? he has fought perhaps as much as any body, though the result we see is not defiantly brilliant.

—According to Russians, Kinglake's vol v.[12] contains severe mistakes, but alone are to blame for it. Kinglake never refused learning the truth, & asked for information. I wrote accordingly to several quarters. No serious result followed! One said "he had no time", another "thought it a shame to look for appreciation abroad" the third would not "commit his comrades" etc, etc... What c[oul][ld] Kinglake do under such circumstances?—

I am not leading an idle life, & never lost any opportunity for doing honestly <u>my best</u> in the largest sense of the word. But, dear me, how little supported I am by others, how few really care for anything beyond their wretched selfs!

—Alois Riehl—from Gras, wrote to me the other day, assuring again, that he'll <u>prove</u> me, that y[ou][r] example has not been lost; & that you'll see proofs of it in his work on I. Kant.[13]

Do let me know <u>whether & when</u> you are likely to publish y[ou][r] present lectures at the R[oyal]. Institution.[14] How I deplore being away from London!!—

But enough scribbling! Good bye!—Ever y[ou][rs] heartily Olga.

RI MS JT/1/N/22

1. *Poste Restante*: see letter 4392, n. 7.
2. *y[ou]r kind, charming & most clever letter of Feb. 19*: letter 4561.
3. *my 2 brothers*: Aleksandr (1833–1910) and Nikolai Kiréeff (d. 1876), who at this time both served in Russia's Imperial Horse Guards Regiment.
4. *y[ou]r "address & Preface"*: *Belfast Address*, 7th thousand and 'Preface to the Seventh Thousand', pp. v–xxxii.
5. *my closest brother*: probably Nikolai, who was later described as 'a splendid pattern of a Christian soldier . . . an upright and zealous Orthodox, and he not only believed, but acted accordingly. If ever practical Christianity shone forth from the life of a man we find it here' (*M.P. for Russia*, vol. 1, pp. 222–3).
6. *Xdom & Xians*: Christendom and Christians.
7. *Réville's article on y[ou]r works*: see letter 4535, n. 1.
8. *hostile intentions of the Contemp[orary]: Review*: this seems an error and Novikoff instead means the *Quarterly Review*, the putatively hostile intentions of whose editor Tyndall discussed in letter 4561, although he himself seems to have meant the editor of the *Edinburgh Review*.
9. *the attacks of the Edinbro' Review (I think December 74)*: [H. Reeve], 'Mill's *Essays on Theism*', *Edinburgh Review*, 141 (1875), pp. 1–31, which was published in January 1875 (see letter 4561, n. 2).
10. *the Ultra "<u>Orthodox</u>"*: the most conservative members of the Russian Orthodox Church.
11. *as you have done already*: Tyndall had added a footnote about Frohschammer's struggles to 'Preface to the Seventh Thousand', p. xxiii (see letter 4469).
12. *Kinglake's vol v.*: volume 5 of A. W. Kinglake, *The Invasion of the Crimea*, 9 vols (Edinburgh: William Blackwood, 1863–87), which was published in 1875 and entitled 'The Battle of Inkerman'. The mistakes Novikoff mentions would have been about the conduct of the defeated Russian troops at the decisive battle against the British and French in November 1854.
13. *his work on I. Kant*: A. Riehl, *Der philosophische Kritizismus und seine Bedeutung für die positive Wissenschaft*, 2 vols (Leipzig: Wilhelm Engelmann, 1876). Tyndall's writings were quoted in the book at vol. 1, pp. 35 and 146–7. On Immanuel Kant, see letter 4455, n. 5.
14. *y[ou]r present lectures at the R[oyal]. Institution*: Tyndall's lecture course on electricity at

the RI (see letter 4557, n. 4). Tyndall had already addressed the question of publication in letter 4564, which Novikoff had not yet received.

To Emily Tyndall 4 March 1875 4566

Royal Institution of Great Britain, 4th. March 1875.

My dear Emma[1]

I have been very very busy of late working at various subjects,[2] and have had therefore little time for letter writing. The winter is particularly severe here—we have had recently for several successive days a covering of snow upon the ground, and the wind has been very bitter indeed. I dare say it is not quite so bitter in Gorey[3] though the difference cannot be great.

I hope the cold agrees with you—I do not dislike it: still I shall be glad when the year advances so far as to give us some of the beauties of spring.

I hope all around you are well, and that Dr Tyndall bears the cold better than usual.[4] If he cared about it I would send him a charcoal and cotton-wool respirator[5]—but I do not know how the instrument works. I have given one to a friend here. I think it would soften the temperature of the air, and therefore be of service to him.

Give my love to all my friends, and my special and affectionate thanks to Georgie[6] for her remembrance of me. If you need anything for your comfort pray let me know.

Your affectionate brother | John

RI MS JT/1/TYP/10/3317
RI MS JT/5/14

1. *Emma*: a diminutive of Emily, which was Tyndall's sister's given name (see 'Deaths', *Enniscorthy Guardian*, 5 December 1896, p. 2).
2. *working at various subjects*: among other things, Tyndall was delivering lecture course on electricity at the RI (see letter 4557, n. 4).
3. *Gorey*: the town in County Wexford, Ireland, where Tyndall's sister was living under the care of her cousin John Tyndall at his house The Lodge on Main Street (known as Charlotte Row).
4. *Dr Tyndall bears the cold better than usual*: John Tyndall was suffering from acute bronchitis from which he would die on 17 April (see letter 4591).
5. *a charcoal and cotton-wool respirator*: a device worn over the mouth to increase the temperature of air as it was breathed. They were expensive, and considered to be helpful for patients with respiratory ailments.
6. *Georgie*: Georgina Mary Tyndall (d. 1939), the youngest daughter of John Tyndall and his wife Catherine.

To Robert Arthur Talbot Gascoyne-Cecil 5 March 1875 4567

British Association for the Advancement of Science, |
22 Albemarle Street, London, March 5, 1875.

MY LORD,—

By the desire of the General Committee and of the Council of the British Association, I beg to lay before your Lordship the accompanying resolutions regarding the continuance of Solar Observations in India.[1]

Researches of the character here contemplated are of comparatively recent date, and have been hitherto pursued with conspicuous ability by independent observers. They may be divided into three distinct groups:—namely, Sun-spot Periodicity;[2] the relation of the Periodicity to Terrestrial and Planetary phenomena; and the Physical and Chemical changes of the sun's visible surface. It is the opinion of the Council of the British Association that observations of the sun conducted under these three heads would furnish results of the highest scientific importance, and that India, presenting as it does every diversity of climate and of atmospheric condition, and every degree of elevation from the sea-level to the greatest mountain heights, is a field eminently suited to the successful prosecution of such observations.

The specific proposal which, on behalf of the British Association, I have the honour to submit for your Lordship's consideration is, that the instruments recently supplied for the Observation of the Transit of Venus[3] should, now that they have served that purpose, be made to contribute to the equipment of a Physical Observatory to be established in the Himalayas,[4] the Nielgherries,[5] or some other fit locality. These instruments are suitable for solar observations, and with the addition of a spectroscope[6] and a few other minor adjuncts would suffice for the present. They would be ready to be brought into practical action the moment the necessary buildings, which might be of the simplest and most inexpensive character, are erected.

But to extract from solar observations their full value it is necessary that they should be continuous. No day ought to pass without observations of the solar surface. This can only be accomplished by establishing, in connexion with the principal observatory, stations in positions selected with the view of rendering it in the highest degree probable that at one or other of them favourable weather would always be found. When, therefore, the results obtained in the proposed observatory shall have justified the extension (and of such justification the Council entertain a confident hope), outlying stations may be added, provided with the moderate equipment needed to multiply the chances of that continuity of observation which it is so desirable to secure.

It is specially agreeable to me, personally, to have the privilege of bringing this important question under the notice of a nobleman whose scientific acquirements render unnecessary any lengthened argument[7] to prove that the proposed observatory is likely to redound to the honour of England, and to materially assist in the advancement of Natural Knowledge.

I have the honour to be, | My Lord, | Your most obedient Servant, | John Tyndall, | *President. The Most Honourable* | *The Marquis of Salisbury,* | *Secretary of State for India.*

Brit. Assoc. Rep. 1875, pp. xlviii–xlix

1. *the continuance of Solar Observations in India*: interest in India as a site for solar observations had been prompted by several expeditions there to observe solar eclipses and the transit of Venus.
2. *Sun-spot Periodicity*: changes in the number of observed sunspots on the solar surface enabled the measurement of alterations in the sun's activity in an eleven-year period.
3. *the instruments recently supplied for the Observation of the Transit of Venus*: this had occurred on 9 December 1874, and had been observed from two stations in the Indian Ocean, on the Kerguelen Archipelago and the island of Rodrigues, with equipment provided by the British colonial government in India.
4. *the Himalayas*: the high altitude and low humidity of this mountain range in Northern India were considered an ideal environment for spectroscopic work.
5. *the Nielgherries*: variant spelling of the Neilgherries, a series of mountains that form part of the Western Ghats mountain range in Southern India. It was in this range, at the hill station of Kodaikanal, that the first formal solar observatory was set up in 1899.
6. *a spectroscope*: an instrument specially designed for the production and examination of spectra (*OED*).
7. *a nobleman whose . . . unnecessary any lengthened argument*: Lord Salisbury had conducted original research in the physical sciences, publishing in the *Phil. Mag.*, and equipping his family seat, Hatfield House in Hertfordshire, with a modern laboratory. He was made FRS in 1869, and on becoming Chancellor of Oxford University in the same year made serious provisions for physical science there.

From Harriet Hooker [7 March 1875][1] 4568

Kew[2] | Sunday Evening

My dear Dr Tyndall

I do not know how to thank you enough for the lovely thimble I found awaiting my arrival this afternoon. I think it was most kind of you to remember it and I value it highly, it fits beautifully too, I assure you I feel quite proud of my possession.

I am very unhappy about Sir Arthur Helps's death.[3] I shall miss him very much, he was always so kind to me, poor Papa[4] feels it very keenly, he has lost so many friends lately; it is a terrible break for the family; one's own sorrows bring our neighbour's nearer to us.[5] I start on Wednesday,[6] I don't like leaving at all, and I suppose when I get comfortably settled I shan't like coming back, however I hope the change will set me up for the summer, and Papa too needs one greatly.

With many thanks again, and much love, | Believe me | Yours aff[ectiona]tely | Harriet Hooker.

RI MS JT/1/TYP/8/2726
LT Typescript Only

1. *[7 March 1875]*: the date is suggested by the relation to letter 4569. 7 March was a Sunday.
2. *Kew*: 49 Kew Green, residence of her father Joseph Hooker and recently late mother Frances Hooker.
3. *Sir Arthur Helps's death*: Arthur Helps (1813–75), a writer and clerk of the Privy Council, died of pleurisy on 7 March (*ODNB*). Helps supported Hooker in his dispute with Acton Ayrton (see letter 4571, n. 2) over the management of the Royal Botanic Gardens, Kew (see letter 4571, n. 3).
4. *poor Papa*: Joseph Hooker.
5. *one's own sorrows bring our neighbour's nearer to us*: Helps and his wife Bessy (née Fuller), married in 1836, lived at Queen Charlotte's Cottage in Kew Gardens, which he had been granted by Queen Victoria in 1867. Frances Hooker, Harriet's mother, died suddenly on 13 November 1874 (see letter 4480, n. 2).
6. *I start on Wednesday*: this was 10 March, but it is unclear where Harriet was going.

From Joseph Dalton Hooker 8 March 1875 4569

Kew,[1] March 8th 1875.

Dear Tyndall

You have enchanted Harriet; the poor little thing is so proud of your kindness and of the token of it.[2]

I thought you would like to see the enclosed.[3] If we could have Thomson[4] for the next succeeding Presidentship[5] it would be nice. Sharpey[6] has I know the very highest opinion of his scientific work. I well remember his contributions to some of the earlier medical Cyclopaedias[7] and which are marked by great originality and breadth of knowledge as well as of views.

Ever yours aff[ectiona]tely | J. D. Hooker.

RI MS JT/1/TYP/8/2727
LT Typescript Only

1. *Kew*: see letter 4568, n. 2.
2. *the token of it*: Tyndall had sent Harriet Hooker a thimble (see letter 4568).
3. *the enclosed*: enclosure missing.
4. *Thomson*: Allen Thomson (1809–84), who was Regius Professor of Anatomy at the University of Glasgow and had been president of the biological section of the BAAS in 1871 (*ODNB*).
5. *the next succeeding Presidentship*: presumably of the BAAS, of which Thomson would serve as president in 1876, succeeding the engineer John Hawkshaw, who was president in 1875.
6. *Sharpey*: William Sharpey (1802–80), an anatomist and physiologist who had been secretary of the RS from 1853–72 (*ODNB*).
7. *his contributions to some of the earlier medical Cyclopaedias*: Thomson contributed the entries on 'Circulation' and 'Generation' to R. B. Todd (ed), *The Cyclopædia of Anatomy and Physiology*, 3 vols (London: Sherwood, Gilbert, and Piper, 1836–47), vol. 1, pp. 638–83; vol. 2, pp. 424–80.

To Vernon Lushington[1, 2] 9 March 1875 4570

British Association for the Advancement of Science, |
22 Albemarle Street, London, March 9, 1875.

SIR,—

The Council of the British Association have had recently referred to them by the General Committee of the Association a question which in various forms has been already under their consideration—the importance, namely, of attaching Naturalists (that is to say, persons specially trained in Natural-history observation) to Surveying-ships generally, and more especially to those engaged in the survey of unfrequented or little-known regions.

The Council have requested me to communicate to Her Majesty's Government their conviction of the importance of making, wherever practicable, this addition to Surveying Expeditions. They believe that such action on the part of the Government would not only be of advantage to Science, but that it would be conducive to the commercial interests of the country to an extent far outbalancing the trifling outlay which such appointments would render necessary.

We are here in reality only asking for a further application of the enlightened policy which enables the Government to utilize the talents of such men as Banks and Solander[3] in the last century, and which has more recently given scope to the abilities of such men as Darwin, Hooker, and Huxley.[4] Even in a commercial point of view the advantages which have flowed from this policy have been quite out of proportion to its cost to the country.

The obvious desirability of associating trained observers with the Surveys of the future is thus strengthened by the experience of the past. The Council of the British Association beg therefore to urge upon the favourable consideration of Her Majesty's Government the question submitted to them by the General Committee.

I have the honour &c.

Vernon Lushington, Esq., Q.C.[5]

Brit. Assoc. Rep. 1875, p. 1–li

1. *Vernon Lushington*: Vernon Lushington (1832–1912), a British lawyer and permanent secretary of the Admiralty (*ODNB*).
2. At this time Tyndall was President of the BAAS.
3. *Banks and Solander*: Joseph Banks (1743–1820) and Daniel Carlsson Solander (1733–82), British naturalists who travelled to the Pacific Ocean with James Cook on HMS *Endeavour* between 1768 and 1771 (*ODNB*).
4. *Darwin, Hooker, and Huxley*: they had sailed on, respectively, HMS *Beagle* (1831–6), HMS *Erebus* (1839–43), and HMS *Rattlesnake* (1846–50).
5. *Q.C.*: Queen's Counsel, a senior trial lawyer technically appointed by the monarch.

To Olga Novikoff 14 March 1875 4571

Royal Institution of Great Britain | 14th. March, 1875

Dear Friend

I have despatched Count Keyserling's memoir[1] to my friend Hooker.

I send you by this post a short sketch of the life of Hooker written at a time when he was suffering from the treatment of Mr. Ayrton,[2] then first Commissioner of works.[3]

This Manifesto[4] created a great commotion in the newspapers: the subject was brought by Lord Derby before the House of Lords.[5] But it was mismanaged in the Commons.[6] However it put a stop to Hooker's annoyances.

Hooker is now President of the Royal Society.[7] He is probably the most famous botanist of the present day.

Yours ever | John Tyndall

[Envelope] Madame Olge de Novikoff | née de Kiréef | Poste Restante[8] | Moscow Russia.

[Postmark] LONDON·W | MA15 | 75]

John Tyndall Correspondence in the Olga Novikoff Correspondence Collection. Courtesy of Special Collections, Kenneth Spencer Research Library, University of Kansas. MS 30:Q11

RI MS JT/1/TYP/3/935

1. *Count Keyserling's memoir:* possibly A. Keyserling and P. T. von Krusenstern, *Wissenschaftliche Beobachtungen auf einer Reise in das Petschora-Land in Jahre 1843* (St Petersburg: Carl Kray, 1846), which Novikoff mentioned in letter 4544.
2. *Mr Ayrton*: Acton Smee Ayrton (1816–86), a lawyer and politician (*ODNB*).
3. *first Commissioner of works*: Ayrton had been appointed as First Commissioner of Works and Public Buildings in 1869, and had engaged in an extended struggle with Joseph Hooker over the management of the Royal Botanic Gardens, Kew. On Tyndall's involvement, see *Tyndall Correspondence*, vol. 13; and *Ascent of John Tyndall*, pp. 295–9.
4. *This Manifesto*: the Memorial, written out by Tyndall, signed by ten others—Darwin and Huxley included—and addressed to William Gladstone, in support of Hooker in his dispute with Acton Ayrton (see letter 3755, *Tyndall Correspondence*, vol. 13).
5. *brought by Lord Derby before the House of Lords*: Edward Henry Stanley (1826–93), 15th Earl of Derby, was the leader of the House of Lords and had addressed his fellow peers on the dispute on 29 July 1872 ('The Royal Gardens, Kew—Dr Hooker and The First Commissioner of Works', HL Deb, 29 July 1872, vol. 213 cc3–23, HANSARD 1803–2005 (online, accessed 30 May 2021); this was also published as 'Dr. Hooker and Mr. Ayrton', *Times*, 30 July 1872, p. 5).
6. *it was mismanaged in the Commons*: seemingly a criticism of John Lubbock, who as an MP had supported Hooker's cause in the House of Commons. The issue was debated in the House of Commons on 8 August 1872 ('The Royal Gardens, Kew—Dr. Hooker, and the First Commissioner of Works', HC Deb, 8 August 1872, vol. 213 cc710–758, HANSARD 1803–2005 (online, accessed 30 May 2021)). In the end, Gladstone essentially evaded the issue, Hooker did not retract any of his complaints and remained Director of the Royal Botanic Gardens, Kew, and Ayrton, despite parliamentary and press support in favour of Hooker, made no apology and received no immediate consequences for his treatment of Hooker. However, the support of the scientific community around Hooker resulted in his election to the presidency of the RS in November 1873, while a few months before, in August, Gladstone placed Ayrton in a new position within his administration and thus no longer in control of public institutions such as the Royal Botanic Gardens, Kew.
7. *Hooker is now President of the Royal Society*: Hooker served as president of the RS from 1873–8.
8. *Poste Restante*: see letter 4392, n. 7.

From John Laing[1] [*c.* mid-March][2] 1875 4572

To Professor John Tyndall, LL.D., F.R.S.:

SIR: The address delivered by you last summer, in Belfast, is an appeal through the British Scientific Association[3] to the world. In it you have not confined yourself to subjects purely scientific, but have dealt with "debatable questions," confessedly "ultra-experimental,"[4] and on these, plain men as well as scientists, are by you expected to form an opinion; you thus admit the right of one, who makes no pretension to scientific learning, to examine your speculations. This I venture to do. Let me assure you, however, I do so in no hostile spirit, for I have received both pleasure and profit from your interesting writings, and rejoice in your ability and success in presenting to plain readers some of the most far-reaching and recondite truths of physical science. I hail you as a fellow-worker in the discovery and elucidation of truth. My only regret is, that one who seems desirous of finding the truth, should have failed to give moral and spiritual truths and their methods that patient and earnest consideration to which they are justly entitled, and thus should be unable to share with Christians the joy and peace which the knowledge and love of God afford. Thus, sir, I am satisfied, and thus only can you find rest and satisfaction for the intellectual faculty, the want of which you so unmistakably declare.

The properly scientific portions of your address I am quite incompetent to review. These have been examined by men competent for the task. Nor will I refer to your historical exposition of the doctrine of atoms, further than to say, that, considering the importance of the subject, the character of the writer, and the assemblage before which it was delivered, I consider it defective, partial, one-sided, and unsatisfactory. The correctness or falseness of that exposition in no way affects the questions on which I am about to animadvert, and, therefore, I may leave it to be dealt with by reviewers who have explored the writings of Greek poets and philosophers, are acquainted with the learning of Arabic savans, and have some knowledge of the subtleties of the school-men.

The object of your paper, professedly, is to vindicate for science the unrestricted right of search, "as regards certain questions affecting religion," and "freedom to discuss them." And inasmuch as you think that "many of the religions of the world have been dangerous, nay, destructive to the dearest privileges of free men. * * * and would, if they could, be again," you essay to show the falseness of ordinary religious conceptions; you also wish to contribute your share in letting the world "know the environment which, with or without our consent, is rapidly surrounding us, and in relation to which some readjustment on our part may be necessary," so that an answer may be formed for those "inevitable fundamental questions" on which the unaided intellect of man has, during all the long past, been in vain exercised. "The problem of problems" which you wish to have solved is "How to yield the religious sentiment reasonable satisfaction," * * * to find some way of satisfying the understanding in its inexorable advance in the path of knowledge, and "finding, consistently with the unquenchable claims of man's emotional nature, which the understanding can never satisfy, a basis for natural phenomena apart from the intervention of Deity."[5]

I cannot, sir, bring myself to believe that you seriously intend to charge true religion – that is, Christianity as delineated in the Scriptures—with hostility to freedom. In the light both of the teachings of scripture and the contendings to the death of Christian martyrs, the charge is utterly absurd, while any unprejudiced reader of history knows that free thought and scientific inquiry have found their widest scope under the sheltering care and fostering influences of Protestant Evangelical Christianity. Why thus malign your best friends, and alienate them from your cause? Why not discriminate between the prosecuting dogmatism of a superstition which claims infallibility, and the opposition which some views advanced by speculative philosophers have encountered in the shape of an appeal to each man's consciousness, but which concedes the right of private judgment? In bringing thus a sweeping charge against religion, without discrimination, you offend your best friends and do your cause an injury. So far, sir, from wishing to restrict the search of science or prevent the discussions of all questions it may raise, even those affecting religion, true Christians cheerfully take part in these discussions and try to help on the search. We only ask that the discussion shall not be all on one side, but that religion shall get a hearing as well as science. And, sir, it does seem passing strange, that, while Mr. Tyndall is permitted to "discuss these questions affecting religion "from his stand-point as a physicist, and in the name "of physical science," Dr. Watts[6] should be denied the liberty of discussing *these same questions* from his stand-point as a theologian, and in the name of theological science. Christians are not afraid to listen to scientists, even when they attack their religion. Are scientists afraid to listen to a theologian when he puts in "a plea for peace and co-operation between science and theology?"[7] Afraid lest theological science be found too much for them: Or is this an attempt to stifle free thought in one direction, and silence any abler thinker who may believe in Christ? Sure we are that scientists cannot approve of such things, and on calm reflection will regret, as we do, that the scientific association denied to theology the "unrestricted right of search" on theological questions which affect science, and freedom "to discuss them,"[8] before the same audience that hailed with applause an attack upon religion, which can be shown to be as unfounded as it is covert and unjust.

I thank you, sir, for your outspokenness as to the environment "which you see surrounding this age, and calling for adjustment."[9] Christians have no fear on that account. We know that truth is one and harmonious in all its parts, that a truth of science cannot contradict, or, when properly understood, conflict with any other truth. Christians are not infallible, and you must know how readily, before the advancing light of science and criticism, theologians have made the "necessary adjustments," by modifying their interpretations of scripture and religious *opinions* on many points. In this respect they have gone as fast as true science warrants them. But you must also be aware, if you are at all candid, that after all these adjustments have been made, not only is the Christian religion unaffected in any essential point, but it has been greatly strengthened by these corrections; and, by reason of the labours and writings of its enemies, to-day shines with a lustre and wields a power unknown in former ages. Other adjustments may be necessary, and will be made as soon as truth call for them, but Christians have not the least apprehension of being required to alter "one jot or tittle" of their faith on "fundamental questions." We are at rest, and with joyous expectation and liveliest interest await and hail every fresh discovery made by men of science; we love all truth and fear none, even though hostile hands may try to turn it against our blessed faith. Your "problem of problems" we have solved; our "religious sentiment has reasonable satisfaction." We have found a "basis for natural phenomena;" a sufficient as well as efficient cause. The solution, however, is not by excluding God from all intervention in the world, but by including him and referring all to him as the first cause. Nor can I see why the exclusion of Deity is demanded by you; certainly *reason* does not make the demand; nay, she calls aloud for God. And, sir, you may rest assured that you weary yourself in vain, as many have done before you, when you try to solve your problem by leaving out a chief factor.

In the treatment of your subject, you are quite aware that you are assailing fondly cherished opinions, and are doing violence to the common sentiments of mankind; that it is "dangerous ground over which you have lead" your reader, when, with "clearness and thoroughness" you propose to change the long received definition of matter, and "abandoning all disguise, make the confession, * * * that you discern in matter the promise" (whatever that may mean in science) "and potency of every form and quality of life." You even adopt an apologetic strain when you "crave the gracious patience" of your reader "to the end," till you explain your novel views so as to be able to say, "There is no very rank materialism here." You seem conscious, that while you profess "to cherish our noble Bible," and at the same time labor to upset its first sentence, you are liable to the charge of "hypocrisy and insincerity,"[10] and I add, in view of our exposition, strange inconsistency. You will, therefore, be quite prepared for it, if, when your views pass under examination, some plain and strong things should be said. I shall assume that in the proper spirit of wholesome controversy, you give me credit for the same love of truth and zeal for it, which you claim as impelling you in your attack upon religion, and will attribute any mistake or misrepresentation of your views, not to an unfair spirit but to an honest misconception of your meaning.

First, then, I direct your attention to the question, is matter created, or is it eternal, that is, without beginning? Does science prove that "God did not in the beginning create the heavens and the earth?"

You admit that matter exists – you assert that its ultimate elements are atoms of inconceivable minuteness; that by the combination of these in the formation of molecules and their association, all forms, organic and inorganic, are produced; that life, sensation, and thought result from this interaction of atoms; but that they themselves are the "foundation stones" of the material universe, "which, amid the ceaseless changes of an immeasurable past, remain unbroken and unworn;"[11] you attribute also to these atoms qualities, by virtue of which combinations are effected and movements produced. These qualities constitute force, which is indestructible, though capable of change of form. Further, these atoms, thus endowed with potential force, you recognize at the beginning, in a "Primordial Fog;"[12] you tell us that science knows nothing before them, and knows nothing of their creation. Elsewhere, you put it thus: "Science knows nothing of the origin or destiny of nature. * * * Who or what made and bestowed upon the ultimate particles of matter their wondrous power of varied interaction, science does not know. The mystery, though pushed back, remains unaltered."—(*Fragments of Science*, page 415.)[13]

Had you left the matter there, no Christian would have had reason to complain; for, as you state it, the whole amount of the teaching of science is, that science knows no creation and no creature, that is, *physical* science knows none. It reaches its utmost limit, when, by your boasted *vorstellung*,[14] it has described atoms in motion in "cosmical mist."

This result of the scientific imagination may be true, or it may be false. True or false, it does not affect the Christian belief of creation. Another question comes in—a question of fact—a question which science cannot answer – Whence came these atoms? Maxwell speaks of them "as prepared materials," but who prepared them? Gassendi,[15] you tell us, does the same, and Darwin has never denied creation. This is scientific caution and candor. But *you* do violence to science—you are tempted to close with Lucretius,[16] when he affirms "that nature is seen to do all things spontaneously without the meddling of the Gods." Or with Bruno, who says, "matter is not that mere *capacity* which philosophers have pictured her to be, but the universal mother who brings forth all things as the fruit of her own womb."[17]

Reason demands a cause for the atoms—a cause for their qualities—a cause for their first movement. This demand science cannot satisfy, for she knows no such cause; but she dares not say "there is no cause." To do so is unscientific. To do so is atheism, thence atheism is unscientific. You have not committed yourself by saying in so many words, there is no cause, no God. Yet, from the way in which you leave the subject, the inference seems inevitable, that, in your opinion, the following is a possible, nay probable answer: "Matter had no beginning and no creator; hence, matter is eternal and self-existent; and there is no God prior to matter or outside of it." I gladly allow your disclaimer, "that it is not in hours of clearness and vigor that this doctrine commends itself to your mind, and that in the presence of stronger and healthier thoughts it even dissolves and disappears, as affording no solution of the mystery in which we dwell and of which we form a part."[18] I do not charge you with *material atheism*, but I appeal to you as a man of truth, honor, and scientific candor, was it fair to thrust in the face of mankind, backed with all the power and weight which your position gave you, and where reply was not permitted, a crude speculation, so unscientific, and, to yourself, unsatisfactory, knowing its tendency to unsettle the mind, and produce discomfort and mischief, while it affords no solution of the mystery? You, sir, may not have yielded to the temptation and adopted the creed of the fool, "who says in his heart, there is no God,"[19] but you know how readily others rush into atheism, goaded by guilty fear of retribution. You know, also, the historical fruits of atheism in anarchy, bloodshed, a carnival of lawless lust, and horrors untold. Others, sir, may slip and be lost in the gloomy abyss, to the brink of which you have led them, and on which you stand trembling as you peer into its impenetrable shades, afraid to launch forth; but, if we may judge from the tenor of your address, regardless of what becomes of those who adopt your unscientific conclusion, that matter is all, and there is nothing besides, and follow out that dogma to its legitimate bitter issues.

Secondly.—Let us look at the question of God's Providence. Does the Deity exercise any influence in the universe? or, are all the changes which take place in the material world, including the phenomena of life, sensation, intelligence, and will, the result of *mere force* acting among the atoms of matter?

You have not said, "There is no God;" nay, you seem to agree with Epicurus, "that the gods are eternal and immortal beings, whose blessedness excludes every thought of care or occupation of any kind." Like him, you think "the idea of Divine power, properly purified, is an elevating one," but that man's "relation to the gods is subjective." In short, that a deity is an "ethical requirement of human nature," and nothing more; a something which is "fluent, and varies as we vary, being gross when we are gross, and becoming, as our capacities widen, more abstract and sublime."[20] Thus, you give us a God (or gods as the case may be, for both are equally unreal) that has no objective existence – that did not create the world, and does not interfere with mundane concerns.

Is this, sir, the Deity of Science? Is this the result of its vaunted exact process of experiment and induction? Do scientific men adore such a God? Away with the baseless superstition! Whatever Mr. Tyndall may say, science disowns the imposture—SCIENCE *knows no* SUCH *God.* Sir, you have no evidence for the existence of such a being; you *cannot* have. That such a God exists is an unscientific assumption, quite as opposed to science as atheism. To worship such a figment of imagination is cultured superstition; you, forsooth, set yourself forth with flourish of trumpets as the champion of free thought—science's called apostle—to do away with "anthropomorphism," and to set poor captives free from the fetters of religious belief; and *you* present to them for their adoration a mere imaginary being, a shadow that passes over the mirror of the soul and abides not, and bid them worship! You take away their infinite, eternal, and unchangeable God, and give them instead a god afar off, but not nigh at hand; a god, everchanging and various as man's views and moods; a god, whose existence cannot be known; a mere subjective ethical requirement, who exists for those that feel that ethical need, and for others is not! Mighty achievement! Let mankind shout Hosannah to this scientific savior! Sir, I reject your religion as an imposture, a superstition, unfounded, unscientific, false, for all practical purposes. Your god is no god, your religion a delusion, your epicureanism is atheism, and your worship an absurdity.

But I have been carried in advance of my argument. Let us see if science "demands the radical extirpation" of Divine interposition, when it reveals the "Reign of Law" in nature, and disproves the idea of "caprice"[21] in the operation of nature's laws.

I grant you all phenomena occur in accordance with law. I acknowledge, also, that scientists have discovered *some* of nature's laws, but not all of them. I will even, for the sake of argument, concede the truth of the laws to which you refer, viz., evolution, variation, natural selection, and inheritance. What, now, is the ultimate finding of science in accordance with these laws? What does she teach as regards the interposition of God in mundane events?

A law of nature, you admit, is only the order in which phenomena occur. In stating a law, you simply state certain antecedents and certain consequents in their mutual relation. The *fact* of this order or relation of succession is all you know. When you have ascertained all the antecedents and all their consequents, you have ascertained the law. The law of evolution, then, is the *fact* that simpler forms preceded the more complex in close connection. The law of variation is the *fact* that change of environment is followed by variation in organisms. The law of natural selection is the *fact* that some organisms perished, while others survived in certain circumstances. The law of inheritance is the *fact* that the qualities of the parent reappears in the progeny. We have here four important generalized *facts*, which have been occurring from time immemorial, although only formulated or known as laws yesterday. These generalized facts are nature's laws, and, in accordance with them, the processes of nature move on.

But science must now be interrogated as to another class of things which are not facts, but causes. Let us ask her whence came the first simple form? Science answers, "I do not know; it was there as far back as I can see." What produced a variation in that form? "I cannot tell; I only know the variation occurred." What destroyed the organisms which have perished? "They were unfit for their new environments." What caused the change in environment? "I cannot tell; I only know there was a change." By what power are qualities transmitted to progeny? "I know nothing of the powers, I only record the fact." Here, then, is the result. The cause which produced the first form is unknown; the cause of the complication in the form is unknown; the cause of the variation is unknown; the cause of the inheritance is unknown. The facts and the order of succession between the occurrences are ascertained; these are the laws of nature. As to the cause of these occurrences science modestly, but firmly and decisively, says, "I know nothing."

Now, sir, what do *you* tell us? That the eternal primordial atoms possess "the potency of every change;" that the *principle* of every "change is in the matter;" that while clashing against each other, the atoms and molecules follow an order fixed and necessary; that they are "self-moved and self-posited;"[22] that they first combine mechanically; that their inorganic forms, by their own potency, adjust themselves to new environments and become living forms; that these living forms, by further adjustment, attain sensation, intelligence, power of will, morality, and religious sentiment. Of a truth you are a leader of science, and are far in advance of your follower! She modestly says, I know nothing of causes; you presumptuously, rashly, unscientifically, and irreverently assert: "The cause I know, it is nature itself." I descry in it the *potency* of every form "and quality of life." Science cautiously states the laws, but is dumb as to the principle. You fearlessly guess at the principle, and jauntily say, "it is in matter, of matter's essence, and it is not God." No wonder, sir, that you fain would screen yourself by saying, "I *feel bound* to make a confession!" When doing violence to all scientific requirements, and recklessly giving utterance to unverified imaginings, you "prolong your vision across the boundary of experimental evidence," and tell us with unblushing effrontery that you have yielded to the "temptation," and in spite of evidence to the contrary, furnished by "the firmly granular character" of the albumen endowed with life, have "discerned the potency of every form and quality of life"[23] in dead matter. Sir, your discovery is a delusion; you have discovered a lie. You cannot *prove* that power is in matter, still less can you prove that the power is not in God. In prolonging your vision you have violated the rules of inductive science, and your assertion is baseless. This, sir, is not science. This is not to follow evidence, and to stop when it stops. You indulge in mere assumptions and hypotheses, and in the name of science I protest. As a man of science you were in no wise bound to make any such humiliating confession.

We reject, then, your conclusion that God is not present in and with his world, and that material changes take place apart from the outpouring of his power. Science tells us clearly that there is a power by which all changes, whether secular or sudden, are effected, but she dare not say that power is in matter, for she knows not where it resides. When you change the definition of matter in order to assert a potency therein, and to do away with a living *Power*, you "cross the boundaries of science," and enter the delusive territory of mere speculation and unscientific conjecture. Your attempt to exclude God, on your own showing, is unscientific, and assuredly it is unsuccessful.

Space, sir, will not allow of other considerations, which are suggested by your idea of law, as excluding the doctrine of Final Cause, on which rests the theistic argument, from skillful contrivance to an intelligent designer. Still, I will try to show how absurd the doctrine you set forth appears to most plain men. Take, then, the exquisite organ of sight, the eye. Its production from "tactual sense" by the action of light "in a mere disturbance of chemical processes," you try to depict. "This action," you tell us, "becomes localized in a few pigment cells, more sensitive to light than the surrounding tissue. The eye is here incipient. Then a kind of anticipatory touch" of closely adjacent objects, is sight—"A bulging out of the *epidermis supervenes.* A lens is incipient, and through the operation of infinite adjustment (*sic*) at length reaches the perfection that it displays in the hawk and eagle."[24]

Here, as before, you state a number of facts in order (which, for the sake of argument, we may allow), and impose upon yourself, and try to impose upon your reader, the belief that you have shown the *cause* for these facts. If now we ask the cause, you will doubtless say, "Matter's innate potency." What, then, is this matter? You reply—"It's real nature we can never know." That is, you give us no *true cause* for one part of the tissue being more sensitive to light than another; none for your infinite series of adjustments or their so-called operations; none for the bulging out or the forming of a lens or an iris. These are facts, you say, and they occurred in accordance with fixed laws controlling the order of action among the molecules. While of the *nature* of these molecules, you say, you never can know anything, you, at the same time, assert that an innate potency belongs to them, in which the eye was produced without any design or contrivance for that end, solely through the fortuitous concourse of atoms.

To say no more about this palpable inconsistency, let us take a telescope. It has its tubes and lenses like the eye; its local distances and inverted images. It also has its nice arrangements for adjusting the various parts so as to produce a perfect image on the retina, according to the laws of optics. It has an iris, too, which, by marvelous adjustments, secures the increase, dimunition, and exclusion of light, as may be required. Further, it, too, has attained its present comparative perfection through continuous improvements on the original clumsy glass. The telescope is an eye for all the purposes of our argument, and we know how it has been made. Now, sir, like the eagle's eye, it is the result of the combination of your atoms. The molecules came into the form of a telescope according to the fixed laws of matter. There was no "caprice" in its production. Will you, sir, tell us that the skillful operations which have produced the telescope, are purely and solely the adjustments of "self-moved and self-posited atoms?" That it had no maker *ab extra?* That there was no contrivance in the arrangement of the molecules, but that the atoms dashed thus together without the intervention of a designing cause? You cannot do so. But, perhaps, you will say that the constructive genius of the designer, and the skill of the maker, were the result of self-moved and self-posited atoms, and therefore the telescope, of a blind, unintelligent potency, which was eternally in the matter, just as you have told us that the "formless fog" of primordial atoms contained potentially "the sadness" and "the thought" of which you were anxious, as you mused upon the weathered point of the Matterhorn in the year of grace 1868.[25]

But, sir, this would be an evasion of the point at issue. Looking at the optical instrument, was there a constructive genius and skillful mechanism concerned in its making? Yes, must be the reply, we *know* there was. Were the constructive genius and mechanician potencies residing in the atoms which constitute the telescope, or powers distinct from and outside of these molecules? The latter alternative must be the answer. Had the astronomer any *design* or contrivance in his operations? Unquestionably, you reply. Was a conscious, intelligent *creative* effort put forth by the maker? Again, yes. Did that design and conscious effort result from the interaction of the molecules *which constitute the telescope?* or were they something outside of these atoms? Undoubtedly, the latter. Let us, then, sum up. You admit in the case of the telescope-making, that there is a constructive and skillful power distinct from the potency of the atoms constructing the instrument, and outside of them; that the power formed a design, according to which he operated; and put forth a conscious, intelligent, creative effort in accordance with that design; and that the design and effort were something outside of the interaction of the molecules which constitute the instrument. In short, you admit that the telescope has as its cause of power—conscious, intelligent, active, designing; without whom the atoms did not clash together in fortuitous order to form that ingenious instrument. *This we know by* EXPERIENCE.

And, sir, when you and I *know* this about man's handiwork, can you seriously propose to mock my common-sense by asking me to believe that the infinitely superior, more perfect organ, the eye—with its matchless contrivances and unerring adjustments—had no analogous cause? No maker that formed it after a wise design? but that the atoms of which it consists moved and posited themselves by chance in accordance with fixed laws? As me, sir, to disbelieve my consciousness, and believe that *I* never design or put forth an effort which determines the order of physical phenomena, and what answer should I give? Tell me that the letters in which your address is printed fell into that order without a designing cause in both author and printer? Sir, I *cannot* believe it. I cannot believe that the eye had no contriving maker or intelligent cause; still less can I believe that the beautiful *cosmos* which science reveals to my astonished view, with its manifest contrivances and endless interdependencies, is the result of a blind potency, residing in the atoms of which you say it is composed. To believe as you suggest would be not only unscientific, as assenting to a thing unproved, but would be contrary to all analogical reasoning and preposterously absurd. Common-sense requires for the orderly universe, so replete with evidences of design, a designing cause, outside of the atoms on which he operates.

I will not dwell longer on your strange speculations as to the origination of life, sensation, intelligence, and emotion, without a cause outside of matter. We believe with you, although in another sense perhaps, "Out of nothing, comes nothing;" life is something, and so is reason and will. We cannot believe that they came out of nothing; that life came out of death without a life-giving cause; and that out of senseless, unintelligent, unemotional atoms, sense, reason, and affection came out without the intervention of a cause equal to the effect. Dead atoms in a primordial fog is not such a cause. But, sir, give us, as a cause, a power-intelligent, self-conscious, designing, moral,—and allow his intervention, and I care not how slow the process, how infinite the number of stages in the progression, how secular the periods of evolution, how stable and fixed the laws, I shall readily accept all of the teachings of science, which will only enlarge my conceptions of that God.

Thirdly. We come now to a third of the "debatable questions," as you are pleased to term them. Is there in man a spirit, distinct from matter? Or is matter all of us? Does science prove that when the material body is decomposed, the spirit ceases to exist; or is the belief that there is an immortal spirit consistent with the teachings of science?

The belief in a spiritual nature in man—a soul; in a hereafter judgment to come, and rewards and punishments after death, has had in all ages a powerful influence upon men. Is that belief unfounded and false? Is it, as you would have us believe, a mere superstition, which the light of science must dispel wherever it shines?

I have carefully noted your concessions such as these, "You cannot satisfy the human understanding in its demand for logical continuity between molecular processes and the phenomena of consciousness. This is the work on which materialism must inevitably split whenever it pretends to be a complete philosophy of life. We can trace the development of a nervous system and correlate with it the parallel phenomena of sensation and thought. We see with undoubting certainty that they go hand in hand. But we try to soar in a vacuum, the moment we seek to comprehend the connection between them* * * There is no fusion possible between the two classes of facts, no motor energy in the intellect of man to carry it without logical rupture from the one to the other."[26]

These concessions are valuable, as evidence that you cannot *intellectually* apprehend spiritual processes. They also show that in uttering your further speculations, you do so with the full consciousness of their unwarrantable assumptions. Had you, as a true disciple of physical science, simply acknowledged the two classes of facts, and then confined your remarks to the modifications of matter, which alone is the subject of physical science, without attempting to account for sensation or thought, it had been well. Your conclusions then could have been tested only by experiment and observation, and thus confirmed or refuted. This Mr. Huxley does. When investigating questions cognate to those you discuss, he says, "Why, in fact, may it not be that the whole of man's *physical* actions are mechanical, his *mind living apart*, like one of the gods of Epicurus, but, unlike them, occasionally *interfering by means of his volition?*"[27] Instead of this, however, you are pleased to go beyond and deny the existence of a living mind apart, and to assert that volition is a physical fact, to be accounted for, like all other phenomena, by the potency of primordial atoms. Speaking of Epicurus, you say approvingly, "One main object of Epicurus was to free the world from superstition and the fear of death. Death he treated with indifference. It merely robs us of sensation. As long as we are, death is not; and *when death is, we are not.*"[28] Again, "I once had the discharge of a Leyden battery passed unexpectedly through me. I felt nothing, but was simply *blotted out of conscious existence* for a sensible interval. Where was my true self during that interval? * * Where is the man himself during the period of insensibility?"[29] Yet, again, you tell us that these themes "will be handled by the loftiest minds ages after you and I, like streaks of morning cloud, shall have melted into the infinite azure of the past."[30]

Now, sir, if you did not intend to deny the immortality of the spirit of man and to teach that death is the end of us, you have been most unfortunate in your use of language. All your reasoning, however, seems to go in that direction, and I doubt not you mean what you say. Let us now examine your doctrine. You admit that I exist, but assert that I exist only when conscious, and no longer. What am I then? What is this Ego? Evidently an animal organism with a nervous system, which is "correlated to certain states of consciousness." Of the existence of the material organism and nervous system, you have not, like Berkeley,[31] any doubt or question. They, you *know*, exist; and they exist, whether the state correlated with them be one of consciousness or unconsciousness. Mr. Tyndall's organism and nervous system continued to exist when *you* "were blotted out of conscious existence" by the Leyden jar, and your *true self* disappeared-was not. Well, then, do I exist, does the Ego exist, when there is no state of consciousness? No, you reply. So Mr. Tyndall ceased to exist and came again into existence after his blotting out! The veritable true self was not and is again! And the French soldier of whom Mr. Huxley tells us, ceases to exist "for a day or two in each month, * * when he passes into an abnormal life * * in which he is not *conscious of anything whatever*," and yet in that state "he is an inveterate thief!"[32] I begin to exist and cease to exist as often as a state of consciousness or unconsciousness is correlated with my nervous system, and I am the same man still! The atoms which constitute my nervous system all change, and my states of consciousness are correlated with new atoms every day, but still I am the same being, existing and ceasing to exist as consciousness comes and goes for three-score years and ten; and yet I am not a being! What man blessed with common-sense will believe such nonsense and call it science.

I do not overlook your solution of this difficulty. "To man as we know him matter is necessary to consciousness, but the matter of any period may be all changed, while consciousness exhibits no solution of continuity. *Like changing sentinels, the oxygen, hydrogen, and carbon that depart seem to whisper their secret to their comrades that arrive*, and thus while the non-ego shifts, the Ego remains intact." (*Fragments of Science* page 414.)[33] I do not overlook this statement, and I acknowledge that it is fine writing, but as science it is preposterous, unsatisfactory to the last degree. I admit what you say of man as we know him, but what of men after death, as we do not know him? Can science tell us whether matter is necessary to consciousness then? You admit that she cannot tell, for *she knows nothing about it.* Therefore, when you venture an opinion you are untrue to science, and add to her testimony your own fancies. Physical science says nothing about the continued existence of the Ego, or the continuity of consciousness, these are questions out of its range. Your idea of whispering sentinels is a pretty conceit, a beautiful poetic conception, to which as *such* I have no objection, but it is not science. But, sir, have you verified the whisper of the oxygen atom as it told its tale to its succeeding comrade? No, it is only a "seem" to whisper, and illustration; not even a presentation of the "vorstellung."

And you, a man of the experimental sciences, deliberately ask me in the name of science to deny my spiritual existence, to ignore my intuition of consciousness, and to disbelieve my immortality, solely on the authority of a poetic fancy of seeming whispers among interacting atoms! No, sir, you may dream away, but serious men will believe all the same in the *reality* of the Ego which is not matter, though, as we now exist, correlated to it: which is the subject of sensation and thought, and which remains *one* and the *same*, while your atoms fly hither and thither, relieve each other's periodic watch and change the non-ego continually; nor will they by your sophistry be cozened out of their intuitions and blessed hopes of immortality.

But, sir, you, physicist as you are, a materialist that scouts Idealism, have your metaphysical system, such as it is – you cannot do without metaphysics. There is something beyond physical nature, and whether you will or no, like other thinking men, you must recognize that *not-physical* something. Accordingly we find you speaking of "intuitions, sentiments, emotions, passions, deep-set feelings, an elemental bias of man's nature, "which the senses cannot know, which "are as old as the understanding, and have claims as ancient and valid," and which you fancy Mr. Spencer has shown to be "the result of the play between organism and environment during cosmic ranges of time;"[34] that is, to have had a material origin, I presume; though really a plain man need not be ashamed to acknowledge that such fine writing is beyond him. Let us then examine your metaphysics. You start with a simple recognition of "*states of consciousness.*" These you divide into sensations and emotions; the former apprehended by the intellect, the latter not cognizable by the intellect, but felt "facts of consciousness over and above man's understanding"—sentiments. Then you introduce the *Ego, or conscious being*, that uses the intellect and is the substratum of the sentiments—the *I* that thinks and feels. Next Mr. Spencer tells us that these "states of consciousness are mere symbols." When interpreted they reveal to us, with undoubting certainty, "an *outside entity* which produces them and determines the order of their succession, but the real nature of this entity we can never know." The entity, outside of the Ego, suggested to it by sensation—*i.e.* through the senses, is the external world, or *matter*, as manifested in sensuous phenomena. Its existence is *intuitively inferred* by the Ego interpreting our sensations; but the Ego is conscious only of the phenomenon, and "knows nothing of the real nature" of this material, sensuous, external world. This matter is known to us only as the producer and modifier of our sensations. All we know is "an organism with a nervous system on which" the image and superscription of the "external world is stamped as states of consciousness; but that external world is not what consciousness represents it to be."[35] (Query, as we pass: Are not the organism and nervous system also part of the external world?)

The sum, then, of your philosophy is, a *conscious Ego*, the subject of the sensation—"I feel"—*sensations* which are merely suggestive symbols. A *non-ego* intuitively suggested by the sensations to the Ego. This non-ego is the *external, sensuous world*. This external world, by its activities, modifies the symbols—*i.e.*, changes the sensations. The Ego is conscious of these *changes*. This much, and no more, we know; but of the "real nature" of the Ego, or of the external, sensuous, material world, or of its modifying activities, we know nothing.

We are now ready for another step. The states of consciousness are said to be two-fold, sensations and emotions. We have seen how you deal with the former, let us in like manner deal with emotions. The *Ego* is conscious of an *emotion*. It is a symbol suggesting something—*a non-ego*—which by its activities modifies the symbols—*i.e.*, the emotions. This non-ego is not the sensuous, material world. It is a *super-sensuous entity*, a being which the senses cannot apprehend in sensation. Not matter but *spirit*. Our knowledge of this spiritual entity is equally certain with our knowledge of the material entity. Both are non-ego, both are intuitively inferred by the Ego interpreting their symbols; both as to their real nature are unknown, both are suggested by phenomena of which the Ego is conscious. The one is sensuous and material, the other super-sensuous and spiritual. We have a non-ego which is matter, and a non-ego which is spirit.

Thus your own metaphysical philosophy demonstrates, first, the existence of a conscious Ego, which is not matter, but outside of it. This is the *spirit in man*. Secondly, the existence of a non-ego, which is not matter. This is the *spirit which we call God*. The existence of both is as undoubtedly certain as the existence of matter, certified by the same process. Thus on your own showing science does not disprove, but postulates the existence of a spirit distinct from matter. A spirit in man and a spirit – God.

But, sir, you acknowledge elsewhere this duality of the non-ego, and try to explain it by saying, "They are two opposite faces of the self-same mystery."[36] That is, matter is spirit and spirit is matter, the same being only looked at from opposite sides. This is a bare assertion of yours, unsupported by a tittle of evidence, and not in accordance with our intuitions. You yourself have admitted that "the real nature of the outside entity (matter) "*can never be known*"[37]. And yet you set forth as a truth of science an affirmation which it is impossible to know or to verify. Is this consistency? Is it scientific caution?

When you tell us of a "Power absolutely inscrutable to the intellect of man,"[38] you speak of a being in whom we believe most firmly—a being outside of ourselves whom we adore. But when you dare to say, "That power is only matter,"[39] I charge you with an unproved assertion and I regret your *dictum*, because it is not confirmed by my consciousness and religious sentiments. Quite as resolutely do I reject your other *dictum* that the Ego is only matter, for that also my consciousness contradicts. As to the continued existence of my true self, from childhood to the present day, I cannot have a doubt, although states of unconsciousness may have at time supervened. I am *sure* I never yet ceased to exist. My belief in immortality also is a "deep-set feeling" of natural intuition, strengthened by revealed teaching. I have no reason from science to give up that belief, nor *can* science disprove its truth. For as she *knows* nothing but "the world as it is," and only *infers* the existence of a spirit now, it is impossible for *her* to say, that in a world which is not now, a spirit does not exist. Hence, as science postulates, but does not prove, the existence of a spirit in man, so she does not by her investigations and discoveries disprove the immortality of that spirit; and we conclude that the belief in immortality is not inconsistent with the teachings of science.

You further speak of a power "whose garment is seen in the visible universe," of a cosmical life, in which the phenomena both of nature and mind have their unsearchable roots."[40] In that Power I believe. "He covereth himself with light as with a garment, and stretcheth out the heavens like a curtain."[41] "In him we live, and move, and have our being."[42] I believe most firmly in that all-providing, all-active, living, life-giving Power. It is only when you most unscientifically assert that, *because* by the methods of physical science you cannot ascertain that that power is distinct from matter, *therefore* you "may cross the boundary" of experimental science, and conclude that he is one and the same as matter, only an opposite face of the mystery, and that there is nothing but matter and space – it is only then that I solemnly dissent from your unfounded assertion – an assertion which alike transgresses the rules of logic, contradicts the intuitions of my nature, and shocks "that deep-set feeling of religious sentiment which has an immovable basis in the emotional nature of man," that force, if you will, "which it is in vain to oppose with a view to its extirpation."[43] And here let me add, that it does seem strange that you should be so complimentary so [sic] this religious sentiment, and yet allow it no weight whatever in your speculations, but treat it a superstition which must disappear in the clear dry light of science.

We have thus examined together three of the "debatable questions," on which, in your address, you so painfully touched. I do not shrink from discussing them, nor will any Christian man; nay, we wish to see them discussed. As a conclusion I think it has been shown:

1. That science does not teach the eternity or self-existence of matter, but leaves the question unsolved and unsolvable by experimental inquiry; also, that science does not disprove creation, or show a belief in a Creator to be inconsistent with the truths of experimental science.

2. That science does not prove that atoms are self-moved and self-posited, or that matter has the potency and principle of any change in itself, but leaves the question as to the cause of motion and change unsolved and unsolvable; also, that as experience proves that in known results there are contrivance and design, so from analogy we may infer a final cause for the universe. Hence, science does not disprove the providence of God, or the interposition of deity in mundane concerns.

3. That science does not disprove the existence of a spiritual being distinct from matter, and outside of the external, sensuous, material world. On the contrary, that we are intuitively led to believe in our own true self and its continued existence, while our material organism continues not; and to believe in a power which is not matter, but which actively modifies matter. Hence, science does not disprove the immortality of the spirit of man, or show that death is the extinction of being.

These conclusions we have reached in perfect accordance with all that on your own showing science has *proved*. Your assertions and implications, that there is no Creator; no power outside of matter by which changes in the external world are effected; no spirit of man distinct from his organism; though at present correlated to it, which may be immortal and survive the decomposition of the body, I utterly reject. They are mere assertions, baseless speculations, wild conjectures, which go beyond the evidence of experience. They, as you yourself admit, leaves the intellect as well as the heart unsatisfied, and afford no solution of the mystery of our being, and no comfort or hope for the deep yearnings of the soul. Your speculation is as disappointing to the mind as sickening to the heart of earnest men.

Now, sir, I ask you in a spirit of friendliness, even if severe, why did you thus attack our religion? You know that men are made better by the religion of Jesus, and that the extinction of it would be ruin to the world. You know that true religion has never cast one rock of offence in your way, and that her ministers in all lands have done science the highest service. Why then seek to bring calamity on mankind? Why seek to wound and destroy your best friend? I would remonstrate with you, also, for the unfairness of your attack. You plume yourself on being one of those "who have escaped from religions in the high and dry light of the understanding;"[44] and you deride it; nay, you misrepresent it, in order to find a butt for your derision, "but, in so doing," I cannot help feeling that you must be aware "you deride *accidents*, and fail to touch its immovable basis." You say "Science claims unrestricted search; the ground which these questions cover is scientific ground."[45] So say I. But, sir, not the exclusive ground of *physical* science; it may not warn all others off as intruders. A super-conscious science must be allowed; a moral experimental induction must be practiced; historical facts must have due weight; theological science must have a fair hearing. Let us bring *all* science to bear on these questions, and ask light from all sources for these highest and infinitely important concerns. It is preposterous folly to attempt to solve them solely by the rush-light of physical science, and arrogant assumption to claim for scientists the sole right to speak authoritatively on them. Let us, sir, work together; let earnest men in every department so place their feeble lights, that, as if reflected from one of your mirrors, all the rays may centre in one focus, and blending together give the strongest light that man's nature can produce, in which we may examine this deepest of problems and darkest of mysteries.

Let me remind you of a fact, which probably you have heard before, although evidently it has made less impression upon you than the facts of Epicurus' teaching. More than eighteen hundred years ago a man of great learning[46] met with the Epicureans and Stoics in the *agora* at Athens. They were the scientists of that day, the materialists and fatalists of the age, and like you they said they were not Atheists, and approved of the worship of an unknown God. This learned Hellenist was taken to Mars' Hill,[47] that the wise men might judge of the new doctrine which he was promulgating. He there in substance told his audience: (1) That the unknown God made the world and all things therein. (2) That he giveth to all life and breath and all things. (3) That the godhead is not like gold or silver, etc., but that we are his offspring is to be known from our consciousness. (4) That this God commands repentance, and has appointed a day in which he shall judge the world. (5) And to give assurance thereof he has raised Jesus from the dead. You will not be surprised to learn that when Paul spake of the resurrection some mocked, for you do the same. Nevertheless, some believed, and Dionysius, The Areopagite,[48] was among the number.

Now, sir, we meet your arguments with just these statements, which are eighteen hundred years old. When you demand that "full weight be given to nature's protest" against the resurrection, because it is a *miracle*, before we give our assent to it as a fact (in which you are perfectly right), we demand in turn that full weight be given to the historical evidence on which it claims our assent, because it is a *fact*. "A knowledge of nature" and the exact sciences is not "the one thing needful" for investigating this question. There is required also a knowledge of the human mind and some acquaintance with *probable evidence*. It a man comes to this inquiry without the latter, he is as unfit for the work as a juryman would be who insisted upon deciding a question of fact by the rules of arithmetic; even more unfit than the man who has no knowledge of science. Such a person, though the first scientist living, would, for the purposes of this inquiry, be as much "a noble savage, and nothing more,"[49] as was Dionysius and the other educated men of Greece and Rome who embraced Christianity. An education exclusively scientific is truly defective, and tends to lead the mind astray. A mere scientist is just as fallible as a jurist or a metaphysician, and, in judging of matter of fact, much more so.

Further, in view of your ridiculous misrepresentations of religion, I wish you to understand that Christians know only one true religion, as there is only one

God. It is the same in all ages, as its God is unchangeable. It consists in *knowing, loving, and serving the true God.* It requires the intellect, the affections, and the will to be engaged. This knowledge of God is within the reach of every reasonable being. "For the invisible things of Him from the creation of a *cosmos*, being apprehended by the intellect by means of the works, are clearly seen, and his essentially enduring (without sensible species) power and Godhead, so that man may be without excuse" – (Rom. I: 20). This *natural* religion, then, Christianity assumes, and requires the reverence and homage due to the God thus apprehended in nature. But it goes further, and assumes the existence of sin or moral evil; and in so far as Christianity goes beyond natural religion, it is the revelation of God in saving men from sin. If professes to be a *remedial* interpretation of God, who is love, to save man from the effects of sin. This claim I do not now care to establish, I merely refer to it, in order to let you see, that if you wish to treat distinctions fairly, you must not overlook this feature in the alleged revelation of God, and ought to give it due weight. That such a revelation of God has been made we next propose to prove by an appeal to *facts*. Christianity rests on facts, and we cheerfully assume the onus probandi.[50] We do not ask any one to believe these facts unless the proof is sufficient. We do not claim authority, but appeal to each man's private judgment. If the fundamental *facts* of revelation cannot be proved, our religion cannot be established. In particular, "if Christ be not risen, then is our preaching vain, and our faith also vain."[51] To the examination of the evidence on which the fact rests we confidently challenge you. Decide for yourself. Did Jesus rise from the dead? Examine the evidence as recorded in Scripture. If it satisfies, you then believe it. Do not go to the examination, saying, I will look at the evidence, but *I know it* CANNOT *be a fact.*

Such a prepossession makes it impossible for you to believe, and would leave you unconvinced if you yourself saw a dead man rising out of the grave. If, having candidly weighed the evidence, you still say Jesus did not rise, but his body was in some way disposed of, there is, then, I fear, no hope of convincing you of the truth of Christianity. But if the result of your examination should be that, convinced of the conclusiveness of the evidence in favor of the fact on the one hand, and pressed by the apparent impossibility of science on the other, you are unable to give your assent to the fact as true, I simply ask you how in scientific inquiries you decide when a stubborn fact is opposed to some law you have previously ascertained? Which must yield, the fact or the law? You know that *you are bound to modify your law so as to embrace the fact.* I ask no more in the result proposed. Modify your law of nature, which says the dead cannot rise, by adding, except in certain cases, of which we have one instance in the resurrection of Jesus Christ, the Son of God, when he was revealed to make an end of sin.

One thing more sir, and I close. You make a great deal of a "*tested experience.*"[52] Even in spiritual matters you think that this is the only ground for scientific knowledge. I remember your proposal to put the efficacy of prayer to the test.[53] Doubtless, you now see clearly the egregious confusion of thought which that proposal implied, and are convinced of its worthlessness as a scientific experiment. Christians saw that at once, and knew it was aside from the question. But, sir, there is a practicable test of which *you* can avail yourself, and if you try it, I am confident you will find out the truth. The God, whom, alas! you do not know, but whose existence you will not deny, has said: "If any of you lack wisdom, let him ask of God, that giveth to all men liberally and upbraideth not, and it shall be given him."[54] Now, sir, this God loves you, and waits to reveal himself to you in mercy, notwithstanding all your treatment of him. Put him to the test – prove him. You sincerely and earnestly wish to know the truth, ask him to show it to you; pray then, if he exists, to show you himself. Stretch out your feeble hands in the darkness to feel after him, lift up your voice in the awful silence and cry after him. If, when you have done this with honest heart, God does not answer, then you may excuse your unbelief. But, sir, until that agonizing prayer has been wrung from your doubting breast, *as a man of science you are inexcusable, for you have left one method of testing truth untried.*

You will not, sir, melt away in the azure of the past, like a streak of morning cloud. You will exist forever. Whether you believe it or not, you will stand before the bar of God for judgment. You will find when the near approach of death gives tension and clearness and vigor to the mind, and produces strong and healthy thought; when the religious sentiment, which has its immovable basis in your very nature, but is now weakened by not being allowed free exercise, becomes intensified – an irresistible force that will stir the depths of your being – you will find in that hour that your recent speculations can afford no comfort or support; and as you gaze with sickening uncertainty and fear into the mystery before you, and hear no voice in the dark valley, and feel no hand to uphold you as this sensuous world fades from your view, and your feet find no rock on which to rest, may you even then by faith lay hold on the living God, and trust his Son Jesus Christ, who is the Resurrection and the Life.

Princeton Review, 4 (1875), pp. 229–53

---

1. *John Laing*: John Laing (1828–1902), the minister of Knox Presbyterian Church in Dundas, Ontario, Canada. His family had emigrated from Scotland in 1843, and he completed his education at King's College and Knox College, Toronto. In 1871 he received a BA degree from Victoria College, Cobourg.
2. *[c. mid-March]*: the date is suggested by this open letter being published in the April number of the *Princeton Review*, which would have gone to press at the end of March, thus suggesting that the letter itself was written a few weeks before.
3. *the British Scientific Association:* the BAAS.
4. *"debatable questions," confessedly "ultra-experimental"*: quoted from *Belfast Address*, pp. 63 and 53, although the latter is a mistranscription of Tyndall's 'ultra-experiential'. It is unclear precisely which published version of Tyndall's address Laing refers to, and in fact it

seems that he uses at least two different versions simultaneously, so references will be given to the above edition.

5. *"as regards certain . . . intervention of Deity"*: quoted from pp. 64, 60, and 63, although, once again, they contain mistranscriptions. Laing's use of quotation marks throughout the letter is inconsistent and does not always indicate direct quotes.
6. *Dr. Watts*: Robert Watts (see letter 4449, n. 4). On 23 August 1874, Watts preached a sermon attacking Tyndall's address at Belfast's Fisherwick Place Church.
7. *"a plea for peace and co-operation between science and theology"*: this was the title of a paper that Watts submitted to the BAAS to be read in the biology section (Section D) of the Belfast meeting. However, it was rejected, and Watts instead read it, on 24 August 1874, in Belfast's Elmwood Presbyterian Church. It was published, along with Watts's sermon at Fisherwick Place Church, in *Atomism: Dr. Tyndall's Atomic Theory of the Universe Examined and Refuted* (Belfast: William Mullan, 1875). Watts and Tyndall are discussed in D. Livingstone, 'Darwinism and Calvinism: The Belfast-Princeton Connection', *Isis*, 83 (1992), pp. 408–28.
8. *"unrestricted right of search"* . . . *"to discuss them"*: quoted from *Belfast Address*, pp. 63–4 and 64.
9. *"which you see surrounding this age, and calling for adjustment"*: quoted from p. 63, although the transcription is inaccurate.
10. *"dangerous ground over . . . "hypocrisy and insincerity"*: quoted from pp. 63, 55, 56, 58, and 34, although as well as the usual mistranscriptions Laing also quotes two passages, 'dangerous ground over which you have lead' and 'abandoning all disguise, make the confession', that only appeared in newspaper reports of Tyndall's address and did not appear in his published version of it.
11. *"foundation stones"* . . . *remain unbroken and unworn"*: a very loose transcription of Tyndall's quotation from James Clerk Maxwell in *Belfast Address*, p. 26.
12. *a "Primordial Fog"*: another mistranscription, this time of 'primordial form', which Tyndall quotes from Charles Darwin (*Belfast Address*, p. 53).
13. *"Science knows nothing . . . Fragments of Science, page 415)*: the page reference suggests Laing is quoting from the American edition of *Fragments of Science for Unscientific People* (New York: D. Appleton, 1871). The lines are from 'Vitality', pp. 410–17, on p. 415.
14. *vorstellung*: mental image, idea, imagination (German).
15. *Gassendi*: Pierre Gassendi (1592–1655), a French philosopher who sought to reconcile Epicurean atomism with Christianity.
16. *Lucretius*: see letter 4413, n. 4.
17. *Maxwell speaks of them "as prepared materials" . . . fruit of her own womb"*: quoted from *Belfast Address*, pp. 25–6 and 55.
18. *"that it is not in hours . . . we form a part"*: quoted from p. viii.
19. *the fool, "who says in his heart, there is no God"*: an allusion to Psalms 53:1 'The fool hath said in his heart, There is no God'.
20. *"that the gods are . . . abstract and sublime"* : quoted from *Belfast Address*, pp. 6, 8, and 9.
21. *"demands the radical . . . "caprice"*: quoted from p. 4, although 'the "Reign of Law"' may be an allusion to the George Campbell's book *The Reign of Law* (London: Alexander Strahan, 1867).

22. *"the potency of every . . . "self-moved and self-posited"*: these are Laing's précis of Tyndall's presumed beliefs rather than actual quotations.
23. *"prolong your vision . . . form and quality of life"*: quoted from *Belfast Address*, p. 55, although again with significant mistranscriptions.
24. *"tactual sense" . . . the hawk and eagle"*: quoted from pp. 47 and 48.
25. *the "formless fog" of primordial atoms . . . the year of grace 1868*: quoted from Tyndall's *Hours of Exercise in the Alps*, p. 292.
26. *"You cannot satisfy . . . from the one to the other"*: quoted from *Belfast Address*, pp. 33–4, with additional passages that only appeared in newspaper reports of the address.
27. *"Why, in fact, may . . . means of volition?"*: T. H. Huxley, 'Are Animals Automatons?', *Popular Science Monthly*, 5 (1874), pp. 724–34, on p. 728. This was the published version of the address Huxley gave at the BAAS meeting at Belfast (see letter 4327, n. 9).
28. *"One main object of Epicurus . . . we are not"*: quoted from *Belfast Address*, p. 8.
29. *"I once had the discharge . . . period of insensibility?"*: quoted from p. 29.
30. *"will be handled by the loftiest . . . azure of the past"*: this passage was omitted from *Belfast Address*, but restored in *Belfast Address*, 7th thousand, p. 65.
31. *Berkeley*: George Berkeley (1685–1753), an Irish philosopher who espoused an idealist theory that he termed 'immaterialism' (*ODNB*).
32. *the French soldier of whom Mr. Huxley . . . an inveterate thief!"*: Huxley, 'Are Animals Automatons?', p. 730.
33. *"To man as we know . . . Fragments of Science page 414)*: again quoted from the 1871 American edition of *Fragments* (see n. 13).
34. *"intuitions, sentiments . . . during cosmic ranges of time"*: quoted from *Belfast Address*, p. 60.
35. *"states of consciousness are . . .what consciousness represents it to be"*: quoted from p. 57 and 51.
36. *"They are two opposite faces of the self-same mystery"*: quoted from J. Tyndall, 'On the Scientific Use of the Imagination', in *Fragments of Science*, pp. 127–63, on p. 160.
37. *"the real nature . . . can never be known"*: quoted from *Belfast Address*, p. 57.
38. *"Power absolutely inscrutable to the intellect of man"*: quoted from pp. 57–8.
39. *"That power is only matter"*: this seems to be Laing's own précis of Tyndall's supposed opinion rather than an actual quotation.
40. *"whose garment is seen . . . unsearchable roots"*: quoted from p. 58–9.
41. *"He covereth himself . . . the heavens like a curtain"*: Psalms 104:2.
42. *"In him we live, and move, and have our being"*: Acts 17:28.
43. *"that deep-set feeling . . . view to its extirpation"*: quoted from *Belfast Address*, p. 60.
44. *"who have escaped . . . of the understanding"*: quoted from p. 60.
45. *"Science claims unrestricted . . . scientific ground"*: a loose paraphrase of the line 'as regards these questions science claims unrestricted right of search' from p. 63–4.
46. *a man of great learning*: Saint Paul (*c.* 5–*c.*64/67 CE). Laing gives an account of his so-called 'Areopagus sermon' delivered to the high court in Athens and described in Acts 17:16–34.
47. *Mars' Hill*: the Latin name for the Athenian Areopagus or high court.
48. *Dionysius, The Areopagite*: Dionysius (*c.* first century CE), a judge at the Areopagus who converted to Christianity after hearing Paul's sermon and became the first Bishop of Athens.

49. *"full weight be given . . . and nothing more"*: quoted from J. Tyndall, 'Miracles and Special Providences,' in *Fragments of Science*, pp. 46–69, on p. 68.
50. *onus probandi*: burden of proof (Latin).
51. *"if Christ be not risen . . . and our faith also vain"*: Corinthians 15:14.
52. "*tested experience*": not an actual quotation, but possibly relating to Tyndall's comments on how the early notions of humanity were 'tested by observation and reflection' (*Belfast Address*, p. 1).
53. *your proposal to put the efficacy of prayer to the test*: see letter 4428, n. 2.
54. *"If any of you lack wisdom . . . shall be given him"*: James 1:5.

To John Hall Gladstone 17 March 1875 4573

17th March | 1875

My dear Gladstone.

Some supervision will I fear be necessary regarding the operations carried on in our chemical laboratory. Last night there was an explosion, which, had not William[1] been present, might have set fire to the Institution.[2] Mr. Williams[3] does not appear to be aware of the danger of leaving operations which involve the use of flames going on in his absence.

Yours faithfully | John Tyndall | Dr. Gladstone

Uncatalogued RI purchase, RI.CG31/35

---

1. *William*: presumably a laboratory servant at the RI (see letter 4659).
2. *the Institution*: the RI, where Gladstone was Fullerian Professor of Chemistry from 1874–7.
3. *Mr. Williams*: Matthew Witley Williams (b. *c.* 1856), who, along with Walter Hibbert, was an assistant in the Chemical Laboratory. In the 1881 Census, he was recorded as a scientific chemist, living at 18 Kempsford Gardens, Brompton.

From William Spottiswoode 18 March 1875 4574

The Royal Society, | Burlington House, London. W. | 18 March 75

My dear Tyndall

I return the papers[1] which I have read, & also shown to Williamson[2] before the RS Council today.

Please consider whether it might not be well to summon the other members of the Eclipse Com[mitt]ee[3] in case a meeting then & there should be convenient. If so you have only to send word in to the clerk (Stewardson) at No. 22.[4]—I will be at the meeting on Saturday.[5]

Ever y[ou]rs | W. Spottiswoode.

RI MS JT/1/S/203

1. *the papers*: not identified.
2. *Williamson*: Alexander William Williamson.
3. *the Eclipse com[mitt]ee*: the Eclipse Committee of the BAAS; this meeting of the committee was likely in reference to the imminent eclipse of 6 April.
4. *the clerk (Stewardson) at No. 22*: H. C. Stewardson (d. 1920) was the clerk of the BAAS, which had an office at 22 Albemarle Street.
5. *the meeting on Saturday*: 20 March; it is not clear if this was a meeting of the Eclipse Committee or of another body.

From Mary Agatha Russell 19 March 1875 4575

Pembroke Lodge.[1] | Richmond Park. | March 19. /75.

Dear Professor Tyndall

Now that the lectures[2] are over I must write one time again to thank you for the very great enjoyment you have given us—my brother[3] & I have been able to go to London every Thursday since you so kindly invited us; & we now mean to go most carefully through the Notes.[4]—I must also thank you for the Notes, which I had not received when we saw you, but which were given to me afterwards;—so that we can now read through them from the beginning.—I hope we may soon see you again here.—

Yours most sincerely | Agatha Russell.

RI MS JT/1/R/51

1. *Pembroke Lodge*: see letter 4237, n. 1.
2. *the lectures*: Tyndall's lecture course on electricity at the RI (see letter 4557, n. 4).
3. *my brother*: Francis Albert Rollo Russell.
4. *the Notes*: Tyndall told Olga Novikoff that the lectures 'are not to be published: for I have not the time necessary to organize them But I am having notes of the course printed' (see letter 4564).

To Robin Allen 23 March 1875 4576

Royal Institution of Great Britain, | 23rd March, 1875.

SIR,

I have read with great pleasure the instructive report of Colonel Campbell, on the recent experiments on sound executed at Woolwich.[1]

With the recommendations contained in the Report[2] I beg to express my entire concurrence.

The proved inferiority of the 18-pounder, and the clear advantage gained with the parabolic muzzle, are practical results of obvious importance.

I am of opinion that the Elder Brethren[3] will act wisely in keeping their attention fixed on gun-cotton. Even the four-ounce charge, on many of the Brethren's lightships, would furnish, in foggy weather, a signal of extreme value to the mariner. The great handiness of gun-cotton is also a recommendation.

In the present state of the question, I should hardly venture on an hypothesis to explain the apparent difference between bronze and cast-iron.[4] Colonel Campbell refers to "a true cause," as far as it goes; but I am not certain that further experiments with other guns would not abolish the observed difference. "Shape," to use the words of Colonel Campbell, "is of more importance than material."[5]

I have the honour to be, | Your obedient Servant, | JOHN TYNDALL | ROBIN ALLEN, ESQ.

National Archives, MT 10/290. Papers re Fog Signals. H8476

---

1. *the instructive report . . . sound executed at Woolwich*: 'Fog Signal Guns. Minute 32,732', *Extracts from the Proceedings of the Department of the Director of Artillery*, 13 (1875), pp. 68–70, which contains 'a report, 2/3/75, by the Superintendent Royal Gun Factories, of the further experiments with fog signal guns and guncotton' (p. 68). The experiments, involving firing charges of guncotton from fog signal guns of various shapes and with different muzzle sizes, had been conducted on 22 February at the Proof Butts at the Royal Arsenal, Woolwich.
2. *the recommendations contained in the Report*: the report proposed that 'Charge of guncotton yield louder reports at all ranges than charges of powder of equal weights, however fired. The addition of a parabolic reflector gives a decided increase to the power of guncotton', and concluded: 'Recommends, before taking steps to manufacture, that he [i.e. Frederick Campbell] be authorized to repeat the third experiment [in which 'comparison charges of dry guncotton were detonated in the open'] on a somewhat extended scale. Thinks it desirable to fire 12 series, a day being selected when little wind is blowing, and when the weather is "muggy"' (pp. 68–70).
3. *Elder Brethren*: the governing body of Trinity House (see letter 4187, n. 2).
4. *the apparent difference between bronze and cast-iron*: in the sound produced by gun cylinders made of these respective metals, with the differential in favour of cast-iron increasing in relation to distance (see 4513, n. 4).
5. *"Shape," to use . . . more importance than material"*: quoted from p. 69 of 'Minute 32,372'.

To William Spottiswoode 23 March 1875 4577

23rd March 1875.

My dear Spottiswoode
I have been up to Osnaburgh Street,[1] and have had a long look at Faraday, both with the electro-magnet and with the Ring.[2]
The former is obvious to the meanest understanding. But I do not hesitate to express my preference for the less obvious, and I would add, less commonplace ring.[3]
Please look at both once more. They will wait for your decision.
I write by this post to Pollock[4] expressing the opinion expressed here.
Du reste[5]—It is a very noble production—in the highest degree dignified, and very life-like.
Yours ever | John Tyndall.

RI MS JT/1/TYP/6/2154
LT Typescript Only

1. *Osnaburgh Street*: the location of sculptor John Henry Foley's studio in Marylebone, close to Regent's Park.
2. *Faraday, both . . . with the Ring*: after the completion of the full-length marble sculpture that Foley had been commissioned to create (see letter 4479), it was still undetermined whether Michael Faraday should be holding an electromagnet or an induction ring, and the statue was evidently shown to Tyndall holding both.
3. *my preference for the . . . less commonplace ring*: it was Tyndall's view that prevailed, and in the completed statue Faraday held the induction ring.
4. *I write by this post to Pollock*: to William Frederick Pollock (see letter 4578).
5. *Du reste*: besides, furthermore (French).

To William Frederick Pollock 23 March 1875 4578

23rd March 1875.

My dear Pollock,
I have been up to Foley's Studio,[1] and have had a long look at Faraday, both with the electro-magnet and the ring.[2]
The former is certainly obvious to the meanest understanding. Still I do not hesitate to express my preference for the less obvious, and I would add, less commonplace ring.[3]

I wish you would see them once more. The Sculptor[4] is waiting for the final word from Spottiswoode and yourself.

When the wires are a little emphasized upon the ring there will be no mistake about its general character It is, moreover, a bit of truth which is some advantage.

I liked the model immensely. The countenance is full of truth and strength, and dignity.

Last night I saw one of the Senate, to whom I had previously written, and he said he would do all in his power to set things in train in Fred's behalf.[5]

Yours ever | John Tyndall.

RI MS JT/1/TYP/6/2153
LT Typescript Only

---

1. *Foley's Studio*: John Henry Foley's studio was in Osnaburgh Street, Marylebone, close to Regent's Park.
2. *Faraday, both with the electro-magnet and the ring*: see letter 4577, n. 2.
3. *my preference for the . . . less commonplace ring*: it was Tyndall's view that prevailed, and in the completed statue Faraday held the induction ring.
4. *The Sculptor*: Thomas Brock (see letter 4479, n. 4).
5. *one of the Senate . . . in Fred's behalf*: Frederick Pollock, Pollock's eldest son, was a Fellow of Trinity College, Cambridge, so Tyndall may be referring to a member of the University's Senate.

To William Frederick Pollock 24 March [1875][1] 4579

24th March.

My dear Pollock,

The ring[2] will bear but little "modification"—the rendering of the wire more pronounced is, in my opinion, all that we ought to venture on

Yours ever | John Tyndall

RI MS JT/1/TYP/6/2157
LT Typescript Only

---

1. *[1875]*: the year is established by the relation to letters 4577 and 4578.
2. *The ring:* the induction ring held by Michael Faraday in the full-length marble statue of him commissioned from John Henry Foley (see letter 4577, n. 2).

From Herbert Spencer 24 March 1875 4580

38, Queen's Gardens[1] | Mar[ch]. 24. 1875

My dear Tyndall,

I send the enclosed[2] thinking that if it reaches you before you leave town for Easter, you may perhaps find time, during the recess, to cast your eye over the more important of the changes I have made; and to add to my already heavy obligations, by telling me whether you think your objections have had the desired effect.

You need not I think trouble yourself to re-read the chapter on the Indestructibility of Matter.[3] I have duly attended to all the points noted in it and have put at the end of it a sufficiently emphatic note concerning the meaning of à priori.[4] The chapter on the Continuity of Motion[5] is most of it quite transformed; and is now, I think, not so far from what it should be. I am very glad you have persisted in making me think it over again, and re-cast it. It is now at any rate very much better than it was.

Respecting the chapter on the Persistence of Force,[6] I still find myself unable to take the view that "conservation" is a good word and that "energy" suffices for all purposes. By an added sentence or two,[7] I have sought to make this point clearer

Ever yours truly | Herbert Spencer

P.S. Keep the proof till we meet.

RI MS JT/1/S/107
RI MS JT/1/TYP/3/1122

---

1. *38, Queen's Gardens*: see letter 4303, n. 1.
2. *the enclosed*: enclosure missing, but presumably manuscript or proof pages from H. Spencer, *First Principles*, 3rd edn (London: Williams and Norgate, 1875).
3. *the chapter on the Indestructibility of Matter*: Chapter IV of Part II 'The Knowable', pp. 172–9.
4. *a sufficiently emphatic note . . . à priori*: 'Lest he should not have observed it, the reader must be warned that the terms "à priori truth" and "necessary truth", as used in this work, are to be interpreted not in the old sense, as implying cognitions wholly independent of experiences, but as implying cognitions that have been rendered organic by immense accumulations of experiences, received partly by the individual, but mainly by all ancestral individuals whose nervous systems he inherits. On referring to the *Principles of Psychology* (§§426–133), it will be seen that the warrant alleged for one of these irreversible ultimate convictions is that, on the hypothesis of Evolution, it represents an immeasurably-greater accumulation of experiences than can be acquired by any single individual' (p. 179).

5. *The chapter on the Continuity of Motion*: Chapter V of Part II 'The Knowable', pp. 180–9.
6. *the chapter on the Persistence of Force*: Chapter VI of Part II 'The Knowable', pp. 190–192*d*.
7. *an added sentence or two*: 'Some explanation of this title seems needful. In the text itself are given the reasons for using the word "force" instead of the word "energy"; and here I must say why I think "persistence" preferable to "conservation". Some two years ago (this was written in 1861) I expressed to my friend Prof. Huxley, my dissatisfaction with the (then) current expression "Conservation of Force": assigning as reasons, first, that the word "conservation" implies a conserver and an act of conserving; and, second, that it does not imply the existence of the force before the particular manifestation of it which is contemplated. And I may now add, as a further fault, the tacit assumption that, without some act of conservation, force would disappear. All these implications are at variance with the conception to be conveyed. In place of "conservation" Prof. Huxley suggested *persistence*. This meets most of the objections; and though it may be urged against it that it does not directly imply pre-existence of the force at any time manifested, yet no other word less faulty in this respect can be found. In the absence of a word specially coined for the purpose, it seems the best; and as such I adopt it' (p. 190). Tyndall and Spencer had been disagreeing about the relation between 'force' and 'energy' for some time (see letter 4234).

To Thomas Archer Hirst 25 March 1875 4581

Folkestone[1] Thursday | March 25 1875

My dear Tom

I must be losing my wits—that a thing so obvious should have escaped me is a proof of extraordinary muddiness. It is rather an illustration of the tendency of clear, but narrow vision, to overstep its own bounds. I was distinct enough in my motion, but it applied only to a fragment of the law—strictly speaking a fragment infinitely small; to see the whole, the eye would have to wander round the curve of intersection of the water surface with the conical surface whose apex is as far below the water as the eye is above it.[2] Thank you much.

Yours aff[ectionate]ly | John

I came down here yesterday with Lady Claud.[3] We have got comfortable rooms.[4] Love to Lilly.[5]

RI MS JT/1/T/716
RI MS JT/1/T/HTYP/630

1. *Folkestone*: a seaside town in Kent.
2. *I was distinct enough . . . the eye is above it*: not identified.
3. *Lady Claud*: Elizabeth Hamilton.

4. *comfortable rooms*: presumably at the West Cliff Hotel, which opened in Folkestone in 1861, and where Tyndall stayed on his following visit (see letter 4599).
5. *Lilly*: Emily Hirst.

To William Spottiswoode 25 March [1875][1] 4582

25th March

My dear Spottiswoode

One last word.

You know that Helmholtz wishes for a bust of Faraday[2] & that the Managers[3] have said that it w[oul]d give them pleasure to present one to him.

Mr. Teniswood (Foley's executor)[4] has been here & he says that the bust of the statue[5] without surroundings w[oul]d not do. But Foley he says produced an excellent bust of Faraday & a copy of it can be made for five guineas.

Now I am off tomorrow morning early, and cannot therefore see the bust, therefore cannot say whether it w[oul]d be suitable. Could you make an appointment with Mr. Teniswood to see it? If you are too busy, Pollock[6] I know w[oul]d do the work.

I send you Mr. Teniswood's address.

Yours ever | John Tyndall

If in the all embracing range of your knowledge you should have an opportunity of being kind to the friend of Helmholtz[7] I am sure you will.

He might have been caught in that Squall yesterday. I have passed through black weather somewhere between Lymington & Alum Bay.[8]

RI MS JT/1/T/1347

1. *[1875]*: the year is established by reference to the statue of Michael Faraday (see n. 5).
2. *Helmholtz wishes for a bust of Faraday*: Hermann Helmholtz wished to install busts of eminent physicists in the Physical Institute that was then under construction at the University of Berlin (see letter 4223), although his request for a bust of Faraday would not be fulfilled until 1878. See D. Cahan, *Helmholtz: A Life in Science* (Chicago: University of Chicago Press, 2018), p. 456.
3. *the Managers*: of the RI (see letter 4549, n. 4).
4. *Mr. Teniswood (Foley's executor)*: see letter 4479, n. 3.
5. *the bust of the statue*: the head and shoulders of the full-length marble statue that was commissioned from John Henry Foley and which had recently been completed.
6. *Pollock*: William Frederick Pollock.
7. *the friend of Helmholtz*: not identified, but Tyndall possibly means himself.
8. *Lymington & Alum Bay*: ports either side of the Solent in England and the Isle of Wight.

From Robert Hall[1] 29 March 1875 4583

Admiralty. | March 29, 1875.

Sir,—

I am commanded by My Lords Commissioners of the Admiralty to thank you for your letter of the 9th instant,[2] conveying the opinion expressed by the Committee of the British Association for the Advancement of Science as to the desirability of trained Naturalists being attached to all Surveying-vessels, more especially to those engaged in the survey of unfrequented regions.

I am, Sir, | Your obedient Servant, | Robert Hall.

The President of the British Association | for the Advancement of Science, | 22 Albemarle Street, S.W.

Brit. Assoc. Rep. 1875, p. li

1. *Robert Hall*: Robert Hall (1817–82), Naval Secretary of the Admiralty (*ODNB*).
2. *your letter of the 9th instant*: letter 4570.

To Charles Garth Colleton Rennie[1] 30 March [1875][2] 4584

Royal Institution of Great Britain | 30th March.

Dear Sir,

Accept my best thanks for the memoir[3] you have been good enough to send me. I have spent many pleasant hours in your father's society, and your account of him will on that account be doubly interesting to me.

Faithfully yours | John Tyndall

I have been away from London,[4] otherwise your kindness should have been more quickly acknowledged

National Library of Scotland, Papers of Sir John Rennie (1875), Ms. 19946, f. 44 r

1. *Charles Garth Colleton Rennie*: Charles Garth Colleton Rennie (1818–91), son of the civil engineer John Rennie (1794–1874).
2. *[1875]*: the year is established by reference to Rennie's memoir of his father (see n. 3).
3. *the memoir*: possibly 'Memoirs: Sir John Rennie', *Minutes of the Proceedings of the Institution of Civil Engineers*, 39 (1875), pp. 273–8, although the author was anonymous so it is unclear whether it was written by his son.
4. *away from London*: in Folkestone (see letter 4581).

From Albert Réville 1 April 1875 4585

Neuville sur Dieppe (Seine-Infér.) | 1 Avril 1875.

Cher Monsieur,

Dans l'espoir que votre congé pascal bien mérité aura retrempé vos vaillantes forces, je vous écris ces lignes pour vous prévenir que l'article de la Revue des D. Mondes qui vous concerne a paru dans le n° du 15 mars dernier. Je crains, par la teneur de votre lettre, que vous ne le sachiez pas. Je n'ai jamais plus regretté qu'à cette occasion la règle absolue à laquelle la direction de ce recueil est obstinément attachée et qui consiste à ne pas envoyer d'exemplaires à part aux hommes éminents dont les œuvres sont l'objet d'une critique détaillée. Je comprends jusqu'à un certain point ses motifs, mais c'est souvent très gênant. Si, comme je m'en flatte, vous lisez mon article, vous verrez que, sans déguiser les quelques points où je serais plus affirmatif que vous en matière de réalité religieuse, je vous tends sincèrement une main d'association que je serai heureux et fier de sentir serrée par une main loyale et courageuse comme la vôtre.

Je connais plutôt de réputation M. James Martineau que je n'ai lu beaucoup de lui. C'est même un reproche que je me fais souvent. Le temps, le temps! Qui inventera le moyen de l'allonger? J'avais même fait venir, en m'occupant de vos travaux, une brochure oubliée par lui, que j'ai eu entre les mains, et qui est si bien cachée quelque part dans mes papiers que je n'ai pu la retrouver. En tout cas il me semble que vous ne pouviez choisir un meilleur representative man. A moins que son titre d'unitaire, si mal porté en Angleterre, ne soit pour beaucoup de vos orthodoxes un motif suffisant de dire : Bah! Ce sont querelles de mécréants entr'eux, cela ne nous regarde pas!

Je suis de retour depuis seulement quelques jours d'une tournée de conférences en Alsace, en Franche Comté et en Suisse. En Alsace mon œuvre était plutôt patriotique (car je ne sais pas si l'on se rend compte en Angleterre que, vu la résistance indomptable des Alsaciens qui ni veulent pas devenir allemands, il couve là un feu qui incendiera l'Europe d'ici à quelques années et que ce n'est pas à nous de l'éteindre); en Franche Comté et en Suisse il s'agissait surtout de la propagande protestante-libérale, c'est-à-dire au fond anti-miraculeuse. J'ai rapporté sous tous ces rapports les meilleures impressions. E pur il spirito si muove. J'ai eu à Genève une curieuse correspondance avec l'ex-père Hyacinthe, bien abandonné par les vieux-Catholiques, parce qu'il est trop conservateur et qui fulmine contre les protestants-libéraux. Telum ni belle sine situ.

Je vous fais un gré infini de m'avoir envoyé votre portait et celui de M. Huxley. Permettez-moi de vous offrir le mien.

Ne repassez-vous pas l'un de ces beaux jours sur le continent? Vous savez qu'il y a ligne directe de Londres à Paris par Newhaven et Dieppe. Il y a près de cette dernière ville un ermitage gentiment situé et dont l'ermite serait enchanté de vous faire les honneurs. Seulement c'est vous qui devriez apporter les indulgences.

Croyez-moi, cher Monsieur, votre tout dévoué, | A. Réville.

Neuville sur Dieppe (Lower-Seine.) | 1 April 1875

Dear Sir,

In the hope that your well-deserved Easter break will bring your valiant strength back, I am writing this to tell you that the article in the Revue des D. Mondes[1] about you appeared in issue 15 of last March. I fear, by the content of your letter,[2] that you did not know that. I have never more regretted the fixed rule, to which this collection stubbornly complies, and which consists of not sending copies to share with eminent men whose works are the subject of a detailed critique. I understand their reasoning to some extent, but it is often very inconvenient. If, as I flatter myself, you read my article, you will see that, without disguising the few points where I would be more assertive than you in the matter of religious reality, I sincerely offer my hand of association, and I would be happy and proud to feel it gripped by a hand as loyal and courageous as yours.

I know more of Mr James Martineau's[3] reputation, rather than having read many books by him.[4] I reproach myself with that often. Time, time! Who will invent a means of extending it? While busying myself with your works, I even brought a brochure that was forgotten by him, which I had in my hands, and which is now so well hidden somewhere in my papers that I could not find it again. In any case, I do not think that you could have chosen a better representative man. Unless his title of a Unitarian,[5] so badly carried in England, would be sufficient grounds for many of your Orthodox to say: Bah! This is the quarrelling of infidels amongst themselves; it does not concern us!

I am back since only a few days from a tour of lectures in Alsace, Franche-Comté[6] and Switzerland. In Alsace my work was more patriotic (since I do not know if you realize in England that given the resolute resistance of the Alsatians who do not want to become German,[7] there is a fire there, which will burn down Europe in a few years from now, and which is not for us to extinguish). In Franche-Comté and Switzerland it was mostly liberal-protestant propaganda, in other words, the bottom line is anti-miraculous. I made my best impressions in all these respects. E pur il spirito si muove.[8] In Geneva I had a curious correspondence with the former Father Hyacinthe,[9] long abandoned by old Catholics,[10] because he is too conservative and argues loudly against liberal-protestants. Telum imbelle sine ictu.[11] I am infinitely grateful to you for sending me yours and Mr Huxley's portrait. Allow me to offer you my own.

Would you not come back to the continent on one of these fine days? You know that there is a direct line from London to Paris by Newhaven and Dieppe.[12] There is near the latter city a hermitage nicely located and whose hermit[13] would be most pleased to do you the honours. Only it is you who should bring indulgences.[14]

Believe me, dear Sir, yours truly, | A. Réville.

RI MS JT/1/R/22
RI MS JT/1/TYP/3/1016–7

1. *the article in the Revue des D. Mondes*: see letter 4535, n. 1.
2. *your letter*: letter missing.
3. *Mr James Martineau*: Martineau had criticized Tyndall's Belfast Address in an address at Manchester New College, London in October 1874 (see letter 4443, n. 4).
4. *books by him*: Tyndall had presumably told Réville of Martineau's *Religion as Affected by Modern Materialism* (London: Williams and Norgate, 1874), in which his address was published.
5. *a Unitarian*: a doctrinal position in which God is a single entity rather than, as in most other branches of Christianity, a Trinity. Martineau was an exponent of the so-called 'new Unitarianism', which encouraged a greater emphasis on orthodox spirituality.
6. *Franche-Comté*: a region in eastern France, on the border with Switzerland.
7. *the Alsatians who do not want to become German*: the formerly French region, on the west bank of the upper Rhine, had been annexed by Germany in 1871 after the Franco-Prussian War.
8. *E pur il spirito si muove*: And yet the spirit moves (Italian), a pun on an apocryphal statement by Galileo.
9. *Father Hyacinthe:* the religious name of Charles Jean Marie Loyson (1827–1912), a French theologian noted for his attempts to reconcile Roman Catholicism with modern ideas.
10. *old Catholics*: see letter 4558, n. 5.
11. *Telum imbelle sine ictu*: 'A feeble weapon without a thrust' (Latin), Virgil, *Aeneid*, II.544. The line is conventionally used to refer to an argument that falls short.
12. *London to Paris by Newhaven and Dieppe*: a railway service offered by the London and Brighton Railway Company, using a steamer across the channel between the latter two ports.
13. *a hermitage . . . and whose hermit*: this was presumably Réville himself; an advertisement in the *Spectator*, offering Réville's services as a teacher, stated that 'he lives in a large house on a hill, about a mile from the town of Dieppe, four hours from Paris, seven-and-half hours by way of Newhaven from London. The situation is extremely healthy, and there is a very fine view over the sea and country' ('Education in France', *Spectator*, no. 2518 (1876), p. 1225).
14. *indulgences*: remissions of the punishment which is still due to sin after sacramental absolution, traditionally dispensed by clergy of the Roman Catholic Church (*OED*).

From Jules Jamin[1] 2 April 1875 4586

Paris, 2 avril 1875

Mon cher ami

Un des élèves de la Sorbonne Mr Kriukov de nationalité Russe et qui travaille avec moi depuis plusieurs années part pour Londres avec le désir de visiter les laboratoires et particulièrement le vôtre. C'est un jeune homme distingué et très laborieux. Je me permets de vous le recommander vous priant de lui faciliter ses visites peut-être passera-t-il jusqu'à Oxford.

Je saisis cette occasion pour me rappeler à votre souvenir tout en regrettant de ne pas vous avoir vu depuis trop longtemps. Veuillez me croire toujours

Votre ami bien dévoué | J. Jamin

Paris, 2 April 1875

My dear friend,

One of the students of the Sorbonne, Mr Kriukov,[2] who is Russian and who has been working with me for many years, is leaving for London in order to visit the laboratories and in particular yours. He is a young distinguished man, and he is very hard-working. I recommend him to you, and I ask you to assist him with his visits. Perhaps he will go up to Oxford.[3]

I seize this opportunity to remind you of myself, while deeply regretting that I was not able to see you for a long time. Believe me, always.

Your devoted friend | J. Jamin

RI MS JT/1/J/17

1. *Jules Jamin*: Jules Célestin Jamin (1818–86), a French physicist, educated at the École Normale Superieur. He was professor of physics at the École Polytechnique (1858–81), and from 1863 he was professor of physics at the Faculté des sciences, Paris. He worked in many branches of physics, including optics, acoustics, electricity, magnetism, and radiation, and in 1858 was awarded the RS's Rumford Medal.
2. *Mr Kriukov*: not identified.
3. *up to Oxford*: presumably to visit the Clarendon Laboratory, which had been completed in 1872 as the first purpose-built physics laboratory in Britain.

From Louis Mallet[1] 2 April 1875 4587

India Office, Westminster, S.W., | April 2, 1875.

SIR,—

I am directed by the Secretary of State for India[2] in Council to acknowledge the receipt of your letter of the 5th March,[3] setting forth the desirability of instituting continuous Solar Observations in India, and, in reply, to transmit to you a copy of a Despatch which his Lordship has addressed to the Government of India[4] on the subject.[5]

I am, Sir, Your obedient Servant, | Louis Mallet. | *Professor Tyndall, F.R.S.*

Brit. Assoc. Rep. 1875, pp. xlix

1. *Louis Mallet*: Louis Mallet (1823–90), Permanent Under-Secretary of State for India (*ODNB*).
2. *the Secretary of State for India*: Robert Arthur Talbot Gascoyne-Cecil.
3. *your letter of the 5th March:* letter 4567.
4. *the Government of India*: the British government in India, which, since 1858, had been led by the Viceroy and the appointed members of his Council. The Viceroy of India at this time was Thomas George Baring (*ODNB*), who served in the role from 1872–6.
5. *a Despatch which . . . India on the subject*: the enclosed letter had been written on 24 March 1875 and read: 'MY LORD,—Para. 1. I have received and considered in Council your Excellency's Despatch, dated 12th February (No. 2, Industry, Science, and Art), 1875, reporting your sanction of an arrangement by which Lieutenant-Colonel Tennant, with a small establishment, will be employed, during the year 1875–76, to make observations, at Roorkee, of the sun and of Jupiter's satellites, and to reduce the transit-observations. The instruments* for use at Roorkee will be ordered in this country, and sent out with as little delay as possible. 2. I observe that your Government, in sanctioning these arrangements, have declined to engage themselves to any thing further at present; and that the suggested establishment of a solar observatory at Simla remains an open question for future consideration. 3. I herewith transmit a copy of a letter, which I have received from the President of the British Association, on the importance of continuous Solar Observations in India; and I would suggest for your consideration whether an observatory on an inexpensive scale might not usefully be established at Simla after the ensuing year, with this object, for which spectroscopes only would be necessary, in addition to the instrument already at Roorkee. *A parallel wire micrometer £ 20 | Solar and stellar spectroscopes 130 | Micrometer for measuring solar photographs 50 | £200'.

To George Gabriel Stokes 7 April 1875 4588

Royal Institution of Great Britain | 7th April 1875.

Dear Stokes,

A telegram just received from Mr. Dewar[1] induces me to say to you, that his lectures here[2] have given evidence of great power as a teacher, and a great promise as an original investigator.[3]

Yours truly | John Tyndall

Prof. Stokes DCL.

CUL SC—add 7656

1. *A telegram just received from Mr. Dewar*: letter missing. James Dewar (1842–1923) was a

Scottish chemist and physicist who would succeed John Hall Gladstone as Fullerian Professor of Chemistry at the RI in 1877 (*ODNB*).

2. *his lectures here*: Dewar delivered a Friday Evening Discourse at the RI, 'On the Physiological Action of Light', on 5 February (*Roy. Inst. Proc.*, 7 (1873–75), pp. 360–7). He had previously delivered a Friday Evening Discourse on 7 March 1873 ('On the Temperature of the Sun, and the Work of Sunlight', *Roy. Inst. Proc.*, 7 (1873–75), p. 57), and would give a lecture course at the RI, 'Four Lectures on the Progress of Physico-Chemical Inquiry', on Thursdays from 13 May to 3 June 1875 (*Roy. Inst. Proc.*, 7 (1873–75), p. 384).
3. *evidence of great power . . . original investigator*: Stokes may have been asking about Dewar's capacity in teaching and research in relation to the Jacksonian Professor of Natural Experimental Philosophy at Cambridge, to which Dewar was elected later in the year.

To Thomas Archer Hirst 14 April [1875][1] 4589

14th April

My dear Tom.

This is a very remarkable letter.[2] I have told the writer[3] I w[oul]d send it to you.

I have offered him money aid should he need it to tide him over a temporary difficulty.

But I have stated that before taking any step both you and I would require to see him. I hope you are in vigour.

Yours aff[ectionatel]y | John Tyndall

Love to Lilly[4]

If you could join Debus[5] and Guthrie[6] here[7] tomorrow at dinner at 7 P.M. precisely it would give me pleasure.

This preface[8] has been hanging about my table for weeks: give me your notion of it.

[Mem.

Letter from E. C. Copas enclosed. The son of a labouring man rose by his intelligence to be a schoolmaster and matriculated at the L.U.[9] changed his views on religious matters & ceased to attend the communion of Church of England. Has read Carlyle, Tyndall Huxley. Lost his situation[10] he fears because of his changed views on Religion asks advice. | T A H][11]

RI MS JT/1/T/712
RI MS JT/1/HTYP/626

---

1. *[1875]*: the year is suggested by the preface Tyndall refers to in the letter (see n. 8).
2. *a very remarkable letter*: enclosure missing.
3. *the writer*: E. C. Copas, a schoolmaster with heterodox religious views.

4. *Lilly*: Emily Hirst.
5. *Debus*: Heinrich Debus.
6. *Guthrie*: Frederick Guthrie (1833–86), an English physicist and chemist who, like Tyndall, had studied under Robert Bunsen in Germany and gained a doctorate from the University of Marburg (*ODNB*). In 1868, Guthrie succeeded Tyndall as the Chair of Natural Philosophy at the Royal School of Mines, a position which Tyndall had held since 1859.
7. *here*: presumably the RI. According to his journal, Hirst was staying at the Athenaeum Club from 13 to 15 April, although he does not mention dining with Tyndall (*Hirst Journals*, p. 2012).
8. *This preface*: probably 'Preface to the Second English Edition' in *Six Lectures on Light* (1875), pp. v–viii. The published preface was dated '*May* 1875' (p. vii), so Tyndall presumably sent Hirst a slightly earlier draft. He asked Hirst to read another section from the same book in letter 4595.
9. *the L.U.*: probably London University, which was established in 1826 as a secular institution, in contrast to the universities of Oxford and Cambridge, who only awarded degrees to members of the Church of England. London University was granted a Royal Charter and renamed University College London in 1836, though the original name remained in use.
10. *Lost his situation*: i.e. lost his position as a schoolmaster. It is likely that Copas's heterodox religious views were the cause of his dismissal.
11. *T A H*: this brief memorandum of the enclosed letter was written by Hirst.

To Ellen Lubbock 17 April [1875][1] 4590

Royal Institution of Great Britain | 17 April

Dear Lady Lubbock,

I was almost startled with pleasure at John's victory,[2] for he seemed to me in the morning to have no hope of victory.

I had been lunching with Hofmann[3] at Dr. Henrys,[4] & met there Mr. Yorke.[5] I enjoined him to vote for John, but this was not necessary as he always sided with him.

Tomorrow I cannot move, but further on in the year, if you give me the chance, I shall be glad to do so.

Yours ever | John Tyndall

RI MS JT/1/T/1043

---

1. *[1875]*: the year is established by reference to John Lubbock's parliamentary victory (see n. 2).
2. *John's victory*: on 8 February Lubbock presented a Bill for the Preservation of Ancient Monuments in the House of Commons, and on 14 April it passed on the second reading by a majority of twenty-two. The bill later became encumbered in parliamentary procedure and was not finally enacted until 1882.
3. *Hofmann*: August Wilhelm Hofmann (see letter 4299, n. 3) delivered the Chemical Society's

Faraday Lecture on 18 March, held at the RI. This was published as A. W. Hofmann, 'The Faraday Lecture. The Life-work of Liebig in Experimental and Philosophic Chemistry; with Allusions to His Influence on the Development of the Collateral Sciences, and of the Useful Arts', *Journal of the Chemical Society*, 28 (1875), pp. 1065–1140; and as a book by Macmillan in 1876.
4. *Dr. Henry*: not identified.
5. *Mr. Yorke*: John Reginald Yorke (1836–1912), the Conservative MP for East Gloucestershire.

To Catherine Tyndall[1] 18 April 1875 4591

Royal Institution of Great Britain | April, 18th, 1875.

My dear Friend

The sad and solemn event which has taken your excellent husband[2] from your side fills me with sorrow. I had no dream of the possibility of such an occurrence; thinking that the difficulties of a severe winter[3] had been completely surmounted, and that he had thus given proof of more than ordinary strength. If you think I can be of service in any way during this emergency, count on my sympathy and willingness to aid you in every possible way. Should you find money needed pray draw upon me. Indeed I wish I could in any practical way show you how deeply I sympathise with you in your great bereavement. Give my affectionate regards to the girls;[4]

and believe me ever | Your faithful friend | John Tyndall

RI MS JT/5/14
LT Typescript Only

1. *Catherine Tyndall*: Catherine Tyndall (née Hartford, 1811–98), who had married John Tyndall in *c*. 1840.
2. *your excellent husband*: Tyndall's cousin, the medical practitioner John Tyndall of Gorey, County Wexford, Ireland, died from acute bronchitis on 17 April ('Obituaries', *Lancet*, 105 (1875), p. 633).
3. *the difficulties of a severe winter*: see letter 4566.
4. *the girls*: John and Catherine Tyndall had three daughters: Catherine (known as Kathleen; d. 1936), Anne Elizabeth, and Georgina Mary (see letter 4566, n. 6).

To Heinrich Rudolf Vieweg 22 April 1875 4592

Royal Institution of Great Britain | 22nd. April 1875

Gentlemen,[1]

I am this moment engaged upon a new edition of the "Fragments"[2] in which considerable changes will be made. It will be ready before I go to Switzerland this year[3]—that is, before the end of June.

A new edition of Sound[4] is also on the point of being published in which considerable additions & alterations occur.

A new edition of Light[5] is also nearly ready. I think the preface,[6] and the frontispiece—a portrait of Dr. Young engraved in steel[7]—might with advantage appear in the German edition.[8]

I have abolished the old frontispiece.[9]

Yours faithfully | John Tyndall

TU Braunschweig/Universitätsbibliothek/UABS V1T:63

---

1. *Gentlemen*: despite the plural, the firm of Friedrich Vieweg und Sohn was at this time managed only by Heinrich Rudolf Vieweg, the grandson of the founder. Heinrich's father, Eduard Vieweg, who had run the firm since the 1830s, had died in 1869, and his own son, also called Eduard, was still a child at this time.
2. *a new edition of the "Fragments"*: J. Tyndall, *Fragments of Science: A Series of Detached Essays, Lectures, and Reviews*, 5th edn (London: Longmans, Green, and Co., 1876).
3. *before I go to Switzerland this year*: Tyndall left for Switzerland on 19 June.
4. *A new edition of Sound*: *Sound* (1875).
5. *A new edition of Light*: *Six Lectures on Light* (1875).
6. *the preface*: 'Preface to the Second English Edition', pp. v–vii.
7. *the frontispiece—a portrait of Dr. Young engraved in steel*: from the portrait of Thomas Young painted by Thomas Lawrence, engraved in stipple by Henry Adlard.
8. *the German edition*: J. Tyndall, *Das Licht: Sechs Vorlesungen gehalten in Amerika im Winter 1872–1873*, ed. G. Wiedemann (Braunschweig: Friedrich Vieweg, 1876), which included both the preface, pp. xiii–xv, and the frontispiece.
9. *the old frontispiece*: in the first edition of *Six Lectures on Light* (1873), the frontispiece was a lithograph entitled 'Plumes Produced by the Crystallization of Water. Photographed by Professor Lockett'. On this image, see letters 4021, 4033, and 4093, *Tyndall Correspondence*, vol. 13.

From George Campbell 23 April 1875 4593

Argyll Lodge, | Kensington.[1] | Ap[ril].: 23/75

Dear Professor Tyndall

I have sent you a copy of a Lecture lately read by me before the young men connected with a college in London.[2] I hope you will find nothing in it inconsistent with the sincere respect I entertain for your love of all discoverable truth.

I have taken no part in the outcry about y[ou]r Belfast Address[3]—because I thought it greatly misunderstood—and that its tendency is rather to spiritualise matter, than to materialise Thought.

But, I need not say, that by this route we may come round very much to the same goal—and I think we must always keep separate in language the two most separate things which can be conceived in Thought.[4]

Y[ou]rs truly | Argyll

RI MS JT/1/A/101
RI MS JT/1/TYP/1/74

---

1. *Argyll Lodge, | Kensington*: previously Bedford Lodge and later Cam House, a home of Argyll's in Kensington, London from 1852 until his death in 1900.
2. *a copy of a Lecture . . . college in London*: 'Anthropomorphism in Theology', which Campbell had given at the Presbyterian College, Queen's Square on 10 March. It was published in *Problems of Faith* (London: Hodder & Stoughton, 1875), pp. 3–56.
3. *y[ou]r Belfast Address*: Tyndall's presidential address to the BAAS at Belfast, delivered on 19 August 1874.
4. *keep separate in language . . . conceived in Thought*: in his lecture, Campbell commented on 'the language of those who habitually ascribe to matter the properties of mind; using this language not metaphorically, like the old Aristotelians whom they despise, but literally,—declaring that mind, as we know it, must be considered as having been contained "potentially" in matter, and was once nothing but a cosmic vapour, or a fiery cloud. Well may Professor Tyndall call upon us "radically to change our notions of matter", if this be a true view of it; for in this view it becomes equivalent to "nature" in that largest and widest interpretation to which I referred at the commencement of this lecture—viz., that in which nature is understood as the "Sum of all Existence"' (p. 42).

From Henry Tracey Coxwell[1] 23 April 1875 4594

Tottenham[2] April 23rd: 1875

Dear Sir,

I wish you could find it in your heart and mind to set the world right about experimental ballooning[3] at great altitudes—It is a delicate matter for me to avow my opinion, lest I might be charged with egotism and a disposition to expose the incompetency of professed aeronauts—The French men who lost their lives,[4] poor fellows, managed badly—I believe their oxygen was not of sufficient use, and that they were subjected to such a dose of carbonated hydrogen combined with rarefied air, as obviously to produce the direful effects we read of—almost similar results are not unfrequently noticed in the immediate vicinity of balloons undergoing an inflation especially in hot weather, and when the process of filling is conducted by ignorant persons—Work men under the silk are too often asphyxiated owing to carelessness and I cannot think that the height attained could possibly be chargeable with causing blood in the mouth and actual blackness of facial contour.—It has been said that I was so at seven miles high, but this is incorrect. I lost the use of my hands because I had taken off a pair of thick gloves in order to empty ballast, and then place them when Mr Glaisher[5] became miserable on the frozen hoop which was the temperature of the air at that elevation—Mr Glaisher was not so much affected as the Parisian voyagers—his face was pale but placid and I merely thought he was resting for awhile until he failed to reply to my entreation to continue his observations.[6] Our own explorations were made under great disadvantages as we had to rise and fall sometimes very rapidly in order to avoid the Lea[7]—once we dropped two miles in four minutes after having attained nearly the height at which the French men were almost dead—Paris should be most favourable as a starting point, but Scientific men and Scientific aeronauts are, as you know, totally distinct personages, and I question whether science is not equally indebted to both, for unless the points aimed at are not well understood and comprehensively dealt with by the helmsman, success is not likely to attend such undertakings as these, where constant meteorological, gaseous and other changes are going on in quick succession and demand skill, judgement and a ready resource to meet them smartly and efficiently as they arise—

I have been hoping too, that some one, who may be above the natural prejudices and jealousies of the professional aeronauts would explore the ignorance and folly of those who pretend to have attained high latitudes with small balloons; a letter recently appeared in the newspaper which adverted most injudiciously to the death of De Grof at Chelsea.[8] The entire allusions to that event were ill advised and uncalled for, since they were made by one who was responsible for the life of that man and who should not in an unseemly manner have alluded to his death—

Much by be said on this subject at the present time and something ought to be advanced to check our so called scientific teachers from taking up and identifying themselves with novices and adventurers. You have already done me the honour and justice of stating that I declined to seek honour and fame in your own fog-signal experiments,[9] because I thought it would be attended with risk to life in the immediate viewing of the channel, and I should be glad if you would give us the benefit of your views on a subject which cannot just now escape your notice and has, I know, occupied a great deal of your attention. You are quite at liberty to make any use you like if these hasty lines—I should have written to the newspaper myself, but I know that in a great and important matter like this speech is sometimes silver, but silence is gold—Still the intelligence and engaging spirit of this age require proper and reasonable explanations and I hope to read of your own views knowing that they will be original and profound.

Please to excuse this opportune appeal and to speak a word in season.[10] It is not too much to say that the scientific world and the public will highly appreciate your utterance.

I remain dear Sir Yours truly | Henry Coxwell | Professor Tyndall FRS / & DC LC[11]

RI MS JT/1/C/58

1. *Henry Tracey Coxwell*: Henry Tracey Coxwell (1819–1900), a dentist and ballooning pioneer who had worked with the BAAS in 1862 to use balloons to explore the upper atmosphere, and helped establish the world's first military ballooning units during the Franco-Prussian War (*ODNB*).
2. *Tottenham*: a town north-east of London.
3. *experimental ballooning*: the use of balloons to conduct scientific experiments. Tyndall had approached Coxwell about using a balloon to conduct experiments on infrared radiation at high altitudes where there would be less absorptive water vapor. From 1858 to 1867, Tyndall had served on the BAAS Balloon Committee, which functioned 'to legitimate the balloon as a "philosophical instrument" for field work in the new science of the Earth's atmosphere' (J. Tucker, 'Voyages of Discovery on Oceans of Air: Scientific Observation and the Image of Science in an Age of "Balloonacy"', *Osiris*, 11 (1996), pp. 144–76, on p. 147). Tyndall never actually made an ascent. See letters 2363 and 2367, *Tyndall Correspondence*, vol. 9.
4. *The French men who lost their lives*: Joseph Crocé-Spinelli (1845–75) and Théodore Sivel (1834–75), who, along with Gaston Tissandier (1843–99), had ascended in the balloon *Zénith* to conduct meteorological experiments on 15 April in Paris. They had brought oxygen with them, though Sivel and Crocé-Spinelli were apparently asphyxiated before they were able to use it. See P. L. S. De Oliveira, 'Martyrs Made in the Sky: The *Zénith* Balloon Tragedy and the Construction of the French Third Republic's First Scientific Heroes', *Notes and Records: The Royal Society Journal of the History of Science*, 74 (2020), pp. 365–86.
5. *Mr Glaisher*: James Glaisher (1809–1903), an English astronomer and meteorologist, who worked at the Royal Observatory, Greenwich (1835–74), managing the meteorological department there from 1841 until his retirement. Glaisher was perhaps best known for his pioneering balloon ascents during the 1860s with Coxwell (*ODNB*). The observations they made during these ascents were published in the *Brit. Assoc. Rep. 1863–6*.
6. *at seven miles high . . . his observations*: on 5 September 1862, Coxwell and Glaisher had made a flight that broke the world record for altitude at *c.* 31,000–35,000 feet, although Glaisher passed out before he could take an accurate reading. After he recovered, he noted that Coxwell had lost the use of his hands, which had turned black, and had to pull the release cord to begin their descent with his teeth.
7. *the Lea*: a river in southern England.
8. *a letter recently appeared . . . death of De Grof at Chelsea*: 'The Fatal Balloon Ascent', *Times*, 19 April 1875, p. 7. The letter, which was written by the balloonist Joseph Simmons, described the death of Vincent de Groff (1850–74), a Belgian shoemaker, who had fallen from Simmons's balloon on 9 July 1874 while testing a kite-like glider above Cremorne Gardens in East London.
9. *stating that I declined . . . own fog-signal experiments*: in relation to the effect of wind on the propagation of sound, Tyndall had noted 'This explanation calls for verification, and I wished much to test it by means of a captive balloon rising high enough to catch the

deflected wave; but on communicating with Mr. Coxwell, who has earned for himself so high a reputation as an aeronaut, I learned with regret that the experiment was too dangerous to be carried out' ('Vehicle of Sound', p. 227).

10. *speak a word in season*: an allusion to Isaiah 50:4, 'The Lord GOD hath given me the tongue of the learned, that I should know how to speak a word in season to *him that is weary*'.
11. *DC LC*: presumably erroneous references to Tyndall's honorary doctorates, a DCL from Oxford in 1873, and an LLD from Cambridge in 1865.

To Thomas Archer Hirst 25 April 1875 4595

25th April / 75

Dear Tom.

Cast please an eye over this "appendix"[1] while enjoying your pipe. I have, as you know, omitted Youngs Reply;[2] but there is something so good and original in the utterances that I think they ought to [be] reproduced.

Yours affe[ctionate]ly | John Tyndall

RI MS JT/1/T/717
RI MS JT/1/HTYP/630

---

1. *this "appendix"*: *Six Lectures on Light* (1875), pp. 229–64.
2. *omitted Youngs Reply*: the first edition of *Six Lectures on Light* (1873) had included 'Lord Brougham's Articles on Dr. Thomas Young in the "Edinburgh Review"' and 'Dr. Young's Reply to the Animadversions of the Edinburgh Reviewers' in its appendix, but these were both omitted in the second edition. In a preface to this edition, Tyndall explained: 'To the first English edition of these Lectures, Dr. Young's "Reply to the Edinburgh Reviewers" was appended. Numbers of scientific men were, to my knowledge, but imperfectly acquainted with this great discussion; while the general public knew nothing whatever about it. The end contemplated having been gained, the "Reply" is here omitted; and in lieu of it a portrait of Dr. Young, engraved with great success by Mr. Adlard, forms the frontispiece of the volume' (pp. vi–vii). The reply had already been omitted from German and French translations of the book's first edition (see letters 4307 and 4350).

To Robert Eli Hooppell[1] [*c*. 25][2] April 1875 4596

Royal Institution

Dear Sir,—

I thank you for the courtesy you have shown, both in sending me your lecture,[3] and in drawing my attention to it by your friendly note.[4] I entirely agree with the statement you have made at page 14, that our difference is simply 'as questions of evidence.'[5] I cannot think the expressions which have given you pain[6]—pain I should not willingly inflict—unjustifiable. You will bear in mind that I am contrasting that notion of the world's origin and governance which referred every new species, separated from its predecessors and its successors by an interval of time more or less vast, to a distinct creative act, with that view which regards the whole as a process of growth. I hardly think it unfair to say of the first notion that it represents the aggregate of species now existing, and which have existed, as the result of separate, or "broken," efforts of creative power.[7] The continuity, which forms the essence of the one conception, is absent from the other; this is what I wished to say. I have been taken much to task by real materialists for introducing the passage from Wordsworth.[8] Push your examination of Wordsworth and me[9] into the least detail, and you would discern differences of the gravest kind. But diverse branches may have a common root, and men whose opinions in some respects widely diverge may have a large area of common ground. I have such with Wordsworth when he recognizes the unscanned and inscrutable power 'which rolls through all things,'[10] and which, I hold, will be the profoundest theme of the poet in ages to come. To not one in fifty of the communications made to me have I ventured the least reply, and if I have done so to yours, it is through the desire to acknowledge the general fairness of spirit which marks your discussion of my address.—

Faithfully yours, | JOHN TYNDALL. | The Rev. R. E. Hooppell.

Morning Post, 26 April 1875, p. 3[11]

---

1. *Robert Eli Hooppell*: Robert Eli Hooppell (1833–95), a clergyman and principal of Winterbottom Nautical College, South Shields, who had interests in mathematics and astronomy (*ODNB*).
2. *[c. 25]*: this date is suggested by the letter having been published in the *Morning Post* on the following day (see n. 11).
3. *your lecture*: R. E. Hooppell, *Materialism: Has it Any Real Foundation in Science?*, 3rd edn (London: Rivingtons, 1874).
4. *your friendly note*: letter missing.
5. *the statement you have . . . simply 'as questions of evidence'*: this is not a direct quotation from Hooppell's lecture, and instead seems to refer to his concluding statement, on p. 16 rather than 14, that 'believers in Spirit, and particularly believers in Revelation, are governed by Evidence, according to the requirements of the scientific laws of Evidence, which govern the reception of physical truths, no less than of moral or historical ones'.
6. *the expressions which have given you pain*: in his lecture, Hooppell stated: 'I do not wish to say one single hard word of Professor Tyndall, nor a single harsh word of the Address at Belfast . . . but I cannot refrain from expressing the pain I felt . . . when I read repeatedly such words as these:—"a manlike creator,"—"anthropomorphism,"—"the adaptations on which this notion of a supernatural artificer has been founded"' (p. 10).

7. *"broken," efforts of creative power*: in the Belfast Address, Tyndall commented on the anthropomorphic notion of the deity as 'an Artificer, fashioned after the human model, and acting by broken efforts, as man is seen to act' (*Belfast Address*, p. 58). In his lecture, Hooppell responded: 'Why does man work by broken efforts? Because he has little strength, because he must rest, and because . . . he has imperfect knowledge. But Christians are taught that God is all powerful, everywhere present, unwearied. How can His efforts be broken efforts then?' (p. 11).
8. *the passage from Wordsworth*: see letter 4423, n. 5.
9. *your examination of Wordsworth and me*: in his lecture, Hooppell stated: 'It is Science that is drawing nearer to Revelation, though many of her votaries know it not . . . A most striking illustration of the truth of this is furnished by Professor Tyndall himself in the quotation with which he closes the recently published reprint of his address . . . listen to the words, they are from Wordsworth . . . Who does not recognise in these beautiful lines the teaching of the 139th psalm? Who does not recognise the doctrine, proclaimed on Mars' Hill, by the Apostle [Paul], to the Epicureans and Stoics of Athens?' (pp. 12–13). In the seventh printing of the Belfast Address, Tyndall removed the quotation from Wordsworth from the conclusion and placed it at the beginning as an epigraph (*Belfast Address*, 7th thousand, p. iv).
10. *the unscanned and inscrutable power 'which rolls through all things'*: a paraphrase and slight misquotation of 'A motion and a spirit, that impels | All thinking things, all objects of all thought, | And rolls through all things' (W. Wordsworth, 'Lines Composed a Few Miles above Tintern Abbey, on Revisiting the Banks of the Wye during a Tour, July 13, 1798' (1798),11. 102–4).
11. *Morning Post, 26 April 1875, p. 3*: Tyndall's letter was published in the newspaper under the heading 'Professor Tyndall and Materialism' and prefaced by an announcement stating: 'The following letter from the pen of Professor Tyndall, is made public by the consent of the writer. It is addressed to the Rev. Dr. Hooppell, of South Shields, a third edition of whose lecture, entitled "Materialism: Has it Any Real Foundation in Science?", has just been published by Messrs. Rivington'.

To William Spottiswoode 25 April [1875][1] 4597

25th April.

My dear Spottiswoode

Our electric battery[2] is complete without a jar to spare. It is of high historic interest, and unless I can have your own guarantee that no harm will come to it I should not like to let it out of the house.[3]

Mr. Ward[4] does not seem to know how highly we prize it.

Yours ever | John Tyndall

RI MS JT/1/T/1353

1. *[1875]*: the year is suggested by reference to the Special Loan Collection of Scientific Apparatus (see n. 3).
2. *Our electric battery*: possibly the chemical battery given to Michael Faraday by Alessandro Volta in 1814 and subsequently held at the RI.
3. *I should not like to let it out of the house*: the battery had presumably been requested for the Special Loan Collection of Scientific Apparatus at South Kensington, which was originally planned for June but postponed until March 1876. Spottiswoode was on the committee that, on 13 February, decided on the formation of the collection. Tyndall lent a number of items, though apparently not the battery. See *Catalogue of the Special Loan Collection of Scientific Apparatus at the South Kensington Museum* (London: Her Majesty's Stationary Office, 1876); and *Handbook to the Special Loan Collection of Scientific Apparatus* (London: Chapman and Hall, 1876). A series of free lectures at the South Kensington Museum based on the collections and sponsored by the Science and Art Department of the Committee of Council on Education ran from June through August 1876 (*Free Evening Lectures, Delivered in Connection with the Special Loan Collection of Scientific Apparatus, 1876* (London: Chapman and Hall, 1876). Tyndall delivered a lecture on 'Faraday's Apparatus' on 1 July 1876 (pp. 118–33), and Spottiswoode, on 'Experimental Illustrations of Polarised Light', on 19 August 1876 (pp. 471–92). See also letter 4618.
4. *M^r^. Ward*: probably William Sykes Ward (1815–85), a Yorkshire lawyer interested in natural philosophy and engineering who lent a number of electrical devices, including a galvanometer and telegraph of his own design, to the Special Loan Collection (*Catalogue of the Special Loan Collection of Scientific Apparatus at the South Kensington Museum*, pp. 167, 335, 370, 563, and 754).

To Olga Novikoff 26 April 1875 4598

Royal Institution of Great Britain | 26th. April. 1875.

Dear Friend

Your note[1] has come to me on the eve of my departure for Folkestone,[2] whither I fly for a day's rest. When your last note[3] reached me I had not had any plans for my vacation: Nor have I any definite plans now. I shall in all likelihood go to Switzerland,[4] and a few days ago had a little conversation with some friends who thought of accompanying me. But all is as yet a dream, and until I have shaken off my present weariness I shall not come to any definite resolution.

I have been thinking for some years past of building a little house in the Alps—I went so far last year as to measure the land for it[5]—But I do not know whether the notion will come to anything.

You have I suppose read Reville's Article in the "Revue".[6] It is excellent.

I purpose when I have when I have a little more time to write to Professor Frohschammer I should like much to learn from time to time how the Ultramontanes[7] behave towards him.

The lecture notes[8] which I promised to send you are still lying in type unfinished, I hope in a few days to have more strength than I possess at present, and then I will finish the notes.

Excuse this very dull production; I thought it, however, better not to give you any further cause of complaint against me; therefore I have not lost a post in replying to your last note.

Ever faithfully Yours | John Tyndall

RI MS JT/1/TYP/3/936
LT Typescript Only

1. *Your note*: letter missing.
2. *Folkestone*: Tyndall also visited the Kent seaside town, with Elizabeth Hamilton, in the previous month (see letter 4581).
3. *your last note*: possibly letter 4565, or a missing letter sent subsequent to that.
4. *Nor have I any definite plans . . . likelihood go to Switzerland*: Tyndall seemed much more certain of his plans four days earlier in letter 4592, and indeed travelled to Switzerland on 19 June. In a response that is now missing, Novikoff seemed to doubt the truth of Tyndall's statement here, and he clarified the nature of his plans, and what parts of them were still uncertain, in letter 4605.
5. *measure the land for it*: see letter 4379.
6. *Reville's Article in the "Revue"*: see letter 4535, n. 1.
7. *the Ultramontanes*: see letter 4558, n. 4.
8. *The lecture notes*: for Tyndall's lecture course on electricity at the RI (see letter 4557, n. 4).

To George Campbell 28 April [1875][1] 4599

West Cliff Hotel[2] | Folkestone | Wednesday 28th | Ap[ril].

My dear Lord Duke.

I came down here[3] on Monday evening to rest a weary head—this accounts for the tardiness of my reply to your note.[4]

I have read through your essay[5] and can take no exception to its tone. It stands out indeed in honourable distinction from many other recent essays on the same subject.

But your kindness in writing to me demands a warmer acknowledgement on my part than this mere recognition of your justice. I will not overload my thanks with words—accept them—they are sincere.

You may remember when I had last the honour of meeting you at the Deanery, Westminster.[6] You were good enough to ask me to come to see you at Inveraray[7] on my return from Belfast. More than once or twice then I have asked myself whether, if I were to present myself at the Duke's door, I should find it closed against me. That question can arise no more.

In fact among them whom I esteem I do not think I can count a single 'averted eye'.[8]

Most faithfully yours | John Tyndall

RI MS JT/1/T/105
RI MS JT/1/TYP/1/75

---

1. *[1875]*: the year is established by the relation to letter 4593.
2. *West Cliff Hotel*: see letter 4581, n. 4.
3. *here*: see letter 4598.
4. *your note*: letter 4593.
5. *your essay*: see letter 4593, n. 2.
6. *meeting you at the Deanery, Westminster*: not identified, but Tyndall regularly attended social gatherings at the residence, in the grounds of Westminster Abbey, of Arthur Penrhyn Stanley, the Dean of Westminster. See *Ascent of John Tyndall*, p. 209 and passim.
7. *Inveraray*: Inveraray Castle in Argyll, western Scotland; a Gothic Revival country house built in the 1740s and since then the ancestral seat of the Dukes of Argyll.
8. *'averted eye'*: possibly an allusion to Lord Byron's poem 'To Belshazzar' (1815), which closes 'But tears in Hope's averted eye | Lament that even thou hadst birth—| Unfit to govern, live, or die'; ll. 22–4.

To Sedley Taylor[1] 3 May 1875 4600

Royal Institution of Great Britain

Mr. Sedley Taylor is thanked for his courtesy[2] by
John Tyndall | 3rd May 1875

Cambridge University Library, Sedley Taylor Correspondence—Add 6260:102

---

1. *Sedley Taylor*: Sedley Taylor (1834–1920), a non-affiliated scholar of Trinity College, Cambridge, with wide-ranging academic interests including the relation between science and music. He was an early advocate of Hermann Helmholtz's theories on the subject, and in 1870 had contended that the version of the theory of musical consonance that Tyndall had presented in *Sound* (1867) was 'radically different from the original, and erroneous' (S. Taylor, 'On Professor Tyndall's Exposition of Helmholtz's Theory of Musical Consonance', *Nature*, 1 (1870), pp. 457–9, on p. 457).

2. *his courtesy*: probably relating to a paper that Robert Holford Macdowall Bosanquet (1841–1912) gave at the Musical Society on the same day—3 May—in which he stated: 'I must allude to a serious defect in Tyndall's exposition of Helmholtz's theory of beats in the ear, to which my attention was drawn by Professor Mayer's paper in the "American Journal of Science", October 1874, though I believe it had been commented on before by Mr. Sedley Taylor' ('On Temperament, or the Division of the Octave', *Phil. Mag.*, 50 n.s. (1875), pp. 164–78, on p. 169). See letters 4454 and 4645.

To William Spottiswoode 4 May [1875][1] 4601

4th May

Dear Friend

If I can be present even at M. Cornu's Lecture[2] I shall be rejoiced at my achievement.

Yours ever | John Tyndall

RI MS JT/1/T/1355

---

1. *[1875]*: the year is established by reference to Marie Alfred Cornu's lecture (see n. 2).
2. *M. Cornu's Lecture*: the French physicist Marie Alfred Cornu (1841–1902) delivered a Friday Evening Discourse, 'On New Determinations of the Velocity of Light', at the RI on 7 May 1875 (*Roy. Inst. Proc.*, 7 (1873–75), pp. 472–5).

To Thomas Carlyle 6 May 1875 4602

6th. May 1875

Well-beloved Chieftain![1]

Deep thanks for your gift.[2]—To no one can you show a similar kindness who is likely to prize it more than

Your ever loyal | John Tyndall | Thomas Carlyle Esquire

I had a conversation some time ago with Lady Derby[3] in regard to the honour which Her Majesty proposed to bestow upon you.[4] The mission of the Countess was to move you:[5] but she rejoiced to find you immovable.

I shall often look at this last autograph.[6] It is full of interest to me. Were I of a fibre and religious tint like those of Sterling,[7] I would say like him God bless you![8]

RI MS JT/1/T/153
RI MS JT/1/TYP/1

1. *Chieftain:* see n. 2.
2. *your gift*: presumably T. Carlyle, *The Early Kings of Norway* (London: Chapman and Hall, 1875), which had just been published. The 'chieftain' figure was an important part of Carlyle's analysis of Norse leadership, so Tyndall's greeting suggests this book.
3. *Lady Derby*: see letter 4561, n. 3.
4. *the honour which Her Majesty proposed to bestow upon you*: on 27 December 1874, Carlyle received an offer to be made a Knight of the Order of the Bath from Queen Victoria (1819–1901) at the behest of the Prime Minister Benjamin Disraeli. He declined it, claiming, as Disraeli reported, that 'titles of honour, of all degrees, are out of keeping with the tenor of my poor life' (W. F. Monypenny and G. E. Buckle, *The Life of Benjamin Disraeli*, 6 vols (London: John Murray, 1910–20), vol. 5, p. 358).
5. *The mission of the Countess was to move you*: Disraeli's biographers noted that it was 'Lady Derby, whom Carlyle credited, perhaps rightly, with the origination of the idea' (vol. 5, p. 358).
6. *this last autograph*: presumably Carlyle's inscription in the book.
7. *Sterling*: John Sterling (1806–44), a Scottish writer about whom Carlyle wrote a celebrated biography, *The Life of John Sterling* (London: Chapman and Hall, 1851). Tyndall had read the book and visited Sterling's grave several times (see *Ascent of John Tyndall*, pp. 94, 120, and 244).
8. *say like him God bless you!*: in Carlyle's biography, several of Sterling's letters to his mother conclude: 'Meanwhile and always, God bless you, is the prayer of Your affectionate son' (p. 314).

To Thomas Archer Hirst 12 May [1875][1] 4603

12th May

My dear Tom.

I have asked Spottiswoode & Co[2] to send you a copy of my preface to Sound.[3] Glance at it please over your pipe, and say whether you think it will do.

I am trying to get off to the Isle of Wight for a day or two.

Yours aff[ectionate][ly] | John Tyndall

RI MS JT/1/T/899
RI MS JT/1/HTYP/630/3

1. *[1875]*: the year is established by reference to Tyndall's preface (see n. 3).
2. *Spottiswoode & Co*: the printers at New Street Square, London.
3. *my preface to Sound*: 'Preface to the Third Edition', in *Sound* (1875), pp. vii–xxv. The published preface was dated '*June* 1875' (p. xxv), so the printers presumably sent Hirst proofs of a slightly earlier draft.

To Fanny Mitchell[1] [16 May 1875][2] 4604

ST. JOHN'S HOUSE | RYDE. I.W.[3]

Dear Mrs. Mitchell

In Saint John's heights, above Ryde, with meadows & flowers, and trees & birds all round; and the blue ocean gleaming beyond, I received your note[4] from the hills of Selkirkshire[5]—Tho' practically pledged against dining out I would join you on Friday, were I not pledged to a young orator at the Royal Institution,[6] who makes his debut on that day—Many thanks to you for your friendly thought of me.

Most faithfully yours | John Tyndall

I have eaten hospitable dinners at 6 Great Stanhope Street[7] before now.[8]

Private Collection #8

---

1. *Fanny Mitchell*: Fanny Mitchell (née Hasler, 1833–1917), widow of the politician Alexander Mitchell (d. 1873), who served as an independent Liberal MP for Berwick-up-on-Tweed from 1865–8. He was a member of the RI.
2. *[16 May 1875]*: this date has been added to the letter in pencil, in a hand other than Tyndall's.
3. *ST. JOHN'S HOUSE | RYDE. I.W.*: the home of the English electrician and wine merchant John Peter Gassiot (1797–1877). St John's House had been built in the 1760s, and was surrounded by an ornamental park designed by Humphry Repton; it overlooked the town of Ryde on the north-east corner of the Isle of Wight.
4. *your note*: letter missing, but presumably an invitation to dine on Friday 21 May.
5. *Selkirkshire*: a region in the lowlands of Scotland, where Alexander Mitchell's family estate was close to the village of Stow.
6. *a young orator at the Royal Institution*: James Baillie-Hamilton (1837–*c.* 1926), a Scottish inventor of an organ called a 'Vocalion' that was played by wind. He delivered a Friday Evening Discourse at the RI on 21 May entitled 'On the Application of Wind on Stringed Instruments' (*Roy. Inst. Proc.*, 7 (1873–75), pp. 488– 95).
7. *6 Great Stanhope Street*: a townhouse near Park Lane in Mayfair. Mitchell had lived there with her husband during the London season, and then stayed on by herself after his death.
8. *eaten hospitable dinners . . . before now*: 6 Great Stanhope Street had previously been shared with Louisa Baring (1827–1903), Lady Ashburton, the widow of William Baring (1799–1864), Lord Ashburton (see *Boyle's Court Guide* (London: Court Guide Office, 1870), p. 243). Baring, a prominent and well-connected society hostess, and art collector and philanthropist, was a close friend and correspondent of Tyndall's, although this volume contains no letters to or from her. See V. Surtees, *The Ludovisi Goddess: The Life of Louisa Lady Ashburton* (Wilby, UK: Michael Russell, 1984).

To Olga Novikoff 17 May 1875 4605

Royal Institution of Great Britain. | 17th. May. 1875.

My dear Friend

On my return from the Isle of Wight last night I received your note.[1] I told you the exact truth regarding Switzerland.[2] I shall certainly go there, but up to the present hour I have not decided as to the part of Switzerland that I shall visit. To my intense perplexity and weariness, my publisher[3] demanded, at the beginning of the year, new editions of four of my books;[4] and I have been labouring at them ever since. It has pulled me down so much that I was first forced to go to Folkestone, and then to the Isle of Wight, from which I returned last night. In most cases it is pleasant to be called upon for new editions; but I have told my publisher that if he repeats such a call I shall simply "strike work."[5]

I am very grateful to you for all your kindness. I hope Frohschammer will join you: your free and vigorous intellect will pass over him like a refreshing breeze, and make him forget his ultramontane foes.[6] The Abbé Moigno has just published a translation of a little book of mine on Light,[7] and has introduced it to the French public by a characteristic preface.[8]

If I knew how to address them I would send you from time to time anything I may happen to write. With regard to the Notes,[9] I am not sure that I shall be able to get them done within a fortnight: but if they should be completed I will address them to the "poste restante"[10] Moscow.

Believe me | most faithfully yours | John Tyndall

RI MS JT/1/TYP/3/936
LT Typescript Only

1. *your note*: letter missing.
2. *I told you the exact truth regarding Switzerland*: in letter 4598 Tyndall told Novikoff that he did not have 'any definite plans now. I shall in all likelihood go to Switzerland', although with other correspondents (see letter 4592) he was more certain about going to Switzerland.
3. *my publisher*: presumably Thomas Longman (see letter 4561, n. 7).
4. *new editions of four of my books*: see letter 4561, n. 8.
5. *"strike work"*: refuse to continue working (*OED*).
6. *his ultramontane foes*: see letter 4558.
7. *a translation of a little book of mine on Light*: see letter 4277, n. 2.
8. *a characteristic preface*: F. Moigno, 'Préface du traducteur', pp. v–xii.
9. *the Notes*: for Tyndall's lecture course on electricity at the RI (see letter 4557, n. 4).
10. *"poste restante"*: see letter 4392, n. 7.

From Hermann Helmholtz 17 May 1875 4606

Berlin 17.5.75

Verehrter Freund

die mit Ihrem letzten Brief gesendeten Blätter der neuen Ausgabe von „On Sound“ habe ich durchgesehen und finde die Darstellung der Theorie der Consonanz und Dissonanz ganz richtig und genau, so weit es in solcher Kürze möglich ist. Es sind alle wesentlichen Bedingungen erwähnt und anschaulich gemacht.

Es hat uns gefreut, dass Ihnen die Übersetzung der Fragmente gut gefällt. Die erste Auflage ist ziemlich schnell verkauft worden. Mit der zweiten wird Vieweg, wie ich voraussetze warten, bis die neue Englische da ist. Ihre Absicht die Rechte des Autors für neue Auflagen der Uebersetzung auf uns ein für alle Mal zu übertragen, ist ausserordentlich freundschaftlich gemeint, sogar so sehr dass es uns in Verlegenheit setzt, ob wir ein solches Opfer Ihrerseits annehmen können. Ist es Ihnen lästig über jede neue Auflage mit Vieweg zu correspondiren, so können wir ja irgend eine Ihnen zusagende Anordnung darüber treffen. Ich meine aber, es wäre besser, wenn Sie Ihr Recht nicht aus den Händen gäben, abgesehen davon, dass, wie ich glaube, Sie wirksamer mit Vieweg verhandeln können als wir.

Im Laufe des Juni kommt der jüngste Bruder meiner Frau, Otmar von Mohl, zur Zeit Cabinets-Secretär der Kaiserinn, nach London, und wird sich Ihnen vorstellen. Wollen Sie mit ihm besprechen, wie Sie es am liebsten eingerichtet haben möchten, so werden wir gern darauf eingehen. Ich habe ihm eine kleine Abhandlung über die Rückführung der anomalen Dispersion auf die Theorie des Mitschwingens mitgegeben für Sie. Er ist ein intelligenter und gewandter junger Mann, der schon als Consul des deutschen Reichs in New York und Singapore war, und ist sehr begierig, Sie kennen zu lernen.

Mit besten Grüssen, auch von meiner Frau

Ihr | H. Helmholtz

Berlin 17.5.75

Dear friend

I read the sheets from the new edition of “On Sound” [1] that you enclosed with your last letter[2] and I find the presentation of the theory of consonance and dissonance entirely correct and precise,[3] as much as is possible with such brevity. All fundamental conditions are mentioned and made clear.

We are glad that the translation of the Fragments[4] pleased you. The first edition sold out quickly. I suppose that Vieweg will wait until the new one in English[5] has come out before publishing the second edition. Your intention to transfer altogether to us the rights for new editions of the translation is extraordinarily kind, so much so that it makes us feel embarrassed and question whether we are entitled to demand such a sacrifice from you. If it were that you do not want to correspond with Vieweg regarding every new edition, we can certainly make an arrangement that satisfies you. But I believe that you should not give away your rights, not to mention that I believe that you would be able to negotiate with Vieweg more effectively.[6]

My wife's[7] youngest brother, Otmar von Mohl,[8] currently a secretary in the empress's cabinet,[9] will be in London during the course of June and will introduce himself to you. If you like to discuss your preferred arrangement with him, we will be happy to accept it. I sent with him a short article for you on the explanation of anomalous dispersion through the theory of resonance.[10] He is an intelligent and articulate young man, who has worked as consul of the German Empire in New York and Singapore, and who is very eager to make your acquaintance.

Kind regards, also from my wife

Your | H. Helmholtz

RI MS JT/1/H/52

1. *the new edition of "On Sound": Sound* (1875). Tyndall was showing proofs of the new edition to trusted friends (see letter 4603).
2. *your last letter*: letter missing.
3. *I find the presentation . . . correct and precise*: on p. 359 (see letters 4454 and 4645).
4. *the translation of the Fragments*: see letter 4192, n. 11.
5. *the new one in English*: see letter 4592, n. 2.
6. *I believe that . . . with Vieweg more effectively*: Tyndall adopted this suggestion in letter 4629.
7. *My wife*: Anna Helmholtz.
8. *Otmar von Mohl*: Ottmar von Mohl (1846–1922), a German diplomat (*NDB*).
9. *a secretary in the empress's cabinet*: Mohl was cabinet secretary to Empress Augusta of Saxe-Weimar-Eisenach (1811–90), the consort of Kaiser Wilhelm I.
10. *a short article . . . through the theory of resonance*: H. Helmholtz, 'Zur Theorie der anomalen Dispersion', *Annalen der Physik und Chemie*, 230 (1875), pp. 582–96.

To Thomas Archer Hirst [*c.* 19 May 1875][1] 4607

My dear Tom.

M[rs]. Tunstal[2] is gone to Cromer[3]—she will be away Thursday: hence we must postpone our little dinner.[4]

Yours ever | John

RI MS JT/1/T/993
RI MS JT/1/HTYP/630/4

1. *[c. 19 May 1875]*: the date is suggested by LT giving the date as 20 May 1875, which was a Thursday, the day mentioned in the letter, so it is likely to have been written a day earlier.
2. *M^rs^. Tunstal*: Jane Tunstall, Tyndall's housekeeper at the RI.
3. *Cromer*: a resort on the coast of Norfolk.
4. *postpone our little dinner*: rather than being postponed, it seems to have been relocated from the RI, where Tyndall's housekeeper was unable to assist, to the nearby Athenaeum Club. Hirst recorded in his journal that on 20 May he 'Dined at Athenaeum with Tyndall and Sir J. Whitworth' (*Hirst Journals*, p. 2013). On the Athenaeum Club, see letter 4186, n. 1.

From John Fryer[1] [*c.* May 1875][2] 4608

One day, soon after the first copy of your work on Sound[3] reached Shanghai,[4] I was reading it in my study, when an intelligent official, named Hsii-chung-hu,[5] noticed some of the engravings and asked me to explain them to him. He became so deeply interested in the subject of Acoustics that nothing would satisfy him but to make a translation. Since, however, engineering and other works were then considered to be of more practical importance by the higher authorities, we agreed to translate your work during our leisure time every evening, and publish it separately ourselves. Our translation,[6] however, when completed, and shown to the higher officials, so much interested them, and pleased them, that they at once ordered it to be published at the expense of the Government, and sold at cost price. The price is four hundred and eighty copper cash[7] per copy, or about one shilling and eightpence.[8] This will give you an idea of the cheapness of native printing.

Sound (1875), p. viii[9]

1. *John Fryer*: John Fryer (1839–1928), an English sinologist and professor of English at Tung-Wen College in Shanghai, China. He translated numerous Western scientific texts as Chief Translator of Scientific Books in the Department for the Translation of Foreign Books at the Kiangnan Arsenal. See D. Wright, 'John Fryer and the Shanghai Polytechnic: Making Space for Science in Nineteenth-Century China', *British Journal for the History of Science*, 29 (1996), pp. 1–16.
2. *[c. May 1875]*: the date is suggested by Tyndall's comments in the preface to *Sound* (1875), which was dated '*June* 1875', where he stated that he was 'favoured' with this letter 'a few weeks ago' (pp. xxv and viii).
3. *the first copy of your work on Sound*: J. Tyndall, *Sound: A Course of Eight Lectures Delivered at the Royal Institution of Great Britain*, 2nd edn (London: Longmans, Green, and Co., 1869).
4. *Shanghai*: a port city in eastern China on the southern estuary of the Yangtze River which, in the second half of the nineteenth century, was largely controlled by Western business interests through the Shanghai Municipal Council.

5. *Hsü-chung-hu*: Xu Yianyin (1845–1901), son of the chemist Xu Shou and a government official at the Kiangnan Arsenal in Shanghai, where he became a close friend of Fryer. See D. Wright, *Translating Science: The Transmission of Western Chemistry into Late Imperial China, 1840–1900* (Leiden: Brill, 2000), pp. 54–71.
6. *Our translation*: J. Tyndall, *Shengxue*, trans. J. Fryer and X. Yianyin (Shanghai: Kiangnan Arsenal, 1874).
7. *copper cash*: a coin used in China since the fourth century BCE with a square hole in the centre.
8. *one shilling and eightpence*: in Britain, the book cost ten shillings and six pence, nearly ten times as much.
9. Sound *(1875), p. viii*: Tyndall introduced the letter by stating: 'Before me . . . lie two volumes of foolscap size, curiously stitched, and printed in characters the meaning of which I am incompetent to penetrate. Here and there, however, I notice the familiar figures of the former editions of "Sound". For these volumes I am indebted to Mr. John Fryer, of Shanghai, who, along with them, favoured me, a few weeks ago, with a letter from which the following is an extract'. At the end of the letter, he noted: 'Mr. Fryer adds that his Chinese friend had no difficulty in grasping every idea in the book'.

To Heinrich Rudolf Vieweg 2 June 1875 4609

Royal Institution of Great Britain | 2nd. June 1875.

Dear Sir,

If the printers[1] permitted the works to which you refer[2] to leave their hands without my express permission they would expose themselves to censure.

I have desired them this day to send you a copy of the lectures on Light.[3] I wish the portrait of D^r^. Young, and the "plumes" of crystallization[4] to be introduced in the German edition,[5] as they are in this English one.

A new edition of "Sound"[6] is also ready, in which considerable changes have been made. This shall be sent to you in a few days.

The Fragments[7] are not so far advanced. 400 pages are however printed.

Yours truly | John Tyndall

TU Braunschweig/Universitätsbibliothek/UABS V1T:63

1. *the printers*: Spottiswoode & Co., New Street Square, London.
2. *the works to which you refer*: not identified; the letter in which Vieweg presumably mentioned them is missing.
3. *the lectures on Light*: *Six Lectures on Light* (1875).
4. *the portrait of Dr. Young, and the "plumes" of crystallization*: see letter 4592, n. 7 and n. 9.
5. *the German edition*: see letter 4592, n. 8.

6. *A new edition of "Sound"*: *Sound* (1875).
7. *The Fragments*: see letter 4592, n. 2.

To Moncure Daniel Conway[1] 5 June [1875][2] 4610

5th. June

My dear Mr Conway.

In sweeping up my papers I have come upon a note of yours[3] which reminds me of my intention to say to you that I read your sermon on the Moody & Sankey commotion[4] with true pleasure. It will endure when they have passed away

Yours faithfully | John Tyndall

Moncure Daniel Conway Papers, Box 21; University Archives, Rare Book & Manuscript Library, Columbia University Libraries

---

1. *Moncure Daniel Conway*: see letter 4467, n. 1.
2. *[1875]*: the year is suggested by reference to the 'Moody & Sankey commotion' (see n. 4).
3. *a note of yours*: letter missing.
4. *your sermon on the Moody & Sankey commotion*: Dwight Lyman Moody (1837–99), an American evangelist (ANB), and Ira David Sankey (1840–1908), a gospel singer and composer (ANB), conducted a popular revival campaign in Britain from 1873 to 1875. Conway's sermon may have used the 'commotion' of this evangelical revivalism to reflect on his own more heterodox beliefs, as a contemporary later recalled: 'About the time of the Moody and Sankey revivals, Mr. Conway gave an account of his own conversion almost unparalleled in its candour: "It was my destiny to be born in a region where this kind of excitement is almost chronic... When the summer came the leading Methodist families—of which my father's was one—went to dwell in the woods in tents. About two weeks were there spent in praying and preaching all the day long, pausing only for meals; and during all that time the enclosure in front of the pulpit was covered over with screaming men and women, and frightened children . . . While I was there women came and wept over me; preachers quoted Scripture to me. No one whispered to me that I should resolve to be better,—more upright, true, and kind. Hundreds were converted by my side, and broke out into wild shouts of joy; but I had no new experience whatever. I was not in the least a sceptic: I believed every word told me. Yet nothing took place at all. On a certain evening I swooned. When I came to myself I was stretched out on the floor with friends singing around me, and the preachers informed me that I had been the subject of the most admirable work of divine grace they had ever witnessed. I took their word for it. All I knew was that I was thoroughly exhausted, and was ill for a week"' (J. M. Davidson, *Eminent Radicals In and Out of Parliament* (London: W. Stewart, 1880), pp. 205–6).

To Hector Tyndale 10 June 1875 4611

10th. June 1875

My dear Hector

Thanks to your kindness the Philadelphia Press has reached me, and given me the great pleasure of reading your marked article.[1] Agassiz, before he died, wrote to me to say something similar.[2] That the little seedling should sprout out into such fruit is surely very gratifying.

I have been long indebted to you as regards writing. Four books[3] were on my hands during the Spring. Three out of the four are now finished; the fourth is nearly so, but I intend to take it with me to Switzerland and finish it there. They were new editions of Light, Sound, Heat and the "Fragments". The last is the unfinished one. Sound I have revised and augmented. To my regret I have been forced to notice the last Report of Prof. Henry to the Lighthouse Board at Washington;[4] and to my still greater regret I learn from Americans who have visited me that he speaks with bitterness of an enquiry[5] which I imagined could give him nothing but pleasure. He is wrong throughout, and time will prove him so.

I have been very low in health; so low that I sought the counsel of a medical friend. He, by the recommendation of a somewhat ascetic mode of life, is bringing me surely round. I am getting steadily stronger, and hope to return from the Alps this year with more than my accustomed vigour. I thought alcohol necessary to my sleep, but I begin to think this a mischievous delusion: even imperfect sleep without it, is better than profound sleep under its influence.

Next week I start for the Alps.[6] The religious world is quiet here for the present. Stray shots are fired at me now and then; and from time to time a battery is converged upon me. A few nights ago, at the Victoria Institute,[7] I am told the orator chosen to give the anniversary Lecture[8] made me and my doings a principal subject of comment and criticism.

"Let them rave."[9]

D^r^. John Tyndall of Gorey died of bronchitis some time ago.[10] He suffered annually from attacks of this malady; but he appeared to have passed the last severe winter with singular success, when he was suddenly smitten and carried away.

Give my kind remembrances to M^r^. Furness[11] when you meet him. I retain a pleasant memory of his genial kindly nature. Also kind remembrances to D^r^. Gross.[12] Give my love to Julia[13] & take my affectionate regards yourself.

John Tyndall

RI MS JT/1/T/1457
RI MS JT/1/TYP/4/1738–9

---

1. *your marked article*: not identified; the *Philadelphia Press* was a daily newspaper, published by John Wien Forney, that ran from 1857–1920.

2. *Agassiz, before he died . . . something similar*: letter missing; Louis Agassiz died on 14 December 1873.
3. *Four books*: see letter 4561, n. 8.
4. *forced to notice . . . Lighthouse Board at Washington*: *Sound* (1875), pp. xi–xxv. The part of Joseph Henry's 'last Report' that Tyndall noticed was *Annual Report of the Light-House Board of the United States* (Washington, DC: Government Printing Office, 1874), pp. 715–21. The main points at issue were Tyndall's apparent reluctance to give appropriate acknowledgement of previous American experiments on fog-signals, and a disagreement over Tyndall's idea of acoustic clouds obstructing sound and creating abnormal sound phenomena.
5. *he speaks with bitterness of an enquiry*: Henry's report noted: 'In May, 1873, Professor Tyndall commenced a series of investigations on the subject of the transmission of sound, under the auspices of the Trinity House, of England, in which whistles, trumpets, guns, and a siren were used, the last-named instrument having been lent by the Light-House Board of the United States to the Trinity House for the purpose of the experiments in question. The results of these investigations were, in most respects, similar to those which we had previously obtained'. On the issue of acoustic clouds, the report also observed: 'A fatal objection, we think, to the truth of the hypothesis Professor Tyndall has advanced is that the obstruction to the sound, whatever may be its nature, is not the same in different directions' (pp. 720 and 720–1).
6. *Next week I start for the Alps*: Tyndall left for Switzerland on 19 June.
7. *the Victoria Institute*: an organization, also known as the Philosophical Society of Great Britain, founded in 1865 to defend evangelical beliefs against science and biblical criticism. See R. L. Numbers, *The Creationists: From Scientific Creationism to Intelligent Design*, 2nd edn (Cambridge, MA: Harvard University Press, 2006), pp. 162–6, 175–9; S. Mathieson, *Evangelicals and the Philosophy of Science: The Victoria Institute, 1865–1939* (New York: Routledge, 2020); and S. Mathieson, 'The Victoria Institute, Biblical Criticism, and the Fundamentals', *Zygon*, 56 (2021), pp. 254–74.
8. *the orator chosen to give the anniversary lecture*: Robert Main (1808–78), an astronomer and Anglian priest (*ODNB*). On 7 June he delivered the Victoria Institute's annual address on 'Modern Philosophic Skepticism Examined', during which he declared: 'I am of opinion that it was a bad day for science (not for science properly so called, but for the popular development of it) when Professor Tyndall composed during a summer holiday, and subsequently delivered at the meeting of the British Association at Liverpool in 1870, his celebrated discourse on the "Use of the Imagination in Science". I heard that eloquent discourse, and I considered at the time that many of the instances adduced from the mathematical sciences were legitimate deductions from established premisses, and implied no use of the imagination properly so called. There has, however, been abundant use made of it since that time, both by the lecturer himself and by others, and I think a note of warning on this head is not out of place'. Later, Main also stated: 'Let us now proceed to devote a few minutes to the study of atomism as understood by the ancients, with the express purpose of offering a few criticisms on the Belfast Address. This would be scarcely necessary if that celebrated Address had been compiled from original sources' (R. Main, *The Annual Address of the Victoria Institute* (London: Robert Hardwicke, 1875), pp. 5 and 23).
9. *"Let them rave"*: A. Tennyson, 'A Dirge' (1830), l. 7.
10. *Dr. John Tyndall . . . some time ago*: on 17 April; see letter 4591.

11. *Mr. Furness*: William Henry Furness.
12. *Dr. Gross*: Samuel David Gross.
13. *Julia*: Julia Tyndale.

To John Leighton[1] 12 June [1875] 4612

12th. June

My dear Mr. Leighton.

I send you with a great many thanks, and a great many apologies for having kept it so long, the book[2] which you were so obliging as to send me last year.

Towards the end of this week new editions of my books on "Sound" and "Light"[3] will, I hope be published by the Messrs. Longman. I have requested them to send you copies of both, and you would do me a favour if you would accept them with my best wishes.

Faithfully yours | John Tyndall

PS Mss.509.L56

1. *John Leighton*: see letter 4479, n. 9.
2. *the book*: not identified, but possibly *Madre Natura Versus the Moloch of Fashion* (London: Chatto & Windus, 1874), which Leighton published under his pseudonym Luke Limner. The book dealt with the harmful effects of fashion on public health.
3. *new editions of my books on "Sound" and "Light"*: *Six Lectures on Light* (1875) and *Sound* (1875).

From Henrietta Huxley 15 June [1875][1] 4613

4 Marlborough Place | Abbey Road N.W.[2] | 15th June

Dear Brother John,

I shall be very pleased to come and see you with some of the children (4)[3] tomorrow at 1 o'clock.

It is a good thing for you that you are going off to the Alps[4]—I wish I were too.

Hal runs down on Wednesday night from Edinboro' and returns on Saturday.[5] He seems uncommonly well and pleased with his work. I shall keep all chit chat till we meet.

Always | Your affectionate sister | Nettie Huxley.

RI MS JT/1/TYP/9/3045
LT Typescript Only

1. *[1875]*: the year is established by reference to Tyndall's trip to Switzerland and Thomas Huxley being in Edinburgh (see n. 4 and n. 5).
2. *4 Marlborough Place | Abbey Road N.W.*: see letter 4213, n. 1.
3. *some of the children (4)*: the Huxleys had seven children at this time: Jessica Oriana, Marian, Leonard, Rachel, Henrietta, Henry, and Ethel.
4. *going off to the Alps*: Tyndall left for Switzerland on 19 June.
5. *Hal runs down . . . returns on Saturday*: 19 June; Thomas Huxley spent much of the summer in Edinburgh giving a course of lectures on zoology at the city's university, in place of those normally given by Charles Wyville Thompson, who was away on the expedition of HMS *Challenger*. Huxley did so again in the summer of 1876. See J. A. Ritchie, 'A Natural History Interlude: Huxley's Teaching at Edinburgh University', *University of Edinburgh Journal*, 10 (1940), pp. 206–12. On HMS *Challenger*, see letter 4489, n. 2.

To Alfred Mayer 16 June 1875 4614

Royal Institution of Great Britain | 16th. June 1875

Dear Prof. Mayer.

Just on the point of starting for the Alps[1] I received your note.[2] I immediately placed it in the hands of our librarian[3] who furnished the enclosed answer to your question.[4] I hope it is what you require.[5]

With best wishes believe me | Ever yours faithfully | John Tyndall

Tyndall, John (1820–1893); Hyatt and Mayer Collection, C0076, Manuscripts Division, Department of Special Collections, Princeton University Library

1. *the point of starting for the Alps*: Tyndall left on 19 June, and stayed in Switzerland until 18 August.
2. *your note*: letter missing.
3. *our librarian*: Benjamin Vincent.
4. *the enclosed answer to your question*: Vincent's answer stated: 'Bakerian Lecture. Nov. 12. 1801 | The sensation of different colours depends on the frequency of vibrations, excited by lights in the Retina. | Phil Trans. vol 92, p. 18. | Youngs Syllabus is dated Jan 1802'. The publications mentioned were T. Young, 'The Bakerian Lecture. On the Theory of Light and Colours', *Phil. Trans.*, 92 (1802), pp. 12–48; and T. Young, *A Syllabus of a Course of Lectures on Natural and Experimental Philosophy* (London: The Press of the Royal Institution, 1802).
5. *what you require*: Mayer had asked for information on Thomas Young for his article 'The History of Young's Discovery of His Theory of Colors', *American Journal of Science*, 9 n.s. (1875), pp. 251–67.

From Charles Wilson Vincent[1] 17 June 1875 4615

Royal Institution of Great Britain | 17 June 1875

Dear Sir,[2]

Dr Young lectured here in 1802, in 1803,[3] & was invited to give 20 lectures in 1804.[4] His syllabus[5] was published Jan 19, 1802 before he began his lectures.

In these notes he refers the perception of all coloured light to three primitive colours, red, yellow, & blue.[6] The undulatory theory itself is set forth with diffidence & only as an alternative for the emission theory.[7] In my opinion these notes were not so much intended to express his own views as to lead his auditors forwards.

The lectures[8] were not published till 1807 owing to delay on the part of the engravers &c.[9] but in the absence of any proof to the contrary I think they should be taken to represent fairly & truly the substance of the courses which he delivered here in 1802 and 1803

Yours faithfully | Chles. W. Vincent

Prof. Tyndall D.C.L.

Tyndall, John (1820–1893); Hyatt and Mayer Collection, C0076, Manuscripts Division, Department of Special Collections, Princeton University Library

---

1. *Charles Wilson Vincent*: Charles Wilson Vincent (1837–1905), the RI's assistant librarian since 1851, assisting his father, Benjamin Vincent, who had become the RI's librarian two years earlier.
2. *Dear Sir*: although the letter is addressed to Tyndall, Vincent was presumably aware that the ultimate recipient of the information would be Alfred Mayer (see letter 4616).
3. *Dr Young lectured here in 1802, in 1803*: after being appointed the RI's professor of natural philosophy in August 1801, Thomas Young gave three series of lectures in 1802, seventeen on mechanics beginning in January, seventeen on hydrodynamics beginning in March, and sixteen on physics and astronomy beginning in January. These were then repeated, with some additional lectures, at the start of the following year.
4. *invited to give 20 lectures in 1804*: Young had resigned his professorship in June 1803 after the RI's managers rejected his request for a pay rise, and when he was subsequently invited to deliver twenty further lectures on his own terms he declined. See G. N. Cantor, 'Thomas Young's Lectures at the Royal Institution', *Notes and Records: The Royal Society Journal of the History of Science*, 25 (1970), pp. 87–112.
5. *His syllabus*: T. Young, *A Syllabus of a Course of Lectures on Natural and Experimental Philosophy* (London: The Press of the Royal Institution, 1802). Mayer (see n. 2) wanted this information for his article 'The History of Young's Discovery of His Theory of Colors',

American Journal of Science, 9 n.s. (1875), pp. 251–67. In the article, Mayer observed that 'Young printed the syllabus of his first course of lectures on January 19th, 1802 . . . I have not been able to procure a copy of this syllabus, but evidently it does not contain even the corrected statement of his theory of color' (p. 266).

6. *In these notes . . . red, yellow, & blue*: 'Light is distinguished by its effect on the sense of vision, into white and coloured light; and coloured light into a great number of various hues; but they may all be referred to three primitive colours, red, yellow, and blue' (Young, *Syllabus*, pp. 96–7). When Mayer's article on Young (see n. 5) was subsequently reprinted in Britain in the *Phil. Mag.*, he added a footnote stating: 'Professor Tyndall has recently informed me that Young's syllabus gives red, yellow, and blue as the three elementary colour-sensations' (A. M. Mayer, 'The History of Young's Discovery of His Theory of Colours', *Phil. Mag.*, 1 n.s. (1876), pp. 111–27, on p. 126).
7. *The undulatory theory . . . for the emission theory*: in his *Syllabus* Young began the account of his new undulatory theory of light by praising the 'ingenuity' of Isaac Newton's proposal that light was emitted from luminous objects, and observing: 'yet few later opticians have been willing to admit the whole even of his essential hypotheses; although scarcely any attempt has been made to substitute more satisfactory ones in the place of those which have been abandoned. It will be sufficient for our present purpose, to enumerate the respective explanations of the principal phenomena of light, as they are furnished by the Newtonian system, and by the theory lately submitted to the Royal Society' (pp. 115–6).
8. *The lectures*: T. Young, *A Course of Lectures on Natural Philosophy and the Mechanical Arts*, 2 vols (London: Joseph Johnson, 1807).
9. *delay on the part of the engravers &c.*: in the 'Preface' to his *Lectures*, Young noted: 'Drawings were also to be made, for representing to the reader the apparatus and experiments exhibited at the time of delivering the lectures, for showing the construction of a variety of machines and instruments connected with the different subjects to be explained, and for illustrating them in many other ways. These figures have been extended to more than forty plates, very closely engraved, and the execution of the engravings has been minutely superintended' (vol. 1, p. vi).

To Alfred Mayer [*c.* 17 June 1875][1] 4616

Dear Prof Mayer,

I have asked our Ass[istan]t Librarian[2] to go a little more thoroughly into the matter about Young:[3]—this is the result.[4]

J. T.

Tyndall, John (1820–1893); Hyatt and Mayer Collection, C0076, Manuscripts Division, Department of Special Collections, Princeton University Library

---

1. *[c. 17 June 1875]*: the text of this letter was added, by Tyndall, in an upward slant at the top left of letter 4615, which is dated 17 June.

2. *our Ass[istan]t Librarian*: Charles Wilson Vincent (see letter 4615, n. 1).
3. *the matter about Young*: Mayer had asked for information about Thomas Young for his article 'The History of Young's Discovery of His Theory of Colors', *American Journal of Science*, 9 n.s. (1875), pp. 251–67 (see letter 4614).
4. *the result*: letter 4615.

To Heinrich Debus 27 June 1875 4617

27th June 1875.

My dear Heins

Your letter[1] has followed me.[2] I looked out for you in the smoking room[3] but you had vanished, and I had got into a talk with Tom[4] which carried me unconsciously to the door. For the last week we have had atrocious weather here. At the present moment the rain splashes through a fog. I have my stakes and instruments all ready for my measurements[5] when the weather lets me begin. I am almost alone—besides Cottrell there is but one other guest here. The bad weather has driven all away. Canon Liddon was here for 3 or 4 days, and I was glad to meet him. I had not known him previously. Probably he had pictured me, before he saw me, as a creature with hoofs and horns. But we parted very cordially, and I am to dine with him when I return. There is something wonderfully kind & sympathetic in the Canon's eye. The world will be better when such men rely upon their natural impulses instead of tacking on to them the tag rag and bobtail of an impossible religion.

I have done a good deal of proof-work[6] within doors: it fills my time pleasantly. Though the weather has been so bad I have still contrived to have a good deal of exercise out of doors. I expect to return very strong to London.

My mind is almost a vacuum, and external nature is nowhere in this fog, so I must end dear Heins by subscribing myself

ever Yours | John Tyndall

Would you kindly post the enclosed two letters[7] for me.

RI MS JT/1/T/272
RI MS JT/1/TYP/7/2376

1. *Your letter*: letter missing.
2. *followed me*: to the Alps, to where Tyndall had set off on 19 June.
3. *the smoking room*: of the Athenaeum Club, where Tyndall dined with Debus, as well as Thomas Hirst and Robert Roupell, on 18 June (*Hirst Journals*, p. 2014). On the Athenaeum Club, see letter 4186, n. 1.
4. *Tom*: Thomas Hirst.
5. *my measurements*: of glacial motion.

6. *a good deal of proof-work*: presumably for *Fragments of Science: A Series of Detached Essays, Lectures, and Reviews*, 5th edn (London: Longmans, Green, and Co., 1876), of which '400 pages' had been printed at the beginning of the month (see letter 4609).
7. *the enclosed two letters*: enclosures missing.

To William Spottiswoode 30 June [1875][1] 4618

30th June.

My dear Spottiswoode.

I have read the memorial, or petition,[2] bearing the signatures of half a dozen friends of mine—among them your own—and therefore commending itself to my attention.

That attention I have bestowed upon it, and my conclusion is that, wisely organised and directed, the proposed museum will be an immediate help to the investigator, and will act mediately to his advantage by the instruction it will furnish to the philosophical instrument maker.

I am not enthusiastic about any scheme of the kind, for if the spirit of investigation be not there, such a collection cannot create it. But given the spirit the contemplated museum will afford it material aid, and on this account I should be glad if you would append my name to the memorial.[3]

ever faithfully Yours | John Tyndall

W^{m}. Spottiswoode Esq | &c. &c. &c.

RI MS JT/1/T/1363
RI MS JT/1/TYP/3/1245

---

1. *[1875]*: the year is suggested by reference to the petition for the Special Loan Collection of Scientific Apparatus (see n. 2). This is a formal letter, on a particular matter of business, which explains why Tyndall wrote again to Spottiswoode two days later, in letter 4619, giving an informal account of his journey to Switzerland.
2. *the memorial, or petition*: to the government, urging the 'importance of establishing a museum of pure and applied science: that is to say a museum containing scientific apparatus, appliances and chemical products illustrating the history and latest developments of science' and also proposing the assimilation of the Patent Office Museum (quoted in R. Budd, 'The 1876 Loan Collection of Scientific Apparatus and the Science Museum', *Science Museum Group Journal*, no. 1 (2014), pp. 1–14, on p. 10). This petition requested that the government purchase the instruments lent to the Loan Collection of Scientific Instruments, which had been scheduled to open at South Kensington in June but was delayed until March 1876 (see letter 4597, n. 3), to create a permanent museum of scientific instruments from which items could be loaned to researchers.
3. *append my name to the memorial*: it eventually had 140 signatories.

To William Spottiswoode 2 July 1875 4619

Bel Alp | Brieg[1] | Canton de Valais 2nd. July. | 1875.

My dear Spottiswoode

The steamer urged us through a calm sea[2] on the 19th. I invested in a coupé[3] from Paris to Neuchatel. Thence to Sierre—being held for three hours at St. Maurice[4]—where I visited a cave,[5] and found the aspect of its excavation still at work. I reached this place on the Monday morning. The weather has been atrocious. Canon Liddon, and his agreeable sister[6] were here for three or four days, but they had an opportunity of seeing the grandeur of the Mountains. I was glad to meet the Canon; for doubtless some of those excellent pious men figure me as a creature with hoofs & horns. We parted very cordially.

Last Sunday Cottrell made his appearance. He is not yet strong in the mind. We have had two days hard work at our measurements;[7] and today I have from him a little rest. "I thought" he remarked at dinner yesterday "that it was all [up] with me, as I ascended the last step zigzag to the hotel."

We are almost alone here as the bad weather has driven every body away. But it cannot be always bad, and I hope to get a good deal of work done before my return. I have proof and other things[8] to occupy me on bad days.

I know you will glance in from time to time at the Institution, and see that all is going on right.

Give my best love to the boys[9] & my very kindest regards to Mrs. Spottiswoode

Yours ever | John Tyndall

Would you kindly have these scraps[10] posted for me?.

BL Add MS 53715, f. 23

---

1. *Bel Alp | Brieg*: see letter 4306, n. 12.
2. *a calm sea*: the English Channel.
3. *a coupé*: an end compartment in a railway carriage, seated on one side only (*OED*).
4. *Neuchatel . . . Sierre . . . St. Maurice*: cities in Switzerland.
5. *a cave*: probably the Grotte aux Fées in the cliffs above Saint-Maurice in the Swiss canton of Valais, which contained an underground waterfall. The excavations Tyndall mentions were led by Chanoine Gard of the Abbey College of Saint-Maurice.
6. *his agreeable sister*: probably Annie Poole King (née Liddon, 1832–1913), with whom Henry Parry Liddon often travelled, although he also had another sister, Louisa Gibson Ambrose (née Liddon, 1836–1909).
7. *our measurements*: of glacial motion.
8. *proof and other things*: presumably for *Fragments of Science: A Series of Detached Essays,*

Lectures, and Reviews, 5th edn (London: Longmans, Green, and Co., 1876), of which '400 pages' had been printed at the beginning of June (see letter 4609).

9. *the boys*: William Hugh Spottiswoode and Cyril Spottiswoode.
10. *these scraps*: enclosures missing.

To Thomas Archer Hirst 8 July 1875 4620

Bel Alp.[1] 8th July 1875

My dear Tom,

Your pleasant little note[2] has just come to me; but it required the qualification which Heinrich's note[3] imparted. He describes you as in blooming health, growing stronger and stronger daily. Is there no way of relieving you of this after clap[4] of work? It must be horribly trying objectively and subjectively. Since I wrote, the weather has brightened here, and I have got a good deal of work done. I have already set out and measured five lines,[5] learning thereby something perfectly definite regarding certain appearances on the Aletsch glacier.[6] Cottrell has been of a good deal of use to me: but I had to warn him somewhat sternly yesterday for losing himself upon the glacier. He is by no means a first rate iceman; Still he wanders about forgetting how little he is and how big the hills and hollows of a glacier are. We had Canon Liddon for three or four days—I aim to dine with him in London on the 24th of August: he and I got on very well together. He is a man of naturally sweet and sympathetic temper. I do not know whether to regard it as a misfortune, or the reverse, when such a nature weds itself to the nonsense of High Churchism, or ritualism.[7] It altogether depends on the answer to the question whether the nonsense is the necessary nutriment of the sweetness and sympathy. I cannot believe it.

My health is growing all it ought to be. I hope to return very strong. Lady Claud[8] is coming out on the 12th. I dare say she is starting about this time. I remain here till my glacier measurements are finished, and then go on to the Æggischhorn to supplement them with a line or two on the higher glacier. Then I purpose pushing up the Rhone valley, possibly climbing the Galenstock; over the Grimsel Pass; down the Imhof: turn aside to Engsteln, and probably go up the Titlis. Thence over the Sheideck to Grindelwald,[9] where in all probability my vacation will come to a close.

I have been reading a little[10] between whiles: and I see with pleasure the hold which the Germ theory of disease[11] has taken upon thoughtful minds. As long as the good work is making progress one need not care to triumph over those who more or less stupidly withstood the theory. Ardent and able workers are engaged upon it mainly in Germany and England. Sanderson and his colleagues are doing a most important work. If the vivisectionists were to succeed in hampering or extinguishing such researches as theirs it would be a calamity to the human race.[12]

When you again look in at the Institution I wish you would put a number of those letters—not the dividend warrants[13]—together and have them posted to me. If the invitation cards be not too weighty I should like to have them. I am in evil odour in relation to such matters, and I might occupy a wet day now & then in disinfecting myself, and rendering my social savour sweet. Besides we can now have fifteen grammes for 25 centimes.[14]

Your wickedness regarding Frankland[15] drew a laugh from me. You have been always wicked upon this score, and I suppose are likely to continue so. I liked the little girl[16] too. I think she will hold her own.

Yesterday was glorious. Today the clouds look more & more silky, and I should not wonder if they dissolved in thunder. I wish you could come out here. It surely would do you good

Best love to Lilly.[17] I hope when you have time you will let me have a word about Lady Lubbock.[18] But I will take the trouble off your hands & write to her myself.

Ever affectiona[te]ly yours | John

BL Add MS 63902, f. 44
RI MS JT/1/T/HTYP/631

---

1. *Bel Alp*: see letter 4306, n. 12.
2. *Your pleasant little note*: letter missing.
3. *Heinrich's note*: letter missing, but see letter 4617 for Tyndall's reply to Heinrich Debus.
4. *after clap*: something occurring after the conclusion of an affair or incident, particularly when unexpected or undesirable (*OED*).
5. *I have already set out and measured five lines*: this was the first time that Tyndall conducted research on glaciers since the summer of 1871 (see *Ascent of John Tyndall*, pp. 325–6).
6. *the Aletsch glacier*: the longest glacier in the Alps, which flows past the Eggishorn in the Swiss canton of Valais.
7. *High Churchism, or ritualism*: a position within the Church of England that emphasized rituals, liturgical ceremony, and tradition.
8. *Lady Claud*: Elizabeth Hamilton.
9. *Æggischhorn . . . Grindelwald*: various mountains and locales in the Swiss Alps.
10. *I have been reading a little*: Tyndall mentions some of the books he brought with him in letter 4621.
11. *the Germ theory of disease*: the conception that microorganisms or 'germs' transmit diseases (see letter 4463).
12. *If the vivisectionists were to succeed . . . the human race*: Tyndall was responding to the Royal Commission on Vivisection, which had begun on 22 June after the suffragist Frances Power Cobbe (1822–1904) and other antivivisectionists successfully lobbied the

government to introduce a bill restricting animal experimentation. He later made his views public, stating: 'while abhorring cruelty of all kinds, while shrinking sympathetically from all animal suffering—suffering which my own pursuits never call upon me to inflict—an unbiased survey of the field of research now opening out before the physiologist causes me to conclude that no greater calamity could befall the human race than the stoppage of experimental inquiry in this direction' (J. Tyndall, 'Fermentation and Its Bearing on the Phenomena of Disease', *Fortnightly Review*, 20 (1876), pp. 547–72, on p. 566). See T. Holmes, 'Science, Sensitivity and the Sociozoological Scale: Constituting and Complicating the Human-Animal Boundary at the 1875 Royal Commission on Vivisection and Beyond', *Studies in History and Philosophy of Science*, 90 (2021), pp. 194–207.

13. *dividend warrants*: the documentary orders or authorities on which a shareholder receives his dividend (*OED*).
14. *fifteen grammes for 25 centimes*: the weight allowable within the standard cost of sending letters to Switzerland.
15. *Your wickedness regarding Frankland*: seemingly relating to Edward Frankland's marriage to a second wife who was more than twenty years younger than him, which had taken place in May. Hirst recorded in his journal that on 3 July he had 'Called . . .on the Franklands (to be introduced to the new Mrs F.)' (*Hirst Journals*, p. 2015).
16. *the little girl*: Ellen Frances Frankland (née Grenside, 1848–99).
17. *Lilly*: Emily Hirst.
18. *a word about Lady Lubbock*: possibly about her state of health, about which Tyndall had been concerned (see letter 4519). According to Hirst's journal, on 1 July he 'Took luncheon with Lady Lubbock at her club in Albemarle Street' (*Hirst Journals*, p. 2015).

To Olga Novikoff 9 July 1875 4621

Bel Alp, Brieg[1] | 9th, July, 1875.

My dear Friend

Your note[2] has found me at the Bel Alp where I now sit swathed in fog, with the rain plashing through it at intervals. I have been here for more than a fortnight, and on the whole, the weather has been as bad as I have ever seen it in the Alps. I brought out some instruments and an assistant[3] with a view of making glacier measurements and observations: but I have been able to accomplish only very little. Only this morning I sent a telegram[4] to a friend[5] who purposed joining me here, declaring the state of the weather to be detestable. Such warnings are necessary, for disappointment is not uncommon, so should my friend come in the face of this warning the responsibility will not rest on my shoulders.

I am perched at a height of about 7000 feet above the sea on an eminence overlooking the great Aletsch Glacier.[6] It is a noble position when the atmosphere permits surrounding objects to be seen. The Aletschhorn is adjacent, and it is the second mountain of the Oberland;[7] but the most majestic masses are the Dom, the Matterhorn and the Weisshorn.[8] But their grandeur is now practically non-existent, all of them being lost in infinite haze.

I brought some indoor work with me which I have been trying to complete. A new edition of the Fragments[9] has been called for, and as the reading of the old one[10] showed me many things capable of improvement, I have been trying to make the new edition better than the old. I brought two or three books with me, which I hope to open by and by. Among them is Lange's History of Materialism,[11] which I need only for purposes of references, as I have already conned[12] it thoroughly. The other books that I have brought are Frohschammer's "Freiheit d. Wissenschaft" and "Das Recht der Eigenen Ueberzeugung"[13] both of which I hope to master before I return to England. Pray, give him my kind wishes.

I remain here for some days, and then wander on to the AEggischhorn, to measure and observe the higher portions of the glacier. Afterwards I hope to go to the Oberland for a week or two I am in duty bound to return to England before the Meeting of the British Association[14] in August, as I have then to hand over the seals of office to my successor. All pious souls will have a calm year; for Sir John Hawkshaw, my successor, is a great Engineer, and instead of talking heresy he will give some account of the achievements of his own profession.

I see by the newspapers that Mr Gladstone is still active I saw him last at the dinner of the Royal Academy,[15] when he seemed very cheerful. <u>You</u> have the credit of spurring him on to the assault of the Vatican.[16]

Canon Liddon was here for a few days. I had never met him, and I was glad to meet him. He appears to be a man of kind and sympathetic nature; and I was glad to be able to prove to him that I am not endowed with hoofs and horns, as many pious people doubtless imagine me to be.

The greater energy of the protestant Cantons of Switzerland is conspicuous. But then comes the question—is the protestantism a cause, or an effect? Are the people energetic because they are protestants; or are they protestants because they are energetic? I think the latter. The remark might have some application to countries where Jesuitism prevails. Doubtless, however, we have action and reaction; and the people that endure the predominance of Jesuitism, is still further debased by the Yoke.

Now I have written you a long letter without the least delay. Your note came to me only a few hours ago, and without the loss of a single post I have replied to it. The Edelweiss, to my knowledge does not grow in this neighbourhood, but if I should see it adorning the rocks I will secure the flower even at some risk

Goodbye | Yours ever | John Tyndall

RI MS JT/1/TYP/3/937–8
LT Typescript Only

---

1. *Bel Alp, Brieg*: another address has been added in hand to the typescript: 'Hotel du Nord | Frankfort am Main'. This seems to have been taken from the envelope in which the letter was sent, indicating where Novikoff was staying at this time (see letter 4625).
2. *Your note*: letter missing.
3. *an assistant*: John Cottrell.
4. *a telegram*: letter missing.
5. *a friend*: not identified.
6. *the great Aletsch Glacier:* see letter 4620, n. 6.
7. *The Aletschhorn . . . of the Oberland*: the mountain above the Aletsch glacier in the Bernese Oberland (or highland).
8. *the Dom, the Matterhorn and the Weisshorn*: three of the tallest mountains in the Swiss Alps.
9. *A new edition of the Fragments*: see letter 4592, n. 2.
10. *the old one*: J. Tyndall, *Fragments of Science: A Series of Detached Essays, Lectures, and Reviews*, 4th edn (London: Longmans, Green, and Co., 1872).
11. *Lange's History of Materialism*: F. A. Lange, *Geschichte des Materialismus und Kritik seiner Bedeutung in der Gegenwart*, 2nd edn, 2 vols (Leipzig: J. Bädeker, 1873–5).
12. *conned*: pored over, perused, committed to memory (*OED*).
13. *Frohschammer's "Freiheit . . . Eigenen Ueberzeugung"*: J. Frohschammer, *Ueber die Freiheit der Wissenschaft* (München: J. J. Lentner, 1861); *Das Recht der eigenen Ueberzeugung* (Leipzig: Fues, 1869).
14. *the Meeting of the British Association*: the BAAS annual meeting was held in Bristol from 25 August to 1 September (*Brit. Assoc. Rep. 1875*).
15. *the dinner of the Royal Academy*: the annual dinner held, since 1770, at the opening of the Royal Academy of Arts' summer exhibition, on 1 May 1875 (the exhibition opened on 3 May). There is no record of Tyndall attending the dinner or seeing William Gladstone there (he was not among the many attendees listed in the *Times* ('Banquet at the Royal Academy', *Times*, 3 May 1875, p. 9).
16. *the assault on the Vatican*: Gladstone published two pamphlets contending that the civil allegiance of British Catholics had been compromised by the Vatican Council's decree of papal infallibility: *The Vatican Decrees in their Bearing on Civil Allegiance* (London: John Murray, 1874) and *Vaticanism: An Answer to Reproofs and Replies* (London: John Murray, 1875). Novikoff, who first met Gladstone in 1873 and became friendly with the former Prime Minister while she was living in England in 1874, later reflected on 'the sensation produced by the great English statesman's pamphlet on *The Vatican*', and commented: 'I will only say that it was the general public, and not Mr. Gladstone's personal friends, who were so astonished at the views expounded in that pamphlet. In his own intimate circle, I constantly heard him repeat his opinion that "Roman Catholicism is the systematic tyranny of the priest over the layman, the Bishop over the priest, and the Pope over the Bishop"' (O. Novikoff, *Russian Memories* (London: Herbert Jenkins, 1917), p. 57).

To Eliza Spottiswoode 13 July 1875 4622

Bel Alp.[1] July, 13th, 1875.

Here is your husband,[2] in the absence of his friend, doing his best to convert that friend into a fat bull of Basan,[3] gorged with the good things of this life.

Well if I know myself it is not the money, but the feeling which prompted the act, that is precious to me. In fact I have begun to think of late that I have enough of money for the evening of my life. What with my books, and my pluralities[4] and the kindness of my friends at the Royal Institution, I have gradually slipped into the position of a millionaire! When I return I must talk to your husband on these points, and he will then be able to judge whether he ought to aid in augmenting a treasury which is already so plethoric.[5]

We have had very changeful weather here: clouds, fogs, and splashing rain. At times glorious sunshine, making the mountains all the more beautiful and more grand, because the sunshine has emerged from such dismal antecedents. Cottrell is wonderfully well—bronzed and hardy—We have contrived notwithstanding the bad weather, to get a fair amount of work done.

I have been finishing the new edition of my fragments,[6] and have sent off a load of 'copy'[7] to the printer to day.

I did not tell Longman's to send your husband copies of Sound and Light;[8] because I hoped to have received them and sent them myself, before my departure for Switzerland. Pray have the enclosed little note[9] posted for me: the book will come immediately afterwards, and I will write his name in them after my return.

Give my warmest love to the boys[10] and believe me always | Your friend | John Tyndall

RI MS JT/1/TYP/3/1249
LT Typescript Only

1. *Bel Alp*: see letter 4306, n. 12.
2. *your husband*: William Spottiswoode.
3. *a fat bull of Basan*: 'Many bulls have compassed me: strong *bulls* of Bashan have beset me round' (Psalm 22:12). Cattle feeding on the fertile plain of Bashan were proverbial for their strength, size, and self-contentment.
4. *pluralities*: the holding of two or more offices or positions concurrently (*OED*).
5. *plethoric*: bulging, swollen, or turgid (*OED*).
6. *the new edition of my fragments*: see letter 4592, n. 2.
7. *'copy'*: manuscript matter prepared for printing (*OED*), so this was presumably handwritten and distinct from the 'proofs', preliminary impressions of a printed text (*OED*), on which Tyndall was working earlier in the trip (see letters 4617 and 4619).

8. *copies of Sound and Light*: *Six Lectures on Light* (1875) and *Sound* (1875).
9. *the enclosed little note*: enclosure missing.
10. *the boys*: William Hugh Spottiswoode and Cyril Spottiswoode.

To Heinrich Debus 14 July 1875 4623

14th. July 1875

My dear Heins

Your pleasant gossiping note[1] reached me in due time, and helped during its reading to make the vile weather here[2] more tolerable.

I am getting on very well, and hope to return to London with more than my usual stock of strength, and furthermore I hope to be able to preserve it better than hitherto. I have had a good deal of outdoor work since I came here, work that did not tax my brain, that compelled me to be about the glaciers, and that in its way is useful work.

Cottrell is with me, and is getting on exceedingly well: he is now hardened and bronzed, but he sometimes complains of fatigue. Yesterday two particular friends of mine came here, and in a day or two I expect two others. I mean the Hamiltons[3]—This will brighten and occupy the evenings; which are rather a difficulty here.

In fact I go to bed soon after 9; but then I rise correspondingly. This morning I was at a chalet, crunching my crust,[4] and drinking my fresh tepid milk at 6 °C. The weather has improved: yesterday was cold on the heights, but altogether glorious; the sun undimmed by cloud. Today is not so promising, but it is still fine.

I probably run some risk in sending you these letters, as you may be out of London before they arrive.

With regard to the hospital,[5] it is a matter not to be decided in a hurry. I think you ought to aim at securing a little competence[6] for yourself. Nothing I think ought to prevent this. But you may see your way to the subsequent realization of this, even should you now give up a portion of your bread-work.[7] If this be so then your course is clear. It is however a serious matter, and ought not to be dealt with lightly.

I continue here for probably another week or so; then I go for a week to the AEggischhorn,[8] and then sweep through the Oberland.[9] I hope to reach London about the 20th of August.

Yours ever dear Heins | John Tyndall

Would you kindly put stamps on the enclosed[10] and post them for me?

RI MS JT/1/T/273
RI MS JT/1/TYP/7/2377

---

1. *Your pleasant gossiping note*: letter missing.
2. *here*: Bel Alp in Switzerland (see letter 4306, n. 12).
3. *the Hamiltons*: Elizabeth Hamilton and her children Louisa, Emma, and Douglas.
4. *crunching my crust*: eating breakfast; the phrase seems to come from a story in Washington Irving's *The Alhambra*, 2 vols (Philadelphia: Carey & Lea, 1832), vol. 2, on p. 161.
5. *the hospital*: presumably Guy's Hospital in London, at whose medical school Debus had worked as a lecturer in chemistry since 1870. In 1873 he was also appointed professor of chemistry at the Royal Naval College, Greenwich, so he may have been considering giving up the former post.
6. *competence*: a sufficiency of means for living comfortably (*OED*).
7. *bread-work*: this appears to be Tyndall's own coinage, seemingly meaning the work required to maintain basic sustenance.
8. *Aeggischhorn*: the Eggishorn, a mountain in the Swiss canton of Valais.
9. *the Oberland*: a highland area in the Swiss canton of Bern.
10. *the enclosed*: enclosures missing.

From Emil du Bois-Reymond 20 July 1875 4624

Berlin, W., 16 Victoria Str[asse]. | July 20th, '75

My dear Tyndall,

You will have been somewhat puzzled by my last Sendung[1] (how strange that you have no corresponding word in English), of course my intention was to write you some explanatory words along with it, but I found no time to do so. The fact is that the Tragedy of Henny-Penny[2] is the work of a little girl of just 13 years (my third daughter),[3] that as such it did appear remarkable enough to her fond parents to have it printed as a memorial of her precocity, and that I thought it might amuse you a moment.

Believe me, my dear Tyndall, | yours faithfully, | E du Bois-Reymond.

RI MS JT/1/D/152
RI MS JT/1/TYP/7/2443

---

1. *my last Sendung*: shipment or parcel (German); letter missing.
2. *the Tragedy of Henny-Penny*: a children's story, originating in Denmark, in which a chicken fears that the world is coming to an end.
3. *a little girl of just 13 years (my third daughter)*: Aimée du Bois-Reymond (1862–1941), who later became a writer of books for children (G. Finkelstein, *Emil du Bois-Reymond: Neuroscience, Self, and Society in Nineteenth-Century Germany* (Cambridge, MA: MIT Press, 2013), p. 197).

To Olga Novikoff 23 July 1875 4625

The Bel Alp, Breig,[1] | Switzerland | 23rd July 1875

Dear Friend

Your letter[2] reached me here where I have been imprisoned for some time by the vilest weather that I have experienced in the Alps. To break the monotony of our lives my friends[3] and myself purpose moving away tomorrow to another mountain, but with no prospect of bettering our condition. It is many years since I have been at Frankfort.[4] I never returned from Switzerland by Frankfort.[5] My way home is by Neuchatel and Pontarlier.[6] All Europe I suppose is swathed by this rainy atmosphere. We have serious accounts from England, of regions inundated and lives lost.[7] So that we, during our holiday, only have our share of a general calamity.

The review of Fels Petri[8] has duly reached me. From the newspapers I infer that the triumph of the Ultramontanes in Bavaria is not so decided as Frohschammer supposed it would be.[9] In fact so long as we keep the schools out of their hands—so long, that is, as we prevent them from extinguishing the light of the mind in childhood, they cannot prosper. Among themselves elements of disintegration and ruin would be found; for humanity, in the long run, must turn its face towards the light.

Excuse a bad pen and cold fingers. The mountains around are covered with freshly fallen hail.

Yours ever | John Tyndall

[Envelope] Madame de Novikoff | Hotel du Nord. | Frankfort am Main

[Postmark] BRIGUE | 23 VII 75

[Postmark] FRANKFURT a. M. | 25·7 | 75 | 7–8N

John Tyndall Correspondence in the Olga Novikoff Correspondence Collection, Courtesy of Special Collection, Kenneth Spencer Research Library, University of Kansas. MS 30:Q 14
RI MS JT/1/TYP/3/938

---

1. *The Bel Alp, Breig*: see letter 4306, n. 12.
2. *Your letter*: letter missing.
3. *my friends*: Elizabeth Hamilton and her children Louisa, Emma, and Douglas.
4. *Frankfort*: Frankfurt am Main, Germany. Novikoff was staying at the Hotel du Nord (see letter 4621, n. 1).
5. *I never returned from Switzerland by Frankfort*: Novikoff seems to have asked if Tyndall could visit her there.

6. *Neuchatel and Pontarlier*: cities in, respectively, western Switzerland and eastern France.
7. *regions inundated and lives lost*: a week earlier, the *Times* had reported that in Wales the 'heavy rains of the last few days have produced . . . very disastrous consequences . . . the damage and loss of life being considerable. The most fatal event is the loss of 13 lives at Cwm Carn, about ten miles from Newport' ('The Floods', *Times*, 16 July 1875, p. 11).
8. *Fels Petri*: J. Frohschammer, *Der Fels Petri in Rom* (Kempten: Tobias Danheimer, 1873). The pamphlet, which argued that Saint Peter never went to Rome and thus was not the first bishop there, was Jakob Frohschammer's response to his excommunication by the Vatican in 1871. It was included, and praised, in a notice of German theological publications in the *Unitarian Review* in March 1875 ('Recent Theological Publications in Germany', *Unitarian Review*, 3 (1875), pp. 314–17, on pp. 315–16), although it is not clear if this was the review to which Tyndall referred.
9. *From the newspapers . . . supposed it would be*: the *Times* reported of the elections to the second chamber of the Bayerische Ständeversammlung (Bavarian States General) that 'notwithstanding the immense efforts made by the Ultramontanes, and the glowing pastorals issued by all the Bishops, the Ultramontanes will have either no majority or, at any rate, that their majority will not be large enough to compel the appointment of an Ultramontane Cabinet' ('Bavaria', *Times*, 19 July 1875, p. 5). The final result gave the Ultramontanes a majority of only two over the Liberals. On Frohschammer's suppositions about the election, see letter 4558.

From Joseph Henry[1] 26 July 1875 4626

Smithsonian Institution | July 26th 1875

My Dear Dr. Tyndall,

Your two books have been received, namely, the report of your lectures in America and the third edition of your work on sound,[2] for which please accept my thanks.

The preface to the latter[3] has been read with interest, but with mingled feelings of gratification and sorrow: with gratification on account of the kind manner with which you have spoken of me,[4] for I value very highly your friendship and good opinion; and with sorrow on account of the annoyance which apparently my dissent from your views has occasioned you.[5] I regret that this difference in opinion should have occurred especially at a time when a discussion was going on in the Light House Board which grew up after the return of Maj[or]. Elliott,[6] and which led to his resignation and subsequent attacks on our Light House system tending to its entire disorganization.

The remark in the Light House report to which you take special exception[7] was intended as an answer to the charge that the Light House Board of the United States had done nothing to improve the system of aids to navigation, and was by no means designed to disparage the independent labor of the Trinity House. Whether the expression was too strong must be left to the judgment of the world: if it has given offense to the Trinity Board[8] I would prefer that it had not been written.

It was not from a spirit of criticism, which in most cases is a manifestation of egotism, but from a sense of justice to myself and to the Light House Board that I published the remarks I have made in the appendix to the Report in reference to the conclusions which you drew from your researches.[9] In these remarks, for which I am alone responsible, I intended to give my views in such a manner as not to interfere with our friendship; if I am in error the error is that of my head, not of my heart. You may recollect that you were invited to attend a reception given by the Washington Philosophical Society in honor of yourself,[10] and that on that occasion I presented a communication, being a brief account of the labors of the Light House Board on Sound in regard to its application to fog signals.[11] In this paper I distinctly stated my opinion that fog did not materially interfere with the propagation of sound, and especially with the practical application of our fog signal to the purposes of navigation. I also stated the fact, which had been determined by our investigations, that the divergence of sound rays in case of powerful instruments was such as to render reflectors of but little or no use. I also gave an account of the abnormal phenomena of sound which had been observed by General Duane,[12] giving an explanation of one of them, in which sound was heard better against the wind during a north east snow storm than in clear and still weather. I referred this to the simultaneous existence of a south west upper currents which, on the principle of Professor Stokes,[13] threw the sound down upon the ears of the observer. The existence of such an upper current does not rest merely on hypothesis, but I think has been abundantly proved by the observations of Prof. Espy[14] and others. Wherever a break occurs in the lower clouds in one of these storms the "send" as it is called, is seen to be moving rapidly in an opposite direction to the wind at the surface. Another case of the abnormal phenomena of sound was mentioned, that which was observed on the visit of Sir Frederick Arrow and Capt. Webb[15] in which the sound was heard at a distance and lost upon approaching an island, this I explained on the principle of the divergence of the sound rays at a distance and the existence of a sound shadow near by; the origin of the sound being on the opposite side of the island. During the reading of this paper, from fatigue or other cause, you appeared to be inattentive, and the remarks which you subsequently made in regard to sound[16] had no definite relation to the points presented in my communication. I felt at the time some little surprise, and this feeling was shared by a number of the members of the Society, because I had long since established something of a reputation in the line of research founded on many original investigations the results of which have stood the test of time. Afterward on reading the account of your researches in the proceedings of the Royal Society[17] I was not much surprised that you did not mention the labors of our Light House Board in the same line, but I beg to assure you that the omission was received by me with no other feeling than that of sorrow, for while your paper gave abundant evidence of your scientific genius in the way of fertility of invention and suggestion, it was, in my opinion, marred by too hasty a conclusion, which might, perhaps, have been modified had you attentively considered the facts which I had stated. I subsequently read the same communication I had presented to the Philosophical Society of Washington, with additional remarks in regard to your experiments, at a meeting of the National Academy of Sciences[18] and it was published in one of the New York papers[19] to which you allude.

After an attentive consideration of your remarks in the preface to your work on sound I am in candor obliged to say that nothing you have said tends to change my opinion as to the efficient cause of the obstruction of loud sound in case of fog signals, nor am I convinced that the clue to all the difficulty is to be found in the aërial echos.[20] It may be possible that this is produced by a reflection from the atmosphere, and yet that the principle cause of the obstruction of the sound may not be due to a flocculent condition of the air. I freely admit that the cause you have assigned is a true one, but

<rest of letter missing>

Smithsonian Institution Archives. RU 7001, Microfilm Reel 6

---

1. *Joseph Henry*: the letter is a draft in the handwriting of a clerk, with corrections by Henry. On the back of one sheet is written 'Draft of letter to Tyndall not sent'. It is unclear if a corrected version was subsequently sent to Tyndall, but if it was, that letter is missing. This draft is included here because of its significance for the dispute between Tyndall and Henry.
2. *Your two books . . . work on sound*: *Six Lectures on Light* (1875) and *Sound* (1875).
3. *The preface to the latter*: 'Preface to the Third Edition', in *Sound* (1875), pp. vii–xxv.
4. *the kind manner . . . have spoken of me*: Tyndall noted that the 'venerable Professor Joseph Henry . . . gives his services gratuitously . . . and I think it will be conceded that he not only deserves well of his own country, but also sets his younger scientific contemporaries, both in his country and ours, an example of highminded devotion' (pp. xi and xiv).
5. *the annoyance which . . . your views has occasioned you*: in the preface Tyndall articulated numerous refutations of Henry's views on Tyndall's apparent reluctance to give appropriate acknowledgement of previous American experiments on fog-signals (pp. xi–xv), and his disagreement with Tyndall's idea of acoustic clouds obstructing sound and creating abnormal sound phenomena (pp. xix–xxv).
6. *Maj[or]. Elliott*: George Henry Elliot (1831–1900), a military engineer in the United States Army who served on the Light House Board of Washington from 1870–4. In 1873 he spent several months inspecting lighthouse systems in Europe, and upon his return wrote a lengthy report that Henry felt unfairly represented the American system as being

behind that of Britain, particularly with regard to fog signaling. On 21 May 1874, Elliot resigned from the Light House Board and published his report through the United States Senate rather than the Board, making his conclusions public and bringing criticism on Henry. See *Annual Report of the Light-House Board of the United States* (Washington, DC: Government Printing Office, 1874), pp. 614–15.

7. *The remark in . . . take special exception*: in the preface to *Sound* (1875), Tyndall stated: 'On this able Report of their own officer [i.e. Elliott] the Lighthouse Board at Washington make the following remark:—"Although this account is interesting in itself and to the public generally, yet, being addressed to the Lighthouse Board of the United States, it would tend to convey the idea that the facts which it states were new to the Board, and that the latter had obtained no results of a similar kind; while a reference to the appendix to this report will show that the researches of our Lighthouse Board have been much more extensive on this subject than those of the Trinity House, and that the latter has established no facts of practical importance which had not been previously observed and used by the former"' (p. xi). The passage is quoted from *Annual Report of the Light-House Board of the United States*, pp. 614–15.
8. *the Trinity Board*: the Elder Brethren of Trinity House (see letter 4187, n. 2).
9. *the remarks I have . . . from your researches*: 'Appendix: Report of the Operations of the Light-House Board Relative to Fog-Signals', *Annual Report of the Light-House Board of the United States*, pp. 687–721.
10. *a reception given by the . . . in honor of yourself*: this had taken place on 11 December 1872.
11. *a communication, being . . . application to fog signals*: J. Henry, 'On Certain Abnormal Phenomena of Sound in Connection with Fog Signals', *Bulletin of the Philosophical Society of Washington*, 1 (1871–4), appendix: pp. 45–52.
12. *General Duane*: James Chatham Duane (1824–97), an officer in the United States Army Corps of Engineers who was responsible for building a number of fortifications along the east coast of the United States.
13. *the principle of Professor Stokes*: probably the principle of the role of wind direction on sound established by George Gabriel Stokes in 'On the Effect of Wind on the Intensity of Sound', *Brit. Assoc. Rep. 1857*, pp. 22–3.
14. *Prof. Espy*: James Pollard Espy (1785–1860), the first official meteorologist of the United States government. He pioneered the use of the telegraph and other instruments for gathering weather data, and his theory of storm formation was highly influential (*ODNB*).
15. *Sir Frederick Arrow and Captain Webb*: on Arrow, see letter 4266, n. 1; on Webb, see letter 4314, n. 7. They were both members of Trinity House's Committee of Elder Brethren who visited the United States in 1872 to observe the American lighthouse system, during which time they both met Tyndall who was in the United States for his lecture tour (see letter 4201, n. 4)
16. *the remarks which . . . made in regard to sound*: it was recorded that 'Prof. Tyndall made some remarks on this subject [i.e. fog signals], citing certain cases of abnormal phenomena which had come under his notice' (*Bulletin of the Philosophical Society of Washington*, 1 (1871–4), p. 65).

17. *the account of your researches in the proceedings of the Royal Society*: J. Tyndall, 'Preliminary Account of an Investigation on the Transmission of Sound by the Atmosphere', *Roy. Soc. Proc.*, 22 (1874), pp. 58–68.
18. *a meeting of the National Academy of Sciences*: this annual meeting of the peripatetic organization, which had been founded in 1863, was held at Columbia College in New York in late October 1873. Henry's communication was entitled 'Sound in Relation to Fog Signals', and was delivered on 28 October.
19. *published in one of the New York papers*: 'Academy of Sciences', *New-York Daily Tribune*, 29 October 1873, p. 3.
20. *the aërial echos*: in the preface to *Sound* (1875), Tyndall proposed of distortions in the sound of fog-signals: 'The clue to all the difficulties and anomalies of this question is to be found in the aerial echoes, the significance of which has been overlooked by General Duane, and misinterpreted by Professor Henry' (pp. xxi–xxii).

To Emil du Bois-Reymond 1 August 1875 4627

My dear DuBois

On the point of starting from the Aeggischhorn[1] I acknowledge the receipt of your little note,[2] which has followed me hither. The Tragedy[3] has not come, but in a fortnight I shall be in London, and then I shall study it with care, and derive from it I doubt not profit & pleasure—Meanwhile give my love to the little one.[4]

Yours ever | John Tyndall. | 1st. Aug[ust]. 1875

RI MS JT/1/T/405
RI MS JT/1/TYP/7/2444

---

1. *Aeggischhorn*: the Eggishorn, a mountain in the Swiss canton of Valais.
2. *your little note*: letter 4624.
3. *The Tragedy*: the Tragedy of Henny-Penny (see letter 4624, n. 2).
4. *the little one*: Aimée du Bois-Reymond (see letter 4624, n. 3).

From André-Prosper-Paul Crova[1] 2 August 1875 4628

A Monsieur Tyndall Professeur à Royal Institution. | Montpellier, 2 Août 1875.

Monsieur,

J'ai lu avec le plus grand plaisir votre ouvrage sur la lumière. Il est impossible d'exposer d'une manière plus claire et plus attrayante une science si ardue pour beaucoup de personnes. Comme il est infiniment probable que cet ouvrage aura comme les précédents un nombre considérable d'édition, je tiens à vous signaler une petite inexactitude qu'il serait bon de corriger. En décrivant (page

64 de l'édition française) mon appareil pour la projection des mouvements vibratoires vous dites que les courbes tracées sur le disque sont des spirales. Ce sont des circonférences excentriques. Koenig m'avait dit depuis longtemps avoir construit pour vous un de mes appareils. Je pensais qu'il vous avait aussi donné une brochure que j'ai publiée depuis longtemps sur ce sujet. Permettez moi, Monsieur, de vous l'offrir, et d'y joindre une publication que j'ai faite il y a quelque temps sur des phénomènes d'interférence par les réseaux. Or la lumière solaire, ces expériences constituent une des plus belles projections de l'Optique. Peut-être jugerez vous à propos d'en illustrer vos savantes leçons.

Veuillez agréer, Monsieur, l'expression de mes sentiments dévoués. | A. Crova | Professeur à la Faculté des Sciences de Montpellier

P.S. Vous trouverez aussi la théorie de l'appareil de projection des mouvements vibratoires, avec quelques additions, dans les Annales de Chimie et de Physique. | 4 Série Tome XII p. 288.

To Mister Tyndall Professor at the Royal Institute. | Montpellier, 2 August 1875.

Sir,

I have read with the greatest pleasure your book on Light.[2] It is impossible to explain in a more clear and attractive manner a science that is so arduous for many people. Since your book, like the previous ones, will very likely have a considerable number of editions, I want to draw your attention on a slight mistake that would be good to correct. While describing (p. 64 of the French edition) my instrument[3] for the projection of vibrating movements, you said that the curves drawn on the disk are spirals. They are eccentric circumferences. Koenig[4] told me a long time ago that he built for you one of my instruments. I thought he also gave you one of my booklets[5] that I published on the matter a while ago. Let me offer it to you, and with it a publication that I made a little while ago on the phenomena of interference by gratings.[6] Apart from the solar light, these experiments constitute one of the most beautiful projections of Optics. Perhaps you will judge it suitable to illustrate your scientific lessons.

Please accept, Sir, the expression of my devoted regards. | A. Crova | Professor at the faculty of Sciences in Montpellier.

P.S. You will also find the theory of the projection device of the vibrating movements, with some additions, in the Annals of Chemistry and Physics. | 4th Series Volume XII p. 288.[7]

RI MS JT/1/C/60

1. *André-Prosper-Paul Crova*: André-Prosper-Paul Crova (1833–1907), professor of experimental physics at the Lycée de Montpellier, known for his work in optics and electricity, particularly radiant energy and the determination of the solar constant.

2. *your book on Light*: see letter 4277, n. 2.
3. *my instrument*: a series of rotating disks in which a beam of light was used to represent different sorts of vibratory motions.
4. *Koenig*: Karl Rudolph Koenig (1832–1901), a German physicist and instrument maker based in Paris who developed new forms of tuning forks and pioneered the use of graphical methods to represent sound (*CDSB*).
5. *one of my booklets*: A. P. Crova, *Description d'un appareil pour la projection mécanique des mouvements vibratoires* (Montpellier: Boehm, 1866).
6. *a publication . . . of interference by gratings*: A. P. Crova, *Sur les phénomènes d'interférence produits par les réseaux parallèles* (Montpellier: Boehm, 1873). The phenomena Crova referred to are interference by diffraction gratings.
7. *the Annals of Chemistry and Physics. | 4th Series Volume XII p. 288*: A. P. Crova, 'Description d'un appareil pour la projection mécanique des movements vibratoires', *Annales de Chimie et de Physique*, 12 (1867), pp. 288–308.

To Heinrich Rudolf Vieweg 13 August 1875 4629

Gentlemen,[1]

I sometime ago requested my American publisher Messrs Appleton of New York[2] to give me an account of the number of copies of each of my books which had been printed and published in the United States. They were good enough to do so; thus making plain to me the fair and honourable manner in which they, as their business, behaved toward me.

I have never had any statement from you of the number of copies of my works which have been circulated in Germany. This I should like to know; and I should be thankful to you if you would give me the information.

In fact I am desirous to make a more definite arrangement than that hitherto existing regarding the German translations of my books.[3] I hope and think that they have proved a source of profit to you. But for future editions I should like to make our arrangement as definite as that existing between the Mess[rs]. Longman[4] and myself. And in the events, which is probable, of my writing a work on a new subject I should like a business-like statement of conditions to exist between you and me.

I always preserve the control of my books in my own hands. The Messrs. Longman purchase each separate edition, and here their control ends. A difficulty has never arisen between us, not do I suppose that a difficulty ever will arise. A written agreement is drawn out for every edition, in which the number of copies to be printed is stated. This is signed by Messrs. Longman and me, each preserving a copy. The total profit is ascertained by deducting the cost of production from the total amount arising from the sale of the book, and two thirds of this profit are paid to me on the day of publication.

Some definite arrangement of this kind I should like to make with you, and it was only because the translations were supervised by my personal friends[5] that I refrained from desiring some such arrangement earlier.

It is not, as you will eventually learn, any desire to increase my own profits that I wish matters to be established on a more business-like footing between you and me.

To repeat—I would beg of you to inform me what is the number of each of my books that has been printed and sold in Germany. I wish also to be informed whether a new edition of any of the books, besides the "Fragments"[6] will soon be required; and for each such editions I should require a definite statement of the number of copies, and of the terms of publication.

I shall be in London in a few days,[7] and would therefore beg of you to be good enough to send your reply[8] as usual, to the Royal Institution.

Your obedient Servant | John Tyndall

Grindelwald[9] 13th. Aug[ust]. 1875.

TU Braunschweig/Universitätsbibliothek/UABS V 1 T : 63

1. *Gentlemen*: see letter 4592, n. 1.
2. *Messrs Appleton of New York*: William Henry Appleton, George Swett Appleton (1821–78), and Daniel Sidney Appleton (1824–90), proprietors of the publishers D. Appleton & Co.
3. *I am desirous to . . . translations of my books*: this had been suggested to Tyndall by Hermann Helmholtz in letter 4606.
4. *Messrs. Longman*: Thomas and William Longman, proprietors of the publishers Longmans, Green, & Co.
5. *the translations were supervised by my personal friends*: by Helmholtz and Gustav Wiedemann, with the actual translations made by their wives Anna Helmholtz and Clara Wiedemann.
6. *the "Fragments"*: see letter 4592, n. 2.
7. *in London in a few days*: Tyndall returned from Switzerland on *c.* 18 August (see letter 4632).
8. *your reply*: Vieweg replied in letter 4639.
9. *Grindelwald*: a village in the canton of Berne in the Swiss Alps.

From Thomas Henry Huxley 13 August 1875 4630

Cragside, Morpeth,[1] | August 13th 1875.

My dear Tyndall,

I find that in the midst of my work in Edinburgh[2] I omitted to write to De Vrij[3] so I have just sent him a letter expressing my pleasure in being able to cooperate in any plan for doing honour to old Benedict,[4] for whom I have a most especial respect.

I am not sure that I won't write something about him to stir up the Philistines.

My work in Edinburgh got itself done very satisfactorily and I cleared about £1000 by the transaction[5]—one of the few examples known of a Southern coming north and pillaging the Scots. However I was not sorry when it was all over as I had been hard at work since October and began to get tired.

The wife and babies from the south and I, from the north, met here a fortnight ago and we have been idling very pleasantly ever since. The place is very pretty and our host[6] kindness itself. Miss Matthaei[7] and five of the bairns are at Cartington—a moorland farm-house three miles off—and in point of rosy cheeks and appetites might compete with any five children of their age and weight. Jess and Mady[8] are here with us and have been doing great execution at a ball at Newcastle. I really don't know myself when I look at these young women, and my hatred of possible sons-in-law is deadly.

All send their love. | Ever yours very faithfully | T. H. Huxley

Wish you joy of Bristol.[9] I send this to Royal Institution not knowing where you are.

IC HP 8.173
RI MS JT/1/TYP/9/3046
Typed Transcript Only

---

1. *Cragside, Morpeth*: see n. 6.
2. *my work in Edinburgh*: see letter 4613, n. 5.
3. *De Vrij*: Hugo de Vries (1848–1935), a Dutch botanist and pioneering geneticist (*CDSB*).
4. *old Benedict*: the Dutch philosopher Baruch Spinoza (1632–77). Benedictus is the Latinized form of his name, and Benedict the anglicized form. The plan for doing honour to him, presumably prompted by the impending bicentenary of his death, was probably for the statue that was erected in the Hague in 1880, following an international subscription scheme.
5. *the transaction*: in the Scottish university system, lecturers collected fees directly from students attending their lectures courses; Huxley's students paid £4 each for the course he gave. See A. Desmond, *Huxley: Evolution's High Priest* (London: Michael Joseph, 1997), p. 78.
6. *our host*: William Armstrong (1810–1900), first Baron Armstrong, was an English arms manufacturer who founded W. G. Armstrong & Co. (1847). Armstrong was well known for his work on hydraulics and electrostatics, service to the BAAS, and involvement in a number of engineering institutions. He was knighted in 1859 for giving the patent of the Armstrong gun to the Crown. In 1875 he was still in the process of completing his large house, Cragside, in Northumberland, having begun in 1869. He was raised to the peerage as Baron Armstrong of Cragside in 1887 (*ODNB*).
7. *Miss Matthaei*: the Huxleys' German governess (see Desmond, *Huxley*, p. 66); possibly Louise Henriette Elizabeth Matthaei (*c.* 1843–1921), who lived in Marylebone at this time,

close to the Huxley's home in St John's Wood. Although born in London, her nationality was German; she was granted British citizenship in 1915, when Britain and Germany were at war. She began working for the family in December 1871, after Leslie Stephen helped interview her (J. W. Bicknell (ed), *Selected Letters of Leslie Stephen*, 2 vols (London: Macmillan, 1996), vol. 1, p. 194).

8. *Jess and Mady*: Jessica Oriana and Marian Huxley.
9. *joy of Bristol*: at the forthcoming meeting of the BAAS (see letter 4621, n. 14), where Tyndall would relinquish the presidency.

To George Wareing Ormerod[1] 19 August [1875][2] 4631

Royal Institution of Great Britain | 19th Aug[ust].

Dear Sir,

I am much obliged to you for the note & paper[3] which you have been kind enough to send me.

yours truly | John Tyndall

Oxford, Bodleian Library, MS. Eng. lett. d. 220, fol. 374

1. *George Wareing Ormerod*: George Wareing Ormerod (1810–91), an English lawyer and amateur geologist who published several papers on the geology of Devonshire and Cheshire (*ODNB*).
2. *[1875]*: the year has been added to the original letter in a hand other than Tyndall's, presumably based on a postmark on the envelope.
3. *the note & paper*: letter missing; the paper is possibly 'On the Murchisonite Beds of the Estuary of the Ex', *Quarterly Journal of the Geological Society*, 31 (1875), pp. 346–54, which was Ormerod's most recent publication, having been read at the Geological Society on 24 February.

To Margaret Ginty[1] 20 August 1875 4632

Royal Institution of Great Britain | 20th. Aug[ust]. 1875.

Dear Mrs Ginty,

A day or two ago I returned from Switzerland, having spent my holiday there, for the most part in the midst of bad weather. I am now about to begin my earnest autumn's work.[2]

I quitted London in low health. In fact I had been in a physician's hands[3] sometime before quitting it, but I have returned much stronger—fit indeed for the weighty duties that now demand my attention.

I do remember Miss Wright[4] and am concerned to read the few lines you have written regarding her.[5] Do you think a little present of two or three pounds would be acceptable to her?

faithfully yours | John Tyndall.

RI MS JT/1/TYP/11/3674
LT Typescript Only

1. *Margaret Ginty*: Margaret Ginty (née Roberts, b. *c.* 1826), the widow of William Ginty (*c.* 1820–66), Tyndall's close friend during his time on the Irish and English Ordnance Surveys. She married William in Ireland in 1846, after which they relocated to Manchester. In 1854 the family moved to Rio de Janeiro, Brazil where William Ginty died in 1866. After William's death, Tyndall served as Margaret's trustee and supervised the education and employment of her children.
2. *my earnest autumn's work*: on 10 September Tyndall began a series of experiments using optically pure air to try to determine whether germs could be found in the air.
3. *in a physician's hands*: Tyndall had recently begun using the services of Andrew Clark.
4. *Miss Wright*: not identified, but possibly either one of the sisters of Tyndall's childhood friend William Wright (1818/9–92) or a relation of Mr and Mrs Wright, with whom Tyndall lodged while working as a railway surveyor in Halifax. It is unclear what difficulty she was experiencing at this time.
5. *the few lines you have written regarding her*: letter missing.

From Mary Agatha Russell 21 August 1875 4633

Pembroke Lodge.[1] | Richmond Park. | Aug[ust]. 21. / 75.

Dear Professor Tyndall

I cannot tell you with what pleasure I yesterday received two of your books[2] "from the Author"—it is most kind of you to have sent them to me, & I am extremely grateful. As I have never read the book on Sound it is a pleasure to come—& the additions about fog-signalling[3] must be very interesting.—I do not know whether you are now in Switzerland enjoying the delightful mountain air, or whether you have come back to England,[4] but I will send this to Albemarle St. in hopes it may reach you before long.—Rollo[5] is going to Bristol for the British Association[6]—we hope he may see you there? Mama[7] begs to be kindly remembered;—& thanking you again very much, believe me,

Yours very sincerely | Agatha Russell

RI MS JT/1/R/52

1. *Pembroke Lodge*: see letter 4237, n. 1.

2. *two of your books*: *Six Lectures on Light* (1875) and *Sound* (1875).
3. *the additions about fog-signalling*: for *Sound* (1875), Tyndall added extensive new sections on fog-signaling (pp. 257–69 and 317–18).
4. *come back to England*: Tyndall returned from Switzerland on *c.* 18 August (see letter 4632).
5. *Rollo*: Francis Albert Rollo Russell.
6. *Bristol for the British Association*: see letter 4621, n. 14.
7. *Mama*: Frances Russell.

From Thomas Archer Hirst 23 August 1875 4634

Paris Aug[u]st 23rd 1875

My dear John,

I send you a line just to say that I am all right,[1] and in full enjoyment of perfect rest and freedom from Greenwich worries. In this freedom consists my happiness, for it enables me to pursue my Geometrical studies undisturbed; and my Geometry has now become my relaxation. It should, it is true, have been my chief object in life, and every thing else have been subordinated to it; it is thus and only thus that these creations eminentes of which Renan speaks[2] would have been rendered possible. But it has been otherwise ordered and I do not complain. Mes creations, en place d'entrainer une rupture d'equilibre me la conservent. Voila tout![3] as an off-hand Frenchman would say. I do not forget, however, that such preservation is either restoration of equilibrium [or] is purchased, to a great extent, at the cost of the eminence of the creations. Chasles and Bertrand,[4] whom I have often seen, have often spoken kindly of you. I have given up my intentions of going to Nantes, Cremona[5] (who has been my companion here for a week) having been called away home suddenly in consequence of the illness of his brother.[6] Tomorrow I leave for Strasburg; hence I repair to the Bergstrasse between Heidelberg and Darmstadt[7] where I am to spend a few days with Sturm.[8] Up to the end of the month a letter addressed Poste Restante[9] Darmstadt will find me. I hope you are in good condition so far as health is concerned and that you will enjoy your visit to Bristol.[10] Remember me kindly to all friends and believe me to be ever yours affectionately.

T. Archer Hirst

P.S. Normandy was of great service to me though my chief occupation there was nursing a very severe cold with which I started from England.

RI MS JT/1/H/259

---

1. *I am all right:* Tyndall had been concerned about Hirst's health for months (see, for example, letter 4538).

2. *these creations eminentes of which Renan speaks*: the French biblical scholar Joseph Ernest Renan (1823–92) proposed: 'Les plus belles choses du monde se sont faites à l'état de fièvre; toute création éminente entraîne une rupture d'équilibre, un état violent pour l'être qui la tire de lui [The most beautiful things in world are done in a state of fever; every great creation involves a breach of equilibrium, a violent state of the being which draws it forth]' (*Vie de Jésus* (Paris: Michel Lèvy, 1863), p. 453).
3. *Mes creations, en . . . la conservent. Voila tout!*: My creations, instead of causing imbalance, preserve it for me. That is all! (French).
4. *Chasles and Bertrand*: Michel Chasles (1793–1880), a French geometer who taught at the Sorbonne (*CDSB*), and Joseph Louis François Bertrand (1822–1900), professor of mathematics at the École Polytechnique and Collège de France (*CDSB*).
5. *Cremona*: Luigi Gaudenzio Giuseppe Cremona (1830–1903), an Italian mathematician who specialized in algebraic curves and surfaces (*CDSB*).
6. *his brother*: Tranquillo Cremona (1837–78), an Italian painter whose health suffered from making his own oil pigments and testing them on the skin of his arms.
7. *the Bergstrasse between Heidelberg and Darmstadt*: the ancient mountain road between the two cities in southwest Germany.
8. *Sturm:* Friedrich Otto Rudolf Sturm (1841–1919), a professor of descriptive geometry and graphic statics at the technical college in Darmstadt (*CDSB*).
9. *Poste Restante*: see letter 4392, n. 7.
10. *your visit to Bristol*: for the meeting of the BAAS (see letter 4621, n. 14), where Tyndall would relinquish his presidency.

To Thomas Archer Hirst 26 August 1875 4635

BRITISH ASSOCIATION FOR THE ADVANCEMENT OF SCIENCE. | BRISTOL MEETING, AUGUST 25TH–SEPTEMBER 2ND, 1875. | BRISTOL, 26th Aug[ust]. 1875.

My dear Tom.

I came here yesterday, handed over the reins last night to Hawkshaw, and am now clear of official responsibility. I read your letter[1] to Cayley[2] in the Committee-room of Section A.[3] Of course it gave me pleasure to hear that you were all right; but how you can get right in Paris is a mystery to me. You and I require very different modes of restoration. I was speaking today to a lady in the Reception Room. After a time I noticed that a gentleman was near us, apparently waiting to speak to me. She moved away & Prof Reynolds[4] was there. They passed to other papers, & I came away. He afterwards brought it[5] forward, but I did not hear it. I wish I had; for I am anxious to set that gentleman right. He is spreading fog over a very clear matter.[6] Spottiswoode comes down tomorrow and he lectures tomorrow night:[7] I spent Sunday, and a portion of Monday[8] with him in the country. I sent you a parcel of cuttings from the local papers & the Daily News,[9] which may amuse you. The reporters seem to have taken down accurately what I said[10]—the word struggling, however, was meant to be wrestling[11]—a small mistake enough, and probably mine. I am in Lodging, having resisted all temptations to tie myself to friends. I returned very strong from Switzerland, but care will be necessary to maintain that strength. From Grindelwald, in company with two Miss Hamiltons and their young brother[12] I climbed the Wetterhorn.[13] We were at the summit at 9 A.M. and had an unspeakably glorious view. The sunset and the sunrise ranked among the finest I have seen in the Alps. Wiedemann was at Grindelwald.

I have been home about 10 days; and shall quit this excitement on Saturday[14] & return to, quiet work. My official life ended prosperously—the audience being very cordial.

ever dear Tom yours affection[at]ely | John Tyndall.

RI MS JT/1/T/718

---

1. *your letter*: letter 4634.
2. *Cayley*: Arthur Cayley (1821–95), an eminent mathematician who specialized in algebra and group theory (*ODNB*).
3. *Section A*: Mathematics and Physics.
4. *Prof Reynolds*: Osborne Reynolds (1842–1912), professor of engineering at Owen's College, Manchester (*ODNB*).
5. *it*: presumably the paper 'On the Steering of Screw-Steamers' that Reynolds delivered at the BAAS meeting (*Brit. Assoc. Rep. 1875*, pp. 141–6).
6. *spreading fog over a very clear matter*: Reynolds challenged the results of Tyndall's experiments on atmospheric sound in his 'On the Refraction of Sound by the Atmosphere', *Roy. Soc. Proc.*, 22 (1874), pp. 531–48.
7. *Spottiswoode comes . . . lectures tomorrow night*: on the evening of 27 August in Colston Hall, William Spottiswoode delivered a lecture on 'The Colours of Polarized Light' (*Brit. Assoc. Rep. 1875*, p. lxvii).
8. *Sunday . . . Monday*: 22 and 23 August.
9. *cuttings from the local papers & the Daily News*: among the reports from local papers that Tyndall sent were 'British Association', *Western Daily Press*, 26 August 1875, p. 5; 'The Inaugural Address', *Bristol Times*, 26 August 1875, p. 2; and 'British Association Inaugural Address', *Bristol Daily Post*, 26 August 1875, p. 2. The report from a national paper was 'The British Association at Bristol', *Daily News*, 26 August 1875, p. 2.
10. *what I said*: on the evening of 25 August in Colston Hall, Tyndall gave a brief speech when he handed over the presidency to John Hawkshaw ahead of the latter's address ('Address of Sir John Hawkshaw, C.E., F.R.S., F.G.S., President', *Brit. Assoc. Rep. 1875*, pp. lxviii–c).
11. *the word struggling . . . meant to be wrestling*: Tyndall was reported as having said that

Hawkshaw was 'one who, struggling Antæus-like with his subject here to-night, will know how to maintain throughout a refreshing contact with his mother earth' ('British Association', *Western Daily Press*, 26 August 1875, p. 5).

12. *two Miss Hamiltons and their young brother*: Louisa, Emma, and Douglas Hamilton.
13. *the Wetterhorn*: the highest of the three peaks of the Wetterhörner mountain in the Swiss Alps, above the town of Grindelwald.
14. *Saturday*: 28 August.

To Mary Adair 28 August 1875 4636

BRITISH ASSOCIATION FOR THE ADVANCEMENT OF SCIENCE. | BRISTOL MEETING, AUGUST 25TH–SEPTEMBER 2ND, 1875. | BRISTOL, 28th. Aug[ust]. 1875.

My dear Miss Adair,

I send you a scrap from the Bristol News,[1] which when the small pencil corrections[2] are made, comes very close to what I said.[3]

It was very brief but I thought it best to err on the side of brevity if at all.

There was nobody in that room[4] to whom it would give me greater pleasure to give pleasure than to you. I am very glad the little introduction pleased you.

This morning I have been over the suspension Bridge[5] and a good way through the lovely country beyond. If Darwin's theory[6] be true I must have had a savage ancestry; for I still retain a stronger bias for outside nature than for the aspect under which she exhibits herself in our section-rooms.[7]

Perhaps it would be truer to say that there are moods in which the one aspect is preferable, and moods again in which the other aspect is preferable.

It has given me great pleasure to meet you here. With best wishes believe me always

Yours faith[ful]ly J. Tyndall

I return today to London.

RI MS JT/1/T/88
RI MS JT/1/TYP/1/14–16

---

1. *a scrap from the Bristol News*: 'British Association', *Western Daily Press*, 26 August 1875, p. 5.
2. *small pencil corrections*: Tyndall was reported as having said that John Hawkshaw, who was about to give his address as the incoming president of the BAAS, was 'struggling Antæus-like with his subject', but Tyndall had used the verb 'wrestling' and seemingly corrected this by hand (see letter 4635).
3. *what I said*: see letter 4635, n. 10.
4. *that room*: the main hall of Colston Hall, a large concert hall opened in Bristol in 1867.

5. *the suspension Bridge:* the Clifton Suspension Bridge, which had opened in Bristol in 1864. It was built by Hawkshaw, the new president of the BAAS whose address Tyndall introduced.
6. *Darwin's theory*: that humans evolved from primitive ape-like ancestors, as articulated in C. Darwin, *The Descent of Man*, 2 vols (London: John Murray, 1871).
7. *our section-rooms*: the BAAS was divided into sections, of which there were seven in 1875, and these met in their own rooms.

To Thomas Henry Huxley 29 August [1875][1] 4637

Royal Institution of Great Britain, | 29th. Aug[ust].

Dear Hal

I hope you may be induced or inspired to say something about Spinoza.[2] A word is needed to put the true figure of the man before the world. He exercises a notorious influence at the present day, and it would be refreshing to the public to know <u>why</u> he exercises it. No man can tell them why so well as yourself.

I was very brief at Bristol[3]—I thought it fair to leave the stage clear to the obviously "safe" man who succeeded me.[4] Last year the parsons had something to handle, but this year their occupation will be entirely "gone". How they are to get over to day's sermons[5] I know not.

I suspect a touch of heresy would have been more to many of their tastes than this dead neutrality. They have now a touch of true practical Materialism,[6] but it appears neither to provoke opposition nor excite jubilation.

Give my best regards to Armstrong[7] and his excellent wife.[8] I hope he was not angry with the effusion about Whitworth.[9]

And give my love to all the children and their dam.[10]

Tell Jess[11] that I intend to call her to account for her doings at Newcastle.[12]

always Yours faithfully | John Tyndall

I kept quiet at my lodgings at Bristol, accepted no invitations, and got home last night.

Foster has been pressing me to go to Leeds on the opening of the Yorkshire College of science[13]—but I do not see my way to it—Such demands are hard upon a worker.

Again we have run neck and neck. My books this year have brought me in about the same sum as your Edinburgh lectures have brought you.[14]

IC HP 8.174
RI MS JT/1/TYP/9/3047
Typed Transcript Only

---

1. *[1875]*: the year is established by the relation to letter 4630 and by reference to the BAAS meeting at Bristol.
2. *say something about Spinoza*: Huxley had agreed to cooperate in a plan to honour the philosopher Baruch Spinoza (see letter 4630, n. 4).
3. *very brief at Bristol*: see letter 4635, n. 10.
4. *the obviously "safe" man who succeeded me*: John Hawkshaw, who took over as president of the BAAS. In his speech introducing Hawkshaw's address, Tyndall called him 'a wise and prudent head, a leader not likely to be caught up into atmospheric vortices of speculation about things organic or inorganic, about mind or matters beyond the reach of the mind' ('British Association', *Western Daily Press*, 26 August 1875, p. 5).
5. *to day's sermons*: presumably Hawkshaw's presidential address given on 25 August, which focused on the benefits of engineering to human civilization and avoided references to religion and other controversial issues ('Address of Sir John Hawkshaw, C.E., F.R.S., F.G.S., President', *Brit. Assoc. Rep. 1875*, pp. lxviii–c). Tyndall may also have had in mind the similarly uncontroversial public lecture, on 'Railway Safety Appliances', that was delivered by the British engineer Frederick Joseph Bramwell (1818–1913) on 30 August in Colston Hall (*Brit. Assoc. Rep. 1875*, p. lxvii).
6. *true practical Materialism*: i.e. engineering.
7. *Armstrong*: William Armstrong (see letter 4630, n. 6), at whose home in Northumberland Huxley was then staying.
8. *his excellent wife*: Margaret Armstrong (née Ramshaw, 1807–93).
9. *the effusion about Whitworth*: the Friday Evening Discourse on 'Whitworth's Planes, Standard Measures, and Guns' that Tyndall had delivered at the RI on 4 June (*Roy. Inst. Proc.*, 7 (1873–75), pp. 524–39). Armstrong and Whitworth were rival industrialists, especially in the manufacture of armaments. At the end of his discourse on Whitworth, Tyndall indicated, albeit somewhat obliquely, that he would be willing to offer the same service to Armstrong (see *Ascent of John Tyndall*, p. 337).
10. *dam*: a female parent (*OED*).
11. *Jess*: Jessica Oriana Huxley.
12. *her doings at Newcastle*: see letter 4630.
13. *the opening of the Yorkshire college of science*: presumably the meeting that was held at Leeds Town Hall on 6 October to mark the opening of the Yorkshire College of Science, which had been established in 1874 with support from individual subscribers and local industry.
14. *My books this . . . lectures have brought you*: Huxley claimed to have 'cleared about £1000' from the lecture series he gave at Edinburgh University (see letter 4630). The books Tyndall refers to were presumably *Six Lectures on Light* (1875) and *Sound* (1875).

From Mary Adair 29 [August 1875][1] 4638

5 Gloucester Row. Clifton, Bristol. | Sunday 29th

My dear Dr. Tyndall,

Thank you many times—I shall value the version (which I had not seen) with its amendment[2] very much; and I do think you most kind to have amended this promptly to my petition.

I am glad you secured a good walk yesterday[3] and trust you do not intend to overtax yourself with work through the autumn.

This sectional work[4] is exhausting and makes me marvel at the strength of mind (and body) of friends who can enjoy it all day and every day. However my present mood is gregarious and I do really like a moderate amount of the wisdom at present accessible here.[5]—maybe it's because of my unknown ancestry,[6] or maybe from my actual environment—or perhaps it is only the reaction after 8 months of continued country calm.[7] Small words and cases believe me your's very faithfully

H. M. Adair.

RI MS JT/1/A/14

1. *[August 1875]*: the month and year are established by the relation to letter 4636.
2. *the version (which I had not seen) with its amendment*: of 'British Association', *Western Daily Press*, 26 August 1875, p. 5, to which Tyndall added a pencil amendment (see letter 4636, n. 2).
3. *a good walk yesterday*: over the Clifton Suspension Bridge (see letter 4636).
4. *sectional work*: the separate meetings of the BAAS's seven sections (see letter 4636, n. 7).
5. *the wisdom at present accessible here*: at the BAAS meeting in Bristol.
6. *my unknown ancestry*: Adair's distinguished family could trace their lineage back to a knight granted an estate in Ireland after being killed at the Battle of Flodden in 1513, so it is unclear to what she is referring.
7. *8 months of continued country calm*: at her family's estate at Heatherton Park in Somerset.

From Heinrich Rudolf Vieweg 7 September 1875 4639

Braunschweig 7 September 1875.

Sehr geehrter Herr!

Wir theilen durchaus Ihre Ansicht, daß es besser und geschäftsmäßiger ist, wenn das Verhältniß, in welches Autor und Verleger für ein literarisches

Unternehmen treten, von vornherein vertragsmäßig festgestellt wird. Daß zu einer solchen Feststellung der gegenseitigen Rechte und Pflichten bis jetzt weder von Ihnen noch von uns die Initiative ergriffen worden ist, liegt zumeist in den besonderen Umständen, unter denen wir zuerst die Ehre hatten, mit Ihnen in Verbindung zu treten. Die von uns hochgeschätzten Übersetzer gaben die Veranlaßung für uns zu den ersten deutschen Ausgaben von Werken Ihrer Feder, sie waren die Mittelspersonen und in der That glaubten wir in ihnen die Repräsentanten des Autors für unsere Sprache zu erblicken. Es schien uns, daß Ihrerseits das Verhältniß ähnlich aufgefaßt würde, denn Sie ertheilten und, ohne Zweifel eben dieser besonderen Umstände wegen, die Berechtigung zur Publication der Übersetzungen durch die berufene Hand Ihrer Freunde, ohne daran eine Beschränkung in Bezug auf die Anzahl der Exemplare oder die Bedingung von Honorar für Sie selbst zu knüpfen. Wir hielten es jedoch für in der Ordnung, Ihnen auch ohne eine solche Stipulation ein Honorar zu übersenden, welches Sie d. Z. mit dem Ausdrucke der Befriedigung in Empfang nahmen, mit dem Sie darin „your own free act and not a consequence of any contract between you & me" erblickten. Später wiesen Sie uns an, solche Zahlungen in Ihrem Auftrage an die Damen zu richten, welche bei der Übersetzung betheiligt waren.

Wenn Sie uns nun den Wunsch nach einem bestimmten Übereinkommen für *[1 word illeg]* Publicationen aussprechen, so finden Sie uns vollkommen damit einverstanden. Der Buchhandel in Deutschland ist jedoch auf ein von dem englischen so durchaus verschiedenes System basiert, so ganz andersartig organisirt, daß es uns unmöglich ist, auf den von Ihnen vorgeschlagenen Modus der Betheiligung an dem Unternehmen einzugehen. Wir können nicht, wie es in England geschieht, gleich nach Vollendung des Druckes die ganze Auflage an die Zwischenhändler verkaufen, sondern müßen die Exemplare den Wiederverkäufern auf unsere eigene Gefahr in Commission geben, und erhalten erst nach einem Jahre das Verkaufte bezahlt, die nicht verkauften Exemplare werden uns aber in natura zurückgeliefert, u. so vergeht oft ein Zeitraum von mehreren Jahren, ehe wir wissen können, ob wir an einem Buch Gewinn oder Verlust gemacht haben. Ihr Verhältniß mit den Herren Longman, wonach Sie schon bei Erscheinen der Bücher den Gewinn nach einem gewissen Verhältniße theilen, setzt einen ganz sichern Gewinn voraus, und wir wollen nicht bezweifeln, daß bei Ihren Werken in England auf einen solchen jedesmal zu rechnen ist. Wie würde sich aber die Sache gestalten, wenn einmal nicht so viele Exemplare zu verkaufen wären, um die Herstellungskosten zu decken? Würden Sie in diesem Falle auch Antheil an dem Verluste nehmen?

Wenn bei uns einige Ihrer Bücher einen Erfolg gehabt haben, worauf bei Büchern, welche aus einer fremden Sprache übersetzt sind, keineswegs immer zu rechnen ist, wie denn z. B. die Schrift über Faraday und „in den Alpen" noch nicht die Kosten gedeckt haben, so ist ein Theil des Erfolges jedenfalls den berühmten Namen und der ausgezeichneten Arbeit der Übersetzer zuzuschreiben. Jedenfalls haben wir bei jedem neuen Buche ein neues Risico zu übernehmen, und zwar das ganze. Dem gegenüber müßen wir im Princip an der bei uns ganz allgemein üblichen Weise der Zahlung eines bestimmten Honorars bei jeder Publication festhalten, womit der Autor abzufinden ist, ohne daß er an dem Erfolge oder Mißerfolge weiter betheiligt ist.

Wir erlauben uns nun, Ihnen den Vorschlag zu machen, daß wir Ihnen in Zukunft für jeden Bogen einer ersten Auflage 20 Mark, für jeden Bogen einer zweiten oder späteren Auflage 10 Mark Honorar zahlen, wobei wir das Recht haben, die Auflage bis 1500 Exemplare stark zu machen; meistens wird die Anzahl der Exemplare eine geringere sein. Sollte der Falle eintreten, daß wir eine Auflage stärker machen wollten, so zahlen wir für die Exemplare, welche die Anzahl von 1500 übersteigen, das Honorar pro rata der angegebenen Sätze. Dabei haben wir natürlich die Kosten der Übersetzung allein zu tragen.

Von Ihren Schriften wurden bisher folgende Auflagen gedruckt:

Wärme,	1[e] Aufl.	1200	Exemplare
"	2. Aufl.	1200	"
"	3. Aufl	1500	"
Schall	1. Aufl.	1200	"
"	2. Aufl.	1500.	"
Faraday	1 Aufl.	1000.	"
in den Alpen	1 Aufl.	1500	"
Fragmente	1 Aufl.	1200	"
Licht	1 Aufl.	2500	"

Das letztere Buch ist gleich in stärkerer Auflage gedruckt, weil wir einen ähnlichen Verkauf wie bei „Wärme“ und „Schall“ davon hoffen u. wieder auf unserem Vorschlage zu honoriren sind; ebenso die dritte Auflage der „Wärme“, welche in diesen Tagen fertig geworden ist (1500 Ex[em]pl[are].)

Wie die neue Auflage „der Fragmente“ ist die Anzahl der neu zu druckenden Exemplare noch nicht bestimmt. Wie die erste Auflage ist noch ein nicht unbedeutender Vorrath vorhanden, theilweise noch bei auswärtigen Buchhändlern lagernd. Wir werden jetzt diese Exemplare zurückverlangen, und nachdem wir die Anzahl verificirt, diejenigen Bogen, welche in Ihrer Neubearbeitung Veränderungen enthalten, neu drucken und gegen die alten vertauschen. Je nach der Anzahl, welche wir so von der alten Auflage noch verwenden können, werden wir die Anzahl der neu herzustellenden Exemplare bemeßen u. Ihnen d. Z. Nachricht davon geben.

Mit den „Alpen“ haben wir eine buchhändlerische Manipulation vorgenommen, um dadurch den bis jetzt langsamen und ungenügenden Absatze vielleicht einen neuen Impuls zu geben. Wir haben nämlich die bis zum Mai d. J. unverkauft gebliebenen Exemplare, etwa noch 900, mit einem neuen Umschlag u. Titel versehen, und als „zweiten Abdruck“ von Neuem an unsere Geschäftsfreunde versandt. Das Buch kommt dadurch dem Publicum noch einmal vor Augen und wir können hoffen, auf diese Weise die vorhandenen noch sehr bedeutenden Vorräthe rascher zu verkaufen u. dann vielleicht zum Druck einer wirklichen neuen Auflage zu gelangen, als wenn wir dieselben in unserem Magazin behielten.

Es ist dieses eine sehr oft vorkommende Procedur und führt zuweilen, allerdings nicht immer, zu dem gewünschten Resultate. Ein Neudruck dieser Bücher hat also factisch nicht stattgefunden.

Wir glauben in Vorstehenden Ihnen, sehr geehrter Herr, die gemischten Aufschlüsse gegeben zu haben, u. sind zu weiteren Aufklärungen jeder Zeit bereit. Wir dürfen wohl einer baldigen Antwort auf unsern Vorschlage entgegen sehen.

Mit ausgezeichneter Hochachtung | ganz ergebenst, | Friedr. Vieweg & Sohn

Braunschweig 7 September 1875.

Dear Sir!

We thoroughly share your view that it is better and more business-like if the relationship into which author and publisher enter for a literary undertaking is determined by contract from the outset. That the initiative to such a determination of mutual rights and duties has until now been taken neither by you nor by us, lies mostly in the special circumstances under which we first had the honour of coming into contact with you.[1] The translators,[2] who were highly valued by us, gave the occasion for us for the first German editions of works from your pen, they were the intermediaries, and in fact we believed that we saw in them the representatives of the author for our language. It seemed to us that the relationship would be understood similarly on your part, for you gave and, without doubt precisely because of these special circumstances, the authorisation for the publication of the translations by the competent hand of your friends, without attaching a limitation with regard to the number of copies or the condition of a royalty for yourself. We thought it in order, however, to send you a royalty without such a stipulation also, which you received at the time with an expression of satisfaction, with which you saw in it "your own free act and not a consequence of any contract between you & me".[3] Later you instructed us to direct such payments on your behalf to the ladies who were involved in the translation.

If you are now expressing to us the wish for a fixed income for *[1 word illeg]* publications, then you will find us in complete agreement with it. The book trade in Germany, however, is based on such a thoroughly different system from the English one, and organised in such an entirely different way, that it is impossible for us to enter into the mode of participation in the undertaking proposed by you.[4] We cannot, as it happens in England, sell the entire edition to the middlemen immediately after completion of printing, but rather have to give the copies to resellers for sale on commission at our own risk, and only receive payment for what has been sold after one year; the unsold copies, though, are delivered back to us in kind, and thus a period of several years will often elapse before we are able to know whether we have made a profit or a loss on a book. Your relationship with Messrs Longman,[5] according to which you already share the profit according to a certain relationship upon appearance of books, presupposes a quite certain profit, and we do not wish to doubt that such a one is to be expected in England with your works every time. How would the matter turn out, though, if not as many copies had to be sold to cover production costs? Would you, in this case, also bear part of the loss?

If some of your books have had success here – which is by no means always to be expected with books which are translated from a foreign language, for e.g. the work on Faraday and "in the Alps"[6] have still not covered costs – then at any rate part of the success has to be attributed to the famous name and the excellent work of the translators. At any rate, with every new book we have to accept a new risk, and indeed the whole risk. On the other hand, we have to hold, in principle, to the quite common customary way here of paying a specific royalty with every publication, whereby the author is to be compensated without being further involved in its success or failure.

We now permit ourselves to make the proposal to you that in future we shall pay you a royalty of 20 Marks for every sheet[7] of a first edition and a royalty of 10 Marks for every sheet of a second or later edition, whereby we have the right to make the edition up to 1500 copies in size; most of the time, the number of copies will be a lesser one. Should the case arise that we wish to make the edition larger, then for the copies which exceed the number of 1500, we shall pay the royalty on a pro rata[8] basis of the stated settings. In doing so we shall naturally have to bear the costs of the translation alone.

Of your works, the following editions have been printed up to now:

Heat,[9]	1st ed.	1200	copies
"	2nd ed.	1200	"
"	3rd ed.	1500	"
Sound[10]	1st ed.	1200	"
"	2nd ed.	1500.	"
Faraday	1st ed.	1000.	"
In the Alps	1st ed.	1500	"
Fragments[11]	1st ed.	1200	"
Light[12]	1st ed.	2500	"

The latter book has just been printed in a larger edition because we are hoping for similar sales from it as with "Heat" and "Sound", and once again have to remunerate according to our proposal; likewise the third edition of "Heat", which has just been finished in the last few days (1500 cop[ies].).

Like the new edition of "Fragments"[13] the number of copies which have to be reprinted has not yet been determined. Like the first edition, a not insignificant supply still exists, partly still in storage with foreign booksellers. We are now going to ask for these copies back, and after we have verified the number, reprint those sheets which contain emendations in your revision and exchange them against the old ones. Depending on the number which we can thus still use from the old edition, we are going to calculate the number of copies to be produced again and shall inform you about this at the time.

With the "Alps" we have carried out a trick of the book trade in order to perhaps thereby give the hitherto slow and unsatisfactory sales a new impulse. We have, you see, provided the copies which remained unsold up to May of this year, some 900 still, with a new cover and title page and sent them as a "second printing" to our business friends again. The book will thereby be seen by the public once again and we can hope to sell the very significant supplies that still exist quicker in this way and then perhaps go to printing a really new edition than if we were to keep them in our warehouse.

This is a procedure which happens very often, and leads occasionally, though not always, to the desired result. A reprint of these books has thus, in a factual sense, not taken place.

We believe we have given you, dear sir, all information, both good and bad, in the above, and are happy to provide further explanation at any time. We think we can look forward to a prompt reply to our proposal.

With excellent esteem | your most humble servants, | Friedr. Vieweg & Son

TU Braunschweig/Universitätsbibliothek/UABS V 1 T: 63

---

1. *the special circumstances . . . into contact with you*: in 1867 Hermann Helmholtz requested that Vieweg publish a translation of *Heat as a Mode of Motion*, having urged his wife to undertake the role of translator (see D. Cahan, *Helmholtz: A Life in Science* (Chicago: University of Chicago Press, 2018), p. 311).
2. *The translators*: Anna Helmholtz and Clara Wiedemann.
3. *an expression of satisfaction . . . you & me"*: letter missing; Tyndall did not use the words quoted in any of his extant correspondence with Vieweg.
4. *the undertaking proposed by you*: in letter 4629.
5. *Messrs Longman*: Thomas and William Longman, proprietors of the publishers Longmans, Green, & Co.
6. *the work on Faraday and "in the Alps"*: J. Tyndall, *Faraday und seine Entdeckungen*, ed. H. Helmholtz (Braunschweig: Friedrich Vieweg, 1870); J. Tyndall, *In den Alpen* (Braunschweig: Friedrich Vieweg, 1872). These were, respectively translations of *Faraday as a Discoverer* and *Hours of Exercise in the Alps.*
7. *sheet*: the original sheet of paper that was printed and then folded the number of times required for the format of the book. As Tyndall's books, and their translations, were generally published in octavo editions, this would equate to eight pages.

8. *pro rata*: proportional (Latin). A pro rata basis is a calculation made according to a rate already determined for a larger total amount.
9. *Heat*: J. Tyndall, *Die Warme Betrachtet als eine Art der Bewegung*, ed. H. Helmholtz and G. Wiedemann (Braunschweig: Friedrich Vieweg, 1867). The second edition was published in 1871, and the third in 1875.
10. *Sound*: J. Tyndall, *Der Schall: Acht Vorlesungen gehalten in der Royal Institution von Grossbritannien*, ed. H. Helmholtz and G. Wiedemann (Braunschweig: Friedrich Vieweg, 1869). The second edition was published in 1874.
11. *Fragments*: J. Tyndall, *Fragmente aus den Naturwissenschaften*, trans. A. Helmholtz., ed. H. Helmholtz (Braunschweig: Friedrich Vieweg, 1874).
12. *Light*: J. Tyndall, *Das Licht: Sechs Vorlesungen gehalten in Amerika im Winter 1872–1873*, ed. G. Wiedemann (Braunschweig: Friedrich Vieweg, 1876).
13. *the new edition of "Fragments"*: Helmholtz suggested that Vieweg was waiting for the new English edition, *Fragments of Science: A Series of Detached Essays, Lectures, and Reviews*, 5th edn (London: Longmans, Green, and Co., 1876), before committing to a new translated edition (see letter 4606).

From Jane Barnard 11 September 1875 4640

Barnsbury Villa, | 320, Liverpool Road, | N.[1]

My dear Dr Tyndall

Had I answered your kind offer of a visit[2] yesterday, I should have said—"We shall be most happy"—but during the night my dear Aunt[3] had one of her attacks from the digestion failing[4] & tho' the attack has not been so severe as some of the previous ones, she is quite prostrated. I hope early next week to ask you to offer us another day. I am very glad you were not disturbed yesterday for me.

Ever very sincerely yours | Jane Barnard | 11 Sept[ember] 1875 | Dr Tyndall | &c &c &c

RI MS JT/1/B/57

---

1. *Barnsbury Villa, | 320, Liverpool Road, | N.*: see letter 4226, n. 1.
2. *your kind offer of a visit*: letter missing.
3. *my dear Aunt*: Sarah Faraday.
4. *one of her attacks from the digestion failing*: presumably intestinal failure; on Faraday's previous attacks, see letters 4418 and 4537.

From George Forbes 13 September 1875 4641

Charing X Hotel,[1] London | 1875 Sept[ember] 13

Sir

Last June I left England for the Sandwich Islands, to observe the Transit of Venus there.[2] I returned only the day before yesterday.

I learn that during my absence you have again reopened the glacier controversy.[3] I have a few weeks to spare before my regular work begins & if you will oblige me so far as to send me a copy of your publication[4] to

Dysart Cottage | Pitlochrie N.B.[5]

I will give it my careful consideration,

I am, Sir, | Y[ou]rs faithfully | Georges Forbes

RI MS JT/1/F/26
RI MS JT/1/TYP/12/3944

---

1. *Charing X Hotel*: Charing Cross Hotel, which was built above the railway station and had opened in 1865.
2. *the Sandwich Islands . . . Transit of Venus there*: Forbes travelled to Hawai'i with Henry Glanville Barnacle (1849–1938) as part of an international expedition to make observations of the December 1874 Transit of Venus, having already published an account of the preparations in *The Transit of Venus* (London: Macmillan, 1874). See also J. Ratcliff, *The Transit of Venus Enterprise in Victorian Britain* (London: Pickering and Chatto, 2008).
3. *reopened the glacier controversy*: in the late 1850s Tyndall had been embroiled in an acrimonious dispute with Forbes's father, James David Forbes, regarding the priority of his work on the motion of glaciers (see the Introduction to *Tyndall Correspondence*, vol. 6). This had been reignited by Tyndall in *Forms of Water*, although for Tyndall and his allies the principal cause was the publication of the *Life of Forbes*. The renewed controversy was then further stoked by a new edition of *Glaciers of Savoy*, which Forbes had published before his departure for Hawai'i. As such, Tyndall would have denied the charge that it was he who had reopened the controversy, and saw himself as responding to provocations from Forbes and his supporters.
4. *a copy of your publication*: J. Tyndall, 'Rendu and His Editors', *Contemporary Review*, 24 (1874), pp. 135–48.
5. *N.B.*: North Britain, a term for Scotland used by supporters of the union with England.

From Paolo Volpicèlli[1] 13 September 1875 4642

LYNCAEI

Cher Monsieur | Rome 13 septembre 1875 | 4 Piazza fiammetta

J'ai l'honneur de vous présenter mon ami le comte Dandini de Sylva qui est professeur du brevet d'invention pour un réflecteur approuvé par une commision nommée par notre gouvernement, dont j'en faisais partie.

J'espère que vous serez si aimable de lui être favorable, et que vous voudrez protéger mon ami.

Agréez Monsieur l'assurance de ma sincère reconnaissance, et des mes hommages distingués.

Votre tout dévoué | Paul Volpicelli

LYNCAEI[2]

Dear Sir | Rome 13 September 1875 | 4 Piazza Fiammetta

I have the honour of introducing to you my friend Count Dandini de Sylva,[3] a professor with a patent for a reflector[4] that was approved by a commission appointed by our government,[5] and in which I took part.

I hope that you will be so kind as to look favourably on him, and that you will wish to become my friend's patron.

Please accept, Sir, the assurance of my sincere gratitude and of my highest respects.

Your most devoted | Paul Volpicelli

RI MS JT/1/V/29

---

1. *Paolo Volpicèlli*: Paolo Volpicèlli (1804–79), an Italian mathematician and physicist who was professor of mathematical physics at the University of Rome.
2. *LYNCAEI*: the Accademia Nazionale Reale dei Lincei (Royal National Lincean Academy). It was founded in 1874 to encourage secular science in the newly-unified Italy, although it adopted the name (literal translation: lynx-eyed) of a papal academy originally founded in 1603 with Galileo Galilei as a member.
3. *my friend Count Dandini de Sylva*: Saverio Dandini de Sylva (1835–1923), an Italian aristocrat who lived in Rome and was a member of the papal Guardia Nobile (Noble Guard).
4. *a patent for a reflector*: Dandini de Sylva was not the inventor of this catoptric device to enhance the propagation of light, and was involved with the patent only as a guarantor for its inventor Pasquale Balestrieri. When the patent was issued in the United States,

Dandini de Sylva was listed (albeit with his name misspelled as its feminine form) as one of Balestrieri's witnesses, along with Giuseppe Clementi. See '162, 263: Apparatus for Collecting the Rays of Light and Heat', *Specifications and Drawings of Patents Issued from the U.S. Patent Office* (Washington, DC: Government Printing Office, 1875), pp. 625–7, on p. 627. The reflector is described by Balestrieri in letter 4669.

5. *a commission appointed by our government*: this was a commission established by the Italian government to evaluate Balestrieri's reflector. In December 1875 it was reported in Italy that 'L'onorevole Ministro dei lavori pubblici, assai provvidamente nominò una commissione, composta dei professori BLASERNA (presidente), KESPIGHI, VOLPICELLI, oltre agli ingegneri signori CORNAGLU e MILESI, con incarico di far conoscere al Ministero stesso, il merito scientifico e pratico dell'indicato istrumento [The honorable Minister of Public Works, very providently appointed a commission, composed of professors BLASERNA (president), KESPIGHI, VOLPICELLI, as well as the engineers CORNAGLU and MILESI, with the task of making known to the Ministry itself, the scientific and practical merits of the indicated instrument]' ('Classe di scienze fisiche, matematiche e naturali', *Atti della Reale Accademia dei Lincei*, 2nd series 3 (1875–6), pp. 14–19, on p. 18–19).

From John Wigham 13 September 1875 4643

33 to 36. Capel Street | Dublin 13th of Sept[embe]r. 1875.

John Tyndall Esq. L.L.D. F.R.S. &.c. | Royal Institution | London.

My dear Sir,

Your favor of the 11th.[1] only reached here this morning. In reply I enclose a "Freeman" containing report of Dr. Moriarty's sermon.[2] I do not find that it has been printed in any other form but if it has & you let me know I will have much pleasure in searching for it.

I only received your very kind note of the 11th of June[3] written on the eve of your departure for Switzerland. I immediately took steps in the important matter to which it refers[4] & with the consent of Capt Morant[5] the new Inspector of Irish Lights I have had the Triform Light[6] flashed upon Kingston Harbour from Howthe Bailey[7] for 15 minutes nearly every night since the receipt of your letter. Special persons have been appointed to look out and report. I will send you their reports when I can get them from the office.

There have not been many occasions on which the weather was so thick during the exhibition of the lights as to show very strikingly the usefulness of the great flash in catching the eye but I am aware of at least 2 very remarkable instances in which the fog was so dense as to obscure, to some observers entirely, and to others nearly so, the ordinary Bailey Lights, and yet the Triform flash shone out with perfect distinction. It was not only quite visible but, if I may use the expression, compelled itself to be seen, lighting up the haze in the striking manner noticed on previous occasions. If I could have the flashing light exhibited for the whole night, instead of only for 15 minutes, and directed over a greater arc, we would obtain very conclusive reports in the matter; but the resources of the Commissioners of Irish Lights at Howth Bailey are limited & I know that they feel a difficulty in asking the Board of Trade to sanction even a trifling expenditure, pending the settlement of the general gas question – I thought your most exhaustive report about Haisbro[8] would have brought this about before now, but the Board of Trade have not yet agreed to the amount of royalty or compensation proposed by the Commissioners of Irish Lights to be paid to me—Now the fact is I care very little about the amount of that Compensation & would be most anxious to meet the news of the Board of Trade. I have offered to the Trinity House to make no charge whatever for royalty or compensation in respect of the two Haisbro Lighthouses[9] in order that they may have full lease to carry out their desire of applying gas to both of them. The difficulty lies in the correspondence which has to go through the Commissioner of Irish Lights to the Board of Trade & then I suppose to the Trinity House. If I could only be brought into personal contact with some person authorised to deal with the matter I think I could settle it in 5 minutes. Perhaps you may have an opportunity of speaking to Mr Farrer[10] in this subject viz you saw no objections & would advise the Board of Trade to have an interview with me, instead of prolonging the correspondence you would do me a great favor & most probably bring the matter to a speedy termination

I am | yours very truly | John R Wigham

National Archives, MT 10/211. Illumination of lighthouses by gas. H6704

---

1. *Your favor of the 11th.*: letter missing.
2. *a "Freeman" containing report of Dr. Moriarty's sermon*: 'The Maynooth Synod', *Freeman's Journal*, 11 September 1875, p. 7. In his sermon at the Maynooth Synod, David Moriarty (1814–77), the Bishop of Kerry (*ODNB*), compared the struggle of Roman Catholic clerics with the civil government in Ireland with that of early Christians with the Roman state. The report of it in *Freeman's Journal* derided the sermon as 'nothing but a monotonous repetition of the sweeping assumptions which constitute the commonplaces of Roman Catholic exhortation and denunciation' (p. 7).
3. *your very kind note of the 11th of June*: letter missing.
4. *the important matter to which it refers*: presumably the continuing controversy over whether to install gas lamps in lighthouses.
5. *Capt Morant*: George Digby Morant (1837–1921), an officer in the British Navy who was promoted to Captain in 1873 and appointed to the post of Inspector of Irish Lights in 1875.
6. *the Triform Light*: a gas-powered light devised by Wigham with three lenses in tiers that produced a flash appearing in three different forms.
7. *Kingston Harbour from Howthe Bailey*: Kingston (now called Dún Laoghaire) was on the Irish coast near Dublin, and Howth Baily was the Baily Lighthouse on the Howth Head

in Dublin Bay. Tyndall conducted experiments here comparing oil and gas burners in June 1869, and the lighthouse was the first site where Wigham's burner was installed.
8. *your most exhaustive report about Haisbro*: letter 4438.
9. *the two Haisbro Lighthouses*: see 4427, n. 4.
10. *an opportunity of speaking to Mr Farrer*: Tyndall wrote to Thomas Henry Farrer nine days later (see letter 4646).

To Heinrich Rudolf Vieweg 14 September 1875 4644

Royal Institution of Great Britain | 14th Sept[ember]. 1875

Gentlemen.[1]

I am very much obliged to you for the letter with which you have favoured me.[2]

It only needs supplementing in one direction. The Messrs. Longman[3] always furnish me with a statement of the cost and profit of my books; and I therefore see immediately both their proportion of the profit and mine. I wish for nothing more than a fair and reasonable division of the profits.

Now your letter does not furnish me with any information of this kind. You leave me in entire ignorance of the value of intellectual property which it has cost me great labour to produce.

I have to request therefore that you will be good enough to supply me with information similar to that supplied to me by the Messrs. Longman.

It would be with extreme regret and reluctance that I should change my German publisher, for I have every reason to be satisfied with the highly creditable form in which you present my books to the public. I therefore sincerely hope that no difference leading to so serious an issue will occur between us.

I am Gentlemen | Your obedient servant | John Tyndall

Messrs. Vieweg & Son

TU Braunschweig/Universitätsbibliothek/UABS V 1 T : 63

1. *Gentlemen*: see letter 4592, n. 1.
2. *the letter with which you have favoured me*: letter 4639.
3. *The Messrs. Longman*: Thomas and William Longman, proprietors of the publishers Longmans, Green, & Co.

To William Francis[1] 15 September 1875 4645

DEAR FRANCIS

Would you kindly permit me, in your next Number,[2] to express my indebtedness to Mr. Sedley Taylor,[3] Professor Mayer of Hoboken, and Mr.

Bosanquet[4] for pointing out an error in the statement of Helmholtz's theory of Consonance in my eighth Lecture upon Sound?[5]

It would be easy, if it were of any use, to show the origin of this mistake. Suffice it to say that it has been long known to me, that it has been corrected in the last edition of my work on Sound,[6] and that the corrected statement of the theory,[7] though necessarily brief, is, I have reason to know, regarded by Helmholtz as "perfectly clear and exact."[8]

With regard to the experimental data referred to in my eighth Lecture, I may have something to say on a future occasion.

I am, dear Francis,

Faithfully yours, | JOHN TYNDALL. Royal Institution, | September 15, 1875.

Phil. Mag., 50 n.s. (1875), p. 336

---

1. *William Francis*: William Francis (1817–1904), a scientific publisher who, with Richard Taylor (his biological father), founded Taylor and Francis & Co. in 1852 (see letter 4368, n. 5). He co-edited the *Phil. Mag.* from 1851 until his death.
2. *your next Number*: no. 331 of the *Phil. Mag.*, published in October 1875.
3. *Mr. Sedley Taylor*: see letter 4600, n. 1.
4. *Mr. Bosanquet*: Robert Holford Macdowall Bosanquet (1841–1912), a musicologist and tutor at the Natural Science School, Oxford, who discussed Taylor's and Alfred Mayer's criticisms of Tyndall's interpretation of Hermann Helmholtz's theory of musical consonance in 'On Temperament, or the Division of the Octave', *Phil. Mag.*, 50 n.s. (1875), pp. 164–78, on p. 169.
5. *an error in . . . eighth Lecture upon Sound*: see letter 4439, n. 2.
6. *the last edition of my work on Sound*: *Sound* (1875).
7. *the corrected statement of the theory*: on p. 359.
8. *regarded by Helmholtz as "perfectly clear and exact"*: see letter 4606.

To Thomas Henry Farrer 22 September [1875][1] 4646

Royal Institution of Great Britain | 22nd Sept[ember].

My dear Farrer,

Before going to Switzerland this year I wrote to M^r^. Wigham,[2] urging upon him the importance of making experiments on the glare of his gaslights, when the light itself is invisible through fog.

He has sent me some data, which are referred to in the enclosed two communications.[3] But they merely whet the desire to see the experiments completed.

The time is now approaching when such observation can be made. I would suggest that the attention of the Board of Irish Lights be directed to the subject.

So important do I deem this point that I should be almost inclined to sacrifice a fortnight in Ireland to its examination and decision.

But this I hope will not be necessary, as the observations require only ordinary powers of observation, and truthfulness of record.

Ever yours | John Tyndall

It would be a great point gained if it were proved that fog flashes could be intercalated with our Sound Signals.

How utterly helpless our navy appears to be in the matter of Sound Signals![4] This ought not to be.

National Archives, MT 10/211. Lighthouses, Illumination by means of gas. H6704

1. *[1875]*: the year is established by an ink stamp added to the letter stating: 'BOARD OF TRADE | SEP 22 1875 | HARBOUR DEPARTMENT'.
2. *I wrote to Mr. Wigham*: letter missing, but see letter 4643.
3. *the enclosed two communications*: presumably one is letter 4643.
4. *How utterly helpless . . . matter of Sound Signals!*: Tyndall objected to the Royal Navy's continued use of horns and whistles and considered steam sirens a much more effective alternative (see *Ascent of John Tyndall*, p. 345).

To Hector Tyndale 22 September 1875 4647

22nd Sept[ember]. 1875.

My dear Hector

Last night before I went to bed your letter[1] reached me. It was pleasant to take the thought of it with me to bed. The fund[2] is becoming a fine child. And the object aimed at, namely that of maintaining two students "perpetually" is already practically realised. We only want the men.

That was a capital investment. I wish I could find as good a one for the little sums that bring me in my driblets of dividends in England.

Tom and Debus have just returned from their brief vacation.[3] They will be rejoiced to hear from you. I say it not conventionally—Both men entertain a profound esteem for you.

I had heard, and was grieved to hear of your illness, from Gorey.[4] Your loving wife[5] must have had a hard and trying time of it. But there are some natures lifted above the consciousness of their own sufferings, when administering to the sufferings of others, and such a nature I always took Julia's to be. Well I hope you will soon be restored to her in health and strength.

I was very sorry to criticise Henry: but he gratuitously provoked it. I was quite taken aback by his remarks.[6] He is about to reply by publishing his

Report,[7] perhaps with additions. But he is wrong—and the worst of it is that error of this kind spreads. Prof. Mayer of Hoboken I am told declares Henry to be right.—Well one must wait, and trust to the force of truth when one has once clearly stated one's case.

I will turn your invitation[8] over in my mind, and communicate it to Hirst and Debus. It would certainly be something to remember.—And still dear Hector—These crowds of men do not edify me so much as my quiet communings with Nature. I had a good time in Switzerland this year, notwithstanding the fact that the weather during the greater part of it was atrocious. I met M[r] Carl Schurz[9] upon the glaciers, and was glad to hear him speak of you as his 'friend'

I am just getting a new edition of the "Fragments"[10] off my hands, and am also deeply entangled in other work.[11] London is now favourable to the worker. If he can only resist invitations on Sundays he can throw himself without disturbance into his work. I have written an introduction to the Fragments[12] which I hope and think you will like. It deals with a few objectors; but more especially with the Rev[eren][d] James Martineau.[13] I hope M[r]. Furness will not be angry with me—but I really have dealt very courteously though very firmly with my opponent. The Introduction will be first published as an article in the first number of the "Fortnightly Review".[14]

Goodbye dear Hector, give Julia my affectionate regards

& believe me always | Your affe[ctionate] Cousin | John Tyndall

RI MS JT/1/T/1458
RI MS JT/1/TYP/4/1740–1

---

1. *your letter*: letter missing.
2. *the fund*: the 'Tyndall Trust for Original Investigation' (see letter 4214, n. 5).
3. *Tom and Debus . . . their brief vacation*: although Thomas Hirst spent much of August and September on the Continent and then stayed on the Sussex coast after his return, there are no entries in his journal from 14 September to 25 December, so it is unclear where he and Heinrich Debus went for their brief vacation.
4. *from Gorey*: presumably either from Tyndall's sister Emily, who was staying in the town in County Wexford, Ireland, with the family of her late cousin John Tyndall, or his widow Catherine Tyndall.
5. *Your loving wife*: Julia Tyndale.
6. *I was very sorry . . . aback by his remarks*: Joseph Henry's remarks related to the experiments on fog signaling which he and Tyndall had both been conducting (see letter 4626).
7. *his Report*: J. Henry. 'Appendix: Report of the Operations of the Light-House Board Relative to Fog-Signals', *Annual Report of the Light-House Board of the United States* (Washington, DC: Government Printing Office, 1874), pp. 687–721.
8. *your invitation*: not identified, but possibly to visit the United States once more (see letter 4201, n. 4).

9. *Mr Carl Schurz*: see letter 4227, n. 8. He met Tyndall while spending several months in Europe following the loss of his senatorial seat, for Missouri, in the elections of November 1874.
10. *a new edition of the "Fragments"*: see letter 4592, n. 2.
11. *other work*: see letter 4632, n. 2.
12. *an introduction to the Fragments*: 'Introduction' to Part II, pp. 325–56.
13. *more especially with the Rev[eren]d James Martineau*: pp. 331–48.
14. *an article in the first number of the "Fortnightly Review"*: J. Tyndall, '"Materialism" and Its Opponents', *Fortnightly Review*, 24 (1875), pp. 579–99.

From Sarah Faraday 24 September 1875 4648

Barnsbury Villa, | 320, Liverpool Road N.[1] | 24th. Sep[tember]. 1875.

Dear Dr Tyndall

I am longing to see you once more and am now pretty well restored to my usual state[2]—can you spare an hour next Monday Tuesday or Thursday[3] I make no apologies for I know to whom I am writing

Affectionately yours | S. Faraday | J. Tyndall Esquire.

The earliest convenient day for you the most suitable for us.

RI MS JT/1/TYP/12/4187
LT Typescript Only

1. *Barnsbury Villa, | 320, Liverpool Road N.*: see letter 4226, n. 1.
2. *restored to my usual state*: following an attack related to intestinal failure (see letter 4640).
3. *next Monday Tuesday or Thursday*: 27, 28, and 30 September.

To Thomas Henry Farrer 25 September [1875][1] 4649

Royal Institution of Great Britain | 25th Sept[ember].

My dear Farrer.

I have a case of Brain-waves for Hutton.[2] Last night in a dream I stood in a hall, where I know not. Mrs. Grote came in, and placing her arm round the waist of a lithe young maiden waltzed with her round the hall. And this morning your letter[3] comes speaking of Mrs. Grote—assuredly a case of "brain-wave" communication.[4]

I promised faithfully that if I went any where at the time you mention it would be down to Gassiot, St. John's House Ryde.[5] But I am utterly entangled and cannot move from London. It would certainly be a delight to me to join you, and is a great cause of regret to me that I cannot.

Matters I think might certainly be so arranged that a fortnights residence in Ireland at the most would be sure to yield opportunity for the desired experiments.[6] They are of the first importance—Could not an Elder Brother[7] be persuaded to go?

If Tom[8] is at home[9]—but he is sure not to be[10]—remember me lovingly to him—also (I had almost put my foot in it) respectfully to Miss Farrer.[11]*

Ever yours | John Tyndall

*Perhaps it would have been forgiven in an Elderly man.

Trinity College Cambridge, Cullum. P6/1

1. *[1875]*: the year is established by Farrer's note on the letter (see n. 4).
2. *a case of Brain-waves for Hutton*: Tyndall possibly mistakes Richard Holt Hutton (1826–1897), the editor of the *Spectator*, for James Thomas Knowles, editor of the *Contemporary Review*, who had written the article 'Brain-Waves: A Theory', *Spectator*, 42 (1869), pp. 135–7. Knowles proposed that thoughts and emotions induced movements in the particles of the brain, which then caused vibrations in the ether.
3. *your letter*: letter missing.
4. *a case of "brain-wave" communication*: Farrer later added a handwritten note to the letter stating: 'What is more. Mrs Grote did actually dance the pas de Zephyre in my hall this week in order to teach it to my daughter. This at 86 is remarkable THF 10 Oct /75'.
5. *Gassiot, St. John's House Ryde*: the home of the physicist John Peter Gassiot on the Isle of Wight (see letter 4604, n. 3).
6. *the desired experiments*: the experiments with gas lamps for lighthouses (see letter 4646).
7. *an Elder Brother*: of the Elder Brethren, the governing body of Trinity House (see letter 4187, n. 2).
8. *Tom*: Thomas Cecil Farrer (1859–1940).
9. *home*: Abinger Hall, near Dorking in Surrey.
10. *he is sure not to be*: Farrer's eldest son was probably away at Eton College, which he attended until 1878.
11. *Miss Farrer*: Emma Cecilia Farrer (1854–1946).

To Louisa Hamilton 25 September 1875 4650

Royal Institution of Great Britain | 25th Sept[ember]. 1875

My dear Miss Hamilton.

Your letter[1] brings vividly to my mind my first visit to the Wartburg.[2] I felt my heart tingling with the associations of the place, as you now feel yours.

When you return put me in mind of sending you a scrap of doggerel that I wrote there.[3] It will at all events show you how similar our thoughts & reflections have been.

It gave me great pleasure to hear from you, and although you will probably be on your way home when this note reaches Thüringen[4] I could [not] resist the desire to tell you the pleasure your letter[5] gave me.

It is Saturday and we near post time. If I do not catch the post today, I am put off till Monday—So I shorten my note in order to give it velocity. Kindest regards to Douglas.[6]

Ever yours | John Tyndall

It would have been a delight to be with you in Marburg.[7] But you chose a good spot. Often in autumn with the yellow leaves flying like living things through the air around me, have I found myself upon the kirchspitze.[8]

I have very good accounts from Mother.[9] Every precaution appears to be taken.

You have given me a lecture to dwell upon in that beautiful Young widow in her schloss[10] in the Thüringer Wald.[11]

RI MS JT/1/T/479

---

1. *Your letter*: letter missing.
2. *my first visit to the Wartburg*: Tyndall had visited this castle, situated on a precipice above the town of Eisenach in central Germany, on 28 May 1849 and wrote a lengthy description of the visit in his journal (RI MS JT/2/13b/433–5). See also letters 0376 and 0377, in M. Baldwin and J. Browne (eds), *The Correspondence of John Tyndall*, vol. 2, *The Correspondence, September 1843–December 1849* (Pittsburgh: University of Pittsburgh Press, 2016); and a letter to William Wright dated 27 August 1850 (courtesy Aaron Sampson), which will appear in the twentieth volume of *The Correspondence of John Tyndall.*
3. *a scrap of doggerel that I wrote there*: Tyndall had written a poem in homage to Martin Luther in the room in the castle in which he had translated the New Testament into German and thrown an inkwell to banish a vision of the devil. It was published as 'A Whitsuntide Ramble', *Preston Chronicle*, 16 June 1849, p. [3], although the version Tyndall proposed sending was probably one of two handwritten copies he made in June 1871 (RI MS JT/8/2/1/55–71). See R. Jackson, N. Jackson, and D. Brown (eds), *The Poetry of John Tyndall* (London: UCL Press, 2020), pp. 177–8. This letter to the *Preston Chronicle* will appear in the twentieth volume of *The Correspondence of John Tyndall.*
4. *Thüringen*: a state in central Germany.
5. *your letter*: letter missing.
6. *Douglas*: Douglas James Hamilton, Louisa's younger brother.
7. *Marburg*: see letter 4216, n. 6.
8. *the kirchspitze*: the summit of a hill with panoramic views overlooking Marburg.
9. *Mother*: Elizabeth Hamilton.

10. *schloss*: castle (German).
11. *a lecture to dwell . . . the Thüringer Wald*: not identified.

To Thomas Archer Hirst 25 September [1875][1] 4651

25th Sept[ember].

Dear Tom.

Be pleased to keep yourself free for Tuesday[2] at seven oClock.

I hope it will be at the Athenaeum,[3] as I have asked Webster[4] to procure an invitation for M. Pitanga.[5]

Yours affec[tionatel]y | John. | Love to Lilly[6]

RI MS JT/1/T/925
RI MS JT/1/HTYP/633

1. *[1875]*: the year is proposed by LT in a handwritten annotation on the MS letter, although there are no entries for this period in Hirst's journal, so it cannot be confirmed.
2. *Tuesday*: 28 September.
3. *the Athenaeum*: the Athenaeum Club (see letter 4186, n. 1).
4. *Webster*: James Claude Webster (1830–1908), the secretary of the Athenaeum Club for 34 years (F. R. Cowell, *The Athenaeum Club and Social Life in London, 1824–1974* (London: Heinemann, 1875), p. 108; M. Wheeler, *The Athenaeum: More Than Just Another London Club* (New Haven: Yale University Press, 2020), p. 115).
5. *M. Pitanga*: probably Epifânio Cândido de Sousa Pitanga (1828–94), a Brazilian engineer who was a member of the Paris Physics Society. See A. G. Hanley, 'Men of Science and Standards: Introducing the Metric System in Nineteenth-Century Brazil', *Business History Review*, 96 (2022), pp. 17–45, on p. 39.
6. *Lilly*: Emily Hirst.

From Sarah Faraday 25 September 1875 4652

Barnsbury Villa, | 320, Liverpool Road, N.[1] | 25th. Sep[tember]. 1875.

No my dear friend! we cannot agree, you must have your chop with us so we shall look for you next Monday[2]

Ever yours | S. Faraday

RI MS JT/1/TYP/12/4187
LT Typescript Only

1. *Barnsbury Villa, | 320, Liverpool Road, N.*: see letter 4226, n. 1.
2. *next Monday*: 27 September.

To Thomas Archer Hirst [28 September 1875][1] 4653

Athenaeum Club, Pall Mall | Tuesday

My dear Tom.

We dine here[2] at seven.

Yours aff[ectionatel]y. | John

RI MS JT/1/T/958
RI MS JT/1/HTYP/633

1. *[28 September 1875]*: the date is suggested by the relation to letter 4651.
2. *here*: the Athenaeum Club (see letter 4651). On the Athenaeum Club, see letter 4186, n. 1.

To Heinrich Rudolf Vieweg 1 October 1875 4654

Royal Institution of Great Britain | Oct[ober] 1st. 1875

Gentlemen[1]

For your private information I beg to send you the original draft of an agreement between Messrs. Longman[2] & myself. Also the statement of expenses & division of profits which accompanied it.[3] Be good enough to send both back to me.

faithfully Yours | John Tyndall

TU Braunschweig/Universitätsbibliothek/UABS V1T:63

1. *Gentlemen*: see letter 4592, n. 1.
2. *Messrs. Longman*: Thomas and William Longman, proprietors of the publishers Longmans, Green, & Co.
3. *the original draft . . . which accompanied it*: enclosures missing. In letter 4644 Tyndall requested that his German publisher send him a 'a statement of the cost and profit of my books' similar to that provided to him by Longmans, Green, & Co., and so was presumably sending the statement to show Vieweg what he wanted.

To Thomas Henry Huxley 6 October [1875][1] 4655

Royal Institution of Great Britain | 6th. Oct[ober].

My dear Huxley

On Tuesday next[2] I hope to be able to show you something that will interest you.

Among other things I hope to reduce to demonstration[3] the practical correctness of your statement that bacteria germs exist "in myriads" in the atmosphere.[4]

Give me a line to day to say that you can join me at luncheon at one o clock

Yours ever | John Tyndall

I hope Sanderson will also be able to come.

IC HP 8.175
RI MS JT/1/TYP/9/3047
Typed Transcript Only

---

1. *[1875]*: the year is established by reference to Tyndall's research on germs in the air (see n. 3).
2. *Tuesday next*: 12 October.
3. *I hope to reduce to demonstration*: on 10 September Tyndall began a series of experiments using optically pure air to try to determine whether germs could be found in the air.
4. *your statement . . . "in myriads" in the atmosphere*: in his presidential address to the BAAS in 1870, Huxley contended of germs that it was 'certain that these living particles are so minute that the assumption of their suspension in ordinary air presents not the slightest difficulty. On the contrary, considering their lightness and the wide diffusion of the organisms which produce them, it is impossible to conceive that they should not be suspended in the atmosphere in myriads' ('Address of Thomas Henry Huxley', *Brit. Assoc. Rep.* 1870, pp. lxxiii–lxxxix, on p. lxxxii; or 'Address of Thomas Henry Huxley, LL.D., F.R.S., President', *Nature*, 2 (1870), pp. 400–6, on p. 403).

From Emma Darwin[1] 7 October [1875][2] 4656

Down, | Beckenham, Kent | Thursday Oct[ober] 7

Dear Professor Tyndall

We are expecting Professor & Mrs. Huxley[3] on Sat[urday]. the 16th to stay over the Sunday with us, & Mr Darwin & I would be so glad if we could persuade you to come for the same time.[4] I am in hopes we may still have a remnant of summer to receive you.

Believe me | very truly yours | Emma Darwin.

CUL GBR/0012/MS DAR 261.8:34[5]
RI MS JT/1/TYP/9/2804

---

1. *Emma Darwin*: Emma Darwin (née Wedgwood, 1808–96), wife of Charles Darwin.
2. *[1875]*: the year is established by the relation to letter 4675.
3. *Mrs Huxley*: Henrietta Huxley.
4. *persuade you to come for the same time*: the Darwin Correspondence Project notes that Tyndall did indeed stay with Darwin at Down House on the weekend of 16 and 17 October 1875, along with Thomas Huxley and his wife Henrietta.
5. *CUL GBR/0012/MS DAR 261.8:34*: published in *Darwin Correspondence*, vol. 23, pp. 385–6; this is letter 10185F in the online edition of the Darwin correspondence.

To Olga Novikoff 9 October 1875 4657

Royal Institution of Great Britain | 9th Oct[ober]. 1875.

My dear Friend

I was with the Trinity Brethren at Blackwall[1] when you called yesterday:[2] I must be with them again tonight. But I will come down to see you[3] tomorrow night—or if I knew the hour of your afternoon walk[4] I would join you, & shield you from cabs, omnibuses[5] and bores.

Yours ever | John Tyndall

Wellcome MS. 7777/18

---

1. *Blackwall*: the location of Trinity House's experimental lighthouse on the Thames in East London. Tyndall was there with the Elder Brethren, the governing body of Trinity House (see letter 4187, n. 2).
2. *you called yesterday*: presumably at the RI; Novikoff, who was in Germany in July (see letters 4621 and 4625), had now come to London for her annual visit.
3. *come down to see you*: at Symonds' Hotel, Brook Street, Mayfair, where Novikoff stayed during her visits to London.
4. *your afternoon walk*: probably in Regent's Park; Novikoff enjoyed taking a 'long healthy stroll in one of the beautiful parks of which every Londoner may be proud', and had a particular 'liking for Regent's Park' (*M.P. for Russia*, pp. 150 and 120).
5. *omnibuses*: large public horse-drawn vehicles carrying passengers by road, running on a fixed route and typically requiring the payment of a fare (*OED*).

From Charles Cecil Trevor 9 October [1875][1] 4658

John Tyndall Esq[ui]r[e]. LLD. FRS. | Royal Institution | Albemarle Street

Sir:—

Referring to your recent communication to Mr Farrer,[2] suggesting the importance of having careful observations taken of the glare of the Triform Gas Light[3] when that light is itself is invisible through fog, I am etc. to transmit herewith copy of a communic[atio]n which on the 29th ulto.[4] the B[oar]d caused to be addressed to the C[ommissione]rs. of Irish Lights on the subject,[5] as also of the Comm[issione]rs.' reply of the 2nd inst.,[6] and I am to request that you will favour the Board of Trade by furnishing the C[ommissione]rs. of Irish Lights with the information they require.[7]

I am | Signed by Mr. Trevor | 9 Oct[ober].

Mr Hamilton[8] | Adm[iral] Bedford[9] | Mr Jennings[10]

National Archives, MT 10/211 H6835 2399–400

---

1. *[1875]*: the year is established by the relation to letter 4646.
2. *your recent communication to Mr Farrer*: letter 4646.
3. *the Triform Gas Light*: a gas-powered light devised by John Wigham with three lenses in tiers that produced a flash appearing in three different forms.
4. *ulto.*: standard abbreviation of ultimo mense, last month (Latin).
5. *a communic[atio]n which . . . on the subject*: enclosure missing.
6. *the Comm[issione]rs.' reply of the 2nd inst.*: enclosure missing; inst. is a standard abbreviation of instante mense, this month (Latin).
7. *the information they require*: not identified.
8. *Mr Hamilton*: Robert George Crookshank Hamilton (1836–96), assistant secretary at the Board of Trade.
9. *Adm[iral] Bedford*: George Augustus Bedford (1809–79), a Vice-Admiral in the Royal Navy who served on the Board of Trade's harbour department.
10. *Mr Jennings*: Frederick Thomas Jennings, first-class clerk at the Board of Trade.

To John Hall Gladstone 11 October 1875 4659

Royal Institution of Great Britain | 11th. Oct[ober]. 1875.

My dear Gladstone,

In consequence of some remarks, made by William,[1] regarding the closing of the chemical Laboratory, which I did not quite understand I went down to the laboratory to enlighten myself by communication with your junior assistant Mr. Williams.[2] The door was locked and I knocked, but could not get in.

I then requested Mr Cottrell to ask Mr. Williams to have the goodness to come to my room for a moment; but having waited for some time and Mr Williams not coming, I was obliged to leave without seeing him.

It is in the highest degree unpleasant to me, to meddle directly or indirectly with the chemical laboratory. But I am responsible for the safety of this house[3] and it is necessary, until I have been relieved of the responsibility by the managers and members, that both myself and the servants[4] should be able to obtain access to every part of it. You are yourself aware that an explosion occurred some time ago[5] in connection with the work of Mr. Williams, which had the laboratory servant not been present might have burnt us all out. It is Mr. Williams, as I understand who now adopts a course, which during the long connection of Faraday,[6] and during my own connection of nearly three and twenty years with the Royal Institution, has never been resorted to.

To avoid these difficulties I would respectfully submit to you whether it would not be better to definitely close the laboratory until you are able to give the work carried on there your personal supervision. I find myself in a very difficult position, my duty as superintendent of this house, and my reluctance to meddle with your department, pulling me in opposite directions.

I am, dear Gladstone | Yours faithfully | John Tyndall.

Dr. J. H. Gladstone Esq.

I intend to send a copy of this note to Mr. Spottiswoode to be laid before the Managers[7]

RI MS JT/1/TYP/1/410
LT Typescript Only

---

1. *William*: a laboratory servant at the RI.
2. *Mr. Williams*: see letter 4573, n. 3.
3. *this house*: 21 Albemarle Street, the building occupied by the RI.
4. *the servants*: in the 1871 Census, the servants recorded as living at the RI were Jane Tunstall, Eliza Moore (b. 1849), Elias Hill (b. 1845), Sarah Bowers (b. 1841), and Elizabeth Bridgman (b. 1849).
5. *an explosion occurred some time ago*: on 16 March (see letter 4573).
6. *the long connection of Faraday*: Michael Faraday had been associated with the RI from 1813, when he had started as a laboratory assistant, until his death in 1867.
7. *the Managers*: of the RI (see letter 4549, n. 4).

From John Hall Gladstone 12 October 1875 4660

Brighton | 12 Oct[ober]. /75.

My dear Tyndall,

I am sorry another difficulty[1] has arisen between our subordinates; and I will try to see you about it on Thursday.[2] It is quite clear that you as Superintendent of the House[3] must always have access to all parts of it, but we must arrange some way of preventing any unnecessary interference with those engaged in the Chemical Laboratory during working hours.

Believe me | Very truly yours | J. H. Gladstone

Prof. Tyndall, F.R.S.

RI MS JT/1/TYP/1/411
LT Typescript Only

1. *another difficulty*: see letter 4659.
2. *Thursday*: 14 October.
3. *the House*: 21 Albemarle Street, the building occupied by the RI.

To John Hall Gladstone 13 October 1875 4661

13th Oct[ober]. 1875.

My dear Gladstone,

I should be very much obliged if you would inform me whether you have hitherto had any reason to complain of "unnecessary interference with those engaged in the chemical laboratory."[1]

If so would you kindly name to me the occasions on which this interference took place?

I ask you for this information so that I may state to the Managers,[2] at their first meeting in November, what my conduct, or the conduct of those under my control, has been on such occasions.

A few words with you from time to time would, I imagine, have prevented a difficulty from ever arising; but you are so rarely here that such opportunity have not been afforded me. Your note to me,[3] for instance, conveys the first hint that I have had from you of "unnecessary interference"; and it excites in me the wish to have my position in this house[4] so clearly defined as to render it impossible for me to interfere where I have no right to do so.

faithfully Yours | John Tyndall

RI MS JT/1/TYP/1/412
LT Typescript Only

1. *"unnecessary interference . . . chemical laboratory"*: Tyndall is quoting Gladstone's own words in letter 4660.
2. *the Managers*: of the RI (see letter 4549, n. 4).
3. *Your note to me*: letter 4660.
4. *this house*: 21 Albemarle Street, the building occupied by the RI; Tyndall's position was Superintendent of the House.

To Olga Novikoff 13 October [1875][1] 4662

13th Oct[ober].

My dear Friend.

I was going to write to you about the "Revue".[2] Our librarian[3] has had London scoured today, but could not find a copy to buy. I will ascertain tomorrow whether the Royal Society take it in and would lend it.

I have been steadily declining invitations from my most intimate friends,[4] the pressure of my work is so great. If I can manage to dine with Charles Darwin,[5] whose health is delicate, and whose years are advancing—and with M^rs^. Grote (wife of the historian)[6] who is 83, it is all that I can accomplish for some time to come.

You and I, I hope will always be perfectly straightforward with each other—and frankly said, if I went to you under my present circumstances[7] it would simply beget gossip which neither of us would like.

Ever faithfully Yours | John Tyndall

RI MS JT/1/T/392
RI MS JT/1/TYP/3/941

1. *[1875]*: the year is established by reference to Tyndall dining with Charles Darwin (see n. 5).
2. *the "Revue"*: presumably the *Revue des Deux Mondes*. Novikoff may have wished to read an article in its most recent bimonthly issue, for September 1875. Given her growing concern at this time with Turkish imperialism and its impact upon Russia, it was possibly É. Burnouf, 'La Grèce et la Turquie en 1875', *Revue des Deux Mondes*, 11 n.s. (1875), pp. 29–58.
3. *Our librarian*: Benjamin Vincent.
4. *I have been steadily . . . most intimate friends*: Novikoff was back in London at this time, and Tyndall, having initially promised to visit her (see letter 4657), seemed now, four days later, to have decided to make no further visits.
5. *dine with Charles Darwin*: Tyndall was invited to dine with Darwin and his family in letter 4656.

6. *the historian*: George Grote (1794–1871), a political theorist and historian of ancient Greece (*ODNB*).
7. *my present circumstances*: presumably his burgeoning friendship with Louisa Hamilton; their engagement would be announced in February 1876.

From Thomas Henry Huxley 13 October 1875 4663

4 Marlborough Place[1] | Oct[ober]. 13 1875.

My dear Tyndall,

Will you bring with you to the "X" to-morrow[2] a little bottle full of fluid containing the Bacteria &c. you have found developed in your infusions?[3] I mean a good characteristic specimen. It will be useful to you I think if I determine the forms with my own microscope[4] and make drawings of them which you can use.

Ever yours | T. H. Huxley.

I can't tell you how delighted I was with the experiments.

IC HP 8.176
RI MS JT/1/TYP/9/3049
Typed Transcript Only

---

1. *4 Marlborough Place*: see letter 4213, n. 1.
2. *the "X" to-morrow*: the monthly meeting of the X Club (see letter 4363, n. 13), which were held on Thursdays at St George's Hotel in Albemarle Street; as they were held during the social season from October to June, this was the first meeting for several months.
3. *a little bottle full . . . in your infusions*: boiled infusions from the experiments Tyndall was then conducting on spontaneous generation (see letters 4655 and 4667).
4. *It will be useful . . . my own microscope*: in a published account of the experiments, Tyndall noted that 'doubtful of my skill as a microscopist I took specimens . . . and sent them to Prof. Huxley, with a request that he would be good enough to examine them' ('The Optical Deportment of the Atmosphere in Relation to the Phenomena of Putrefaction and Infection', *Phil. Trans.*, 166 (1876), pp. 27–74, on p. 36).

To James Chambers[1] 14 October 1875 4664

Royal Institution of Great Britain

To Mr. James Chambers, in compliance with his request,[2] from
John Tyndall | 14th October 1875.

Private Collection #52

1. *James Chambers*: not identified.
2. *his request*: not identified, but possibly a request for an autograph.

To Heinrich Rudolf Vieweg 14 October 1875 4665

Royal Institution of Great Britain | 14th. October 1875

Gentlemen,[1]

I thought my last letter[2] would make my position clear to you; but it does not seem to have done so. I do not complain of your terms. What you offer me may be liberal—even more liberal than I should demand. What I want to know is the ratio that my profits bear to yours.

I have already stated to you that no matter what terms you may offer me I shall reap no personal advantage from them. It is not on my own account that I make this reasonable request. I do not want you to pay me as M^r^. Longman[3] pays me. I want no money in advance. I repeat, what I want to know is the share I receive of the profits arising from my books. Do I receive half the profits? Do I receive a third? Do I receive a tenth? If you give me the information I desire the upshot may be that I shall decline to accept any money whatever from you. I may consider that your profits are small enough without paying me. What I object to is being kept in total darkness as to the value of my property. As far as money is concerned I have no interest whatever in this matter, but I do feel an interest in seeing a reasonable demand complied with, and a plain principle of justice carried out.

I am Gentlemen | Your obed[ien]t servant | John Tyndall

Messrs Vieweg & Sohn

TU Braunschweig/Universitätsbibliothek/UABS V1T:63

1. *Gentlemen*: see letter 4592, n. 1.
2. *my last letter*: letter 4644; Tyndall evidently considered letter 4654 as merely a supplement to this.
3. *M^r^. Longman*: probably Thomas Longman.

From George Forbes 14 October 1875 4666

Andersonian University[1] | Glasgow | 1875 Oct[ober] 14

Sir,

About a month ago I wrote to ask you for a copy of an article on the glacier question[2] which I believed you had written within the last 18 months. I have received no answer to that letter.[3] Will you be so good as to inform me where I may see the article alluded to or in what form it is published.

Yours faithfully | George Forbes

Prof Tyndall | Royal Institution

RI MS JT/1/F/27
RI MS JT/1/TYP/12/3945

1. *Andersonian University*: an institution founded in 1796 to provide scientific instruction with particular reference to the practical application of scientific ideas. It was named after John Anderson (1726–96), who left a bequest that helped fund the opening of the initial Andersonian Institute (*ODNB*).
2. *an article on the glacier question*: see letter 4641, n. 4.
3. *that letter*: letter 4641.

To Thomas Henry Huxley 15 October [1875][1] 4667

Royal Institution of Great Britain | 15th. Oct[ober]

My dear Huxley

I send you some specimens of the infusions.[2]

Would it not be well when the matter is further advanced to look in for an hour or so, and choose from the infusions those that you consider "characteristic?"

The "article"[3] as it now stands will I think meet your views. Indeed the small alterations you suggested did not sensibly interfere with the substance of it. The matter, if necessary, can be further developed at any time.

I must live the life of an ascetic, to clear away this Bastian fog. It is amazing what audacity can do in England; and his audacity has powerfully influenced numbers of intelligent people. Without entering into controversy with him I hope to set him in his true light.

I have been re-reading his criticisms on your Liverpool Address[4]—They almost take one's breath away—but these go down with numbers. Something, I think, may be done to dispel this illusion.[5]

Ever faithfully Yours | John Tyndall

IC HP 8.177
RI MS JT/1/TYP/9/3050
Typed Transcript Only

1. *[1875]*: the year is established by reference to Tyndall's research on germs in the air (see n. 2).

2. *specimens of the infusions*: from the experiments Tyndall was then conducting to refute the theory of spontaneous generation (see letter 4655).
3. *The "article"*: probably a draft version of J. Tyndall, 'The Optical Deportment of the Atmosphere in Relation to the Phenomena of Putrefaction and Infection', *Phil. Trans.*, 166 (1876), pp. 27–74.
4. *his criticisms on your Liverpool Address*: Henry Charlton Bastian's principal critique of Huxley's 1870 presidential address to the BAAS was 'The Evolution of Life: Professor Huxley's Address at Liverpool', *Nature*, 2 (1870), p. 492. He continued his criticisms in his books *The Modes of Origin of Lowest Organisms* (London: Macmillan, 1871) and *The Beginnings of Life* (London: Macmillan, 1872). For Bastian, see letter 4229, n. 2; and for Huxley's address, see letter 4655, n. 4.
5. *this illusion*: presumably that living organisms spontaneously generate from non-living matter. See J. Strick, *Sparks of Life: Darwinism and the Victorian Debates over Spontaneous Generation* (Cambridge, MA: Harvard University Press, 2000).

To Olga Novikoff 15 October 1875 4668

15th Oct[ober]. 1875.

Dear Friend.

No man knows when the supreme hour comes what his courage may be. I have sometimes felt the touch of courage in danger, coming upon me like a kind of anger, but I have also felt the touch of fear.

At present I fear anything and everything that interferes with my work,[1] which demands the most calm and concentrated attention: and I am quite determined for a good while to come to hold on loyally to it and to set my face like a flint against all Social temptations.[2] This is almost an annual habit with me at this season of the year when I come to London with the express object of working.

I am conscious of no other fear than this,[3] and it is not a fear which a brave man need shrink from avowing.

Ever faithfully yours | John Tyndall

Dear Friend

This was written before the your last note[4] reached me. I have broken the envelope to say that you place far too much emphasis on my "fear". What I fear is that which has been stated on the other side. At the present time I care little for any other danger—But my work must go on without interruption. Pitfalls and enemies are before me and around me, and I am resolved not to be tripped up by the one, nor overcome by the other. But failure can only be avoided by making every part of my investigation unassailable. This God helping me, as old Luther would say,[5] I intend to do, and thus can only be done by solitary thought and severe experiment.

I saw the name of your friend M[r]. Meyrick in a late number of the Guardian newspaper.[6]

Goodbye.

RI MS JT/1/T/388

1. *my work*: the experiments on germ infusions (see letter 4667).
2. *set my face like a flint against all Social temptations*: see letter 4662, n. 4.
3. *I am conscious of no other fear than this*: possibly an allusion to the 'gossip' that, as Tyndall told Novikoff in letter 4662, 'neither of us would like'; this letter suggests that she may have responded by accusing him of being fearful of it.
4. *your last note*: letter missing.
5. *This God helping me, as old Luther would say*: see letter 4384, n. 3.
6. *M[r]. Meyrick in a late number of the Guardian newspaper*: the activities of Frederick Meyrick (1827–1906), the dean of Trinity College, Oxford (*ODNB*), were mentioned regularly in the Anglican weekly journal the *Guardian*. His participation in the Old Catholic Conference in Bonn in August received particular coverage.

From Pasquale Balestrieri[1] 16 October 1875 4669

Rome ce 16 octobre 1875

Monsieur et très honorable Professeur Tyndall

Quoique je n'ai pas l'honneur d'être connu personnellement de vous, j'ai été bien content qu'un homme si illustre, un savant aussi consciencieux que vous êtes, soit parmi les membres de la Commission qui doivent juger mon invention. Ces membres doivent en juger principalement par l'appareil qu'ils ont sous leurs yeux; mais vous Monsieur, devez en jugez autant par ce que l'on voit par les yeux du corps que par ceux de la raison scientifique. Aussi osé-je vous diriger ce peu de lignes pour vous développer moins ce que l'on voit dans la machine que vous voyez que le principe, j'ose dire nouveau sur lequel elle se fonde, et ce que l'on en peut obtenir.

D'abord le modèle que vous avez sous vos yeux n'est point la limite de ce que l'on peut obtenir par mon principe: ce modèle n'a pas été fabriqué dans le but d'éclipser ce qu'il y a de mieux dans l'industrie moderne en ce genre; car, pour cela, j'aurai dû avoir des modèles de Phares Fresnel pour me proposer le but de surpasser sa puissance; je n'ai fait fabriquer qu'un modèle quelconque, et c'est ce modèle quelconque que la Commission compare maintenant avec ce qu'il y a de mieux dans l'industrie.

Mais, malgré que votre Fresnel soit le nec plus ultra de l'industrie contemporaine et mon Collecteur soit un modèlc quelconque qui est d'une échelle d'autres innombrables au dessus de lui, il peut bien en soutenir la comparaison: sa supériorité doit être incontestable moins par ce que l'on voit que par ce que la raison scientifique démontre. D'abord, dans mon modèle, une grande quantité de rayons est dispersée par l'irrégularité de la surface des segments de Cône réfléchisseurs qui forment les armilles. Car ces armilles ont été construises à la main et sans être dressées au tour, comme il était nécessaire pour la parfaite régularité. Mais l'on sent bien que ce défaut de mon appareil qui est un défaut de construction, faute de machines et d'ouvriers habiles, doit être considéré comme une circonstance plaidoyant en faveur de mon appareil dans la comparaison avec un autre que l'on doit considérer comme la plus parfaite construction dont l'industrie humaine soit capable! Pour apprécier tout ce que ce défaut de construction ôte d'effet à mon appareil, il suffit de placer à son foyer une très petite lampe, comme celle d'une bougie; on apercevra alors une quantité innombrable de taches obscures qui surgissent de la dispersion opérée par les irrégularités dont j'ai parlé. En faisant autant avec le Fresnel, on y voit bien une diminution correspondant dans l'intensité de la lumière mais point de taches et point d'interruptions.

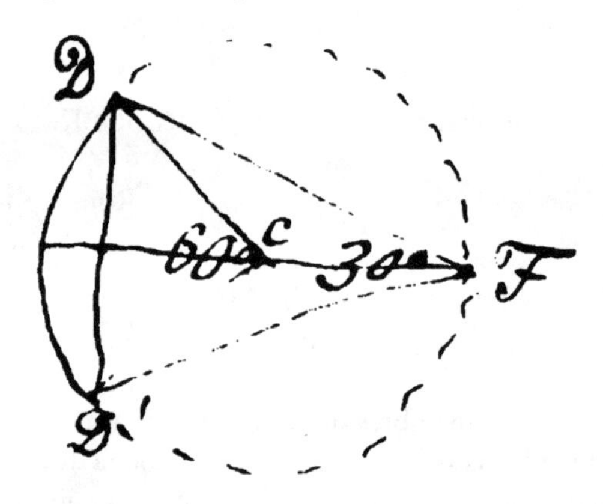

Après ces considérations préliminaires, occupons-nous du système lui-même considéré dans sa perfection. Sous d'égales circonstances, mon appareil recueille trois, quatre, cinq, etc fois plus de lumière que la lentille la plus étendue que l'on puisse fabriquer. On ne peut pas construire une lentille dont le diamètre soit plus grand que le rayon de sa courbure. Alors ce diamètre de lentille est représenté par le côté D D' du triangle équilatéral DD' C. Dans ce cas l'angle DCD' est de 600 degrés; et par conséquent l'angle en F extrémité du diamètre de la sphère, n'a que 300 degrés. Donc la Lentille d'un Fresnel ne peut pas recueillir plus de 300 degrés de lumière, degrés comptés sur DD' F dont le plan passe par le centre de l'atmosphère lumineuse, parce que le Foyer d'une lentille plano-sphérique est à peu près à l'extrémité du diamètre de la sphère dont elle a été détachée.

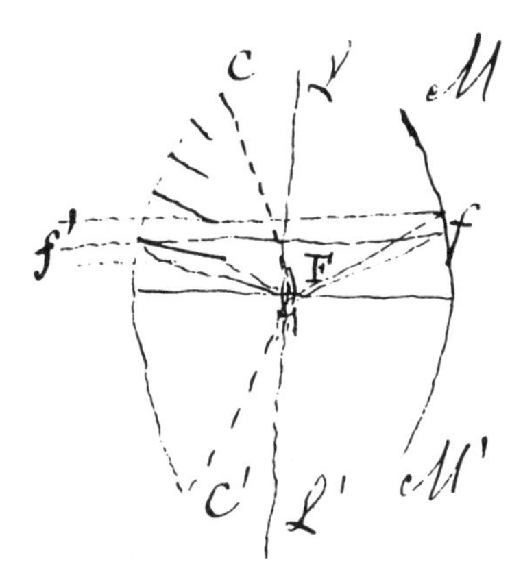

La chose procède bien autrement dans mon Collecteur; car on le peut fabriquer de manière à avoir la distance focale que l'on désire. Pour l'objet de projeter la lumière à de grandes distances, il convient de placer ce foyer très près du Collecteur. Alors le nombre des degrés de lumière est représenté par l'angle CFC' qui est d'autant plus proche de 1800 degrés que le foyer f est plus proche du Collecteur.

Tout cela ne regarde que l'émisphère lumineux antérieur qui est borné en avant par le collecteur et en arrière par la ligne LL' qui est la représentation du plan qui sépare l'émisphère antérieur d'avec le postérieur. La lumière de cet émisphère postérieur est irrévocablement perdue dans l'appareil Fresnel mais elle n'est pas perdue dans notre Collecteur. Car, en plaçant un miroir concave ou parabolique derrière la flamme en MM', le faisceau postérieur Fff, se reflétant en f, va sortir parallèlement entre une armille, et l'autre. Cette récolte des rayons de l'émisphère postérieur est d'autant plus grande que les armilles sont plus disposées sur un plan au lieu d'être disposées sur une surface sphérique. Aussi dans le système plan du Collecteur, nous y ajoutons, presque toujours, un miroir sphérique pour recueillir les rayons de l'Emisphère postérieur. Tel est le grand Collecteur système plan, qui a été exposé sur la Piazza del Popolo à Rome. Tant le Collecteur que le miroir sphérique ont deux mètres de diamètre: l'appareil antérieur recueille 140 degrés et le postérieur 70, somme totale 210 degrés de lumière: son effet a été prodigieux; c'est ce grand appareil de l'effet duquel la presse du monde entier a parlé aux mois de mars et d'avril de cette année.

Pour obvier à l'inconvénient de la récolte de très peu de degrés de rayons que les lentilles Fresnel peuvent faire à cause de la longue distance focale, on a imaginé d'entourer les lentilles à échelons d'autres échelons de prismes disposés sur une surface courbe en dôme supérieurement et inférieurement en panneau. Mais l'on sent toute la différence qu'il y a entre des prismes réflecteurs et des armilles. D'abord la coupole ou dôme ne peut pas être prolongée autant que la série des armilles jusqu'au tirage de la lampe, car l'angle-limite de la réflexion totale ne le permet pas. Puis, ces échelons de prismes offrent des interruptions dans la réflexion pour une quantité de rayons perdus que le Collecteur ne présente pas, où nul rayon n'est intercepté, car le rayon qui rase le limbe intérieur d'une armille va battre sur le limbe antérieur de l'armille suivante où il est reflété. Ajoutez à cela la double réflexion d'immersion et d'émersion du rayon dans le moyen transparent, l'absorption de la masse du verre, et la dispersion par les bulles et turbinations de la pâte du verre même; et vous verrez que c'est de trop si nous admettons que la quatrième partie à peine des rayons qui entrent dans ces prismes est refléctée parallèlement ou quasi. Ajoutez encore pour la lentille la double aberration de sphéricité et de réfrangibilité, et vous verrez encore quelle grande diminution de lumière le Fresnel souffre en comparaison du Collecteur Balestrieri.

La grande aberration chromatique que l'appareil Fresnel fait subir aux rayons est révélée par la teinte bleuâtre que la lumière du Fresnel montre en comparaison de la lumière du Collecteur, laquelle est absolûment la même que celle de la lampe; en sorte que la lumière du Collecteur apparait toujours plus rouge et plus vive que celle du Fresnel, dont la teinte est toujours bleuâtre. Cette teinte provenant des rayons bleus, indigos et violets qui sont plus fortement réfractés que les rouges, orangés et jaunes, se voyent bien de près et à de petites distances, mais sont absolûment perdus à de grandes distances; car si on règle la distance focale en sorte de donner le parallélisme *aux* rayons les moins refrangibles, les rayons les plus réfrangibles seront dispersés en cônes très élargis dans l'espace. Rien de tout cela, pas un de ces inconvénients n'a lieu dans le Collecteur. Même, la très grande obliquité des rayons incidents rend presque nul ce peu d'absorption que les surfaces ont de leur nature.

C'est à cause de tout cela, que dans les expériences à Civitavecchia entre notre Collecteur et le Fresnel de ce port, Fresnel fabriqué par LePaute de Paris, la lumière du Phare paraissait presque éclipsée en comparaison du Collecteur. Si cela n'est pas arrivé dans les expériences à Blackwel c'est sans doute pour le déplacement que l'on a opéré sur la lampe. La Lentille, ayant des foyers conjugués, peu de pouces de déplacement de la lampe n'altèrent pas sensiblement l'effet; mais le Collecteur qui est un Instrument tout-à-fait sui-generis, et qui n'a pas de foyers conjugués, exige une rigoureuse centralisation de la lampe dans son foyer, tel que l'on a établi dans la construction: un ou deux pouces de déplacement de la lampe en quelque sens que ce soit font manquer tout l'effet. Or dans le modèle qui est sous les yeux de la Commission de Trinity-House, la lampe doit être placée en sorte que le tube de son tirage doit toucher la dernière armille, c'est à dire la plus éloignée de l'axe. Le placement était bien réglé sur la machine; mais puisque cette machine a été démontée pour appliquer le Collecteur à la même lampe qui anime le Fresnel, je crains bien que la distance de la lampe à mon Collecteur soit plus grande qu'elle ne doit être.

Sans cela, la lumière du Collecteur aurait dû apparaitre cinq ou six fois plus vive que celle du Fresnel.

J'ai dit que l'on peut donner au Collecteur la distance focale que l'on veut, mais qu'il est bon de la faire aussi courte que l'on peut pour embrasser un plus grand nombre de rayons. Or c'est un très grand avantage que mon Collecteur a sur la Lentille Fresnel, laquelle ayant une distance focale d'au moins un mètre et demi ou deux mètres, pour tourner sur l'axe qui passe par le centre de la flamme a besoin d'un tour de presque six mètre de développée. Ce tour, ne pouvant s'accomplir qu'en un temps très long, si l'on veut un éclair tous les 40 ou 50 secondes, on est obligé de disposer 10 ou 12 appareils sur un même tambour tournant. Or, puisque on peut donner à un Collecteur 25 centimètres de distance focale, son tour peut n'avoir qu'un mètre et demi de développée; et alors un seul appareil, ou au plus deux; peuvent donner cette fréquence d'éclairs que les Fresnel ne peuvent donner qu'en 10 ou 12 appareils. Comprend-on bien tout l'avantage qui en résulte dans les frais d'installation d'un Phare?

Outre cela, considérons ce que peut coûter cet assemblage d'armilles en feuilles métalliques argentées qui constituent le Collecteur et cet assemblage de lentilles, échelons sphériques et prismatiques dont les Fresnel sont composés! Ajoutez enfin que le Collecteur peut être fermé hermétiquement dans un tambour métallique entre deux glaces; et alors la manutention ne se réduit à autre chose qu'à un polissage à l'éponge mouillée de ces glaces mêmes comme sur celles d'une boîte.

Agréez Monsieur et très honorable Professeur le témoignage de ma plus profonde estime.

Naples etc | Prof. Pasquale Balestrieri

Rome, 16 October 1875

To the highly respected Professor Tyndall,[2]

Although I do not have the honour of being personally acquainted with you, I was very glad that such a celebrated man, a scholar as conscientious as you are, would be among the members of the Commission[3] in charge of judging my invention.[4] These members are to judge it primarily by the device before their eyes; but you, Sir, ought to judge it as much from a physical perspective as from that of the scientific reason. Hence, I have the audacity to send you these few lines not so much to elaborate on what is seen in the machine presented before you as to elaborate on the principle, I dare say novel, upon which it is based, and what can be obtained from it.

First, the model in front of you is not the limit of what can be achieved through my principle: this model was not built with the aim of outshining the best of its kind in the modern industry ; because for that I should have had some replicas of the Fresnel light,[5] to allow myself to exceed its power. I only had an ordinary prototype built, and it is this ordinary prototype that the Commission now compares to what is best in the industry.

However, despite the fact that your Fresnel is the nec-plus-ultra[6] in the modern industry and my Collecteur an ordinary design that has many others measuring above it on the scale, it can well bear comparison with it: its superiority should be undeniable less from what is seen than from what the scientific reason demonstrates. First, in my design, a large quantity of rays are dispersed by the unevenness in the surface of the reflecting cones segments forming the armils.[7] For these armils have been hand-manufactured and without being trued[8] on a lathe, as is necessary to achieve perfect evenness. But it is clear that the flaw in my device, which is a flaw in its construction, for lack of machinery and skilful workers, should be regarded as a favourable circumstance in defence of my device when comparing it to another which must be considered as the most perfect construction that the human industry is capable of! Placing a really small light on the focal center of my device, such as a candle, is sufficient to appreciate the full scale of the device's impact taken away by this flaw in construction; a countless quantity of dark spots arising from the dispersion made by the irregularities I mentioned is then visible. By doing the same with the Fresnel, a clear and equivalent reduction in the intensity of the light is visible, but there are no spots nor interruptions.

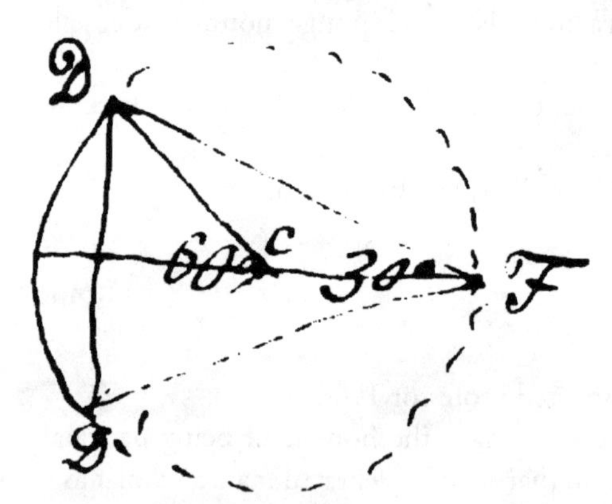

Following from these introductive observations, let us turn to the system judged in its own perfection. Under equal circumstances, my device collects three, four, five, etc. times more light than the largest lens that can be manufactured. It is not possible to produce a lens with a diameter larger than the radius of its curvature. Then, this diameter of the lens is represented by the side DD' of the equilateral triangle DD'C. In this case, the angle DCD' is 600 degrees; and consequently the angle at F, at the extremity of the diameter of the sphere, is only 300 degrees. Thus, the lens of a Fresnel cannot collect more than 300 degrees of light, which are counted on DD'F in which the plane passes through the centre of the luminous atmosphere, because the focus of a plano-spherical[9] lens is close to the edge of the diameter of the sphere from where it was detached.

The process is very different in my Collecteur: the reason is that it can be manufactured in such a way as to obtain the specified focal distance. In order to project light over great distances, this focus must be placed very close to the Collecteur. Then the quantity of degrees of light is represented by the angle CfC' which is all the closer to 180° degrees that the focus f[10] is brought closer to the Collecteur.

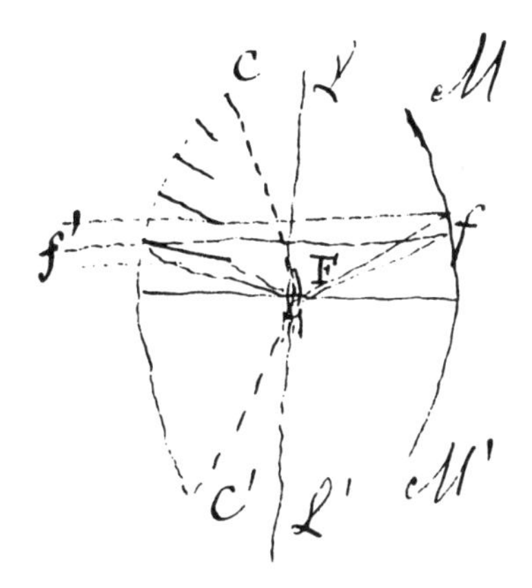

This only concerns the illuminated anterior hemisphere that is marked out in front by the Collecteur and behind by the line LL', which is the representation of the plane that separates the anterior hemisphere from the posterior one. The light of that posterior hemisphere is irreversibly lost in the Fresnel device, but it is not lost in our[11] Collecteur. For, by placing a concave or parabolic mirror behind the flame in MM', the posterior beam Fff', reflected in f, will emerge in parallel between one armil and the other. This quantity of beams hence collected by the posterior hemisphere is even larger when the armils are more arranged on a plane instead of being arranged on a spherical surface.[12] Also, we almost always add a spherical mirror to the plane system of the Collecteur to collect the beams from the posterior hemisphere. Such is the great Collecteur that was exhibited in the Piazza del Popolo in Rome. Both the Collecteur and the spherical mirror are two meters in diameter: the anterior device collects 140 degrees and the posterior collects 70, for an added total of 210 degrees of light: its effect was phenomenal: it is the effect from the same large device that was the talk of the worldwide press in the months of March and April last year.

To avoid the Fresnel lenses' drawback in that they can collect very few degrees of rays because of their lengthy focal distance, it was conceived that the stepped lenses would be surrounded by other steps of prisms arranged on a curved surface with its upper side in the shape of a dome and the lower as a panel. Yet, there is a noticeable difference between a reflective prism and an armil. First, the cupola or dome[13] cannot be extended as much as the series of armils towards the draught of the lamp, because the furthest angle of the total reflection does not allow it. Then, these steps of prisms allow some interruption in the reflection resulting in the loss of a quantity of rays which the Collecteur does not allow, as no ray is blocked, because the ray that grazes the inner limb of an armil will be projected on the anterior limb of the next armil where it is reflected. Add to that the double reflection of immersion and emersion of the ray in the transparent medium, the absorption of the glass mass, and the dispersion by the bubbles and whirls contained in the glass paste itself, and you will see that it is in the way if we assume that hardly a fourth part of rays that enter the prisms is reflected in parallel or quasi. Regarding the lens, add to this the double aberration of sphericity and refrangibility, and you will see again how significantly the light is reduced in the Fresnel in comparison to the Balestrieri Collecteur. The bluish shade shown by the light from the Fresnel reveals the large chromatic distortion that the rays are subjected to in the Fresnel device while, in comparison, the light of the Collecteur is absolutely similar to that of the lamp; as a result, the light of the Collecteur always appears redder and brighter than that of the Fresnel, the shade of which is always bluish. This shade, originating from the blue, indigo, and purple rays which are refracted more strongly that the red, orangy, and yellow ones, are clearly visible up close and at short distances, but are entirely lost in long distances; because, if the focal distance is adjusted in order to give parallelism to the less refrangible rays, the most refrangible rays will be dispersed in very widened cones in the space. There is nothing like that, none of these drawbacks, happening with the Collecteur. Even, the very large obliquity of the incident rays almost entirely nullifies the little absorption that the surfaces naturally have.

This is the reason why, in the experiments conducted in Civitavecchia[14] between our Collecteur and the Fresnel in this port, a Fresnel manufactured by LePaute[15] from Paris, the light from the lighthouse paled in comparison to the Collecteur. If that did not occur during the experiments in Blackwell,[16] it is most likely because of the displacements that were made on the lamp. As for the lens, with its combined foci, a few inches of displacements in its lamps do not substantially distort its effect; but the Collecteur, which is a completely sui generis[17] device and does not have combined lenses, requires a strict centralisation of the lamp in its focus point, as was established in the construction: the lamp is moved by one of two inches, in whichever direction, and the whole effect is missed. However, in the model presented to the Commission of the Trinity-House, the lamp must be placed in such a manner that the draft pipe must touch the last armil, i.e. the armil the furthest from the axis. The position was well adjusted on the machine; but since this machine was disassembled to install on the Collecteur the same lamp operating on the Fresnel, I fear that the distance between the lamp and my Collecteur is greater than it should be. Had this not happened, the light of the Collecteur should have appeared five to six times brighter than that of the Fresnel.

I said we can give the Collecteur the focal distance we want, but it is desirable to shorten it as much as possible to capture a larger amount of rays. But, it is a really great advantage that my Collecteur has over the Fresnel lens, with its focal distance of a least a meter and a half or two meters and which requires an evolute of a curve[18] of almost six meters to turn on the axis that goes through the centre of the flame. Because this turn can only be accomplished in a very long time, 10 or 12 devices must be placed on a single rotating cylinder in order to obtain a flash every 40 to 50 seconds. But, since the Collecteur can be adjusted to a focal distance of 25 centimetres, its rotation would require not more than an evolute of a meter and a half; and then, a single device, or rather two, could provide with the frequency in flashes that the Fresnel can only achieve with 10 or 12 devices. Are all the ensuing benefits in the costs of installing a lighthouse clearly understood?

Besides that, let us consider the price of this assemblage of armils in silver-plated leaves, which forms the Collecteur, and that of this assemblage of lenses, spherical and prismatic steps, which the Fresnel consists of! Finally, add that the Collecteur can be hermetically sealed in a metallic cylinder between two glass screens; and that the maintenance comes down to no more than polishing with a wet sponge these same screens as is done on a box's glass screens.

Please accept, highly respected Professor, the assurance of my deepest esteem.

Naples etc.| Prof. Pasquale Balestrieri

RI MS JT/1/B/10

---

1. *Pasquale Balestrieri*: Pasquale Balestrieri (*fl.* 1850s–70s), an Italian physicist originally from Naples who, on 9 January 1875, was granted a British patent for his invention of an 'Improved Apparatus for Collecting Rays of Light and Heat' (*English Patents of Inventions*, no. 89 (1875), p. 1). In Italy, the patent seems to have been held by Balestrieri's aristocratic guarantor Saverio Dandini de Sylva (see letter 4642).
2. *To the highly respected Professor Tyndall*: Balestrieri's language throughout the letter is somewhat stilted, as French was evidently not his first language. We have tried to retain this for the English translation.
3. *the Commission*: a board established by Trinity House to oversee a trial of Balestrieri's invention. In December 1875 it was reported in Italy that his invention was 'ortato a Londra, ove si sperimento ripetutamente sul Tamigi dalla corporazione della Trinity House, sotto la direzione del dotto fisico inglese signor Tyndall. Dal rapporto fatto alla corporazione medesima da tale scienziato risulta, che la forza illuminante del collettore fototermico armillare, supera quella dei fari alla Fresnel, e che può riescire molto utile pei bastimenti, per le coste, pei segnali, ed egualmente per altri usi, che sapranno bene indicare gli esperti [brought to London, where it was subjected to repeated experiments on the Thames by the Trinity House corporation, under the direction of the learned English physicist Mr Tyndall. From the report made to the corporation itself by this scientist, it appears that the illuminating force of the armillary photothermal collector exceeds that of the Fresnel

lighthouses, and that it can be very useful for ships, for coasts, for signals, and equally for other uses, which they will be able to point out the experts]' ('Classe di scienze fisiche, matematiche e naturali', *Atti della Reale Accademia dei Lincei*, 2nd series 3 (1875–6), pp. 14–19, on p. 19). For Trinity House, see letter 4187, n. 2.

4. *my invention*: a photo-thermic hollow sphere collector (collecteur photo-thermique armillaire) that was also known as the 'Balestrieri Reflector'. It was a catoptric device to enhance the propagation of light, of which the *Nautical Magazine* proposed: 'It will, perhaps, be better understood by our readers, if we state that the principle of the new arrangement consists chiefly in anterior reflection, the reflectors being a series of metallic annular belts, or conical segments, so adjusted as to receive at a certain angle on their polished surfaces, 180° of the light radiating from a luminous body placed in focus behind it. The rays of light are therefore sent out at the same angle as that at which they strike the reflecting surface, and each ring is so adjusted, that all the impinging rays are reflected outwards in a direction parallel to each other, thus forming a compact horizontal shaft of light to be sent out to the sea horizon, similar to the beam which issues from a lenticular apparatus' ('Light and Heat Collector', *Nautical Magazine*, 44 (1875), pp. 1033–4, on p. 1034).
5. *the Fresnel light*: a light for use in lighthouses containing a catadioptric lens, facilitating both refraction and reflection, invented by Augustin-Jean Fresnel (1788–1827), a French physicist.
6. *nec-plus-ultra*: ultimate (Latin)
7. *the armils*: the metal bands in Balestrieri's device.
8. *trued*: made straight, level, round to the desired degree (*OED*).
9. *plano-spherical*: the convex side of a plane having spherical curvature (*OED*).
10. *the focus f*: while the original handwritten letter clearly states 'f', it seems that Balestrieri means 'F'.
11. *our*: it is unclear whether Balestrieri's use of 'notre' or 'nous' in the original indicates a plural subject (possibly himself and the manufacturer of the device) or simply a generic subject.
12. *This quantity of beams . . . arranged on a spherical surface*: the original sentence is awkwardly expressed (see n. 2), but it seems to mean that there are more armils on a plane surface than there are on a spherical one.
13. *the cupola or dome*: while both terms are regarded as interchangeable, the French 'coupol' refers to a concave shape, while dome refers to the convex shape.
14. *Civitavecchia*: an ancient area of Rome featuring a harbour and a lighthouse.
15. *LePaute*: Henry-Lepaute, a lens-manufacturing firm in Paris started by Augustin Michael Henry (1800–85), who later changed his last name to Henry-Lapaute (adding his mother's maiden name).
16. *Blackwell*: Trinity House's experimental lighthouse in East London.
17. *sui generis*: unique (Latin).
18. *an evolute of a curve*: the single point at the centre of a curve that is the locus of all its centres of curvature.

To Olga Novikoff 18 October [1875][1] 4670

Royal Institution of Great Britain | 18th. Oct[ober].

My dear Friend

Late last night on my return from Mr Darwin's[2] I found your note[3] here. I thank you for it much.

I expect the proof[4] sometime to day, or at the latest tomorrow—I will, if you will allow me—bring it myself

Yesterday I picked up a number of the Theological Review in Darwin's drawing room, and read a sweet and affectionate article by our friend Reville on a friend of his I think named Caqueril[5]—a young minister of the French protestant church.

Yours ever faithfully | John Tyndall

RI MS JT/1/TYP/3/946
LT Typescript Only

1. *[1875]*: the year is established by reference to Tyndall's visit to the home of Charles Darwin (see n. 2).
2. *my return from Mr Darwin's*: Tyndall accepted an invitation to stay with Charles Darwin and his family at their home Down House in Down, Kent, on the weekend of 16 and 17 October (see letter 4656).
3. *your note*: letter missing.
4. *the proof*: possibly of 'Introduction' to Part II, in J. Tyndall, *Fragments of Science: A Series of Detached Essays, Lectures, and Reviews*, 5th edn (London: Longmans, Green, and Co., 1876), pp. 325–56, or the version of it that was published, in November, as J. Tyndall, '"Materialism" and Its Opponents', *Fortnightly Review*, 24 (1875), pp. 579–99. See letter 4647.
5. *a sweet and affectionate article . . . named Caqueril*: A. Réville, 'Athanase Josué Coquerel. In Memoriam', *Theological Review*, 12 (1875), pp. 500–14. Coquerel (1820–75) was a French Protestant theologian whose liberal views on religious freedom had aroused controversy.

To William Lees[1] 20 October 1875 4671

Copy[2] | Royal Institution, London | 20th. Oct[ober]. 1875.

Dear Sir,

I am exceedingly anxious to ascertain for the information of the Board of Trade, whether the gas-light in its most powerful form, namely, the

Triform[3]—is capable of illuminating fogs by a momentary thrill, after the steady light has ceased to be distinguishable.

I have been in communication with Mr. Wigham[4] with reference to this question, and have given him definite information as to the knowledge I desire to obtain. I will only add to these instructions the expression of an opinion that a steamer of some kind will be an indispensible adjunct. The steamer to be placed at a point from which the Bailey Light[5] is invisible, and at a second point from which the steady Triform is invisible, the flashing of the Triform upon the fog being observed from both points, and its performance carefully recorded.

On board the steamer one of the observers appointed by Professor Jellett[6] might take his place.

It will add greatly to our resources if the Triform should be found capable of affording effectual warning by its glare upon the fog, after it has ceased to be useful as a steady light.

I need hardly press upon you the importance of this subject. The guidance of the mariner in fogs, or rather the absence of guidance, constitutes the most serious gap in our system of coast signals, and I trust that you will make such efforts as lie in your power to render that system complete.

Even after the establishment of sound signals the intercalation of flashing lights should they be able to act in a manner approaching the thrills of lightning would be of inestimable value.

Yours etc. | (signed) John Tyndall—| The Secretary | Board of Irish Lights.

National Archives, MT 10/211. Lighthouses, Illumination by means of gas. H8585

---

1. *William Lees*: the secretary of the Board of Irish Lights since the 1850s.
2. *Copy*: the letter has been copied in a hand other than Tyndall's, possibly by Lees himself.
3. *the Triform*: see letter 4643, n. 6.
4. *in communication with Mr. Wigham*: see letter 4643.
5. *the Bailey Light*: the dioptric light of the Baily Lighthouse on the Howth Head in Dublin Bay.
6. *Professor Jellett*: John Hewitt Jellett (1817–88), a mathematician at Trinity College, Dublin who specialized in optics (*ODNB*).

To Olga Novikoff 20 October [1875][1] 4672

Royal Institution of Great Britain | 20th. Oct[ober].

Many thanks to you my dear friend. It is however but little reading[2] that I can accomplish now. I am just preparing for my visit of piety to Mrs Grote[3]

Yours ever | John Tyndall

RI MS JT/1/TYP/3/946
LT Typescript Only

1. *[1875]*: the year is established by reference to Tyndall's visit to Harriet Grote (see n. 3).
2. *little reading*: it is unclear what reading material Novikoff had either recommended or sent to Tyndall.
3. *my visit of piety to Mrs Grote*: Tyndall previously mentioned this visit in letter 4662.

From Charles Darwin 20 October [1875][1] 4673

Down, | Beckenham, Kent. | Railway Station | Orpington. S.E.R.[2] | Oct[ober]. 20th

My dear Tyndall

Your tube[3] with the solution dated Oct[ober]. 16th was quite clean at 8°30' AM on the 19th, but at 4°30' P.M was slightly muddy. At 8 AM today (20th) it was more muddy & contained many Bacteria (& apparently vibrios)[4] in lively movement. I believe that this is all that you wanted to hear.—.

I very greatly enjoyed your visit[5] & the talking had done me no harm.—

Yours very sincerely | Ch. Darwin

Gerald and Sue Friedman Manuscript Collection (MC 72), Folder 22, Box 1, Institute Archives and Special Collections, Rensselaer Polytechnic Institute, Troy, NY[6]

1. *[1875]*: the year is established by reference to Tyndall's research on boiled infusions (see n. 3).
2. *Orpington. S.E.R.*: a station on the South Eastern Railway opened in 1868.
3. *Your tube*: a test tube containing a boiled infusion from Tyndall's experiments on spontaneous generation. He probably brought it with him when he visited Darwin the previous weekend (see n. 5).
4. *vibrios*: aquatic microorganisms.
5. *your visit*: Tyndall, along with Thomas Huxley and his wife Henrietta, stayed with Darwin at his home Down House in Down, Kent, on the weekend of 16 and 17 October (for Emma Darwin's invitation, see letter 4656).
6. *Gerald and Sue . . . Troy, NY*: published in *Darwin Correspondence*, vol. 23, p. 408; this is letter 10207 in the online edition of the Darwin correspondence.

To Thomas Henry Huxley 21 October [1875][1] 4674

Royal Institution of Great Britain | 21st. Oct[ober].

My dear Huxley

If you possibly can would you examine these[2] closely to day?

Yours ever | John Tyndall

Prof. Huxley

IC HP 8.178
RI MS JT/1/TYP/9/3050
Typed Transcript Only

1. *[1875]*: the year is established by the relation to letter 4663.
2. *these*: presumably boiled infusions from his experiments on spontaneous generation (see letter 4663).

To Charles Darwin 23 October [1875][1] 4675

Royal Institution of Great Britain | 23rd Oct[ober].

My dear Darwin

I am very much obliged to you for your note.[2]

I wonder if I send two or three other tubes[3] down whether Mr Henry[4] (he lives near you, I believe) would kindly place them somewhere in the open and observe them for me?

Trust me, your friends were delighted to see you so joyful on Sunday,[5] and it is a doubled pleasure to learn that you have not suffered from the dissipation.[6]

Ever yours | John Tyndall

CUL GBR/0012/MS DAR 106:C19[7]
RI MS JT/1/TYP/9/2837

1. *[1875]*: the year is established by reference to Tyndall's research on boiled infusions (see n. 3).
2. *your note*: letter 4673.
3. *two or three other tubes*: test tubes containing boiled infusions from Tyndall experiments on spontaneous generation (see letter 4663).
4. *Mr. Henry*: not identified.
5. *your friends were . . . joyful on Sunday*: Tyndall, along with Thomas Huxley and his wife Henrietta, stayed with Darwin at his home Down House in Down, Kent, on the weekend of 16 and 17 October (for Emma Darwin's invitation, see letter 4656).

6. *dissipation*: waste of the moral and physical powers by indulgence in pleasure (*OED*).
7. *CUL GBR/0012/MS DAR 106:C19*: published in *Darwin Correspondence*, vol. 23, pp. 417–8; this is letter 10218 in the online edition of the Darwin correspondence.

To Thomas Henry Huxley 23 October [1875][1] 4676

Royal Institution of Great Britain | 23rd. Oct[ober].

My dear Huxley

I am extremely obliged to you for the trouble you have taken.[2] The result is what I expected, and is (to me) in the highest degree instructive.

I have three test tubes of turnip infusion on the same stand. Two of them showed what I have called the white muddiness. These two are thickly covered with mould. For some days the liquid has been clearing; the bacteria dying and falling to the bottom. In fact the mould has cut off their oxygen and stifled them.[3]

In the same stand is a third test tube (they were all filled with the same infusion) of a bright yellow green colour. Not a trace of mould is on its surface, and the liquid continues as muddy as ever it was. These pigment-producing bacteria have by some means or other overcome the mould spores and prevented their development.[4]

Six other tubes, placed in a different position, all show the white muddiness. They are all covered with mould: they have been standing now more than a month—the liquid underneath the mould is clear, after having been long muddy.

The subject seems to have no boundaries, to my short vision.

Darwin and Lubbock have both written to me[5]—Bacteria have developed both at Down and High Elms[6]

Yours ever | John Tyndall

IC HP 8.179
RI MS JT/1/TYP/9/3051
Typed Transcript Only

1. *[1875]*: the year is established by reference to Tyndall's research on boiled infusions (see n. 2).
2. *the trouble you have taken*: Huxley was providing microscopical analysis of the boiled infusions from Tyndall's experiments on spontaneous generation (see letter 4663).
3. *the mould has cut off their oxygen and stifled them*: in a published account of this experiment, Tyndall noted: 'The turnip-infusion, after developing in the first instance its myriadfold Bacterial life, frequently rapidly contracts mould, which stifles the Bacteria and clears the liquid all the way between the sediment and the scum' ('The Optical Deportment of the Atmosphere in Relation to the Phenomena of Putrefaction and Infection',

Phil. Trans., 166 (1876), pp. 27–74, on p. 66). Tyndall failed to draw the conclusion that the mould, *Penicillium*, secreted an antibacterial chemical that could have important medical implications (see *Ascent of John Tyndall*, p. 344).

4. *These pigment-producing bacteria . . . prevented their development*: in his published account of the experiment, Tyndall reported that 'The third tube, the middle one of the three, contained a bright yellow-green pigment, and on its surface no trace of mould was to be seen. It never cleared, but maintained its turbidity and its Bacterial life for months after the other tubes had ceased to show either. It cannot be doubted that the mould-spores fell into this tube also, but in the fight for existence the colour-producing Bacteria had the upper hand' (p. 35).
5. *Darwin and Lubbock have both written to me*: letter 4673; John Lubbock's letter is missing.
6. *Down and High Elms*: Down House, Charles Darwin's home, and High Elms, Lubbock's estate, were close to each other in Kent.

From Charles Darwin 25 October [1875][1] 4677

Railway Station | Orpington. S.E.R.[2] | Oct[ober]. 25th

My dear Tyndall

My son Frank[3] has been in Cambridge & only returned this evening so that I c[oul]d. not answer your note[4] by return of Post. He will be very glad to do his best in what you require.—Give us full instructions.—Direct the tubes[5] as above. If it is necessary that they sh[oul]d reach us quickly, please write outside "to be forwarded by foot-messenger".

Otherwise parcels always wait till carriage goes to station, which is never long.—

Ever yours | C. Darwin

CUL GBR/0012/MS DAR 261.8:23[6]
RI MS JT/1/TYP/9/2838

1. *[1875]*: the year is established by reference to Tyndall's research on boiled infusions (see n. 5).
2. *Railway Station | Orpington. S.E.R.*: in the MS letter, Darwin's home address of 'Down, Beckenham, Kent' has been crossed out, and a hand symbol drawn pointing to this part of the address. Orpington. S.E.R was a station on the South Eastern Railway, opened in 1868.
3. *My son Frank*: Francis Darwin (1848–1925), who had been visiting his elder brother George Darwin in Cambridge (*ODNB*).
4. *your note*: letter 4675.
5. *the tubes*: test tubes containing boiled infusions from Tyndall experiments on spontaneous generation. Tyndall was sending them across England to sample the effect of air in different locations (see letter 4663).
6. *CUL GBR/0012/MS DAR 261.8:23*: published in *Darwin Correspondence*, vol. 23, p. 420; this is letter 10222 in the online edition of the Darwin correspondence.

From Gustav Wiedemann 25 October 1875 4678

Leipzig, 25 Oct[ober]. 75

My dear Tyndall,

Many thanks for all your troubles. I have just received your correspondence with Vieweg,[1] will think carefully about it and then consult with Helmholtz. It seems to us that it would not be inappropriate to arrange provisionally a fixed honorarium in proportion to the copies, if it is not possible to arrange for the English mode.[2]

Please answer a question for me, if possible immediately. In Vieweg's letter of September 7 1875,[3] the number of copies for the German edition of Heat[4] are given for the 3 editions[5] resp. as 1200, 1200, 1500, for Sound[6] as 1200, 1500 for Light[7] as 2500! The honorarium for the 3 translations amounted to 8 taler 7 pf;[8] 10 taler, 15 taler per sheet. Thus, without even asking, Herr Vieweg has combined an increase in the honorarium with a completely disproportionate increase in the number of copies, which makes the former naturally more than illusionary. What is even worse than this pecuniary question, however, and closely concerns you as well as us, is that it takes much longer until an edition with a large number of copies is sold out and the public will accordingly judge your works and our translations of earlier publications as of lesser value.

I would be very grateful, if you would authorise me to confront Herr Vieweg with the above mentioned numbers, without discussing the other content of your correspondence. I would like to first direct the attention of the director of the house, Mr Heinrich Vieweg, privately to the consequences of such actions. One must not put up with everything.

Please send me a few lines immediately to let me know.

With fond regards and many thanks again, also from my wife,[9]

Your | faithfully devoted | G. Wiedemann

RI MS JT/1/W/55

1. *your correspondence with Vieweg*: presumably letters 4629, 4639, 4644, 4654, and 4665.
2. *the English mode*: where, on publication, authors received an advance of the publisher's anticipated profits from a book, as described by Heinrich Rudolf Vieweg in letter 4639.
3. *Vieweg's letter of September 7 1875*: letter 4639.
4. *the German edition of Heat*: J. Tyndall, *Die Warme Betrachtet als eine Art der Bewegung*, ed. H. Helmholtz and G. Wiedemann, 3rd edn (Braunschweig: Friedrich Vieweg, 1875).
5. *the 3 editions*: the first edition of the translated *Die Warme* had been published in 1867, the second in 1871.
6. *Sound*: J. Tyndall, *Der Schall: Acht Vorlesungen gehalten in der Royal Institution von Grossbritannien*, ed. H. Helmholtz and G. Wiedemann, 2nd edn (Braunschweig: Friedrich Vieweg, 1874). The first edition had been published in 1869.

7. *Light*: J. Tyndall, *Das Licht: Sechs Vorlesungen gehalten in Amerika im Winter 1872–1873*, ed. G. Wiedemann (Braunschweig: Friedrich Vieweg, 1876).
8. *8 taler 7 pf*: a taler, or thaler, was a silver coin, formally called a Vereinsthaler, that remained legal tender in the newly-unified Germany even after its official replacement by the mark. Pf. is an abbreviation of pfennig, a low denomination coin equivalent to a British penny, that was worth 1/100th of the new mark (see Note on Money).
9. *my wife*: Clara Wiedemann, who had been involved in the translation of several of Tyndall's books.

To Olga Novikoff 26 October [1875][1] 4679

Royal Institution of Great Britain | 26th Oct[ober].

Dear Friend.

Thank you very much for the correction.[2] I like to spell names accurately.

I was just going out to Guildhall[3] when your note[4] reached me. I thought for once in my life I would see the pageant there.[5]

I know the affair[6] you mention only in a general way. But I thank the Gods, in my own way, that the King was found in a courageous mood.[7]

To day I go down to Shoeburyness[8]—and for the next two or three days I shall be cruising about at the mouth of the Thames, making fog signal observations.[9]

Ever Yours | John Tyndall

[Envelope] Madame de Novikoff | Symonds's Hotel | Brook Street W.

[Postmark] LONDON·W | *[10]* | OC26 | 75

John Tyndall Correspondence in the Olga Novikoff Correspondence Collection, Courtesy of Special Collection, Kenneth Spencer Research Library, University of Kansas. MS 30:Q 16
RI MS JT/1/TYP/3/947

---

1. *[1875]*: the year is established by reference to the 'pageant' at the Guildhall (see n. 5).
2. *the correction*: of the surname of the French theologian Athanase Josué Coquerel, the spelling of which Tyndall had mistakenly surmised in letter 4670.
3. *Guildhall*: a gothic building in Moorgate that had served as the ceremonial and administrative centre of the City of London and its Corporation since the thirteenth century.
4. *your note*: letter missing, but it was received, and presumably written, on 25 October.
5. *the pageant there*: a ceremony to mark the award of the freedom of the City of London to Prince Leopold, Queen Victoria's youngest son, held at the Guildhall on 25 October. The *Times* reported: 'The day was fine and the ceremony excited considerable interest, among not only the members of the [City of London] Corporation, but among the citizens generally. The Lord Mayor and Lady Mayoress, escorted by the sword and mace bearers of the

Corporation, went in state from the Mansion-house to take part in the proceedings of the day'. It was also noted that the 'Civic dignitaries had . . . mustered in considerable force . . . Among the company were . . . Professor Tyndall . . . and many more' ('Prince Leopold at Guildhall', *Times*, 26 October 1875, p. 6).

6. *the affair*: a standoff between the Liberals in the Bayerischen Abgeordnetenkammer (Bavarian Chamber of Deputies) and the conservative Ultramontanes who were the largest party in the Bayerische Ständeversammlung (Bavarian States General). Tyndall and Novikoff had earlier discussed the election in which the Ultramontanes achieved their small majority (see letter 4625, n. 9).
7. *the King was found in a courageous mood*: King Ludwig II of Bavaria (1845–86), who on 20 October issued a public letter, which was posted on noticeboards, expressing his support for the Liberals in their opposition to the Ultramontanes. See 'Bavaria', *Times*, 25 October 1875, p. 5.
8. *Shoeburyness*: a town in Essex at the mouth of the Thames Estuary, close to the Maplin lighthouse, which was constructed from screw-piles in 1838.
9. *making fog signal observations*: of the sound produced by different guns firing charges of gun-cotton. Tyndall gave an account of the guns 'tested at Shoeburyness' in 'Recent Experiments on Fog-Signals', *Roy. Soc. Proc.*, 27 (1878), pp. 245–58, on p. 248.

To William Spottiswoode 30 October 1875 4680

Royal Institution of Great Britain | 30th Oct[ober]. 1875

My dear Spottiswoode.

I shall need your aid and countenance[1] on Monday.[2]

Two things are surely profoundly gratifying to me. It is gratifying to find the Institution after so serious an outlay upon its laboratories,[3] able to afford an addition of £200 a year to my salary.[4] It is more than I can express to you gratifying to find the disposition of the managers[5] towards such as to cause them to contemplate this addition.

But having thus gratified me they will stamp and ratify their friendship for me by not pressing me to accept any addition to my salary. What they allow me is liberal; and my present salary added to that which I receive from the Trinity House and Board of Trade,[6] and swelled by the proceeds of my books,[7] places me in perfectly easy, not to say affluent circumstances.

What I wish you to aid me in is this. Cottrell has now been with us for ten years. Others have been with us and have quitted us, and are now receiving elsewhere far better pay than Cottrell. Mr. Wills,[8] for example, receives £150 a year from Greenwich, and he adds to this income by other employments. Cottrell's income here ought to be equal to that of Mr. Wills at Greenwich. It is now £115 a year: I want you to help me to get £35 a year added to it.

The managers were kind enough to vote, on whose proposition I know not, though I can guess, that some funds were to be placed at my disposal for the payment of my second assistant.[9] Hitherto I have paid him myself, and I have not drawn upon the fund placed at my disposal. But as he is employed almost exclusively in the laboratory I should not object to accepting the salary which I have paid him during the past year; nor should I object to his being paid £50 a year in the future by the Institution. I would therefore draw upon the Institution for £85 instead of £200. I should feel especially grateful to the managers if they would consent to this arrangement.

Do not try to alter the wish here expressed. Your not doing so will make your friend a happier man.

Warm love to Cyril[10] & kindest regard to your wife.[11]

Ever faithfully yours | John Tyndall

RI MS JT/1/T/1303
RI MS JT/1/TYP/3/1250
RI Manager's Minutes (1874–1903), XIII, pp. 64–5

---

1. *countenance*: sanction, encouragement or 'back up' (*OED*).
2. *on Monday*: the RI's business meeting, to be held on 1 November; the meetings were held on the first of every month.
3. *so serious an outlay upon its laboratories*: the construction, in 1872, of new chemical and physics laboratories at the RI had been funded by a bequest of £2,000. See F. A. J. L. James and A. Peers, 'Constructing Space for Science at the Royal Institution of Great Britain', *Physics in Perspective*, 9 (2007), pp. 130–85, on p. 162.
4. *my salary*: Tyndall's salary from the RI at this time was £450 per annum, having been raised to this level, from £300 per annum, in 1868 (see *Ascent of John Tyndall*, p. 133).
5. *the managers*: see letter 4549, n. 4.
6. *that which I receive from the Trinity House and Board of Trade*: when Tyndall took over from Michael Faraday in 1867 as scientific adviser to Trinity House and the Board of Trade he was awarded a salary of £400 per annum (see letter 2614, *Tyndall Correspondence*, vol. 10; and *Ascent of John Tyndall*, p. 283). On Tyndall and Trinity House, see letter 4187, n. 2.
7. *the proceeds of my books*: in letter 4637 Tyndall told Thomas Huxley 'My books this year have brought me in about the same sum as your Edinburgh lectures have brought you'; in letter 4630 Huxley claimed to have 'cleared about £1000' from the lecture series.
8. *Mr. Wills*: Thomas Wills (see letter 4372, n. 10), who had come to the RI in 1868 as the assistant of the new Fullerian Professor of Chemistry, William Odling (1829–1921). He resigned his post in 1873 to become demonstrator in chemistry at the Royal Naval College, Greenwich, and in 1874 also became secretary to the Chemical Section of the Society of Arts. See 'Report of the Council', *Proceedings of the Physical Society*, 3 (1880), pp. 1–14, on pp. 12–13.
9. *my second assistant*: not identified. Until now, Tyndall paid for a second assistant, who he had used since 1862, out of his own pocket (see *Ascent of John Tyndall*, p. 172).
10. *Cyril*: Cyril Spottiswoode.
11. *your wife*: Eliza Spottiswoode.

BIOGRAPHICAL REGISTER

This register contains the names and biographical details of people who are mentioned three or more times in the letters included in this volume. Biographical information about people mentioned only once or twice is contained in the notes appended to the appropriate letter(s).

Conventions

The sources for the biographical entries are standard biographical dictionaries, such as the *Complete Dictionary of Scientific Biography*, the *Oxford Dictionary of National Biography*, the *American National Biography*, Poggendorff's *Biographisch-literarisches Handwörter-buch der exacten Wissenschaften*, *Historisches Lexikon der Schweiz*, *Neue Deutsche Biographie*, and the *Dictionary of Nineteenth-Century British Scientists*. These have not been specified in the entries unless necessary. Additionally, the editors have found material from the Tyndall correspondence and from the journals of Tyndall and Thomas Hirst very useful. Where additional sources add important material, they are given in parentheses at the end of the entry.

Adair, Henrietta Mary (1843–1921), was the youngest of the seven children of Alexander (1791–1863) and Harriet Eliza (née Atkinson, 1811–78) Adair, of Heatherton Park in Somerset. She attended Tyndall's lectures at the RI and they discussed scientific matters. Tyndall fell in love with Adair and proposed to her in late 1869; she declined. They nevertheless continued to correspond. In 1879 Adair was admitted as a member of the RI. She never married.

Adams-Reilly, Anthony (1836–85), was an Irish mountaineer and cartographer. With Peter Guthrie Tait and John Campbell Shairp, he coedited *Life of Forbes* in 1873. This book sought to posthumously vindicate the scientific reputation of James David Forbes, which had been tarnished in a longstanding dispute with Tyndall regarding the motion of glaciers.

Agassiz, Alexander (1835–1910), was an American zoologist, oceanographer, and mining engineer. He was the eldest son of Louis Agassiz, from his first marriage, in Switzerland, to Cecile Braun (d. 1848). In 1874 Agassiz *fils* corresponded with Tyndall regarding the renewed controversy over his father's work on glaciers and the accusations of plagiarism leveled at James David Forbes.

Agassiz, Louis (1807–73), was a Swiss-born American naturalist. He studied medicine, botany, geology, and zoology in Germany and France before being appointed to a position at the University of Neuchâtel. In 1847 he was appointed professor of zoology and geology at Harvard in the United States and founded the University's Museum of Comparative Zoology in 1859. He was a key figure in developing the study of the earth's history, proposed the theory of ice ages, and made a close study of glaciers in the Alps. In 1840 he published *Études sur les glaciers*, which helped establish the scientific study of glaciers. It was on this subject that Agassiz became involved in a dispute over priority of discovery with James David Forbes. Later, in the late 1850s and early 1860s, when Forbes and Tyndall entered a prolonged and acrimonious debate regarding theories of glacial motion, the unsettled question of Forbes's claim to originality would lend weight to highly damaging accusations of plagiarism made by Tyndall.

Allen, Robin (1820–98), was a poet who was appointed as clerk in the Secretary's Office at Trinity House in 1837. From 1867 to 1881 he was the secretary to the Corporation of Trinity House, to which Tyndall served as scientific adviser between 1865 and 1883.

Appleton, William Henry (1814–99), was an American publisher who joined the family business at the age of sixteen. After the death of his father, Daniel Appleton (1785–1849), William expanded the interests of D. Appleton & Co. to include the fields of travel, poetry, and science, becoming the first American publisher to release editions of works on evolution by Charles Darwin and Herbert Spencer. In 1844 he married Mary Moody Worthen (1824–84) and together they had four children. At their Wave Hill estate in Hudson Hill, New York, they hosted guests such as Darwin and Thomas Huxley. Appleton was the proprietor of *Popular Science Monthly*, edited by Edward L. Youmans, and also with Youmans he played a crucial role in launching the International Scientific Series in 1871. William and Mary Appleton visited England in the spring of 1874 to recuperate following the death of their daughter Kate Geary (née Appleton, 1848–73). He also advised Tyndall on investing the surplus generated by his lecture tour of the United States; this resulted in a loss.

Argyll, Duke of: see Campbell, George.

Bacon, Francis (1561–1626), was a renowned statesman and natural philosopher. He is widely credited with developing the standard scientific method through his espousal of inductive reasoning based upon careful experiment and observation. Bacon therefore served as a role model for Tyndall and many other nineteenth-century men of science.

Barnard, Frank (1828–95), was a watercolor painter and nephew of Michael Faraday. He was born into the Sandemanian Church, but formally joined only in 1877, becoming a deacon by the mid-1880s. From the late 1860s, he lived with his sister Jane in Islington.

Barnard, Frederick Augustus Porter (1809–89), was an American academic and university president. In the family tradition, Barnard studied at Yale and began his own career in education following his graduation in 1828. He held several academic positions over the years, professor of mathematics and natural philosophy at the University of Alabama (1837–54), professor of physics, astronomy, and civil engineering at the University of Mississippi (1854–61), and president of Columbia College (1864–89). It was at Columbia that Barnard was most active in educational reform, which included support for coeducation.

Barnard, Jane (1832–1911), was the daughter of Michael Faraday's younger sister Margaret (1802–62), who had married John Barnard (1797–1880) in 1826. Jane Barnard was baptized at the Sandemanian Church in Islington in 1844, her mother having become a church member a week earlier. As an adult, she formally joined the church in 1857. Barnard was close to her uncle, and lived with him at the RI from the early 1860s, performing the duties of secretary and nurse. Following Faraday's death in 1867, his will described her as "my dear niece Jane who has now for many years been our affectionate companion and support" (quoted in Geoffrey Cantor, *Michael Faraday: Sandemanian and Scientist* [London: Macmillan, 1991], 84). She continued to care for Faraday's widow, Sarah, and also endeavored to maintain her uncle's posthumous reputation.

Barnard, Margaret (née McMurray, 1820–91), was the wife of Frederick Augustus Porter Barnard. She was born in England before moving with her parents to Ohio, and first met her husband while staying in Alabama with a cousin, who was a friend of Barnard's. They married in 1847. Although she dedicated her life to managing her husband's affairs, her own strong concern with temperance reform led him to take up the same cause.

Bence Jones, Henry (1813–73), was a physician and medical chemist. He was educated at Harrow School and Trinity College, Cambridge, before studying medicine at St George's Hospital in London. Attracted by the potential applications of chemistry to clinical medicine, he studied organic chemistry with

Bence Jones, Millicent (née Acheson, 1812–87), was the wife and cousin of Henry Bence Jones. She was the daughter of Archibald Acheson (1776–1849),

2nd Earl of Gosford, and Mary Acheson (née Sparrow, 1777–1841). Following their marriage in 1842, the Bence Joneses had three sons and four daughters. They lived first in Lower Grosvenor Street, and later on in Brook Street, Grosvenor Square, both in fashionable Mayfair, although Millicent was compelled to relocate to Marylebone following Henry's death.

Bruno, Giordano (1548–1600), was an Italian philosopher. Bruno entered the Dominican Order of monks as a young man and he soon distinguished himself as a brilliant scholar. However, his increasingly unorthodox views forced him to flee his native city of Naples and seek sanctuary in various European cities, including Paris and London. During his time in England, Bruno published key works on cosmology, which expanded on the theories of Nicolaus Copernicus (1473–1543). Following Copernicus, Bruno proposed that the Earth orbited the sun, but he took this proposition further, speculating that the stars were also suns with systems of planets that might harbor life. After almost two decades in exile, Bruno returned to Italy, where he was arrested by the Catholic Inquisition, imprisoned for seven years, and finally burned at the stake for his heretical beliefs. Bruno has subsequently been remembered as a martyr to science, although these interpretations are challenged by some scholars. His theological stance has also been identified with pantheism, and it was for this reason that Bruno was a key historical figure referenced by Tyndall in his Belfast Address.

Bunsen, Robert (1811–99), was a German chemist. A pioneer in the fields of spectroscopy and photochemistry, and perhaps best remembered for developing the Bunsen burner, a gas-fueled burner widely used in laboratories. Among his numerous achievements was the discovery of the elements cesium and rubidium (with the physicist Gustav Kirchoff). Bunsen was greatly admired by Tyndall, who chose to study at the University of Marburg in 1848, where Bunsen was a professor at the time. The two subsequently maintained a friendly correspondence.

Burdon-Sanderson, John Scott (1828–1905), was a physiologist who investigated outbreaks of diphtheria and cholera. In 1871 he was appointed principal of the newly founded Brown Animal Sanatory Institution, a veterinary research laboratory in London, and three years later became Jodrell Professor of Physiology at UCL.

Campbell, Frederick Alexander (1819–93), was a colonel in the British army and the superintendent of the Gun Factories, Royal Arsenal, Woolwich.

Campbell, George (1823–1900), 8th Duke of Argyll, was a Liberal statesman who served in several government posts under three different prime ministers. Campbell also wrote and published on a diverse range of subjects, including ornithology, geology, economics, and evolution. His influential book *The*

Reign of Law (1867) argued that the recognition of design in nature elevated humans above animals. He was elected FRS in 1851, and served terms as president of the BAAS (1855) and the Geological Society (1872–74).

Carlyle, Thomas (1795–1881), was a Scottish essayist and historian, and one of the most influential intellectual figures of the nineteenth century. His marriage in 1826 to Jane Baillie Welsh (1801–66) was not happy, as their letters indicate, but they were inimitable hosts who entertained an impressive circle of friends, particularly during their time in London. Carlyle was a major literary influence on Tyndall, who avidly read his publications. After reading *Past and Present* (1843), Tyndall commented that Carlyle "must be a true hero. My feelings towards him are those of worship 'transcendental wonder' as he defines it" (Journal, 18 July 1847, RI MS JT/2/13a/231). From the 1860s, Tyndall and Carlyle became close friends, and Carlyle's authoritative advice helped Tyndall during the renewal of the glacier controversy in 1873. In the following year, he was awarded the Prussian Order of Merit, but declined an offer to be made a Knight of the Order of the Bath.

Clark, Andrew (1826–93), 1st baronet, was a Scottish physician who had many prominent patients, including William Gladstone, Thomas Huxley, and Charles Darwin. Regarded as one of the leading society doctors, Clark had the largest London consulting practice of his day. He was married twice, first in 1851 to Seton Mary Percy (d. 1858) and second in 1862 to Helen Annette Doxat (1838–1922), with whom he had a son and two daughters. Clark was created baronet in 1883, elected FRS in 1885, and became president of the Royal College of Physicians in 1888.

Clausius, Adelheid (née Rimpham, 1833–75), was the first wife of Rudolf Clausius. Born in Brunswick, she was an orphan and lived with her sister until in 1859 she married Clausius, with whom she had six children. Their son Rudolf John was named after Tyndall, who was a close friend.

Clausius, Rudolf John "Johnny" (1864–1915?), was the son of Adelheid and Rudolf Clausius. Known affectionately as "Johnny," he was named after Tyndall, who was a close friend of his parents. He later studied law at the University of Bonn, and practiced as a lawyer in Cologne.

Clausius, Rudolf Julius Emanuel (1822–88), was a German mathematician and physicist and one of the founders of modern thermodynamics. His famous paper on the theory of heat, "Ueber die bewegende Kraft der Wärme," was published in the *Annalen der Physik* in 1850. He held a professorship in mathematical physics at the Polytechnikum in Zurich from 1855, moved to the University of Würzburg in 1867, and finally to the University of Bonn in 1869. He was injured working as a volunteer ambulance officer at the Battle of Gravelotte on 18 August 1870 during the Franco-Prussian War, and suffered with mobility problems for the rest of his life. Tyndall translated many of his major works into English for publication in the *Phil. Mag.* and *Scientific Memoirs* in the early 1850s. He was very close to Tyndall, as was his wife, Adelheid, and they named their son Rudolf John after Tyndall. Following Adelheid's death in 1875, Rudolf married Sophie Sack (1856–1941) in 1886, with whom he had another child.

Cottrell, John (1843–90), was Tyndall's senior laboratory assistant from ca. 1866. Cottrell served at the RI from 1866 until 1885. In 1872 he traveled to the United States with Tyndall and in 1875 assisted with observations at the Aletsch glacier. Cottrell also helped to develop the apparatus through which Tyndall demonstrated his research on sound waves, and had a paper read (by Tyndall) to the RS in February 1874 ("On the Division of a Sound-Wave by a Layer of Flame or Heated Gas into a Reflected and a Transmitted Wave," *Roy. Soc. Proc.* 22 [1874]: 190–91). In the 1871 census, he was recorded as a laboratory assistant, living at 35 Upper Charlton Street, Marylebone.

Coxe, James (1811–78), was a Scottish physician who specialized in the causes and treatment of mental illness. As commissioner in lunacy for Scotland from 1857 until his death, he did much to reform the institutions responsible for the care of mentally ill people, arguing against the practice of restraining patients. Coxe, who was knighted in 1863, was a friend of Tyndall, and in 1874 advised him during the renewed glacier controversy, in addition to writing supportive articles about it for the *Scotsman* newspaper.

Coxe, Mary Anne (née Cumming, 1806–75), was the wife of James Coxe. She collected portraits of Tyndall, and assisted her husband in his efforts on Tyndall's behalf during the renewed glacier controversy.

Darwin, Charles (1809–82), was a naturalist and geologist, best known for his book *On the Origin of Species* (1859). In it, Darwin introduced his theory of evolution by natural selection, born out of his five-year research trip aboard HMS *Beagle* from 1831 to 1836. Two years after returning, he married his first cousin Emma Wedgwood (1808–96). Though a devout Christian, she supported her husband's evolutionary research. Together they had ten children and made their family home at Down in Kent. In 1864 he was awarded the RS's Copley Medal but only after being nominated three years in a row. When *The Descent of Man* was published in 1871, "Darwinism" became a catchword among the public. His evolutionary writing cycle was completed in 1872 with the publication of *The Expression of the Emotions in Man and Animals*. Earlier in Tyndall's career, Darwin had encouraged his research on glaciers and supported him in his disagreement with James David Forbes, although he did not become involved when the controversy was revived in 1873. In the following year, Darwin read proofs of Tyndall's Belfast Address, which praised his evolutionary theories, and in 1875 assisted in Tyndall's experiments on germ bacteria that refuted spontaneous generation.

Debus, Heinrich (1824–1915), was a German-born chemist who studied at the University of Marburg (1845–48), where he met Tyndall, Edward Frankland, and Thomas Hirst. Debus moved to England in 1851 and taught chemistry at Queenwood College until 1867, when he began teaching at Clifton College, Bristol. Later, Debus lectured at Guy's Hospital Medical School (1870–88) and became professor of chemistry at the Royal Naval College, Greenwich (1873–88).

Delane, John Thadeus (1817–79), was the editor of the *Times* from 1841 to 1877. Delane wielded considerable influence through his position, and became the confidant of many cabinet ministers and other powerful politicians. Privately, he generally favored Liberal policies, but attempted to maintain the newspaper's impartiality.

Disraeli, Benjamin (1804–81), was a British statesman of the Conservative Party and novelist who served twice as prime minister, in 1868 and from 1874 to 1880.

Douglass, James Nicholas (1826–98), was a civil engineer who specialized in the construction of lighthouses, and worked extensively with Tyndall, and earlier with Michael Faraday, on lighthouse illumination and fog signals. He contended that experiments at Haisborough lighthouse in Norfolk, in which Tyndall was involved, validated his own plan for oil burners, rather than gas burners.

Draper, John William (1811–82), was an American chemist and historian. Born in northern England, he emigrated in 1832 after studying chemistry at UCL. An early experimenter with photography, he took the first known photographs of the moon in 1839–40, as well as some of the first photomicrographs, and demonstrated that only absorbed rays were effective on daguerreotypes (which became known as "Draper's law"). As a historian, Draper is known for a number of works, including *A History of Intellectual Development in Europe* (1862), *History of the American Civil War* (1867–70), and, most notably, *History of the Conflict between Religion and Science* (1874). The latter, which argued that Christianity had suppressed the advance of science, went through multiple editions and was translated into several languages.

du Bois-Reymond, Emil Heinrich (1818–96), was a German physiologist. While studying anatomy and physiology at the University of Berlin, du Bois-Raymond became interested in fish that were able to produce electrical currents. After completing his thesis on this subject in 1843, du Bois-Reymond spent the rest of his career researching the conduction of electricity by nerve and muscle tissue, producing foundational works in the field of electrophysiology. In 1858 he was appointed professor of physiology at the University of Berlin, becoming an influential figure in German science. He was one of the first and most ardent supporters of Darwinism in the German-speaking world.

Duppa, Adeline Frances Mary (née Dart, 1845–95), was the wife of Baldwin Francis Duppa. After her marriage in 1869, she became a botanical artist, making drawings of plants native to Italy and Madeira, as well as of English fungi.

Duppa, Baldwin Francis (1828–1873), was a chemist at the Royal College of Chemistry. He was close friends with Edward Frankland, with whom he worked at the Royal College of Chemistry, having previously done so at the RI. One of Duppa's final acts before his death in November 1873 was to dictate a letter addressed to Tyndall regarding the refraction of light in a glass of water (letter 4212).

Egerton, Mary "May" Alice (1848–1924), was the eldest daughter of Mary and Edward Egerton. She married Beauchamp Tower (1845–1904) in 1902 ("Births," *Times*, 19 February 1848, 8; "Wills and Bequests," *Times*, 11 June 1924, 15).

Egerton, Mary Frances (née Pierrepont, 1819–1905), was the daughter of Charles Pierrepont (1778–1860), 2nd Earl Manvers. In 1845 she married Edward Christopher Egerton (1816–69); they had six children, Charles Augustus (1846–1912), Mary Alice, Charlotte (1849–1926), Emily Margaret (1852–94), Georgina Renira (1854–1930), and Hugh Edward (1855–1927). Egerton was a frequent correspondent of Tyndall's and regularly attended his lectures at the RI, as well as discussing philosophical issues with him.

Emerson, Ralph Waldo (1803–82), was an American author, poet, and lecturer. He was a leading figure in the transcendentalist movement, which he heavily influenced through such essays as "Nature" (1836), "The American Scholar" (1837), and "Self-Reliance" (1841). Emerson's trip to Egypt and Britain in 1872–73, accompanied by his daughter Ellen (1850–1920), included many social engagements with London's literary and scientific elite. Tyndall was deeply influenced by reading Emerson's work as a young man, and they met in person during his lecture tour of the United States.

Epicurus (341–270 BCE), was an ancient Greek philosopher who established a school of philosophy known as Epicureanism. Central to this school of thought was the principle that pleasure, defined as the absence of suffering, was the primary aim in life. Furthermore, Epicurus was an atomist, believing that the physical universe consisted of indivisible components known as atoms, and that all phenomena could be explained by the motion of these fundamental particles. This materialistic theory of the universe, which argued that the gods played no part in determining the course of events, led Epicurus to be branded, incorrectly, as an atheist by later commentators. When Tyndall chose to discuss the work of Epicurus in his Belfast Address, he consequently laid himself open to accusations of materialism and atheism.

Faraday, Michael (1791–1867), was a natural philosopher, Fullerian Professor of Chemistry (1833–67) at the RI, and a member of the Sandemanian Church. His extensive research focused on electricity and encompassed electrochemistry, electromagnetism, diamagnetism, and magneto-optics. Faraday also held the positions of director of the laboratory (1825–67) and superintendent of the house (1852–67) at the RI. He was Tyndall's patron and mentor and helped him to obtain the position of professor of natural philosophy at the RI in 1853. Despite significant differences in their views on religion and on the nature of magnetism, they held the highest mutual regard for each other. In 1867 Tyndall succeeded Faraday as superintendent of the house and director of the laboratory at the RI. Tyndall's *Faraday as a Discoverer* (1868) was his tribute to his mentor, and in 1874–75 he oversaw the creation of a marble statue of Faraday, although the process proved problematic.

Faraday, Sarah (née Barnard, 1800–79), was the wife of Michael Faraday. She was the daughter of Edward Barnard (1767–1855), a successful London silversmith and Sandemanian, and Mary Barnard (née Boosey, ca. 1769–1847). She likely met Faraday at the Sandemanian meeting house in Paul's Alley in the City of London. After they married in 1821, she moved into his rooms at the RI. They had no children, but two nieces, Jane Barnard and Margery Ann Reid (1815–88), who lived with them at different times for extended periods. In later years, she took on many of her husband's administrative responsibilities at the RI. From 1858 onward, the Faradays lived at Hampton Court Palace, in rooms provided by Queen Victoria (1819–1901). After her husband's death in 1867, Sarah lived with Jane Barnard at Barnsbury Villa in Islington.

Farrer, Thomas Henry (1815–99), 1st baronet, was a civil servant and statistician. On the recommendation of his brother-in-law Stafford Northcote (1818–87), he started work at the Board of Trade, first as an assistant but soon moving up the ranks, eventually becoming the board's first sole permanent secretary. He served in this role from 1867 to 1886, and during that time had significant influence over policy. Farrer was married twice, first to Frances Erskine (1825–70), with whom he had four children, and second to Katherine Euphemia Wedgwood (1839–1931), Charles Darwin's niece.

Fichte, Johann Gottlieb (1762–1814), was a German philosopher and leading figure in the movement known as German idealism. Although Fichte was an original thinker in his own right, much of his work built on that of Immanuel Kant (1724–1804). Tyndall was deeply influenced by Fichte's work, which he first encountered through an 1847 English translation of *Einige Vorlesungen über die Bestimmung des Gelehrten* (*The Vocation of a Scholar*, 1794). Tyndall regularly discussed Fichte in letters with Mary Egerton, and drew on his ideas in the Belfast Address.

Foley, John Henry (1818–74), was an Irish sculptor who spent his working life in London. Born in Dublin, his talent earned him a place at the Royal Academy in London. Foley's works include numerous statues of public figures, most notably the bronze figure of Prince Albert at the center of the Albert Memorial in Kensington. He died suddenly in August 1874, leaving unfinished a marble statue of Michael Faraday, a commission funded by public subscription and overseen by Tyndall.

Forbes, George (1849–1936), was a Scottish electrical engineer and astronomer, and the second son of James David Forbes. He was appointed professor of natural philosophy at Glasgow's Andersonian University in 1873. In the following year, he edited a new translation of Louis Rendu's *Glaciers of Savoy*, hoping to refute accusations, made by Tyndall and Thomas Huxley, that his father had plagiarized from the book. Before Tyndall responded in June 1874, Forbes had embarked on the British expedition to observe the transit of Venus from the Hawai'ian Islands, leading a team of astronomers. Forbes's subsequent scientific work focused primarily on the applications of electricity, notably developing the use of carbon brushes in electric motors, and he served as a consultant on pioneering hydroelectricity schemes around the world. This innovative research did not make him a wealthy man, and Forbes lived in relative poverty toward the end of his life.

Forbes, James David (1809–68), was a Scottish physicist and geologist. He was elected FRS in 1832, and the following year was made professor of natural philosophy at Edinburgh University. Forbes received the RS's prestigious Rumford Medal in 1838 for his experimental investigations on the polarization of heat, which he had been the first to demonstrate in 1834. He was also a pioneering mountaineer, and while in the Alps studied the motion of glaciers, which, at the end of the 1850s, brought him into controversy with Tyndall and Thomas Huxley. Tyndall was particularly critical of Forbes's apparent misuse of terminology, and for taking credit for discoveries that Tyndall and Huxley believed had originally been made by Louis Rendu. The dispute would be revived by Tyndall in 1872 with *Forms of Water*. It was then intensified in 1873 by the publication of the *Life of Forbes*, which sought to vindicate its subject's posthumous reputation, and continued in the following year when Forbes's son George edited a new translation of Rendu's *Glaciers of Savoy*.

Foster, Michael (1836–1907), was a physiologist. He studied medicine at UCL, and returned there, after a brief period in medical practice, to teach practical physiology, becoming an assistant to and close friend of Thomas Huxley. Foster was made professor in 1869, but a year later was elected as praelector in physiology at Trinity College, Cambridge, and would later become the first chair in this subject at the university. He was an influential figure in the organization of the Cambridge biological school.

Fox, Wilson (1831–87), was a physician. He became professor of clinical medicine at UCL in 1866, and was appointed as physician extraordinary to the queen in 1870. During 1873–74 he treated Thomas Hirst for various ailments.

Frankland, Edward (1825–99), was a chemist and one of Tyndall's closest friends. They had met at Queenwood College in 1847, before leaving together in the following year to study at the University of Marburg in Germany. After holding a professorship at Owens College, Manchester (1856–57), Frankland moved to London and held a number of positions including posts at St Bartholomew's Hospital and the East India Company's military college. He was professor of chemistry at the RI (1863–68) and afterward at the Royal College of Chemistry (1868–85). In 1864 Frankland joined Tyndall and seven others as members of the X Club, a monthly dining club dedicated to the promotion of science in British society.

Frohschammer, Jakob (1821–93), was a German theologian and philosopher. He became a Catholic priest in 1847, but quickly resented the need to submit to Church authority, especially on intellectual issues. In 1855 Frohschammer switched from theology to philosophy, becoming professor of the subject at Ludwig-Maximilians-Universität in Munich, and publishing a series of books, including *Einleitung in die Philosophie und Grundriss der Metaphysik* (1858) and *Über die Freiheit der Wissenschaft* (1861), in which he disputed traditional Catholic doctrines and affirmed that science should be independent of all religious authority. His views were denounced by Pope Pius IX (1792–1878) in 1862 and students were prohibited from attending his lectures, although he received support both from them, with five hundred signing an address to him, and from the Bavarian king Maximilian II (1811–64). In the following year, Munich's liberal Catholics began a schismatic Old Catholic movement, protesting at the recent assertion of papal infallibility. However, Frohschammer refused to join the movement, considering that their liberalism did not go far enough, and instead he founded his own journal, the *Athenäum*, in which he expressed support for Darwinian evolution. He was excommunicated from the Catholic Church in 1871. In subsequent works such as *Das neue Wissen und der neue Glaube* (1873), Frohschammer opposed materialism and insisted that a creative power was necessary in nature. Tyndall was introduced to his work by

Olga Novikoff, and read his books with considerable interest. In particular, he sympathized with Frohschammer's persecution by Jesuits, which he mentioned in his "Preface to the Seventh Thousand" of the *Belfast Address*. They corresponded briefly in early 1875, when Frohschammer published an account of Tyndall's career ("Professor John Tyndall und die Freiheit der Wissenschaft in England," *Beilage zur Allgemeine Zeitung*, 9 January 1875, 121–23).

Furness, William Henry (1802–96), was an American Unitarian minister and author. As the minister of the First Unitarian Church of Philadelphia from 1825 until his retirement in 1875, he was known for asserting that biblical miracles had natural explanations.

Gascoyne-Cecil, Robert Arthur Talbot (1830–1903), 3rd Marquess of Salisbury, was a Conservative politician who would serve as prime minister three times in the 1880s, 1890s, and 1900s. Educated at Eton and Christ Church, Oxford, he was elected to the House of Commons in 1854, serving as secretary of state for India from 1866 to 1867. He returned to the role in 1874 in the newly elected Conservative government, before becoming foreign secretary in 1878. From 1881 he led the Conservative Party from the House of Lords, which he had joined in 1868, and was the last British prime minister to serve from the upper chamber. He also conducted original research in the physical sciences, publishing in the *Phil. Mag.*, and equipping his family seat, Hatfield House in Hertfordshire, with a modern laboratory. He was elected FRS in 1869, and on becoming chancellor of Oxford University in the same year made serious provisions for physical science there. Tyndall visited him at Hatfield House on several occasions, and in the period covered by this volume they corresponded in an official capacity regarding solar observations in India.

Gauthier-Villars, Jean-Albert (1828–98), was a French engineer and publisher. Born to a family of printers in Lons-le-Saunier, he studied at the École Polytechnique from 1848 to 1850 before becoming a telegraph engineer and then serving in the military. In 1864 he purchased the Mallet-Bachelier printing house in Paris, which published the *Comptes rendus hebdomadaires des séances de l'Académie des sciences*. Under his direction, the firm, renamed Gauthier-Villars, specialized in mathematical and scientific books and magazines, and published French translations of Tyndall's works.

Gladstone, John Hall (1827–1902), was a chemist who worked on the chemical relations of light, and also on batteries. He was elected FRS in 1853, and was active in many other scientific societies, becoming a founding member of the Physical Society of London in 1874. Gladstone met Michael Faraday during his time serving on the Committee for Lighthouses, and would later become his biographer. From 1873 to 1877 Gladstone was Fullerian Professor of Chemistry at the RI, during which time he clashed with Tyndall about the safety of experiments conducted in the Chemical Laboratory.

Gladstone, William Ewart (1809–98), was a Liberal politician and author born into a wealthy evangelical family. Educated at Eton and then Christ Church, Oxford, Gladstone was encouraged by his family and connections to pursue a career in politics. This was assisted by his marriage, in 1839, to Catherine Glynne (1812–1900), whose aristocratic family owned the Hawarden estate in northern Wales. After entering the House of Commons in 1832 as a Conservative, he joined Robert Peel's breakaway faction in 1846. With the Peelites merging with the Liberals, Gladstone served as chancellor in three Liberal governments between 1852 and 1866. In 1868 he became prime minister, serving until January 1874, the first of his four terms in the post. A general election in early February 1874 resulted in a Conservative victory, with Benjamin Disraeli taking on the role of prime minister for a second time. As a result, Gladstone stepped down as leader of the Liberal Party, but retained his seat in the House of Commons. In 1880 the Liberals defeated the Conservatives, and Gladstone resumed his role as the party's leader, serving in his second premiership until 1885. He would hold this position twice more: for a brief period in 1886, and from 1892 to 1894.

Glaisher, James (1809–1903), was a meteorologist and aeronaut. From 1836 to 1874 he was superintendent of the Department of Meteorology and Magnetism at the Royal Greenwich Observatory. He was also a pioneering balloonist, frequently making ascents to high altitude with his copilot Henry Tracy Coxwell (1819–1900).

Grattan, Richard (1790–1886), was an Irish physician and writer. A fellow of the King and Queen's College of Physicians in Ireland from 1817, he first became a writer while serving as physician to Cork Street Fever Hospital in Dublin, composing that institution's *Annual Medical Report*. Related to Henry Grattan (1746–1820), an Irish politician and vocal opponent of the 1800 Act of Union between Britain and Ireland, he was an ardent supporter of Irish independence and gained notoriety for publishing on this subject and many other controversial issues, including freethought in relation to religion. As an octogenarian, he offered vociferous support to Tyndall during the controversy over the Belfast Address, although it was not accepted.

Gray, Elisha (1835–1901), was an American electrical engineer who visited Europe in August and September 1874 to test his system for telegraphically transmitting musical tones on British submarine cables. He would subsequently become embroiled in a famous legal dispute with Alexander Graham Bell (1844–1922) over the invention of the telephone, which Gray lost.

Griffith, George (1833–1902), was a science master at Harrow School, and assistant general secretary to the BAAS from 1862 to 1878 and 1890 to 1902.

Gross, Samuel David (1805–84), was an American surgeon. He trained at Jefferson Medical College in Philadelphia, where he would eventually return as professor of surgery from 1856. Gross became one of the most prominent men in his field, and is notably depicted in the *Gross Clinic*, an 1875 painting by Thomas Eakins (1844–1916). He met Tyndall in Philadelphia during the latter's lecture tour of the United States in 1872–73.

Grote, Harriet (née Lewin, 1792–1878), was a writer and the wife of the political radical and historian George Grote (1794–1871). She was a friend of Tyndall's.

Hamilton, Claud (1813–84), was a Conservative politician and the father of Louisa Hamilton. He was MP for County Tyrone, Ireland, from 1835 to 1837 and 1839 to 1874, and served as treasurer of the household and vice-chamberlain of the household in four Conservative governments. Hamilton married Elizabeth Proby in 1844, and they had three daughters and one son.

Hamilton, Douglas James (1856–1931), was a soldier and the son of Elizabeth and Claud Hamilton, and brother of Louisa Hamilton. He served in Egypt and Sudan, rising to the rank of brevet-colonel before his retirement in 1908. He also had a brief political career, serving as an MP for less than a year in 1910.

Hamilton, Elizabeth Emma (née Proby, 1821–1900), was a translator and the mother of Louisa Hamilton. She was the daughter of Admiral Granville Leveson Proby (ca. 1782–1868), 3rd Earl of Carysfort, and Isabella Proby (née Howard, 1783–1836). In 1844 she married Claud Hamilton, with whom she had four children. In the early 1870s she introduced her children into London scientific society, socializing with the families of John Lubbock, John Birkbeck (1817–90), and John Herschel. Through these connections, Lady Claud, as she was known, became acquainted with Tyndall in early 1873, and she encouraged his friendship with her eldest daughter, Louisa, even ensuring that their summer holidays in Switzerland that year coincided. Following her daughter's marriage to Tyndall in February 1876, she assisted her son-in-law by translating a biography of Louis Pasteur by René Vallery-Radot (1853–1933). With Pasteur himself requesting that the book be published in English, the translation, as Tyndall explained to its readers, "was confided, at my suggestion, to Lady Claud Hamilton" (*Louis Pasteur: His Life and Labours* [London: Longmans, Green, and Co., 1885], xi).

Hamilton, Emma (1847–1924), was the second daughter of Elizabeth and Claud Hamilton, and sister of Louisa Hamilton.

Hamilton, Louisa (1845–1940), was the eldest daughter of Elizabeth and Claud Hamilton. In the early 1870s, her family, led by her mother, began socializing with the families of John Lubbock, John Birkbeck (1817–90), and John Herschel, all friends of Tyndall. He first met the Hamiltons in early 1873, soon after his return from the United States, and they began attending his lectures and related social events at the RI. Tyndall quickly became a close friend of the family, visiting their country house, Heathfield Park, in Sussex, and, during August and September 1873, traveling with them in the Swiss Alps. By the fall of 1875, Tyndall and Hamilton began corresponding, sharing a mutual enthusiasm for German landscape and history in their letters. Hamilton married Tyndall at Westminster Abbey in February 1876, and she spent the next decade acting as hostess for dinners and other gatherings at the RI. She also worked closely with Tyndall on his research and writing, and assisted with experiments and note taking in the laboratory. They did not have children. After John Tyndall's death in 1893, which Louisa Tyndall inadvertently helped bring about with an accidental overdose of chloral hydrate, she began the task of collecting and editing her husband's papers and correspondence, intending to write his biography, but this work was not completed.

Hamilton, Mary (1851–1939), was the third daughter of Elizabeth and Claud Hamilton, and sister of Louisa Hamilton.

Hastings, Charles Sheldon (1848–1932), was an American physicist who, in 1874, was a recipient of the Tyndall Fund, which used the surplus earnings from Tyndall's lecture tour of the United States in 1872–73 to support American students of physics. In 1873 he completed a PhD under Chester Smith Lyman (1814–90) at Yale's Sheffield School of Science and was made a demonstrator in physics. With support from the Tyndall Fund, he left this post and spent the next three years studying in Europe, in Heidelberg, Berlin, Munich, and Paris. Hastings joined the new Johns Hopkins University when he came back to the United States, then returned to the Sheffield School of Science as professor of physics and director of its physics laboratory. His most significant research was in optics, especially the design of large telescopes that were integral to the development of astronomy in the United States.

Hawkshaw, John (1811–91), was a civil engineer. He distinguished himself in the construction of railways and canals, most notably the Charing Cross and Cannon Street railways, and the Severn Tunnel between southern Wales and southwestern England. Hawkshaw was knighted in 1873, and in September 1875 succeeded Tyndall as president of the BAAS.

Helmholtz, Anna (née von Mohl, 1834–99), was a German translator, and the wife of Hermann Helmholtz. Born into an academic family, she became well known in Berlin as a hostess of intellectual salons. She married Hermann

Helmholtz in 1861, and together they had three children. With Clara Wiedemann, Anna Helmholtz translated the German editions of Tyndall's *Sound* (*Der Schall*, 1869) and *Heat Considered as a Mode of Motion* (*Wärme betrachtet als eine Art der Bewegung*, 1875).

Helmholtz, Hermann Ludwig Ferdinand (1821–94), was a German physicist and physician. Educated in military medicine at the Friedrich-Wilhelms-Universität in Berlin, he was first a teacher of anatomy at the Berlin Academy of Arts in 1848. In the following year, he became professor of physiology at the University of Königsberg, and then professor for anatomy and physiology at the University of Bonn in 1855, where he turned his attention to physics. From 1858 Helmholtz was professor of physiology at Heidelberg, but in 1870 he accepted an offer to become professor of physics at the University of Berlin on the condition that the Prussian government provide funds to build a new physics laboratory. These funds were still not agreed by the end of 1870, largely because of the Franco-Prussian War, and in January 1871 Helmholtz was offered a new professorship of experimental physics at Cambridge and the directorship of the university's new Cavendish Laboratory. However, he declined the offer and came to Berlin in April 1871, where his physical institute was finally completed five years later. Helmholtz made several contributions to numerous fields, including thermodynamics, electrodynamics, and the conservation of energy, as well as physiological studies of sound and optics. He also arranged for Tyndall's books to be published in German, urging his wife, Anna, to undertake the role of translator. In 1883 he was raised to the German nobility, meaning that he and his family were thereafter styled "von" Helmholtz.

Henry, Caroline (1839–1920), was the youngest daughter of Harriet and Joseph Henry.

Henry, Harriet (née Alexander, 1808–82), was the wife and cousin of Joseph Henry. They married in 1830 and had four children.

Henry, Helen Louisa (1836–1912), was the middle daughter of Harriet and Joseph Henry.

Henry, Joseph (1797–1878), was an American physicist. After attending Albany Academy from 1812 to 1822, Henry held various roles there, including laboratory assistant and teacher, eventually becoming professor of mathematics and natural philosophy in 1826. His early research focused on electromagnetism. In 1832 Henry was appointed to the chair of natural philosophy at the College of New Jersey (subsequently Princeton University) and, three years later, was elected to the APS. A trip to Europe in 1837 brought an opportunity to meet many leading men of science, Michael Faraday among them. In 1846 Henry was appointed the first secretary of the new Smithsonian Institution in Washington, DC, serving in this role until 1878. In 1870 he took a second trip to Europe, and attended the international conference on the metric standard in Paris. He was a friend of Tyndall's, whom he met during the latter's lecture tour of the United States in 1872–73, and was a trustee of the Tyndall Fund, which used the surplus earnings from the tour to support American students of physics. However, the two conducted competing experiments on fog signals, and in 1875 they clashed over the idea of acoustic clouds obstructing sound.

Henry, Mary Anna (1834–1903), was the eldest daughter of Harriet and Joseph Henry. She kept an extensive diary that recorded her experiences living at the Smithsonian Institution in Washington, DC, during the American Civil War, which was later published as *The Civil War out My Window: Diary of Mary Henry* (2014).

Herschel, John (1837–1921), was a military engineer. The son of the celebrated astronomer John Frederick William Hershel, he was born in Cape Town, South Africa, while his father was working on an astronomical survey. He entered the East India Company as an engineer, and took part in the Great Trigonometrical Survey of India from 1864 to 1872, also using this time to conduct astronomical observations. Returning to England in 1873, he began work on collecting his father's letters. Herschel eventually became deputy superintendent of the Trigonometrical Survey before retiring from the army in 1886.

Herschel, John Frederick William (1792–1871), was a polymathic man of science, particularly noted in the field of astronomy. The son of astronomer William Herschel (1738–1822), he began working with his father after graduating from St John's College, Cambridge, where he was senior wrangler in 1813. He made extensive observations of stars and nebulae, particularly while living in South Africa from 1834 to 1838, and also named moons of Saturn and Uranus, the latter planet having been discovered by his father. Among his numerous other scientific investigations were experiments in photography and the chemical power of ultraviolet rays. In addition, Herschel's *Preliminary Discourse on the Study of Natural Philosophy* (1830) was a hugely influential treatise on scientific method.

Hirst, Emily "Lilly" Anna (1853–83), was the niece and housekeeper of Thomas Hirst. Her parents were William and Rachel Hirst, and she studied at Bedford College, London, which had been founded in 1849 as the first higher education college for women in Britain (*Hirst Journals*, Index, 50).

Hirst, Thomas Archer (1830–92), was a mathematician, educator, and Tyndall's closest friend and correspondent. His friendship with Tyndall began in Halifax, Yorkshire, in 1845, when they both worked for the land agent surveyor Richard

Carter (1818–95). Encouraged by Tyndall, who acted as a mentor to him, Hirst studied at the universities of Marburg, Göttingen, and Berlin between 1849 and 1852. In 1853 he replaced Tyndall as mathematics teacher at Queenwood College in Hampshire, then resigned in 1856 to look after his sick wife, Anna Hirst (née Martin, 1831–57), who died from tuberculosis in Paris in the following year. During the 1860s Hirst established himself as an active and respected member of the London scientific community. He taught mathematics at UCL from 1860 to 1864, and subsequently held professorships of physics (1865) and mathematics (1867–70). Hirst was elected to the RS council in 1864, and in the same year was, with Tyndall, a founding member of the X Club, a monthly dining club dedicated to the promotion of science in British society. He also served as general secretary of the BAAS from 1866 to 1870. In 1873 Hirst was appointed director of studies at the Royal Naval College, Greenwich, although persistent ill health interfered with his duties. For much of his life, he kept a detailed journal, later published as *Hirst Journals*, which provides invaluable information on his close and enduring friendship with Tyndall.

Hooker, Frances Harriet (née Henslow, 1825–74), was a translator and botanist, and the wife of Joseph Hooker. She was the daughter of John Stevens Henslow (1796–1861), professor of botany at Cambridge, where he taught and mentored Charles Darwin. After a long engagement, she married Hooker, a close friend of Darwin's, in 1851, and together they had three daughters and four sons. In 1873 she translated *Traité général de botanique* (1867), by Emmanuel Le Maout (1799–1877) and Joseph Decaisne (1807–82), into English as *A General System of Botany, Descriptive and Analytical*. Her sudden death a year later was a profound shock both to her husband and to Tyndall.

Hooker, Harriet Anne (1854–1945), was a botanical illustrator and the eldest daughter of Frances and Joseph Hooker. She married the botanist William Turner Thiselton-Dyer (1843–1928) in 1877, the same year in which she began contributing illustrations to *Curtis's Botanical Magazine*, a journal edited by her father. She went on to contribute hundreds of illustrations to various botanical publications.

Hooker, Joseph Dalton (1817–1911), was a botanist who specialized in taxonomy and the geographical distribution of plants. He was the youngest son of William Jackson Hooker (1785–1865), the director of the Royal Botanic Gardens, Kew. After studying medicine at the University of Glasgow, he became a naval surgeon and spent the years 1839–43 on HMS *Erebus* on a geomagnetic survey of the southern oceans, where he collected plants alongside his medical duties. Subsequently Hooker explored and collected plants in India (1847–51), achieving considerable mountaineering success, before joining his father as deputy director of Kew in 1855. He was a close friend of Charles Darwin, and in

1859 was the first prominent man of science to publicly endorse Darwin's theory of evolution by natural selection. In 1864 Hooker joined Tyndall and seven others in founding the X Club, a monthly dining club dedicated to promoting science in British society. He succeeded his father as director of Kew in 1865, and became embroiled in a long-running controversy over the management and independence of the gardens. Hooker served as president of the BAAS in 1868, and advised Tyndall, a close friend, when he took on the same position in 1874. He was also president of the RS from 1873 to 1877.

Huxley, Ethel (1866–1941), was the youngest daughter of Henrietta and Thomas Huxley. In 1889 she married the artist John Collier (1850–1934), who was the widower of her elder sister Marian, and in later life became the formidable matriarch of the Huxley family.

Huxley, Henrietta (1863–1940), was a daughter of Henrietta and Thomas Huxley, who became a singer and illustrator. In 1889 she married Harold Roller (1859–1934), who owned a firm of picture restorers, but she spent most of her time traveling in Europe, supporting herself by singing.

Huxley, Henrietta Anne (née Heathorn, 1825–1915), was born in the West Indies to Henry (1790–1879) and Sarah (née Thomas, 1793–1878) Heathorn, and was educated in Neuwied, Germany, for two years. In 1843 she, her mother, and her half-sister Oriana Richardson (1816–1900) moved to Australia to join her father. She kept house for Richardson and her brother-in-law William Fanning (1816–87) in Sydney, where she met Thomas Huxley in 1847, who was then a surgeon and zoologist on HMS *Rattlesnake*. After an eight-year engagement, Henrietta arrived in London in 1855 and married Thomas; they had eight children. Huxley assisted her husband in his work by translating German and drawing diagrams for his lectures. She was also a poet and later in life published *Poems of Henrietta A. Huxley* (1913). Huxley was very much a part of the scientific community in which her husband was involved; she corresponded with Charles Darwin and his family as well as with Tyndall.

Huxley, Henry (1865–1946), was a son of Henrietta and Thomas Huxley, who later became a general practitioner. In 1890 he married Sophy Wylde Stobart (1865–1927), and they had five children.

Huxley, Jessica "Jessie" Oriana (1858–1927), was the eldest daughter of Henrietta and Thomas Huxley. In 1877 she married the architect Frederick William Waller (1846–1933), and they had two children.

Huxley, Leonard (1860–1933), was the eldest surviving son of Henrietta and Thomas Huxley, and later became a biographer, poet, and editor. After graduating from Balliol College, Oxford, he took a teaching post at Charterhouse School and then joined the publishing firm of Smith, Elder & Co. in 1884. He had four children from his first marriage, to Julia Frances Arnold (1862–1908), including the biologist Julian Huxley (1887–1975) and the novelist Aldous Huxley (1894–1963). He had two further children from his second marriage, to Rosalind Bruce (1890–1994), including the physiologist Andrew Huxley (1917–2012). Leonard's publications included the *Life and Letters* of his father (1900) and Joseph Hooker (1918) and a biography of Darwin (1920), as well as a volume of poems (1920) and an account of the publishing firm where he worked (1923). He also edited the *Cornhill Magazine* from 1917 until his death.

Huxley, Marian "Mady" (1859–87), was the second daughter of Henrietta and Thomas Huxley. As a child she showed artistic promise and attended drawing classes at the Slade School of Art, where she was a prize-winning student; in addition, she had a dedicated "studio" in the Huxleys' family home at 4 Marlborough Place (see letter 4272). She subsequently exhibited paintings at the Royal Academy and the Grosvenor Gallery. In 1879 she married the artist John Collier (1850–1934). She suffered with mental illness following the birth of their daughter Joyce (1884–1972), and died of pneumonia in Paris, where she was being treated by Jean-Martin Charcot (1825–93).

Huxley, Rachel (1862–1934), was the third daughter of Henrietta and Thomas Huxley. In 1884 she married the civil engineer William Alfred Eckersley (1856–95), with whom she had three children. Following her husband's death, she married Harold Shawcross (1855–1930) and had a further two children.

Huxley, Thomas Henry (1825–95), was a zoologist, educator, science popularizer, and one of Tyndall's closest friends. Following medical training at Charing Cross Hospital in London, he was a ship's surgeon aboard HMS *Rattlesnake*'s expedition to Australia and New Guinea from 1846 to 1850. While in Australia he met Henrietta Anne Heathorn, whom he married in 1855, after returning to Britain, following an eight-year engagement. He became professor of natural history at the Royal School of Mines (1854–71) and moved with the school to South Kensington where he taught in a new laboratory until 1885. He joined Tyndall twice at the RI as Fullerian Professor of Natural History (1855–58; 1866–69). He was a vociferous public advocate of evolution, and his major publications on the topic include *Evidence as to Man's Place in Nature* (1863) and *Evolution and Ethics* (1893). In 1864 Huxley joined Tyndall, Edward Frankland, Thomas Hirst, Joseph Hooker, and four others as members of the X Club, a monthly dining club dedicated to the promotion of science in British society. In the period covered by this volume, Huxley was still recovering from a lengthy period of ill health, both physical and mental, but he maintained a regular correspondence with Tyndall, discussing issues such as the Belfast meeting of the BAAS, where both delivered controversial public addresses, and experiments with bacterial growth aimed at disproving spontaneous generation.

Jamin, Jules (1818–86), was a French physicist who conducted research on a wide range of subjects, including optics, electricity, and magnetism. Among his most notable achievements was the development of the Jamin interferometer, which accurately measured the refraction and dispersion of light by gases. A renowned public lecturer, Jamin was professor of physics at the École Polytechnique in Paris from 1852 to 1881, and his name is one of the seventy-two scientists inscribed on the Eiffel Tower. Tyndall met him in Paris in 1855. During the Franco-Prussian War in 1870, Tyndall offered to care for Jamin's wife and children in England, but they went to Brittany while Jamin remained in Paris.

Key, Astley Cooper (1821–88), was a royal naval officer. He saw service in the Crimean War and the Second Opium War, rising to the rank of rear admiral in 1866. Key was director of naval ordnance from 1866 to 1869 at a time when the Royal Navy underwent considerable reorganization and began using armor-plated ships. His expertise in the science of gunnery earned him election as FRS in 1868.

Keyserling, Alexander von (1815–91), was a Baltic German geologist and paleontologist born into a noble family. He also carried out work in the fields of zoology and botany, and theorized about the transmutation of species, although he was initially skeptical about natural selection. In 1861 Charles Darwin referenced Keyserling's hypothesis about the impact of alien molecules on an embryo in the formation of new species in the preface to the third edition of *On the Origin of Species* (see chapter 14 in C. N. Johnson, *Darwin's Historical Sketch: An Examination of the 'Preface' to the* Origin of Species [Oxford: Oxford University Press, 2020]).

Kinglake, Alexander William (1809–91), was a travel writer and historian, who published a nine-volume account titled *The Invasion of the Crimea* (1863–87). He was a close friend of Olga Novikoff, who provided him with information about the Russian armies during the Crimean conflict, and he may have introduced her to Tyndall (see letter 4411, n. 4). Kinglake and Tyndall met regularly at Novikoff's salon during her stay in London in 1874, and occasionally thereafter at the Athenaeum Club. A gossipy letter that Kinglake wrote to Novikoff in 1876 about Tyndall's marriage was later published in *M.P. for Russia* (vol. 1, 157).

Kirchhoff, Gustav (1824–87), was a German physicist. He developed a number of scientific laws, all of which are named after him, relating to electrical circuit laws, spectroscopy, thermal radiation, and thermochemistry. His circuit laws, fundamental to electrical engineering, were developed while he was still a student at the University of Königsberg. In 1854 Kirchhoff joined the University of Heidelberg, where he began a fruitful collaboration with Robert Bunsen. Together they jointly discovered the elements cesium and rubidium, and also invented the spectroscope, which Kirchoff used to identify the elements of the sun. However, William Thompson wrote to Kirchhoff claiming that George Gabriel Stokes had already published much of the work for which Kirchhoff claimed priority. Stokes himself did not pursue these claims, but Kirchhoff felt it necessary to defend himself, sparking a controversy that Tyndall joined in support of Kirchhoff. In 1875 he was appointed to the first professorship of theoretical physics at the University of Berlin. Following an accident that injured his foot in 1868, Kirchoff was forced to walk with crutches or use a wheelchair. Although it lessened, the injury never healed fully.

Klein, Emanuel Edward (1844–1925), was a British microbiologist of Croatian origin. A pioneer in the study of bacteria, Klein had his career mired in controversy due to his outspoken support of vivisection in physiological and medical experiments. He became particularly infamous for his apparently callous statements when called to defend these practices before a Royal Commission, which would lead to the establishment of the 1876 Cruelty to Animals Act. His work did much to establish the role of bacteria in numerous diseases in humans and animals.

Knoblauch, Karl Hermann (1820–95), was a German physicist. He established his reputation through experiments on the properties of radiant heat performed while completing a doctorate at the University of Berlin in 1847. As professor of physics at the University of Marburg from 1849 to 1853, Knoblauch carried out research on diamagnetism in collaboration with Tyndall. In 1853 he moved to the University of Halle, where he remained for the rest of his career. At Tyndall's invitation, Knoblauch came to Belfast for the BAAS meeting in 1874, where Tyndall delivered the presidential address.

Knowles, James Thomas (1831–1908), was a writer, periodical editor, and architect. In the latter role he designed hundreds of houses, three churches, and numerous other commissions for wealthy clients. Knowles also pursued a literary career, publishing his own version of the Arthurian legends, and contributing articles to numerous journals. In 1866 he met Alfred Tennyson, and formed a close friendship with the poet. In 1869, with the help of Tennyson, Knowles founded the Metaphysical Society, an invitational debating society that brought together prominent clergymen, philosophers, politicians, and men of science. Notable members included Tyndall, Thomas Huxley, and John Ruskin. The lofty aim of the society's discussions was to determine the bases of morality, but ultimately the group was dissolved in 1881. In 1870

Knowles became editor of the *Contemporary Review* and held this position until 1877, when a disagreement with the periodical's owners over the treatment of theological matters led him to resign and establish the *Nineteenth Century*. Both publications were highly influential under his editorship, and in them he published several important articles by Tyndall.

Lange, Friedrich Albert (1828–75), was a German philosopher. Among his most notable works was *Geschichte des Materialismus und Kritik seiner Bedeutung in der Gegenwart* (*A History of Materialism and a Critique of Its Present Importance*), first published in 1866, then revised and expanded into two volumes in 1873–75. This work, of which Tyndall was a keen reader, discussed and developed the writings of Immanuel Kant. In 1872 Lange accepted a chair in philosophy at the University of Marburg, and is credited with reviving interest in Kantian philosophy at that institution.

Lesley, J. Peter (1819–1903), was an American geologist and director of the second Geological Survey of Pennsylvania. He met Tyndall during his lecture tour of the United States in 1872–73, of which Lesley was one of the principal organizers. At this time, Lesley was editor of the *United States Railroad and Mining Register*, professor of geology, mining, and engineering, and dean of the science faculty at the University of Pennsylvania. The Second Pennsylvania Survey (1874–89), which expanded the work of the first and detailed the economic resources of the state, is considered his most significant contribution to science. Indefatigable in his work ethic, Lesley suffered from recurrent illnesses and ultimately a breakdown in 1893, after which point he became a semi-invalid and was cared for by his wife, Susan Lesley (née Lyman, 1823–1904), until his death.

Liddon, Henry Parry (1829–90), was an Anglican clergyman. Through his preaching and writing, Liddon did much to reestablish the influence of Tractarianism—the philosophy of the Oxford movement—within the Church of England. He was appointed canon of St Paul's Cathedral in 1870, and in the same year was made professor of the exegesis of Holy Scripture at Oxford. In these two prestigious positions, Liddon was able to exert considerable social and political influence.

Liebig, Justus von (1803–73), at Giessen in the early 1840s, before becoming a successful private physician and physician at St George's. In 1842 he married a cousin, Millicent Acheson, with whom he had seven children. He published important research on protein chemistry, identifying the protein excreted by patients with multiple myeloma, and in 1852 made an abstract in English of Emil du Bois-Reymond's work on electrophysiology that annoyed Carlo Matteucci (1811–68), the Italian pioneer of the subject. In 1853 Bence Jones became a manager of the RI, and was elected secretary in 1860. His *Life and Letters of Faraday* was published in two volumes in 1870. He frequently entertained Tyndall at both his London home and at Folkestone, where he kept a summer house. Bence Jones resigned his position at St George's on health grounds in 1862, and he continued to suffer from rheumatism and heart disease for the next decade. His condition worsened while Tyndall was on his lecture tour of the United States in 1872, hastening the latter's return shortly before Bence Jones died in April 1873.

Lockyer, Joseph Norman (1836–1920), was an astronomer and the founding editor of the science journal *Nature.* Privately educated, he wrote his first scientific paper for the Royal Astronomical Society in 1863. Lockyer was an early astrophysicist and used a spectroscope to make solar observations. In 1868 he demonstrated that the protuberances seen during a solar eclipse were from the sun. He confirmed that there was an atmosphere around the sun and concluded that it contained an as yet unknown element, which he labeled helium, a discovery that was not verified until the 1890s. After establishing *Nature* in 1869, Lockyer's editorial approach encouraged controversy, as is seen in the letters in this volume with the bitter exchanges, over glacial motion, between Tyndall and Peter Guthrie Tait. *Nature*'s provocative role in this dispute caused Tyndall to distrust Lockyer, and Tyndall called him "a man whose conceit has rendered him intolerable to his best friends, and from whom I never disguised my opinion of his conceit" (letter 4300). Lockyer also used *Nature* to support his own causes, particularly in relation to public funding of science, and he remained editor for more than fifty years.

Longman, Thomas (1804–79), was a publisher. With his younger brother William, he was proprietor of the firm Longmans, Green, & Co. This family business was originally established in 1724, and the siblings inherited from it from their father, Thomas Norton Longman (1771–1842).

Longman, William (1813–77), was a publisher. With his elder brother Thomas, he was proprietor of the firm Longmans, Green, & Co.

Lubbock, Ellen "Nelly" Frances (née Horden, 1835–79), was the first wife of John Lubbock, whom she married in 1856. They had six children. Along with her husband, she became one of Tyndall's closest friends and confidants. She helped him with the proofs of *Heat as a Mode of Motion* (1863).

Lubbock, John (1834–1913), 1st Baron Avebury, was a banker, anthropologist, and politician. In his youth he befriended Charles Darwin and other men of science, including Tyndall, and he later became part of Darwin's inner circle. In 1865 Lubbock published the influential *Pre-historic Times,* in which he coined the terms *paleolithic* and *neolithic.* He also published books on insects and plants. After the death of his first wife, Ellen, in 1879, Lubbock married Alice Fox-Pitt (ca. 1862–1947), daughter of the ethnologist Augustus Pitt Rivers (1827–1900), with whom he had five children. Lubbock was the youngest member of the X Club, and regularly joined Tyndall on his climbing adventures in the Alps. He was also a member of the RI, the Geological Society, and the RS, having been elected FRS in 1858.

Luther, Martin (1483–1546), was a German priest and theologian. He was the central figure of the Protestant Reformation, and the founder of Lutheranism. He publicly objected to the teachings and practices of the Roman Catholic Church, leading to his excommunication. Tyndall closely identified with Luther, and frequently invoked his dauntless resistance of established authority as a model for his own conduct.

Lyell, Charles (1797–1875), was a highly influential Scottish geologist. Trained as a lawyer, he became professor of geology at King's College, London. Lyell was most noted for his controversial three-volume work *Principles of Geology* (1830–33), in which he argued that the only legitimate type of causal explanation for the past is that seen in operation in the present; natural laws were nothing if not consistent. In consequence of this philosophical commitment, Lyell argued that the earth was many millions of years older than even the most generous of his contemporaries had hitherto been willing to admit. Though controversial, Lyell was highly regarded, both as a philosopher of science and as a geologist. He had mentored Charles Darwin after his return from the voyage of HMS *Beagle* in 1836, and in 1858, along with Joseph Hooker, pressed Darwin to publish an abstract of his big species book, which became *On the Origin of Species* (1859). He entered the debate on human evolution with *Geological Evidences of the Antiquity of Man* (1863), and by the 1870s was one of the elder statesmen of British science.

Macmillan, Alexander (1818–96), was a Scottish publisher. He was the cofounder, with his brother Daniel (1813–57), of Macmillan & Co., which they formed in London in 1843 before moving to Cambridge a few months later. In addition to books, the firm published *Macmillan's Magazine*, which began in 1859, and, from 1869, the science journal *Nature*, which was edited by Norman Lockyer.

Martineau, James (1805–1900), was a theologian and philosopher. His work was primarily concerned with questions of religion, and he came to be deeply influenced by the transcendental movement. Martineau was a Unitarian, although he was an exponent of the so-called new Unitarianism, which encouraged a greater emphasis on orthodox spirituality. From 1840 to 1885 he was professor of mental and moral philosophy and political economy at Manchester New College in London, an institution for Unitarian students. In an address at the college in October 1874, he criticized the theological implications of Tyndall's Belfast Address, publishing his lecture as *Religion as Affected by Modern Materialism* (1874).

Mattress, John Henry (1828–96), was a printer who worked for Taylor and Francis. He had begun working for the firm by March 1852 (Weekly Wages Pay Book 1852–55, 1841B.14, Taylor & Francis Archive, St Bride Library, London), and may have become overseer of the company's printing office by the mid-1860s (see "Photographic Society of London," *Photographic Journal* 10 [1866]: 79). In the 1871 census, he was recorded as a printer compositor, living at 5 Goswell Street (now Road), Holborn, which was within walking distance of Red Lion Court in Fleet Street, the location of Taylor and Francis's printing works. He married Mary Ann Hall (1828–94) in 1849, and they had eight children.

Maxwell, James Clerk (1831–79), was a Scottish physicist and mathematician. He graduated in the Cambridge mathematics Tripos as second wrangler in 1854, after which he held professorships at Marischal College, Aberdeen (1856–60), and King's College, London (1860–65), and, from 1871, the Cavendish Professorship at Cambridge. Maxwell is best known for his contributions to the study of electromagnetic radiation. He developed Michael Faraday's field theory and demonstrated that electricity, magnetism, and light were instances of the same phenomenon. Maxwell's research culminated in his 1873 publication *A Treatise on Electricity and Magnetism.*

Mayer, Alfred Marshall (1836–97), was an American physicist. He became professor of physics and astronomy at the newly established Lehigh University in 1867, where he worked extensively on electromagnetism and astronomy. His groundbreaking study of the 1869 solar eclipse, funded by the United States Nautical Almanac Office, used photographs he had taken of the phenomenon. In 1871 he became professor of physics at Stevens Institute of Technology in Hoboken, New Jersey, known for its state-of-the-art facilities. In his first decade there, Mayer published thirty-five articles and three books, and became the leading authority on sound research, in addition to his continued work in electromagnetism and heat conduction. Mayer was involved in the organization of Tyndall's lecture tour of the United States, especially the farewell dinner held in New York on February 4, 1873. Later that same year he traveled to Britain, where he was received by Tyndall and other eminent men of science.

Mayer, Julius Robert (1814–78), was a German physician and physicist. He practiced medicine as a profession, but his principal interest was in physics. Notably, he first hypothesized the mechanical equivalent of heat, but the priority of his discovery was overlooked in favor of James Prescott Joule (1818–89). Mayer's achievement would later gain some recognition, with Tyndall championing his cause. As a consequence, Tyndall was drawn into a public dispute with Joule's collaborator, William Thomson, during the 1860s, and their argument was returned to during the glacier controversy, in which Tyndall again made assertions about priority of discovery, in the mid-1870s. In 1867 he was raised to the German nobility, meaning that he and his family were thereafter styled "von" Mayer.

Miller, William Hallowes (1801–80), was a Welsh mineralogist. In 1832 he succeeded William Whewell (1794–1866) as professor of mineralogy at Cambridge, a position he held until 1870. Miller's most important publication was *A Treatise on Crystallography* (1839). He served as foreign secretary of the RS from 1856 to 1873.

Moigno, François-Napoléon-Marie (1804–84), was a French Catholic priest, physicist, writer, and editor. Educated at a Jesuit seminary, he studied science alongside theology and in 1836 was given the chair in mathematics at the Jesuit College of Sainte-Geneviève in Paris. After leaving the order in 1843, he became the science editor of the newspaper *L'Epoque*, and later became editor of the scientific journal *Cosmos* (1852–62). He founded the journal *Les Mondes* in 1863 and undertook the translation and publication of many of Tyndall's works in French, including *Radiation* (1867), *Le Son* (1869), *La Chaleur* (1864), and *Faraday Inventeur* (1868) ("L'Abbé Moigno," *Nature* 30 [1884]: 291–92).

Murchison, Roderick Impey (1792–1871), was a geologist and influential director of scientific institutions. After serving in the army as a young man, he became fascinated by the burgeoning science of geology. Over the course of his distinguished scientific career, Murchison investigated and described the Silurian, Devonian, and Permian stratigraphic systems. In 1855 he was appointed director-general of the British Geological Survey, a position he held until his death. He also became director of both the Royal School of Mines and the Museum of Practical Geology.

Novikoff, Alexander (1861–1913), was the only child of Olga and Ivan Novikoff (d. 1890). He later became a land captain, and wrote essays on rural life in the Russian provinces.

Novikoff, Olga (née Kiréeff, 1840–1925), was a cosmopolitan Russian aristocrat and socialite who developed a brief but intense friendship with Tyndall in the final months of 1874. Born into a noble family in Moscow, in 1860 she married Ivan Novikoff (d. 1890), a general on the general staff of Grand Duke Nicholas (1831–91), brother of Tsar Alexander II (1818–81). The couple moved to Saint Petersburg, where Novikoff became a member of the salon of Grand Duchess Elena Pavlovna (1807–73), cultivating friendships with Russian intellectuals, particularly the geologist Alexander Keyserling. Having developed her own salon, through which she became acquainted with politicians and diplomats, Novikoff corresponded with several prominent statesmen across Europe. She first came to London in 1868, after which she seems to have visited annually, and she also traveled extensively on the Continent. She took a strong interest in theological and ethical questions, privately printing pamphlets that she sent to her correspondents. In September 1874 she sent one to Tyndall (see letter 4401) shortly before coming to London, where she stayed until late November. During this time, Novikoff transferred her salon to Symonds' Hotel in Mayfair, where Tyndall visited her on several occasions. They developed an intimate, potentially even romantic, friendship, often seeming to hanker for each other's company and counsel. When Novikoff left to return to Russia, Tyndall, seeking to avoid an emotional farewell, thought it "on the whole best I should not be here" (letter 4492). She returned to London the following October, but by then Tyndall's relationship with Louisa Hamilton had grown closer, and he worried that meeting with Novikoff again might "beget gossip which neither of us would like" (letter 4662). Through Novikoff, Tyndall was introduced to a new circle of European philosophers and theologians, including Albert Réville, Alois Riehl, and Jakob Frohschammer, who influenced his thinking when he was responding to criticisms of the alleged materialism of the Belfast Address. By 1876 Novikoff had become an ardent advocate of Russian imperial interests, especially during the Russo-Turkish War of 1877–78, and her journalism, in collaboration with the campaigning editor William Thomas Stead (1849–1912), helped determine the foreign policy of William Gladstone's Liberal opposition. With these new interests preoccupying her, the close friendship with Tyndall seems not to have been resumed, and Novikoff's later published reminiscences of him do not always correspond with what is recorded in their letters. See *M.P. for Russia*, vol. 1, 149–60.

O'Brien, Lucy Harriet de Vere (née Wynne, 1843–1932), was the daughter of Anne and George Wynne. In 1871 she married Aubrey Stephen Vere O'Brien (1837–98), with whom she had two children.

Pasteur, Louis (1822–95), was a French chemist and pioneering bacteriologist. His discoveries have become central to medical science, as he did much to demonstrate the germ theory of disease, and it was his advocacy that first convinced Tyndall of it (see letter 4463). Pasteur is perhaps most famous for the process of pasteurization, the method he developed to prevent bacterial contamination of milk and wine. Over the course of his career, Pasteur held various professorships and other academic positions in Strasbourg, Lille, and Paris, and he founded the Pasteur Institute in 1887. From the mid-1860s to 1870, he carried out work on diseases in silkworms, saving France's silk industry. Pasteur's significant research on spontaneous generation (which he disproved), immunology, and vaccination would be carried out from the 1870s onward. In 1874 he was awarded the RS's Copley Medal for his work on fermentation.

Peel, Emily (née Hay, 1836–1924), was the seventh daughter of George Hay (1787–1876), 8th Marquess of Tweeddale, and Susan Montagu (1797–1870). In 1856 she married the politician Robert Peel (1822–95), son of the former prime minister Robert Peel (1788–1850). They had four children. The Peels divided their time between London and Drayton Manor in Staffordshire. In later years, she spent most of her time at Lammermoor, the family's villa on Lake Geneva, Switzerland, from where she also engaged in climbing in the Swiss Alps. She was a regular attendee of Tyndall's lectures, and was regarded as a great hostess and philanthropist ("Births, Deaths, Marriages and Obituaries," *Tamworth Herald*, 12 April 1924, 5).

Pollock, Frederick (1845–1937), was a lawyer, academic, and the eldest son of Juliet and William Frederick Pollock. He was called to the bar in 1871, and went on to write influential textbooks on English law and to teach the subject at Oxford.

Pollock, Juliet (née Creed, 1819–99), was an author and the wife of William Frederick Pollock. The Pollocks married in 1844 and had three sons, Frederick, Walter Herries (1850–1926), and Maurice Emilius (1857–1932). Pollock was a staunch supporter of the RI, of which her husband was manager, and frequently had Tyndall to dinner at the family's London home on Montagu Square. She also organized home theatricals, in which she insisted Tyndall perform, and the two shared a close and affectionate friendship. Pollock also wrote several books on literature and theater, including *Julian and his Playfellows* (1852), *New Friends* (1858), and *Macready As I Knew Him* (1885).

Pollock, William Frederick (1815–88), 2nd baronet, was a jurist, author, and manager of the RI, and the eldest son of Frances Rivers (d. 1827) and Jonathan Frederick Pollock (1783–1870). In 1832 he went to Trinity College, Cambridge, where he obtained his MA in 1840. Four years later, he married Juliet Creed, with whom he had three sons. Having been admitted to the Inner Temple in 1833, Pollock was called to the bar in 1838, and then became a master of the Court of Exchequer in 1846. In 1874 he was appointed queen's remembrancer, the most prestigious judicial role in the English legal system. Pollock's *Personal Remembrances* (1887) illustrates how he and his wife circulated among London's elite and entertained at their Montagu Square home, where Tyndall was a frequent visitor.

Priestley, Joseph (1733–1804), was a renowned chemist and theologian. Born into a Dissenting family, Priestley had a varied career as an educator and minister, frequently relocating to different parts of England, before religious and political persecution forced him in 1794 to emigrate to the United States, where he died. He was a prolific writer on varied subjects, ranging from history to politics, theology, and science. Priestley's reputation as a chemist rests largely on his discovery of ten gases, including oxygen (which he termed "dephlogisticated air"), carbon monoxide, and nitrous oxide (laughing gas). He also designed the apparatus for producing carbonated water, and experimented with electricity.

Rendu, Louis (1789–1859), was a French theologian and the Roman Catholic bishop of Annecy. He also produced an important work on glacial motion, *Théorie des glaciers de la Savoie* (1840). In 1859, during the course of a lengthy dispute regarding priority of discovery for theories of glacial motion, Tyndall and Thomas Huxley accused James David Forbes of plagiarizing from Rendu. Although this serious charge was never proven, the mere suggestion of wrongdoing was enough to seriously damage Forbes's reputation. In an attempt to clear his father's name, George Forbes edited the first English translation of Rendu's book, published as *Glaciers of Savoy* in 1874.

Réville, Albert (1826–1906), was a French Protestant theologian and journalist who articulated his liberal and often heterodox views in the *Revue des deux Mondes*, where he published an account of Tyndall's career ("Les Sciences naturelles et l'orthodoxie en Angleterre," *Revue des deux Mondes*, n.s., 8 [1875]: 283–318). Tyndall was introduced to him by Olga Novikoff, and they corresponded briefly in 1875. Réville lived near Dieppe at this time, but later moved to Paris, where he taught at the Collège de France, and became embroiled in the Dreyfus Affair in the 1890s as a leading supporter of the Dreyfusard cause.

Riehl, Alois (1844–1924), was an Austrian philosopher at the University of Graz who put forward a neo-Kantian critique of perception. In November 1874 he published a two-part review of Tyndall's *Fragments of Science* in the *Wiener Abendpost*. Tyndall was introduced to his work, particularly *Moral und Dogma* (1871), by Olga Novikoff.

Robinson, Thomas Romney (1792–1882), was an Irish astronomer and physicist. In 1823 he was appointed director of the Armagh Observatory, a post he held until his death.

Ruskin, John (1819–1900), was an art critic and writer. Alongside his friend and mentor Thomas Carlyle, Ruskin was one of the most prominent intellectual figures of the nineteenth century. He first came to prominence writing in defense of the painter J. M. W. Turner (1775–1851), and his views on art gained considerable influence, notably inspiring the paintings of the Pre-Raphaelite Brotherhood, whose work he championed. Ruskin went on to publish works on architecture that contributed to the Gothic revival of the period. As his reputation as a public intellectual became more secure, Ruskin's writings encompassed social critique, political economy, science, and a diverse array of other subjects.

In 1848 he married Euphemia Gray (1828–97), but the marriage was annulled after six years on the grounds that it was allegedly never consummated, and in 1855 Gray married the artist John Everett Millais (1829–96). The controversy fueled speculation regarding Ruskin's sexuality, which was subsequently exacerbated by gossip about his infatuation with a former student, Rose La Touche (1848–75), almost thirty years his junior. Ruskin's interest in geology and glaciers led him to befriend and correspond with James David Forbes, whose scientific reputation was damaged in a dispute with Tyndall regarding glacial motion. Consequently, Ruskin joined other supporters of Forbes in a concerted attack on Tyndall in 1873, and he devoted an installment of his serialized work *Fors Clavigera* to criticizing Tyndall's theories and conduct.

Russell, Frances Anna Maria (née Elliot-Murray-Kynynmound, 1815–98), was the second wife of former prime minister Lord John Russell, whom she married in 1841. She was the daughter of Gilbert (1782–1859) and Mary (née Brydone, 1786–1853) Elliot-Murray-Kynynmound. Lord Russell brought six children to his second marriage, and he and Lady Russell had four more between 1842 and 1853. They lived at Pembroke Lodge in Richmond Park, a home provided for them by Queen Victoria, where they entertained the political, social, and scientific elite, including Tyndall. Lady Russell regularly attended Tyndall's lectures, and she also had a keen interest in politics as well as strong religious beliefs.

Russell, Francis Albert Rollo (1849–1914), was a meteorologist and the son of former prime minister Lord John Russell and his second wife, Frances. As with other members of his family, he was a friend of Tyndall's, and like him was interested in public health and contagious diseases, and particularly the relation between atmospheric conditions and human health.

Russell, John (1792–1878), 1st Earl Russell, was a politician and first entered Parliament in 1814. He was the son of John Russell (1766–1839), the 6th Duke of Bedford, and Georgiana Elizabeth Russell (née Byng, ca. 1768–1801). Russell became a leader of the Whigs and was a crucial figure in the passage of the First Reform Act of 1832. He was prime minister from 1846 to 1852 and from 1865 to 1866. In 1856 he became the first Earl Russell. John Russell and his wife, Frances, supported the RI and regularly attended Tyndall's lectures.

Russell, Mary Agatha (1853–1933), was the daughter of the former prime minister Lord John Russell and his second wife, Frances. Agatha, as she was called, never married and lived at the family's grace-and-favor home, Pembroke Lodge, in Richmond Park, until her mother's death in 1898. She subsequently coedited a book about her mother, *Lady John Russell: A Memoir* (1910).

Rutherford, Elizabeth (née Bunyan, b. ca. 1816), was the mother of the Scottish physiologist William Rutherford (1839–99), Fullerian Professor of Physiology at the RI from 1872 to 1874. It is unclear how she and Tyndall came to be acquainted, but Tyndall clearly considered her a trusted correspondent, sending her an extract of a letter from Thomas Carlyle in October 1873 (see letter 4197). She had married Thomas Rutherford, a farmer from Ancrum, Roxburghshire, and William was the youngest of their seven children. He never married, and remained close to his mother, who from 1872 lived at 13 Shandwick Place in Edinburgh. Her age was recorded as twenty-five in the 1841 census, but given the number of children she had had by this time it may have been entered wrongly.

Sabine, Edward (1788–1883), was an Irish military officer, astronomer, and geodesist. He served in the Royal Artillery, rising to the rank of general. As astronomer to the arctic expeditions of John Ross (1777–1856) and William Parry (1790–1855), Sabine began a series of experiments to determine the shape of the earth, and also carried out observations of the earth's magnetic field, for which he received the RS's Copley Medal in 1821. He spent a further two years traveling around the globe in order to collect sufficient measurements to accurately calculate the precise figure of the earth. As scientific advisor to the Admiralty, he was a primary instigator in the so-called Magnetic Crusade, which sought to establish magnetic observatories across the British empire. Among his numerous other scientific interests, Sabine was a keen ornithologist, giving his name to three bird species.

Salisbury, Lord: see Gascoyne-Cecil, Robert Arthur Talbot.

Shairp, John Campbell (1819–85), was a Scottish poet, critic, and academic. After studying at Glasgow University and Balliol College, Oxford, he held various posts at the University of St Andrews. Having won the Newdigate Prize for his poetry as a student, Shairp returned to Oxford as the professor of poetry in 1877. His publications included *Kilmahoe: A Highland Pastoral* (1864), *Studies in Poetry and Philosophy* (1868), and *Culture and Religion* (1870). In 1868 Shairp had become principal of St Andrews's United College, and five years later coedited the *Life of Forbes*, commemorating his predecessor in the role. This posthumous tribute helped reignite the bitter controversy in which James David Forbes had clashed with Tyndall over glacial motion.

Sharpey, William (1802–80), was a Scottish anatomist and physiologist. He was known as the father of British physiology. Educated at Edinburgh, he took his MD in 1823 with a thesis on stomach cancer. After practicing with his stepfather, in 1836 he became the new chair of anatomy and physiology at UCL, a position he held until 1874. He became FRS in 1839 and a member of the RS's council in 1844. He was also the secretary of the RS from 1854 to

1872. Some of his pupils at UCL included Joseph Lister (1827–1912), John Marshall (1818–91), and John Boon Hayes (1826–56).

Siemens, Carl Wilhelm (1823–83), was a German-British electrical engineer and inventor. He was born into an old farming family in what was then the Kingdom of Hanover, and was educated at the University of Göttingen. He came to London in 1843 as an agent for his elder brother, the electrical engineer Ernst Werner Siemens (1816–92), who had patented several important inventions in telegraphy. By 1858 Siemens had established a separate branch of his brother's company in London, and over the next two decades Siemens Brothers & Co. was commissioned to build long-distance telegraph networks across the globe. While developing his business, Siemens continued to design his own new products, including, in 1874, an innovative telegraph cable-laying ship that he named the CS *Faraday*. He was granted British citizenship when he married Anne Gordon (1821–1902) in 1859.

Somerville, Martha (1813–79), was the eldest daughter of Mary Somerville and her second husband, William Somerville. She spent most of her life caring for her mother, and also edited her autobiography, published posthumously as *Personal Recollections from Early Life to Old Age of Mary Somerville* (1873).

Somerville, Mary (née Fairfax, 1780–1872), was a Scottish polymath, with a particular expertise in astronomy and mathematics. Her earliest scientific work, which was published by the RS, concerned the relationship between light and magnetism. Somerville's fame was established by *Mechanism of the Heavens* (1831), a mathematical exposition of the solar system that remained a standard reference work throughout the nineteenth century. It was followed by *On the Connexion of the Physical Sciences* (1834), and *Physical Geography* (1848), both of which proved commercially successful and secured Somerville's reputation as both an original thinker and a gifted writer of "popular" works that synthesized various branches of science. Along with Caroline Herschel (1750–1848), Somerville became the first woman to be elected an honorary member of the Royal Astronomical Society. She was married twice, first to Samuel Greig (1778–1807), who did not encourage his wife's scientific studies, and then to William Somerville (1771–1860), who supported her research.

Somerville, Mary Charlotte (1817–75), was the younger daughter of Mary Somerville and her second husband, William Somerville. Like her sister Martha, she spent much of her life caring for her mother.

Spencer, Herbert (1820–1903), was a journalist, philosopher, biologist, and sociologist. He spent his early life in Derby, but at the age of twenty-eight he eventually secured, through family connections, a post as subeditor of the

Economist, which put him at the heart of literary London. He published his first book, *Social Statics*, in 1850 and his second, *Principles of Psychology*, in 1855. Spencer's "Synthetic Philosophy" used evolution to explain both organic and social change, and in his *Principles of Biology* (1864) he coined the phrase "survival of the fittest." Along with Tyndall and seven others, Spencer was a member of the X Club, a monthly dining club dedicated to promoting science in British society. Spencer was the only member who was not an FRS, turning down a nomination in 1873 because he felt it came too late. He was a close friend of Tyndall's, and they regularly corresponded and read proofs of each other's work, although Tyndall and Thomas Huxley often shared jokes at Spencer's expense in their own correspondence.

Spottiswoode, Cyril Andrew (1867–1915), was the younger son of Eliza and William Spottiswoode. With his elder brother Hugh, he became a partner in the family printing firm.

Spottiswoode, Eliza Taylor (née Arbuthnot, 1837–94),was born in Madras, India, the eldest daughter of William Urquhart Arbuthnot (1807–74), a colonial administrator, and his wife, Eliza Jane Arbuthnot (née Taylor, 1815–92). She married William Spottiswoode in 1861 and they lived in Grosvenor Place, London. They had two sons: Hugh, who became a partner and later director in his father's publishing firm, and Cyril, who also joined the family business.

Spottiswoode, William (1825–83), was a mathematician and physicist, as well as a partner in the printing firm Eyre and Spottiswoode & Co., the queen's printer. He married Eliza Taylor Arbuthnot in 1861, and parties at the couple's home were considered the height of the London season, attracting men of science, politicians, and aristocratic socialites. These events often featured scientific demonstrations in their home laboratory. Spottiswoode was a consummate administrator, and worked closely with Tyndall as the treasurer (1865–73) and then secretary (1873–9) of the RI. Among other posts, he also served as president of the BAAS in 1878 and of the RS between 1878 and 1883. His influence and social status were invaluable for the X Club, the monthly dining club dedicated to promoting science in British society, of which he and Tyndall were both members.

Spottiswoode, William Hugh (1864–1915), was the eldest son of Eliza and William Spottiswoode. After attending Eton and then Balliol College, Oxford, he was made a partner in his father's printing firm, later becoming its manager. He married Sylvia Tomlin (1870–1922) in 1893, and they had two children.

Steuart, Elizabeth Dawson (née Duckett, 1802–93), was born into a wealthy Irish family with an estate of twelve thousand acres and a gothic mansion—Duckett's Grove—situated northeast of Carlow. In 1843 she married William

Richard Steuart (ca. 1798–1852), formerly the high sheriff of County Carlow, and they lived at Steuart's Lodge, Leighlinbridge. Tyndall's father worked as land agent for the Steuart estate, and she took a keen interest in Tyndall's career, especially following her husband's death, and was probably his most consistent and long-lived patron and correspondent. Letters in this volume show that Steuart frequently helped Tyndall arrange financial support for his relatives and acquaintances in Ireland, and she also advised him on taking legal action against an accusation in the Irish press, following the Belfast Address, that he was a degenerate.

Stokes, George Gabriel (1819–1903), was a mathematician and physicist. He graduated from Cambridge as senior wrangler in 1841, and became the university's Lucasian Professor of Mathematics eight years later, continuing in the role until his death. He served as secretary of the RS for more than thirty years (1854–85), during which time he was responsible for refereeing papers for the *Phil. Trans.* He often sought Tyndall's assistance in this onerous task, and also, as seen in the letters in this volume, provided detailed reports on Tyndall's own papers. In the 1850s Stokes turned his attention to optics, and in 1852 published on fluorescence, a term he coined, showing that the phenomenon was the result of changes in the wavelength of light. For this work he was awarded the RS's Rumford Medal (1852), and was later awarded the Copley Medal (1893) for his contributions to science.

Sylvester, James Joseph (1814–97), was a mathematician. He excelled in his studies at Cambridge, but did not receive a degree until 1872 on account of being Jewish. Among his most significant contributions to the field of mathematics was his work on invariant theory, number theory, and matrix theory (coining the term *matrix*). He held various academic positions in Britain and the United States, and also offered his mathematical expertise to the burgeoning insurance industry. In 1869 he was compelled to retire as professor of mathematics at the Royal Military Academy, Woolwich, because of the institution's statutory retirement age of fifty-five, and a lengthy controversy over the payment of his pension prompted Sylvester, who had a fiery temper, to write to the *Times*. His financial situation remained parlous in the mid-1870s, and Tyndall, a longstanding friend, seems to have provided philanthropic support (see letter 4274).

Tait, Peter Guthrie (1831–1901), was a Scottish mathematician and physicist. He graduated as senior wrangler in the mathematics Tripos at Cambridge in 1852, and two years later became professor of mathematics at Queen's College, Belfast. He left in 1860 to take the chair of natural philosophy at Edinburgh University, which he held until shortly before his death. During the 1860s Tait forged a friendship with William Thomson that resulted in their *Treatise on Natural Philosophy* (1867). Tait and Thomson also began a dispute with Tyndall over his claims about Julius Mayer's priority over James Prescott Joule in developing the mechanical equivalent of heat. This controversy spilled over into the press before their differences were resolved at the 1867 meeting of the BAAS, where Tyndall reflected that they "seemed like brothers together" (letter 2646, *Tyndall Correspondence*, vol. 10). Six years later, Tyndall's feelings for Tait had soured again after he coedited the *Life of Forbes*, which helped reignite the dispute over glacial motion that James David Forbes had had with Tyndall in the 1850s. As the letters in this volume show, this led to a number of bitter exchanges between Tait and Tyndall, in *Nature* as well as several other publications.

Tennyson, Alfred (1809–92), 1st baronet, was a celebrated poet. He served as poet laureate from 1850 until his death. Notable poems include "The Lady of Shalott" (1833), *In Memoriam A. H. H.* (1850), and "The Charge of the Light Brigade" (1854). Tyndall was an ardent admirer of Tennyson's verse, and they became friends. Tyndall visited Tennyson's home on the Isle of Wight in April 1874 (see letter 4318).

Thomson, William (1824–1907), 1st Baron Kelvin, was a distinguished Irish-Scottish mathematician, physicist, and inventor. He was educated at the University of Glasgow and Peterhouse College, Cambridge, where he was second wrangler and Smith's Prizeman in 1841. He was professor of natural philosophy at Glasgow University from 1846, while also paying frequent visits to London. He first met Tyndall at the Edinburgh meeting of the BAAS in 1850, and from the start the two physicists had a strained relationship. For much of his career, Tyndall regarded Thomson as a scientific rival; however, they had largely resolved their differences by the late 1860s. In 1867 Thomson published *Treatise on Natural Philosophy* with Peter Guthrie Tait, and his connection with the latter drew him into new controversies, although he did not participate in the dispute with Tyndall over glacial motion that Tait helped reignite in 1873.

Thorp, Charles (ca. 1818–96), was an Irish solicitor who practiced from 26 William Street, Bagenalstown, and also had an office at 60 Lower Dominick Street in Dublin. In 1876 he became the sessional crown solicitor for County Carlow. His father, also Charles Thorp, had begun the family's legal practice in Bagenalstown, and he worked there, initially known as Charles Thorp Jr., from the mid-1840s (*The Dublin Almanac, and General Register of Ireland* [Dublin: Pettigrew and Oulton, 1847], 587). He was later succeeded by his own son, Charles Henley Thorp (1852–1937). In late 1874 Tyndall consulted him over a potential action for libel against the *Tuam Herald*.

Trevor, Charles Cecil (1830–1921), was assistant secretary to the Board of Trade's Harbour Department between 1867 and 1895. After training in law, he was called to the bar in 1855, and soon afterward published *A Treatise on the Taxes on Succession* (1856), which remained in print for several decades. Trevor married

Mary Weston (d. 1892) in 1863. Four years later, soon after assuming the role of assistant secretary, he wrote to Tyndall inviting him to take over Michael Faraday's role as scientific adviser to the Board of Trade. In this volume, the letters between Trevor and Tyndall principally concern the use of gas in lighthouses.

Tunstall, Jane (1811–92), was a widowed housekeeper who worked for Tyndall and Thomas Hirst when they lived at 14 Waverley Place in London. Tunstall continued as Tyndall's housekeeper when he moved into Michael Faraday's rooms at the RI in January 1868, following Faraday's death the previous summer. From 1871 to 1875, Tunstall was also employed as the housekeeper at the RS, while her husband, Thomas Tunstall, was the porter (RS CMP/4/21; RS CMP/4/64).

Tyndale, Hector (1821–80), was an American businessman, Union general, and Tyndall's cousin. He was the son of Tyndall's father's brother, Robinson Tyndall (1775–1842), who had emigrated from Leighlinbridge to Philadelphia, where he married Sarah Biles (née Thorn; ca. 1792–ca. 1859) in 1812. He also founded a glass and ceramics import business, which Tyndale inherited. Tyndale first met his cousin in 1854 on one of his frequent business trips to Europe, and they immediately struck up a close rapport and began a longstanding correspondence. During the American Civil War, Tyndale rose to the rank of brigadier general in the Union Army; after his resignation in 1864 due to ill health, he was promoted to brevet major general in acknowledgment of his service. Tyndall spent time with Tyndale during his lecture tour of the United States in 1872–73, and Tyndale became a trustee of the Tyndall Fund, which used surplus earnings from the tour to support American students of physics.

Tyndale, Julia (née Nowlen, 1823–97), was the wife of Hector Tyndale. She was born in New York to Julia and Edward Nowlen (1786–1843). She married Hector in Philadelphia in 1842, and although they did not have children they raised one of their nieces. Upon Tyndale's death, her husband's significant collection of ceramics was bequeathed to the Philadelphia Museum of Art, where it is known as the General Hector Tyndale Memorial Collection.

Tyndall, Caleb (1800–79), was Tyndall's paternal uncle. He appears to have had various jobs, including maltster and farmer, while from 1819 he served as the weighmaster of the butter crane in Leighlinbridge. During the 1841 general election, he achieved notoriety by shooting Mary McAssey while facing a pro-repeal mob. At his trial, he was cleared of malicious wounding. He subsequently served as high constable for the Barony of Idrone West and overseer of the poor. For reasons that are far from clear, Tyndall's father had an intense dislike of Caleb, and his son continued to have a difficult relationship with his uncle and his children, especially when, as in letters in this volume, they requested financial assistance.

Tyndall, Dorothea (née Shirley, 1802–87), was the wife of Tyndall's paternal uncle Caleb, with whom she had ten children, at least seven of whom survived into adulthood.

Tyndall, Emily "Emma" (1817–96), was Tyndall's sister. She does not appear to have married, and later in life suffered from a mental illness that manifested as an extreme form of religiosity. At this time, she lived in Gorey, County Wexford, with the family of her cousin John Tyndall. Although she was generally known as Emma, this was a diminutive of Emily, her given name ("Deaths," *Enniscorthy Guardian*, 5 December 1896, 2).

Tyndall, John (ca. 1815–75), was an Irish physician and Tyndall's cousin. He trained in medicine in Dublin, where he worked at the city's Lying-In Hospital, and in Edinburgh, before becoming a physician in Gorey, County Wexford. In ca. 1840 he married Catherine Hartford (1811–98), and they had five children. They also cared for Tyndall's sister Emily at their home The Lodge. As the letters in this volume show, Tyndall suffered from acute bronchitis, for which his cousin offered to send expensive medical devices, although to no avail.

Valentin, William George (1829–79), was a German chemist. He came to Britain in 1855 to study at the Royal College of Chemistry, where he soon became senior assistant in the laboratory. Having also taught at the Science Schools, South Kensington, Valentin was made gas examiner for the Great Western Gas Company, and finally chemical adviser to Trinity House, in which role he assisted Tyndall with the testing of gas illuminants in lighthouses. In 1871 he published *A Laboratory Text Book of Practical Chemistry*, which remained in print long after his sudden death from apoplexy.

Vieweg, Friedrich (1808–88), was a German publisher and the second son of Friedrich Vieweg (1761–1835), founder of the firm Vieweg Verlag. When his elder brother Eduard inherited the firm after their father's death, he moved to Paris and began his own publishing company.

Vieweg, Heinrich Rudolf (1826–90), was a German publisher who was the third generation of Vieweg Verlag's proprietors. He was the son of Eduard Vieweg (1797–1869), and grandson of Friedrich Vieweg, who had founded the family publishing business in Berlin in 1786 and later Brunswick in 1799. Vieweg worked and studied natural sciences in Heidelberg before joining his father as a partner in the family firm in 1853, and assuming full control in 1866. He continued his father's emphasis on science and published German translations of a number of Tyndall's books, although, as letters in this volume show, there were wrangles over the financial arrangements.

Villiers, Charles Pelham (1802–98), was a Liberal politician and MP for Wolverhampton. He had first been elected in 1835 and held the seat until 1898, making him the longest-serving MP in British parliamentary history. He also served as president of the Poor Law Board from 1859 to 1866. He was a friend of Olga Novikoff, to whom he expressed his admiration for Tyndall.

Vincent, Benjamin (1818–99), was the RI's assistant secretary from 1848 to 1889, and librarian from 1849 to 1889. He had previously worked as a translator and editor for the publishers Gilbert and Rivington. Like Michael Faraday, Vincent was a Sandemanian, having made his confession of faith in 1832. He was a deacon from 1844 and an elder in the Church from 1849 to 1864.

Wheatstone, Charles (1802–75), was an experimental physicist and inventor. The son of a music teacher, he was apprenticed to an uncle who made musical instruments. After an intensive self-education, during which he used musical instruments to experiment with sound, in 1834 he became professor of experimental philosophy at King's College, London, where he was a reluctant lecturer. There he developed the electric telegraph with William Fothergill Cooke (1806–79) as well as a number of optical instruments, including the stereoscope that led him into controversy with David Brewster (1781–68). He married Emma West (1813–65) in 1847 and together they had five children by 1855. The family settled in Hammersmith, where Wheatstone frequently entertained Tyndall, whom he first met in 1852.

White, Walter (1811–93), was assistant secretary of the RS. He left school at age fourteen to work as an upholsterer and cabinetmaker, but pursued self-improvement through reading and attending lectures. In 1844 White took on the position of attendant at the RS's library, undertaking the extensive task of cataloging their collections, and became librarian and assistant secretary in 1861. He held this position until his retirement in 1884. Through his role at the RS, White became acquainted with many of its most eminent members, and his private record of his activities were posthumously published as *Journals of Walter White* (1898).

Whitworth, Joseph (1803–87), 1st baronet, was an engineer, inventor, and industrialist. He devised the first nationally standardized system for screw threads, known as the British standard Whitworth. He also designed the Whitworth rifle, a firearm renowned for its accuracy, and a larger, rifled artillery piece, although both were rejected by the British army. He was renowned for the enlightened treatment of the workers at his manufactory in Manchester, and Thomas Carlyle considered him a model captain of industry. He also contributed to philanthropic schemes such as the Manchester Mechanics' Institute, and in 1868 he founded the Whitworth Scholarship to fund the academic studies of promising young engineers and chemists. Whitworth was friends with Tyndall, and, in the period covered in this volume, offered assistance with his experiments on the effectiveness of various guns as fog signals. Tyndall returned the favor, in June 1875, with a Friday Evening Discourse at the RI that he jokingly called an "effusion about Whitworth" (letter 4637).

Wiedemann, Clara Louise (née Mitscherlich, 1827–1914), was a German translator, daughter of the chemist Eilhard Mitscherlich (1794–1863), and wife of Gustav Wiedemann. They married in 1851 and had two children. She translated many of Tyndall's books, working in collaboration with Anna Helmholtz.

Wiedemann, Gustav Heinrich (1826–99), was a German physicist and editor. In 1847 he completed a doctorate in organic chemistry at the University of Berlin; he maintained that chemistry was a necessary foundation of physics. In 1854 Wiedemann became professor of physics at the University of Basel. A year earlier, he had developed, together with Rudolph Franz (1826–1902), the so-called Wiedemann-Franz law, a fundamental premise of physics concerning the conditions in which the ratio of thermal and electrical conductivity are constant. In 1877 Wiedemann succeeded Johann Christian Poggendorff (1796–1877) as editor of *Annalen der Physik und Chemie*, with his wife, Clara Wiedemann, assisting with the translation of foreign scientific papers. Clara had performed the same role with German translations of Tyndall's books, which Gustav himself edited and helped to facilitate.

Wigham, John Richardson (1829–1906), was a Scottish lighthouse engineer. Born in Edinburgh, in 1843 Wigham went to Dublin, where he served an apprenticeship in a hardware and manufacturing business under his brother-in-law Joshua Edmundson (1806–48). After Edmundson's death, the Edmundson's Electricity Corporation grew rapidly under Wigham's stewardship and focused on the provision of gas lighting to private and public buildings. Wigham was engineer to the Commercial Gas Company of Ireland and served as director of the Alliance and Dublin Consumers' Gas Company from 1866, director and vice-chairman of the Dublin United Tramways Company from 1881, and was a member of the council of the Dublin chamber of commerce, as well as serving as its secretary (1881–93) and president (1894–96). He was best known as the inventor of powerful gas burners that were adopted by the Commissioners for Irish Lights. A system invented by Wigham, with an on-site gas works, was installed in Baily Lighthouse in Howth, and underwent testing by Tyndall in his work for Trinity House.

Williamson, Alexander William (1824–1904), was a chemist who worked on the formation of ethers. In 1849 he became professor of chemistry at UCL, a position he held until his retirement in 1887. Williamson was Tyndall's predecessor as president of the BAAS, handing over at the annual meeting in

Belfast in August 1874. While president he also became the RS's foreign secretary, serving in this influential role from 1873 to 1889.

Wills, Alfred (1828–1912), was a judge and mountaineer. Aside from his distinguished legal career, Wills also gained notoriety for his numerous ascents of the Swiss Alps, and served as president of the Alpine Club in the mid-1860s. Despite having climbed alongside Tyndall, Wills chose to take James David Forbes's side in his dispute with Tyndall regarding theories of glacial motion. In the early 1870s, Wills completed an English translation of Louis Rendu's *Théorie des glaciers de la Savoie*, seeking to settle the question of whether Forbes was guilty of plagiarizing Rendu's work. Published as *Glaciers of Savoy*, it was edited by Forbes's son George.

Wynne, Francis George (1846–1929), was a civil engineer and the youngest son of Anne and George Wynne. After spending time working in Germany, he married Dora Frome (1847–1930) in 1887.

Wynne, George (1804–90), was an Irish military engineer. Born in Dublin, he enlisted in the Royal Engineers in ca. 1825, becoming 1st lieutenant by 1831. He joined the Irish Ordnance Survey in 1835, and three years later was stationed at Leighlinbridge, where he was in charge of the 5th Division, C District, and first met Tyndall as a young surveyor in the division. Wynne left the Irish Survey in 1840, apparently because his wife, Anne Wynne (née Osborne, 1808–64), was ill. George Wynne later served in Greece, the West Indies, and China, where he was wounded during the Second Opium War in 1858. He also served in the relief works during the Irish famine in 1846, and in the following year was made government inspector of railways, serving in this role for ten years. He was made colonel in 1862 and rose to general by the time he retired in 1877. In 1874 he married Henrietta Jane Darrah (1841–1917). He maintained a longstanding friendship with Tyndall, and in a letter to the *Times* in 1890, the latter recorded: "In 1839, I being fresh from school, Lieutenant Wynne became my chief. A few years afterwards he was sufficiently my friend to offer me the use of his purse for the prosecution of my studies in Germany. Our friendship, afterwards, was of the most intimate and cordial character; and notwithstanding my heterodoxy, and his Christian piety, it was never for a moment shaken. Let me briefly say that a spirit more erect and pure I have never known. . . . The earth's noblest ones are sometimes found among its noiseless ones. A conspicuous illustration of this truth is furnished by the life and character of General George Wynne" ("The Late General Wynne," *Times*, 12 July 1890, 11).

Wynne, Henry Le Poer (1836–74), was a civil servant and the eldest son of Anne and George Wynne. He served in the Royal Engineers before joining the Bengal civil service, and became acting foreign secretary to the Government of India. In 1869 he married Emily Marianne Goold (d. 1898). He died suddenly of cholera in Calcutta following a short visit to Britain.

Youmans, Edward Livingston (1821–87), was an American science writer and editor. As a child he contracted ophthalmia, which impaired his vision for the rest of his life, and he was practically blind for much of his early adulthood. Youmans nevertheless trained in chemistry, and forged a successful career as a lecturer and science journalist. In 1872 he founded the journal *Popular Science Monthly*, which he edited until his death. Through the magazine and his advisory role with the publisher D. Appleton & Co., Youmans was a crucial conduit for publishing British science in the United States, and he arranged American editions of books by Tyndall, Thomas Huxley, Herbert Spencer, and several others. Many of these works were published in the International Scientific Series, which Youmans began, with William Henry Appleton and other publishers, in 1871. He was also involved with Tyndall's lecture tour of the United States in 1872–73, and became a trustee of the Tyndall Fund, which used surplus earnings from the tour to support American students of physics.

Young, Thomas (1773–1829), was a renowned polymath who made important contributions to several fields, especially physics and the study of Egyptian hieroglyphics. He originally trained as a physician, but in 1801 was appointed professor of physics at the RI. He resigned the post two years later, as it interfered with his medical practice. Among his most significant scientific achievements was a series of experiments that demonstrated the wave theory of light. Tyndall included a discussion of Young's theory, and the criticisms it aroused, in an appendix to *Six Lectures on Light* (1873), although, as shown in the letters in this volume, it was removed in French and German editions of the book. Young also carried out pioneering work on physiological optics, studying how light interacts with the human eye, and his theory of color vision was later developed by Hermann Helmholtz.

Zöllner, Johann Karl Friedrich (1834–82), was a German astrophysicist. He studied with Heinrich Gustav Magnus (1802–70) and Gustav Heinrich Wiedemann, receiving his PhD in 1857 for a thesis on photometry. Zöllner became professor of astrophysics at the University of Leipzig in 1872. His research included the development of several instruments for astrophysical research as well as more theoretical work, including a theory of comets that he elaborated in *Über die Natur der Cometen* (1872). From the early 1870s, he began using his published work to attack other scientific practitioners, including Hermann Helmholtz, William Thomson, Peter Guthrie Tait, and Tyndall himself. This, along with a growing commitment to spiritualism, led to his increasing isolation from the scientific community, with some considering that he suffered from mental illness.

INDEX

The index refers to the letters using their serial numbers. The letter 'n' denotes a footnote reference and appears if an item is not otherwise mentioned in the letter to which the footnote pertains. A number in **bold** signifies a letter to or from a correspondent.